EVOKED POTENTIALS

ANNALS OF THE NEW YORK ACADEMY OF SCIENCES
Volume 388

EVOKED POTENTIALS

Edited by Ivan Bodis-Wollner

The New York Academy of Sciences
New York, New York
1982

Library of Congress Cataloging in Publication Data

Main entry under title:

Evoked potentials.

(Annals of the New York Academy of Sciences; v. 388)
"Result of a conference . . . held by The New York Academy of Sciences on June 16–19, 1981"—P. v.
Bibliography: p.
Includes index.
1. Evoked potentials (Electrophysiology)—Congresses. I. Bodis-Wollner, Ivan, 1937– II. Series.
Q11.N5 vol. 388 500s [616.8'047] 82-8171
[RC386.6.E86]
ISBN 0-89766-166-4
ISBN 0-89766-167-2 (pbk.)

SP

Printed in the United States of America

ISBN 0–89766–166–4 (cloth)

ISBN 0–89766–167–2 (paper)

ANNALS OF THE NEW YORK ACADEMY OF SCIENCES

VOLUME 388

June 18, 1982

EVOKED POTENTIALS*

Editor and Conference Chairman

IVAN BODIS-WOLLNER

Advisory Committee

ROGER Q. CRACCO, FLOYD RATLIFF, JEROME SHERMAN, AND HERBERT G. VAUGHAN, JR.

CONTENTS

*This series of papers is the result of a conference entitled *Evoked Potentials,* held by The New York Academy of Sciences on June 16–19, 1981.

Part IV. Visual Evoked Potentials in Ophthalmic Practice

Part V. Visual Evoked Potentials in Neurology

Part VI. Somatosensory Evoked Potentials in Clinical Practice

Part VII. Brainstem Auditory Evoked Responses

Part VIII. Ontogeny of the Evoked Potentials

Part IX. Simultaneous Electroretinograms and Evoked Potentials

Part X. Roundtable Session: The True Binocular Visual Evoked Potential

Financial assistance was received from:

- THE DAGAN CORPORATION
- E.I. DU PONT DE NEMOURS & COMPANY
- MERCK SHARP & DOHME RESEARCH LABORATORIES
- THE NATIONAL EYE INSTITUTE
- NICOLET BIOMEDICAL INSTRUMENTS
- SANDOZ, INC.
- THE TECA CORPORATION

INTRODUCTION

Ivan Bodis-Wollner

*Mount Sinai School of Medicine
of the City University of New York
New York, New York 10029*

This volume is the result of an international conference on evoked potentials held under the auspices of the New York Academy of Sciences from June 16 through June 19, 1981. Several previous conferences have addressed the analysis of surface potentials from a formal viewpoint, without much interest in the underlying biology. Others have tried to relate cognitive functions and evoked potentials. Our aim was to provide a common platform for physiologists and clinicians who use evoked potential techniques. Although we have powerful new instruments and improved capabilities for analyzing signals, even the most sophisticated or comprehensive analysis can be useless without considering the structures and mechanisms of neural function. We hope that these proceedings will establish useful links between sensory physiology and clinical practice.

Over the last decade a branch of the neurosciences, visual psychophysics, has helped bridge single cell electrophysiology and human sensory physiology. From these studies a much clearer understanding of neuronal mechanisms of vision emerged, and by referring to psychophysics we are now better able to discuss the relationship between visual physiology and evoked potentials. For this reason the first sections on vision serve as a model for other sensory modalities. It will become apparent that in order to select a particular stimulus or decide between linear and nonlinear analytical techniques, researchers and clinicians need to understand the physiological foundations of evoked potential measurements. In the pages concerning stimulus-specific evoked potentials, the applicability of systems analytical approaches in sensory studies will be discussed: several "basic" and clinical studies reveal that there is an interdependence between stimulus selection and analytical technique, depending on what kind of neuronal ensembles one wishes to investigate. This type of analytical approach can lead not just to abstract mathematical models of brain circuitry, but also to experimentally verifiable, quantitative and qualitative predictions about the response properties of particular neuronal structures, and in turn to clinically relevant questions concerning the vulnerability of parallel pathways.

Evoked potential measurements represent a promising technology—not an end in themselves. They can and should be used in the future to gain *de novo* information, not merely for confirmation of facts known from anatomical studies or clinical testing. One promising area of evoked potential research in this respect is the possibility of studying *in vivo* synaptic malfunction and neurotransmitter abnormalities in human diseases.

The central themes mentioned above are reflected in the organization of this volume. The first chapters are entirely addressed to physiological aspects of evoked potentials; however, these concepts recur in the section on clinical applications, and should be especially evident in the chapters on the roundtable discussions, which summarize the main topics presented at the conference. The roundtable sessions were organized by posing to each participant, prior to the meeting, a question geared to his own research. All questions were formulated so that they would mesh with questions asked of others in the same roundtable session. We asked for an imaginative reply, not necessarily a statement which had already appeared in print, but rather a consideration of the intuitive aspects of the topic so that the answers would lead to a dialogue with other members of the panel. The papers resulting from these sessions reflect the fact that physiologists and clinicians met at the same table and these chapters will, I hope, suggest directions for further research.

It should become evident that by building on the physiological foundation, one can make more astute clinical diagnoses. However, it was not our sole aim to improve evoked potential diagnosis. We also believe that clinical scientists can perform invaluable research by properly applying evoked potential studies to patients: their data may define the territory and generate questions for scientists who can investigate their cellular or behavioral mechanisms of the nervous system using technologies other than evoked potentials. In many ways, EP technology links these ends of biological studies as the light microscope provides a useful step between unaided inspection and electronmicroscopy. Thus we hope that both clinicians and physiologists will find useful information in this volume.

Acknowledgments

This conference follows by 17 years a symposium organized by Herbert Vaughan and sponsored by the New York Academy of Sciences in 1964. It is a pleasure to acknowledge his help and participation in 1981. He, along with Drs. Roger Cracco, Floyd Ratliff, and Jerome Sherman, made up the Advisory Committee who helped to design this conference. Another important conference was held in Brussels in 1974 and was organized by John Desmedt, who also participated in this symposium. We were happy to follow in the tradition of these international conferences.

I thank the New York Academy of Sciences and its staff, who were helpful in staging this symposium. I thank the National Eye Institute for partial support of the conference, and the following companies: Dagan Corporation, Nicolet Instrument Co., Teca Corporation, Traicor Analytic, DuPont, Merck Sharp & Dohme, and Sandoz Pharmaceuticals. I also thank the participants for their contribution to the success of this meeting.

ELECTRORETINOGRAPHIC AND VISUAL CORTICAL POTENTIALS IN RESPONSE TO ALTERNATING GRATINGS

L. Maffei

Istituto di Neurofisiologia
del Consiglio Nazionale delle Ricerche
56100 Pisa
Italy

It is a general law of sensory physiology, and therefore also of visual physiology, that some dimensions of the physical stimulus are transmitted and others are rejected. The central nervous system, with its economic plan oriented to survival, cannot afford, for instance, the distraction of analytically processing stimuli that are homogeneous either in time or in space.

It is the surprise—that which is new—that is particularly attractive and stimulating: *Variatio delectat.*

Changes in space and in time are most effective in attracting our visual interest and in eliciting the discharge of visual neurons. Contrast seems to be the variable that matters whereas a slow background flux of light either in time or in space is progressively discarded up to its almost complete neglect at the level of the visual cortex.

Starting from the stage of receptors (cones) a particular organization of visual receptive fields becomes apparent, i.e., a spatial distribution of excitatory and inhibitory influences apt to remove the DC component of light distribution. This is progressively achieved through various stages along the visual pathway.

The study of physiology of visual neurons from retina to cortex has long been hindered by the stubborn determination of investigators to use flashes of diffuse light. Similarly the study of visual evoked potentials (VEP) in man has meandered rather inconclusively.

It was following the lead of physiologists of single neurons of the visual system and in particular the work of Hubel and Wiesel that visual evoked potentials in man had a fresh start.

Campbell and I[1] more than ten years ago recorded evoked potentials in response to sinusoidal gratings, the phases of which were alternated in time. A property of the grating stimulus is that luminance, considered over a sufficiently large summation area, remains constant while contrast and spatial frequency can easily be changed. The most important finding was that from visual evoked potentials it was possible to predict related psychophysical data. We found, for instance, that for each spatial frequency the amplitude of VEP was linearly related to the logarithm of contrast of the grating. In addition, we found that extrapolation of the regression line between the evoked potential amplitude and the logarithm of contrast to zero amplitude predicts the psychophysical contrast threshold. The physiological astigmatism that exhibits itself as higher contrast sensitivity in vertical and horizontal meridians than in oblique ones was also present in evoked potentials.[2] These showed an amplitude twice as much for the vertical and horizontal as for the oblique meridians.

This difference was not present in the electroretinographic responses. Campbell, Piccolino, and I[3] used in the cat the same technique of evoked potentials as for the human subjects and found again a linear relationship between the amplitude of evoked

0077–8923/82/0388–0001 $1.75/0 © 1982, NYAS

potentials and the logarithm of contrast. Also in the cat extrapolation of the regression line predicted the behavioral contrast threshold of the cat.[4]

Visual evoked potentials to alternating gratings have now become a standard easy technique to investigate contrast sensitivity of animals in which behavioral methods can often be difficult and time consuming.

Since the relation between VEP amplitude and log contrast held for the experimental animal, it became possible to investigate the question of how and where in the visual pathway contrast is elaborated at the cellular level. It was found that a linear relation between the amplitude of cell response and the log of grating contrast holds only in visual cortical cells and is particularly strong in simple cells.[5]

The strong correlation between VEP in response to alternating gratings and psychophysical data suggested the use of this technique in human ophthalmological or neurological disease so as to permit a differentiation of impairment of visual performance caused by the optics of the eye from that caused by nervous alteration of the visual pathway.

At this stage it was not possible to discriminate in the visual pathway what was caused by retinal disease from what was caused by more central alterations, for instance, at the cortical level.

The following series of experiments, which shows that the ERG in response to alternating gratings is correlated with ganglion cell activity, offers a method of disentangling what is of retinal origin from electrical activity of the visual pathway that originates more centrally.

Electroretinographic Response to Alternating Gratings

The ERG response to diffuse light flashes consists of several components, the electrical sources of which are located in the receptor layer or in the inner nuclear layer of the retina. It is a well-established finding, as classical as it is surprising, that ganglion cells do not contribute to the flash ERG.[6] A possible explanation of this fact is that the flash is much more powerful in exciting the electrical sources of the *a* and *b* waves than the ganglion cell activity, whose contribution remains negligible. Periodical patterns, such as alternating gratings of suitable spatial frequency and contrast, are a more appropriate stimulus for retinal ganglion cells, which are detectors of contrast rather than of luminance. Indeed, the average luminance of an alternating grating remains constant. One may expect therefore that the ERG response to alternating grating exibits some sign of ganglion cell activity.

This question has been investigated[7] by recording pattern reversal ERG in an animal before and after chronic section of the optic nerve, which causes retrograde degeneration of ganglion cells.

Experiments were performed on 8 adult cats. After tracheal intubation the animal was mounted in a stereotactic apparatus and paralyzed with Pavulon. Artificial ventilation was supplied with a mixture N_2O (75%) + O_2 (25%). Pupils were dilated with atropine. Platinum ring electrodes for ERG recording were applied to the corneas with contact lenses having an artificial pupil of 4 mm diameter. Refraction was corrected with additional lenses (usually +2 to +4 diopters). ERGs were recorded alternately from either eye while the nonrecording eye was patched.

The papillae of both eyes were projected by inverted ophthalmoscopy on a tangent screen at 40 cm from the eye. The screen was then replaced by a video display (HP 1300A, Hewlett-Packard, (20 × 25 cm) positioned at 40 cm in such a way as to cover the central part of the visual field of either eye. Sinusoidal gratings of various spatial frequencies were generated on the display and shifted in spatial phase by 180° (pattern reversal) at the rate of 8 Hz (16 reversals per sec). The contrast was 30% and the mean luminance was 10 cd/m^2). The responses were filtered with a band-pass filter between 6 and 60 Hz with a slope of 6 dB/octave. ERGs were conventionally amplified, fed into an averaging computer and recorded by an *xy* plotter. At the end of the recording session the cat was allowed to recover from paralysis. A few days after the preliminary experiment, the cat was anesthetized with sodium pentobarbital (35 mg/kg) and the right optic nerve sectioned at the level of the chiasm by an oral approach. The first recording session was performed either in the same day as the

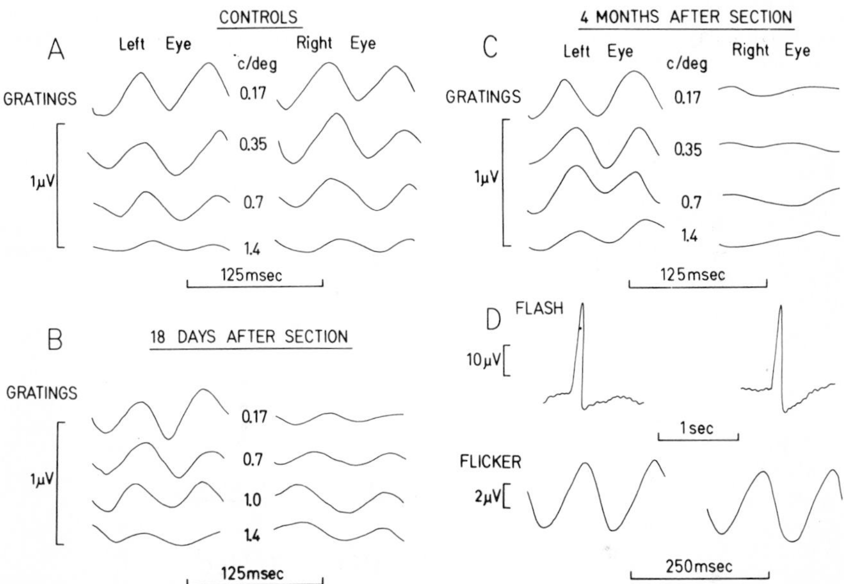

FIGURE 1. Examples of pattern reversal ERGs recorded from one cat before and after the section of the right optic nerve. (A) Control records obtained from both eyes before the section of the optic nerve. (B) and (C) Records obtained 18 days and 4 months after the optic nerve section, respectively. Each record is the average of 500 responses. Stimulus: vertical sinusoidal grating reversed in contrast 16 times/sec, contrast 30%, mean luminance 10 cd/m^2. The spatial frequency is indicated next to each record. (D) ERG in response to 50-msec light flashes (250 td) and to light flickering at 8 Hz (mean luminance, 10 cd/m^2; amplitude of square-wave modulation, 40%) recorded from both eyes 4 months after the right optic nerve section. (From Maffei & Fiorentini.[7] By permission of *Science*.)

operation (2 cats) or 6 days after (1 cat) or later. ERGs were recorded from each cat at various times during the first 4 months after the operation.

In the first session after optic nerve section, pattern reversal cortical evoked potentials were also recorded with monocular stimulus presentation to make sure that no cortical response was present from the right eye.

The ERGs recorded in control sessions preceding the optic nerve section, have approximately sinusoidal waveform (FIGURE 1).

The ERG amplitude is typically independent of spatial frequency from 0.17 to 0.5 cycles/deg and then progressively decreases. The highest spatial frequency at which a pattern reversal ERG can be obtained is similar to that at which a cortical evoked potential can be elicited and of the same order as the behavioral visual acuity of the cat. The ERG waveform in response to a grating alternating at 8 Hz is similar to that of cortical evoked potentials recorded in the same conditions. Both responses, at this temporal frequency have a waveform corresponding to the second harmonic of the stimulus.

At variance with the flash ERG, the alternating-grating ERG has a limited summation area over the retina which is a function of spatial frequency. For spatial frequencies of 0.5–2 cycles/deg the summation area is on the order of 7–5 deg, respectively.

The pattern reversal ERG responses recorded either soon after the optic nerve section or 6 days later had the same amplitude in the two eyes and were comparable to the controls recorded before the operation.

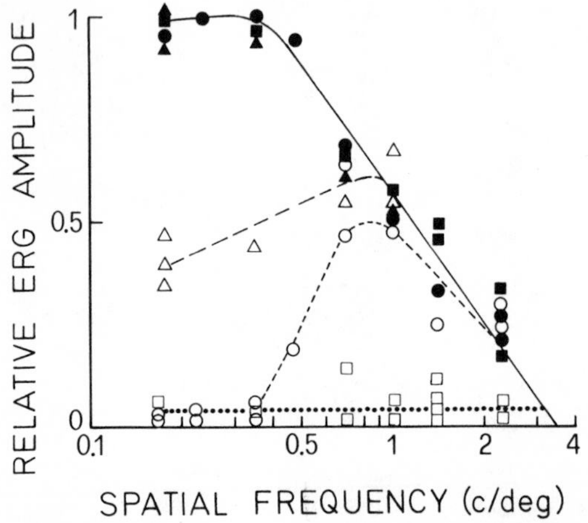

FIGURE 2. Relative amplitude of the pattern reversal ERG plotted as a function of spatial frequency obtained from one cat for the eye ipsilateral to the optic nerve section. Stimulus conditions: vertical sinusoidal gratings reversed in contrast at 8 Hz; contrast, 30%; mean luminance, 10 cd/m^2. Data obtained in 4 experimental sessions: control session, before optic nerve section (filled circles); session performed 18 days (open triangles); and 70 days (open circles) and 4 months (open squares) after optic nerve section. Each point is the average of 2 to 4 records.

The earliest unequivocal effects of optic nerve section on the pattern reversal ERG of the ipsilateral eye were observed in one cat 18 days after the operation (FIGURE 1) and in other cats between the fourth and sixth week. This effect consisted of a considerable reduction in the amplitude of the ERG responses at low spatial frequencies (FIGURES 1 & 2). ERGs at or above 1 cycle/deg were unaffected. No clear response to the lowest spatial frequencies could be obtained any longer about 2 months later. Even the ERGs at 1 cycle/deg were attenuated (FIGURE 2 left, open triangles).

Four months after the operation we could not obtain pattern reversal ERG from the right eye at any spatial frequency (FIGURE 1).

These findings were repeated in seven cats.

In all cats the ERG in response to light flashes or flicker remained normal and had comparable amplitude in both eyes throughout the 4-month period following the operation (FIGURES 1 & 2, open squares).

In agreement with what has previously been reported in the literature,[8] the whole mounted retinae of our animals showed degeneration of the ganglion cell layer 4 months after the optic nerve section. According to these previous reports[8] there is no clear sign of ganglion cell degeneration in preparations obtained earlier than 3 months after the section.

The disappearance of the pattern reversal ERG 4 months after the section is therefore likely to be causally correlated with ganglion cell degeneration.

It is tempting to conclude that the activity of the ganglion cells is the main source of the pattern reversal ERG. Obviously we cannot exclude that the degeneration of the ganglion cells may have secondary effects capable of influencing other possible generators of the ERG, although this is rather unlikely since the flicker ERG remains totally unaffected even after complete degeneration of the ganglion cells.

The finding that the flash ERG and the pattern reversal ERG have different electrical sources can be demonstrated in other ways also. The most dramatic and clear way is a transient block of the retinal artery, obtained by clamping the optic nerve at the exit of the orbit. A block of the retinal circulation for 30 min abolishes the pattern reversal ERG, while the flash ERG returns to normal 30–40 min after the block of the retinal circulation is terminated (FIGURE 3).

Prompted by these results in cats, we have recorded the ERG in response to alternating gratings in man in the attempt to find evidence for an electrical retinal response correlated with ganglion cell activity.[9] The ERG evoked by alternating gratings has been recorded in normal subjects and in a number of patients with retinal or optic nerve diseases causing either degeneration or dysfunction of the ganglion cells. We have found that the ERG in response to alternating gratings is dramatically affected in patients with diseases of the ganglion cells and is absent in cases in which the ganglion cells are expected to be degenerated. In all these patients the flash-evoked ERG was normal.

Sinusoidal vertical gratings of variable spatial frequency and contrast were electronically generated on a circular TV screen (mean luminance 20 cd/m^2, 25 cm dia). The gratings were alternated in phase at the rate of 8 Hz (16 reversals/sec) and were viewed monocularly from a distance of 57 or 114 cm with natural pupils. The same display was used to generate homogeneous sinusoidally modulated light (0–40 cd/m^2, 8 Hz).

The ERG was recorded with a variation of the electrode of Maffei and Campbell[2] consisting of a small silver plate introduced between the lower eyelid and the eyebulb after application of novacaine. An similar electrode placed in the other eye was used as a reference. This was done to avoid interference with cortical evoked potentials. The ground electrode was placed on the forehead. The cortical visual evoked potentials (VEP) were simultaneously recorded with silver plate electrodes placed on the midline 2 cm above the inion and at the vertex. The ERG and VEP signals were amplified and filtered by band-pass filters (12 dB/octave). Filtering was between 5 and 50 Hz for alternating gratings and flickering light and between 2 and 50 Hz for light flashes.

The filtered signals were fed into two analog input channels of a PDP 11/10 computer allowing rejection of single sweeps disturbed by artifacts, on-line averaging of signals and storage of signals for further off-line analysis. Averaged responses were recorded by an *xy* plotter. The nonrecording eye was either covered or presented with a fixation spot optically superimposed on the display that was made invisible to this eye by means of crossed polaroid filters.

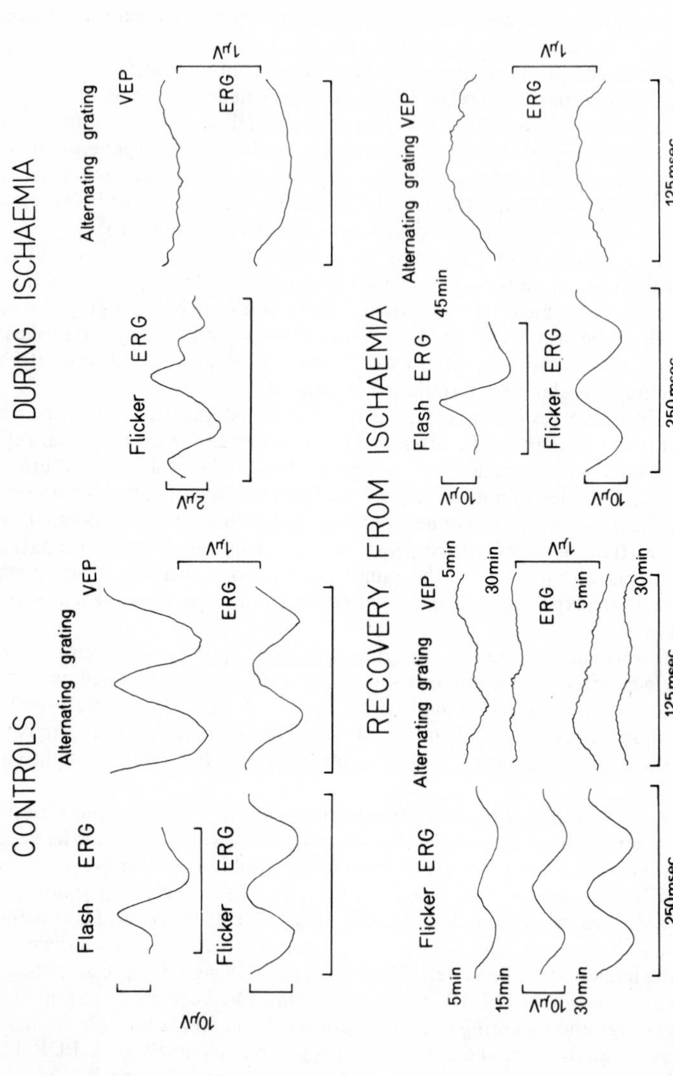

FIGURE 3. Flash, flicker, and alternating grating ERG and VEP in the cat before (controls) during and after 10 min of ischemia of the retinal artery (see text). Flash and flicker responses are the average of 20 responses, the alternating grating ERG responses are the average of 200 responses.

In normal subjects the ERG in response to gratings alternating at 8 Hz has, as in the case of the cats, an approximately sinusoidal waveform with a temporal period of about 60 msec, corresponding to the second harmonic of the stimulus alternation (FIGURE 4). For a fixed contrast of the stimulus (around 30%), the amplitude of the ERG is a function of the grating spatial frequency, with a maximum around 2–3 cycles/deg and a high frequency cut-off about 10 cycles/deg.

Among the patients we have recorded, the most dramatic case was that of a woman who had suffered from a transient occlusion of the left retinal artery, a condition similar to what we had caused artificially in the cat.

FIGURE 4 illustrates the records obtained from this patient. At the time of recording, 8 days after the appearance of the symptoms, the left eye was practically blind, while at the same time retinal circulation had recovered as indicated by

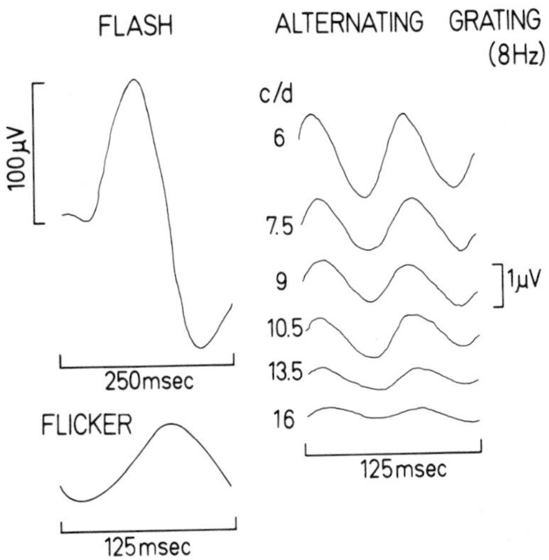

FIGURE 4. Examples of flash, flicker, and alternating grating ERG at various spatial frequencies in a normal subject. The spatial frequencies are reported on the left of the alternating grating ERG response. The flash and the flicker responses are the average of 50 responses. The alternating grating ERG responses are the average of 500 responses.

fluoroangiography. The ERG responses to flashes of light and to diffuse flickering light were normal and of comparable amplitude in the two eyes (FIGURE 5).

The ERG and VEP responses to alternating grating from the right eye were also normal (FIGURE 5). No measurable ERG or VEP in response to alternating gratings could be recorded from stimulation of the left eye (FIGURE 1, bottom right) at any spatial frequency between 0.2 and 1.4 cycles/deg.

A similar case was kindly pointed out to us by Dr. Dodt.[10] His description runs as follows: In a 55-year-old female patient 3 weeks after an accident causing a *contusio nervi optici* of her left eye, all electroretinal recordings including pattern ERG were normal. Reinvestigation of the same eye 3 months later had the same result except for the pattern ERG, which was completely extinguished.

We have also recorded 10 patients suffering from optic neuritis, and in all cases we have found that the alternating grating ERG was severely impaired, while the flash ERG was normal.

ERG Responses to Alternating Gratings in Patients Affected by Retinitis Pigmentosa

It is well known that flash ERG patients affected by retinitis pigmentosa is severely impaired or absent even if patients often maintain a relatively normal central

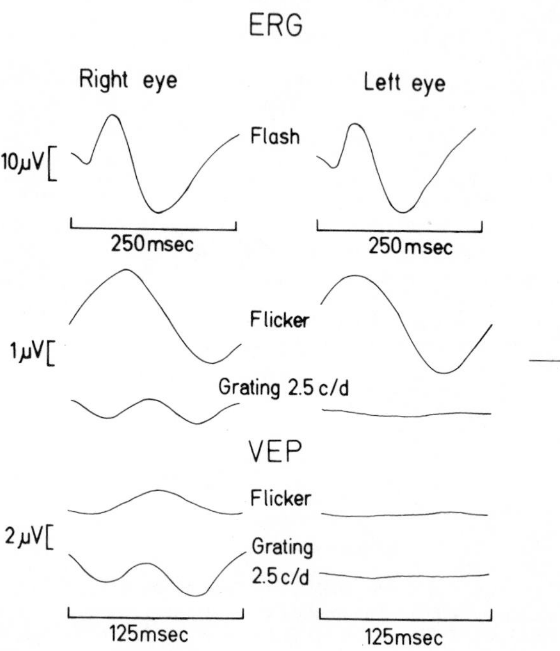

FIGURE 5. Examples of averaged ERG and VEP responses to light flashes, flickering light, and alternating gratings recorded from the right and left eye of a patient who had suffered from temporary occlusion of her left retinal artery. Responses were averaged over 200 to 1200 stimulus sweeps (modulated light and alternating gratings) or 40 stimulus sweeps (flashes).

vision. The anatomical localization of this retinal disease is mainly in the receptor layer and does not involve ganglion cells. Ergo, if our previous results and considerations were correct, pattern reversal ERG should be present in these patients. We (Fiorentini, Maffei, Pirchio, and Porciatti; manuscript in preparation) have examined 7 patients with retinitis pigmentosa. In all of these patients the flash ERG was dramatically reduced or absent while the ERG response to alternating gratings was present, even if often with a reduced amplitude with respect to responses obtained from normal subjects.

References

1. Campbell, F. W. & L. Maffei. 1970. Electrophysiological evidence for the existence of orientation and size detectors in the human visual system. J. Physiol. (London) **207:** 635–52.
2. Maffei, L. & F. W. Campbell. 1970. Neurophysiological localization of the vertical and horizontal visual coordinates in man. Science **176:** 386.
3. Campbell, F. W., L. Maffei & M. Piccolino. 1973. The contrast sensitivity of the cat. J. Physiol. (London) **229:** 719–731.
4. Bisti, S. & L. Maffei. 1974. Behavioural contrast sensitivity of the cat in various visual meridians. J. Physiol. (London) **241:** 201–210.
5. Maffei, L. & A. Fiorentini. 1973. The visual cortex as a spatial frequency analyzer. Vision Res. **13:** 1255–67.
6. Granit, R. 1962. The visual pathway. *In* The Eye. Davson M., Ed. Vol. 2: Academic Press, New York, N.Y.
7. Maffei, L. & A. Fiorentini. 1981. Electroretinographic responses to alternating gratings before and after section of the optic nerve. Science **211:** 953–955.
8. Stone, J. 1966. The naso temporal division of the cat's retina. J. Comp. Neurol. **126:** 585–600.
9. Stone, J. 1978. The number and distribution of ganglion cells in the cat's retina. J. Comp. Neurol. **180:** 753–772.
10. Fiorentini, A., L. Maffei, M. Pirchio, V. Porciatti & D. Spinelli. 1981. The ERG in response to alternating grating in patients with diseases of the peripheral visual pathway. Invest. Ophthalmol. Vis. Sci. **21:** 490–493.
11. Groneberg, A. & C. Teping. 1980. Topodiagnostik von Sehstörungen durch Ableitung retinaler und Korticaler Antworten auf Umkehr. Kontrastmuster. Ber. Deutsch. Ophthalmol. Ges. **77:** 409–415.

Discussion of the Paper

G. Celesia: In order to stimulate preferentially ganglion cells from the fovea versus the peripheral retina, could one vary spatial frequency?

L. Maffei: For cats and for the human, probably yes. We have not done selective stimulation of the peripheral retina, but I would say that most of the pattern ERG waves come from the central part. When we stimulate the peripheral part, the ERG becomes a very small wave of the order of one microvolt or less.

H. Spekreijse; I am particularly intrigued by your observation that, after cutting the optic nerve, the reduction of the ERG was particularly pronounced for, or most evident in the beginning for, the low spatial frequencies. If you extend the spatial frequency scale, you find that there are still lower spatial frequencies, and you would expect thus to find a reduced luminance (flicker) response. Yet, that one was not affected. So if you plotted the ERG amplitude curve as a function of decreasing spatial frequency, first it should decline, and then it has to go up as you observed. Have you done this?

Maffei: I agree with you that if you go to very low spatial frequencies you run the risk of stimulating luminance detection. If you go below one cycle in the human or 0.1 cycle per degree in the cat, you begin to see again that the response is going up.

However, we never go below that frequency because we think that is a poor experiment.

SPEKREIJSE: If you plot the amplitude of the VEP as a function of contrast in man, then it extrapolates to the psychophysical threshold. At the same time you also said in your presentation that the PERG curve crosses the abscissa at a much higher contrast than the VEP. Thus the ERG disappears before the VEP. Well, if the receptors do not give any response, of course, the ganglion cells should not give the response either. How do you explain this?

MAFFEI: That is a very good observation, and it is correct. The PERG signal is a very small signal and so we think it must be present if the evoked potential is present, but the response is not clear if the contrast is very low.

SPEKREIJSE: It is not the absolute amplitude which is of importance here—otherwise I would agree with you. It is the extrapolated value and the slope which matter. The absolute value has no meaning here and if the PERG disappears and extrapolates at a much higher contrast, and if you believe that the same mechanism accounts for both, then there is a paradox.

PARALLEL PATHWAYS IN THE MAMMALIAN VISUAL SYSTEM*

Robert Shapley

Rockefeller University
New York, New York 10021

The study of visual physiology with single-unit techniques is complementary to the use of the evoked potential. In the former technique, the microscopic elements of the visual system are accessible to study; in the latter, large populations are sampled and fine detail is lost. Both techniques have been useful in understanding the neural basis of visual perception.

The excitement in single-unit studies of visual physiology recently has been the discovery of the importance of parallel processing. Many retinal pathways exist, not just one or two. There are many central targets for retinal axons. My paper concerns the properties of two of these main parallel pathways, the X and Y cells. The X and Y pathways are elaborated first in the retina, project in parallel to the lateral geniculate nucleus (LGN), and thence project in parallel to the visual cortex. I will first discuss the defining characteristics of the X and Y cells, then describe some of their visual properties, and finally discuss recent work done in our laboratory on the existence and location of X and Y cells in the LGN of primates.

The stimuli we used were produced on the face of a cathode ray tube (CRT) monitor (Tektronix 606 or 608). The electronic circuitry used has been previously described.[1] The stimulus display was a raster, 500 lines/frame at 200 frames per second. Drifting and contrast reversal gratings were produced by synchronizing the raster to an oscillator that produced the spatial grating pattern. The visual stimulus could be represented formally as

Contrast reversal

$$L(x, t) = L_0 + L_1 \sin(2\pi kx + \phi)\, M(t), \qquad \textbf{(1a)}$$

Drift

$$L(x, t) = L_0 + L_1 \sin[2\pi(kx - ft)]. \qquad \textbf{(1b)}$$

In these equations L_0 is the mean luminance, L_1 is the contrast (or depth of modulation), k is spatial frequency, x is the position in degrees, ϕ is the spatial phase, $M(t)$ is the temporal modulation signal, usually a temporal sinusoid or temporal square wave, and f is the temporal rate of drift in Hz. The mean luminance was constant throughout an experiment.

The basic experiment we performed was measurement of the spatial phase dependence of the response to a grating undergoing contrast reversal as described by Equation 1a. The procedure is illustrated in FIGURE 1. The grating was first positioned to produce a maximal response as in FIGURE 1A. Then, it was moved to a new location 90° away in spatial phase as in FIGURE 1B. In comparing FIGURE 1A and 1B, notice that at positions at which the stimulus was modulated the most in A, it is modulated the least in B. The idea behind the experiment is that if the neural pathway leading up

*This work was supported by research grants (EY1472, EY188, and EY1428) and a Research Career Development Award from the National Eye Institute.

0077-8923/82/0388-0011 $01.75/0 © 1982, NYAS

to the neurons we were studying were linear, then the neural response should behave like the stimulus. If the response is maximal in position A, it should be null in position B.

Rationale for Classification

Understanding the rationale for this experimental procedure is essential for the interpretation of our results. The stimuli used for classifying the cells were contrast reversal sine gratings. For each cell, gratings of several different spatial frequencies

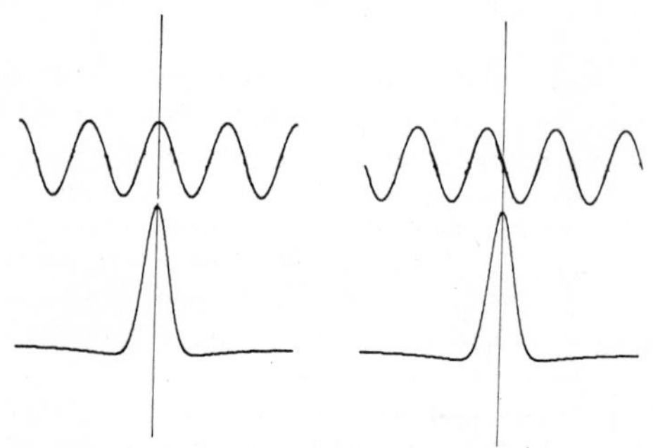

FIGURE 1. Grating patterns at different spatial phases. The curve at the top of each side of the figure represents the one-dimensional luminance profile of the grating. The luminance profile of the stimulus in space is a sinusoid. Its amplitude is modulated by a slow sinusoidal temporal modulation signal [$M(t)$ in Equation 1a]. The upper curve therefore is the luminance profile of the grating at the crest of the temporal modulation signal. The solid curve below the stimulus represents the sensitivity profile of a typical center–surround visual neuron in the retina or LGN. The vertical line represents the midpoint of the receptive field. In the stimulus situation on the left-hand side, denoted 0° Spatial Phase, the crest of the spatial sinusoid is lined up with the peak of the sensitivity profile. This will produce a maximal modulated response. In the stimulus condition on the right, denoted 90° Spatial Phase, a zero-crossing of the grating is lined up with the peak of the sensitivity profile. In a linear cell, responses to modulation to the left of the vertical line will be precisely canceled by neural responses to modulation to the right of the vertical line. In a linear visual neuron which can be described by a single sensitivity profile, the position for maximal response will always be 90° in spatial phase away from the null position.

were used. Modulation was generally at 2–4 Hz. The space-averaged illuminance on the retina did not vary with time; rather, the spatial pattern was time modulated. The illuminance profile on the retina was as given in Equation 1a. The spatial phase (or position) dependence of the optic fiber's response to the contrast reversal grating determined whether we classed it as an X cell or as a Y cell, according to the following reasoning.

Suppose there were a single spatial mechanism in which light-evoked neural signals were added linearly. Then such a mechanism would have sensitivity for the contrast reversal grating, which would be a sinusoidal function of the spatial phase of the grating.[2] In particular, there would be two positions or spatial phases of the grating which evoked zero response. The same grating placed a quarter cycle away in either direction from one of these null positions[3] would give a maximal response for that grafting. Furthermore, the responses of a linear spatial mechanism to a sinusoidally modulated grating would be sinusoidal in time, and these responses would be sinusoids at the modulation frequency only.

If the ganglion cell receives input from several spatial mechanisms within which and between which signals are pooled in a linear manner, and if these mechanisms produce responses that are in phase or exactly 180° out of phase with one another, then the ganglion cell will respond as if it received input from a single linear spatial mechanism. This equivalent single linear mechanism would be the algebraic sum of the several separate linear mechanisms. In particular, center and surround mechanisms which combine in a linear manner and produce responses exactly 180° out of phase with one another will be equivalent to a single linear spatial mechanism. Therefore, if a visual neuron does produce responses that vary in magnitude sinusoidally with the position of a contrast reversal sine grating, and if the responses of the cell are at the modulation frequency of the contrast reversal, one may conclude that the cell is being driven by a single linear spatial mechanism or its equivalent. Such a cell we call an X cell.

There are several different kinds of departures from the ideal linear case presented above. The first departure occurs when there are two (or more) linear mechanisms, but the responses from these separate mechanisms are neither exactly in phase nor exactly 180° out of phase with each other. This will occur if there are differences in the dynamics within the different mechanisms. In this case the responses of the neuron receiving these two (or more) inputs might not vary sinusoidally with spatial phase. In particular, there might be no null positions. Nevertheless, the responses would be at the contrast modulation frequency. We have never observed this kind of departure from the ideal linear neuron in the mammalian retina.

It is possible to imagine receptive field models that are in some sense intermediate between the single linear mechanism and the dispersed ensemble of nonlinear subunits. For instance, one might imagine a receptive field with one or only a few nonlinear subunits, or with many subunits which overlapped in one small retinal region. In this case, one would expect responses to be, to some extent, spatial phase dependent, but one would also expect responses to contain a large component at the second harmonic of the modulation frequency. Such cells have been observed in the retinas of eels and frogs.[4]

In practice the two main properties we used to identify X cells in the cat optic tract and LGN were response at the modulation frequency of contrast reversal and positions for a null response separated by 90° in spatial phase from positions for a peak response. Both of these criteria are what one would expect from the response of a single linear receptive field mechanism to stimulation by contrast reversal of a grating pattern (as stated above). In most cases we performed this identification procedure rapidly by listening to the cell's discharge on an audio monitor while varying the spatial phase of the stimulus pattern.

X Cells

The spatial phase dependence of the responses of X cells is illustrated in Figure 2. The cell was stimulated by a 2 cycle/deg sine grating which was undergoing square-wave contrast reversal at a rate of 4 Hz. One of the cycle histograms displayed in Figure 2 was elicited by the grating at a spatial phase denoted 0° for the position of maximal response. At a spatial phase 90° away from this (or one quarter of a spatial

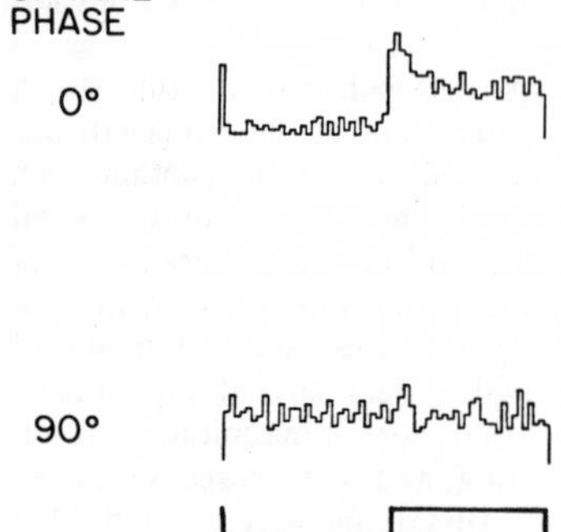

FIGURE 2. X cell response to contrast reversal of a sine grating pattern. The spatial frequency of the grating was 2 cycles/deg. The contrast was 0.36. The temporal frequency was 4 Hz. The contrast reversal was a square wave in time. The peak position, 0° spatial phase, was exactly 90° away from the null position. Parvocellular X cell in macaque monkey LGN.

period, equivalent to 0.125° of visual angle for a 2 c/deg grating) the cell produced no modulated response to grating contrast reversal. In other words, displacement of the grating by 0.125° visual angle changed the response from a peak to a null. This is the basis for saying that an X cell's response is highly dependent on spatial phase.

Strictly speaking, the identification of a cell as an X cell on the basis of linear summation requires more than a strong dependence on spatial phase and response at odd harmonics of the modulation frequency. It is known that Y cells, in the cat retina and lateral geniculate nucleus, have a strong spatial phase dependence and respond primarily at the fundamental modulation frequency *if a sine grating of low spatial frequency is used.*[2,6,7] It is crucial to test spatial phase dependence over a range of spatial frequencies and especially at high spatial frequencies for the cell. For X cells, the strong spatial phase dependence (with peaks and nulls), and the dominance of the response by the fundamental modulation frequency, persisted as spatial frequency was increased up to the highest value that the cell would respond to. This indicates that the smallest neural summing area within the receptive field—presumably the receptive field center—is part of a linear receptive field mechanism in X cells (compare References 2, 5, and 6).

Y CELLS

The primary characteristic we have used to identify Y cells in the cat is the behavior these cells exhibit in response to contrast reversal of *fine* grating patterns.[2,6] The Y cells respond to such patterns predominantly at twice the modulation frequency; the frequency-doubled response of the Y cells does not change as the spatial phase of the grating is varied. An example is given in FIGURE 3. One can see that the cell's responses were dominated by a component at twice the frequency of stimulus modulation, and also that this frequency-doubled response was the same at two spatial phases separated by 90°.

An explanation for the duplex behavior of Y cells in terms of a receptive field model has been offered previously.[7] The essence of the model is that in the Y cell's receptive field there must be many dispersed subunits which excite the Y cell only after their signals have passed through a nonlinear transducer like a threshold or a rectifier. This array of "nonlinear subunits" appears to be arranged in parallel with a conventional, linear receptive field center and surround.[7,8] This explains why Y cells appear most nonlinear in response to gratings of high spatial frequency, gratings which are too fine to be resolved by the Y cell's large linear center but are still resolvable by the cell's nonlinear subunits.

A summary diagram that illustrates our views of the X/Y distinction in the cat retina is given in FIGURE 4. This is a block diagram of the retinal circuitry that leads to X and Y retinal ganglion cells. The X cell pathway is simple and involves linear addition of similar signals separately within center and surround receptive field mechanisms and then linear pooling of center and surround signals. The Y cell gets two inputs. One is a linear center–surround signal like that sent to the X cells. Another input to the Y cell is the ensemble of nonlinear excitatory subunits.

Yuen Tat So and I showed that these receptive field properties were preserved in X and Y cells of the cat's LGN.[6] The spatial phase dependence and spatial frequency of response of LGN relay cells were basically determined by their retinal inputs.

PARALLEL PATHWAYS IN MACAQUE MONKEYS

Recently we have turned our attention to X and Y cells in the visual pathway of the macaque monkey and have concentrated our attention on the monkey's lateral geniculate nucleus.[9]

One of the mysteries of the monkey's geniculate has been the reason for the layering in this structure. One recent explanation is that the purpose of the layering is to keep the X and Y cells segregated—the X cells in the small-cell (parvocellular) laminae and the Y cells in the large-cell (magnocellular) laminae. Several groups have come to this conclusion based on experiments in which response time course and axon conduction latency were measured. However, the important question of whether measurement of these properties truly showed X and Y cell segregation has remained

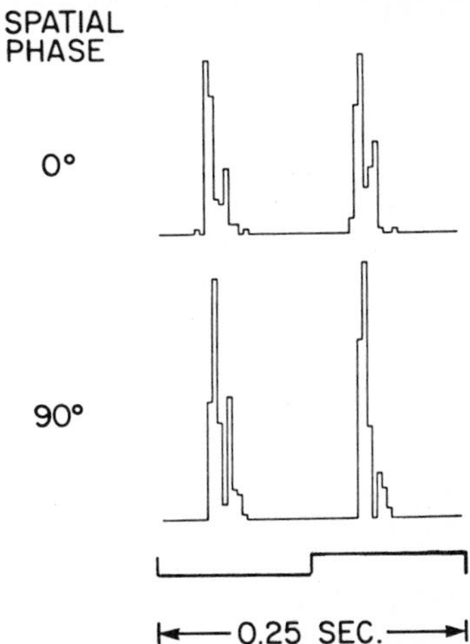

FIGURE 3. Y cell response to contrast reversal of a sine grating pattern. The grating had a spatial frequency of 2.5 cycles/deg, a contrast of 0.2, and was contrast-reversed with a square wave time course at 4 Hz. The response is at twice the modulation frequency, and is more or less unaffected by a 90° change in spatial phase.

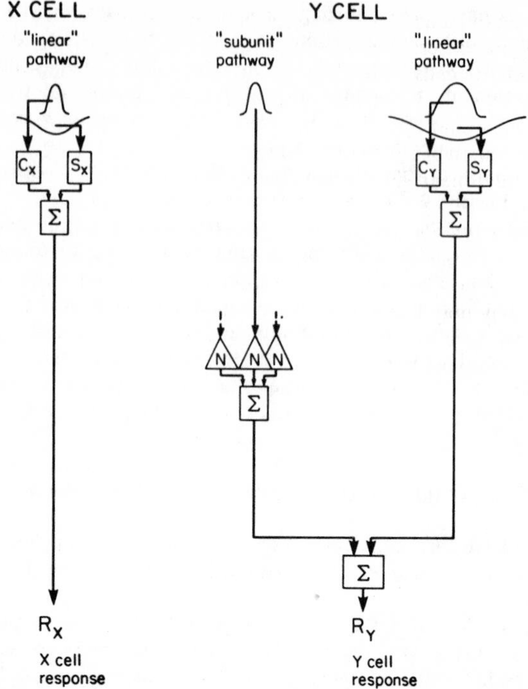

FIGURE 4. A simplified heuristic model of the receptive fields of X and Y cells. X cell responses are qualitatively explained by the superposition of two approximately linear mechanisms, a center and a concentric antagonistic surround. The first-order components of a Y cell's response are also explained by similar mechanisms, although the sizes of Y cell centers and surrounds are somewhat larger than those of X cells in the same region of retina. In addition, the Y cell response contains nonlinear components due to an array of subunits, each of which contains a center–surround organization of approximately the same size as the X cell center and surround. The responses of these subunits undergo transformations similar to rectification before being pooled. This pooled response is the excitatory nonlinear component of the Y cell response.

open because spatial receptive field properties that distinguish X and Y cells in the cat have not been used up to now in the monkey.

Our first objective was to determine if monkey geniculate cells could be classified as X or Y on the basis of linear or nonlinear spatial summation. A second goal was the determination of the positions of these X and Y cells within the cellular laminae of the geniculate. During the course of our experiments, a third objective became important: the correlation of spatial and spectral properties of geniculate cells.

Experiments were done on anesthetized, paralyzed monkeys. To illustrate the LGN of the macaque and our track reconstruction procedure, FIGURE 5 shows a pair of dots left on one track through the geniculate. The four dorsal layers are called the parvocellular layers, and the two more ventral layers are referred to as the magnocellular layers. From layers 6 through 1, the eye changes, which we also used as a clue to reconstructing tracks, are contra, ipsi, contra, ipsi, ipsi, and contra. The most difficult decision is the layer 3–layer 2 assignment, since both are driven from the ipsilateral eye. That is why leaving a green dot at layer 2 was important on this track, to prove we were in layer 2.

Cells were classified as X or Y by means of their response to contrast reversal and drift of sine grating patterns as described above. The results from applying these tests to monkey geniculate cells were given in FIGURES 2 and 3, which were in fact results from monkey LGN cells. Nearly all of the parvocellular cells are X by the criterion of linear summation. About three-fourths of the cells in the magnocellular laminae are X cells. About one-fourth of the magnocellular cells are like Y cells in the cat, according to this linearity/nonlinearity criterion. Thus, we do not confirm that all magnocellular cells are Y cells.

In the cat there is a correlation, though not a perfect correlation, between linearity of spatial summation, chiasm latency, and response time course. In the monkey this loose correlation is broken. The crucial point is that X cells may be "sustained" or "transient"; or they may have fast or slow chiasm latencies, depending on what layer they are in. Parvo-X cells have long latencies to stimulation of the optic chiasm while magno-X cells have short latencies.

CONTRAST SENSITIVITY

Finally, we come to the correlation of spatial and spectral properties. Here we were somewhat surprised by the outcome that the parvocellular X cells had relatively poor spatial contrast sensitivity. Many parvo-cells are clearly color-opponent. In the others

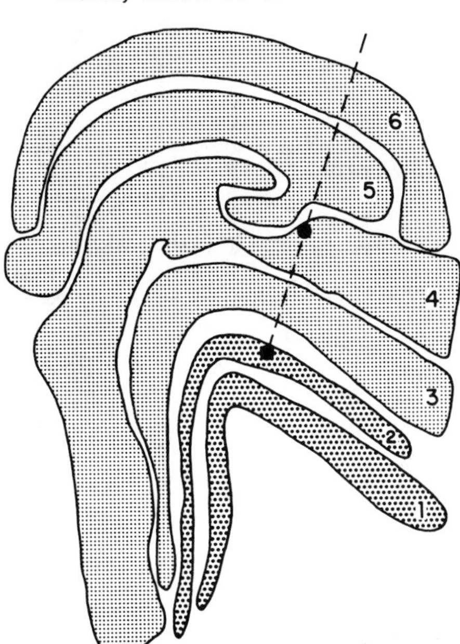

FIGURE 5. Tracking through the monkey's LGN. There are six main layers of cell bodies in the LGN of *Macaca fascicularis,* the monkey species we used. The layers are numbered 1–6, ventral to dorsal. The dorsalmost four layers are parvocellular. One track through such a geniculate is represented here by a charting of 100-μm thick histological sections from one monkey's brain. Fast Green dye marks were left at the locations marked with dark dots. Typical parvocellular and magnocellular neurons were recorded at these locations.

there is also color-opponency but it is concealed. The spectral sensitivities of parvo-cells revealed this hidden color opponency by having narrower spectral bandwidths than would single cones, and by having peak responses at wavelengths displaced from the cone maxima. An example of the spectral sensitivity and the luminance responses of a parvo-X cell are given in FIGURE 6. This cell was typical of many cells in having weak color opponency, which was only revealed by averaging. Nevertheless, its responses to drifting luminance gratings were poor, giving only about 12 impulses/second at best at a contrast of 0.24 or 24%.

This behavior may be compared with that of a magnocellular X cell illustrated in FIGURE 7. As in all our magno-cells, this cell was nonopponent and it had a broad spectral sensitivity, broader than that of individual cone receptors in the macaque. The contrast sensitivity was high—the cell gave more than a 14-impulse/sec response to a contrast of 0.06. At 0.24 contrast, the response was above 30 impulses/sec. In terms of sensitivity the magno-cells were three to ten times more sensitive than parvocellular cells. None of the parvocellular cells that were recorded had a contrast sensitivity as high as the lowest magnocellular sensitivity. The highest contrast sensitivity of magnocellular X cells was approximately the same as that of magno-Y cells.

Therefore, our conclusion is that cells in the geniculate of the monkey are segregated for color and luminance contrast. The color cells are in the parvo-layers and the brightness cells are in the magno-layers. This is completely consistent with the earlier proposals of Wiesel and Hubel[10] and others. Our new proposal is that luminance contrast vision at low to medium contrasts must proceed via the magnocellular route, and that the parvocellular pathway cannot "see" luminance contrast until

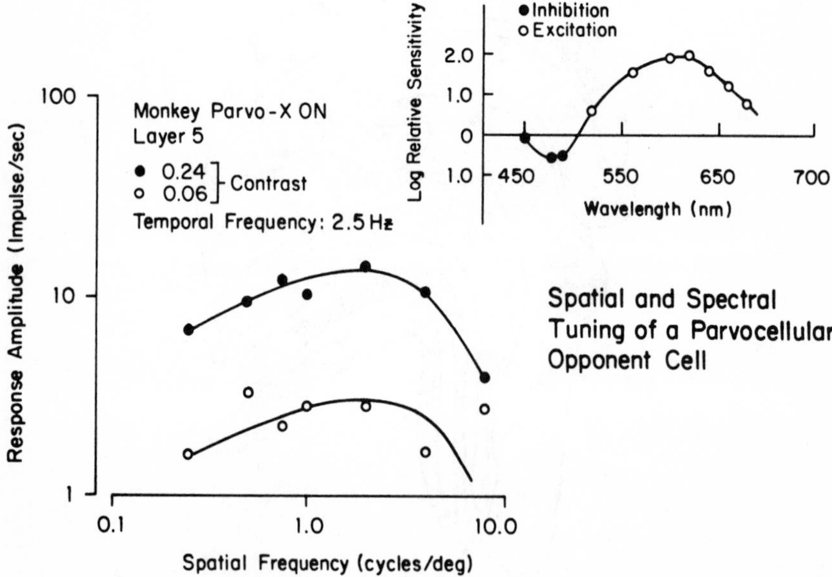

FIGURE 6. Parvocellular X cell. Response amplitude of the neural response at the fundamental frequency of drift of a drifting grating (f in Equation 1b) is plotted versus spatial frequency. Sets of responses at two contrasts, 0.06 and 0.24, are shown. The responses at 0.06 contrast are in the noise. The inset shows the spectral sensitivity of the cell which had a neutral point around 500 nm.

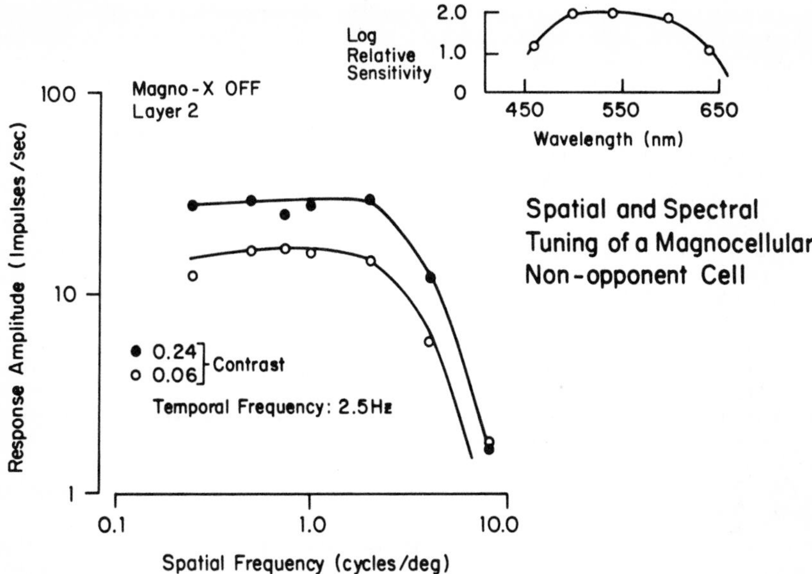

FIGURE 7. Magnocellular X cell. Responses to drifting gratings at two contrasts, 0.06 and 0.24. The responses at 0.06 contrast are as large or larger than the parvocellular X cell's responses at 0.24 contrast. The broad spectral sensitivity of the magnocellular neuron is indicated in the inset.

rather high contrast levels. It suggests further experiments to test whether the parvocellular X cells contribute to brightness contrast at all, or whether all brightness contrast information at all levels of contrast goes to cortex via the magnocellular path. We have shown also that magnocellular X cells are capable of carrying undistorted high spatial frequency information to the cortex.

Since the human lateral geniculate nucleus is structurally similar to the monkey's, it is reasonable to speculate that brightness contrast is mediated by the magnocellular layers in people, while the parvo-layers are specialized for color vision. There ought to be consequences for the evoked potential: amplitude of response ought to be a two-branched function of contrast.

ACKNOWLEDGMENTS

This work has been a cooperative venture with Shaul Hochstein, Jonathan Victor, Yuen Tat So, and Ehud Kaplan.

REFERENCES

1. SHAPLEY, R. M. & M. ROSSETTO. 1976. An electronic visual stimulator. Behav. Res. Meth. & Instru. **8:** 15–20.
2. HOCHSTEIN, S. & R. M. SHAPLEY. 1976. Quantitative analysis of retinal ganglion cell classifications. J. Physiol. **262:** 237–264.

3. Enroth-Cugell, C. & J. G. Robson. 1966. The contrast sensitivity of retinal ganglion cells of the cat. J. Physiol. **187:** 517–552.
4. Gordon, J. & R. M. Shapley. 1978. Contrast sensitivity and spatial summation in frog and eel retinal ganglion cells. Visual Psychophysics and Physiology. J. C. Armington, J. Krauskopf & B. R. Wooten, Eds. pp. 315–329. Academic Press, New York, N.Y.
5. Shapley, R. M. & J. Gordon. 1978. The eel retina. Ganglion cell classes and spatial mechanisms. J. Gen. Physiol. **71:** 139–155.
6. So, Y. T. & R. M. Shapley. 1979. Spatial properties of X and Y cells in the lateral geniculate nucleus of the cat and conduction velocities of their inputs. Exp. Brain Res. **36:** 533–550.
7. Hochstein, S. & R. M. Shapley. 1976. Linear and nonlinear subunits in Y cat retinal ganglion cells. J. Physiol. **262:** 265–284.
8. Victor, J. D. & R. M. Shapley. 1979. The nonlinear pathway of Y ganglion cells in the cat retina. J. Gen. Physiol. **74:** 671–689.
9. Shapley, R. M., E. Kaplan & R. E. Soodak. 1981. Spatial summation and contrast sensitivity of X and Y cells in the lateral geniculate nucleus of the macaque. Nature **292:** 543–545.
10. Wiesel, T. N. & D. H. Hubel. 1966. Spatial and chromatic interactions in the lateral geniculate body of the rhesus monkey. J. Neurophysiol. **29:** 1115–1156.

Discussion of the Paper

O. Braddick: Dr. Shapley, do you know if the difference you find through the magno- and parvocellular layers in the LGN is reflected in a difference in the optic nerve input to those layers?

R. Shapley: Yes. It happens that the LGN is almost an electrophysiologist's dream, and it is possible to record from a geniculate cell and simultaneously pick up extracellularly evoked synaptic potentials which are closely tied to the optic axon and optic tract, and so one can look simultaneously at the retinal input and the geniculate cell activity. And in all cases that we have studied, the geniculate cell is certainly in its main excitatory drive a reflection of its retinal input. For instance, the wave length discrimination property, the contrast sensitivity properties, the spatial resolution, and so on, are all closely tied to the retinal input. This is true both in the cat and in the monkey.

Question: Did you say that with low contrast stimuli we could expect a better response from the magnocellular or the parvocellular layer?

Shapley: Magnocellular. Because its cells have higher contrast sensitivity, at lower contrast they would be the ones to predominate.

THE RELATIONSHIP OF VISUAL EVOKED POTENTIALS TO CORTICAL PHYSIOLOGY

Ken Nakayama

Smith-Kettlewell Institute of Visual Sciences San Francisco, California 94115

The last 15 years have seen a rapid expansion in studies on human visual evoked potentials (VEP), both at the fundamental and at the clinical level. A wide variety of stimulus conditions have been explored,[1] new types of mathematical tools have been introduced,[2,3] and there has been an increasing appreciation of possible VEP correlates with underlying physiological mechanisms.[4] Variability of the folding and size of the human striate cortex[5] as well as the remote location of human scalp electrodes, however, limits the accurate localization of human VEP components to particular cortical structures. In this regard it would seem desirable to explore the VEP of a laboratory animal closely related to the human.

In this paper I discuss two approaches that may prove both useful and complementary in deciphering visual evoked potentials in monkey, and by extrapolation, in man. The first approach involves the use of fine stimulus control, noting the degree to which the VEP amplitude varies with small changes in stimulus characteristics. The second approach, still in the preliminary stages, involves the use of current source density analysis, a method to localize the generating processes of the VEP to specific cortical laminae and sublaminae.

STIMULUS CONTROL OF SPATIAL AND TEMPORAL FREQUENCY

The ability of the human observer to see variations in brightness, either in time or in space, has been measured psychophysically by obtaining the temporal or the spatial modulation sensitivity functions, respectively. In time, this involves the plotting of the reciprocal of threshold modulation (the sensitivity) required to see flicker at various temporal frequencies.[6] In space, it involves the measurement of sensitivity for different spatial frequencies of sine wave gratings. The curve relating contrast sensitivity to spatial frequency is a broad unimodal one (the dashed line in FIGURE 1), peaking around 5 cycles/deg.[7] Rather than being determined by a single mechanism, there is good evidence to suggest that this contrast sensitivity function reflects the upper envelope of sensitivities of many different subsystems, each tuned to a much more narrow range of spatial frequencies.

Psychophysical evidence indicates that one can adapt a spatial frequency "channel," having a tuning function approximating a bandwidth of about 1 octave, much more narrow than the overall contrast sensitivity function.[8] Electrophysiological evidence provides a biological substrate for such channels. Cortical cells have a narrow range of preferred spatial frequencies, much narrower than that reflected by the psychophysical contrast sensitivity function. The center frequency of each cell can

*This work was supported in part by Grants 5RO1 EY-01582, 5RO1 EY-02124, 5P30 EY-01186, and 2SO RR 05566 from the National Institutes of Health and by the Smith-Kettlewell Eye Research Foundation.

0077-8923/82/0388-0021 $1.75/0 © 1982, NYAS

vary over quite a range. FIGURE 1 shows the relationship between the tuning function of individual cortical cells as obtained by DeValois, *et al.*[9] in relation to the psychophysical curve. The range of spatial frequency tunings of individual cortical cells is consistent with the overall sensitivity curve obtained psychophysically.

What should be expected if we were to record from the surface of the skull of the human or monkey, plotting VEP amplitude as a function of spatial frequency? If there were many narrowly tuned mechanisms, each tuned to a slightly different spatial frequency and if each mechanism or cell type could be recorded with approximately equal weighting on the surface of the scalp, then one might expect that the contrast evoked potential results would mirror those obtained by psychophysics.

In fact, the first systematic investigation relating VEP to spatial frequency showed such a relation.[10] By presenting counterphase-modulated sinusoidal gratings to human observers (essentially equivalent to exchanging the bright and dark bars of the grating) at a reversal rate of 16 Hz, Campbell and Maffei noted a linear relationship between VEP amplitude and log contrast of the grating. Extrapolating this linear function to zero μV, they obtained a VEP estimate of the contrast threshold, one that appeared rather close to that obtained by psychophysics. These results suggested that the VEP indeed reflects the psychophysics and that by making VEP measurements one could hope to tap the envelope of activities of highly representative sets of visual cortical neurons. As such, the VEP could be considered as an alternative technique to obtain results that could be obtained psychophysically.

Upon closer examination, using much more resolution and over a wider range of conditions, however, we have shown that the human steady-state VEP has perhaps the opposite characteristic.[11] Rather than being broadly tuned to spatial frequency, it was generally very sharply tuned (FIGURE 2) and with great reproducibility. Furthermore, these spatial frequency tuning functions were highly dependent on temporal frequen-

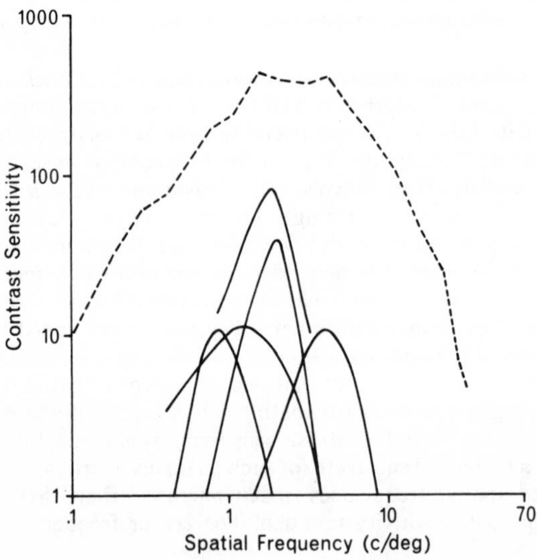

FIGURE 1. Upper dashed curve is the human psychophysical contrast sensitivity function. Sensitivity (defined as the reciprocal of theshold) is plotted as a function of spatial frequency for vertical sine wave gratings. (After Campbell and Robson.[7]) Lower solid curves represent contrast sensitivity functions obtained from visual cortex single units. (After DeValois *et al.*[9])

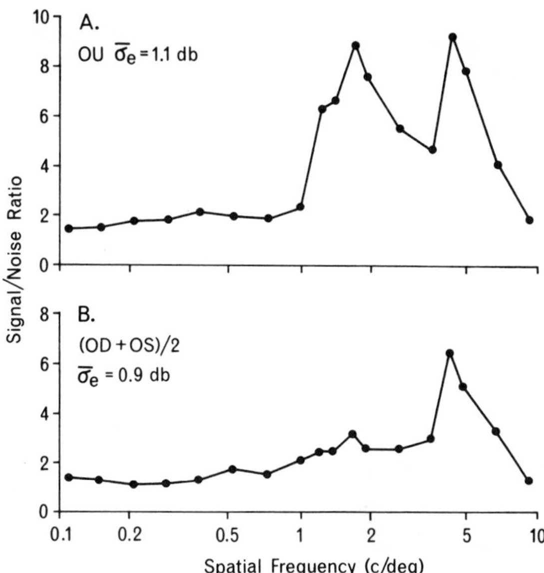

FIGURE 2. (A) Human VEP amplitude versus spatial frequency for binocularly presented gratings. Counterphase reversal rate and recording frequency was 30 Hz. Contrast was 68%. (B) Same as above except for monocular viewing. (After Apkarian, Nakayama & Tyler.[29])

cy. Thus, rather than speaking of a simple spatial or temporal frequency function, it is more accurate to speak of a multiple peaked spatial/temporal frequency surface (FIGURE 3).

Steady-state surface VEPs recorded in alert cats and alert monkeys revealed a qualitatively similar picture.[12,13] For example, in the alert monkey, trained to fixate a spot to receive a water reward, the spatial and temporal frequency tuning was as sharp or sharper than that seen for humans (FIGURE 4). Note that in this figure, the amplitude versus spatial frequency tuning function is a very sharply peaked one, centered on 10 cycles/deg, and that, for at least 4 electrode placements, there is essentially no response below about 3 cycles/deg. Thus, for surface recording from this monkey, the spatial frequency tuning function is as narrow as that seen for individual cortical neurons in the same species. They compare well with the results of DeValois *et al.*,[9] as seen in FIGURE 1 and more directly with the results of Schiller *et al.*,[14] who plotted spike counts as a function of spatial frequency. Tuning in the temporal frequency domain was also very sharp for the monkey VEP. FIGURE 5 shows the relation between temporal frequency and the VEP amplitude for 4 spatial frequencies, also showing the progressive variation of phase lag with increasing temporal frequency.

The existence of very narrow spatial frequency tuning in the VEP raises a wide range of questions. Why was it not seen by Campbell and Maffei[10]? Is it the only response that can be obtained? What does it mean in terms of the genesis of these steady-state visually evoked potentials?

Campbell and Maffei's results were restricted to very few electrode placements and the results of only one temporal frequency were reported. Our own results indicate that results like that found by Campbell and Maffei can be obtained both in the

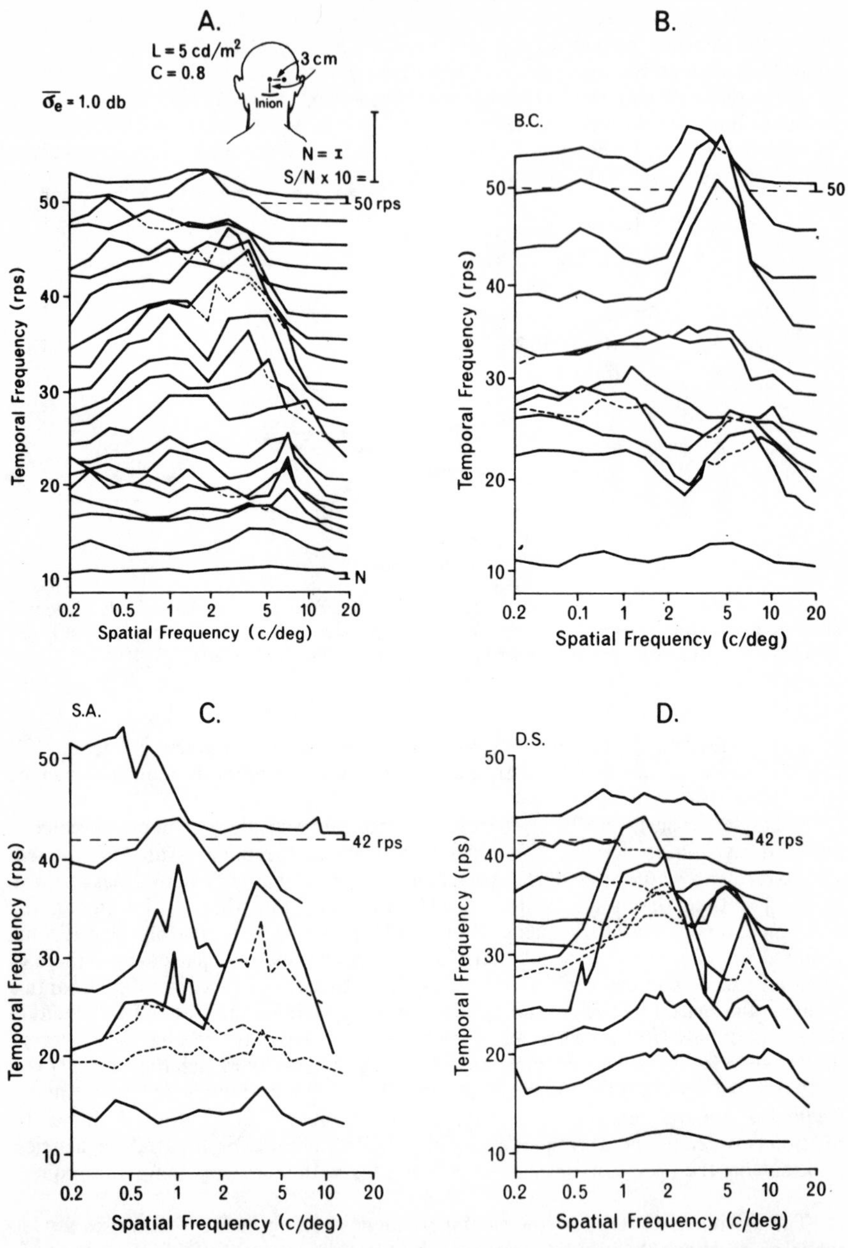

FIGURE 3. Spatial frequency tuning as a function of temporal frequency (rps) for four different human observers (A–D). Each curve is plotted from a baseline at its frequency specified on the ordinate with VEP amplitude plotted as an EEG signal/noise ratio at the recording frequency. Note the high degree of stimulus specificity in both the temporal and the spatial frequency domains. Although the details differ for each observer at the same electrode position, multiple narrow spatial frequency tuning is prevalent. (After Tyler, Apkarian & Nakayama.[11])

human and in the monkey, but they are far from the rule. For example, FIGURE 6A shows the contrast responses of a single electrode (electrode 2 seen in FIGURE 4), associated with an extrapolated VEP sensitivity curve for the monkey. First it should be noted that although the contrast functions are linear over quite a range, they often have two limbs, a relatively shallow section at low contrast and a steeper sloped section at higher contrast. Either function can occur in combination or alone (see below). Although the steep high contrast function has been reported previously,[10] it has

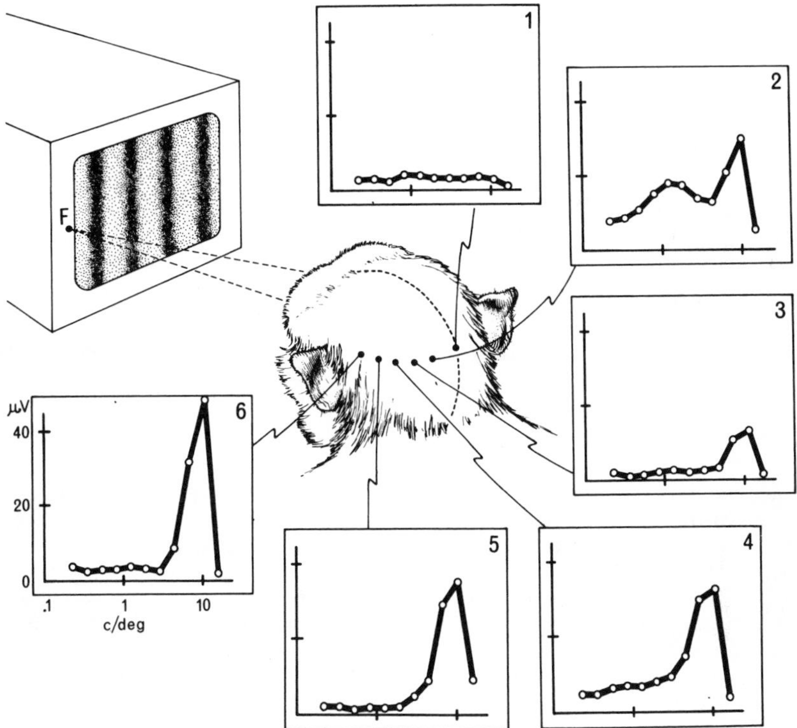

FIGURE 4. Spatial frequency tuning of the steady-state VEP in the alert primate, trained to fixate on a spot (point F). Responses are obtained from 6 epidural electrode placements, localized along the occipital region of monkey cortex. Foveal representation of the striate cortex is represented in the more anterior placements. Note the high degree of spatial frequency tuning for nearly all of the electrode channels. Contrast is 33%. Counterphase reversal rate and recording frequency are maintained at 16 Hz. (From Nakayama, Mackeben & Sutter.[13])

elicited little interest. Its existence, however, indicates a separate pattern sensitive mechanism as yet to be revealed by single unit or psychophysical techniques.

Extrapolating the lower contrast threshold curve to zero μV (arrows in FIGURE 6A) yields a set of extrapolated thresholds (dots) which can be compared with the monkey psychophysical data. This can be seen in FIGURE 6B. The fit of the extrapolated monkey VEP to monkey psychophysics is rather good, both in the shape and the absolute magnitude of the sensitivity curves.

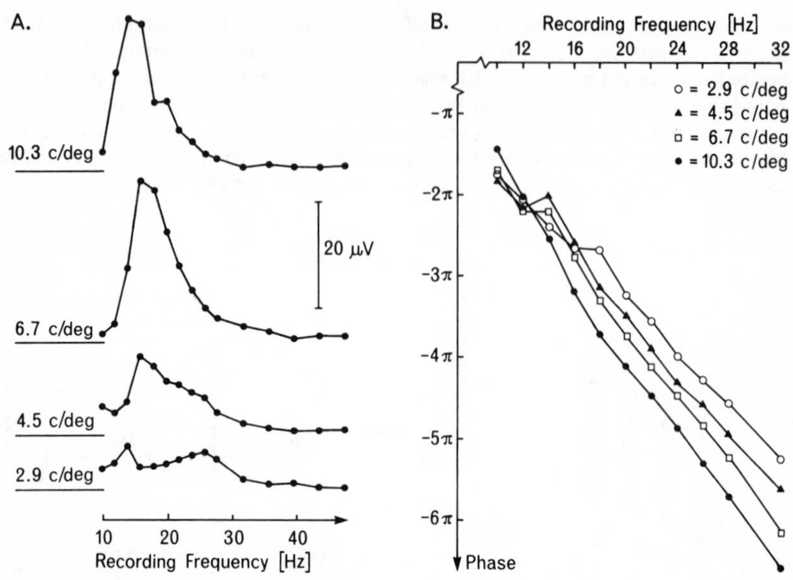

FIGURE 5. (A) Temporal frequency tuning for 4 spatial frequencies in an alert monkey obtained from electrode 6 (as shown in FIGURE 4). (B) Corresponding phase lag at different temporal frequencies. (From Nakayama, Mackeben & Sutter.[13])

This is a rare example, however, of such a close match. Under other circumstances the agreement is much worse. Note that in FIGURE 7, the extrapolated VEP curve bears no similarity to the psychophysics. The reason for this poor fit is important to note. It does not fit because at electrode 3, 4, 5, and 6, (as depicted in FIGURE 4) there is no low-frequency contrast portion of the VEP versus log contrast function, leaving only the steep higher contrast curve. Extrapolating this remaining steep higher contrast to zero μV leads to a high VEP threshold which does not fit the psychophysics. Thus, it should be of no surprise that the VEP versus spatial frequency curve will show some large departures from the psychophysics. The fit will depend heavily on which contrast function is present and obviously on the contrast that is used to obtain the spatial frequency function. FIGURE 6C (bottom right panel), for example, shows the change in spatial frequency tuning obtained from a single electrode placement for 3 different contrast levels. As contrast is increased, one reaches the steeper higher contrast portion of just some of the functions shown on the left of FIGURE 6, and, as a consequence, the spatial frequency tuning is much more sharply tuned.

Before suggesting a neurophysiological framework to interpret the above results, one needs to establish that such a framework is indeed necessary; that the responses, for example, are not due to an electrical artifact, namely, the cancellation of opposing polarity signals from more broadly tuned mechanisms. This purely electrical interpretation seems unlikely for several reasons. First, the temporal phases of the two peaks in the human spatial frequency response were often very close and there was no indication that many of the multiple peaks could be explained by opposite polarity responses canceling.[11] Second, under some conditions individual peaks could disappear almost completely without affecting other peaks. This was especially prominent for some cases of monocular versus binocular stimulation as shown in FIGURE 2. Finally, the monkey results often showed only one very sharp spatial frequency peak (as seen in FIGURE 4) at relatively high spatial frequencies. Thus, there was no second mechanism that could electrically cancel to result in the narrowly tuned peak.

The similarity between the steady-state pattern VEP seen in the human and that in the monkey is noteworthy, especially in light of the very large differences in the gross layout of the striate and parastriate cortex in each species. In humans the

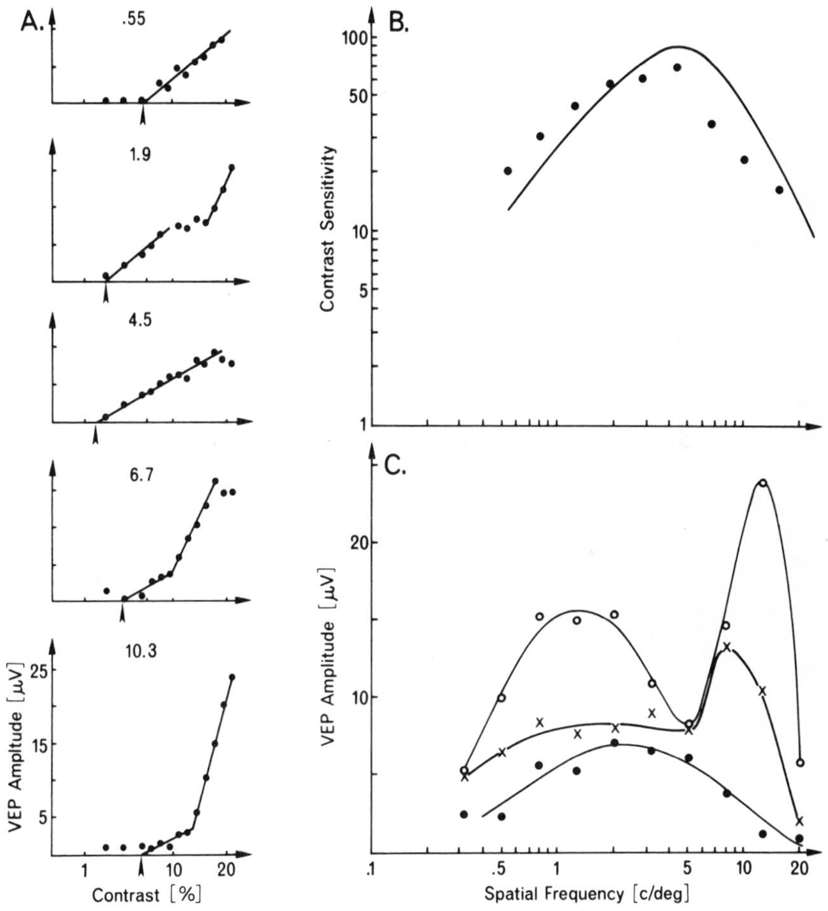

FIGURE 6. (A) Monkey VEP amplitude (plotted on a linear scale) vs grating contrast (plotted on a logarithmic scale) for 5 different spatial frequencies as noted by numerals above each function. Arrows denote the position of the extrapolated VEP threshold. (B) Contrast sensitivity of the VEP as derived from VEP thresholds (solid dots). VEP recordings were obtained from electrode 2 as shown in FIGURE 4 (from Nakayama *et al.*[31]). As a comparison, corresponding monkey psychophysical data is presented as the smoothed solid curve (from DeValois *et al.*[32]). (C) VEP amplitude vs spatial frequency function for 3 different contrasts (dots = 9%; crosses = 18%; open circles = 33%). (From Nakayama & Mackeben.[33])

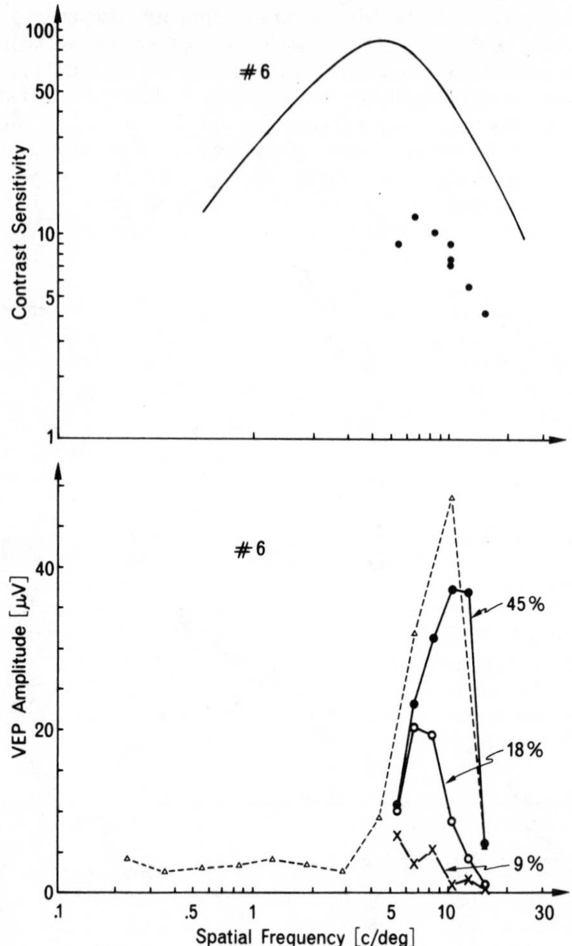

FIGURE 7. Same conditions as shown in FIGURE 6 except for a different electrode in the same monkey. Upper graph: VEP contrast sensitivity (dots) as compared with monkey psychophysical results (solid line). Note the very poor match of VEP and psychophysics. Lower graphs show the VEP amplitude vs spatial frequency for 3 levels of contrast (From Nakayama & Mackeben.[33])

primary cortex is mostly buried within the mesial walls of the posterior lobe and within the calcarine fissure. In contrast, the secondary cortex is exposed. In the macaque, it is almost the reverse, with much of the striate cortex exposed over the posterior cortical region and the parastriate fields buried.[15] The similarities of the steady-state pattern VEP in both man and monkey, especially in light of these gross anatomical differences, indicates that it must be the microanatomical or microphysiological aspects of the cortex that lead to this degree of VEP specificity.

Although many interpretations are possible at this microanatomical level, one broad conclusion seems indicated, namely, that the surface VEP records only a fraction of the cortical neurons which might be thought to respond to the stimulus and which could be expected to be involved in perception. For example, consider the human curves seen in FIGURES 2 and 3 and the monkey data seen in FIGURE 4. For these electrode placements, no VEP is evident over at least some range of spatial/temporal frequencies that are visible to the person or the animal. One hypothesis to explain such a result is that of neural "resonance."[11] This hypothesis suggests that the repetitive nature of the stimulus required to obtain a steady-state VEP brings out narrow tuning characteristics in certain sets of cortical neurons. For example, neurons having certain spatial characteristics (namely, a given receptive field size) might also share similar temporal characteristics, similar time constants in feedback loops, cable constants, and so on. As such, they may be extremely selective to the temporal frequency of the stimulus. By using the term "resonance" rather loosely, it is not assumed that it is identical to the concept of resonance used in linear theory. In fact, many of the narrow temporal peaks are likely to be due to nonlinear processes. Before considering this hypothesis in further detail, however, it would be best to consider the results of another relevant neurophysiological technique, that provided by the measurement of the VEP at different cortical depths.

CURRENT SOURCE DENSITY ANALYSIS

A second possible approach in the search for a physiological interpretation of evoked potentials is that of current source density analysis, a technique which has had a long history,[16] but which has only recently become applicable to the VEP. This endeavor has been furthered most recently by a pair of tutorial papers[17,18] followed by an important set of papers elucidating the electrically elicited field potentials obtained in various visual structures by Mitzdorf and Singer.[19–21]

Because the mathematical reasoning, the assumptions, and the description of the CSD technique are well described in the above papers, the reader is urged to consult them for details. In essence, the approach can be seen as having several aspects. First is the application of electromagnetic theory to the field potential distribution in neural tissue. If one assumes homogeneity and horizontal translation symmetry in two dimensions, a one-dimensional current source/sink distribution can be obtained by estimating the spatial second derivative of potential with respect to cortical depth. As such, this first part of the endeavor is an exercise in applied physics and is relatively straightforward. It estimates the net sink or source contribution for particular cortical depths. A more difficult problem is to provide an interpretation of the sink/source distributions in terms of neuronal activity. This latter task is not straightforward, and, as has been pointed out by others,[22] it requires a wealth of additional knowledge, including the detailed anatomy and physiology of cortical tissue.

The measurement of field potentials at different depths of cortex can be obtained in several ways. Early experiments advanced the electrode in small increments to the cortex, measuring the field potentials at successive depths. More recently, many laboratories are beginning to use multiple electrode arrays, either using metal electrodes or multiple pipettes, as reported by Vaughan in this volume.[23] The primary work in visual structures was collected by using electrical stimulation of the afferent tracks.[19–21] Of significance is the fact that an X, Y system as inferred from conduction velocities could be segregated at the level of cortex in cat and monkey. Furthermore, new X, Y subdivisions, not previously described, were noted in the LGN of cat[19] and now confirmed by intracellular HRP injection techniques.[24] Of particular importance to those interested in the interpretation of the surface VEP is the relevance of the "open" versus "closed" field current generators (as originally described by Lorente de No.[25]) It determines whether a given set of synaptic currents in the depths will be visible as a surface VEP. For example, electrical stimulation of area-18 afferents in the cat leads to a prominent surface positivity which is the result of sinks in layer IV and adjacent sources above.[20] This dipolar sheet configuration generates large potentials even as far away as the cortical surface. With the electrically evoked activity in the monkey striate cortex, however, it appears that the potential deflections seen at the cortical surface is but a pale reflection of the voltages seen in the depths.[21] Current source density analysis of these monkey field potentials suggests a reason for this difference, showing the predominance of "closed field" sink/source distributions in the striate cortex of the monkey, at least for electrical stimulation. Closed field source/sink distributions are not dipolar, but are "sandwich" like, showing a sink bounded by sources above *and* below. As a result, these currents do not generate appreciable potential deflections at distant sites.

Our preliminary studies of the monkey striate cortex in response to *visual* rather than electrical stimulation[26] reveal additional features. In brief, the recording electrode consisted of a set of 16 insulated stainless steel wires (50 μm in diameter) arranged in a parallel array. Ends are cut so as to provide a vertical contact separation of 150 μm between adjacent recording poles. With our electronically matched 8-channel amplifier system, every other contact could be recorded simultaneously and the stimulus was then repeated for the recording from the 8 other channels. Because the differentiation grid used for the current source density calculations was 300 μm, all CSDs were calculated from field potentials obtained simultaneously, removing one of the most important sources of extraneous variability.

In FIGURE 8 we see the field potentials obtained for two intensities of stroboscopic flash for different cortical depths. FIGURE 9 shows the corresponding CSD profiles. Several aspects should be noted. First is the fact that the predominantly positive surface potential is very small compared to the large negative potentials seen in the depths. Note the degree to which the sinks and sources are spatially localized by the CSD analysis and the fact that there is great consistency of this sink source distribution when comparing high versus low intensity flash (compare the right versus left column in FIGURE 9). In both cases, the earliest response is a set of two prominent sinks, the lowest most probably in layer IV (see later), each bordered by a prominent source above. This is then followed by a sequence of current source densities of opposite polarity at the same depth. Small though consistent indications of an early and later sink can also be seen in the bottom CSD tracings, possibly reflecting excitatory activity of layer VI afferents. The main difference between the two intensities of flash is the duration and latency of the deflections. Although the response of the dimmer flash is smaller and slower, it has the same spatial structure. Thus the neuronal elements activated by dim flash appear as nearly identical to those activated by the higher intensity flash.

Two general features of the data deserve further emphasis. First is the fact that in comparison to the depths, the potentials seen at the surface are very small. Second is the rather complex, and yet reliable, pattern of sources and sinks that can be derived from these potentials. This indicates the existence of a "closed field" situation mentioned earlier although it does not prove that the individual neurons themselves have closed field characteristics. In particular, the synaptic currents of several neuronal classes might summate to generate a net cancellation of current when seen from the surface. These results indicate that at least for strobe flash in monkey striate cortex, the surface evoked potential is a poor indicator of cortical activity immediately below the electrode.

So far we have confined ourselves to the localization of net transmembrane current sources and sinks from a CSD analysis of cortical field potentials. This leaves unanswered the question as to the physiological interpretation of such CSD profiles. To start, Mitzdorf and Singer have made a bold hypothesis.[19] They argue that

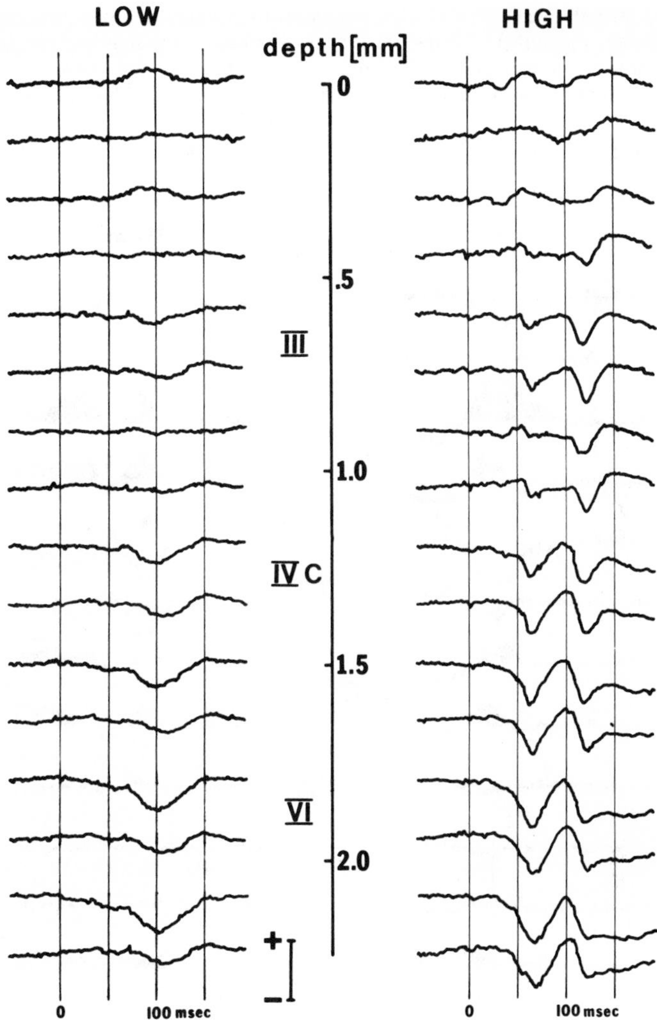

FIGURE 8. Transient VEP profiles in monkey striate cortex for 2 intensities of strobe flash. Potential functions were obtained for 16 depths in cortex spaced at 150 μm. Each time-averaged response is the mean of 64 sweeps. Roman numerals represent the estimated location of key cortical laminae. Voltage calibration is 100 μV. Flash begins at vertical line marked by zero. Distance between time markers is 50 msec. Animal is sedated with light Nembutal anesthesia. (From Mackeben, Mitzdorf & Nakayama.[26])

essentially all electrically elicited field potentials in the visual system, including surface potentials, are reflections of excitatory synaptic currents. Thus the distribution of sinks indicates the relative strength of excitatory synapses whereas the distribution of sources primarily reflects the associated outward loop closing currents adjacent to these excitatory synapses. As such, inhibition in the form of IPSPs makes a negligible contribution of field potentials and associated CSDs in the visual cortex.

In brief, Mitzdorf and Singer's argument runs as follows:[19] Extracellular voltage differences are due to extracellular current flow, which in turn originates from transmembrane current flow. Most current flow in neural tissue is capacitative. Thus, transmembrane current is roughly proportional to the first time derivative of membrane potential. Because the slope of the rising phase of EPSPs of visual neurons is about ten times greater than for IPSPs, EPSPs contribute much more to the extracellular current flow. This difference is further exaggerated by the fact that the

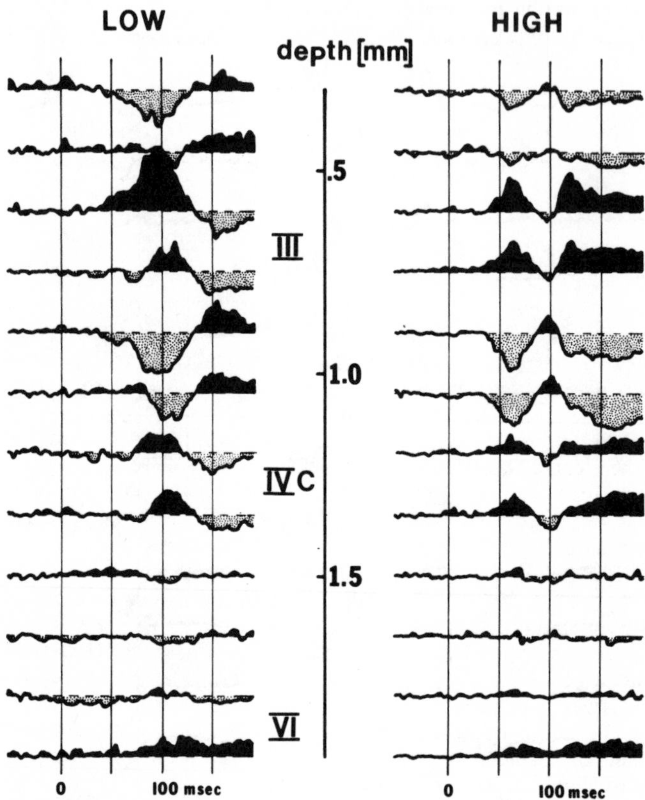

FIGURE 9. Corresponding current source density profiles obtained from the monkey field potentials presented in FIGURE 8. Current source density is plotted as a function of time for 14 depths in cortex. Sinks (net inward transmembrane current density) are depicted by upward deflections and are identified as solid areas. Sources (net outward transmembrane current density) are represented by downward deflections and are filled in as dotted areas. The magnification of the low-intensity flash CSD profiles is twice that as pictured for the higher-intensity flash. (From Mackeben, Mitzdorf & Nakayama.[26])

slope of the rising part of the EPSP is likely to be underestimated by conventional intracellular recording. Excitatory synapses are generally thought to reside mainly on the dendrites of cells (for example, on dendritic spines), whereas inhibitory synapses are closer to the somata. Therefore, the waveforms of EPSPs will show disproportionate low-pass filtering at the somatic intracellular recording site and are even faster relative to IPSPs then intracellular records might indicate. In addition to this biophysical reasoning, preliminary pharmacological results with nembutal and picrotoxin (presumed agonists and antagonists of cortical inhibition) appear consistent with this conclusion.[27]

Assuming such a hypothesis regarding the origins of field potentials, Mitzdorf and Singer[19–21] were able to trace out a plausible flow of excitation through the laminae of the cat and monkey cortex with an interpretation that is consistent with many aspects of anatomy. As such, it lends further credence to their assumption regarding the centrality of excitation in generating visual field potentials. Because of its obvious importance in the interpretation of VEP origins, this hypothesis deserves greater attention and demands more experimental support. In particular, it needs to be seen whether it can be generalized to visual stimulation in addition to electrical stimulation.

Concluding Discussion

So far, I have described two approaches in the study of the VEP. Our long-range objective is to combine both approaches: to use fine stimulus control to isolate different cortical mechanisms and then to make current source density measurements to determine the anatomical location of such mechanisms. Although we have not reached this goal as yet, observations with each technique permit some tentative conclusions.

First is the finding that although there may be little in the way of a surface potential, there can be large and complex patterns of synaptic currents in the depths. The converse can also be true, namely, that a very simple set of sink/source distributions (especially if it has an "open field" configuration) can lead to large potentials on the surface.[20]

Given these considerations regarding the depth distribution of synaptic currents and given the "neural resonance" hypothesis mentioned earlier, one can begin to see how the steady-state VEP could be so narrowly tuned with respect to spatial and temporal frequency. Narrow spatial frequency tuning might result from a restricted set of cells which have "open field" dipole characteristics and which are also particularly sensitive to a small range of temporal and spatial frequencies.

If true, the above hypothesis presents itself as either a set of problems to be overcome and/or opportunities to be exploited. It is a problem if we wish to use the visual evoked potential as a simple one-to-one substitution for psychophysics. For example, if we wished to obtain the full contrast sensitivity function from a human observer, a single VEP amplitude versus spatial frequency function would be too biased. We could overcome this problem, however, by recording from several areas of the human scalp, and over a range of temporal frequencies, thereby obtaining an envelope of many curves which would more closely approximate the psychophysics. In fact, if we were interested in only one aspect of the contrast sensitivity curve, the upper acuity limit, for example, the technique of obtaining only one curve at one temporal frequency and at one electrode location (if properly chosen) is surprisingly good.[28] As such, the selective nature of VEP recording can pose problems but they are surmountable.

What can be said for the specific opportunity afforded by VEP selectivity? Because the spatial frequency tuning of the gross VEP recorded from the scalp can be as sharp as the tuning seen for single neurons, we are presented with the tantalizing possibility that these spatial and temporal frequency peaks represent discrete neural entities sharing the same specific trigger features as well as a common neuronal geometry and anatomical classification.

Easy isolation of such "entities" for a given observer at a given electrode site does not come without some additional effort or without some luck, however. Field potentials, in contrast to single unit recordings, have the property of long distance propagation and summation. Thus, at any electrode site one can record voltages from anywhere else in the brain. Therefore, it is highly likely that, unless specific care is taken, many mechanisms will be "visible" and it will be difficult to isolate each hypothetical mechanism if the overlap is too great. In favorable cases (as shown in FIGURE 2) it appears that the isolation is rather good, but in other cases it is not.[29] One possible approach to achieve a greater isolation in humans is to supplement the fine stimulus control afforded by spatial and temporal frequencies with an additional type of stimulus control, namely, by carefully restricting the area of retinal stimulation, as reported by Tyler and Apkarian in this volume.[30] Another approach is to combine the use of fine stimulus control with the technique of current source density analysis in a suitable animal model. With such experiments we hope to begin the process of correlating specific surface potentials to the activity of identifiable cortical laminae and to particular cell groups.

REFERENCES

1. DESMEDT, J. E., Ed. 1977. Visual Evoked Potentials in Man: New Developments. Clarendon Press, Oxford.
2. SPEKREIJSE, H. 1966. Analysis of EEG Responses in Man. Thesis. Univ. of Amsterdam, The Hague, Junk, The Netherlands.
3. RATLIFF, F., J. D. VICTOR & R. M. SHAPLEY. 1978. Nonlinear analysis of visual evoked potentials in the human. J. Opt. Soc. Am. **68:** 1427.
4. KULIKOWSKI, J. 1977. Separation of occipital potentials related to the detection of pattern and movement. *In* Visual Evoked Potentials in Man: New Developments. J. E. Desmedt, Ed. Clarendon, Oxford.
5. STENSAAS, S., D. K. EDDINGTON & W. H. DOBELLE. 1974. The topography and variability of the primary visual cortex in man. J. Neurosurgery **40:** 747–755.
6. DELANGE, H. 1952. Experiments on flicker and some calculations on an electrical analogue of the foveal systems. Physica **18:** 389–398.
7. CAMPBELL, F. W. & J. G. ROBSON. 1968. Application of Fourier analysis to the visibility of gratings. J. Physiol. **197:** 551–566.
8. BLAKEMORE, C. & F. W. CAMPBELL. 1969. On the existence of neurons in the visual system selectively sensitive to the orientation and size of retinal images. J. Physiol. **203:** 237–260.
9. DEVALOIS, K., D. ALBRECHT & L. G. THORELL. 1978. Cortical cells: Bar and edge detectors or spatial frequency filters? *In* Frontiers of Visual Science. S. Cool, Ed. Springer Verlag, New York, N.Y.
10. CAMPBELL, F. W. & L. MAFFEI. 1970. Electrophysiological evidence for the existence of orientation and size detectors in the human visual system. J. Physiol. **207:** 635–652.
11. TYLER, C. W., P. APKARIAN & K. NAKAYAMA. 1978. Multiple spatial frequency tuning of electrical responses from the human visual cortex. Exp. Brain Res. **33:** 535–550.
12. APKARIAN, P. & J. ADAMS. 1980. Spatio-temporal frequency tuning in alert cat VEP. Invest. Ophthal. Vis. Sci. (Suppl.): 277.
13. NAKAYAMA, K., M. MACKEBEN & E. SUTTER. 1980. Narrow spatial and temporal frequency tuning in the alert monkey VEP. Brain Res. **193:** 263–267.
14. SCHILLER, P. H., B. L. FINLAY & S. F. VOLMAN. 1976. Quantitative studies of single-cell properties in monkey striate cortex. III. Spatial frequency. J. Neurophysiol. **39:** 1334–1351.

15. VAN ESSEN, D. C. & S. M. ZEKI. 1978. The topographic organization of rhesus monkey prestriate cortex. J. Physiol. **277:** 193–226.
16. PITTS, W. 1952. Investigations on synaptic transmission. *In* Cybernetics Trans. 9th Conf. H. von Foerster, Ed., pp. 159–163. Josiah Macy Foundation, New York, N.Y.
17. NICHOLSON, C. & J. FREEMAN. 1975. Theory of current source density analysis and determination of conductivity tensor for anuran cerebellum. J. Neurophysiol. **38:** 356–368.
18. FREEMAN, J. A. & C. NICHOLSON. 1975. Experimental optimization of current source density techniques for anuran cerebellum. J. Neurophysiol. **38:** 369–382.
19. MITZDORF, U. & W. SINGER. 1977. Laminar segregation of afferents to lateral geniculate nucleus of the cat. An analysis of current source density. J. Neurophysiol. **40:** 1227–1244.
20. MITZDORF, U. & W. SINGER. 1978. Prominent excitatory pathways in the cat visual cortex (A17 and A18): A current source density analysis of electrically evoked potentials. Exp. Brain Res. **33:**371–394.
21. MITZDORF, U. & W. SINGER. 1979. Excitatory synaptic ensemble properties in the visual cortex of the macaque monkey: A current source density analysis of electrically evoked potentials. J. Comp. Neurol. **187:** 71–84.
22. NICHOLSON, C. & R. LLINAS. 1975. Field potentials in the alligator cerebellum and theory of their relationship to Purkinje cell dendritic spikes. J. Neurophysiol. **38:** 356–368.
23. VAUGHAN, H. G. 1982. Ann. N. Y. Acad. Sci. This volume.
24. BOWLING, D. & C. MICHAEL. 1980. Projection patterns of single physiologically characterized optic tract fibers in the cat. Nature **286:** 899–902.
25. LORENTE DE NO, R. 1947. A study of nerve physiology, Part 2. Rockefeller Inst. Med. Res. **132.**
26. MACKEBEN, M., U. MITZDORF & K. NAKAYAMA. (Unpublished observations).
27. MITZDORF. U. (Personal communication).
28. TYLER, C. W., P. APKARIAN, D. LEVI & K. NAKAYAMA. 1979. Rapid assessment of visual function: An electronic sweep technique for the pattern visual evoked potential. Invest. Ophthal. **18:** 703–713.
29. APKARIAN, P., K. NAKAYAMA & C. W. TYLER. 1981. Binocularity in the human visual evoked potential: Facilitation, summation, and suppression. EEG Clin. Neurophysiol. **51:** 32–48.
30. TYLER, C. W. & P. APKARIAN. 1982. Ann. N.Y. Acad. Sci. (this volume).
31. NAKAYAMA, K., P. APKARIAN & C. W. TYLER. 1982. Human visual evoked potentials. Isolation of cortical populations tuned to spatial frequency. *In* Neurophysiology and Psychology. E. Donchin, Ed. Academic Press, New York, N.Y. (in press).
32. DEVALOIS, R. L., H. MORGAN & D. M. SNODDERLY. 1974. Psychophysical studies of monkey vision. III. Spatial luminance contrast sensitivity tests of macaque and human observers. Vis. Res. **14:** 75–81.
33. NAKAYAMA, K. & M. MACKEBEN. Steady state visual evoked potentials in the alert primate Vis. Res. (submitted).

DISCUSSION OF THE PAPER

I. BODIS-WOLLNER: You have shown that VEP amplitude as a function of contrast is a complicated function. It seems to me that it may represent not one, but two or even more mechanisms, such as different ganglion cell type responses as they are reflected at the cortex. Along this line of thinking, have you attempted to relate your data to the mechanism Dr. Shapley described in the LGN? It occurs to me that he also described a "speeding up of the Y ganglion cell responses of the retina as a function of contrast" which leads me to the question of whether or not you considered VEP phase versus contrast curves.

K. NAKAYAMA: We have done that, and we do see a speeding up right off. At the moment though, I do not know quantitatively how much the response speeds up.

BINOCULAR NEURONS AND CYCLOPEAN VISUALLY EVOKED POTENTIALS IN MONKEY AND MAN

Bela Julesz and Walter Kropfl

Bell Laboratories
Murray Hill, New Jersey 07974

INTRODUCTION

Interest in visually evoked potentials (VEP) elicited by dynamic random-dot stereograms (RDS) and correlograms (RDC), also called cyclopean stimuli,[1] is twofold. First, it permits study of the nature of VEP under the purest conditions, since cyclopean stimuli operationally bypass several early processing stages in the retinal-thalamic-cortical pathway. Second, regardless of how cyclopean stimuli elicit robust VEP, they provide an unfakeable test for functional binocularity or stereopsis in animals, human infants, and patients with optical pathway defects in the absence of verbal communication. This article tries to give a brief view of both of these diverse aspects.

LINKING BETWEEN NEUROPHYSIOLOGY AND PSYCHOLOGY

One basic problem of brain research is the linking of neurophysiological findings with psychologically observed phenomena. In most cases this linking is indirect and speculative. In the case of global stereopsis using random-dot stereograms (RDS), at least one difficulty is eliminated. Since for RDS the binocular correlation cannot exist prior to the cyclopean retina,[1] one can safely assume that the cyclopean depth phenomena must be the result of cortical neuron pools selectively tuned to certain binocular disparities. Indeed, all neuroanatomical evidence indicates that in the monkey's visual system the first stage where the left and right optical pathways combine is in Area 17 of the striate cortex. So the cyclopean retina in primates and man must be in Area 17 or beyond.

Indirect evidence for binocular-disparity-detecting neuron pools has been provided by Blakemore and Julesz,[2] who could produce a shift in perceived depth by prolonged adaptation to a static random-dot stereogram.

Dramatic direct neurophysiological evidence for neural detectors that participate in global stereopsis comes from Gian Poggio.[3] He used the technique of dynamic RDS,[1] by portraying cyclopean bars. In spite of the fact that such cyclopean bars do not differ in contrast (only in binocular disparity) from their surrounding areas, several neurons were found that fired for such stimuli. This exemplifies the neurological end of the linking hypothesis. Here cyclopean neural detectors were found that fired optimally for elongated cyclopean bars of specific orientations. Psychological studies indicate that global stereopsis detectors might be insensitive to orientation of textural blobs.[4,5] Obviously, global stereopsis must be a complex process whose first processing stage might be as early as Area 17, but will be further processed by higher stages. For instance, Cowey and Porter[6] reported a loss of global stereopsis in medial-temporal-cortex-ablated monkeys without impairment of local stereopsis (e.g., stereoacuity for single bars).

In these psychology–neurophysiology linking hypotheses, VEP research has an

0077-8923/82/0388-0037 $1.75/0 © 1982, NYAS

important role. After all, VEP is regarded as an indicator of the activity of cortical neuron pools.

VEP and Cyclopean Research

Among the many visual stimuli that can elicit VEP, dynamic random-dot stereograms (RDS) and correlograms (RDC) as described by Julesz[1] have a special role. As previously discussed, these cyclopean stimuli directly prime the cortex, and thus the activity of lower stages (retina, LGN) does not confound the VEP. Furthermore, global stereopsis of RDS is the result of a large number of disparity detectors cooperatively coupled[7,8] and the maximum disparity for fusion increases monotonically with the area of the target.[9] This essential globality of the perceptual phenomenon parallels the essential globality of the VEP. Finally, in RDS and RDC only one parameter is varied (binocular disparity or binocular correlation), stimulating thousands of neural detectors in a similar fashion. Therefore, the interpretation of the VEP should be simplified and one should expect rather robust effects.

Several investigators used *static* RDS;[10,11] however, such stimuli usually produce some unwanted monocular cues (e.g., variations in local density) since two static RDS must be alternated. More recently, Lehmann and Julesz[12] succeeded in recording VEP to *dynamic* RDS, in which all monocular cues were absent. Indeed, when dynamic noise arrays of high spatial resolution are presented at fast rates (60 arrays per second or faster) any local density difference is averaged out for the order of magnitude slower pulsating disparity rates.

The study by Lehmann and Julesz[12] addressed another important methodological question concerning VEP. They presented a dynamic RDS having a cyclopean center square pulsating in depth at 1 Hz rate in a steady surround. This depth pulsation synchronized the averaging of VEP, and yielded usable signal strength. The fact that, without monocular depth cues, observers could perceive the square pulsating in depth, and that this pulsating also yielded significant VEP amplitudes at a 1-Hz rate, indicated that VEP to cyclopean stimulation could be used to detect stereopsis (or functional binocularity) in observers without having to communicate with them. While this finding has obvious appeal in accessing stereopsis in animals or human infants, it did not directly identify the mechanism by which stereoscopic depth perception would contribute to VEP. For instance, the observer could have "counted" each time the cyclopean square emerged from its surround, and the VEP merely indicated this counting activity. While Lehmann and Julesz instructed their observers to refrain from counting, some other side-effects, such as convergence movement of the eyes, could also have accounted for the VEP. Therefore, it is very important that in any given experiment the cyclopean square was presented either left or right of a center fixation marker, and that larger VEP activity was always found contralaterally (on the scalp) with respect to the site of the cyclopean target. This finding suggests that VEP are directly generated by neural pools tuned to disparity changes. Of course, it cannot be ruled out that these disparity detectors stimulate some other more central site, which in turn generates the VEP. [It should be noted that Julesz, Kropfl, and Petrig[13] obtained robust VEP to depth pulsations of 6 Hz rate, which is much beyond the rate of convergence motions.]

The apparatus in the Lehmann and Julesz study consisted of relatively small display oscilloscopes with electrostatic deflection, and so the extent of the cyclopean targets was limited. In the following two years, technical efforts were spent to generate dynamic RDS and RDC using projection TV equipment with magnetic deflections at a 5-MHz video information rate.

With the help of special digital hardware developed by us, we were able to obtain large VEP to dynamic RDS and RDC that were presented on 6-foot-large color projection TV screens as red-green anaglyphs.[13] Typical results are reprinted in FIGURE 1 for an adult with good stereopsis. The figure legend explains the details. Here we mention only a few interesting findings. First, VEPs to dynamic RDS and RDC are quite different. Particularly interesting is the low-frequency component of the VEP to RDC. The difference between the VEPs to RDS and RDC is particularly apparent when the periodic disparity or correlation changes are increased to 3 Hz as

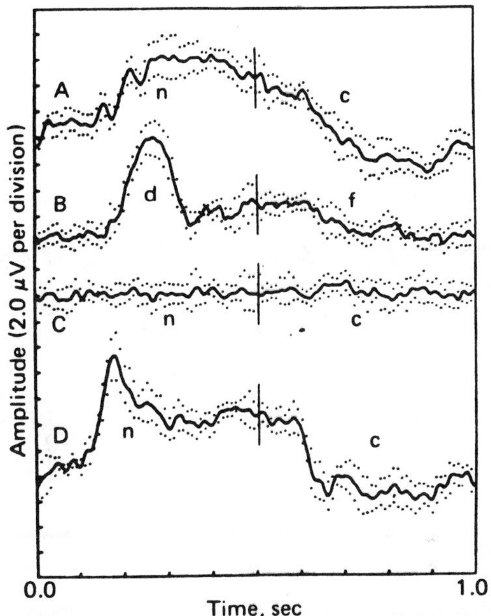

FIGURE 1. Evoked potentials for 2 reversal/sec stimuli for subject W.K. under the following stimulus conditions: (A) Dynamic random-dot correlograms (RDC). Negative correlated (anti-correlated) intervals denoted by *n*; correlated intervals by *c*. (B) Dynamic RDC portraying a cyclopean checkerboard pulsating in depth from a flat background. The depth interval is denoted by *d*, the flat interval by *f*. (C) Monocular control. Same as (A) but with one eye closed, while other eye saw through a red (or green) filter. (D) Same as (A) but anaglyph was viewed without red-green goggles, giving rise to monocular color cues. Solid line is mean of 6 EP ($N = 50$). Dotted lines are ± 1 SD of the 6 EP. Positive polarity, measured on the anterior electrode, is plotted upward.

shown in FIGURE 2. Here the VEP to RDS has a 3-Hz period, while the VEP to RDC has a 6-Hz doubled period. Second, there is a large low-frequency component in the VEP to RDC, almost a square-wave shaped response, as FIGURE 1 shows. This is not necessarily connected with binocular correlation, since the monocular control stimuli exhibit the same squarish shape, but with a shorter latency. It is probably based on suddenly changing a property (e.g., the color or intensity) of many small dots. Finally, the amplitude of the VEP is large (in excess of 10 μV) and is robust (i.e., reproducible for the same subject throughout several months).

These properties can be exploited in many ways. The finding that RDCs elicit VEPs with doubled frequencies might permit the differentiation between functional binocularity and stereopsis in animals and human infants. The robustness of VEP to cyclopean stimuli might be advantageous to locate the site of VEP generators. For instance, using metabolic markers in unison with cyclopean stimulation might be as successful as deoxyglucose methods were in tracing binocular dominance columns.[14]

Cyclopean VEP in Determining the Critical Period of Stereopsis and Clinical Applications

In the previous section we discussed how pools of binocular-disparity-tuned neurons may directly or indirectly contribute to VEP. Here, we are not interested in how cyclopean VEP are generated, but accept the existence of robust cyclopean VEP as a fact, and use it as a tool in diagnosing functional binocularity or stereopsis.

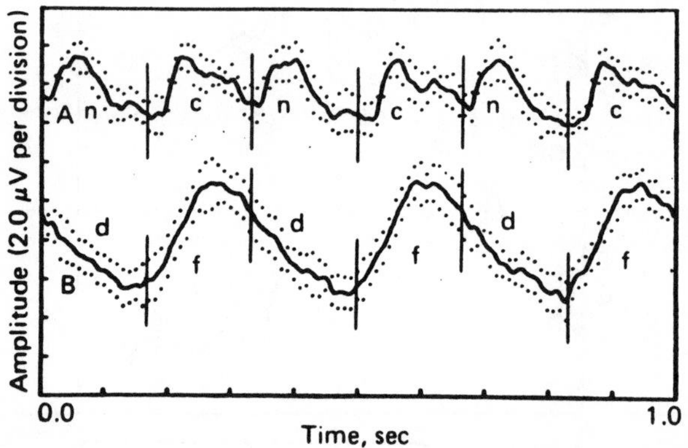

FIGURE 2. Evoked potentials for 6 reversal/sec stimuli for subject W.K. under stimulus conditions (A) and (B), respectively, as described for FIGURE 1. Solid line is mean of 12 EP (whose $N = 75$) for a total of 900 individual responses. Dotted lines are ± 1 SD of the 12 EP.

The first step in this direction was the work (during a visit by one of us to Caltech in 1979) of Miezin *et al.*[15] who elicited large VEPs in a behaving monkey using dynamic RDCs. These VEPs in a macaque monkey are very similar to those obtained in human adults. The adult monkey studied by Miezin *et al.*[15] was alert during the experiments and flexed her muscles. The monkey was surrounded by the stimulus back-projected onto a 6-foot-large screen. The monkey had a mount glued on her head that served as constraint, and the animal was accustomed to vision experiments in which her head was held firm. The fact that robust VEPs could be obtained by cyclopean stimuli shows that the animal kept her eyes open and that eye movements did not affect the results. This finding was instrumental in trying to generalize this technique for human infants.

Work with human infants was carried out at two places using similar techniques. The first was at Mount Sinai Hospital in collaboration with Drs. Bodis-Wollner and Raab. We invited Drs. Braddick and Atkinson to carry out a pilot study using dynamic RDCs on a large number of human infants.[16] We found that between 4 weeks and 2 months of age 4 infants out of 9 had cyclopean VEP, while in the age range of 3 to 5 months 9 out of 9 had cyclopean VEP. We concluded that at least by the age of 12 weeks the normally developing human visual cortex contains neurons that are activated by binocularly correlated stimuli.

The finding is similar to that of Fox *et al.*[17] who found stereoscopic discrimination in dynamic RDS in 3½-month-old infants, but not in younger infants. Their psychophysical techniques were based on stereopsis, while ours required only functional binocularity, and this could explain why we found VEP at a somewhat earlier age.

The other place where our technique was tried on human infants was at the Eidgenoessische Technishe Hochschule (ETH) in Zurich, by Benno Petrig.[18] He used both dynamic RDC and RDS, and found cyclopean VEP in 10–19-week-old infants, several weeks after showing classical checkerboard evoked potentials. We also found doubling of VEP periodicity for dynamic RDC, but only for infants 6 months of age or older. This suggests that the onset of cortical binocularity precedes stereopsis.

So far we have not tried to elicit cyclopean VEP in strabistic infants by compensating for the nonaligned axes of the eyes before amblyopia set in. It is this type of test that might be of greatest potential benefit. To determine the onset of amblyopia ex anopsia, in order to decide on the timing of surgical correction of strabismus, seems an important ophthalmological task.

With Bodis-Wollner we also used cyclopean VEP to study binocular pathway delays, as observed in retrobulbar neuritis. We do not want to go into details here. It suffices that dynamic RDC and RDS at 60 frame/sec rates (i.e., 16-msec frame duration) cannot be fused by patients whose intraocular delays exceed 16 msec. Only if the "leading" eye's view is delayed by adequate amounts is fusion restored. For this study we do not need VEP, since the patient's report of fusion suffices. But again, with patients who cannot verbally communicate, a VEP-based cyclopean method is of potential advantage.

Conclusion

The results of research using cyclopean visually evoked potential methods have been reviewed. It seems that the cyclopean VEP methods not only shed light on some properties underlying VEP, but also and more importantly, yield an unfakeable method to determine functional binocularity or stereopsis in animals and human infants.

References

1. Julesz, B. 1971. Foundations of Cyclopean Perception. Univ. of Chicago Press, Chicago, Ill.
2. Blakemore, C. & B. Julesz. 1971. Stereoscopic depth aftereffect produced without monocular cues. Science **171:** 386–388.
3. Poggio, G. 1980 Neurons sensitive to random-dot stereograms in areas 17 and 18 of rhesus monkey cortex. Soc. Neurosci. Abstracts (Nov.) 6.
4. Julesz, B. 1965. Texture and visual perception. Sci. Am. **212:** 38–48.
5. Mayhew, J. E. W. & J. P. Frisby. 1978. Stereopsis in humans is not orientationally tuned. Perception **7:** 431–436.
6. Cowey, A. & J. Porter. 1979. Brain damage and global stereopsis. Proc. R. Soc. Lond. Ser. B **204:** 399–407.
7. Fender, D. H. & B. Julesz. 1967. Extension of Panum's fusional area in binocularly stabilized vision. J. Opt. Soc. Am. **57**(6): 819–830.

8. JULESZ, B. & J. J. CHANG. 1976. Interaction between pools of binocular disparity detectors tuned to different disparities. Biol. Cybernetics **23:** 33–37.
9. TYLER, C. W. & B. JULESZ. 1980. On the depth of the cyclopean retina. Exp. Brain Res. **40:** 196–202.
10. REGAN, D. H. SPEKREIJSE. 1970. Electrophysiological correlate of binocular depth perception in man. Nature **225:** 92–94.
11. REGAN, D. & K. I. BEVERLEY. 1973. Electrophysiological evidence for existence of neurones sensitive to direction of depth movement. Nature **246:** 504–506.
12. LEHMANN, D. & B. JULESZ. 1978. Lateralized cortical potentials evoked in humans by dynamic random-dot stereograms. Vision Res. **18:** 1265–1271.
13. JULESZ, B., W. KROPFL & B. PETRIG. 1980. Large evoked potentials to dynamic random-dot correlograms and stereograms permit quick determination of stereopsis. Proc. Natl. Acad. Sci. USA **77**(4): 2348–2351.
14. HUBEL, D. H., T. N. WIESEL & M. P. STRYKER. 1978. Anatomical demonstration of orientation columns in macaque monkey. J. Comp. Neurol. **177:** 361–80.
15. MIEZIN, F. M., J. MYERSON, B. JULESZ & J. M. ALLMAN. 1981. Evoked potentials to dynamic random-dot correlograms in monkey and man: A test for cyclopean perception. Vision Res. **21:** 177–179.
16. BRADDICK, O., J. ATKINSON, B. JULESZ, W. KROPFL, I. BODIS-WOLLNER & E. RAAB. 1980. Cortical binocularity in infants. Nature **288:** 363–365.
17. FOX, R., R. N. ASLIN, S. L. SHEA & S. T. DUMAIS. 1980. Stereopsis in human infants. Science **207:** 323–324.
18. PETRIG, B., B. JULESZ, W. KROPFL, G. BAUMGARTNER & M. ANLIKER. 1981. Development of stereopsis and cortical binocularity in human infants: Electrophysiological evidence. Science **213:** 1402–1405.

DISCUSSION OF THE PAPER

H. SPEKREIJSE: Is it possible that your observation that checkboard stimuli give responses in very young infants, while random-dot correlograms do not, just reflects the fact that babies do not focus too well, and your fine grain is projected at a closely placed screen?

B. JULESZ: What is particularly nice about random-dot stereograms is that they can be blurred to a large extent without affecting the evoked response, since the correlation between the left and right images remains constant with blurring. Furthermore, even very young infants yield large square-shaped evoked responses when the random-dot correlograms are viewed without the red-green goggles. Obviously, their monocular resolution is adequate to yield evoked potentials, and only when the same stimulus is cyclopeanly presented (i.e., with the red-green goggles) does the response vanish. Of course, as we have shown (Petrig *et al.* 1981. Science **213**:1402–1405) normal infants yield evoked responses to cyclopean correlograms starting from 10–19 weeks.

C. TYLER: I wanted to address the question of orientation selection in the cylopean single unit work. It seemed that you suggested that this was antithetical to some orientation insensitivity found in psychophysics. I always thought that psychophysics also found orientational sensitivity for contrast.

JULESZ: The cyclopean single units in area 17 of the monkey are rather recent discoveries by Gian Poggio at Johns-Hopkins University. These units are orientationally sensitive when an elongated and randomly dotted rectangle has different binocular disparity from its randomly dotted surround. The firing of these units is really a cyclopean event since the elongated rectangle and its surrounding area are composed of the same dynamic noise, so there is zero contrast. This "cyclopean orientation sensitivity" is very different from the sensitivity to the orientation of elongated grains in monocular filtered noise as studied by Mayhew and Frisby. Psychophysics revealed that global stereopsis (of random-dot stereograms) is independent of the orientation of the grain size. However, as your cyclopean adaptation studies demonstrated, using corrugated depth gratings, the cyclopean process is orientationally sensitive.

QUESTION: You showed that introducing vertical disparity removed evoked response. Without red-green glasses, would the same elimination of response occur with vertical shifts?

JULESZ: The red-green glasses and the red-green filters over the red and green guns of the projection TV system are well matched and with one eye closed no perceptional alternation can be perceived. With both eyes open and wearing the red-green goggles the perceived cyclopean pulsation *slowly* disappears as the vertical disparity exceeds Panum's fusional area (for normal adult observers). Without the red-green goggles a vertical shift in excess of one picture element can cause a sudden disappearance of changes in correlation (within Panum's area). Vertical shift is only one of the controls to test the absence of all monocular cues. We also used the same colored filter over both eyes for control, that yielded no response.

I. BODIS-WOLLNER: The random-dot correlogram itself is not completely flat in its spatial spectrum; it has a dominant spatial frequency. Do we know anything about the spatial transfer properties of those single cells in the visual cortex that would perform the correlation that responds to binocular correlation versus those neurons that would respond to random dot correlograms?

JULESZ: The first question is about the spectrum of a random-dot stereogram. A random-dot stereogram can be regarded as flat (white) noise, until the dimension of the dots is reached. So the spectrum is flat until a rather high spatial frequency. Of course, studying evoked potentials with filtered dynamic random-dot stereograms would be an interesting understanding. We have studied such low- and high-pass random-dot stereograms with Dr. Robert Schumer recently in psychophysical experiments (Julesz & Schumer. 1981. Ann. Rev. Psychol.) With respect to your second question, we found doubling of VEP's temporal frequency for random-dot correlograms, but did not find such doubling for random-dot stereograms if the pulsating depth plane had either crossed or uncrossed disparity, so there was no change in correlation. This recent study was conducted with Ms. Nivian Sanchez and Mr. Walter Kropfl and suggests that functional binocularity (of correlation detection) and stereopsis could be told apart by evoked potential studies.

L. MAFFEI: Would you like to comment on the fact that single cells do not respond to random-dot stereograms?

JULESZ: Of course, the essence of cyclopean methodology is the fact that one can bypass several processing stages beyond the lateral geniculate and the 4th layer of area 17, since the binocular correlation is only extracted after these stages. However, the "cyclopean retina" seems at an earlier level then believed by most, somewhere in area 17. I mentioned that Gian Poggio used some of our techniques and found real cyclopean single units in the monkey cortex tuned to specific binocular disparities. First, he searched for the so-called Poggio-Fischer units, which are tuned to specific binocular disparities, but also require elongated contours of high contrast and specific orientations. After he found such a neural unit, he changed the white bar in black surround (100%) contrast to a white bar with 10% randomly speckled black dots, while the black surround was also speckled with 10% white dots. Usually these units still fired. But, at a 30%/70% bar/surround contrast ratio, most of these units stopped their activities, except for a few. These few continued to fire, even with 50%/50% bar/surround contrast ratio, i.e., when all monocular contrast cues vanished, as long as the binocular disparity of the bar was kept at an optional value. So, these units in area 17 can be regarded as real cyclopean detectors.

COMPARISON OF TRANSIENT AND STEADY-STATE METHODS*

D. Regan

Departments of Physiology and Ophthalmology
Dalhousie University, Gerard Hall
Halifax, Canada B3J 1B6

Distinction Between Steady-State and Transient Stimulation

Transient stimuli follow each other at sufficiently long intervals that the visual system returns to its initial state before the next stimulus occurs. Steady-state stimuli are delivered at a greater rate, so that the response to one stimulus has not died away before the next stimulus is delivered. Transient stimulation gradually changes into steady-state stimulation over a range of stimulus repetition rates. The minimum stimulation rate for steady-state stimulation is the lowest rate for which appreciable overlap occurs between the response to one stimulus and the occurrence of the next stimulus. This rate is different for different types of stimuli.

For a linear system the transient response has a fixed relationship to the steady-state response. Consequently, transient and steady-state descriptions of a linear system's behavior are equivalent, and can be regarded as alternative formulations of the same data. However, the visual and auditory pathways commonly show several types of nonlinear behavior. For example, when successive responses run into each other they may not summate even approximately linearly. In principle, therefore, transient and steady-state stimulation can produce responses that provide complementary information about the sensory system under test. However, this possibility has so far been little exploited. At the present time, the chief practical distinction between transient and steady-state evoked potentials lies in the trade-off of speed versus information. As discussed below, recording steady-state evoked potentials can be much speedier than recording transient evoked potentials, and can be accomplished for much lower signal levels even in adverse environments. For example, mains interference is usually no problem when recording steady-state evoked potentials, but care must be taken to minimize it when recording transient evoked potentials. The price that must be paid for these advantages is that, over a given recording time, steady-state evoked potentials provide less information than transient evoked potentials.

As described below, steady-state evoked potentials can be quantified in a less arbitrary manner than transient evoked potentials. Steady-state evoked potentials can be split up into a small number of harmonic frequency components, each recorded by a separate Fourier analyzer. A Fourier analyzer's output is equivalent to two numbers (namely, amplitude and phase), and these are quite unequivocal. In contrast, describing and measuring a transient evoked potential may not be straightforward, especially when the waveform is unusual or complex (FIGURE 2 illustrates one of the many problems). In such situations different experimenters may measure transient evoked potentials quite differently, may disagree as to whether an evoked potential is normal or abnormal, and may even disagree as to whether an evoked potential is

*This work was supported by the National Eye Institute, the Multiple Sclerosis Society of Canada, and the Medical Research Council of Canada.

0077–8923/82/0388–0045 $1.75/0 © 1982, NYAS

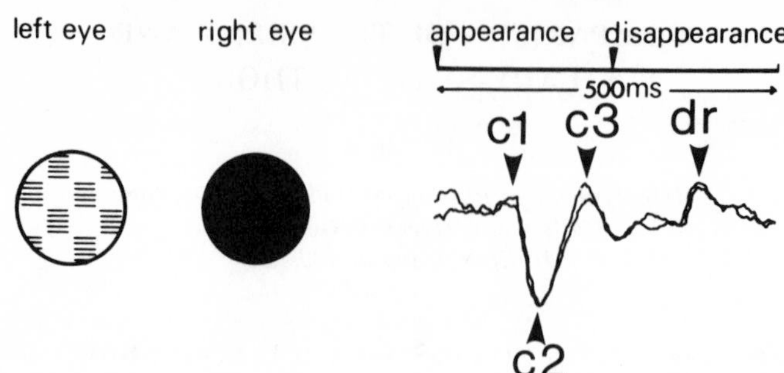

FIGURE 1. The appearance and subsequent disappearance of a stimulus pattern produce two quite distinguishable responses that have different cortical origins, different dynamics, different binocular summation and are differently affected by spatial frequency or check size. The appearance response consists of three components of which only Cl originates in striate cortex. (After Spekreijse, Tweel & Regan.[50])

present at all. The output of a Fourier analyzer leaves comparatively little scope for disagreement between different experimenters and some authors find this an advantage of steady-state recording, though again greater certainty may be offset by less information.

Evoked Potential Recording

Investigators who wish to record evoked potentials must grapple with two problems. The first is that the evoked potentials are commonly invisible in recordings taken directly from the scalp. In other words, the signal (i.e., the evoked potential) is buried in noise, and the experimenter's first problem is to improve the signal-to-noise ratio. In general, the device that improves the signal-to-noise ratio, whether an averager of a Fourier analyzer, effectively defines "signal," and any response that falls outside this definition is rejected.[1] The experimenter's second problem is that the evoked potential may not be the same from moment to moment. It varies with time about some mean value, and this mean value may progressively change with time also. Progressive changes with time are particularly difficult to deal with since the commonly used methods for improving signal-to-noise ratios assume that the evoked potential does not progressively change with time.

Methods for Recording and Analyzing Transient Evoked Potentials

As displayed by a conventional averager, a transient evoked potential is a graph of voltage versus time.† FIGURE 1 illustrates that an averaged transient evoked potential consists of one or more positive and negative deflections. Analysis of such a transient evoked potential is commonly aimed at decomposing the waveform into elementary

†By calculation it is possible to transform such an evoked potential into a plot of volts versus frequency rather than volts versus time. Note, however, that this does not necessarily predict the result of an experiment using repetitive stimulation at the various frequencies. This can only be obtained by practically performing the experiment.

constituents of "components" in the hope that different components will eventually prove to originate in different parts of the brain, or at least in different neural organizations, and/or that the different components will indicate the activities of different functional subunits within the brain.[2] How to analyze the evoked potential into such physiologically meaningful "components" is, however, a problem with no agreed solution. Unfortuantely, different authors define "component" to mean different things, so that the topic of transient evoked potential analysis is confusing.

The chief difficulties in defining an evoked potential component are as follows: (a) Commonly, successive positive and negative deflections overlap considerably, so that it is difficult to identify the peak latency of the several individual deflections. For example, when a small deflection of one polarity rides on top of a larger deflection of opposite polarity, then the only indication of the small deflection's presence may be a shoulder on the rising or falling slope of the larger deflections (FIGURE 2), (b) Many authors define a component by its peak latency, assuming that a deflection of a given peak latency is composed of only one component. However, this is not necessarily so. Deflections originating in different cortical regions might have exactly the same latencies. A second problem is that, when a deflection's latency depends on some stimulus parameter (for example, the latency of a visual evoked potential may decrease as light intensity increases), then it is difficult to decide whether the balance between several fixed-latency components is changing or whether the latency of a single component is changing. (c) Many authors include polarity in the definition of a component (for example, the P100 component). A weakness of this, however, is that

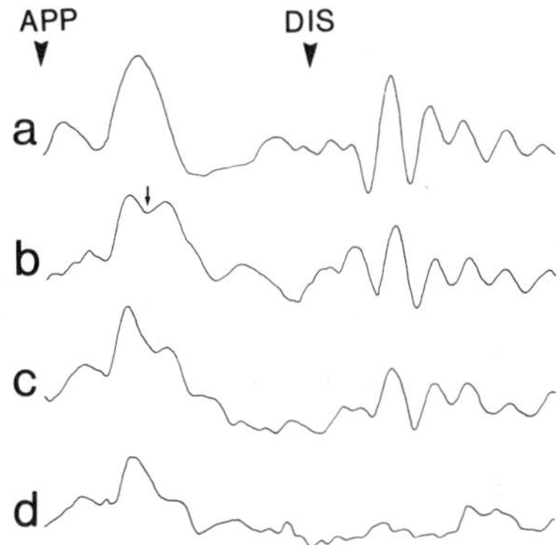

FIGURE 2. Difficulties of interpreting transient evoked potential waveforms. (A) At 80% contrast, the pattern appearance response in this subject chiefly consists of a broad positive deflection peaking at 154 msec. (B) At 40% contrast the broad positive deflection shows itself to consist of two subdeflections, the latency of the second one increasing at lower contrast. An alternative explanation is that a small negative-going deflection (arrowed) rides on the larger positive deflection. (C) In a repeat of (B), for the fellow eye the second subdeflection has the same latency, but relatively lower amplitude. An alternative explanation is that the small negative-going peak has a slightly longer latency than in (B) (40% contrast). (D) The shoulder on the falling flank of the main positive deflection can be interpreted as in (B) and (C) (10% contrast).

These recordings were all taken in one session from a 10-year-old child with normal vision. Responses are to the appearance (APP) and disappearance (DIS) of 10-min checks at constant luminance. Viewing distance, 290 cm; field size, 5°; stimulus cycle rate, 1.2 Hz; sweep time, 0.8 sec; 400 repetitions; inion to linked ears. These records were made by Dr. D. M. Regal using a visual stimulator designed by Dr. K. I. Beverley.

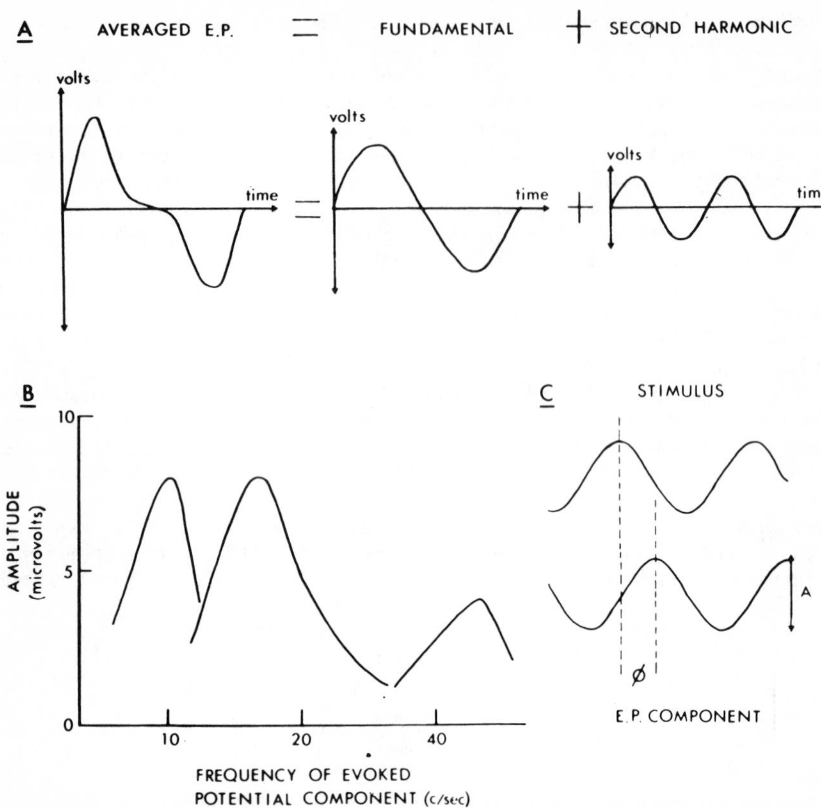

FIGURE 3. (A) Averaged evoked potential waveform elicited by a light flickering at F Hz. The waveform shown can be analyzed into two harmonics. These harmonics are sine waves, the fundamental of frequency F and the second harmonic of frequency $2F$ Hz. (B) Graph of the amplitude of the fundamental (F) harmonic component of evoked potentials to F Hz flicker. Ordinates are evoked potential amplitudes, abscissae are flicker frequencies. Amplitude is maximal at about 10, 16, and 50 Hz. The three curves define three frequency regions. Note that the relative heights of the peaks may vary with subject, electrode position, stimulus field size, and so on. (C) Measurements made on the fundamental component of the evoked response stimulation. A = peak-to-peak amplitude; ϕ = phase lag between stimulus and response. (D) Illustrates how a delay of, for example, 100 msec in the retina-cortex system causes a phase lag of the evoked potential and how this phase lag increases with increasing frequency of stimulation. The plot of phase lag against stimulus frequency is a straight line, the slope of which gives a value for the delay in the system. Note that, although an increase in delay *necessarily* increases the slope of the phase plot, an increase in the slope of the phase plot does not *necessarily* mean an increased delay. (From Milner, Regan & Heron.[51] By permission of *Brain*.)

the polarity of a deflection recorded at the scalp depends on the orientation of the cortical generator. For example, moving a patterned stimulus from one point of the visual field to another can invert the surface polarity of a component. To get over this problem, Jeffreys and Axford[3] proposed that a "component" should be defined in terms of topographical distribution of amplitude over the scalp rather than by latency or polarity.

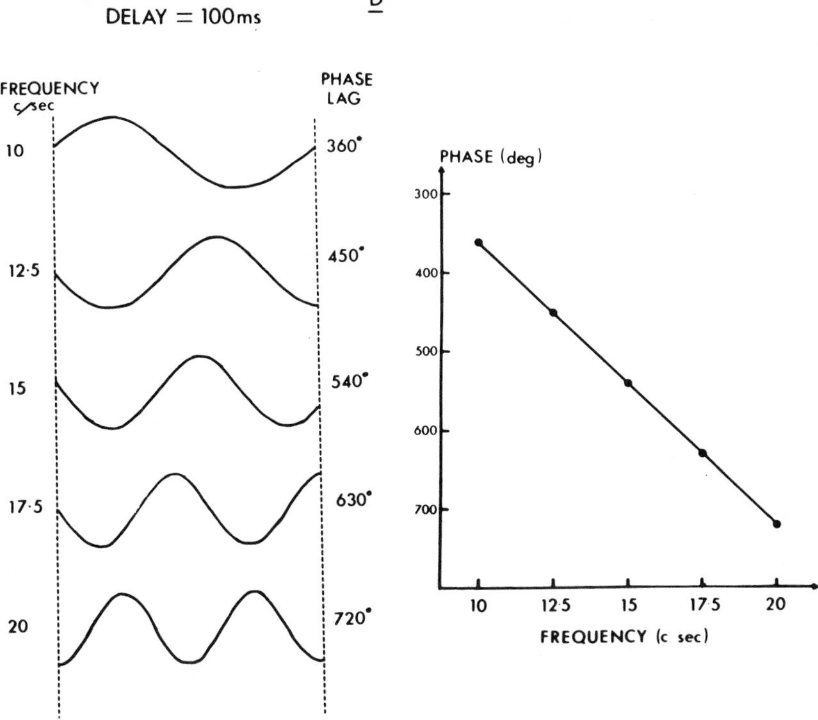

FIGURE 3 (*Continued*).

An alternative way of analyzing transient evoked potential waveforms is by principal factor analysis.[4-9] The evoked potential waveform is described as an array of points in multidimensional space, and the latencies at which evoked potential waveforms differ most significantly is found by statistical analysis. These latencies do not necessarily coincide with a peak latency of a deflection obvious to eyeball inspection. According to John *et al.*,[10] statistical analysis of this general type is effective in detecting brain abnormalities that are not detected by other techniques in children with learning difficulties.

METHODS FOR RECORDING AND ANALYZING STEADY-STATE EVOKED POTENTIALS

Any complex, repetitive evoked potential waveform can be analyzed into an (often small) number of pure repetitive sine waves called harmonic or frequency components, much as a complex note can be analyzed into harmonic tones. If the stimulus repetition frequency is F (in Hz), then the harmonic components have frequencies of F, $2F$, $3F$, $4F$, and so on. For example, FIGURE 3A illustrates how one particular complex waveform of frequency F is composed of two harmonic components, one of frequency F and the other of frequency $2F$. The harmonic components of steady-state evoked potentials can be regarded as analogous to the temporal components of a transient evoked potential.

Evoked potentials can be extracted from noise by on-line Fourier analysis.[11,12] This procedure can be regarded as frequency filtering with a very narrow bandwidth. Harmonic components can be extracted in this way. The technique is also applicable for detecting subharmonic evoked potential frequency components. FIGURES 4 illustrates how steady-state evoked potentials can be analyzed into harmonic components

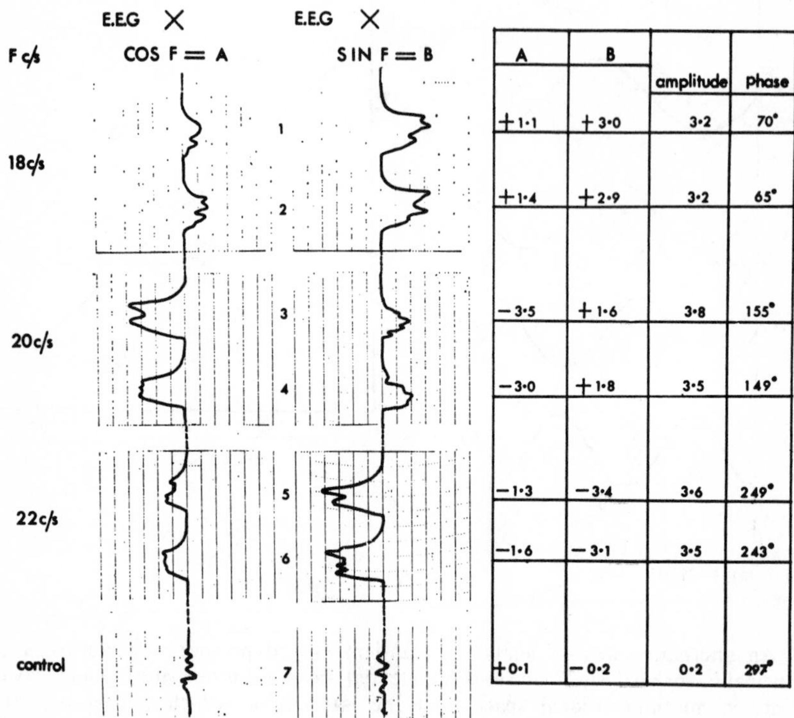

A	B	amplitude	phase
+1·1	+3·0	3·2	70°
+1·4	+2·9	3·2	65°
−3·5	+1·6	3·8	155°
−3·0	+1·8	3·5	149°
−1·3	−3·4	3·6	249°
−1·6	−3·1	3·5	243°
+0·1	−0·2	0·2	297°

FIGURE 4. The subject looked at a light flickering at frequency F Hz while the Fourier analyzer recorded the frequency component at frequency F Hz. The outputs of the Fourier analyzer's multipliers were filtered by three-stage low-pass passive RC filters of two-second time constants for each stage. This figure shows typical raw data. The pair of traces marked (1) are filtered outputs of the sine and cosine multipliers, recorded while the subject viewed 18-Hz flicker. The lines show the mean deflections $+A$ and $+B$. The traces marked (2) are a repeat of (1) and illustrate variability. Evoked potential amplitude and phase were calculated in the way described previously.[2,11,12] Phase (ϕ) is given by [$\phi = \tan^{-1}(A/B)$], and amplitude is proportional to $(A^2 + B^2)^{1/2}$. Thus, traces (1) indicate a phase of 70° and an amplitude of 3.2 divisions, while traces (2) indicate a phase of 65° and an amplitude of 3.2 divisions. Traces (3) and (4) were recorded while the subject viewed 20-Hz flicker and illustrate the increase in phase lag caused by increasing flicker frequency. Traces (3) indicate a phase of 155° and an amplitude of 3.8 divisions and traces (4) indicate a phase of 149° and an amplitude of 3.9 divisions. Traces (5) and (6), recorded at 22 Hz, show a phase of 249° and an amplitude of 3.6 divisions; a phase of 243° and an amplitude of 3.5 divisions, respectively. Trace (7) was recorded with the light stimulus occluded and shows a typical noise level for medium-frequency evoked potentials. The noise amplitude here was 0.2 divisions (Calibration: 1 division = 1.8 μV). (From Milner, Regan & Heron.[51] By permission of *Brain*.)

by means of a simple analog Fourier analyzer that can be constructed by any competent electronics technician.

FIGURE 3 illustrates how amplitude and phase data can be treated. It is clear from FIGURE 3C that the value of C in the equation below is multiply ambiguous by some integral number of whole cycles; it might be ϕ, $(\phi + 360°)$, $(\phi + 720°)$, and so on, but one's uncertainty as to the zero of phase need not matter as shown in FIGURE 3D. For, in general,

$$\phi = 360\, F\gamma + C + K,$$

where γ is the real conduction time, F is the frequency, K is a constant, and C is a function of F. Therefore,

$$\frac{d\phi}{dF} = 360\gamma + \frac{dC}{dF},$$

where $d\phi/dF$ is the slope of the phase-versus-frequency plot.

In general, we cannot be certain that dC/dF is zero. The term "apparent latency" was introduced for this reason [References 2 (p. 77), 11, 12], where

$$\text{Apparent Latency} = \frac{1}{360} \cdot \frac{d\phi}{dF} = \frac{1}{360}(\text{slope of phase plot}).$$

The interpretation of evoked potential phase is not always straightforward. Although a time delay causes a phase lag, it is not necessarily true that any phase lag is caused by a time delay. FIGURE 5 gives two examples where, although evoked potentials for the two eyes differ in phase, the delays nevertheless do not differ since the phase shift is frequency independent. FIGURE 5 illustrates the point that it is the steepness of the phase-versus-frequency plot that is related to delay, not the absolute phase lag.

A second reason why time delay does not directly follow from phase lag is that evoked potential amplitude must vary with stimulus frequency, and this *in itself* can be expected to produce phase shifts even when there is no delay at all.[13] In practical terms, this means that it may be unreliable to estimate evoked potential latency from measurements made at a single frequency. A more subtle point is that estimates of evoked potential latency should be corrected for curvature of the amplitude-versus-frequency characteristic, and this correction can be substantial.[14] It follows that a change in apparent latency need not necessarily be due to a change in real delay. Consider the case of delayed evoked potentials in multiple sclerosis. If we assume that the visual pathway acts as though a real delay stage is in series with a bandpass filter, then a change in the bandpass filter could change the apparent latency of the evoked potential even if the real delay remains constant. For example, if evoked potentials to high temporal frequencies are preferentially attenuated in multiple sclerosis (compatible with reductions in the critical flicker fusion frequency), then this could produce an increase in measured evoked potential latency even though retina-cortex transmission time was unaffected. It would probably be as well to allow for this before interpreting evoked potential delays observed in multiple sclerosis patients as being entirely due to increased transmission time caused by demyelination.

An important and troublesome point is that the method used to record evoked potentials strongly affects the temporal tuning curve obtained. This can be understood as follows. One way to estimate the amplitude of a steady-state evoked potential is first to record the evoked potential by means of an averaging device, and then to measure

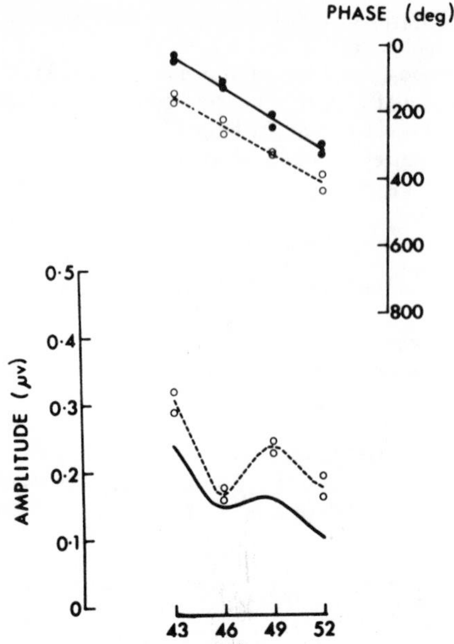

FIGURE 5A. High frequency flicker evoked potentials. Amplitude and phase plots for the left eye are shown by the full line; those for the right eye by the dotted line. This patient had suffered an attack of RBN in the right eye. The slopes of the phase plots (and hence delays) are similar for affected and unaffected eyes. A frequency-independent phase displacement is evident. (From Milner, Regan & Heron.[51] By permission of *Brain.*)

the maximum peak-to-trough deflection. This method confounds different frequency components of the evoked potential (e.g., a 5 μV waveform with the same frequency as the stimulus would measure exactly the same as a 5 μV waveform with twice the stimulus frequency), and this is a disadvantage in cases where harmonic components of different frequencies have different properties and arise in different coritcal sites, as with flicker evoked potentials.[2,15] A second way to estimate the amplitude of a steady-state evoked potential is by means of a Fourier analyzer (see above). The amplitude of the F (first harmonic) component may then be plotted versus stimulus frequency F. This has been done in FIGURE 6 for flicker evoked potentials and pattern evoked potentials. This procedure has been employed in many studies (e.g., References 12 & 16). Alternatively, the amplitudes of several harmonic components may be recorded simultaneously and all plotted on the same graph (e.g., References 14 & 17). In practice, an approximately similar result can usually be obtained by multiplying the EEG, not by quadrature sine waves of frequency F or $2F$, and so on, but by quadrature square waves of frequency F or $2F$, and so on.[18] This is because third harmonic ($3F$) components of evoked potentials are commonly weak.[19] For example, this procedure of square wave multiplication was used by Tyler *et al.*[20] A harmonic component may also be extracted by passing the EEG through a narrow-band tuned filter before averaging.[21] This is the procedure used by Campbell and Maffei[22] and in many subsequent studies.

The main point here is that there may well be considerable differences between these various plots of evoked potential amplitude versus stimulus repetition frequency. For example, when contrast (pattern) reversals are abrupt (square-wave modulation of contrast) and of low frequency, then the evoked potential waveform resembles a brief transient oscillation rather than a sine wave, so that much of the power is at high harmonics. Consequently, comparatively little power is left at the pattern reversal frequency, so that the amplitude of the pattern evoked potential component in FIGURE

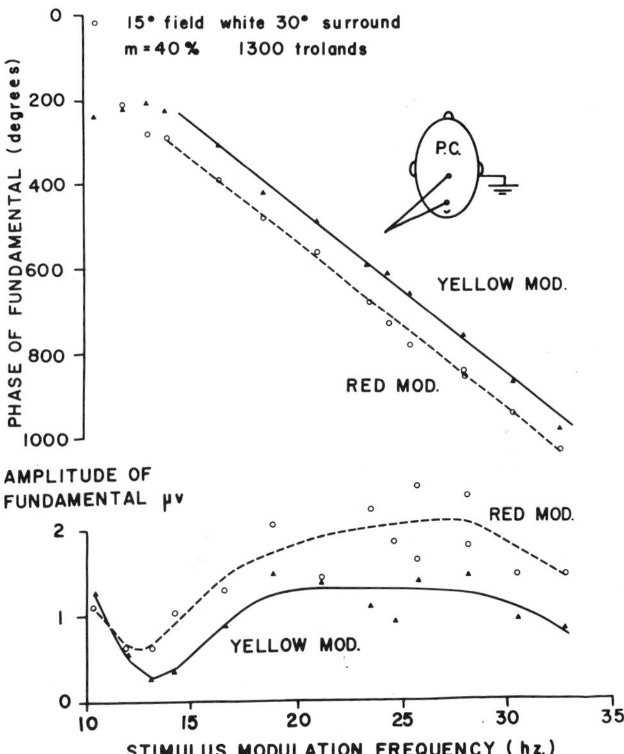

FIGURE 5B. Shows frequency independence of the phase shift between evoked potentials elicited by flickering red light and yellow light. "Yellow mod." means that the red light was of constant intensity, while the yellow light was sinusoidally modulated to ±40% about its mean value. "Red mod." has the opposite meaning. Thus, the net luminance of the stimulus never departed more than 20% from its mean value. The phase difference between the fundamental components of the evoked potentials to red and yellow flicker is roughly 80° and does not vary by more than ±10° even though evoked potential phase was changed by some 800° by increasing flicker frequency from 13 to 33 Hz. The zero of phase is arbitrary. The slopes of the phase plots indicate "apparent latencies" of 110 msec (after correction for amplitude vs frequency plot, the apparent latency was 92 msec with less than 1 msec difference between responses to red and yellow flicker). Over most of the frequency range the subject could not see any difference between red and yellow modulation. (From Regan.[52] By permission of *Vision Research*.)

6 is much less than the amplitudes of the averaged waveform at low frequencies.[19, 23] (They are more similar for sine-wave modulation of contrast.) Again, the flicker evoked potential data of FIGURE 6 are quite different (and less revealing) if the amplitude of the averaged evoked potential is plotted rather than the amplitude of the F (or $2F$) evoked potential component.

CONDITIONS FOR VALIDITY OF THE FOURIER ANALYSIS METHOD

One of the conditions that must be met if Fourier analysis is to be an appropriate or even a valid means of analyzing visual evoked potentials is that an unvarying

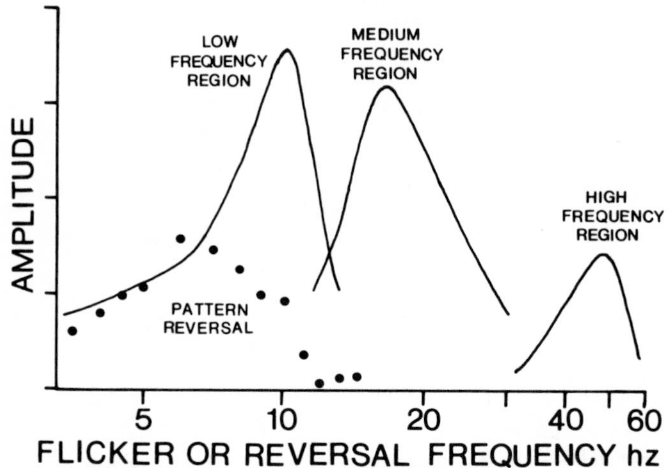

FIGURE 6. Low-, medium-, and high-frequency response to flicker. The amplitudes of evoked potentials produced by flickering an unpatterned patch of light show resonantlike peaks in three frequency regions. The properties of evoked potentials are fairly uniform within a single frequency region, but are quite different in different frequency regions. Evoked potentials in the different frequency regions are generated in different areas of the brain, have different latencies, and different color properties, and, furthermore, demyelinating disease slows medium-frequency but not high-frequency responses. The three types of signal may be separate from an early stage in the visual system. Note that these evoked potentials to unpatterned flicker have quite different properties from evoked potentials to fine patterns (whose frequency dependence is shown by dots). (From Regan.[48] By permission of *Nature,* **253:** 401–407, © 1975, Macmillan Journals Limited.)

repetitive stimulus must give a repetitive visual evoked potential waveform whose phase and amplitude remain constant from moment to moment. It should be noted that the steady-state condition is not established until some time has elapsed after the onset of repetitive stimulation.[11,12,24] By using one form of Fourier analyzer, this "stationarity" of the response can be experimentally checked, since the analyzer outputs visual evoked potential amplitude and phase in the form of a "running average." By careful stimulus control, the moment-to-moment values of visual evoked potential latency can be held constant to about ±1 msec in 100 msec and amplitude to better than ±6% over some minutes.[2,11,12] Conversely, the running-average display makes any deficiencies in stimulus control very evident (e.g., variations of the eye's fixation or accommodation).

Relative Advantages of Fourier Analysis and Averaging Methods

Advantages of Fourier Analysis Compared with Straightforward Averaging

(1) When rapidly repetitive stimulation is used to generate steady-state evoked potentials, the resulting averaged visual evoked potential waveform may be uninterpretable (or almost so) since it is made up of, for example, two or more harmonic components whose properties are quite different.[25]

Two such components add to give an averaged waveform whose shape and peak-to-peak amplitude depends on the phase difference between the two components. Since this relative phase may be arbitrary, the amplitude and shape of the resulting averaged waveform may have little meaning. For the same reason, the latency to peak of the averaged waveform may also be meaningless. Furthermore, one harmonic may be masked by another larger (but possibly less interesting) harmonic. In general, therefore, measurements of the amplitudes and latencies of averaged steady-state visual evoked potential waveforms have little value unless visual evoked potential changes are very large.

However, in some situations the average waveform may be approximately sinusoidal, and useful estimates of phase and amplitude may be obtained.[26] Again, waveforms may approximate sine waves when stimulus frequency is sufficiently high since the high-frequency attenuation of the brain itself ensures a sine wave response. However, to confine measurements to high stimulus frequencies is unnecessarily restricting, and may even give misleading results. This is because steady-state visual evoked potentials of low frequency can have quite different properties than those of high-frequency visual evoked potentials.[14,15,25]

In contrast, evoked potential interpretation may be straightforward when a Fourier analyzer is used instead of an averager. The Fourier analyzer isolates and separately measures the different harmonic components so that their different properties can easily be distinguished. FIGURE 1 in Regan[25] illustrates this point. With ordinary care, and with a reasonable signal-to-noise ratio, latency can be measured to an accuracy of roughly 4 msec standard deviation[12] (though latency *differences* can be measured to within 1 msec).[25] Given reasonable signal-to-noise ratios, amplitude can be measured to about 10%–20% standard deviation.[14,25] This precision compares favorably with the everyday precision of measuring averaged transient visual evoked potentials.

(2) A second advantage of Fourier analysis is that the "running-average" presentation of visual evoked potential data can provide feedback to the experimenter so that when he changes the stimulus he knows almost immediately how the visual evoked potential is affected.[11,12,18] Additionally, it is the running-average output of data that makes possible the methods based on Fourier analysis described below.

(3) A Fourier analyzer can often record a visual evoked potential in much less time than an averager.

(4) A Fourier analyzer can measure smaller signals with precision (e.g., 0.2 μV in 60 μV of noise within a 30-sec recording period (Reference 25, FIGURE 3).

(5) A Fourier analyzer is less disturbed by certain kinds of noise (e.g., alpha activity). This is because the stimulus-repetition frequency (F) can be chosen so that

all harmonics (F, $2F$, etc.) are more than about 1 Hz removed from the alpha frequency. The narrow bandwidth of the analyzer then renders it almost immune to alpha noise (FIGURE 6 of Reference 12).

(6) The relation between spontaneous EEG activity and the steady-state visual evoked potential can be investigated moment by moment. For example, it is not difficult to investigate whether the occurrence of a burst of alpha activity causes the visual evoked potential amplitude to change momentarily.[12]

Disadvantages of Fourier Analysis

(1) A Fourier analyzer is not appropriate for recording transient evoked potentials.

(2) The reason that a Fourier analyzer can be faster than an averager is that it "sees" less information than an averager. In other words, it exchanges bandwidth for speed. This carries the danger that a Fourier analyzer, operated alone, might be blind to important information.‡

(3) Latency cannot be measured unequivocally by recording the visual evoked potential for a single stimulus frequency, since evoked potential phase can be markedly affected by factors other than latency. In other words, the simple equation, $\tau = \phi/360F$ (where τ = latency, ϕ = phase, and F = frequency) is not generally valid for evoked potentials (see above).

METHODS FOR MINIMIZING THE IMPORTANCE OF EVOKED POTENTIAL ANALYSIS

If the method of evoked potential analysis is called into question, this may throw doubt on the conclusions of a whole experiment, even though the experimenter was only using evoked potential recording as a tool to study, for example, color vision or pattern vision. One method for minimizing this danger is to regard the scalp-recorded evoked potential as no more than an objective sign that a visual signal has reached the brain.[27] The evoked potential can then be treated in much the same way as a psychophysicist treats a subject's yes/no response ("yes, I detect the stimulus"/"no, I do not detect the stimulus"). Just as the psychophysicist manipulates the stimulus so as to "dissect" the visual or auditory system, so the evoked potential experimenter can manipulate the stimulus so as to "dissect" a sensory pathway comparatively unfettered by his lack of understanding of how or where the evoked potential is generated.

‡Information missed by a Fourier analyzer might be either stimulus-locked, or not locked to the stimulus. Stimulus-locked visual evoked potential components might be located at harmonic frequencies other than that to which the Fourier analyzer is set. This possibility can be guarded against by using an averager in parallel with a Fourier analyzer; a practical procedure is to rely on an averager for coarse scans, reserving the Fourier analyzer for precise quantitative measurements.

Information not locked to the stimulus would be missed by both averager and Fourier analyzer but could be picked up by a method that examined a broad frequency range. This can be done in a number of ways. For example, banks of narrow-band filters have long been used to carry out such a wide-ranging (but necessarily coarse) analysis of the EEG. Alternatively, a small digital computer using the fast Fourier transform would serve this purpose. It should be noted, however, that these methods give amplitude information only and do not indicate whether the stimulus organizes the relative phasing of EEG frequency components. There is, in any case, little evidence so far available that repetitive visual stimuli have any consistent effect on EEG power at other than harmonic frequencies (apart from some reduction of alpha activity).

Methods for Dealing with Variability and Progressive Changes in Transient Evoked Potentials

Interleaving stimuli is a relatively simple method for minimizing the effects of slow changes in transient evoked potentials.[28] Figure 7 shows an example of transient evoked potentials recorded for an 8-year-old amblyopic child by an interleaving technique. The evoked potentials illustrated are to 30-msec constant-luminance presentations of sine-wave gratings of a single spatial frequency but of 7 different contrasts interleaved in groups of 4. At the time of testing, visual acuity was 6/6 left eye, 6/12 right eye. As expected, EPs at 10 c/deg and 5 c/deg were smaller in the right eye than in the left. Unexpectedly, however, EPs were also considerably asymmetric at 1 c/deg. Two years previously acuity had been 6/7 left, 6/60 right.

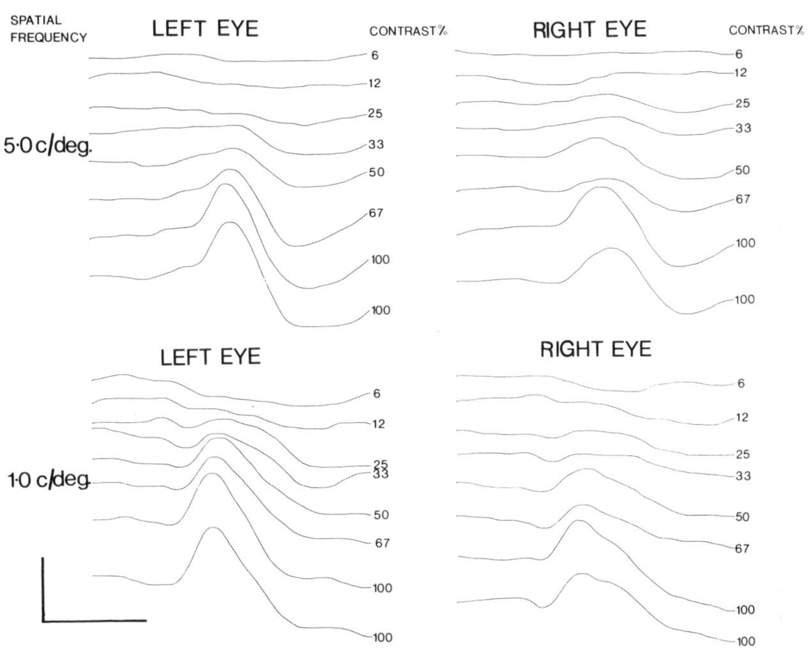

Figure 7. Transient evoked potentials in an amblyopic child. (Top) Averaged transient evoked potentials elicited by the 30-msec presentation of a sine-wave grating pattern at constant luminance. Spatial frequency 5.0 cycles/deg. Reading downwards from the top, contrasts were 6.2%, 12.5%, 25%, 33%, 50%, 67%, 100% and 100%. Two hundred presentations of each contrast were recorded, interleaving in blocks of four. Stimulus cycle time, 320 sec; sweep time, 270 msec. (Bottom) Similar to (Top) except that the spatial frequency was 1.0 cycles/deg. The repeated 100% traces indicate variability. Calibration: 5 μV and 100 msec. (From Regan *et al.*[53] By permission of *Investigative Ophthalmology and Visual Science.*)

Adaptive filtering techniques are capable of dealing with evoked potentials whose latencies vary between successive responses.[30–32] These techniques, however, are quite demanding on computing facilities.

SPEEDY EVOKED POTENTIAL TECHNIQUES BASED ON FOURIER ANALYSIS

The following three techniques are based on the same principle, namely, that an experimenter may obtain a more reliable estimate of the *relationship* between responses to several stimulus parameters if he directly measures the relationship than

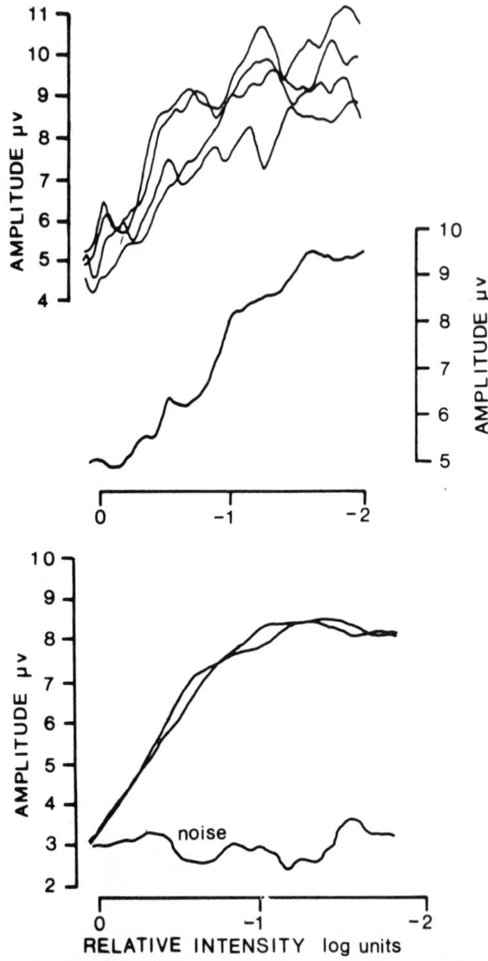

FIGURE 8. The method of averaging graphs. This shows how a graph is made smoother by averaging several rapidly recorded samples of the whole graph. The stimulus was a pattern of 9-min red checks that reversed 6 times per sec. The Fourier analyzer was locked to the 6-Hz rate. The running average of evoked potential amplitudes (ordinates) was continuously recorded while stimulus contrast was slowly changed (abscissae). The upper four traces are individual samples of the whole graph. The middle trace is the average of the upper four samples and shows some smoothing. The lower graph shows two traces, each of which is the average of 16 samples. The noise trace is the average of 16 samples made with the stimulus occluded. All filters were three-stage passive RC low-pass filters, each stage with a 0.3-sec time constant.

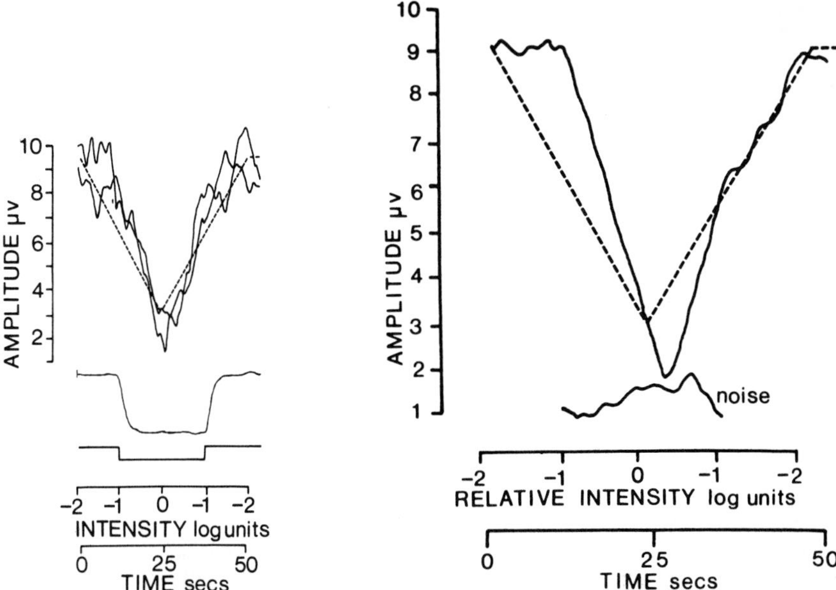

FIGURE 9. Averaging of graphs: distortions caused by evoked potential and by apparatus. (Left) Graph showing that hysteresis and dynamic nonlinearities are negligible using the sweep method for human pattern evoked potentials. The running average of evoked potential amplitudes (ordinates) was continuously recorded while log stimulus intensity was changed with a triangular waveform (shown dotted). The evoked potential traces are approximately similar for increasing and decreasing intensities, and minimum evoked potential amplitude approximately coincides with minimum stimulus intensity. Each trace is the average of 16 samples. The middle trace shows the speed of response to switching a 6-Hz sine-wave calibration signal on and off (onset shown by lowermost trace). All filters were three-stage passive RC low-pass filters, each stage with a 0.15-sec time constant. (Right) Graph showing how increasing the time constants of the Fourier analyzer's filters smooths the plot, but also causes distortion. Filter time constants on the left were doubled on the right. Compared with the evoked potential traces on the left, the trace on the right is smoother, but the minimum is delayed, and the plot is asymmetric. Each trace is the average of 16 samples. (From Regan.[34] By permission of *Vision Research.*)

if he separately measures the evoked potentials to each stimulus parameter and then computes the relationship.

The Method of Averaging Graphs, Also Called the "Sweep" or "Zooming" Method

The technique consists of two operations. The first is to sweep the stimulus parameter of interest while outputing moment-to-moment values of evoked potential amplitude (e.g., by means of the running average facility of a Fourier analyzer). This sweep operation was introduced as a means of speeding evoked potential refraction.[18] The second operation is to average the successive sweep responses.[33] FIGURE 8 shows an application of this technique.

As discussed previously,[29,34,35] this technique has a number of pitfalls and should be used with caution. These pitfalls include the following: (a) As shown in FIGURE 8, averaging many sweeps can improve the signal-to-noise ratio. However, averaging

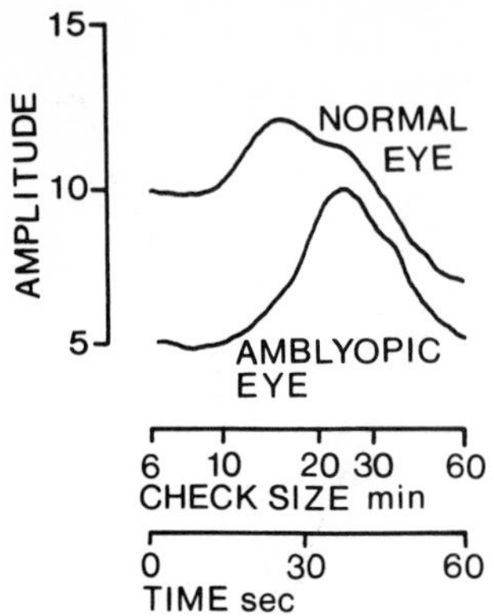

FIGURE 10. Plots of evoked potential amplitude (in μV) versus check size (in min arc) recorded by continuously zooming check size. Amplitude is preferentially attenuated for small check sizes. A movie cartoon was superposed on the checkerboard. Subject was an amblyopic child. (From Regan.[39] By permission of the *Journal of the Optical Society of America.*)

must be used with discretion, for when too many traces are averaged, the experiment is unduly prolonged, variability increases, and the procedure becomes self-defeating, (b) Rather than averaging a greater number of sweeps, it is possible to improve the signal-to-noise ratio by increasing the filtering time constants of the Fourier analyzer or cross-correlation device. Unfortunately, this can introduce substantial errors as excessively high time constants cause the running average output to lag during a sweep. This point is illustrated in FIGURE 9, (c) A change in response phase can produce a spurious change in output amplitude; if filters are placed before the $(x^2 + y^2)^{1/2}$ stage of the analyzer. On the other hand, if filters are placed after the $(x^2 + y^2)^{1/2}$ stage, then the noise rejection may be poor.[34] (d) It is not necessarily true that the sweep procedure is valid in all stimulus situations. This can only be checked empirically, for example, as illustrated in FIGURE 9. Gross dynamic nonlinearities have been observed in some situations that completely invalidate the technique.[36] A practical compromise between filtering and averaging has been described previously.[34]

FIGURES 10–12 give examples of different stimulus parameters that may be swept. FIGURE 10 shows how check size (spatial frequency) may be swept in order to plot evoked potential amplitude versus check size—here in the normal and amblyopic eyes of a child.[39] (See also Tyler *et al.*[37]) FIGURE 10 confirms a previous finding[38] that amblyopia can preferentially attenuate evoked potentials to high spatial frequencies (small checks), while leaving evoked potentials to low spatial frequencies (large checks) comparatively unaffected. The temporal frequency of the stimulus (i.e., the number of pattern reversals per second) can also be zoomed, and the choice of temporal frequency is important in recording from children and infants, since the temporal tuning of pattern evoked potentials changes markedly with age, at least over the age range 5–12 years.[39,40]

FIGURE 11 shows how the orientation of a rapidly-reversing grating pattern can be swept (i.e., rotated) so as to speedily find the preferred orientation. Further to this application, FIGURE 12 shows how the orientation of a slit placed in front of the cornea can be swept so as to distinguish between the neural contribution and the corneal curvature contribution to astigmatism. In FIGURE 12 sweeps are plotted in polar coordinates. By reducing the sweep time to as little as 10 sec (FIGURE 12E) the quality of the recording was improved rather than degraded.

In FIGURE 13 the optical power of a lens placed in front of the eye was slowly swept up and down through a range of 2.5 diopters while the subject looked at a checkerboard pattern that was reversing at a comparatively rapid rate of 6 reversals per sec. The tracing in FIGURE 13 shows moment-to-moment amplitudes of evoked potentials elicited by the checkerboard pattern. Evoked potential amplitude was maximal at the moments when lens power passed through the value that gave the sharpest image on the retina.

The Method of Simultaneous Stimulation: Multiple Sites of Stimulation

One method of comparing evoked potentials from different sites of stimulation (for example, for different locations in the visual field) is to make several recordings on different occasions. A drawback to this method is that, if the evoked potential changes with time, these changes limit the accuracy with which the different evoked potentials

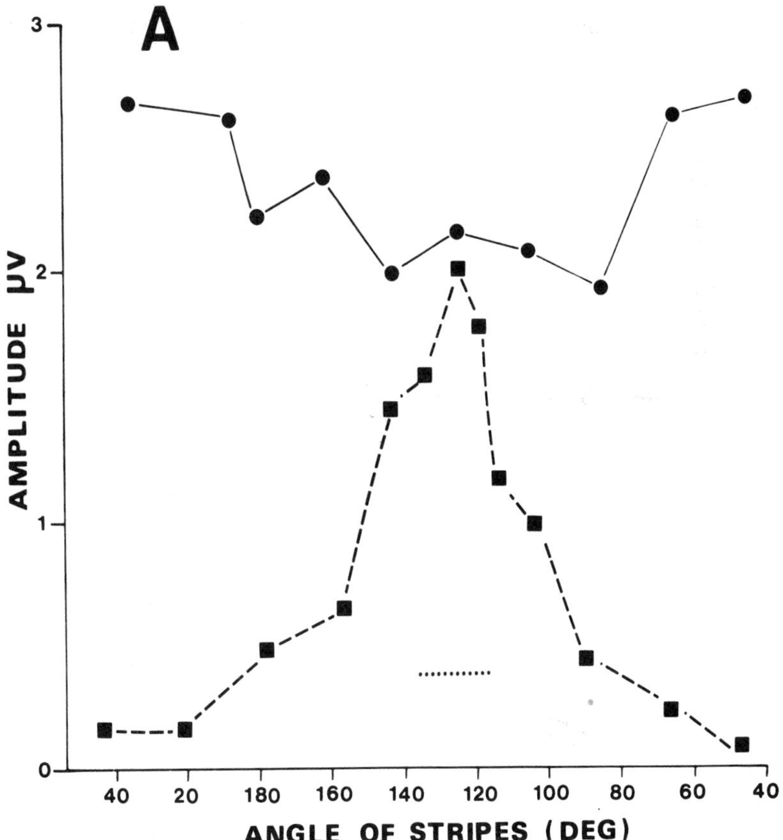

FIGURE 11A. Astigmatism shown by evoked potentials elicited by a pattern of stripes. Static method: Stripe orientation constant during the recording of any point on the graph. Dashed line: astigmatism uncorrected, but −2.5 D sphere in front of eye. Evoked potential amplitude peaks when stripes oriented at approximately 130°. Continuous line: astigmatism corrected by prescription lens. Noise level shown by horizontal dotted line (light occluded).

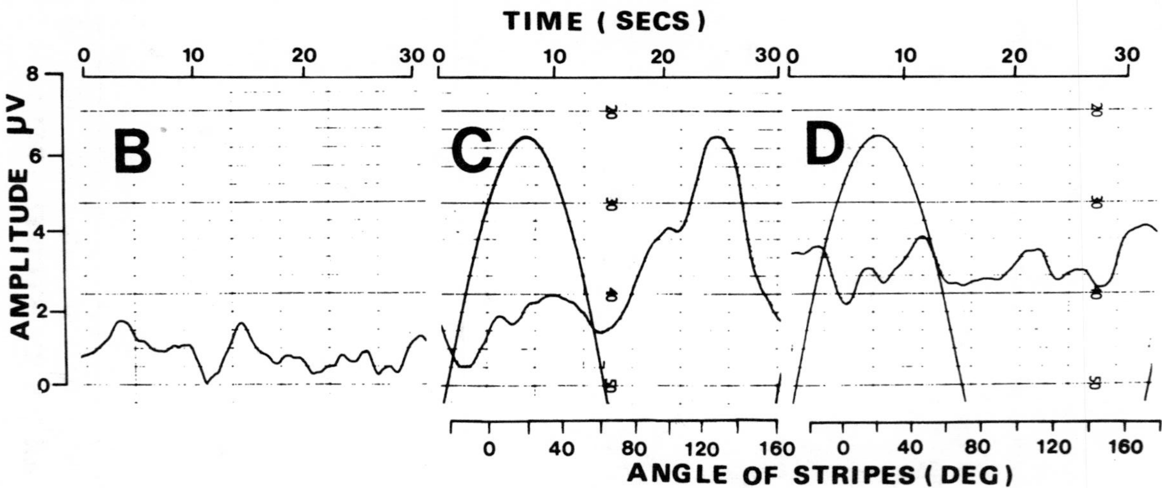

FIGURE 11B–D. Rotating-stripe method: Orientation of stripes continuously rotated through 180° every 30 sec. (B) Noise level, stimulus pattern occluded. (C) Astignatism uncorrected, viewing through −2.5 D sphere. Evoked potential amplitude peaked when stripe orientation was approximately 130°. (D) Astigmatism corrected by prescription lens. Evoked potential amplitude now much less affected by stripe orientation than in (C). Bar width, 10 min. Accommodation not paralyzed. FIGURE 12H gives key to angular measures of orientation. (From Regan.[18] By permission of *Investigative Ophthalmology and Visual Science.*)

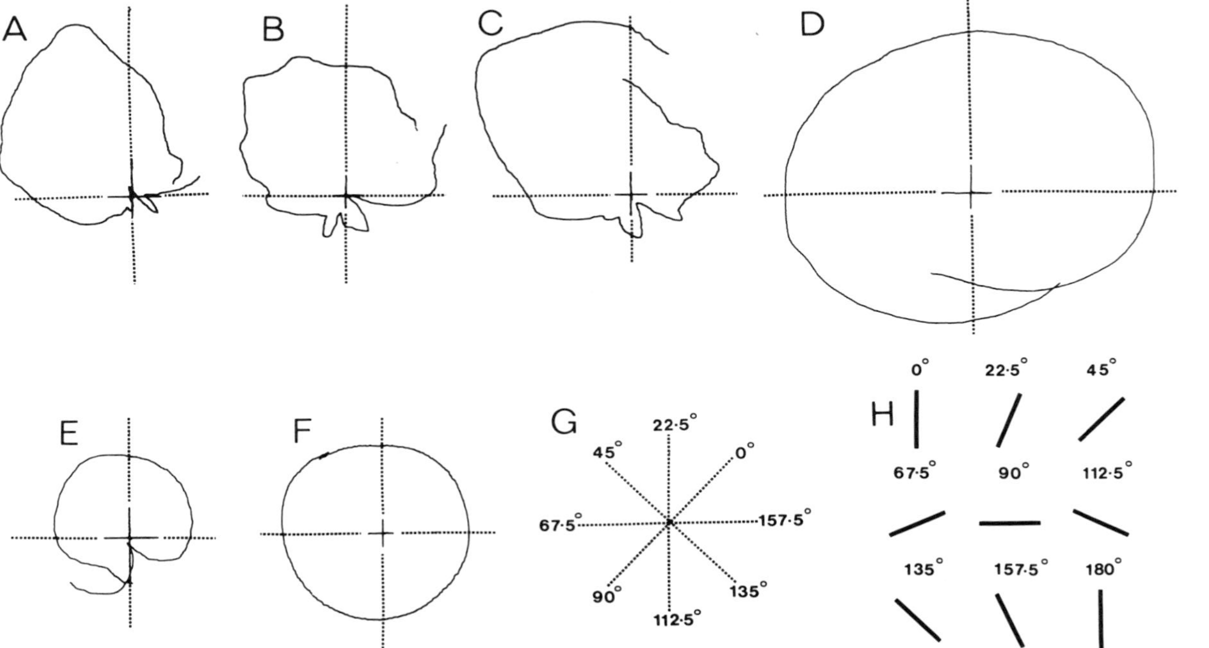

FIGURE 12. Axes of astigmatism displayed by polar plot of evoked potentials elicited when a pattern of checks was viewed through a rotating stenopaeic slit. (A), (B), and (C) Three replications of evoked potentials recorded while the stenopaeic slit rotated through 180° in 18 sec. The instantaneous value of evoked potential amplitude was given by the length of the line joining the origin of the axes to any given point on the trace, and the corresponding slit orientation was given by the orientation of this line. Thus, evoked potential amplitude is largest in the upper left quadrants of (A), (B), and (C), and this corresponds to a slit angle of 30° to 50° as shown in (G). The slit orientation corresponding to 30° to 50° is shown in (H). Similarly, (A), (B), and (C) show that minimum evoked potential magnitude was recorded for slit angles of 110° to 140°. (D) Evoked potentials recorded during one slit rotation after astigmatism had been corrected. (E) Similar to (A), (B), and (C), but recorded in 10 sec. (F) 5μV calibration. Accommodation not paralyzed. (From Regan.[18] By permission of *Investigative Ophthalmology and Visual Science*.)

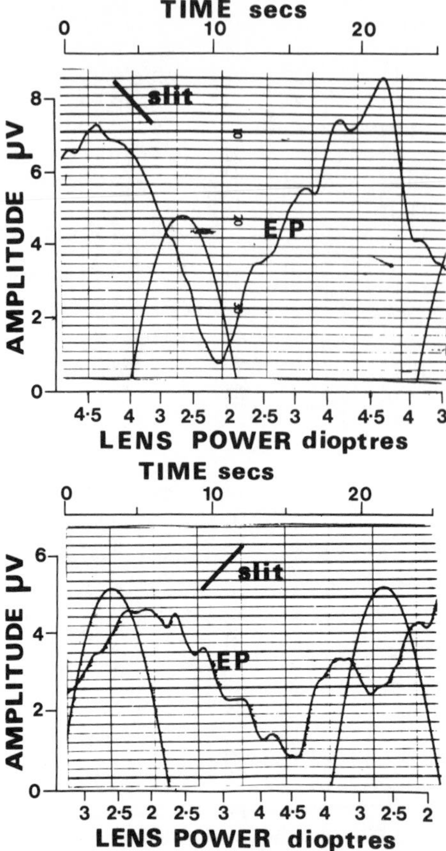

FIGURE 13. Determination of lens prescription for optimal correction of refractive errors using variable-power lens. The axes of astigmatism were first found as shown in FIGURE 12. The stenopaeic slit was then set parallel to one axis of astigmatism and the subject viewed a checkerboard pattern through a lens whose power oscillated from −2 D to −4.5 D once every 18 sec. The stenopaeic slit was then rotated through 90° and another 18-sec recording made. For one slit position evoked potential amplitude peaked near −4.5 D, and for the other slit position evoked potential amplitude peaked at −2 to −2.5 D. Accommodation was not paralyzed. (From Regan.[18] By permission of *Investigative Ophthalmology and Visual Science*.)

can be compared. The "method of simultaneous stimulation" was designed to counter this problem.[41] This method is to stimulate several sites simultaneously. Each stimulus has a slightly different modulation rate, and there is a separate Fourier analyzer locked to each stimulus. This method takes advantage of the very sharp temporal tuning of a Fourier analyzer, since a Fourier analyzer can easily separate temporal modulation rates that are so closely spaced as to be effectively identical to the central nervous system (e.g., 0.15 Hz apart). The success of this method depends on the empirical observation that the *relationship* between evoked potentials elicited from different parts of the visual field remains fairly constant, even when individual evoked potentials change appreciably.

The simultaneous stimulation method has so far only been used to record pattern and flicker evoked potentials from multiple sites in the visual field (although the technique is also applicable to recording somatosensory evoked potentials from multiple sites on the body surface and to recording auditory evoked potentials for several simultaneous carrier frequencies). Two sites of stimulation have been used to investigate hemi-field effects for pattern and for flicker in patients with cortical lesions,[41] and migraine. Four simultaneous sites of stimulation in the four retinal quadrants have been employed to study visual field defects using pattern stimulation and flicker stimulation in patients with cortical damage due to stroke, space-occupying lesions, and after neurosurgery.[43] In these studies, flicker evoked potentials were recorded in both the "high-frequency" region (40–60 Hz), "medium-frequency" region (13–25 Hz), and "low-frequency" region (near 10 Hz) (FIGURE 6). As discussed previously,[2,39] evoked potentials in these three frequency regions seem to be generated in different areas of the brain, so it was envisaged that combining different retinal sites of stimulation with different types of visual stimulus might given further information as to the site of cortical damage.

Although the simultaneous stimulation technique did achieve its object of reducing the effects of slow changes in the evoked potential, by improving accuracy it revealed an inherent problem with studies involving multiple sites of stimulation. This problem limits the value of evoked potential recording as an objective equivalent of perimetry. The problem is that, in control subjects, intersubject variations are considerably greater for stimuli of small diameter than for large stimulus fields. Thus, for example, in two control subjects with fairly similar evoked potentials to full-field pattern stimulation, one subject had a quadrantic absence of pattern evoked potentials that could not be distinguished from patients with a genuine quadrantic field defect caused by a cortical lesion, whereas the other control subject had roughly similar evoked potentials from each quadrant of the visual field.[44] Again, some control subjects who show a polarity reversal for upper versus lower half-field stimulation with a 6° field do not show a polarity reversal with a 3° field, though other control subjects give a polarity reversal for both field sizes. One possible explanation for this finding is in terms of neuroanatomical variability from subject to subject.[45]

The Method of Simultaneous Stimulation: Multiple Sine-Wave Stimuli

One way of obtaining the apparent latency is to record the phase of steady-state evoked potentials to several different frequencies of stimulation on several different occasions. A drawback to this method is that, if the evoked potential phases change with time, these changes limit the accuracy with which the slope of the graph can be estimated. A variant of the "simultaneous stimulation" method designed to counter this problem (FIGURE 14) is to simultaneously stimulate a single retinal site at several widely different temporal frequencies, and to record evoked potentials with several Fourier analyzers, one for each stimulus frequency.[46] FIGURE 14A shows how stimulus luminance could be modulated by a mixture of three sine waves. FIGURE 14B illustrates that, although the absolute phase of the evoked potential changed considerably between runs, nevertheless the slope of the plot (and hence the apparent latency) was comparatively stable. As illustrated, apparent latency was 173 msec in one run and 165 msec in the other, a difference of only about 6%.

This technique has so far been used with flicker evoked potentials and with pattern evoked potentials only, though it has wider applicability. An illustration of its use was to carry out serial measurements of apparent latency in multiple sclerosis (MS) patients while the patients' body temperature was being changed.[47]

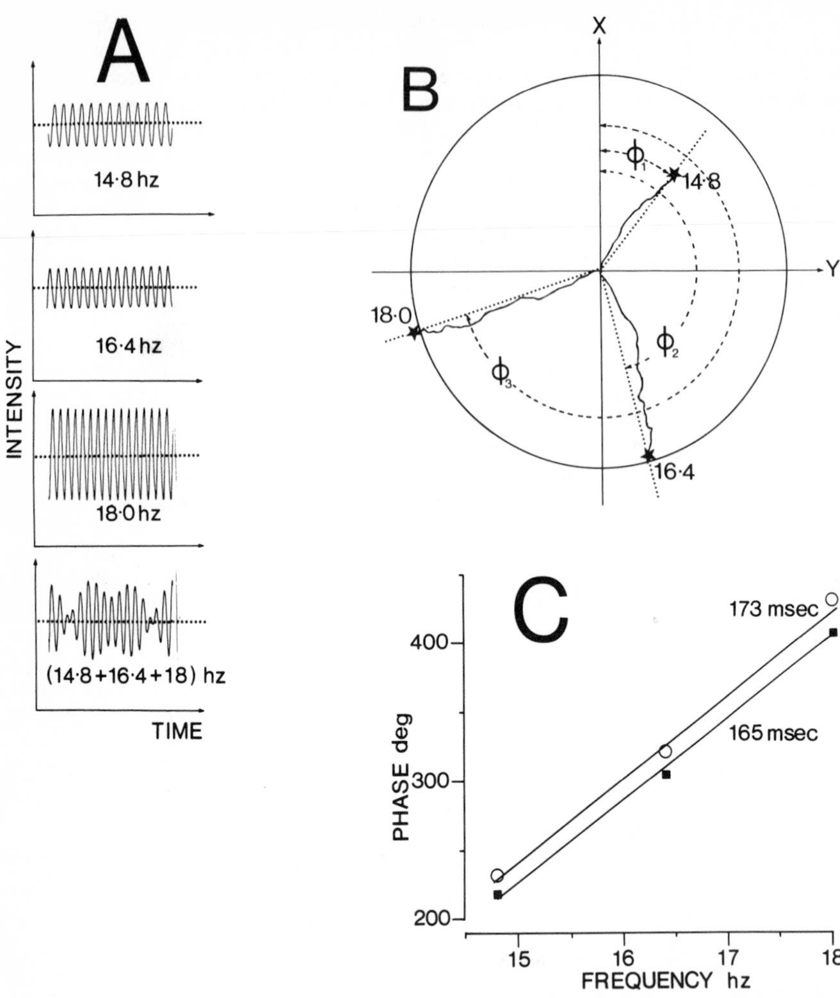

FIGURE 14. (A) Modulating sine waves of frequencies of 14.8, 16.4, and 18.0 Hz and (lowermost) the complex modulating waveform created by adding the three sinewaves. (B) Display of three simultaneously recorded evoked potentials produced by viewing the irregularly flickering light whose timecourse is shown in the lowermost graph of (A). The continuous lines plot the time integrals of the three evoked potentials over 1 min. The radial distance of the superposed star marked 14.8 Hz gives the amplitude of the 14.8 Hz frequency component of the evoked potential while the angle ϕ_1 gives its phase. The stars marked 16.4 and 18.0 Hz similarly give the amplitudes and phases of the 16.4 and 18.0 Hz frequency components of the evoked potential. The circle (continuous line) shows the radial distance corresponding to 5 μV. (C) Plots of evoked potential phase versus frequency for two 1-min stimulations similar to that illustrated in (B). Regression lines are drawn through each set of three points. Note that the slopes of the lines (and hence the apparent latency of the evoked potential) is here little affected by bodily displacement of the plot. (From Regan.[46] By permission of *Electroencephalography and Clinical Neurophysiology*.)

The Method of Evoked Potential Feedback

The method of evoked potential feedback was designed as an efficient means of performing "constant criterion" experiments electrophysiologically. Suppose that we wish to plot the contrast of the stimulus grating that gives a constant response, for example, the stimulus contrast that gave some criterion amplitude or some criterion latency. One way of reducing the degrading effect upon accuracy of slow changes in the evoked potential is to arrange that moment-to-moment values of evoked potential amplitude (or latency) control the stimulus so as to maintain a constant criterion response. In our example, this can be done by arranging that moment-to-moment values of evoked potential amplitude exert continuous control on the contrast of the

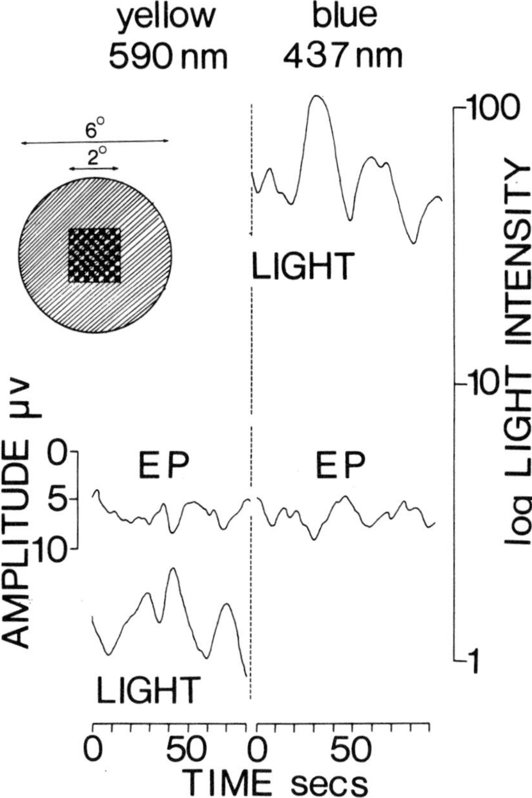

FIGURE 15. Evoked potential feedback when brain signals control the stimulus. The visual stimulus shown in the insert was a red pattern of bright and dark checks that exchanged places 6 times per sec. Superposed on the pattern was a desensitizing patch of light. The pattern elicited a 6-Hz evoked potential, whose amplitude continuously controlled the intensity of the desensitizing light by means of a neutral density wedge in the desensitizing beam. This feedback was set to maintain evoked potential amplitude constant at 6 μV. When the color of the desensitizing light was changed from yellow to blue, the evoked potential immediately moved the wedge so as to increase the intensity of the desensitizing light by about 40 times. This means that the red pattern response is about 40 times more sensitive to yellow than to blue light. (From Regan.[48] By permission of *Nature,* **253:** 401–408, © 1975, Macmillan Journals Limited.)

stimulus grating so that contrast continually "hunts" about the value that gives a criterion response of, say, 2.0 μV amplitude. If, then, spatial frequency is slowly swept over the range of interest, taking something between 0.5 and 3 min per sweep, then stimulus contrast will always assume the value required to give a criterion 2.0 μV response, so that an automatic plot of stimulus contrast versus spatial frequency would constitute the subject's sensitivity curve for a criterion response of 2.0 μV. In order to improve the signal-to-noise ratio, several sweeps can be averaged.

FIGURE 15 illustrates the application of this technique for which it was first developed.[33,34,48]

Some care was needed in selecting a criterion level for evoked potential amplitude that was not too high. If this were set too near the "knee" (FIGURE 8, lower half), a fluctuation might take the ND wedge to a position where evoked potential amplitude was *attenuated* by increasing contrast so that feedback became ineffective (one way of reducing this problem was to use patterns of lower contrast, e.g., 30%).

Estévez, Spekreijse, Berg, and Cavonius[49] combined evoked potential feedback with the "silencing" method to measure spectral sensitivities of color channels in normal and color-blind subjects, and were able to carry out these measurements within quite short recording durations.

The technique illustrated in FIGURE 17 can be regarded as an objective equivalent of Stiles' two-color increment threshold technique with the advantage that Stiles'

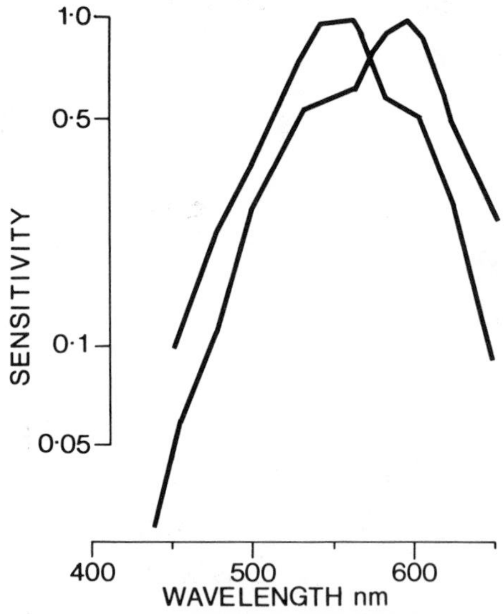

FIGURE 16. Spectral sensitivities for the red and green channels of pattern vision. The red channel (predominantly) was stimulated by a foveal red (676 nm) pattern of fine checks as in FIGURE 15 and a desensitizing patch of light was superposed on the red pattern. The intensities of the desensitizing patch needed for a given attenuation of the red pattern evoked potential were plotted (reciprocally) as ordinates against desensitizing wavelength, giving the right-hand curve. The experiment was repeated with a green (544 nm) pattern to obtain an approximation to the spectral sensitivity of the green channel (left-hand curve). Note: because of the overlap of red and green spectral sensitivities, the color channels are not completely separated by this technique. (From Regan.[48] By permission of *Nature,* **253:** 401–408, © 1975, Macmillan Journal Limited.)

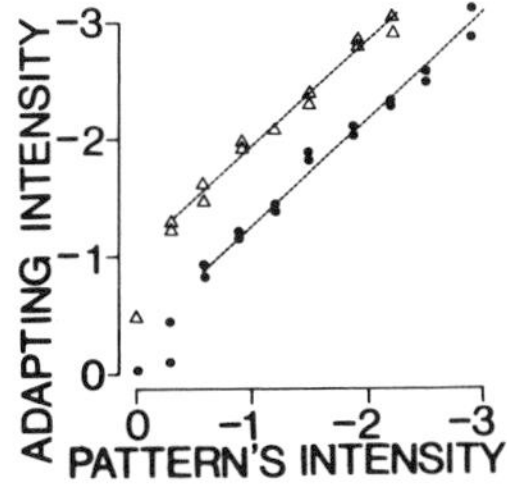

FIGURE 17. Evoked potential analogue of Stiles' two-color increment threshold technique. Abscissa are checkerboard pattern's intensity. Ordinates plot the way in which the intensity of the adapting light was changed in order to maintain evoked potential amplitude constant. Filled circles: both pattern and adapting light at a wavelength of 676 nm. Open triangles: pattern, 544 nm; adapting light, 619 nm. (From Regan.[34] By permission of *Vision Research*.)

psychophysical technique is restricted to threshold levels, whereas the evoked potential equivalent can operate at high superthreshold levels as well as at near threshold levels. However, the evoked potential method is not without disadvantages. One of these is that pattern evoked potentials are quite small when stimulus luminance is reduced below a critical level. Consequently, the plots of FIGURE 17 cannot be obtained over as wide a range of intensities as the psychophysical curves, and this limits the effectiveness of the evoked potential technique in separating different color mechanisms.

REFERENCES

1. DAWSON, G. D. 1953. Autocorrelation and automatic integration. Electroenceph. Clin. Neurophysiol. (Suppl.) **4:** 26–37.
2. REGAN, D. 1972. Evoked Potentials in Psychology, Sensory Physiology and Clinical Medicine. Chapman & Hall, London; Wiley, New York, N.Y.
3. JEFFREYS, D. A. & J. G. AXFORD. 1972. Source locations of pattern-specific components of human visual evoked potentials. Exp. Brain Res. **16:** 1–40.
4. JOHN, E. R., D. S. RUCHKIN & J. VILLEGAS. 1964. Experimental background: Signal analysis and behavioral correlates of evoked potential configurations in cats. Ann. N.Y. Acad. Sci. **112:** 362–420.
5. RUCHKIN, D. S., J. VILLEGAS & E. R. JOHN. 1964. An analysis of averaged evoked potentials making use of least mean squares techniques. Ann. N.Y. Acad. Sec. **115:** 799–826.
6. JOHN, E. R., P. WALKER, D. CAWOOD, M. RUSH & J. GEHRMANN. 1973. Factor analysis of evoked potentials. Electroenceph. Clin. Neurophysiol. **34:** 33–43.
7. DONCHIN, E. 1966. A multivariate approach to the analysis of average evoked potentials. IEEE Trans. Biomed. Eng. BME-**13:** 131–139.
8. DONCHIN, E. 1969. Data analysis techniques in average evoked potential research. *In* Average Evoked Potentials: Methods, Results and Evaluations. E. Donchin & D. B. Lindsley, Eds. U.S. Govt. Printing Office, Washington, D.C.
9. DONCHIN, E. 1979. Event related potentials: A tool in the study of human information processing. *In* Evoked Potentials in Psychiatry. H. Begleiter, Ed. Plenum Press, New York, N.Y.
10. JOHN, E. R., B. Z. KARMEL, W. C. CORNING, P. EASTON, D. BROWN, H. AHN, T. HARMONY, L. PRICHEP, A. TORO, I. GERSON, F. BARTLETT, R. THATCHER, H. KAYE, P. VALDES & E. SCHWARTZ. 1977. Neurometrics. Science **196:** 1393–1410.
11. REGAN, D. 1964. A study of the visual system by the correlation of visual stimuli and evoked electrical responses. Thesis, University of London.
12. REGAN, D. 1966. Some characteristics of average steady-state and transient responses evoked by modulated light. Electroenceph. Clin. Neurophysiol. **20:** 238–248.
13. BODE, H. W. 1945. Network Analysis and Feedback Amplifier Design. Van Nostrand. Englewood Cliffs, N.J.
14. REGAN, D. 1968. A high frequency mechanism which underlies visual evoked potentials. Electroenceph. Clin. Neurophysiol. **25:** 231–237.

15. REGAN, D. 1970. Evoked potential and psychophysical correlates of changes of stimulus colour and intensity. Vision Res. **10:** 163–178.
16. FREEMAN, F. D. & L. N. THIBOS. 1975. Visual evoked responses in human with abnormal visual experience. J. Physiol. (Lond.) **247:** 711–724.
17. REGAN, D. 1968. Chromatic adaption and steady-state evoked potentials. Vision Res. **8:** 149–158.
18. REGAN, D. 1973. Rapid objective refraction using evoked brain potentials. Invest, Ophthalmol. Vis. Sci. **19:** 669–679.
19. SPEKREIJSE, H. 1966. Analysis of EEG Responses in Man. Thesis, Amsterdam. W. Junk, Publ. The Hague, Netherlands.
20. TYLER, C. W., P. APKARIAN & K. NAKAYAMA. 1978. Multiple spatial frequency tuning of electrical responses from human visual cortex. Exp. Brain Res. **27:** 203–209.
21. TWEEL, L. H. VAN DER, C. W. SEM-JACOBSEN, A. KAMP, W. S. VAN LEEUWEN & F. T. H. VERINGA. 1958. Objective determination of responses to modulated light. Acta Physiol. Pharmacol. Neerl. **7:** 528.
22. CAMPBELL, F. W. & L. MAFFEI. 1970. Electrophysiological evidence for the existence of orientation and size detectors in the human visual system. J. Physiol. (Lond.) **207:** 635–652.
23. COBB, W. A., H. B. MORTON & G. ETTLINGER. 1967. Cerebral potentials evoked by pattern reversal and their suppression in visual rivalry. Nature **216:** 1123–1125.
24. TWEEL, L. H. VAN DER. 1964. Relation between psychophysics and electrophysiology of flicker. Doc. Ophthal. (Den Haag) **18:** 287–304.
25. REGAN, D. 1970. An objective method of measuring the relative spectral luminosity curve in man. J. Opt. Soc. Am. **60:** 856–859.
26. TWEEL, L. H. VAN DER & H. F. E. VERDUYN LUNEL. 1965. Human visual responses to sinusoidally modulated light. Electroenceph. Clin. Neurophysiol. **18:** 587–598.
27. REGAN, D. 1973. Parallel and sequential processing of visual information in man investigated by evoked potential recording. *In* Photophysiology, Vol. 8: 185–208. Academic Press, New York, N.Y.
28. DIAMOND, S. P. 1964. A simple programming technique for comparing average responses. Electroenceph. Clin. Neurophysiol. **16:** 305–307.
29. REGAN, D. 1980. Speedy evoked potential methods for assessing vision in normal and amblyopic eyes: Pros and cons. Vision Res. **20:** 265–269.
30. WOODY, C. D. 1967. Characterization of an adaptive filter for the analysis of variable latency neuroelectric signals. Med. Biol. Eng. **5:** 539–556.
31. DAVISSON, L.D. 1966. A theory of adaptive filtering. IEEE Trans. Inf. Theory IT-**12:** 97–102.
32. GROGINSKY, H. L., L. R. WILSON & D. MIDDLETON. 1966. Adaptive detection of statistical signals in noise. IEEE Trans. Inf. Theory IT-**12:** 337–348.
33. REGAN, D. 1974. Colour channels in pattern vision: Electrophysiological evidence in man. Nature (Lond.) **250:** 437–439.
34. REGAN, D. 1975a. Colour coding of pattern responses in man investigated by evoked potential feedback and direct plot techniques. Vision Res. **15:** 175–183.
35. TYLER, C. W., K. NAKAYAMA, P. APKARIAN & D. M. LEVI. 1981. VEP assessment of visual function. Vision Res. **21:** 607–609.
36. REGAN, D., N. A. M. SCHELLART, H. SPEKREIJSE & T. J. T. P. VAN DER BERG. 1975. Photometry in goldfish by electrophysiological recording. Vision Res. **15:** 799–807.
37. TYLER, C. W., P. APKARIAN, D. M. LEVI & K. NAKAYAMA. 1979. Rapid assessment of visual function: an electronic sweep technique for the pattern visual evoked potential. Invest. Ophthalmol. Vis. Sci. **18:** 703–713.
38. SPEKREIJSE, H., L. H. KHOE & L. H. VAN DER TWEEL. 1972. A case of amblyopia. *In* The Visual System. G. B. Arden, Ed. Plenum Press, New York, N.Y.
39. REGAN, D. 1977. Steady state evoked potentials in man. J. Opt. Soc. Am. **67:** 1475–1489.
40. MOSKOWITZ-COOK, A. & S. SOKOL. 1980. Spatial and temporal interaction of pattern-evoked cortical potentials in human infants. Vision Res. **20:** 699–709.
41. REGAN, D. & J. R. HERON. 1969. Clinical investigations of lesions of the visual pathway: A new objective technique. J. Neurol. Neurosurg. Psychiatry **32:** 479–483.

42. REGAN, D. & J. R. HERON. 1970. Simultaneous recording of visual evoked potentials from the left and right hemispheres in migraine. *In* Background to Migraine. A. L. Cochrane, Ed. pp. 66–77. Heinemann, London.
43. MILNER, B. A., D. REGAN & J. R. HERON. 1971. Theoretical models of the generation of steady-state evoked potentials, their relation to neuroanatomy and their relevance to certain clinical problems. *In* The Visual System. G. B. Arden, Ed. Plenum Press, New York, N.Y.
44. REGAN, D. & B. A. MILNER. 1978. Objective perimetry by evoked potential recording limitations. Electroenceph. Clin. Neurophysiol. **44:** 393–397.
45. REGAN, D. 1973. Evoked potentials specific to spatial patterns of luminance and colour. Vision Res. **13:** 2381—2401.
46. REGAN, D. 1976. Latencies of evoked potentials to flicker and to pattern speedily estimated by simultaneous stimulation method. Electroenceph. Clin. Neurophysiol. **40:** 654–660.
47. REGAN, D., T. J. MURRAY & R. SILVER. 1977. Effect of body temperature on visual evoked potential delay and visual perception in multiple sclerosis. J. Neurol. Neurosurg. Psychiatry. **40:** 1083–1091.
48. REGAN, D. 1975. Recent advances in electrical recording from the human brain. Nature (review article) **253:** 401–407.
49. ESTEVEZ, O., H. SPEKREIJSE, T. J. T. P. VAN DER BERG & C. R. CAVONIUS. 1975. The spectral sensitivities of isolated colour mechanisms determined from contrast EP measurements. Vision Res. **15:** 1205–1212.
50. SPEKREIJSE, H., L. H. VAN DER TWEEL & D. REGAN. 1972. Interocular sustained suppression: Correlations with evoked potential amplitude and distribution. Vision Res. **12:** 521–526.
51. MILNER, B. A., D. REGAN & J. R. HERON. 1974. Differential diagnosis of multiple sclerosis by visual evoked potential recording. Brain **97:** 755–772.
52. REGAN, D. 1973. An evoked potential correlate of colour: evoked potential findings and single cell speculations. Vision Res. **13:** 1933–1941.
53. REGAN, D., K. I. BEVERLEY & H. MACPHERSON. 1982. Sinewave grating evoked potentials in amblyopic children and in control subjects. Invest. Ophthalmol. Vis. Sci. (in preparation).

SEQUENTIAL ANALYSIS OF THE VISUAL EVOKED POTENTIAL SYSTEM IN MAN; NONLINEAR ANALYSIS OF A SANDWICH SYSTEM*

Henk Spekreijse† and Dick Reits

The Netherlands Ophthalmic Research Institute and The Laboratory of Medical Physics University of Amsterdam Amsterdam, the Netherlands

Although linear systems exist only in theoretical models, the application of linear techniques in the analysis of physical systems is well established. The reason for this is that nonlinear systems can frequently be investigated in such a way that their behavior is approximately linear, for example, by using "small" signals. In that situation the full power of a linear analysis with its predictive potential can be employed. It should be emphasized that a "perturbation" analysis holds only for signals that are weak at the input of the (smooth) nonlinear elements in a system. Therefore, a constant response criterion, just differing from noise, fulfills this condition only if the threshold "detector" is the last and sole nonlinearity in the chain of transformations. Without this consideration, erroneous conclusions can be reached about the system elements. Suppose, for example, that the physiological system responsible for flicker perception consists of the following three sequential stages: (a) a static square root (saturating) element, (b) a first-order, low-pass linear filter, and (c) a threshold detector. If this system is tested with a sine-wave-modulated light of constant mean intensity, then for stimulus frequencies exceeding the cut-off frequency of the linear filter, increasingly higher modulation depths are required for threshold. To maintain a constant amplitude at the output of the linear filter for increasing frequencies, the input amplitude of the filter must increase proportionally to frequency. Because of the square root element, to obtain a doubling of amplitude at the input of the filter with a doubling of frequency, the stimulus amplitude must quadruple. This results in a high-frequency attenuation that is twice as steep as that actually present. Thus, a constant response analysis indicates a second-order low-pass filter followed by a threshold detector. This erroneous conclusion could have been avoided with the consideration that a constant response approach is applicable only if, in addition to amplitude, all other aspects of the response, like latency, waveform, and so forth, are kept constant! These restrictions are frequently ignored in psychophysics.

When nonlinear aspects of a system are taken into account, the results of an input–output analysis have by definition little predictive value. As the superposition principle holds only for linear systems, knowledge of the input–output relation of a nonlinear system is only valid for the actual stimulus employed. One may argue that this is not the case for a nonlinear system that can be expanded in a Volterra series.[1,2] However, for a complete description *all* Wiener kernels of such a system must be determined. This is not a realistic proposition. In general, a complete description of a nonlinear system is not feasible; the class of input signals is always restricted and all

*This work was supported by grants from the Netherlands Organization for the Advancement of Pure Research (Z.W.O.) and from the Netherlands Organization for Health Research (T.N.O.).

†Address for correspondence: Lab. of Medical Physics, Herengracht 196, 1016 BS Amsterdam, The Netherlands.

0077–8923/82/0388–0072 $1.75/0 © 1982, NYAS

aspects of the output signal cannot be evaluated. Thus, each nonlinear system presents therefore a fresh problem to be solved in its own terms with a "tailored" approach.

Yet there is a paradox here. When a system is nonlinear, sometimes more information, such as organization of successive stages, can be extracted. This approach is especially advantageous in the analysis of biological systems, which in general are accessible to measurement only at a limited number of sites. Furthermore, in a linear system additive noise (e.g., "spontaneous" activity) is no more than an annoyance (the superposition principle holds!), whereas in a nonlinear system noise can modify the input–output relationship.

This paper is restricted to the analysis of a simple nonlinear system consisting of a zero-memory nonlinearity $f(v)$ sandwiched between two linear processes $h_1(\tau)$ and $h_2(\tau)$ (FIGURE 1). Sandwich models have proven to be useful for the analysis of sensory pathways.

Relevant papers on the analysis of the peripheral hearing organ are Weiss,[3] Duifhuis,[4] de Boer and de Jongh,[5] and the Ph.D. thesis of de Jongh,[6] where an exhaustive discussion of the several quantitative models on cochlear encoding can be found. The work of Jones and Milsum[7] is a good introduction to nonlinear analysis of the vestibular system.

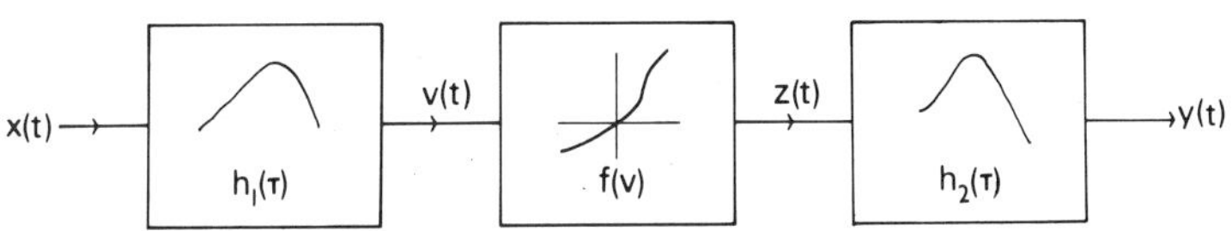

FIGURE 1. A sandwich nonlinear system. The two linear processes $h_1(\tau)$ and $h_2(\tau)$ are separated by the static nonlinearity $f(v)$.

For the visual system the following references may serve as examples of nonlinear analysis:

(1) The pupillary reflex system: Varju,[8] Sandberg and Stark,[9] Kohn and Clynes,[10] Watanabe and Stark.[11]
(2) The direct visual pathway in vertebrates:
  (a) At retinal single cell level: Spekreijse,[12] Schellart and Spekreijse,[13] Marmarelis and Naka,[14] Hochstein and Shapley,[15] Victor and Shapley;[16]
  (b) At peripheral (ERG) level: van der Tweel,[17] Spekreijse *et al.*,[18] Reits,[19] Koblasz;[20]
  (c) At intermediate (subcortical and LGN) level: Lopes da Silva,[21] Gielen;[22]
  (d) At central (visual evoked potentials) level: Spekreijse,[23] van der Tweel and Spekreijse,[24] Dummermuth *et al.*,[25] Reits.[19]
(3) The invertebrate visual system: Reichardt,[26] Marmarelis and McCann,[27] Gemperlein and McCann.[28]

Our purpose is to introduce three different analysis methods for dissecting a sandwich system. These analyses of input–output relationship dissect the system into its various elements and place them in proper sequence. The three methods are: (a) the Gaussian white noise approach resulting in a description of the system based on ("Wiener") kernels, (b) a two sinusoidal approach resulting in a description of the system on sum and difference frequencies, and (c) a linearizing approach in which the harmonic response to a sinusoidal stimulus is modified by an auxiliary signal.

These three methods are not the only ones available for the study of a sandwich system—for example, Victor and Knight[29] present an approach based on N sinusoids—nor are they mutually exclusive. They are, however, optimal from a practical point of view as will be shown in the discussion section. The method selected for an actual analysis depends on the computing facilities available as well as the dynamic characteristics of the two linear elements in the sandwich system. For example, if both behave like a low-pass filter and if the cut-off frequency of the second $h_2(\tau)$ filter lies far below that of the first $h_1(\tau)$ filter, then an analysis on the basis of the difference frequency is indicated. Since this paper was written for an evoked potential (EP) conference, the three methods will be illustrated with examples derived from EP studies on luminance and spatial contrast processing in man.

Wiener Kernel Approach

Wiener[30] introduced the concept that the input–output relation of nonlinear systems can be characterized by analyzing their response to Gaussian (white) noise stimulation. Such an analysis results in a description of the input–output relation on the basis of what is now generally known as Wiener kernels. These kernels are estimated by polycorrelation of input and output signals.[31] In the next paragraphs the mathematical tools will be presented before applying this approach to the analysis of a sandwich system. A thorough presentation written for the analysis of physiological systems can be found in the textbook by Marmarelis and Marmarelis.[34]

Poly Correlation Functions

Higher-order correlation functions exist only if the stationary random processes studied fulfill several conditions. For more details the reader is referred to the article of Rosenblatt and Van Ness.[35] The present paper concerns *physical* processes that, in general, automatically fulfill the required conditions.

For simplicity we assume that the mean value of all input signals $x(t)$ considered equals zero.

The moment functions of such a signal are defined as:

$$m_1(t) = E\{x(t)\} = \lim_{T\to\infty} \frac{1}{2T}\int_{-T}^{+T} x(t)\,\mathrm{d}t = \overline{x(t)} = 0$$

$$m_2(t, t+\tau) = E\{x(t)x(t+\tau)\} = \lim_{T\to\infty} \frac{1}{2T}\int_{-T}^{+T} x(t)x(t+\tau)\,\mathrm{d}t$$

$$m_3(t, t+\tau_1, t+\tau_2) = E\{x(t)x(t+\tau_1)x(t+\tau_2)\}$$

$$= \lim_{T\to\infty} \frac{1}{2T}\int_{-T}^{+T} x(t)x(t+\tau_1)x(t+\tau_2)\,\mathrm{d}t,$$

etc.

Since higher-order moments are related to each other, it is advantageous to define the poly correlation functions $R_{x\ldots x}(\tau_1 \ldots \tau_n)$ of a signal as the independent cumulant functions $S_n(t, t+\tau_1, \ldots, t+\tau_{n-1})$ rather than as the moment functions $m_n(t, t+\tau_1, \ldots, t+\tau_{n-1})$.[36]

For random physically realizable signals $x(t)$ with mean zero, the first three cumulant and moment functions are equal:

$$s_1(t) = m_1(t) = 0$$

$$s_2(t, t+\tau) = m_2(t, t+\tau) = R_{xx}(\tau)$$

$$s_3(t, t+\tau_1, t+\tau_2) = m_3(t, t+\tau_1, t+\tau_2) = R_{xxx}(\tau_1, \tau_2).$$

Higher-order cumulant and moment functions differ. For example:

$$\begin{aligned} R_{xxxx}(\tau_1, \tau_2, \tau_3) &= s_4(t, t+\tau_1, t+\tau_2, t+\tau_3) \\ &= m_4(t, t+\tau_1, t+\tau_2, t+\tau_3) \\ &\quad - m_2(t, t+\tau_1)m_2(t+\tau_2, t+\tau_3) \\ &\quad - m_2(t, t+\tau_2)m_2(t+\tau_2, t+\tau_3) \\ &\quad - m_2(t, t+\tau_3)m_2(t+\tau_1, t+\tau_2). \end{aligned}$$

Since, for a Gaussian noise signal $x(t)$, only the second cumulant differs from zero, $R_{xx}(\tau)$ is the sole autocorrelation function which differs from zero. All other poly correlation functions of this signal equal zero.

Similarly one can define the cross poly correlation functions between an input signal $x(t)$ and the output signal $y(t)$ of a system. In our treatment we restrict our analysis to those systems for which the input–output relation can be described by a Volterra expansion:

$$y(t) = \overline{y(t)} + \sum_{n=1}^{\infty} \int\int_0^{\infty} \cdots \int h_n(\tau_1, \ldots, \tau_n) \prod_{k=1}^{n} \{x(t-\tau_k)\, d\tau_k\}.$$

The most informative cross-correlation functions are:

$$R_{yx}(\tau) = \lim_{T\to\infty} \frac{1}{2T} \int_{-T}^{+T} \{y(t) - \overline{y(t)}\}x(t-\tau)\, dt$$

$$R_{yxx}(\tau_1, \tau_2) = \lim_{T\to\infty} \int_{-T}^{+T} \{y(t) - \overline{y(t)}\}x(t-\tau_1)x(t-\tau_2)\, dt.$$

For a Volterra system these functions become:

$$\overline{y(t)} = \int\int_0^{\infty} h_2(\tau_1, \tau_2)R_{xx}(\tau_1 - \tau_2)\, d\tau_1\, d\tau_2 + \cdots$$

$$\begin{aligned} R_{yx}(\tau) &= \int\int_0^{\infty} h_1(\tau_1)x(t-\tau_1)x(t-\tau)\, dt d\tau_1 \\ &\quad + \int\int_0^{\infty}\int h_2(\tau_1, \tau_2)x(t-\tau_1)x(t-\tau_2)x(t-\tau)\, d\tau_1 d\tau_2 dt + \cdots \\ &= \int_0^{\infty} h_1(\tau_1)R_{xx}(\tau-\tau_1)\, d\tau_1 + \int\int_0^{\infty} h_2(\tau_1, \tau_2)R_{xxx}(\tau-\tau_1, \tau-\tau_2)\, d\tau_1\, d\tau_2 \\ &\quad + \int\int_0^{\infty}\int h_3(\tau_1, \tau_2, \tau_3)[R_{xxxx}(\tau-\tau_1, \tau-\tau_2, \tau-\tau_3) \\ &\quad + R_{xx}(\tau-\tau_1)R_{xx}(\tau_2-\tau_3) + R_{xx}(\tau-\tau_2)R_{xx}(\tau_1-\tau_3) \\ &\quad + R_{xx}(\tau-\tau_3)R_{xx}(\tau_1-\tau_2)]\, d\tau_1 d\tau_2 d\tau_3 + \cdots \end{aligned}$$

If the input signal $x(t)$ has a symmetric amplitude distribution then all even autopoly correlation functions are zero. As a result the even terms in the above expression vanish. Furthermore, if $x(t)$ is a Gaussian noise, then all higher-order autopoly correlation functions are zero and the cross-correlation functions of the Volterra system simplify into:

$$R_{yx}(\tau) = \int_0^\infty h_1(\tau_1)R_{xx}(\tau - \tau_1)\,d\tau_1$$

$$+ \int\int_0^\infty\int h_3(\tau_1, \tau_2, \tau_3)[R_{xx}(\tau - \tau_1)R_{xx}(\tau_2 - \tau_3) + R_{xx}(\tau - \tau_2)R_{xx}(\tau_1 - \tau_3)$$

$$+ R_{xx}(\tau - \tau_3)R_{xx}(\tau_1 - \tau_2)]\,d\tau_1 d\tau_2 d\tau_3 + \cdots$$

$$R_{yxx}(\tau_1, \tau_2) = \int_0^\infty\int h_2(\tau_3, \tau_4)[R_{xx}(\tau_1 - \tau_3)R_{xx}(\tau_2 - \tau_4)$$

$$+ R_{xx}(\tau_1 - \tau_4)R_{xx}(\tau_2 - \tau_3)]\,d\tau_3 d\tau_4 + \cdots$$

Polyspectra

Polyspectra are defined as the Fourier transforms of the poly correlation functions. For example, the automonospectrum is:

$$S_{xx}(\omega) = \int_{-\infty}^{+\infty} R_{xx}(\tau)e^{-j\omega\tau}\,d\tau,$$

and the autobispectrum is:

$$S_{xxx}(\omega_1, \omega_2) = \int_{-\infty}^{+\infty} R_{xxx}(\tau_1, \tau_2)e^{-j(\omega_1\tau_1 + \omega_2\tau_2)}\,d\tau_1 d\tau_2.$$

Therefore, if the autobispectrum of a signal differs from zero, then its amplitude distribution is asymmetric. The skewness α and the kurtosis β of the amplitude distribution are

$$\alpha = \frac{R_{xxx}(0, 0)}{R_{xx}^{3/2}(0)}, \qquad \beta = \frac{R_{xxxx}(0, 0, 0)}{R_{xx}^2(0)}.$$

For a Gaussian noise feeding into a Volterra system one finds for the cross-monospectrum and cross-bispectrum, respectively:

$$S_{yx}(\omega) = H_1(j\omega)S_{xx}(\omega) + 3\int_{-\infty}^{\infty} H_3(j\omega_1, -j\omega_1, j\omega)S_{xx}(\omega)S_{xx}(\omega_1)\frac{d\omega_1}{2\pi} + \cdots$$

$$S_{yxx}(\omega_1, \omega_2) = 2H_2(j\omega_1, j\omega_2)S_{xx}(\omega_1)S_{xx}(\omega_2) + \cdots$$

with $H_n(j\omega_1, \ldots, j\omega_n)$ being the Fourier transform of $h_n(\tau_1, \ldots, \tau_n)$. For the simplest case of a second-order Volterra system:

$$y(t) = \overline{y(t)} + \int_0^\infty h_1(\tau)x(t - \tau)\,d\tau$$

$$+ \int\int_0^\infty h_2(\tau_1, \tau_2)x(t - \tau_1)x(t - \tau_2)\,d\tau_1 d\tau_2.$$

The cross-spectra become:

$$S_{yx}(\omega) = H_1(j\omega)S_{xx}(\omega) \text{ and}$$

$$S_{yxx}(\omega_1, \omega_2) = 2H_2(j\omega_1, j\omega_2)S_{xx}(\omega_1)S_{xx}(\omega_2).$$

Therefore the two kernels of this system can be estimated from:

$$H_1(j\omega) = \frac{S_{yx}(\omega)}{S_{xx}(\omega)} \quad \text{and} \quad H_2(j\omega_1, j\omega_2) = \frac{S_{yxx}(\omega_1, \omega_2)}{2S_{xx}(\omega_1)S_{xx}(\omega_2)}.$$

The following expression can be derived from the autospectrum of the output signal of a second-order Volterra system:

$$S_{yy}(\omega) = |H_1(j\omega)|^2 S_{xx}(\omega) + \int_{-\infty}^{+\infty} |H_2\{j(\omega - \lambda), j\lambda\}|^2 S_{xx}(\omega - \lambda)$$

$$S_{xx}(\lambda)\frac{d\lambda}{2\pi} = \frac{|S_{yx}(\omega)|^2}{S_{xx}(\omega)} + \frac{1}{2}\int_{-\infty}^{+\infty} \frac{|S_{yxx}(\omega - \lambda, \lambda)|^2}{S_{xx}(\omega - \lambda)S_{xx}(\lambda)}\frac{d\lambda}{2\pi}.$$

In this expression the first term represents the amount of power in $y(t)$ due to the first kernel, and the second term that due to the second kernel.

If one wants to estimate in the frequency domain $H_1(j\omega)$ and $H_2(j\omega_1, j\omega_2)$ on the basis of a least-square estimate, the best estimates are[37]

$$H_1(j\omega) = \frac{E[X^*(j\omega)Y(j\omega)]}{E[|X(j\omega)|^2]} \quad \text{and}$$

$$H_2(j\omega_1, j\omega_2) = \frac{E[X^*(j\omega_1)X^*(j\omega_2)Y\{j(\omega_1 + \omega_2)\}]}{2E[|X(j\omega_1)X(j\omega_2)|^2]},$$

with $X(j\omega)$ and $Y(j\omega)$ being the Fourier transform of input and output signals, respectively.

Coherence

To obtain a dimensionless value for the relative contribution of the different kernels to the output of a system, the coherence function has been introduced. The quadratic squared coherence, as introduced by Tick,[38] has the form:

$$C_{yx}^2(\omega) = \frac{|S_{yx}(\omega)|^2}{S_{xx}(\omega)S_{yy}(\omega)} + \frac{1}{2S_{yy}(\omega)}\int_{-\infty}^{+\infty} \frac{|S_{yxx}(\omega - \lambda, \lambda)|^2}{S_{xx}(\omega - \lambda)S_{xx}(\lambda)}\frac{d\lambda}{2\pi}.$$

If the system is linear, only the first term (called linear squared coherence) differs from zero and has the value 1 if $x(t)$ passes a noise-free linear stationary system, and is between 0 and 1 for all other conditions. If only the second-order kernel exists for a noise-free stationary system, the second term equals 1, and is between 0 and 1 in all other conditions. Goodman[39] extensively details the statistics of estimation of the linear squared coherence; such an analysis has not yet been published for higher-order coherences.

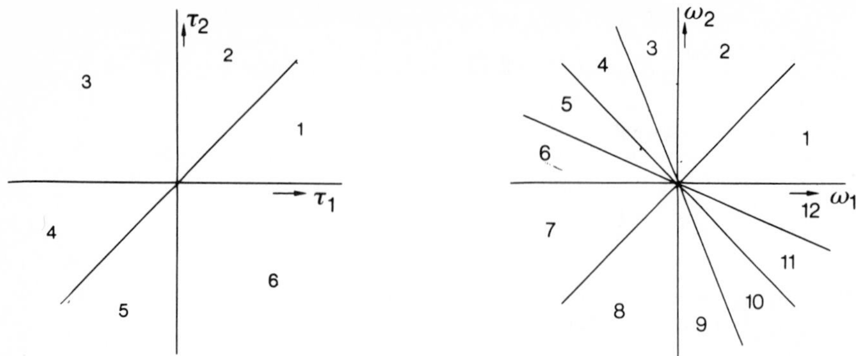

FIGURE 2. This figure shows the 6 sectors to which, because of the symmetry properties, calculation of the autobicorrelation function can be restricted. The right-hand side shows similarly the 12 sectors of the autobispectrum.

Symmetry Properties of Poly Correlation Functions and Spectra

For the analysis of real physical processes, one can use the symmetry properties of these functions to reduce calculation time.[35] For the mono- and bi-functions these properties are:

$$\begin{cases} R_{xx}(\tau) = R_{xx}(-\tau) \quad \text{and} \\ R_{xxx}(\tau_1, \tau_2) = R_{xxx}(\tau_2, \tau_1) = R_{xxx}(-\tau_1, \tau_2 - \tau_1) \end{cases}$$

and

$$\begin{cases} S_{xx}(\omega) = S^*_{xx}(-\omega) = S_{xx}(-\omega) = S^*_{xx}(\omega) \\ S_{xxx}(\omega_1, \omega_2) = S_{xxx}(\omega_2, \omega_1) = S_{xxx}(\omega_1, -\omega_1 - \omega_2) \\ \qquad = S^*_{xxx}(-\omega_1, -\omega_2) = S^*_{xxx}(-\omega_2, -\omega_1) \\ \qquad = S^*_{xxx}(-\omega_1, \omega_1 + \omega_2). \end{cases}$$

As shown in FIGURE 2, symmetry properties restrict the calculation to one of 6 sections for the bicorrelation function and to one of 12 (2×6 since phase information is not considered) for the bispectrum.

The sampling rate (ω_s) used for estimation of the bispectrum from real data determines the highest values of ω_1 and ω_2 to avoid aliasing: $\omega_{max} < \omega_s/2$.

For the cross-correlation functions and cross-spectra, the symmetry properties are:

$$\begin{cases} R_{yx}(\tau) \text{ none} \\ R_{yxx}(\tau_1, \tau_2) = R_{yxx}(\tau_2, \tau_1) \end{cases}$$

and

$$\begin{cases} S_{yx}(\omega) = S^*_{yx}(-\omega) \\ S_{yxx}(\omega_1, \omega_2) = S_{yxx}(\omega_2, \omega_1) = S^*_{yxx}(-\omega_1, -\omega_2) = S^*_{yxx}(-\omega_2, -\omega_1). \end{cases}$$

FIGURE 3 shows the quadrants in which the cross-bispectrum needs to be calculated. In the octants denoted by A and A^*, the sum frequencies $(\omega_1 + \omega_2)$ in the output signal contribute to the bispectrum; in those denoted by B and B^* the difference frequencies $(\omega_1 - \omega_2)$ contribute.

The line $\omega_1 = \omega_2$ is a line of real symmetry; the line $\omega_1 = -\omega_2$ is a line of complex conjugate symmetry. For a treatment in which the Gaussian noise input is approximated by a sum of sinusoids, see Victor and Knight.[29]

If there is a pure transport time in the system studied and if the autospectrum of the input signal is broad-band noise, then the cross-mono- and cross-bicorrelation functions will differ from zero only for positive time values. In such a situation only half of the cross-mono- and a single octant of the cross-bicorrelation function have to be estimated.

The sampling rate (ω_s) chosen determines the highest frequencies in the ω_1, ω_2 plane that contain independent information. In the sum frequency octant, $(\omega_1 + \omega_2)$ has to be smaller than $\omega_s/2$; in the difference octant, $\omega < \omega_s/2$ holds, restricting estimation of the cross-bispectrum to the dashed region in FIGURE 3.

Gaussian Noise Analysis of a Sandwich Nonlinear System

In this section we will apply the polyspectral analysis to a nonlinear system consisting of a zero-memory nonlinearity $f(v)$, preceded by a linear system with impulse response $h_1(t)$ and followed by a linear system with impulse response $h_2(t)$ (FIGURE 1). With $x(t)$ as the input signal, $v(t)$ as the output signal of the first linear process, $z(t)$ as the output signal of the static nonlinearity, and $y(t)$ as the output signal of the last linear process and thus of the complete nonlinear system, the following equations hold:

$$v(t) = \int_0^\infty h_1(\tau_1)x(t - \tau_1)\, d\tau_1$$

$$z(t) = f\left\{\int_0^\infty h_1(\tau_1)x(t - \tau_1)d\tau_1\right\}$$

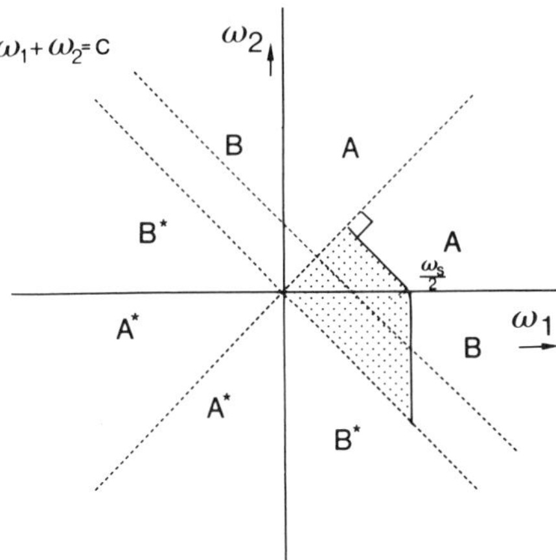

FIGURE 3. The dashed region is the segment of the cross-bispectrum to which the calculation can be restricted if the input and output signals are sampled at a rate of ω_s.

and

$$y(t) = \int_0^\infty h_2(\tau_2 f\left\{\int_0^\infty h_1(\tau_1)x(t-\tau_1-\tau_2)\,d\tau_1\right\}d\tau_2.$$

The different correlation functions between output and input signals $y(t)$ and $x(t)$ can be calculated easily by using the theorem of Price,[40] which was modified by Pawula[41] and extended by McGraw and Wagner[42] to include signals with elliptically symmetric amplitude distributions $P(k,l)$. Elliptically symmetric distributions are characterized by second-order probability densities whose iso-contour lines in the k,l plane are ellipses of the form: $[k^2 + l^2 - 2kl\rho(\tau)]$ with the correlation-coefficient $\rho(\tau) = R_{kk}(\tau)/R_{kk}(0)$.

Signals with a Pearson II (beta) amplitude density fulfill this property. Sinusoids and Gaussian noise belong to this class of signals. Since the amplitude distribution of the latter two signals is invariant after passing through a linear system, Price's theorem can be applied for the analysis of a sandwich system if $x(t)$ is a Gaussian noise signal. This theorem states that:

$$R_{zv}(\tau) = C_1 R_{vv}(\tau)$$

and

$$R_{zvv}(\tau_1,\tau_2) = C_2 R_{vv}(\tau_1) R_{vv}(\tau_2)$$

with

$$C_n = \int_{-\infty}^{+\infty} H_n\left(\frac{v}{\sigma_v}\right) f(v)p(v)\,dv \left(= \int_{-\infty}^{+\infty} \frac{d^n f(v)}{dv^n}\,p(v)\,dv\right),$$

$$p(v) = \frac{1}{\sigma_v(2\pi)^{1/2}} \exp(-v^2/2\sigma_v^2),$$

$$\sigma_v^2 = \frac{1}{2\pi}\int_{-\infty}^{+\infty} |H_1(j\omega)|^2 S_{xx}(\omega)\,d\omega,$$

and $H_n(v/\sigma_v)$ is the nth Chebychev-Hermite polynomial.

Since $p(v)$ is an even function of v, C_1 differs from zero if $df(v)/dv$ is even, and thus if $f(v)$ contains odd terms in v. Similarly, C_2 differs from zero if $f(v)$ contains even terms.

With this theorem and the above general expressions for $v(t)$, $z(t)$, and $y(t)$, one finds for the correlation functions between output and input signal of a sandwich system:

$$R_{yx}(\tau) = C_1 \int\int_0^\infty h_2(\tau_2)h_1(\tau_1)R_{xx}(\tau-\tau_1-\tau_2)\,d\tau_1 d\tau_2$$

$$R_{yxx}(\tau_1,\tau_2) = C_2 \int\int_0^\infty\int h_2(\tau)h_1(\tau_4)h_1(\tau_5)R_{xx}(\tau_1-\tau-\tau_4) \times R_{xx}(\tau_2-\tau-\tau_5)\,d\tau d\tau_4 d\tau_5.$$

and

$$S_{yx}(j\omega) = C_1 S_{xx}(\omega)H_1(j\omega)H_2(j\omega)$$

$$S_{yxx}(j\omega_1, j\omega_2) = C_2 S_{xx}(\omega_1)S_{xx}(\omega_2)H_2\{j(\omega_1+\omega_2)\}H_1(j\omega_1)H_1(j\omega_2).$$

From these two spectra one can calculate the amplitude characteristics of the two linear processes in the sandwich system in case both $C_1 \neq 0$ and $C_2 \neq 0$:[43]

$$|H_1(j\omega)| = \frac{C_1 |S_{yxx}(-\omega/2, \omega)|}{C_2 |S_{yx}(\omega/2) S_{xx}(\omega)|}$$

$$|H_2(j\omega)| = \frac{C_2 |S_{yx}(\omega) S_{yx}(\omega/2)|}{C_1^2 |S_{yxx}(-\omega/2, \omega)|}.$$

By varying the amplitude of the Gaussian noise signal $x(t)$, C_1 and C_2 can be estimated and hence the shape of the zero memory nonlinearity derived. This means that in practice [e.g., the nonlinearity meets the condition of expansion in Chebychev-Hermite polynomials, i.e., $|f(v)| < \exp(v^2)$] a *full* description of the three processes in the sandwich system can be obtained from the mono- and bispectra and that higher-order spectra do not provide independent information.

When even terms are lacking, $C_2 = 0$ and a higher-order spectrum like the cross-trispectrum is needed for sequential analysis. It is also possible to derive sequential information about a sandwich system from the autopolyspectra of $y(t)$.[44]

FIGURE 4A shows a cross-bispectrum for the brain potentials generated in man by stimulation with a Gaussian-noise-modulated light. The spectrum is depicted in the form of equipotential lines expressed in relative values of the maximal value (=100) of the spectrum. Along the line $\omega_1 + \omega_2 =$ constant, the values of the cross-bispectrum do not depend on $H_2(j\omega)$ but only on $H_1(j\omega)$. Furthermore, along this line the bispectrum has a derivative which is zero for $\omega_1 = \omega_2$. This can easily be understood by substituting in the above formula for S_{yxx} ($\omega_1 = \omega + h$ and $\omega_2 = \omega - h$) and taking the derivative d/dh. As seen in the figure, a maximum is nearly always found for $\omega_1 = \omega_2$. This means that $H_1(j\omega)$ has at most one clear maximum. This maximum can be read off by determining the line, $\omega_1 + \omega_2 =$ constant, along which the curvature of the bispectrum is the steepest. This can be understood if one realizes that, for $h \rightarrow 0$, the product $H_1\{j(\omega - h)\} H_1\{j(\omega + h)\}$ grows the fastest for values of $(\omega - h)$ and $(\omega + h)$ below and above the peak frequency of $H_1(j\omega)$. This is the region of the amplitude characteristic where the curvature is the strongest. If, on the other hand, the frequencies $(\omega - h)$ and $(\omega + h)$ were on those regions of the slopes of the amplitude characteristic where the attenuation per octave is constant, then the product would have been constant. For the bispectrum depicted in FIGURE 4A, this analysis results in an estimate for the maximum of $H_1(j\omega)$ between 15 Hz and 20 Hz. An indication of the peak frequences of the late frequency selective $H_2(j\omega)$ processes can be obtained by estimating the peak values along the dotted line in the figure. Peaks can be seen at around 10, 20, and 30 Hz, corresponding to, respectively, the low-, medium-, and high-frequency selective cortical processes.[45,46]

FIGURE 4B shows a cross-bispectrum evoked by Gaussian noise pattern reversal stimulation. The spectrum is depicted in the form of dots whose diameters represent the relative power. In contrast to the luminance bispectrum, *two* optimal regions can be distinguished, one peaking around 3 Hz and the other between 13 and 20 Hz. The latter reflects the same distal process as found with noise-modulated light.

A Two-Sinusoid Approach

Gaussian noise can also be described as a sum of an *infinite* number of sinusoids.[47] If a *finite* number of sinusoids yields a satisfactory description of a Volterra system,

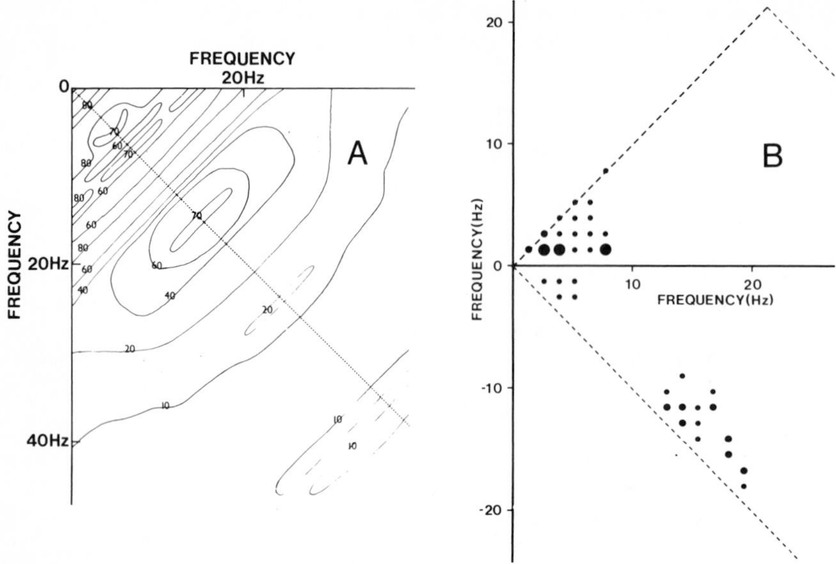

FIGURE 4. (A) Cross-bipowerspectrum to light modulated with Gaussian noise. Only the quadrant with sum frequencies is shown. The values along the equipotential lines are the relative amplitudes in percent of peak amplitude. (For stimulus details see Reits,[19] or Spekreijse *et al.*[45]). (B) Cross-bipowerspectrum to Gaussian-noise pattern reversal stimulation. Only the sector in the sum and difference frequency quadrants that contains independent information is shown (see also FIGURE 3). Two regions of maximal amplitude can be distinguished, one peaking around 3 Hz, the other around 16 Hz. These two regions represent the two distal processes which can also be estimated by a two sinusoidal approach (FIGURE 5). Stimulus details: mean luminance, 2000 asb [1 asb = $(1/\pi)$ cd/m^2]; field size, 3°; check size, 18 minutes of arc; bandwidth of Gaussian noise, 0–30 Hz; modulation depth, $\sigma = 17\%$.

then a sum of sinusoids analysis may be preferred above a noise analysis for computational speed. In this section we will demonstrate that, for the analysis of a sandwich system, a two-sinusoidal input can be advantageous. To use the characteristic function method, the Fourier transform $F(ju)$ of the nonlinearity $f(v)$ must exist:

$$|F(ju)| = \left|\int_{-\infty}^{+\infty} f(v)e^{-juv}\mathrm{d}v\right| < \infty.$$

With the input signal to the sandwich system as the sum of N sinusoids,

$$v(t) = \sum_{k=1}^{N} A_k \cos(\omega_k t + \phi_k),$$

one finds

$$z(t) = f\{v(t)\} = \frac{1}{2\pi}\int_C F(ju)e^{+juv(t)}\,\mathrm{d}u$$

$$= \frac{1}{2\pi}\int_C F(ju)\exp\left\{ju\sum_{k=1}^{N} A_k\cos(\omega_k t + \phi_k)\right\}\mathrm{d}u.$$

By the use of the Jacobi-Anger formula,

$$\exp[juA_k \cos(\omega_k t + \phi_k)] = \sum_{k=0}^{\infty} j^n \epsilon_n J_n(uA_k) \cos n(\omega_k t + \phi_k),$$

where ϵ_n is the Neumann factor ($\epsilon_0 = 1$, $\epsilon_n = 2$ for $n \neq 0$) and J_n is the Bessel function of the first kind, one finds finally that

$$z(t) = \frac{1}{2\pi} \int_C F(ju) \prod_{k=1}^{N} \sum_{n=0}^{\infty} j^n \epsilon_n J_n(uA_k) \cos n(\omega_k t + \phi_k)\, du,$$

where C is a contour extending from $-\infty$ to $+\infty$ selected to exclude singularities.

The purpose of the present study is to determine the smallest number of input sinusoids needed to analyze the sandwich system. If the linear elements in such a system are minimum phase filters, there is an unique relation between amplitude and phase:[48]

$$\ln|H(j\omega)| = -\frac{1}{\pi} \int_{-\infty}^{+\infty} \frac{\phi(u)}{u-\omega} du; \quad \phi(\omega) = \frac{1}{\pi} \int_{-\infty}^{+\infty} \frac{\ln|H(ju)|}{u-\omega} du.$$

Under this condition a sequential analysis can, in principle, be performed with a *single* sinusoid as input. If the linear elements do not fulfill the minimum phase restriction, as is generally the case for biological systems with transport times, then more input signals are required. In the next paragraph we will show that if only the amplitude information of the output signal is considered, a complete description can be obtained with two sinusoids as input.[4]

For $N = 2$ the above expression for $z(t)$ becomes

$$z(t) = \sum_{n=0}^{\infty} \sum_{m=0}^{\infty} B_{nm} \{\cos[n\omega_1 - m\omega_2)t + n\phi_1 - m\phi_2] + \cos[(n\omega_1 + m\omega_2)t + n\phi_1 + m\phi_2]\}$$

with

$$B_{nm} = \frac{1}{4\pi} \epsilon_m \epsilon_n j^{n+m} \int_C F(ju)\, J_n(uA_1)\, J_m(uA_2)\, du.$$

For an easy sequential analysis a number of conditions must be fulfilled:

(a) To exclude the influence of the last linear filter, $h_2(\tau)$, the *frequency* of the component considered in the output signal $y(t)$ must be constant.
(b) To exclude the influence of the nonlinearity the *amplitude* of the frequency component chosen for analysis must be constant and independent of the relative values of A_1 and A_2. This means that for the particular frequency component $m = n \neq 0$ and $\omega_1 \neq \omega_2$.

The frequency components in $z(t)$ of interest for this analysis are:

$$B_{nn}\{\cos[n(\omega_1 - \omega_2)t + n(\phi_1 - \phi_2)]\}$$

and

$$B_{nn}\{\cos[n(\omega_1 + \omega_2)t + n(\phi_1 + \phi_2)]\}, \qquad n = 1, \ldots, \infty.$$

When $f(v)$ can be expanded in a Taylor series

$$f(v) = \sum_{k=1}^{\infty} a_k v^k,$$

then for the sum and difference frequencies one finds:

$$B_{11} = \sum_{m=0}^{\infty} \frac{a_{2m+2}}{2^{2m+1}} \sum_{l=0}^{m} \binom{2m+2}{2l+1} \binom{2l}{l} \binom{2m-2l}{m-l} A_1^{2l} A_2^{2m-2l}.$$

In practice, higher harmonics of the sum and difference frequencies are of little interest for the analysis of a sandwich system, since they depend on higher-order terms of $f(v)$; e.g., a_4 is the lowest term contributing to B_{22}.

If the last process is a low-pass filter, then a straightforward analysis of the first filter is possible on the basis of the *difference* frequency ($\omega_1 - \omega_2$). With $\omega_1 = \omega_2 + \Delta\omega$ and $\Delta\omega$ small, identical amplitudes at the input of $h_1(\tau)$ will yield nearly identical amplitudes at the input of the nonlinearity: $A_1 \approx A_2 = A\,|H_1(j\omega)|$. Substitution in the above expression of B_{11} shows that the amplitude A at the input of the sandwich system will vary in an inversely proportional way to $|H_1(j\omega)|^2$ if one keeps the amplitude of the difference frequency $\Delta\omega$ constant at the output, regardless of ω. If, on the other hand, the last element is a band-pass filter, and if, as is typically the case, the system studied is noisy, the $\Delta\omega$ should be chosen near the peak frequency of the proximal element. In this situation, $|H_1(j\omega)|$ can be estimated reasonably well by the above procedure. Estimation improves the more the input frequencies differ from the cut-off frequencies of $H_1(j\omega)$.

An example of this procedure is given in FIGURE 5, which shows the amplitude characteristics of the distal $h_1(\tau)$ processes for both the luminance and the pattern reversal EPs to sinusoidal stimulation. The frequency component of 7.8 Hz was selected as difference frequency in the pattern reversal response since the signal-to-noise ratio of this component was the highest. The amplitude characteristic of the transient reversal EP drops rather steeply for higher frequencies.

On the other hand, a much lower difference frequency could not be selected in the contrast EP since we intended to use for comparison the same difference frequency for sequential analysis of the luminance EP. It is well known that those EPs can be generated up to much higher repetition rates than the reversal EPs but they decay much faster at lower rates.[46] So the 7.8 Hz is in essence a compromise.

Inspection of FIGURE 5 shows that sinusoidal pattern reversal and luminance stimulation yield different distal amplitude characteristics for low stimulus frequencies, but rather similar ones for high stimulus frequencies. If the frequency of the lowest component in the two input sinusoids f and f + 7.8 Hz becomes higher than 8 Hz, the two amplitude characteristics are similar and quite resemble the distal $h_1(\tau)$ process derived from the bispectrum to Gaussian noise stimulation (FIGURE 4A). Thus, the pattern reversal EP mediates partly through the same element that also yields the luminance EP. Further analysis indicated that this luminance component in the transient reversal EP is the first component in the typical NPN waveform of this response.[49] The other two components in this complex are mediated by a distal low-pass filter that, as can be seen in FIGURE 5, peaks around 3 Hz. Since this cut-off frequency is so low, sequential analysis on the basis of the output difference frequency seems indicated for the contrast EP. In the bottom half of FIGURE 5 also the phase of the difference frequency is given as the function of the input frequency f. Note that below and above 8 Hz two rather constant, slightly different branches can be distinguished, supporting our conclusion that two parallel distal processes contribute to the pattern reversal responses: a low-pass (sustained) one for the contrast and a band-pass (transient) one for the luminance component.

The Linearizing Approach

Instead of an analysis based on two sinusoids with a fixed phase relation, a single sinusoid together with another signal can be used for sequential analysis. The second signal may also be a sine wave but from an unrelated source. Inspection of the

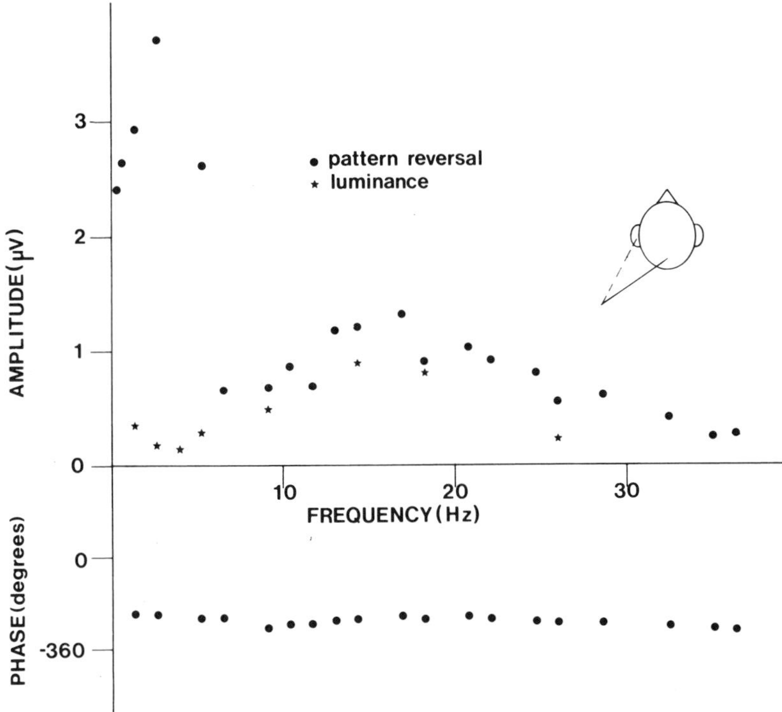

FIGURE 5. Amplitude characteristics of the distal $h_1(\tau)$ processes in the luminance and pattern reversal EPs, determined with a difference frequency component of 7.8 Hz in the EP. The bottom half gives the phase of this component in the pattern reversal EP as a function of the lowest frequency of the two sinusoids F and $F + 7.8$ Hz in the input.

expression of B_{nm} in the preceding section shows that the B_{n0} terms, which are the harmonics of the first sinusoid, depend on the strength of the added sine wave at the input of the nonlinearity. This means that, in principle, the amplitude *variation* of a harmonic component of constant frequency at the output of the sandwich system can be used for sequential analysis. This is the linearizing approach.

This sequential analysis method of a sandwich system is especially suitable if one wants to perform the analysis with an averager. With such a device, only the harmonic components in the response upon periodic stimulation are available. As the periodic stimulus, only the sinusoid can be employed since the amplitude distribution of such a stimulus is invariant after passing through a linear system. Thus, a sinusoid at the input of the sandwich system yields a sinusoid at the input of the zero-memory nonlinearity $f(v)$. As shown in the preceding section with $v(t) = A \cos \omega t$, the output signal $z(t)$ of $f(v)$ has the form:

$$z(t) = \sum_{n=0}^{\infty} \left\{ \frac{\epsilon_n j^n}{2\pi} \int_c F(ju)\, J_n(uA)\, du \right\} \cos n\omega t$$

$$= \sum_{n=0}^{\infty} A_n \cos n\omega t.$$

The linearizing method takes advantage of the fact that for a nonlinear system the superposition principle does not hold. This means that the amplitudes A_n of the harmonic components will be influenced by the presence of another signal at the input of the nonlinearity. When, for example, noise feeds in at a stage preceding of nonlinearity $f(v)$, then the nonlinearity expresses itself differently in the output of the system. Noise and nonlinearity together can be treated as a novel nonlinearity with the characteristic:

$$l(v) = \int_{-\infty}^{+\infty} f(k + v) p(k)\, dk,$$

where $p(k)$ is the amplitude distribution of the noise at the input of the nonlinearity $f(v)$. The effect of the noise is a smoothing of the original nonlinearity, such that the functional characteristic typically behaves more linearly. This is probably the reason why physiological systems, which are intrinsically noisy, behave frequently in a linear way under small signal analysis conditions.

As an extreme example, FIGURE 6 shows the amplitude characteristic of a hard limiter [$f(v) = 1$ for $v > 0$; $f(v) = -1$ for $v < 0$; $f(0) = 0$] under the influence of a Gaussian noise with amplitude density

$$p(k) = \frac{1}{\sigma(2\pi)^{1/2}} \exp(-k^2/2\sigma^2).$$

Substitution of $p(k)$ in the above formula gives:

$$l(v) = \left(\frac{2}{\pi}\right)^{1/2} \int_0^{v/\sigma} \exp(-z^2/2)\, dz,$$

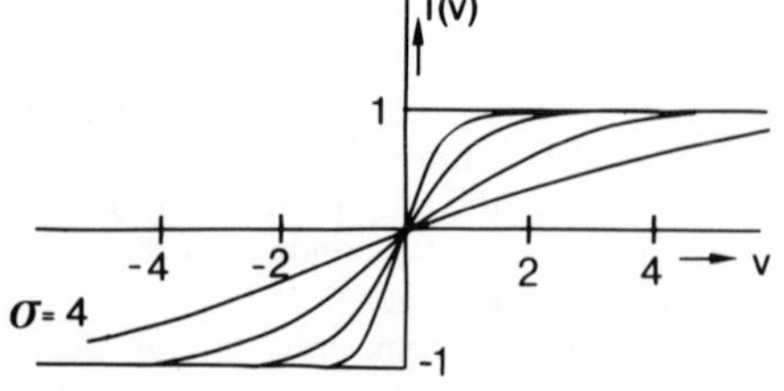

FIGURE 6. Characteristic of a hard limiter in the presence of Gaussian noise with variances $\sigma = 0, 0.5, 1, 2,$ and 4. For increasing variance the characteristic becomes more linear.

which for small signal-to-noise ratios, $|v/\sigma| \ll 1$, simplifies into

$$f(v) \sim \left(\frac{2}{\pi}\right)^{1/2} \cdot \frac{v}{\sigma}.$$

For small amplitudes of the input sinusoid in comparison to the noise, the output signal after averaging is a rather undistorted sinusoid whose amplitude varies proportionally with the input.

For the calculation of the effect upon the harmonic components in the output, the approach based on characteristic function yields an immediate answer. With $L(ju)$ the Fourier transform of $l(v)$ and

$$M_k(ju) = \int_{-\infty}^{+\infty} p(k)e^{+juk}\mathrm{d}k,$$

the characteristic function, one finds

$$L(ju) = F(ju) \cdot M_k(ju).$$

Substitution of this expression in the above formula for A_n gives for the amplitudes $A_{n,k}$ of the harmonic components in the presence of noise (or any other signal):

$$A_{n,k} = \frac{j^n}{\pi} \int_C F(ju)M_k(ju)\, J_n(Au)\mathrm{d}u, \quad n = 1, \ldots, \infty.$$

This modification of the harmonic content is illustrated in FIGURE 7 for the first three harmonic components in the output of a hard limiter in the presence of a Gaussian noise, sinusoidal or triangular linearizing signal. The triangular signal is included in this example since for signal-to-noise ratios $N < 3/2$ (on a power basis), only the fundamental component is present at the output.

The addition of an auxiliary signal also introduces intermodulation frequencies, which cannot be seen with an averaging technique but become apparent if, e.g., the autospectrum of the output signal is determined. In the presence of a Gaussian noise with autocorrelation coefficient $\rho_{vv}(t)$ and variance σ_n^2 at the input of a static nonlinearity, and autocorrelation function becomes:[50]

$$R_{zz}(\tau) = \sum_{n=0}^{\infty}\sum_{m=0}^{\infty} \frac{\epsilon_n}{m!} \alpha_{nm}^2 \rho_{vv}^m(\tau) \cos(n\omega\tau) - \alpha_{00}^2.$$

For static nonlinearities that can be expanded in a Taylor series

$$f(v) = \sum_{n=1}^{\infty} a_n v^n,$$

one finds

$$\alpha_{nm} = \sum_{s=0}^{\infty}\sum_{p=0}^{\infty} \frac{a_{m+n+2s+2p}(m+n+2s+2p)!(A/2)^{2p+n} \cdot \sigma_n^{2s+m}}{(p+n)!p!2^s s!}$$

The amplitude of the nth harmonic component in the response equals $\epsilon_n\, \alpha_{n0}$ ($n \geqq 1$; $m = 0$).

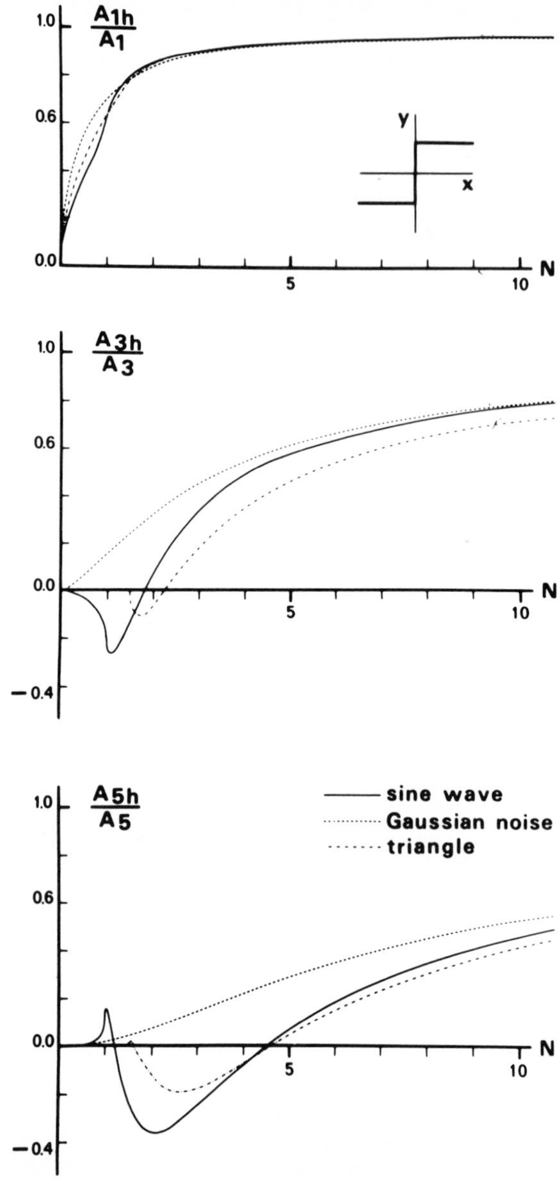

FIGURE 7. Linearizing effect of three auxiliary signals—Gaussian noise, a triangular wave, and a sine wave—upon the amplitudes of the harmonics in the output of a hard limiter to a sinusoidal input signal. A_{nh} is the amplitude of the nth harmonic in the presence of an auxiliary signal; A_n, the same without auxiliary signal; N is the ratio of signal to auxiliary signal, defined on the basis of power. (From Spekreijse & Oosting.[52] By permission of *Kybernetik*.)

The above formula shows that in the autospectrum side-bands appear around the harmonic responses because of intermodulation between the sine wave stimulus and the noise ($n \geqq 1$, $m \geqq 1$). Of course, the noise is also enhanced ($n = 0$, $m \geqq 1$). So, when side-bands are observed in the analysis of a nonlinear system, this can indicate the existence of a noise source at a site preceding the nonlinearity. Another possibility is that the system is nonstationary, i.e., that either the amplitude of the sine wave signal varies in time (multiplicative noise) or that the transport time (which cannot be ignored in biological systems) fluctuates (phase variation).[51] The modification of the nonlinear characteristic by the addition of noise can also be used to determine the transfer characteristic of the linear process preceding this nonlinearity. This can easily be understood by considering the effect of an auxiliary sine wave, $B \sin \omega' t$, upon the system response to an input sine wave, $A \sin \omega t$. The analysis is based on the fact that the incoming signal to the nonlinearity will consist of two sine waves with, however,

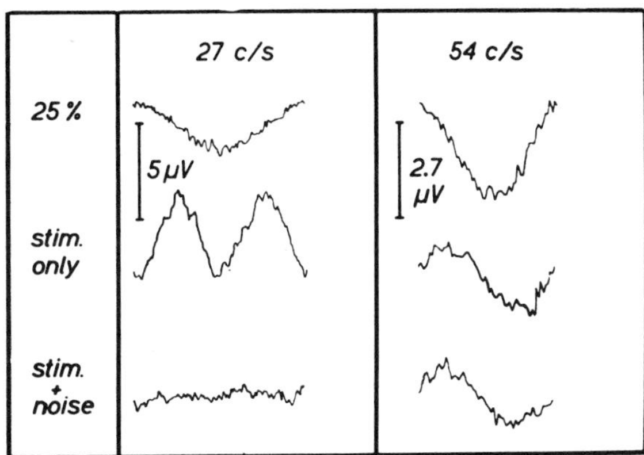

FIGURE 8. Amplitude of the second harmonic (left) and fundamental component (right) dominated responses to sinusoidally modulated light of 27 and 54 Hz, respectively. The addition of Gaussian noise to this stimulus reduces the second harmonic severely, leaving the fundamental unaffected (bottom row). (For details see Spekreijse.[23])

amplitudes A^* and B^*. Suppose that the linear process before the nonlinearity has a high-frequency cut-off at $\omega < \omega'$. For increasing frequencies ω', the amplitude B^* will be progressively reduced and a more distorted response will be obtained. From the amount by which the amplitude B must be increased to obtain a fixed amount of linearizing, one can determine the characteristic of the filter preceding the nonlinearity.

Further details regarding also the determination of the phase characteristic can be found in References 23, 52, and 53. An example of the effect of noise on the evoked potential to sine-wave-modulated light is given in FIGURE 8. Addition of noise reduces the amplitude of the second harmonic but not that of the fundamental component in the response. On the basis of this kind of differential behavior, we reached the conclusion that in the human evoked potential system to sine-wave-modulated light, an asymmetric linear rectifier separates the distal and proximal processes.

EPILOGUE

In this paper we have presented three methods to analyze a sandwich system. There are, of course, other approaches which yield the same result. One of them is, for example, the approach of Victor and Knight[29] based on an input signal consisting of N sinusoids.

If the nonlinearity $f(v)$ can be expanded into a Taylor series

$$f(v) = \sum_{n=1}^{\infty} a_n v^n \quad \text{and} \quad v(t) = \sum_{k=1}^{N} A_k \cos(\omega_k t + \phi_k),$$

then

$$\begin{aligned} z(t) = {} & a_1 \sum_{k=1}^{N} A_k \cos(\omega_k t + \phi_k) \\ & + \frac{a_2}{2} \sum_{k=1}^{N} \sum_{l=1}^{N} A_k A_l [\cos\{(\omega_k + \omega_l)t + \phi_k + \phi_l\} + \cos\{\omega_k - \omega_l)t + \phi_k - \phi_l\}] \\ & + \frac{a_3}{4} \sum_{k=1}^{N} \sum_{l=1}^{N} \sum_{m=1}^{N} A_k A_l A_m [\cos\{(\omega_k + \omega_l + \omega_m)t + \phi_k + \phi_l + \phi_m\} \\ & \qquad + \cos\{(\omega_k + \omega_l - \omega_m)t + \phi_k + \phi_l - \phi_m\} \\ & \qquad + \cos\{(\omega_k - \omega_l - \omega_m)t + \phi_k - \phi_l - \phi_m\} \\ & \qquad + \cos\{(\omega_k - \omega_l + \omega_m)t + \phi_k - \phi_l + \phi_m\}] \\ & \qquad + \text{higher order terms.} \end{aligned}$$

The sinusoids in the input signal can only have a fixed phase relation if they are generated by a single source such as a computer. The repetition cycle of this source sets the frequency resolution. The sample frequency by which the input and output signals are digitized determines the highest frequency available. Thus, the number of possible input frequencies is limited. This set is further restricted by the condition that each frequency component in the output signal (see the above formula) must be generated in a unique way. For instance, the maximal number of sine waves in an input with a period length of 512 points used for the analysis of a quadratic system amounts to 10, for a third-order system $v(t)$ may not contain more than 4 sinusoids.

The estimation of the transfer functions of the linear filters in a sandwich system places another condition on the actual input frequencies selected. For example, the easiest way to determine the amplitude characteristic of the first linear filter, $H_1(j\omega)$, is to follow in a bispectrum the line, $\omega_1 + \omega_2 =$ constant. Thus the frequencies at the input must be selected at preferably equal distances. In the above example with a period length of 512 points, the 10 optimal harmonics are: 8, 40, 61, 92, 116, 143, 163, 197, 220, and 242. In general, even sandwich systems have a nonlinearity in which higher-order terms cannot be ignored. The N-sinusoid approach rapidly reduces to one or two sinusoids of fixed phase relation or to greater than two sinusoids of variable phase in successive records. This is, in a sense, a Gaussian noise approach. The sandwich system is the simplest sequential nonlinear system. If the nonlinear system consists of a cascade of linear and nonlinear elements, which may show up in relatively strong harmonics of sum and difference frequencies at the output, then only under severe restrictions can a sequential analysis be performed.[54] In linear system analysis, the Dirac function is favored as input signal. This is not true for a nonlinear system. For example, the impulse to a sandwich system becomes $h_1(t)$ at the input of the nonlinearity. With a Taylor series expansion for this nonlinearity, the output signal of $h_2(\tau)$ becomes:

$$y(t) = \sum_{n=1}^{\infty} a_n \int_0^{\infty} h_2(\tau)\, h_1^n\,(t - \tau)\mathrm{d}\tau,$$

a complicated expression to solve.

Calculation Procedures for Digital Computers

Averaging Methods

To analyze responses to periodic stimuli like the sum of N sine waves, the standard averaging procedure can be used to find the coherent response. There is another method to estimate an average which consists of shifting the bits of the digital samples.[55] The latter requires less computing time, but the result has a larger variance. To obtain a smaller variance, one has to calculate at which number of averages one has to switch from $(m - 1)$ to m bit shifts. The optimal point lies around the entire function of $1 + 2^m \ln 2$ (Figure 9). Figure 9 shows that the maximal variance increase occurs at 26 averages: This variance is 4.8% larger than the variance of the standard averaging procedure. The figure shows further that the bit shift procedure employed in the averager (Hewlett-Packard, model 5481) is less optimal the results from a procedure proposed by White[56] lie in between.

Methods to Compute Correlation Functions

In the next paragraph it is assumed that the test signal $x(t)$ is an aperiodic signal. For simplicity, it is further assumed that $x(t)$ is a Gaussian noise and that the system studied is the sandwich system depicted in Figure 1.

The Classic Correlation Function Method

The sampled cross-correlation function of signals $x(t)$ and $y(t)$ are calculated according to the formula:

$$R_{yx}(l) = \frac{1}{N}\sum_{k=1}^{N} y(k)\, x(k - l),$$

where $l = \tau/f_s$, f_s being the sample frequency of the signal. The cross-bicorrelation function of $y(t)$ and $x(t)$ is calculated as:

$$R_{yxx}(l, m) = \frac{1}{N}\sum_{k=1}^{N} y(k)\, x(k - l)\, x(k - m).$$

This method gives the best estimate of correlation functions. It is based on multiplication. The variances of the multiplication-free methods in the time domain, which will be treated next, will be expressed in the variance of the classic (analog) method. The effect of digital-sampling on the correlation functions is neglected.

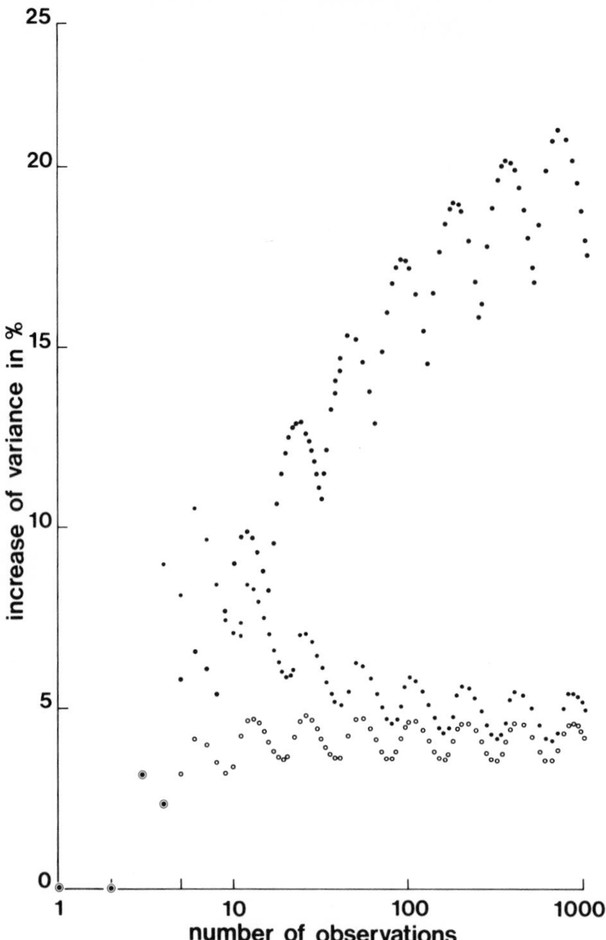

FIGURE 9. The percent increase in variance as a function of the number of averages for three different approaches: (★) the procedure of White; (●) the algorithm used in the HP averager; (O) the optimal procedure described in this paper.

The Ternary Correlation Function

The input signal is transformed into a signal which is $+1$ for $x(t) > \sigma_x$, 0 for $|x(t)| < \sigma_x$, and -1 for $x(t) < -\sigma_x$.[57] This gives for the cross-monocorrelation function of the sandwich system:

$$R_{zx}(\tau) = \left(\frac{2}{\pi e}\right)^{1/2} \sum_{n=0}^{\infty} (-1)^n C_{2n+1} H_{2n}(1) \frac{\rho_{vx}^{2n+1}(\tau)}{(2n+1)!},$$

and for the cross-bicorrelation function:

$$R_{zxx}(\tau_1, \tau_2) = \frac{2}{\pi e} \sum_{m=1}^{\infty} \sum_{k=[m/2]}^{\infty} \frac{C_{2m} H_{2k}^2(1)}{(m!)^2 (2k+1-m)!} \rho_{vx}^m(\tau_1) \rho_{vx}^m(\tau_2) \cdot \rho_{xx}^{2k+1-m}(\tau_2 - \tau_1)$$

with

$$H_{2p}(1) = \sum_{s=0}^{p} \frac{(-1)^s(2p)!}{(2p-2s)!2^s s!}.$$

The variances of the cross-mono- and cross-bicorrelation functions are 1.37 and 1.88 times the variances of these functions calculated with the classic method. Each transformation of a Gaussian signal in the above form increases the variance by a factor of 1.37 (for small correlation coefficients).

The Polarity Correlation Function

The output signal is correlated with the sign of the input signal.[58] For the cross-correlation function of the sandwich system one finds:

$$R_{zx}(\tau) = \left(\frac{2}{\pi}\right)^{1/2} \sum_{n=0}^{\infty} (-1)^n C_{2n+1} \frac{\rho_{vx}^{2n+1}(\tau)}{2^n\, n!(2n+1)}$$

and

$$R_{zxx}(\tau_1, \tau_2) = \frac{2}{\pi} \sum_{m=1}^{\infty} \sum_{k=[m/2]}^{\infty} \frac{C_{2m}\{(2k)!\}^2}{(m!)^2(2k+1-m)!2^{2k}(k!)^2} \rho_{vx}^{m}(\tau_1)$$
$$\times\, \rho_{vx}^{m}(\tau_2) \rho_{xx}^{2k+1-m}(\tau_1 - \tau_2).$$

The variances of the mono- and bicorrelation functions are 1.57 and 2.47 times the variances of the classic correlation procedures. Each time a Gaussian signal is transformed into its sign, the variance is enhanced by a factor $\pi/2 = 1.57$ (this holds for small correlation coefficients, as commonly found in biological systems).

The Triggered Correlation Function

The triggered correlation function is defined as a function obtained by averaging one signal on the synchronization or trigger pulses generated by the same or a different signal.[59] If the output signal $y(t)$ is averaged whenever the input signal $x(t)$ crosses a pre-set amplitude level x_a, then the following triggered correlation function will be obtained:[19]

$$\langle z(t)\,|\,x(0) = x_a \rangle = \int_{-\infty}^{+\infty} \frac{f(v)p_2(v, x_a)\, dv}{p_1(x_a)}$$
$$= \sum_{n=0}^{\infty} H_n\left(\frac{x_a}{\sigma_x}\right) \rho_{vx}^{n}(t) \frac{C_n}{n!},$$

in which $H_n(t)$ are the Chebyshev-Hermite polynomials. If $f(v)$ can be expanded in these polynomials, then the integrals can be easily solved, since the functions are orthogonal. By choosing the trigger level at $x = \sigma_x$, the ρ^2 term in the above formula is canceled. By adding and subtracting $\langle y(t)\,|\,x(0) = x_a \rangle$ and $\langle y(t)\,|\,x(0) = -x_a \rangle$, one can separate the contribution of the even and odd terms of the nonlinearity. Under this condition the optimal trigger levels are for the subtracted triggered correlation function $\sigma_x\sqrt{3}$ and $-\sigma_x\sqrt{3}$ and for the added function $\sigma_x\sqrt{3+\sqrt{6}}$ and $-\sigma_x\sqrt{3+\sqrt{6}}$. The variances are, respectively, 1.85 and 2.1 times the variances of the classic correlation procedures. This correlation procedure can be performed with an averager at the cost of an increase in variance since not all subsequent triggers can be used.

The FFT Method

Instead of calculating the convolution of the two signals $x(t)$ and $y(t)$, one calculates the product of their Fourier transforms. With this method the number of multiplications is drastically reduced. Because of the finite memory of a computer, the signals must be divided into a number of records, each of which is handled independently.

Errors may occur in the estimated cross-spectra, because of this procedure, since the stimulus can be in a different record than the response. The variance increases with decreasing bandwidth of the system studied.

Comparison of the Various Methods for On-Line Applicability

In general, if a (mini)computer is available, the FFT method is the fastest, unless hardware facilities for fast correlation (like array processors) are implemented. For example, for a microprogrammed computer (Hewlett-Packard, model HP 2100), the computation times for each of the above outlined five correlation methods are 130, 30, 65, 26, and 14 msec, respectively, when 128 correlation values are calculated from 128 sample points. For microprocessors without a multiplication instruction, one of the other multiplication-free methods can be chosen. A good candidate is the polarity correlation function method in which both the input and output signals are clipped.

It can be shown that when $y(t)$ is generated by a linear system, the dual polarity cross-monocorrelation function is

$$R_{yx}(\tau) = \frac{2}{\pi} \arcsin \rho_{yx}(\tau).$$

This relation holds for the wider class of elliptically symmetric amplitude distributions.[33] The variance of this expression has been calculated by Wolf.[32] By addition of signals with a uniform amplitude distribution, the polarity method can provide a result that is linearly proportional to the genuine correlation function $R_{yx}(\tau) = C\rho_{yx}(\tau)$.

If the clipped input and output signals are fully linearized by uncorrelated, uniformly distributed auxiliary signals (see also Figure 7), then the polarity cross-correlation functions are proportional to the classically derived functions even when the system studied is nonlinear. Auxiliary signals have the additional advantage that low-frequency artifacts in the signals (like drift) are less troublesome. The disadvantage of this procedure is the increase in variance, which for the cross-mono function is about 2.5 times (and for the cross-bi function about 4 times) the variance of the classically determined correlation functions.

For the determination of the cross-bicorrelation functions of a sandwich system on the basis of a polarity correlation procedure, the addition of auxiliary signals to the Gaussian noise employed for calculation (and of course not for stimulation) is essential, since, for $\tau_1 = \tau_2$, the series expansion of $R_{zxx}(\tau_1, \tau_1)$ does not converge. For a reasonable (error <10%) estimate, ρ_{xx} must be smaller than 0.42. The dual polarity correlation procedure reduces the computation time on the HP 2100 from 65 to 8 msec.

REFERENCES

1. BEDROSIAN, E. & S. O. RICE. 1971. The output properties of Volterra systems (nonlinear systems with memory) driven by harmonic and Gaussian inputs. Proc. IEEE **59**(12): 1688–1707.
2. BUSGANG, J. J., L. EHRMAN & J. W. GRAHAM. 1974. Analysis of nonlinear systems with multiple inputs. Proc. IEEE **62**(8): 1088–1119.
3. WEISS, T. F. 1966. A model of the peripheral auditory system. Kybernetik **3:** 153–175.
4. DUIFHUIS, H. 1976. Cochlear nonlinearity and second filter: Possible mechanisms and implications. J. Acoust. Soc. Am. **59:** 408–423.
5. DE BOER, E. & H. R. DE JONGH. 1978. On cochlear encoding: Potentialities and limitations of the reverse-correlation technique. J. Acoust. Soc. Am. **63**(1): 115–135.
6. DE JONGH, H. R. 1978. Modelling the peripheral auditory system. Ph.D. thesis. University of Amsterdam.
7. JONES, G. M. & J. H. MILSUM. 1969. Neural response of the vestibular system to translational acceleration. *In* Systems Analysis in Neurophysiology. C. A. Terzuolo, Ed. pp. 8–20. Univ. of Minnesota Press, Bloomington, Minn.
8. VARJÚ, D. 1967. Nervöse Wechselwirkung in der pupilanatomischen Bahn des Menschen Kybernetik **3:** 202–226.
9. SANDBERG, A. & L. STARK. 1968. Wiener G-function analysis as an approach to nonlinear characteristics of human pupil light reflex. Brain Res. **11:** 194–211.
10. KOHN, M. & M. CLYNES. 1969. Color dynamics of the pupil. Ann. N.Y. Acad. Sci. **156**(2): 931–951.
11. WATANABE, A. & L. STARK. 1975. Kernel method for nonlinear analysis identification of a biological control system. Math. Biosc. **27:** 99–108.
12. SPEKREIJSE, H. 1969. Rectification in the goldfish retina: analysis by sinusoidal and auxiliary stimulation. Vision Res. **9:** 1461–1472.
13. SCHELLART, N. A. M. & H. SPEKREIJSE. 1972. Dynamic characteristics of retinal ganglion cell responses in goldfish. J. Gen. Physiol. **59:** 1–21.
14. MARMARELIS, P. Z. & K. I. NAKA. 1973. Nonlinear analysis and synthesis of receptive field responses in catfish retina. Two-input white-noise analysis. J. Neurophysiol. **36:** 605–648.
15. HOCHSTEIN, S. & R. SHAPLEY. 1976. Linear and nonlinear spatial subunits in Y cat retinal ganglion cells. J. Physiol. **262:** 265–284.
16. VICTOR, J. D. & R. M. SHAPLEY. 1979. Receptive field mechanisms of Cat X and Y retinal ganglion cells. J. Gen. Physiol. **74:** 275–298.
17. VAN DER TWEEL, L. H. 1961. Some problems in vision regarded with respect to linearity and frequency response. Ann. N.Y. Acad. Sci. **89:** 829–856.
18. SPEKREIJSE, H., O. ESTEVEZ & L. H. VAN DER TWEEL. 1973. Luminance Responses to pattern reversal. Doc. Ophthalmol. **2:** 205–211.
19. REITS, D. 1975. Cortical potentials in man evoked by noise modulated light. Ph.D. thesis, University of Utrecht.
20. KOBLASZ, A. J. 1978. Nonlinearities of the human ERG reflected by Wiener Kernels. Biol. Cybernetics **31:** 187–191.
21. LOPES DA SILVA, F. H. 1970. Dynamic characteristics of visual evoked potentials. Ph.D. thesis, University of Utrecht.
22. GIELEN, C. C. A. M. 1980. Spatio-temporal and chromatic properties of visual neurones in the rhesus monkey geniculate nucleus. Ph.D. thesis, University of Nijmegen.
23. SPEKREIJSE, H. 1966. Analysis of EEG responses to diffuse and to patterned light in man. W. Junk Publ., The Hague.
24. VAN DER TWEEL, L. H. & H. SPEKREIJSE. 1969. Signal transport and rectification of the human evoked response system. Ann. N.Y. Acad. Sci. **156**(2): 678–696.

25. DUMMERMUTH, G., P. J. HUBER, B. KLEINER & T. GASSER. 1971. Analysis of the interrelations between frequency bands of the EEG by means of the bispectrum. A preliminary study. Electroenceph. Clin. Neurophysiol. **31:** 137–148.
26. REICHARDT, W. 1961. Autocorrelation, a principle for the evaluation of sensory information by the central nervous system. *In* Sensory Communication. W. A. Rosenblith, Ed. pp. 303–317. Wiley, New York, N.Y.
27. MARMARELIS, P. Z. & G. D. MCCANN. 1973. Development and application of white-noise modeling techniques for studies of insect visual nervous system. Kybernetik **12:** 74–89.
28. GEMPERLEIN, R. & G. D. MCCANN. 1975. A study of the response properties of retinula cells of flies using nonlinear identification theory. Biol. Cybernetics **19:** 147–158.
29. VICTOR, J. D. & B. W. KNIGHT. 1979. Nonlinear analysis with an arbitrary stimulus ensemble. Q. Appl. Math. **37:** 113–136.
30. WIENER, N. 1958. Nonlinear Problems in Random Theory. MIT Press, Cambridge, Mass. and John Wiley, New York, N.Y.
31. LEE, Y. W. & M. SCHETZEN. 1961. Measurement of the kernels of a nonlinear system by cross correlation. Quarterly Progress Report no. 60, Research Laboratory of Electronics, MIT, (Jan. 15): 118–130.
32. WOLF, D. 1973. Zur Genauigkeit der Bestimmung der Autokorrelations funktion aus der Polaritäts korrelations funktion. AEÜ **27**(6): 279–284.
33. VELTMAN, B. P. T. 1966. Quantisierung, Abtastfrequenz und statistische Streuung bei Korrelationsmessungen. Regelungstechnik **14:** 151–158.
34. MARMARELIS P. Z. & V. Z. MARMARELIS. 1978. Analysis of Physiological Systems. Plenum Press, New York, N.Y.
35. ROSENBLATT, M. & J. W. VAN NESS. 1965. Estimation of the bispectrum. Ann. Math. Stat. **36:** 1120–1136.
36. BRILLINGER, D. R. 1965. An introduction to polyspectra. Ann. Math. Stat. **36:** 1351–1374.
37. KIM, Y. C., W. F. WONG, E. J. POWERS & J. R. ROTH. 1979. Extension of the coherence function to quadratic models. Proc. IEEE. **67**(3): 428–429.
38. TICK, L. J. 1961. The estimation of "transfer functions" of quadratic systems. Technometrics **3:** 563–367.
39. GOODMAN, N. R. 1963. Statistical analysis based on a certain multivariate complex Gaussian distribution (an introduction). Ann. Math. Stat. **34:** 152–177.
40. PRICE, R. 1958. A useful theorem for nonlinear devices having Gaussian inputs. IRE Trans. IT-**4:** 69–76.
41. PAWULA, R. F. 1967. A modified version of Price's theorem. IEEE-IT-**13:** 285–288.
42. MCGRAW, D. K. & J. F. WAGNER. 1968. Elliptically symmetric distributions. IEEE-IT-**14:** 110–120.
43. KORENBERG, M. J. 1973. Obtaining differential equation, functional expansion or cascade representation for nonlinear biological systems. Proc. N. England Bioeng. Conf.
44. REITS, D. & H. SPEKREIJSE. 1980. Sequential analysis of a lumped nonlinear system; A model for visual evoked brain potentials. *In* Signal Processing: Theories and Applications. M. Kunt & F. de Coulon, Eds. North-Holland Publishing Company, New York, N.Y.
45. SPEKREIJSE H., O. ESTEVEZ & D. REITS. 1977. The physiological analysis of visual processes in man. *In* Visual Evoked Potentials: New Developments. J. E. Desmedt, Ed. pp. 16–89. Clarendon Press, Oxford, England.
46. REGAN, D. 1972. Evoked Potentials in Psychology, Sensory Physiology and Clinical Medicine. Chapman and Hall, London.
47. RICE, S. O. 1944. Mathematical analysis of random noise. Bell Sys. Tech. J. **23:** 282–332; 1945. **24:** 46–156.
48. BODE, H. W. 1945. Network Analysis and Feedback Amplifier Design. D. van Nostrand & Company, Inc., New York, N.Y.
49. HALLIDAY, A. M., W. I. MCDONALD & J. MUSHIN. 1973. The visual evoked response in the diagnosis of multiple sclerosis. Br. Med. J. (4): 661–664.
50. MIDDLETON, D. 1960. Statistical Communication Theory. McGraw Hill Book Company, New York, N.Y.
51. DAGNELIE, G., T. J. T. P. VAN DEN BERG & D. REITS. 1977. Unfamiliar effects of flicker on the human EEG. Doc. Ophthalmol. Proc. Series **15:** 173–178.

52. SPEKREIJSE, H. & J. OOSTING. 1970. Linearizing: A method for analysing and synthesizing nonlinear systems. Kybernetik **7:** 23–31.
53. OOSTING, J. 1971. Linearizing of nonlinear systems by means of auxiliary signals. Ph.D. thesis (in Dutch). University of Amsterdam.
54. KORENBERG, M. J. 1973. Cross-correlation analysis of neural cascades. Proc. 10th Ann. Rocky Mountain Bioeng. Symp. pp. 47–52.
55. DEARDORF, J. E. & C. R. TRIMBLE. 1968. Calibrated real-time signal averaging. Hewlett-Packard J. **19**(8): 8–13.
56. WHITE, R. C. 1970. A fast digital computer method for recursive estimation of the mean. IEEE Trans. Computers **19**(9): 847–849.
57. KLEIN, S. & S. YASUI. 1979. Nonlinear systems analysis with non-Gaussian white stimuli: General basis functionals and kernels. IEEE-IT-**25:** 495–500.
58. VAN VLECK, J. H. 1973. The spectrum of clipped noise. Rep. 51, Harvard Univ., Cambridge, Mass.
59. DE BOER, E. & P. KUYPER. 1968. Triggered correlation. IEEE-BME-**15:** 169–179.

PSEUDORANDOM SEQUENCES IN THE STUDY OF EVOKED POTENTIALS*

Richard Srebro† and Weldon Wright

Department of Ophthalmology
Southwestern Medical School
Dallas, Texas 75235

Pseudorandom sequences are deterministic discrete level periodic signals that are useful in system identification. A periodic sequence consists of successive terms, where each term is one of the discrete levels. Term values are constrained to change only at integral multiples of a basic time unit, Δt. The periodic sequence of terms is specified recursively by a weighted modulo-m sum of the previous n terms, where m is the number of discrete levels that a term may assume. Since the signal is periodic, the initial n term values may be selected arbitrarily provided that they are not all zero. However, the weighting coefficients cannot be selected arbitrarily. They are specified by selecting those coefficients that make the characteristic polynomial of the periodic sequence both irreducible and primitive.[1] When correctly selected, the coefficients specify a sequence period, $(m^n - 1)\Delta t$ (maximum length sequence); otherwise the sequence repeats prematurely and its useful properties are destroyed. It is customary to specify the terms of the periodic sequence as logic levels (0, 1, 2, . . .) and to map them into symmetrical stimulus units when constructing a test stimulus. FIGURE 1 illustrates two examples of periodic sequences, i.e., a 15-term binary periodic sequence ($m = 2, n = 4$) and a 26-term ternary periodic sequence ($m = 3, n = 3$).

Some important properties of binary periodic sequences are also illustrated in FIGURE 1. These properties characterize all sequences for which m is an even number. The first-order autocorrelation function $\Phi_{xx}(\tau)$ is an even periodic function with period $(2^n - 1)\,\Delta t$, equal to the period of the sequence, and consists of positive unit triangles with bases $2\Delta t$. It is offset below zero by an amount, $1/(2^n - 1)$, which reflects that there are an odd number of terms contained in one cycle of the sequence. In practice, n is chosen sufficiently large to make the offset negligible. The higher order autocorrelation functions have a complicated structure. The frequency domain properties are illustrated by the power spectrum of the sequence. Because the signal is periodic, the power spectrum is discrete with nonzero values occurring only at integral multiples of the fundamental frequency, $1/(2^n - 1)\Delta t$. The envelope of the power spectrum follows a $(\sin[x]/x)^2$ function with the 3-dB point occurring at approximately $1/3\Delta t$. Thus the bandwidth of the signal is adjustable by selecting n and Δt.

Similar properties for the ternary periodic sequence are also shown in FIGURE 1. These properties characterize all sequences for which m is an odd number. Ternary sequences are symmetrical about the half period point such that the second half period is the negative of the first half period (inverse repeat property). The first-order autocorrelation function, $\Phi_{xx}(\tau)$, is an even periodic function with period $(3^n - 1)\Delta t$, equal to the period of the sequence, and consists of positive unit triangles with bases $2\Delta t$ centered at 0, $(3^n - 1)\Delta t$, $2(3^n - 1)\Delta t$, and so forth, and negative unit triangles with bases centered at $(3^n - 1)\Delta t/2$, $3(3^n - 1)\Delta t/2$, and so forth. The second-order

*This work was supported by Grant EY 03068 from the National Eye Institute.

†Address for correspondence: University of Texas Health Science Center, Department of Ophthalmology, 5323 Harry Hines Boulevard, Dallas, Texas 75235.

0077–8923/82/0388–0098 $1.75/0 © 1982, NYAS

autocorrelation function, $\Phi_{xx}(\tau_1, \tau_2)$, is zero for all arguments. All even-order higher autocorrelation functions share this property. However, the higher odd-order autocorrelation functions have a complicated structure. The frequency domain properties are illustrated by the power spectrum of the sequence. Since the sequence is periodic, the power spectrum is discrete with nonzero values occurring only at integral multiples of the fundamental frequency, $1/(3^n - 1)\Delta t$. However, power is present only at odd harmonics. The envelope of the power spectrum follows a $(\sin[x]/x)^2$ function with the 3-dB point occurring at approximately $1/3\Delta t$. Thus, the bandwidth of the signal is adjustable by selecting n and Δt.

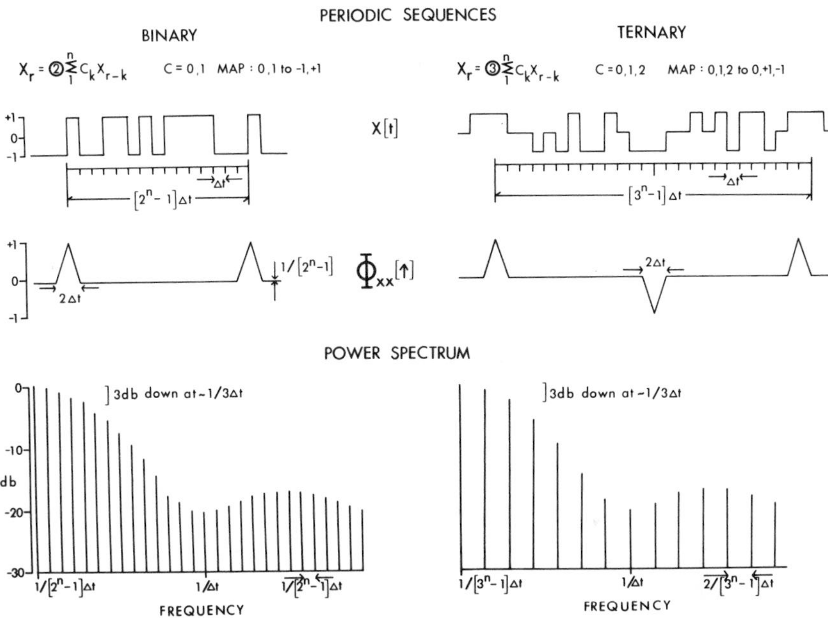

FIGURE 1. Some important properties of binary and ternary periodic sequences. A 15-term binary and a 26-term ternary periodic sequence ($X[t]$) are illustrated. The first-order autocorrelation function $\Phi_{xx}(\tau)$ and the power spectrum are shown for each type of sequence. See text for discussion.

Binary periodic sequences are useful in the identification of linear systems.[2-4] They have several advantages over Gaussian white noise for this purpose in that the signals are physically realizable, they have well-defined properties, and they permit signal averaging. They have been used to a limited extent to study auditory[5] and visually[6,7] evoked potentials and we will not comment further on this use here.

Ternary periodic sequences can be useful in the identification of some nonlinear systems.[8,9] If a nonlinear system can be described by a functional expansion of second order, the first-order kernel can be estimated unambiguously and the second-order kernel can be approximated using ternary periodic sequences. Consider a nonlinear system with input $X(t)$, a ternary periodic sequence with period T, and output $Y(t)$, a periodic function also with period T. It is assumed that the system can be represented by the functional expansion,

$$Y(t) = \int_T h_1(\tau)x(t-\tau)d\tau + \int\int_T h_2(\tau_1, \tau_2)x(t-\tau_1)x(t-\tau_2)d\tau_1 d\tau_2, \quad (1)$$

where $h_1(t)$ and $h_2(t_1, t_2)$ are, respectively, the first- and second-order kernels. The integrations are carried out over one cycle of stimulus and response. The first order cross-covariance function can be expressed as,

$$\phi_{xy}(\tau) = \int_T h_1(\lambda)\phi_{xx}(\tau-\lambda)d\lambda + \int\int_T h_2(\lambda_1, \lambda_2)\phi_{xx}([\tau-\lambda_1], [\tau-\lambda_2])d\lambda_1 d\lambda_2, \quad (2)$$

where $\phi_{xx}(\tau)$, and $\phi_{xx}(\tau_1, \tau_2)$ are, respectively, the first- and second-order autocovariance functions of the input. Since $\phi_{xx}(\tau_1, \tau_2)$ is zero for all arguments, the second integral in Equation 2 is zero, and since

$$\begin{aligned} \phi_{xx}(\tau) &= \sigma^2, \text{ for } \tau = 0, T, 2T, \ldots \\ &= -\sigma^2, \text{ for } \tau = T/2, 3T/2, \ldots \\ &= 0, \text{ otherwise,} \end{aligned} \quad (3)$$

where $\sigma^2 = 1/T \int_T x(t)^2 dt$, it follows that

$$\phi_{xy}(\tau) = \sigma^2 h_1(\tau); \text{ for } h_1(\geq T/2) = 0. \quad (4)$$

Thus, if the period of the ternary sequence is twice as long as the settling time of the first-order kernel, that kernel can be estimated unambiguously.

The second-order cross-covariance function, $\phi_{xy}(\tau_1, \tau_2)$, can be expressed as,

$$\begin{aligned} \phi_{xy}(\tau_1, \tau_2) = &\int_T h_1(\lambda)\phi_{xx}([\tau_1-\tau_2], [\tau_1-\lambda])d\lambda \\ &+ \int\int_T h_2(\lambda_1, \lambda_2)\phi_{xx}([\lambda_1-\lambda_2], [\lambda_1-\tau_1], [\lambda_1-\tau_2])d\lambda_1 d\lambda_2, \end{aligned} \quad (5)$$

where $\phi_{xx}(\tau_1, \tau_2, \tau_3)$ is the third-order autocovariance function of the input. The first integral in Equation 5 is zero, and it can be shown that,

$$\begin{aligned} \phi_{xy}(\tau_1, \tau_2) &= \frac{\sigma^2}{3} h_2(\tau_1, \tau_2) + \text{constant} + \text{contribution from anomalies, for } \tau_1 = \tau_2 \\ &= \sigma^2 h_2(\tau_1, \tau_2) + \text{constant} + \text{contributions from anomalies, for } \tau_1 \neq \tau_2 \end{aligned} \quad (6)$$

The term anomaly is used to label nonzero values of $\phi_{xx}(\tau_1, \tau_2, \tau_3)$ which occur when $\tau_1 \neq \tau_2 \neq \tau_3$ and take values $\pm\ \sigma^2/3$. The anomalies contribute to $\phi_{xy}(\tau_1, \tau_2)$ when both $h_2(\lambda_1, \lambda_2)$ and $\phi_{xx}([\lambda_1-\lambda_2], [\lambda_1-\tau_1], [\lambda_1-\tau_2])$ are nonzero in the second integral of Equation 5. The anomalies are completely determined by the structure of the ternary periodic sequence.[10] Thus, in principle, $h_2(\tau_1, \tau_2)$ can be estimated unambiguously, but this requires solving a set of $(3^n - 1)^2$ simultaneous linear equations, which is impractical. No two sequences, even if their lengths are identical, have the same set of anomalies and there is no convenient symmetry of the anomalies amongst different sequences that cancels their contributions to the cross-covariance. The number of anomalies increases linearly with the sequence length while the space of the second integral in Equation 5 increases as the third power of the sequence length. Thus, since $h_2(\tau_1, \tau_2)$ is bounded, the contributions to $\phi_{xy}(\tau_1, \tau_2)$ due to anomalies decreases within the boundary of $h_2(\tau_1, \tau_2)$ as the sequence length increases. Thus the estimate of $h_2(\tau_1, \tau_2)$ approaches an unambiguous estimate as the period of the ternary sequence increases to values much longer than the settling time of the second-order kernel.

METHODS

Ternary periodic sequences were generated using a computer, the output of which was put through a digital-to-analog converter, buffered by a voltage to current converter, and supplied to a light-emitting diode (LED, Monsanto, MV5352, 40 millicandles at 20 mA, yellow). The LED was viewed monocularly with a natural pupil through a Maxwellian optical system (field size, 24°; aperture size, 3 mm). A cover glass deflected a small amount of light that was focused to fill the aperture of a photomultiplier whose output provided continuous monitoring of the stimulus. Two subjects were used in this study. The subjects used a dental bite plate to stabilize head position and adjusted the position of the optical path so that they saw a homogeneous field of maximum brightness. They were instructed to fixate the center of the field. The highest luminance available from the stimulator was approximately 1100 footlamberts (ft-L). The three levels of the ternary periodic sequence were mapped symmetrically as three luminance levels, L_0, $L_0 + \Delta L$, and $L_0 - \Delta L$. Modulation ($\Delta L \times 100/L_0$, %) was adjusted under computer control. Mean luminance (L_0) was adjusted using a neutral density filter. Stimuli were coded and stored as disk files in advance of an experiment.

The visually evoked potential (VEP) was recorded using a differential electrode derivation (negative electrode on the midline 5 cm above the inion, positive electrode 5 cm lateral to negative electrode ipsilateral to stimulated eye). The VEP was amplified by 10^5 and band-pass filtered between 1 and 100 Hz (3-dB points, single pole filters), and was digitized at 1 kHz.

An experimental run began with 40 sec of steady light (luminance L_0) followed by approximately 4 min containing a variable number of continuously presented cycles of a ternary periodic sequence (depending on the sequence period). In any single run the parameters of the ternary periodic sequence, the mean luminance, and the modulation were constant. Zero modulation runs were used to establish a "baseline" condition for rough signal to noise estimation. During each run, the VEP was digitized and stored on disk for later processing. These data were used to calculate input–output cross-covariance functions. Generally, the entire record was used in making these calculations and no signal averaging was applied. For some purposes it was useful to calculate cross-covariance functions for segments of the record that were coincident with one or more cycles of the stimulus. Segments containing several cycles of the stimulus were continuous. Estimates of h_1 and h_2 were made from input–output cross-covariance functions using Equations 4 and 6, respectively. The estimates of h_1 together with the estimate of the major diagonal of h_2 were used to compute predictions of the VEP to increments and decrements of luminance. These were compared to observed VEPs obtained when the subject viewed a steady field interrupted alternately at 1.21-sec intervals by luminance increments and decrements. The observed VEPs were averaged over 70 repetitions of each stimulus type. Both the estimates of h_1 and h_2 and the observed average VEPs were based on data collected in one recording session. The increment and decrements were of the same duration (Δt), magnitude (ΔL), and applied at the same mean luminance (L_0) as used to make the estimates of h_1 and h_2.

Power spectra were obtained from average VEPs. A three-term Blackmann–

Harris sliding window[11] was applied to the average VEP with overlap and configured to yield a spectral resolution of approximately 1.5 Hz.

RESULTS

FIGURE 2 shows VEPs recorded from subject 1 in response to a stimulus based on a 242-term ternary periodic sequence with a Δt of 10 msecs, a mean luminance of 100 ft-L, and a modulation of 40%. Although 90 cycles of the stimulus were presented continuously, two groups of four VEPs were selected to illustrate the range of VEPs that were recorded. The four VEPs in each group are continuous in time. Each VEP is coincident with one cycle of the stimulus. No clearly consistant waveform is evident in each VEP (presumably) because the signal-to-noise ratio is too low to permit resolution of stimulus-dependent activity. The two groups of VEPs differ in the amplitude of a quasiperiodic component with frequency near 10 Hz, presumably the alpha rhythm. Alpha rhythms waxed and waned through the record with no obvious relationship to the stimulus. The lower part of FIGURE 2 shows three input–output cross-correlation functions, one for each of the groups of four VEPs, and the third for 88 continuous VEPs that include the VEPs shown in the first two columns. The input–output cross-correlation function for the group of four VEPs containing prominent alpha shows more ringing than does that for the group of four VEPs

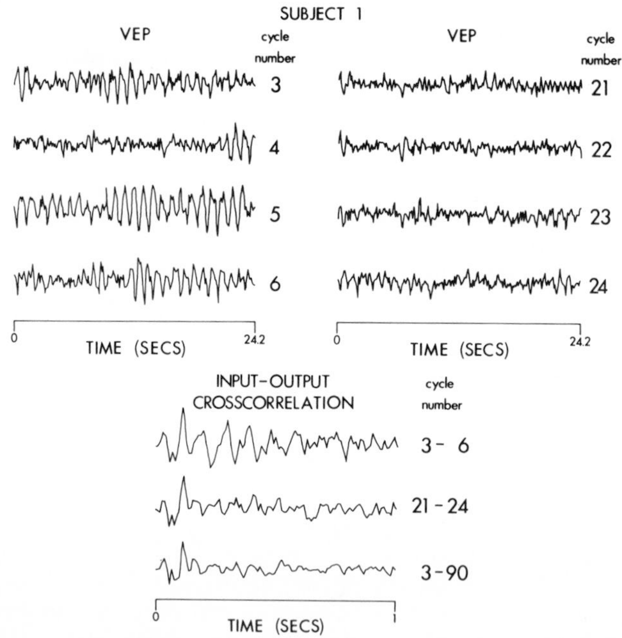

FIGURE 2. VEPs and input–output cross-correlations. Stimulus: 242-term ternary periodic sequence, Δt = 10 msec, mean luminance = 100 ft-L, modulation = 40%. Upper columns from left: continuous VEPs, each coincident with one cycle of the stimulus, selected to show extremes of alpha activity. Lower column: input–output cross-correlations for the groups of VEPs in upper columns and for the entire experiment consisting of 88 continuous cycles of the stimulus.

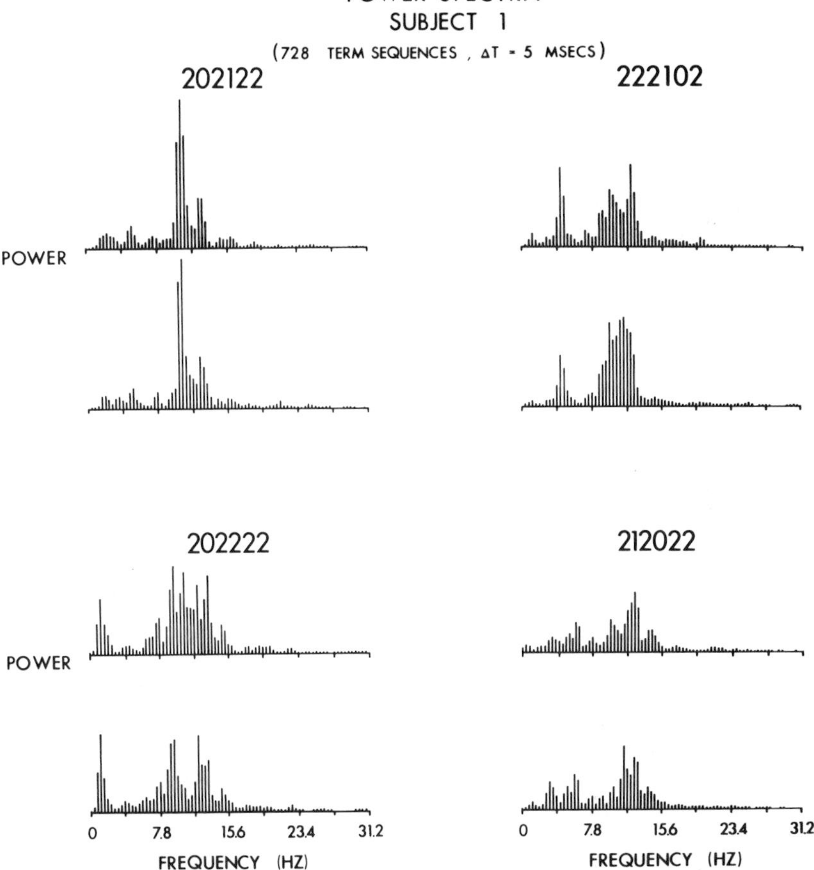

FIGURE 3. Power spectra of average VEPs (70 stimulus cycles). Stimulus: 728-term ternary periodic sequence, $\Delta t = 5$ msec, mean luminance = 10 ft-L, modulation = 20%. Three term Blackmann-Harris sliding window. Spectral resolution is approximately 1.5 Hz. Numbers above each group of two spectra: coefficients in the recursion formula used to generate the sequence (see text). The two spectra in each group are repeated runs. All spectra plotted on the same scale. All data from subject 1 taken in one recording session.

containing less alpha, but there is a consistent waveform that characterizes the first 165 msec of both. The input–output cross-correlation function for the 88 continuous VEPs resembles that for the group of four continuous VEPs with little alpha. It was necessary to use at least four continuous VEPs to obtain a reliable input–output cross-correlation. When all possible groups of four continuous nonoverlapping VEPs were examined, the input–output cross-correlations had consistent waveforms for the first 165 msec and variable amounts of ringing that increased as alpha increased in the VEPs. Subject 2 had less alpha than did subject 1, and there was less ringing in all the input–output cross-correlations for subject 2.

FIGURE 3 shows power spectra calculated from average VEPs to 70 continuously presented cycles of stimuli based on 728-term ternary sequences with a Δt of 5 msecs, a modulation of 20%, and a mean luminance of 10 ft-L. The spectra are based on data from subject 1, all recorded in a single session. Four different sequences were used that differed only in the order of the three discrete luminance levels determined by the six coefficients in the recursion formula used to generate the sequence. Two power spectra based on independent data are shown for each sequence. The power spectra are consistent for any one sequence and similar but not identical for different sequences. In all cases, resolvable power is absent at frequencies higher than 20 Hz. (We verified that there was no resolvable power to 100 Hz in each case.)

FIGURE 4 shows estimates of h_1 and the major diagonal of h_2. Four different

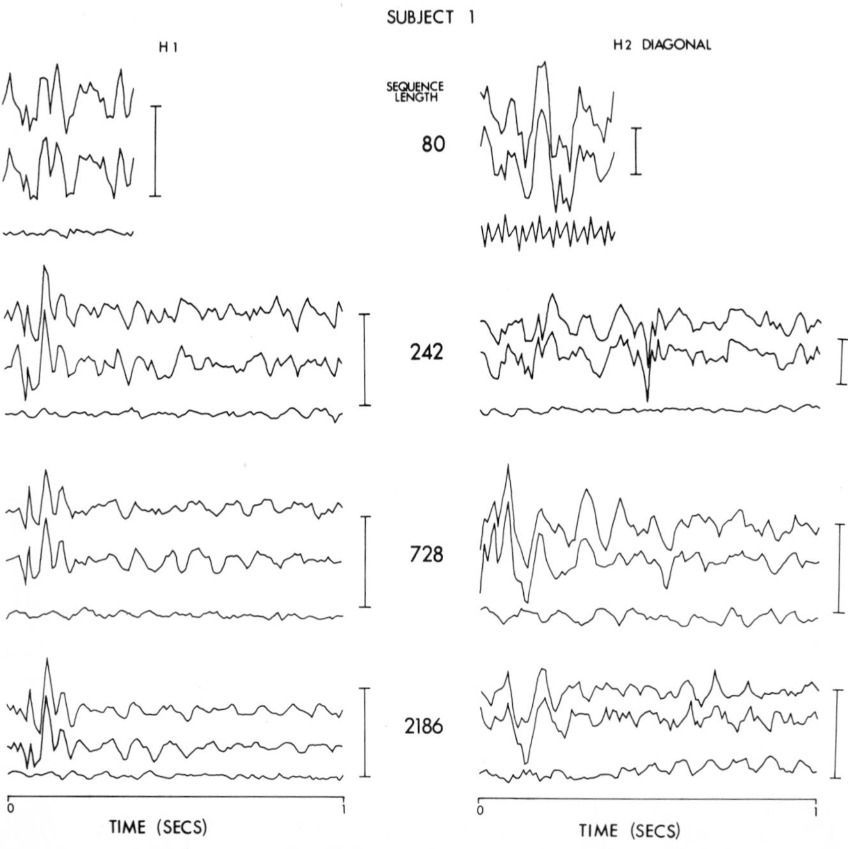

FIGURE 4. Estimates of h_1 (left column) using ternary periodic sequences of lengths specified. Each estimate results from that number of cycles of the stimulus that is contained in approximately four minutes. Three estimates of h_1 and three estimates of the major diagonal of h_2 are shown for each sequence length. Upper two estimates for each sequence: repeated runs, Δt = 10 msec, mean luminance = 100 ft-L, modulation = 40%. Lowest estimate for each sequence: 0% modulation. Scale marks to right of each group of records shows relative magnitudes. All data for subject 1 taken in one recording session.

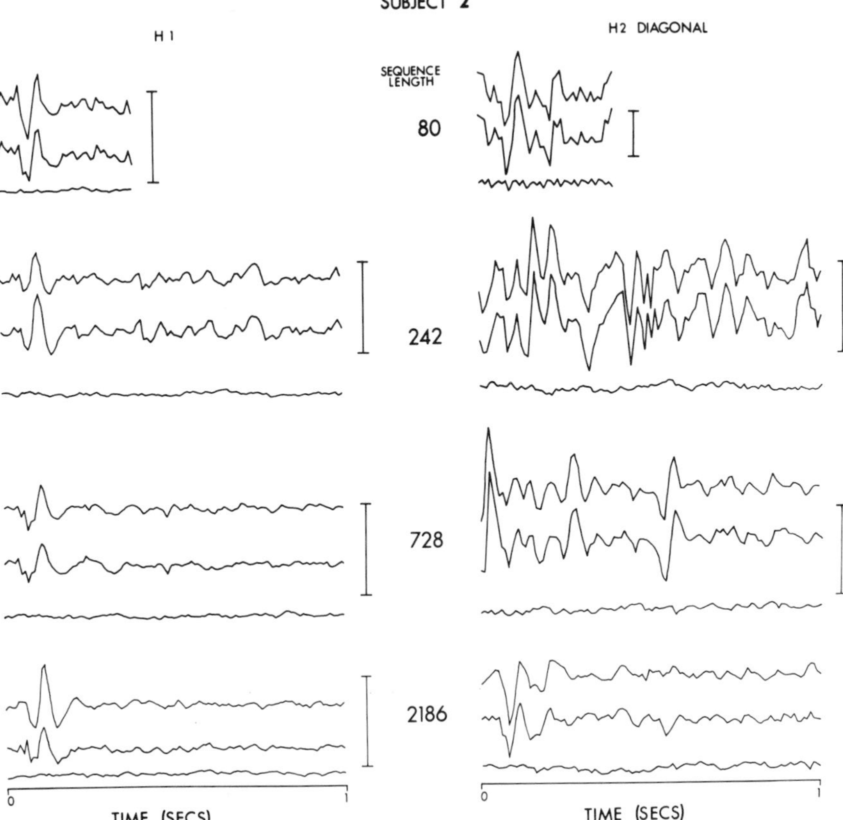

FIGURE 5. Estimates of h_1 and major diagonal of h_2 using ternary periodic sequences of lengths specified. Format and sequences the same as in FIGURE 4. All data from subject 2.

sequences were used with lengths from 80 to 2186 terms. The mean luminance was 100 ft-L, the modulation was 40%, and Δt was 10 msec for all sequence lengths. The estimates are based on data from subject 1 all taken in a single recording session. A group of three estimates is shown for each sequence length. The upper two estimates of each group are for identical conditions but from independent data recorded at two different times during the recording session. The lower estimate in each group is from data recorded under identical experimental conditions except that the modulation was 0% (baseline). For each sequence length, the two estimates based on the 40% modulation stimuli are consistent and different from the baseline estimates. The h_1 estimates are poorly defined for the 80-term sequence. For the 242-, 728-, and 2186-term sequences, the h_1 estimates are much better defined and similar. The h_2 diagonal estimates are poorly defined for both the 80- and 242-term sequences. For the 728- and 2186-term sequences, the estimates are better defined and consistent for the first 250 msec. However, the h_2 diagonal estimates based on the 2186-term sequence settle to values comparable to the baseline at times longer than 300 msec whereas the h_2 diagonal estimates based on the 728-term sequence do not. FIGURE 5 is identical to FIGURE 4 in format and represents the data for subject 2. The results are generally the same as for subject 1.

FIGURE 6 shows estimates of h_1 and the major diagonal of h_2 based on four different sequences with lengths of 2186 terms, a Δt of 10 msec, a mean luminance of 100 ft-L, and a modulation of 40%. The sequences differed only in the order of their terms determined by the seven coefficients in the recursion formula used to generate the sequences. In each case, two estimates, based on independent data, are shown. The data used to make the estimates were all obtained from subject 1. The two groups of estimates at the top of the figure were made from data obtained on two different days using identical sequences. The lower four groups of estimates were made from data obtained in a single recording session. All duplicated runs yielded consistent estimates and the same sequence produced consistent estimates on the two different days. All

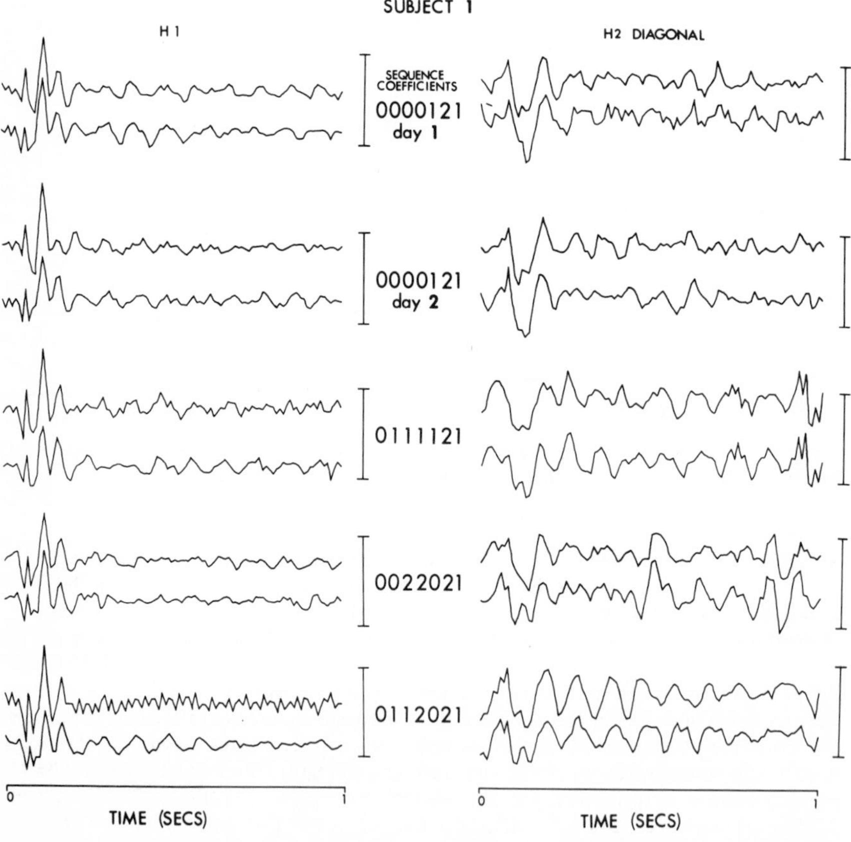

FIGURE 6. Estimates of h_1 and the major diagonal of h_2 based on four different ternary periodic sequences of length 2186, specified by coefficients in recursion formula (see text), Δt = 10 msec, mean luminance = 100 ft-L; modulation = 40%. Repeated runs shown for each sequence. Top two rows: same sequence used on different days. Bottom four rows: all data taken in one recording session. All data from subject 1. Plotting scale the same for all estimates.

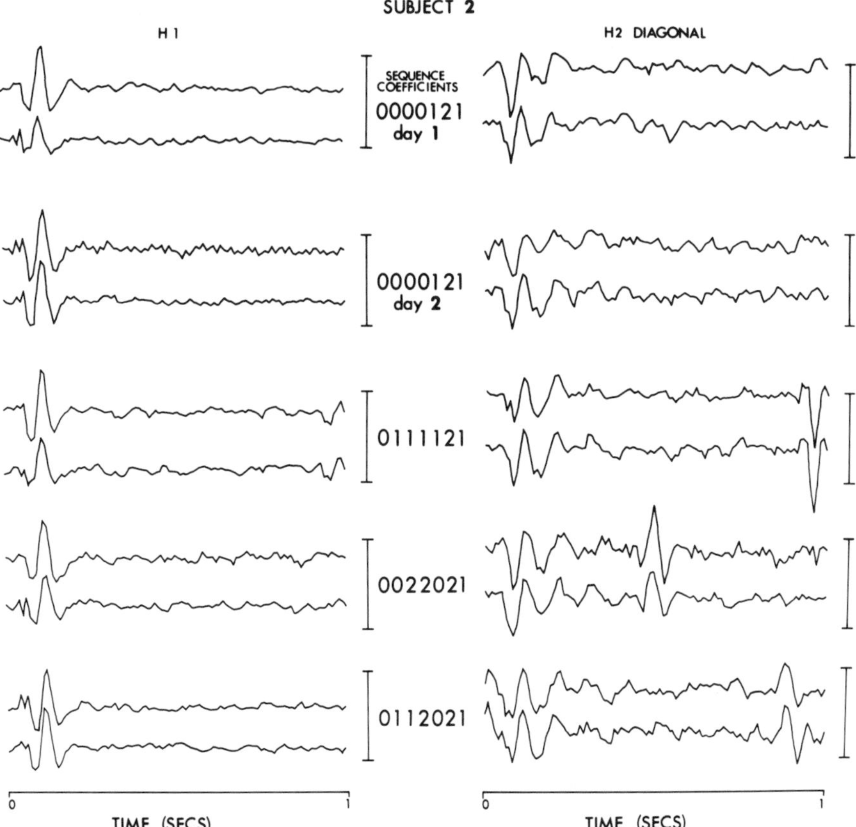

FIGURE 7. Estimates of h_1 and the major diagonal of h_2 based on four different ternary periodic sequences of 2186-term length, $\Delta t = 10$ msec, mean luminance = 100 ft-L, modulation = 40%. Format and plotting scale the same as FIGURE 6. All data from subject 2.

estimates of h_1 are consistent. The estimates of the h_2 diagonal are consistent for times up to approximately 250 msec. At longer times there are differences amongst the h_2 diagonal estimates that are sequence specific. These differences take the form of "transients" and are seen at about 950 msec for the sequence 0111121, and at about 475 and 875 msec for the sequence 0022021. FIGURE 7 is identical to FIGURE 6 in format, but is based on data from subject 2. The results are similar to those of subject 1. In particular, note the transients in the h_2 diagonal estimates for exactly the same sequences and at exactly the same times as for subject 1. (The oscillatory behavior of the h_2 diagonal estimate for subject 1 for the sequence 0112021 is not present for subject 2.)

FIGURE 8 shows nine estimates of h_1 and the major diagonal of h_2 for subject 1, each calculated from the VEP coincident with one cycle of a stimulus based on a ternary sequence with a length of 2186 terms, a Δt of 10 msec, a mean luminance of 100 ft-L, and a modulation of 40%. The coefficients used in the recursion formula to generate the sequence were 0000121, and the VEPs used to make the estimates are

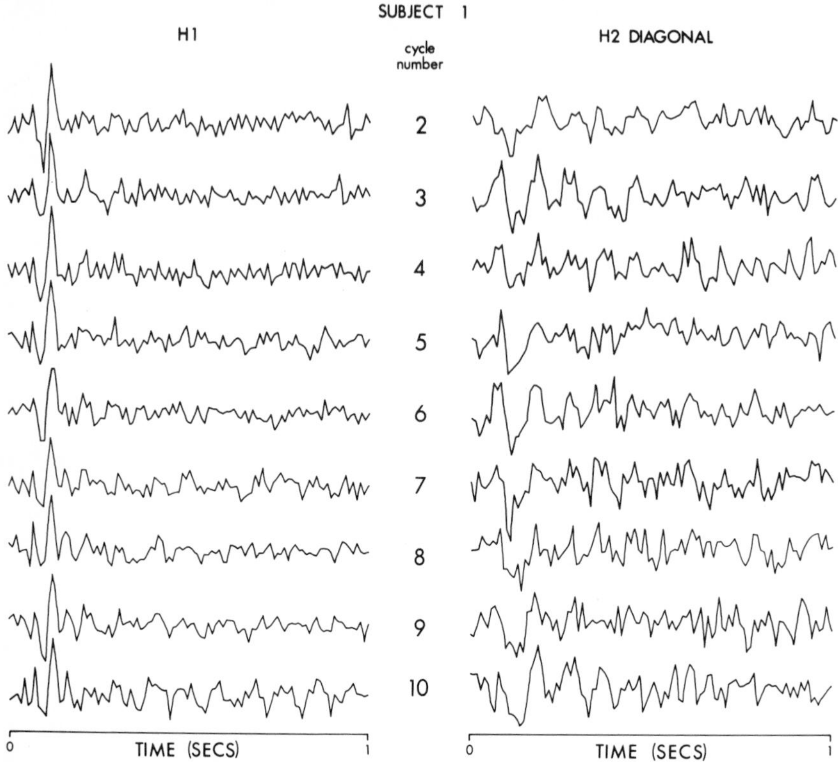

FIGURE 8. Estimates of h_1 and the major diagonal of h_2. Stimulus: 2186-term ternary periodic sequence, Δt = 10 msec, mean luminance = 100 ft-L, modulation = 40%. Coefficients of recursion formula: 0000121. Each estimate results from VEP data coincident with one cycle of the stimulus. Data from subject 1. See FIGURE 6 for comparable estimates based on nine continuous cycles.

identical to those used to make one of the pairs of estimates in the second line of FIGURE 6. These estimates of h_1 and h_2 diagonal are consistent amongst themselves and similar to the estimates shown in FIGURE 6, line 2, that were based on nine (continuous) VEPs due to nine continuous cycles of the same stimulus.

FIGURE 9 compares observed and predicted VEPs for subject 1. The top line of records shows three (superimposed) average evoked potentials to increments of luminance and three (superimposed) average evoked potentials to decrements of luminance. The lower four lines show predictions of the average VEPs. Each line shows predictions based on estimates of h_1 and the major diagonal of h_2 made using a different ternary series stimulus with a length of 2186 terms (the coefficients used in the recursion formula to generate the sequence are indicated) with Δt being 10 msec; mean luminance, 100 ft-L; and modulation, 40%. The estimates of h_1 and h_2 that were used are shown in FIGURE 6. In each case, two independent predictions are shown superimposed. All the data used to construct FIGURE 8 were obtained in a single recording session. Predicted and observed VEPs are similar in waveform. The predicted VEPs are somewhat more oscillatory than the observed VEPs and exaggerate small inflections in the observed VEPs. This tendency, although present in both increment and decrement predictions, is more apparent in the increment predictions. The observed VEPs and the predicted VEPs are *not* plotted on the same scale. The observed VEPs are larger than the predicted VEPs by a factor of approximately 3.

FIGURE 10 compares observed and predicted VEPs for subject 2. The format of this figure is identical to that of FIGURE 9 and the results are similar to those described for subject 1. However, for subject 2 the decrement predictions are not so good as the increment predictions, whereas for subject 1 the reverse is true.

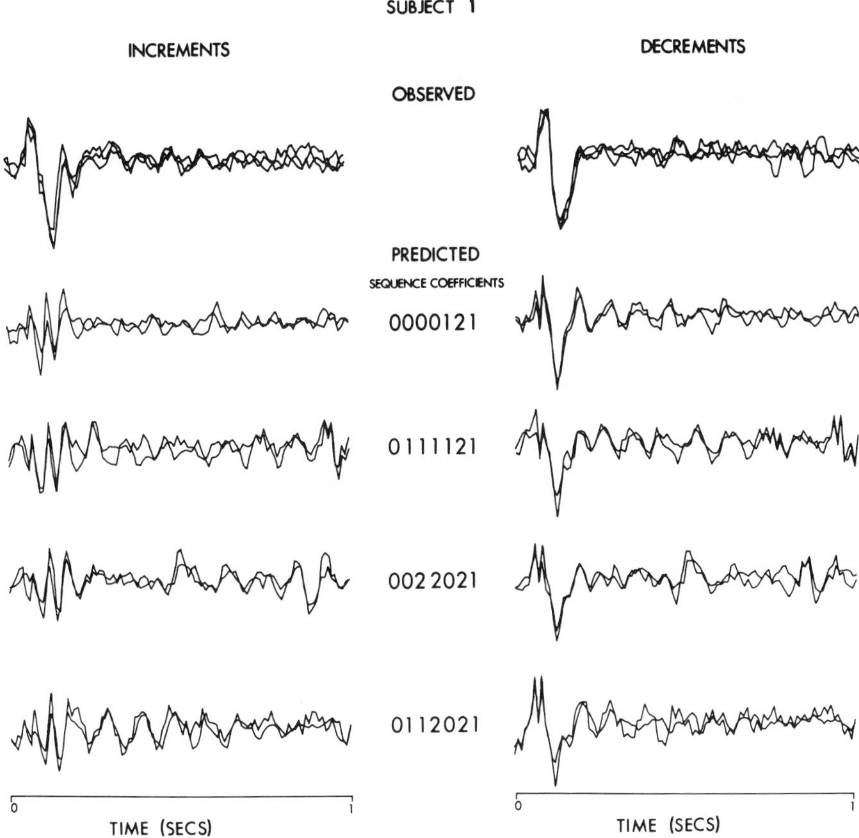

FIGURE 9. Observed and predicted VEPs to increments and decrements of luminance. Observed VEPs: average of 70 stimulus presentation. Subject viewed steady field, luminance = 100 ft-L interrupted at 1.21-sec intervals by 40% increments or decrements of luminance; duration = 10 msec. Three repeat measurements shown as superimposed VEPs. Predicted VEPs based on estimates of h_1 and the major diagonal of h_2 from four different ternary periodic sequences, length = 2186 terms, Δt = 10 msec, mean luminance = 100 ft-L, modulation = 40%. (See FIGURE 6 for estimates.) Predictions based on repeated runs are shown as superimposed traces. Plotting scale for predicted VEPs 3.33 times larger than for observed VEPs. All data from subject 1 taken in one recording session.

DISCUSSION

Our results suggest that ternary periodic sequences may be useful in the identification of the nonlinear properties of the VEP to luminance modulation for the following reasons.

(1) The behavior of the kernel estimates conforms to that anticipated by the theory outlined above. In particular, the estimate of h_1 is well behaved and consistent for stimuli based on sequence lengths of 242 terms or greater, and is independent of the

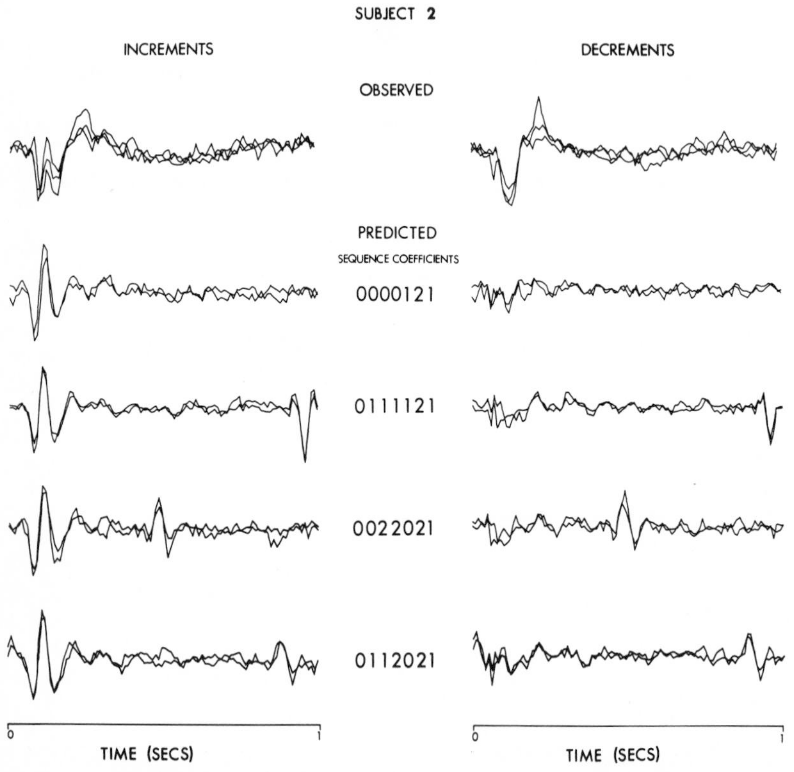

FIGURE 10. Observed and predicted VEPs to increments and decrements of luminance. Format the same as FIGURE 9. (See FIGURE 7 for estimates of h_1 and the major diagonal of h_2 used to calculate predicted VEPs.) All data from subject 2 taken in one recording session.

particular sequence used to make it. The theory anticipates this behavior in that the sufficient condition for a consistent estimate of h_1 is that the period of the sequence is greater than twice the settling time of h_1. Thus, our results suggest that the settling time of h_1 is greater than 0.4 sec and less than 1.21 sec. By contrast, the estimate of the major diagonal of h_2 is well behaved when stimuli based on sequences with lengths of 2186 terms are used to make it, but do not settle to consistently low values when shorter sequence lengths are used. The theory anticipates this behavior, in that longer sequences are required to estimate h_2 than to estimate h_1 because contributions from anomalies distort the estimate of h_2 and are minimized by increasing the sequence length well beyond the settling time of h_2. The differences that we found amongst the estimates of the major diagonal of h_2 based on four different sequences of 2186-term length, appeared as "transients" at times greater than 250 msec, were specific to the sequence, and were similar for a given sequence in both subjects. This behavior conforms to the theory in that residual contributions of the anomalies to the estimate of h_2 are also sequence specific and isolated to relatively discrete regions. Conformity of the behavior of the kernel estimates to the theory suggests the validity of a second-order functional expansion model as a representation of the nonlinear characteristics of the VEP to luminance modulation.

(2) The waveforms of the VEPs to short pulses of luminance increment or decrement are similar to those predicted from the theory using the estimates of h_1 and the major diagonal of h_2. We recognize the deficiencies of the predictions obvious in FIGURES 8 and 9, but these deficiencies must be viewed against the notorious difficulties that have characterized the ability of nonlinear systems models to predict responses to any stimuli other than the test stimuli themselves.

(3) Useful kernel estimates can be made using record lengths as short as 21.86 sec. This result speaks for the relative efficiency of the method and is important because biological systems are often nonstationary.

However, our results also uncover problems that need further study:

(1) It is not clear how the variation in alpha affects the analysis and what implications it has with regard to system stationarity. At the simplest level, our results show that alpha contributes some amount of ringing to the input–output cross-correlation functions and hence to the estimates of the kernels. This result is compatible with the hypothesis that alpha is independent of stimulus-related activity, but our testing of this hypothesis is superficial. Even if alpha is independent of the stimulus in a direct sense, it does not preclude the possibility that its waxing and waning may be correlated to changes in system parameters, or that the amount of alpha over times comparable to the analysis time (i.e., several minutes) may be correlated to changes in system parameters.

(2) Our results are primarily based on stimuli that are bandwidth limited to approximately 33 Hz. The choice was expedient and was justified on the result that stimuli based on sequences with Δt equal to 5 msec (i.e., a bandwidth limited to 66.67 Hz) did not produce resolvable power at frequencies greater than 20 Hz. But it is not impossible that important features of system performance could be contained by low power components at greater than 20 Hz.

(3) The estimates of h_2 made with stimuli based on sequences of 2186-term length contain identifiable contributions from anomalies. Since we have examined only four sequences of this length, and no sequences of greater length, we are not as sure of the validity of the estimate of h_2 as we would like to be. For example, it is possible that there are sequences that result in significantly different estimates of h_2 than those we have found so far. However, we feel that this is a remote possibility for the following reasons. First, the initial 250 msec of the estimate of the major diagonal of h_2 is consistent for the four sequences we used. Second, when we examined all possible valid sequences of 242-term length (there are 21 such sequences) we found that the first 250 msec of the estimate of the major diagonal of h_2 were similar. This suggests that there are unlikely to be major differences in the first 250 msec of the h_2 estimate amongst the different possible valid sequences of the 2186-term length. In any event, it would be impractical to survey all possible 2186-term sequences because there are an enormous number of them.

(4) The predicted VEPs to brief increments and decrements of luminance are

about three times smaller than the observed VEPs. We have no explanation for this result. It implies that there are features of the system nonlinearity that we may not have addressed. For example, system parameters might change in response to the overall pattern of stimulation. A stimulus consisting of an occasional perturbation on a steady background may result in a different system characteristic than a stimulus consisting of a continuous perturbation without any obvious discontinuity. Phenomena such as these fall under the rubric of habituation and for the moment are beyond the scope of our study.

References

1. Davies, W. D. T. 1970. Systems Identification for Self-Adaptive Control. Wiley-Interscience, London.
2. O'Leary, D. P. & V. Honrubia. 1975. On-line identification of sensory systems using pseudorandom binary noise perturbations. Biophys. J. **15:** 505–532.
3. Poussart, D., L. E. Moore & H. M. Fishman. 1977. Ion movements and kinetics in squid axon. Ann. N.Y. Acad. Sci. **303:** 355–379.
4. Poussart, D. & V. S. Ganguly. 1977. Rapid measurement of system kinetics—An instrument for real time transfer function analysis. Proc. IEEE **65:** 741–747.
5. Reddy, S. N. & R. L. Kirlin. 1979. Spectral analysis of auditory evoked potentials with pseudorandom noise excitation. IEEE Trans. Biomed. Eng. BME-**26:** 479–487.
6. Srebro, R. & W. W. Wright. 1980. Visually evoked potentials to pseudorandom binary sequence stimulation. Arch. Ophthalmol. **98:** 296–298.
7. Srebro, R., B. Sokol & W. Wright. 1981. The power spectra of visually evoked potentials to pseudorandom contrast reversal of gratings. Electroenceph. Clin. Neurophysiol. **51:** 63–68.
8. Hooper, R. J. & E. P. Gyftopoulos. 1967. On the measurement of characteristic kernels of a class of nonlinear systems. U.S.A.E.C. Conf. 660206:343–356.
9. Ream, N. 1970. Nonlinear identification using inverse-repeat m sequences. Proc. Inst. Electr. Eng. **117:** 213–218.
10. Barker, H. A. & T. Pradisthayon. 1970. High-order autocorrelation functions of pseudorandom signals based on m sequences. Proc. Inst. Electr. Eng. **117:** 1857–1863.
11. Harris, F. J. 1978. On the use of windows for harmonic analysis with discrete Fourier transform. Proc. IEEE **66:** 51–83.

SOME NEW METHODS FOR THE ANALYSIS OF LATERAL INTERACTIONS THAT INFLUENCE THE VISUAL EVOKED POTENTIAL*

Floyd Ratliff and Vance Zemon

Rockefeller University
New York, New York 10021

The use of patterned stimuli that vary periodically in either space or time, or both, has a very long history in vision research.[1] Applications in the study of visual evoked potentials, however, are relatively recent.[2,3] The rationale behind the use of such stimuli is that analysis of the responses to the periodicity of the pattern can reveal significant information about the form and function of underlying mechanisms—in particular, about the important space constants and time constants of the system. This approach is based on sound physical and physiological principles concerning the transfer of information through complex systems, and well-established mathematical methods of systems analysis can be applied.[4]

Commonly used one-dimensional patterns such as periodic bar gratings or sinusoidal gratings are relatively easy to generate and to display with modern electronic equipment. Waveform generators can be bought "off the shelf" and by using them to control the raster of an oscilloscope one can display a wide variety of simple one-dimensional spatial grating patterns. The technical problems rapidly become more complex, however, as more variables such as contrast reversal, drift, and rotation of the patterns are introduced. The generation and control of two-dimensional patterns, which can be modulated in both space and time, are even more complex and approach the limits of our present affordable technology.

Nevertheless, the problem must be faced. Indeed, the development of methods for the generation and control of complex two-dimensional stimulus patterns is essential for the advancement of vision research. The visual system contains a network of heterogeneous parallel channels that interact with one another in complex ways. To reach a full understanding of these interactions requires methods of stimulus control capable of eliciting the interactions and methods of data analysis capable of treating them.

This paper describes our recent efforts to develop methods of stimulus control and data analysis that have these capabilities. First we describe new methods of generating and displaying a wide variety of two-dimensional patterns that can be varied in both space and time. We then discuss new applications of some of these patterns in the analysis of the influence of lateral interactions on the visual evoked potential.

A Two-Dimensional Computer-Controlled Visual Stimulator

The basic strategy followed in the design of our visual display was to create a television-type raster on a CRT in which, within specifiable regions on the raster, we can produce any of four one-dimensional spatiotemporal functions.[5] This raster differs from an actual TV raster in two major respects. First, it has a frame rate that is

*This work was supported by the National Institutes of Health and by the Esther A. and Joseph Klingenstein Fund.

0077–8923/82/0388–0113 $1.75/0 © 1982, NYAS

approximately an order of magnitude greater (270 Hz vs 30 Hz). This permits us to probe the visual system in a significant temporal frequency region that TV rates fail to include. Second, the raster is rotatable. This feature permits us to probe for directional and orientational selectivity.

A raster of 256 lines is generated from time intervals provided by two counters counting 20-MHz clock ticks. The count in the first counter defines the position of a segment in a line containing 256 segments. The count in the second counter defines the position of a line in a 256-line raster. Thus, the X coordinate is given by the line number and the Y coordinate is given by the segment number. The 256 line by 256 segment coordinates define the positions of 65,536 picture elements (pixels) in the

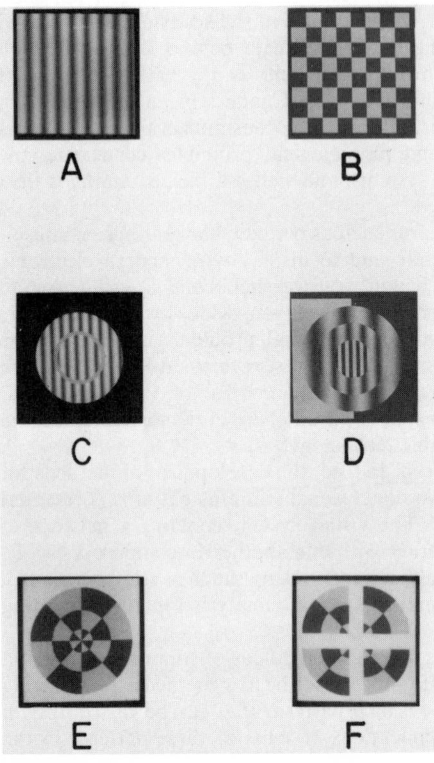

FIGURE 1. Examples of one- and two-dimensional stimulus patterns generated by a computer-controlled stimulus generator (see text).

plane. The idea is to provide a means by which each and every pixel can be assigned to any one of four spatiotemporal functions. The assignments are provided by a memory bit map that contains a 2-bit code for each of the 65,536 pixels. In each frame, the CRT beam traverses all pixels. As the beam moves from one pixel to the next, a unique 2-bit code supplied by the bit map causes a fast electronic switch to select the appropriate spatiotemporal function. In this way, the memory bit map enables us to map arbitrary regions on the raster in which desired spatiotemporal functions are selected.

In FIGURE 1, frames A through D show, respectively, examples of patterns that require one, two, three, and four of the spatiotemporal functions.

Frame A shows an ordinary one-dimensional sinusoidal grating which requires only one of the four spatiotemporal functions.

Frame B shows a two-dimensional checkerboard which requires two of the spatiotemporal functions, one for each set of checks.

Frame C shows a pattern consisting of a central disc surrounded by two annuli. This pattern requires three of the spatiotemporal functions. The outer annulus contains a sinusoidal grating that is 180° out of phase with the sinusoidal grating in the central disc. The annulus surrounding the disc is of a uniform luminance throughout its whole extent. The rectangular surround is not illuminated in this figure (but could be illuminated and controlled independently by the fourth spatiotemporal function).

Frame D shows a stimulus that requires all four of the spatiotemporal functions. The central disc is filled with a square-wave grating. The first (inner) annulus, is filled with a sinusoidal grating. The second (outer) annulus is also filled with a sinusoidal grating but with a different spatial frequency. The surround is filled with a single cycle of a square-wave grating.

Frame E shows a dartboard pattern in which the checks are segments of radial vanes.

Frame F shows a pattern similar to the one in E, except that it has been divided into four quadrants, separated from one another on the uniform background. Each quadrant may be displayed separately.

These still photographs can show only static conditions. A wide variety of dynamic changes can be produced using the temporal controls in each of the four spatiotemporal function generators. Thus, pattern A might be contrast-reversed and simultaneously drifted at some arbitrary velocity across the field. In pattern B the luminance of one set of checks could be modulated at one temporal frequency while the other set was held constant, or modulated at a different frequency. In the pattern shown in frame C, the central grating could be contrast-reversed at one temporal frequency and the outer grating at a different frequency, and so on. Also, each pattern may be set at any chosen rotation angle, or rotated continuously. Patterns with different orientations may be interleaved at a high frame rate to produce composite overlapping patterns, or patterns with different orientations in different parts of the field.

The two-dimensional patterns are not limited to the rectilinear or circular forms shown here. The only limit on the shape of a two-dimensional pattern is that imposed by the "grain" of the individual pixels. Any arbitrary shape can be generated, within those limits, by writing the appropriate program to create the necessary memory bit map.

Effects of Spatial Interactions on the Visual Evoked Potentials

It is well known that the visual evoked potential elicited by a patterned stimulus has a different form depending on whether the pattern is presented by contrast reversal or by appearance and disappearance.[6–8] The contrast-reversal VEP contains only even harmonic components of the frequency of contrast reversal. For example, an 8-Hz contrast reversal produces a predominant response at 16 Hz.[9] The even symmetry of the contrast-reversal VEP is a consequence of the periodicity of the spatial pattern used to evoke it, and the fact that the VEP as usually recorded is a sum of signals over a large fraction of the cortex. Summed over the whole ensemble of neurons that make up the VEP, any odd components of the response to a cycle of contrast reversal must come to naught.

In order to study "contrast" responses, two groups of investigators have devised two independent strategies to bring out the fundamental and other odd symmetric responses in the VEP. Spekreijse and colleagues[7,8] have used pattern appearance-disappearance (*no* contrast reversal) to isolate and characterize the "contrast" component of the VEP. With pattern appearance-disappearance, the odd symmetric components of the VEP do not go to zero, because the "disappearance" phase of the response is not elicited by a pattern that is completely equivalent to that presented in the "appearance" phase (plus a small translation). Rather, during the "disappearance" phase of the stimulus there is no pattern at all. This asymmetry in the stimulus leads to the possibility of asymmetry in the response. In fact, both even and odd harmonics are observed in the appearance-disappearance VEP.

A related but independent approach has been used by Bodis-Wollner and his colleagues.[6] These workers used modulation of the contrast of the whole pattern around a mean level of contrast (*no* contrast reversal) to elicit "contrast" VEPs. At low to moderate depths of contrast modulation, the VEP is clearly dominated by responses at the fundamental and other odd harmonics of the modulation frequency. At higher modulation depths, even harmonics appear to contribute. At very high depths of contrast modulation, this stimulus becomes equivalent to pattern appearance-disappearance, and the VEP elicited by contrast modulation becomes equivalent to the VEP elicited by pattern appearance-disappearance.

Using a new and different technique, unrelated to those described above, we have observed strong fundamental responses in the VEP which appear to be generated by lateral spatial interactions across the boundary between two contiguous patterns.[10,11] The frequency of the major component of the response is determined primarily by the relation between the frequencies of contrast reversal of the two adjacent patterns. If only one pattern is contrast reversed, the major component of the response is at the fundamental frequency of contrast reversal. If both patterns are contrast reversed, however, the major components of the response are at the sum and difference frequencies of the two contrast reversals.

Methods and Procedures

Gold cup EEG electrodes, filled with Grass electrode paste, were attached to the scalps of the subjects. Electrode placement was O_z–C_z, with a ground at P_z according to the 10–20 Electrode System of the International Federation.[12]

The VEP was amplified and filtered (bandwidth: 0.5–100 Hz), and then digitized by computer (270 samples per second). Two types of temporal stimuli were used to evoke responses: 2 min of 1.02-Hz square-wave temporal modulation of certain features of the stimulus pattern, as described below, and 1 min of sinusoidal temporal modulation at a fixed frequency, typically at 4.19 Hz. Averaged responses were displayed on a CRT screen to monitor the progress of the experiment. All data were stored on a disc for later analysis.

The patterned stimuli were generated on the screen of a Tektronix 608 Monitor oscilloscope (P31 phosphor) by means of the visual stimulator described above. The subject viewed the screen binocularly at a distance of 1 meter. Subjects wore corrective lenses if required. At the viewing distance of 1 meter, the raster lines could not be resolved by most subjects.

Three basic spatial patterns were used in these experiments; they are shown in FIGURE 2. A small dark fixation point (a dot 2 minutes in diameter) was displayed in the center of each pattern. One pattern, frame A, consisted of a central disc filled with a sinusoidal grating. The central disc was surrounded by a uniform annulus with luminance the same as the mean luminance of the disc. In the experiment, the grating in the central disc was modulated (contrast reversed) at various temporal frequencies.

Another pattern, frame B, was the same as the one in frame A except that the surrounding disc was filled with a sinusoidal grating of the same contrast and spatial frequency as the grating in the central disc. This grating could either be fixed in a static position or contrast reversed as required in the experiment. Also, both the spatial frequency and the spatial phase of the grating in the annulus relative to the grating in the central disc could be adjusted.

The third pattern, frame C, was the same as the pattern in frame B, except that a uniform annulus, of the same luminance as the mean luminances of the two gratings, separated the grating in the outer annulus from the grating in the central disc. The width of this uniform annulus could be varied as required.

In all patterns, the diameter of the inner disc was 2° of visual angle. The outside diameter of the outer annulus subtended 4°. The mean luminance of the pattern was

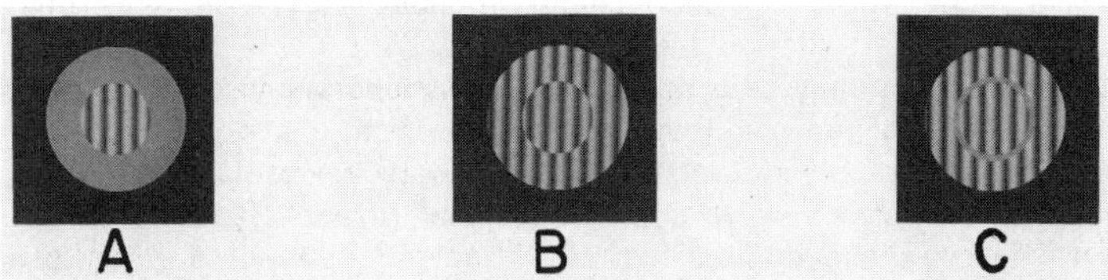

FIGURE 2. Two-dimensional patterns used in the experiments (see text).

30 footlamberts (ft-L). The contrast of the sinusoidal grating was 25%. Contrast was defined in terms of luminance (L) by the following equation:

$$\text{Contrast} = (L_{max} - L_{min})/(L_{max} + L_{min}).$$

Spatial frequency, spatial phase, and temporal frequency of contrast reversal were set as required in the experiments.

The basic phenomena investigated in this series of experiments are illustrated in the set of averaged responses to contrast reversal shown in FIGURE 3. The two stimulus patterns used to obtain these responses are shown at the left of the figure. The pattern at the upper left consisted of a 2° central disc filled with a sinusoidal grating, spatial frequency of 2.4 cycles per degree of visual angle. The disc was surrounded by a uniform annulus, outside diameter of 4°, with a luminance equal to the mean luminance of the central disc. The sinusoidal grating in the central disc was contrast reversed with a 1.02-Hz square wave (left) or with a 4.19-Hz sine wave (right). The length of each record is equal to one period of the square-wave or sine-wave contrast reversal of the stimulus.

The stimulus pattern at the lower left was the same as the one just described except that the outer annulus was filled with a sinusoidal grating of the same spatial frequency, contrast, orientation, and mean luminance as the one in the central disc. The grating in the surrounding annulus was not contrast reversed. It remained unchanged throughout the experiment.

Square-Wave Versus Sine-Wave Contrast Reversal

Contrast reversal of the sinusoidal grating in the central disc when surrounded by the uniform annulus produces only a very small response, with both square-wave and sine-wave reversal—the so-called transient and steady-state conditions (upper left and upper right graphs, FIGURE 3). The response to square-wave reversal shows a small positive wave as the pattern is contrast reversed. Similarly, the response to sine-wave reversal is barely visible in the averaged response. Under these latter conditions, Fourier analysis shows that the amplitude of the response at the fundamental frequency (first harmonic) and at the second harmonic is typically less than 0.5 μV. The absence of a significant response at the fundamental and at other odd harmonics was expected, as explained above: Contrast reversal of a periodic spatial pattern elicits nearly equal out-of-phase responses that cancel one another when summed over a wide area. Absence of a significant response at the even second harmonic results from the particular spatial frequency used, 2.4 cycles per degree. This spatial frequency happens to be at or near the point where there is a pronounced minimum in the response at the second harmonic[13] and in the whole second-order frequency response.[14] In any event, under these particular conditions both the responses to square-wave reversal and to sine-wave reversal are very weak.

When the uniform annulus is replaced by a stationary sinusoidal spatial grating, contrast reversal of the sinusoidal grating in the central disc produces a very strong response, both to square-wave reversal and to sine-wave reversal.[10,11]

The response to square-wave contrast reversal (lower left) shows a very large negative wave when the modulated central grating goes 180° out of spatial phase with the stationary surrounding grating. There is little response when the contrast-reversed central grating goes back in spatial phase with the stationary surrounding grating.

The response to sine-wave contrast reversal of the central grating in the presence

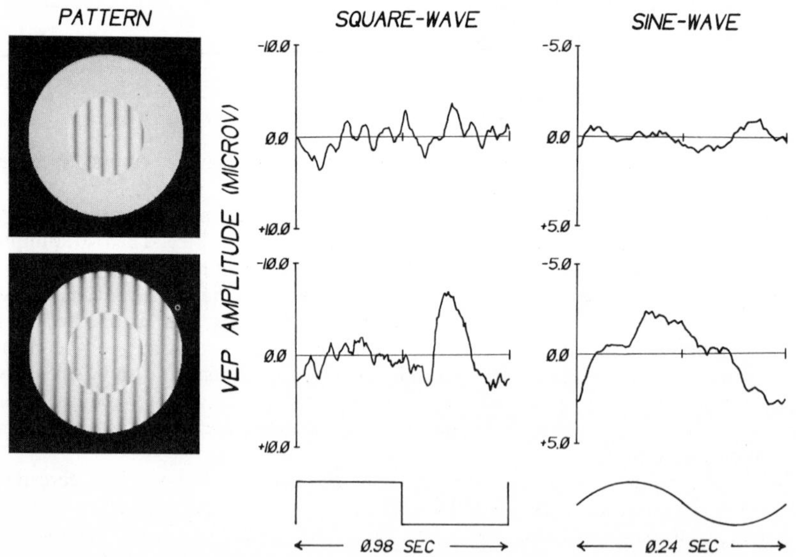

FIGURE 3. Visual evoked potentials elicited by square-wave and sine-wave contrast reversal of a sinusoidal grating in the central disc, with and without a sinusoidal grating in the surround.

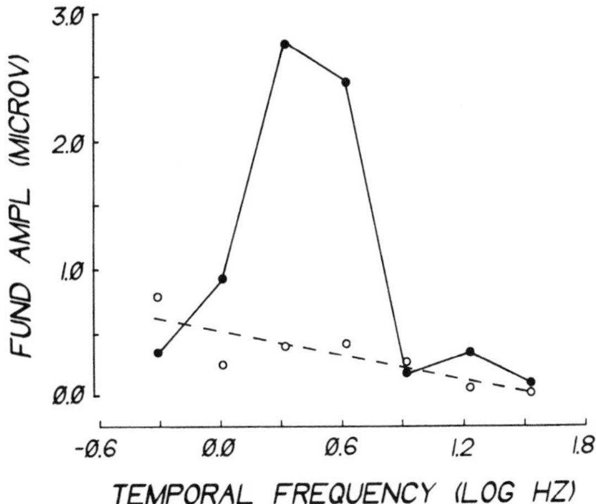

FIGURE 4. Amplitude of response at the fundamental frequency of contrast reversal as a function of temporal frequency.

of a surrounding stationary grating (FIGURE 3, lower right) is equally dramatic. There is a large response at the fundamental frequency. The amplitude of the Fourier component at this frequency, neglible when the surround was uniform, is typically about 2.0 μV for this subject. The response at the second harmonic is about the same as in the condition with the uniform surround.

Some subjects show much larger positive waves in their responses to square-wave contrast reversal of the grating in the central disc with uniform surround than those shown here (upper left panel). The fundamental and second harmonic responses to sine-wave contrast reversal, however, are small in those cases (as they are in the upper right panel). Therefore, the larger positive waves must be composed of responses to temporal frequencies in the square-wave stimulus other than the fundamental 4.19 Hz frequency used for the sine-wave contrast reversal.

Temporal Frequency

The marked effect of the stationary grating in the outer annulus on the evoked potential elicited in response to contrast reversal of the grating in the central disc depends strongly upon the temporal frequency of the contrast reversal. This is shown in FIGURE 4. The open circles represent the response at the fundamental frequency to the contrast reversal of the grating in the central disc when surrounded by a uniform annulus. Amplitude as a function of frequency is essentially flat. (The gradual decrease in amplitude as temporal frequency increases is probably, in part at least, a measure of the power spectrum of the underlying noise.) The filled circles represent the amplitude of the response at the fundamental frequency of contrast reversal of the grating in the central disc when surrounded by a stationary grating. It depends strongly upon the frequency of contrast reversal. For this particular spatial pattern, the maximum response for most subjects is somewhere in the range of 2 to 4 Hz. The response falls off fairly rapidly on either side of this maximum. It cannot be distinguished from the noise at 0.5 Hz and at 8 Hz.

Spatial Phase

The evoked response at the fundamental frequency of the contrast reversal of the grating in the central disc when surrounded by a stationary grating varies with the relative spatial phase of the two gratings. The amplitude of the response at the fundamental temporal frequency (4.19 Hz) is greatest when the surrounding static grating is in spatial phase with the contrast-reversing central grating. This dependence on spatial phase is shown in FIGURE 5. Because the one grating is being contrast reversed and the other is stationary, the contrast *difference* between the two (and the magnitude of the response) goes through a full cycle with a phase shift of the surrounding static grating of only 180° rather than 360°.

Separation of the Two Gratings

The strong dependence of the response at the fundamental frequency on spatial phase of the stationary grating with respect to the contrast reversed grating suggests that the effect may be local, that is, that it may result from the change in contrast at or near the border between the stationary grating and the reversing grating. If so, the response should be reduced by separation of the two gratings. This was done as shown in FIGURE 6. A narrow uniform annulus, of the same mean luminance as that of the two gratings, was inserted between the central disc and the surrounding annulus. The inner diameter of this uniform annulus (the same as the diameter of the inner disc) was held constant. The outer diameter of the uniform annulus could be varied so as to vary its width. (The inner diameter of the outer annulus, containing the stationary grating, varied in the same way.) Thus, as the uniform annulus increased in outer diameter (and width) the outer annulus with the stationary grating increased in inner diameter (and decreased correspondingly in width).

When there was no separation between inner disc and outer annulus, the stimulus conditions and the responses were the same as before: there was a strong response at the fundamental frequency of contrast reversal. As the separation of the gratings in

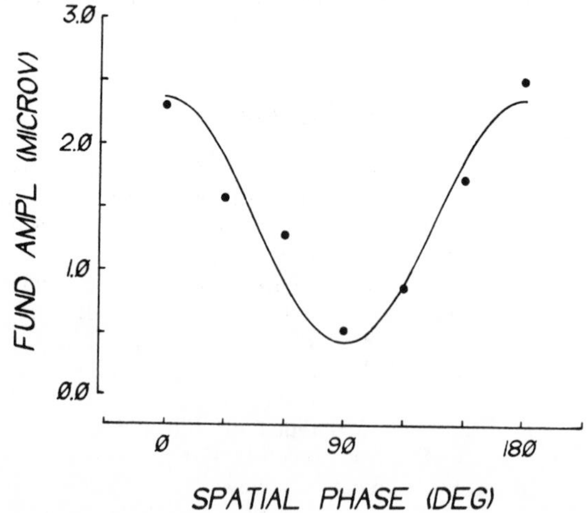

FIGURE 5. Amplitude of response as a function of spatial phase.

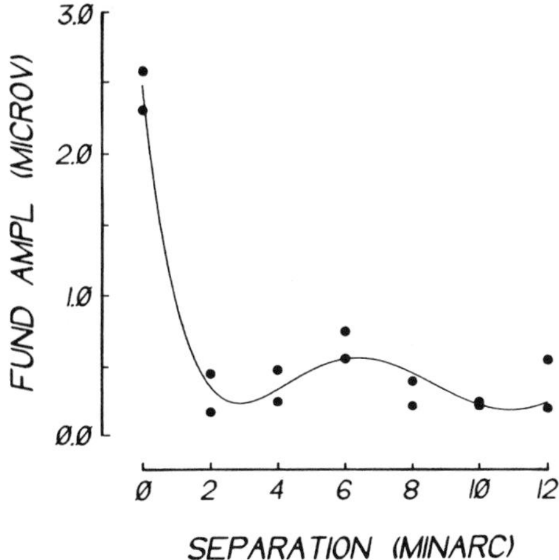

FIGURE 6. Amplitude of response as a function of separation of the two gratings.

the inner disc and in the outer annulus increased, the amplitude of the response at the fundamental decreased. For this subject, the amplitude of the response at the fundamental dropped to the noise level with a separation of only 2.0 minutes of visual angle. Contiguity of the two gratings is crucial.

Intermodulation

The stimuli in all of the experiments described above which produced significant responses at the fundamental frequency consisted of two contiguous sinusoidal gratings of the same spatial frequency. The grating in the central disc was always contrast reversed, usually at a temporal frequency of 4.19 Hz. The grating in the surrounding annulus was always stationary (that is, at a frequency of 0.0 Hz). The experiments on spatial phase indicate that the response depends upon the changing difference in contrast between contiguous points of the two gratings. Since one grating was stationary and one was contrast reversed, the frequency and amplitude of change in relative contrast between the two depended solely on the frequency and amplitude of contrast reversal of the one dynamic grating. If both gratings are contrast reversed, however, the temporal variation in contrast between the two spatial gratings will be some function of the combination of these two frequencies—therefore, the sum and difference frequencies should be significant.

We have conducted an exploratory intermodulation experiment involving two frequencies.

First, the grating in the central disc was contrast reversed at a frequency (F_1) of 4.19 Hz, with no grating in the surrounding (uniform) annulus. Next a grating in the annulus was contrast reversed at a frequency (F_2) of 2.08 Hz, with no grating in the central (uniform) disc. As before, there was no significant response at either of these fundamental frequencies.

For the subsequent measurements (TABLE 1) there was a grating present in both the central disc and the surrounding annulus. As shown in the first row, the grating in the central disc was contrast reversed at a frequency (F_1) of 4.19 Hz and the grating in the surrounding disc was held stationary—that is at a frequency (F_2) of 0.0 Hz. Next the situation was reversed and, as shown in the second row, the grating in the surrounding disc was modulated at a frequency (F_2) of 2.08 Hz and the grating in the central disc was held stationary—that is, at a frequency (F_1) of 0.0 Hz. Finally, as shown in the third row, both gratings were contrast reversed simultaneously—the frequency (F_1) of the contrast reversal of the grating in the central disc was 4.19 Hz and the frequency (F_2) of reversal of the grating in the surrounding annulus was 2.08 Hz. The *difference* ($F_1 - F_2$) between these two frequencies is 2.11 Hz; the *sum* ($F_1 + F_2$) of these two frequencies is 6.27 Hz.

The results shown in the first two rows of TABLE 1 verify those shown in FIGURES 3 and 4. They also show that the lateral interactions can take place in either direction. When the stationary pattern is contiguous with the contrast-reversed pattern, a significant response appears at the fundamental F_1 (when the stationary pattern is in the surrounding annulus) and at the fundamental F_2 (when the stationary pattern is in the central disc). The lateral effect is independent of the relative positions of the contiguous stationary and reversing gratings.

TABLE 1

AMPLITUDES OF RESPONSES (μV IN PARENTHESES) AT VARIOUS FREQUENCIES OF MODULATION (CONTRAST REVERSAL) OF THE GRATING IN THE CENTRAL DISC (F_1), IN THE SURROUNDING ANNULUS (F_2), OR BOTH*

Test	F_1	F_2	$F_1 - F_2$	$F_1 + F_2$
1 (F_1 alone)	4.19	0.0	—	—
	(1.28)	—	—	—
2 (F_2 alone)	0.0	2.08	—	—
	—	(1.68)	—	—
3 (F_1 and F_2)	4.19	2.08	2.11	6.27
	(0.28)	(0.59)	(1.69)	(0.10)

*In test 1, the surrounding grating was held stationary. In test 2, the central grating was stationary. For the difference ($F_1 - F_2$) and sum ($F_1 + F_2$) both gratings were modulated simultaneously (test 3).

When the grating in the central disc and the grating in the surround are modulated simultaneously at frequencies F_1 and F_2, respectively, then the intermodulation (combination) frequencies ($F_1 - F_2$ and $F_1 + F_2$) are the principal determinants of the variation in contrast between the grating in the disc and the grating in the surround. Under these particular conditions, as shown in the third row of TABLE 1, the responses at F_1 and at F_2 when presented simultaneously are neglible or in the noise. The response at the difference frequency ($F_1 - F_2 = 2.11$ Hz) is about the same as the response at the comparable single fundamental ($F_2 = 2.08$ Hz). The sum frequency ($F_1 + F_2 = 6.27$ Hz) is near the upper limit of the effective "window" for single frequencies, as shown in FIGURE 4, and yields little or no response.

SUMMARY AND DISCUSSION

Temporal sinusoidal contrast reversal of a high-contrast spatial sinusoidal grating within a central disc, surrounded by a uniform annulus of constant luminance equal to the mean luminance of the disc, generally evokes a larger cortical potential at the second harmonic than at the fundamental frequency. At certain spatial and temporal frequencies, however, if the surrounding annulus contains a static grating with contrast, orientation, spatial frequency, and spatial phase the same as in the central disc, contrast reversal of the grating in the disc produces a larger evoked potential at the fundamental. The effect varies with the relative spatial phases of the two gratings. (It also falls off rapidly if the spatial frequency of the static grating in the surround is increased or decreased with respect to the dynamic grating in the disc.) Furthermore, the effect depends on the proximity of the static grating. If a narrow blank annulus of the same mean luminance is introduced so as to separate the inner disc (with the dynamic grating) and the outer annulus (with the static grating), the response drops to zero with a separation of a few minutes of visual angle. The effect is bi-directional. Modulation of the grating in the central disc with a static grating in the surrounding annulus, or *vice versa,* produces similar effects. Simultaneous contrast reversal of both the grating in the central disc and the grating in the surrounding annulus by two different frequencies (F_1 and F_2) shows that the response is not strictly dependent upon the fundamental frequency or frequencies *per se* but upon the intermodulation frequencies ($F_1 - F_2$ and $F_1 + F_2$). The initial experiments in which only a single frequency (F_1) was used to contrast-reverse one grating and the other grating was held stationary ($F_2 = 0.0$ Hz) was only a special limiting case in which one frequency (F_2) went to zero and therefore F_1, $F_1 - F_2$, and $F_1 + F_2$ were all the same. The subsequent exploratory experiments using two nonzero frequencies indicate that it is the *intermodulation* frequencies across the boundary of the two patterns that are significant.

The visual evoked potential, as recorded in these experiments, is summed over a large area of visual cortex which contains a very large number of visual neurons. Nevertheless, the techniques of stimulus control and the methods of data analysis that we have applied have enabled us to evoke large fundamental components of the visual evoked potential from the boundary between two contiguous grating patterns. The critical region for evoking these responses has been shown to extend over a visual angle of only about 2.0 or 3.0 minutes of arc. This is equal to a distance of about 0.5 mm in the striate cortex of the human,[15] about the same order of magnitude as the width of one hypercolumn.[16]

Acknowledgments

We thank Norman Milkman and Gary Schick for technical assistance.

References

1. Campbell, F. W. & J. G. Robson. 1968. Application of Fourier analysis to the visibility of gratings. J. Physiol. Lond. **197:** 551–566.
2. Arden, G., I. Bodis-Wollner, A. M. Halliday, J. J. Kulikowski, H. Spekreijse & D. Regan. 1977. Methodology of patterned visual stimulation (Report of the Brussels Symposium ad-hoc committee) Visual Evoked Potentials in Man: New Developments. J. E. Desmedt, Ed. pp. 3–15. Clarendon Press, Oxford.
3. Jeffreys, D. 1977. The physiological significance of pattern visual evoked potentials. Visual Evoked Potentials in Man: New Developments. J. E. Desmedt, Ed. pp. 134–167. Clarendon Press, Oxford.
4. Ratliff, F. 1980. Form and function: Linear and nonlinear analyses of neural networks in the visual system. D. McFadden, Ed. pp. 73–142. Neural Mechanisms in Behavior. Springer-Verlag, New York, N.Y.

5. MILKMAN, N., G. SCHICK, M. ROSSETTO, F. RATLIFF, R. SHAPLEY & J. VICTOR. 1980. A two-dimensional computer-controlled visual stimulator. Behav. Res. Meth. Instr. **12:** 283–292.
6. BODIS-WOLLNER, I., C. D. HENDLEY & J. J. KULIKOWSKI. 1972. Electrophysiological and psychophysical responses to modulation of contrast of a grating pattern. Perception **1:** 341–349.
7. SPEKREIJSE, H., L. KHOE & L. H. VAN DER TWEEL. 1972. A case of amblyopia: electrophysiology and psychophysics of luminance and contrast. *In* The Visual System: Clinical Applications of Neurophysiology and Biophysics, Arden, Ed. pp. 141–157. Plenum, New York, N.Y.
8. SPEKREIJSE, H., L. H. VAN DER TWEEL & T. ZUIDEMA. 1973. Contrast evoked responses in man. Vision Res. **13:** 1577–1601.
9. CAMPBELL, F. W. & L. MAFFEI. 1970. Electrophysiological evidence for the existence of orientation and size detectors in the human visual system. J. Physiol. **197:** 635–652.
10. RATLIFF, F., V. ZEMON & J. D. VICTOR. 1980. Visual evoked potentials: Spatial interaction of dynamic and static sinusoidal gratings. J. Opt. Soc. Am. **70:** 1598.
11. ZEMON, V. & F. RATLIFF. 1982. Visual evoked potentials: Evidence for lateral interactions. In preparation.
12. JASPER, H. H. 1958. The ten twenty electrode system of the International Federation. EEG. J. **10:** 371–375.
13. TYLER, C. W., P. APKARIAN & K. NAKAYAMA. 1978. Multiple spatial-frequency tuning of electrical responses from the human visual cortex. Exp. Brain Res. **33:** 535–550.
14. RATLIFF, F., J. D. VICTOR & R. M. SHAPLEY. 1978. Nonlinear analysis of visual evoked potentials in the human. J. Opt. Soc. Am. **68:** 1427.
15. COWEY, A. & E. T. ROLLS. 1974. Human cortical magnification factor and its relation to visual acuity. Exp. Brain Res. **21:** 447–454.
16. HITCHCOCK, P. F. & T. L. HICKEY. 1979. Banding pattern in human striate cortex as demonstrated by reduced silver stain. Invest. Ophthal. Vis. Sci. (Suppl.) **17:** 157–158.

THE NEURAL ORIGINS OF HUMAN EVENT-RELATED POTENTIALS*

Herbert G. Vaughan, Jr.

Departments of Neuroscience and Neurology
Rose F. Kennedy Center for Research
in Mental Retardation and Human Development
Albert Einstein College of Medicine
Bronx, New York 10461

INTRODUCTION

An understanding of the neural origins of scalp-recorded event-related potentials (ERP) is an essential requirement for their full utilization as clinical tools. In neurological applications, the localization of pathological processes requires that the anatomical source of absent or altered components be identified. Characterization of more subtle pathophysiological disorders demands a more detailed knowledge of the neural mechanisms underlying the generation of the relevant brain potentials. Such information will also clarify fundamental questions concerning the relationship between human perceptual and cognitive processes and underlying brain processes. Thus, investigation of the neural basis of human ERP has two closely related but distinct objectives: (1) identification of the gross anatomical location and configuration of the sources of each ERP component, and (2) definition of the specific cellular processes that generate a particular surface-recorded potential. It is evident that the first objective forms a subset of the second, but as will be discussed later, it is not necessary to have a full knowledge of the cellular mechanisms of ERP generation in order to localize their component sources.

There are three principal methods that can contribute knowledge concerning ERP generators, namely: (1) analysis of the surface topography of electrical and magnetic fields; (2) observation of the effects of selective brain lesions on surface distribution; and (3) intracranial mapping of the potential field. These methods differ both in the information each can provide and in their ease of application to human investigation. Thus, the analysis of scalp topography provides a generally applicable method for estimation of intracranial activity, but it is limited in the kind of information it can provide and requires certain assumptions regarding the intracranial sources. Lesion data are important not only in delineating the consequences of generator pathology on surface ERP distribution, but are necessary to evaluate the rather complex consequences of intracranial pathology that interrupts afferent pathways or destroys central structures in various combinations. As will be seen later, interpretation of lesion effects is not nearly so straightforward as it might appear. Intracranial recordings provide data that are required for definitive characterization of ERP generators, both of their gross anatomical configuration and of their neural mechanism. It is evident, however, that opportunities for human intracranial recording are rare and their scope is limited by specific clinical indications, so the detailed recordings that are necessary to provide definitive information on ERP generation must be carried out in experimental animals. This circumstance imposes the additional requirement that cross-species

*This work was supported by Grants HD 01799 and MH 06723 from the U.S. Public Health Service.

0077–8923/82/0388–0125 $1.75/0 © 1982, NYAS

equivalence of ERP generators and components be established. Therefore, care must be exercised in the selection of an appropriate animal model.

In the following sections, some of our current knowledge of the neural origins of human ERP will be briefly reviewed and the applications of the three main approaches to ERP investigation will be discussed in the light of our own experience with them.

Analysis of the Scalp Topography of Event Related Potentials

The scalp distribution of ERP components provides the principal generally available source of information on the location of their intracranial generators. These data can be obtained in normal human subjects without recourse to invasive recording methods. The effective utilization of topographic data depends upon a quantitative determination of the relationship between the electrical and geometrical characteristics of the intracranial generators and the field potential distribution they set up within the head. This can be done if the generator parameters and the impedance characteristics of the volume conductor which comprises the brain and its coverings are known. For practical purposes, a number of simplifying assumptions are required to permit straightforward computation of the surface potential fields. The generators are assumed to be current dipoles arrayed in a plane so as to represent surface and sulcal cortex.[1] The brain is considered to be spherical and covered by shells of uniform thickness representing the scalp, skull, and subaracnoid space. The volume impedance of the brain and its coverings is essentially resistive and constant throughout each kind of tissue. Although the actual physical situation departs from these assumptions, it can be shown that the computed fields at the surface of the scalp are relatively insensitive to variations in conductivity and relative thickness of the brain's coverings, as well as to inhomogeneities within the brain itself.[1,2]

There are two methods for utilizing the empirical scalp potential distribution to estimate its intracranial generators. In the so-called forward method, assumptions are made regarding the electrical characteristics and geometry of the sources and the field potentials are calculated from these known source configurations. These theoretically derived distributions are compared to the observed topography to determine the adequacy of the assumed source configuration. Alternative source geometries can be compared with one another to determine the best fit with the empirical data. By contrast, the inverse method endeavors to directly compute an intracranial generator configuration from the scalp potential distribution.[3] Although estimation of intracranial generators by the inverse method would be convenient, the approach is of limited utility since it cannot effectively deal with the problem of multiple anatomically extended generators.

Therefore, the forward method of defining relationships between surface potential topography and intracranial generators represents the more useful approach, since any desired generator properties can be specified and the resultant intracranial and surface potential distribution uniquely computed. As more empirical data on the electrical properties of intracranial generators become available from detailed intracerebral recordings, increasingly accurate models of the consequences of specific patterns of neural activation within specific structures can be constructed. Klee and Rall[4] have provided an example of this approach, in which simple model neurons, arranged as pyramidal cells within the surface cerebral cortex, provided the basis for computing the external field distribution. This model differed from other applications of the method[1,5] in the use of a specific neuronal model generator, rather than an equivalent dipolar layer of arbitrary strength. The rough validity of this model is attested by the fact that the computed transcortical voltage generated by a cortical cap (about 400 μV) is of the same order as empirically observed transcortical evoked potential (EP) values.

However, as will be seen later, this simple generator model, based upon somatic depolarization of pyramidal cells, does not accurately depict the actual intracortical pattern of neural activity.

The analysis of scalp topography is further complicated by the presence of multiple, concurrently active generators. In the auditory and visual modalities, the primary projection cortex is bilaterally activated. Responses within secondary areas and the corpus callosum are generated within a few milliseconds after initiation of the primary cortical response. Inasmuch as primary and secondary activity is temporally overlapping and generated in contiguous areas, it is difficult to resolve their relative contributions from the scalp topography alone.

From a series of topographic studies of human ERP,[6-10] we have concluded that the obligatory EP of cortical origin arise from the primary and secondary cortical areas of each sensory modality. Studies of movement related potentials[11] indicate an origin in the pre- and post-central cortex; potentials preceding saccadic eye movements[12] in the posterior frontal and parietal region and potentials associated with visual and auditory and visual discriminative tasks[9] in the secondary cortex of each modality and in modality unspecific loci believed to involve the parietal association cortex. The spatiotemporal overlap of the various ERP components observed in these studies was manifested by complex peaks, suggesting that the intracranial generators we inferred from measurements of individual peaks are not homogeneous. Only when ERP components are substantially different in their timing and spatial distribution can they be readily differentiated and mapped in surface recordings. The use of experimental methods to differentially activate or to selectively inactive specific brain areas can contribute to a clarification of these overlapping surface ERP distributions. For example, selective activation of visual cortex can be achieved by stimulation of restricted portions of the visual field to elicit responses within different regions of primary and secondary visual cortex. Although this approach has yielded data that imply differential sources in striate and extrastriate cortex for certain visual EP components,[13,14] the circumstance that concurrent activity is generated in primary and secondary regions suggests that current interpretations of human VEP generation are oversimple.

A complementary approach endeavors to selectively activate a particular type of processing mechanism such as pattern-specific or luminance-specific processes. Unfortunately, too little is known as yet concerning the specific cellular mechanisms and their anatomical distribution to use this method to localize particular intracranial sources. Nevertheless, experimental manipulation of stimulus variables represents an important tool for differential activation of cellular systems that underlie specific EP components.

Another method, which is particularly valuable for examining potentials associated with higher-order perceptual and cognitive processing, involves the introduction of specific discriminative task requirements. Under these circumstances additional components emerge, some of which may be concurrent with and some following the obligatory nonsignal EP. These task-related or "process-contingent" potentials may be isolated and mapped by subtracting the nonsignal ERP waveforms from those associated with the discriminative response. This "difference waveform" provides an index of the change in neural activity associated with the additional processing requirements.

Although it is convenient to map the topography of scalp ERP at a few points in time, the usual practice of identifying components by the presence of peaks and mapping them at the times corresponding to these peaks, risks the loss of important topographic information and errors because of temporally overlapping but independently generated components. Thus, a more detailed chronotopographic analysis[3,15] is of great value. Time domains in which the field topography is constant can be identified with activity in an anatomically stable configuration of generators, whereas a shifting topography implies the presence of a changing configuration of intracranial sources.

Effects of Brain Lesions

Observation of the impact of selective lesions on EP waveshape and distribution can provide valuable evidence on the structures that generate specific EP components. There are, however, several difficulties in the interpretation of such data. Early studies of the impact of local brain lesions on the EP were impeded by the lack of generally applicable methods for defining the location and extent of intracranial pathology. This difficulty has, to some extent, been resolved by the introduction of CT scan technology, providing a noninvasive visualization of cerebral structures, which unlike postmortem anatomical evidence, can be obtained at the time of EP recording. Despite the undoubted value of the CT scan, these data can be misleading, as shown by two recent case reports[16,17] on the effect of bitemporal lesions on the auditory EP. In both cases the CT scans were considered to demonstrate bilateral destruction of auditory cortex. In one, the AEP was unobtainable, whereas in the other case, it was normal in configuration. There is no clear explanation for this striking discrepancy inasmuch as the location and extent of the lesions disclosed by CT scans were quite similar. Evidently, radiographic evidence should be regarded with caution in the definitive localization of intracranial pathology with regard to ERP generators.

More fundamental complexities in the interpretation of lesion data are illustrated by an experimental study of selective destruction of portions of the monkey's visual system.[18] In Figure 1, the effects of unilateral geniculocortical deafferentation and striate cortex resection are compared. The cortical ablation totally eliminates the VEP ipsilaterally while diminishing its amplitude contralateral to the lesions. By contrast, optic tract section, although eliminating the initial oscillatory potentials and altering the VEP waveshape, leaves a significant response ipsilateral to the lesion. As with the striate ablation, the contralateral VEP was attenuated. These differences in lesion effect may be explained by recurrent connections from area 19 and 18 to area 17 which apparently partially activate the intact but deafferented striate cortex. Thus, human VEP studies that employ the functional criterion of homonomous field defects as evidence of unilateral geniculocalcarine pathology must take into consideration the differential impact of cortical versus subcortical pathology.

Further complications may be expected when lesions effect more than one structure. We observed only slight effects of unilateral striate ablation on the response recorded from inferotemporal cortex. In contrast, a *combination* of unilateral striate and inferotemporal lesions was associated with a marked *increase* in the response of the remaining inferior temporal region.

Given the complexity of connections among the cortical and subcortical structures of each sensory system, as well as the capricious and often uncertain extent of human cerebral pathology, it is unlikely that lesion effects on scalp ERP will be simple. Much additional work, utilizing experimental lesions in animals will be required to define the ERP alterations that are associated with selective interruption of afferent pathways and destruction of cortical generators.

INTRACRANIAL RECORDING

It is evident from the foregoing remarks that direct intracranial recordings are required, both to establish definitively the anatomical location of the structures that generate the surface-recorded EP and to delineate the patterns of neuronal activity within these generators. In the absence of opportunities in humans for the detailed mapping of intracerebral field potentials required to define the location and cellular activity of their generators, we have developed a program of investigation over the past decade that has begun to relate the components and topography of the human

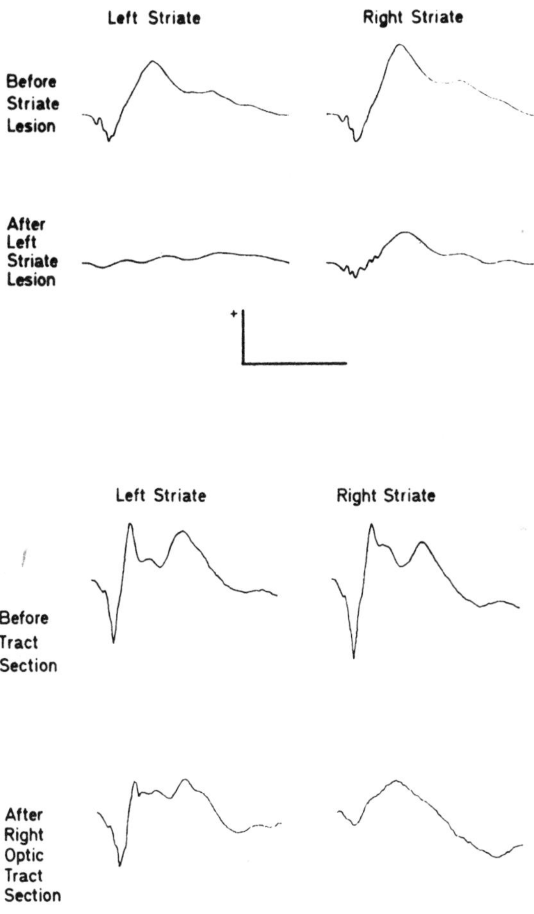

FIGURE 1. VEP to flash before and after unilateral striate cortex removal in a rhesus monkey are depicted in the upper portion of the figure. Note the elimination of the VEP on the side of the lesion and the reduction in amplitude of the slow components of the response contralaterally. The lower traces illustrate the effect of unilateral optic tract section in another monkey. VEP is present, although markedly reduced in amplitude and altered in waveshape on the side of the geniculocortical deafferentation. Calibration: 100 μV, 200 msec. (After Vaughan and Gross.[18])

scalp-recorded ERP to the equivalent potentials and their intracranial generators in experimental animals.

Selection of an appropriate animal model for the comparative analysis of EP generators involves several important considerations. First, the gross anatomy of brain structures and their relationship to the cranial surface should be similar to the human configurations to permit identification of the macroscopic relationship between source geometry and surface potential distributions. Further, the organization of afferent pathways and their central projections, as well as the cellular architecture of cortical regions and the patterns of their intrinsic and extrinsic connections must be comparable to insure that similar patterns of physiological activity occur in both species. Finally, it is desirable that both the human and the experimental animal exhibit similar psychophysical responses and discriminative behavior so that correspondence of EP components can be assessed on functional as well as on anatomical and physiological grounds. Although small mammals such as cats and rats have been extensively employed in neurophysiological investigation, only primate species adequately meet the above requirements for comparative studies. It is evident that there are important differences in the gross configuration and cellular organization of sensorimotor systems between primates and nonprimate experimental animals, rendering the identification of equivalent cross-species ERP components problematic. Among available primate species, the Old World monkeys provide the closest practical experimental model for human ERP generation.

Our methods for detailed intracranial mapping of concurrently recorded field potentials and unit activity in alert monkeys have evolved over the past several years. We employ chronically implanted matrices of stainless steel guide tubes as depicted in FIGURE 2. Two configurations are used. In one, the guides also serve as epidural recording electrodes and are insulated except for the end touching the dura. They are implanted through holes drilled in the skull and placed over the cranial surface to provide both a detailed mapping of the surface potentials and to guide electrode passes for intracortical laminar recording of field potentials and unit activity. The second configuration comprises a dense stereotaxically placed matrix of guides. This array permits the targeting of deep structures within the cerebrum and brainstem with electrode passes at spacings of less than 2 mm. The intracerebral recording electrode comprises up to 16 contacts spaced as closely as 50 μm and mounted within a rigid stainless steel tubing 300 μm in diameter. The recording system has sufficient band-pass to permit the concurrent recording of field potentials and unit data. The latter are derived by high-pass filtering the neuroelectric signals above 500 Hz which are then digitized at 10 kHz, rectified, and averaged to provide an estimate of moment-to-moment net firing within the neuronal sample. Single units may also be isolated with this electrode and their properties examined by post-stimulus time histograms. The field potentials are conventionally averaged and the intracranial distribution of each identifiable component is mapped as follows: after defining and mapping the surface-recorded EP components, depth probes are passed to explore the intracerebral field distributions from the surface toward the possible intracranial generators, as appropriate for each type of ERP. Our initial studies have focused upon the EP to click,[19] to median nerve shock,[20,21] and to flash stimulation and on the movement-related potentials associated with self-initiated wrist extensions.[22]

In considering the EP generators, the short-latency components of subcortical origin must be distinguished from those arising from cortical sources, both surface and deep. Based upon timing considerations alone, the short-latency components of the click and shock EP can be attributed to activity within portions of the afferent pathways. In tracing these components intracranially, one finds that they arise exclusively from the activity generated by ascending volleys of depolarization in afferent fiber tracts. Each of these components, which are surface positive in polarity, increases in amplitude, at first gradually and then at an accelerating rate as an active fiber tract is approached. Components that are generated at earlier portions of the afferent pathway do not exhibit a major increment in amplitude until the electrode approaches the region of their origin. As the electrode moves along an active tract, large polyphasic potentials are recorded with the latency of the initial positive component gradually diminishing as the nucleus giving rise to each fiber tract is approached. When the waveshape of the surface-recorded positive potential is compared with the comparable potential in the incremental depth recordings, it is seen

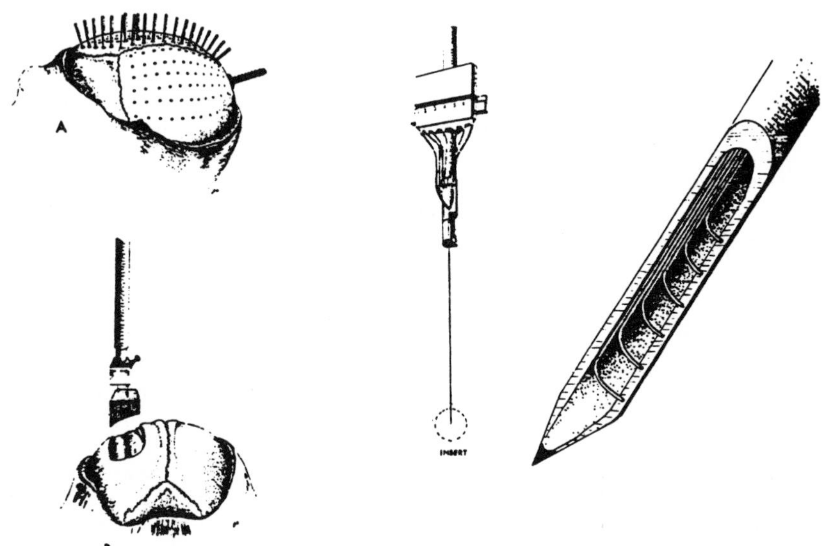

FIGURE 2. Two types of guide matrices employed for intracranial recording. In (A), stainless steel tubing is implanted through holes drilled in the skull, the uninsulated end, which touches the dura, serving as an epidural recording electrode. Intracortical electrodes are introduced through the lumen of each guide for laminar intracortical recording. In (B), a stereotaxically positioned dense matrix of guides is being placed over the exposed dura. This type of matrix is employed for systematic recordings from deep structures. On the right of the figure a multicontact depth probe is depicted. The detail illustrates the position of 25 μm wires in the tip of the probe shaft. The uninsulated tips of these wires provide for concurrent recording of field potentials and multiunit activity.

that the onset of the surface potential corresponds to the shorter latency response from the initial segment of the relevant fiber tract whereas the peak of the surface potential corresponds to the peak latency at the termination of the tract. Thus, the surface-recorded potentials represent the distance-weighted sum of potentials generated by the initial advancing depolarization within a fiber tract. When the electrode enters the nucleus whose cells contribute to a particular tract, large slow potentials, presumed to represent post-synaptic activity, are observed. However, these potentials sharply diminish in amplitude with distance from the nucleus and cannot be traced into the far field, in contrast to the phasic potentials identified with tract activity which can be followed to the surface of the brain. Evidently, the geometrical arrangement of the cells within these subcortical nuclei does not possess the configuration necessary for the generation of far-field potentials.

Despite the clear association of the short-latency SEP and AEP components with synchronized action potentials within afferent pathways, it is not the case that these potentials can be identified with the serial activation of individual tracts. Both click and shock initiate a repetitive sequence of afferent volleys, associated both with repeated neuronal firing at various levels of the afferent pathways and with different conduction velocities in fibers of various diameter. Further complexity is introduced by the presence of parallel afferent pathways, which is of particular importance in the auditory system because of the presence of crossed and uncrossed pathways, as well as several sites of synaptic interruption.

It is important to recognize, therefore, that a simple serial activation conceptualization of the origins of the short-latency auditory and somatosensory EP is not in accord with the observations derived from detailed intracranial mapping. It is particularly important to caution against identifying a surface potential component with an intracranial source based upon correspondence in timing of surface and depth recordings at fixed locations. Our data show many such correspondences in timing of components, most of which are adventitious, since the depth-recorded potential cannot be traced into the far field to the surface recording site. This circumstance is to be expected inasmuch as the components of the short-latency phasic activity within fiber tracts are 1 msec or less in duration and exhibit multiple peaks with similar interpeak intervals. Since these potentials are propagated and thus shift in latency at different recording sites, it is easy to line up peaks in surface and depth recordings that do not represent activity within the same generator.

Short-latency subcortical potentials have not been identified in surface recordings of flash VEP. This is probably due to the relatively prolonged duration of afferent volleys within the retinocortical pathways. The initial repetitive oscillatory potentials, each have a duration of about 10 msec, making it difficult to differentiate the cortically generated oscillatory response from the activity within the afferent pathways. It has further been shown in the visual system, as in the auditory and somesthetic systems, that postsynaptic activity within the thalamic relay nucleus, the lateral geniculate body, makes no detectable contribution to the surface-recorded VEP to flash.[18]

Considering now the ERP components of cortical origin, we encounter a degree of complexity that severely challenges our ability to interpret the surface recorded ERP distributions in terms of their intracranial sources. It is instructive to examine the human and simian cortical anatomy to grasp the origin of a portion of this complexity. Although models of cortical ERP sources have utilized simple geometrical configurations of surface and sulcal cortex, the actual anatomical pattern of cortical convolution is far from simple and it is quite variable from individual to individual. Furthermore, individual differences in location and size of cytoarchitechtonic fields in relation to the gross cortical anatomy introduce a further source of variability in the relationships between intracranial generators and surface potential distributions.

In addressing these complexities it is important to distinguish the field potentials recorded *within* an active structure, to be called *internal* fields, from the *external* fields that are volume-conducted beyond the region that generates them. The fields within active cortex are generally complex and their laminar distribution varies with time in association with the changing intracortical patterns of current sources and sinks. These regions correspond to the laminar distribution of inward and outward transmembrane current flows due to synaptic activity. As depicted in FIGURE 3, the dynamic source-sink configuration within an active cortical region can be estimated by a technique known as current source density (CSD) analysis, employing formulas that

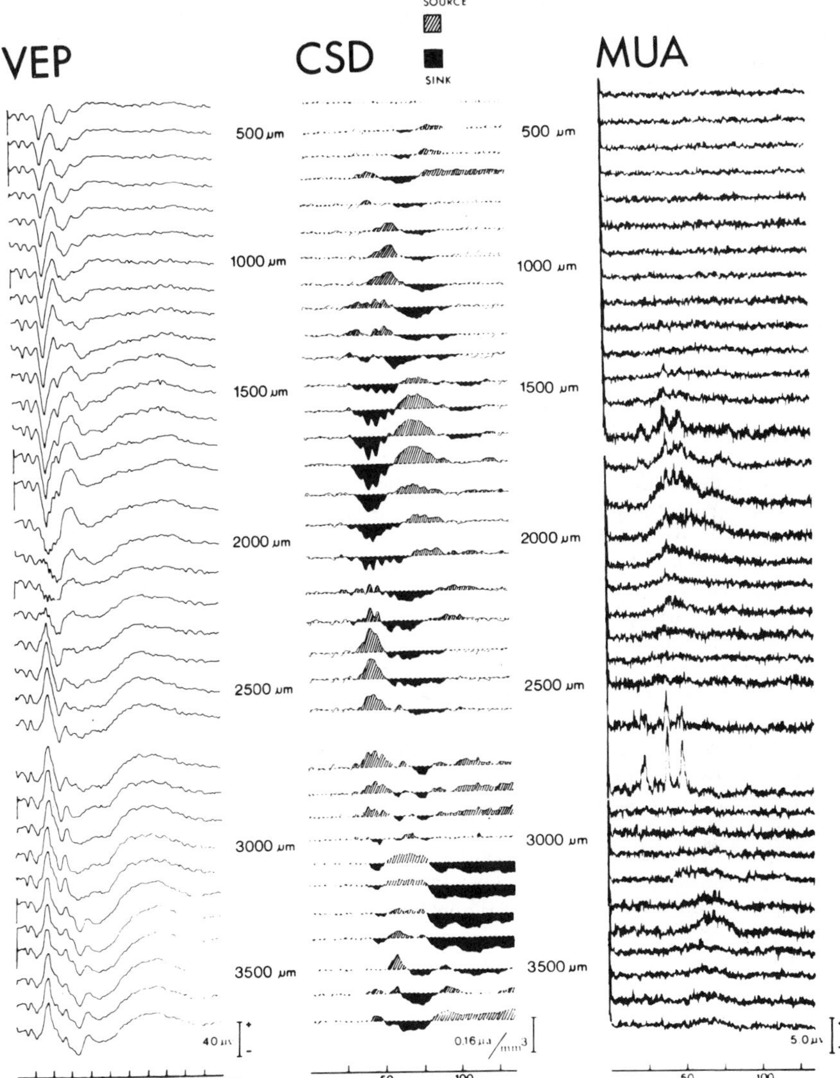

FIGURE 3. Laminar recordings of VEP to flashes taken at 100 μm increments in a pass through the lateral striate cortex of a monkey (left column). Averaged MUA derived by high-pass filtering, rectifying and averaging the activity recorded at each intracortical position is depicted at the right. The center column illustrates source current density obtained from approximation to the second spatial derivative of voltage. Note polarity inversion of early negative VEP component. Current sinks representing excitatory post-synaptic activation are seen immediately above the inversion of VEP, and at the same levels bursts of MUA are present. These events mark the level of geniculocortical fiber termination.

approximate the second spatial derivative of the field potential across the cortical thickness. In this example, the VEP is seen to undergo complex changes with incremental cortical depth, the early components undergoing an inversion in polarity from surface to depth. In the CSD traces a prominent current sink corresponding to the initial excitatory postsynaptic activation is seen superficial to the level of EP inversion. Concurrent increases in MUA are evident at similar cortical depths. It is noteworthy that the sources associated with the return current are deep to the sink. This observation conflicts with the commonly employed pyramidal cell model of evoked potential generation which assumes somatic depolarization with current return by way of the apical dendrites. Although we have just begun to use the CSD technique to characterize the intracortical patterns of current sources and sinks associated with ERP in alert animals, it is evident that intracortical electrogenesis is quite complex, involving temporally overlapping activation of cellular aggregates at several cortical levels.

Despite the complex internal field pattern, the potentials recorded beyond the limits of the active tissue are generated by external current flows that reflect the weighted sum of the internal current sources and sinks. Because of the conservation of charge, any external current flow must derive from a net imbalance of positive and negative charges that represent an equivalent intracortical dipole. This dipolar equivalent generator will vary in strength and intracortical depth over time as the patterns of sources and sinks change in strength and location. In general, the equivalent dipole will be oriented perpendicular to the cortical surface because of the radial symmetry of activity in cortical neurons, so that the only asymmetry of charge distribution occurs across the cortical laminae. Thus, active cortex can be accurately represented by an equivalent time-varying dipolar layer in models of cortical generators. It must be noted, however, that a good deal of information concerning intracortical activity is unavailable in recordings of the external fields because of this algebraic superposition.

In considering the external fields set up by the spatially discrete generators represented by primary and secondary sensory cortical areas, the principle of superposition also applies, so the field potential at any point within the brain or at the surface represents the algebraic sum of the fields produced by each generator. Generally, the primary and secondary cortical regions that contribute to externally recorded EP fields exhibit complex patterns of summation because of the presence of evoked activity that differs somewhat in its timing and waveshape, as well as in the orientation of the respective generators.

In FIGURES 4 and 5, the complex surface topography of the auditory evoked potential to clicks is illustrated, the isopotential mapping of individual peaks is shown, and the correspondence between the surface projections of each component and the intracerebral generators located in different regions of the primary and secondary auditory cortex within the superior temporal plane is depicted. Similar data have been obtained for the cortical somatosensory evoked potentials.[21] These are generated within both surface and sulcal cortex comprising postcentral areas 3, 1, 2, 5, and 7b and from the precentral motor cortex as well. As in auditory cortex the potentials generated within each cytoarchitechtonic area differ in their timing and waveshape. This, together with the different orientation of fields generated within surface and sulcal cortex, produces a complicated resultant surface distribution that is difficult to interpret in the absence of information on the anatomical configuration of each active cortical region and on the waveshape generated therein.

The detailed information on intracranial source anatomy and physiology obtainable from experimental primates should serve as the benchmark for interpretation of human EP studies. Presently, caution must be exercised in the cross-species identifica-

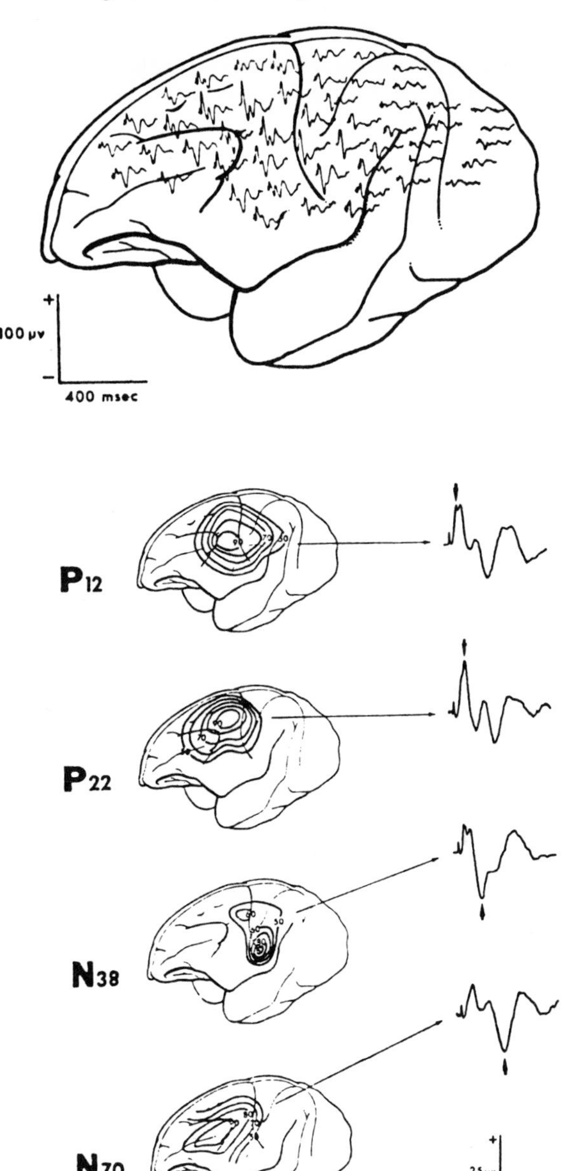

FIGURE 4. Evoked potentials to clicks recorded over the dorsolateral surface of the cerebral hemisphere of a monkey. In the lower diagrams, isopotential maps of the surface distribution of four peaks are shown. Their different topographies imply that these components are generated by sources that differ in anatomical configuration. (After Arezzo *et al.*[19])

tion of specific components, since detailed human scalp topographic data, capable of disclosing the complex component structure that would be anticipated from the simian findings has not yet been obtained. Computerized methods for chronotopographic analysis of scalp potentials will greatly facilitate this endeavor. The complementary information provided by topographic analysis of magnetic field recordings will undoubtedly contribute to more definitive localization of the intracranial generators of human ERP. We can anticipate in the near future a convergence of human topographic analysis and animal investigation, combined with increasingly sophisticated and detailed models of cerebral electrogenesis, that will make it possible to resolve

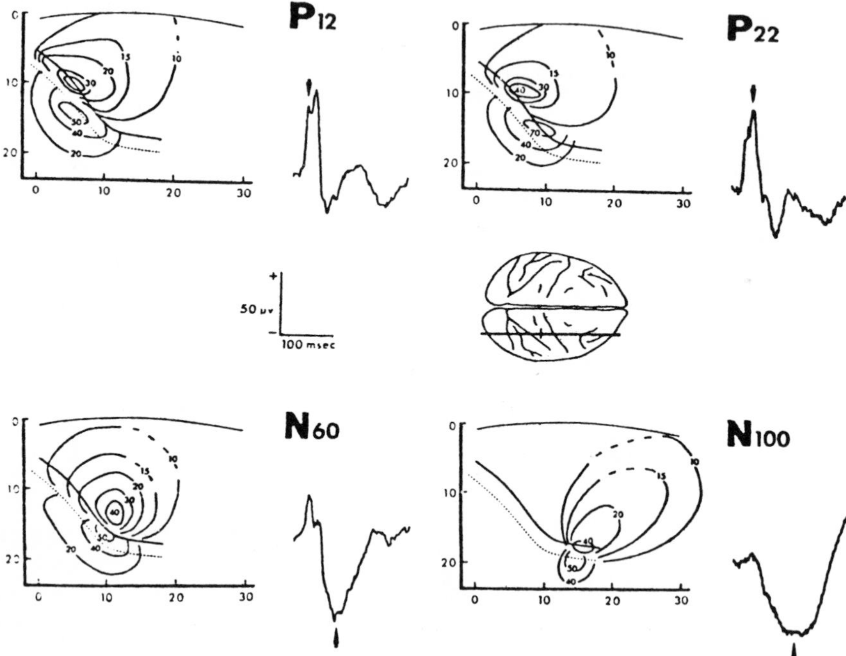

FIGURE 5. Parasaggital intracerebral isopotential maps of four auditory evoked potential peaks. Location of the sylvian fissure which forms the superior boundary of the superior temporal plane is indicated by the curved line. Sources of each component lie in different portions of the superior temporal plane and the fields project toward the surface of the brain with varied sites of maximum amplitude. (After Arezzo *et al.*[19])

surface-recorded human EP waveforms into components associated with specific intracranial generators and to relate these field potentials to patterns of neural activity within the active structures.

ACKNOWLEDGMENT

I would like to acknowledge the essential collaboration of several colleagues in this work: Joseph Arezzo, who has played a principal role in the implementation of the animal studies along with Alan Legatt, Mitchell Steinschneider, and Carl Rosenkilde. Walter Ritter, Diane Kurtzberg, and Richard Simson have collaborated in the topographic studies of human event-related potentials.

REFERENCES

1. VAUGHAN, H. G., JR. 1974. The analysis of scalp-recorded brain potentials. *In* Bioelectric Recording Techniques, Part B: Electroencephalography and Human Brain Potentials. R. F. Thompson & M. M. Patterson, Eds. pp. 157–207. Academic Press, New York, N.Y.
2. WITWER, J. G., G. J. TREZEK & D. L. JEWETT. 1972. The effect of media inhomogeneities upon intracranial electrical fields. IEEE Trans. Biomed. Eng. BME-**19**(5): 352–362.
3. WOOD, C. 1982. Ann. N. Y. Acad. Sci. This volume.
4. KLEE, M. & W. Rall. 1977. Computed potentials of cortically arranged populations of neurons. J. Neurophysiol. **40:** 647–666.
5. DARCEY, T. M., J. P. ARY & D. H. FENDER. 1980. Methods for the localization of electrical sources in the human brain. Progr. Brain Res. **54:** 128–134.
6. VAUGHAN, H. G., JR. 1969. The relationship of brain activity to scalp recordings of event-related potentials. *In* Averaged Evoked Potentials: Methods, Results, Evaluations. E. Donchin & D. B. Lindsley, Eds. NASA Vol. SP-1911: 45–94. National Aeronautics and Space Administration, Washington, D.C.
7. VAUGHAN, H. G., JR. & W. RITTER. 1970. The sources of auditory evoked responses recorded from the human scalp. Electroenceph. Clin. Neurophysiol. **28:** 360–367.
8. SIMSON, R., H. G. VAUGHAN, JR. & W. RITTER. 1976. The scalp topography of potentials associated with missing visual or auditory stimuli. Electroenceph. Clin. Neurophysiol. **40:** 33–42.
9. SIMSON, R., H. G. VAUGHAN, JR. & W. RITTER. 1977. The scalp topography of potentials in auditory and visual discrimination tasks. Electroenceph. Clin. Neurophysiol. **42:** 528–535.
10. SIMSON, R., H. G. VAUGHAN, JR. & W. RITTER. 1977. The scalp topography of potentials in auditory and visual go/nogo tasks. Electroenceph. Clin. Neurophysiol. **43:** 864–875.
11. VAUGHAN, H. G., JR., L. D. COSTA & W. RITTER. 1968. Topography of the human motor potential. Electroenceph. Clin. Neurophysiol. **25:** 1–10
12. KURTZBERG, D. & H. G. VAUGHAN, JR. 1973. Electrocortical potentials associated with eye movements. *In* The Oculomotor System and Brain Functions. V. Zikmund, Ed. pp. 137–145. Buttersworth, London; Publishing House of the Slovak Academy of Sciences, Bratislava.
13. JEFFREYS, D. A. & J. G. AXFORD. 1972. Source locations of pattern-specific components of human visual evoked potentials. I. Component of striate cortical origin. Exp. Brain Res. **16:** 1–21.
14. JEFFREYS, D. A. & J. G. AXFORD. 1972. Source locations of pattern-specific components of human visual evoked potentials. II. Component of extrastriate cortical origin. Exp. Brain Res. **16:** 22–40.
15. LESEVRE, N. 1982. Ann. N. Y. Acad. Sci. This volume.
16. MICHEL, F., F. PERONNET & B. SCHOTT. 1980. A case of cortical deafness: clinical and electrophysiological data. Brain Lang. **10:** 367–377.
17. WOODS, D. L. 1981. The effects of brain lesions on the auditory vertex potential. 6th Int. Symp. on Electrical Potentials Related to Motivation, Motor and Sensory Processes of the Brain, Lake Forest, Ill., June, 1981.
18. VAUGHAN, H. G., JR. & E. G. GROSS. 1969. Cortical responses to light in unanesthetized monkeys and their alteration by visual system lesions. Exp. Brain Res.**8:** 19–36.
19. AREZZO, J., A. PICKOFF & H. G. VAUGHAN, JR. 1975. The sources and intracerebral distribution of auditory evoked potentials in the alert rhesus monkey. Brain Res. **90:** 57–73.
20. AREZZO, J., A. D. LEGATT & H. G. VUGHAN, JR. 1979. Topography and intracranial

sources of somatosensory evoked potentials in the monkey I. Early components. Electroenceph. Clin. Neurophysiol. **46:** 155–172.

21. AREZZO, J. C., H. G. VAUGHAN, JR. & A. D. LEGATT. 1981. Topography and intracranial sources of somatosensory evoked potentials in the monkey II. Cortical components. Electroenceph. Clin. Neurophysiol. **51:** 1–18.
22. AREZZO, J. C. & H. G. VAUGHAN, JR. 1980. Intracortical sources and surface topography of the motor potential and somatosensory evoked potential in the monkey. Prog. Brain Res. **54:** 77–83.

APPLICATION OF DIPOLE LOCALIZATION METHODS TO SOURCE IDENTIFICATION OF HUMAN EVOKED POTENTIALS*

Charles C. Wood

Neuropsychology Laboratory
Veterans Administration Medical Center
West Haven, Connecticut 06516

Departments of Neurology and Psychology
Yale University
New Haven, Connecticut 06520

INTRODUCTION

Attempts to specify the sources of evoked potentials (EPs) recorded from the human scalp have relied upon evidence from a variety of approaches, including studies of scalp topography, analyses of lesion effects, cortical surface and intracranial recordings in humans, and similar experiments in animals.[1-3] In this paper I review the application of quantitative dipole localization methods derived from electrical field theory to the problem of EP source identification. My main objective is not to present the technical details of such methods,[4-7] but to illustrate their value and limitations by applying them to empirical EP data.

Applications of electrical field theory to the identification of electrical sources in the human body can be divided into two broad categories termed the "forward" and "inverse" problems. The forward problem is to calculate the potential field distribution over the surface of a volume conductor generated by electrical sources of known location and configuration in the volume conductor. Based on the work of Helmholtz,[8] Wilson and Bayley[9] presented a method for calculating the potential field on the surface of a homogeneous spherical volume conductor generated by a dipole source of arbitrary location and orientation in the sphere.† A number of EP and EEG investigators have used forward-problem approaches related to that of Wilson and Bayley to calculate the surface potential fields generated by theoretical sources proposed to account for various EEG and EP phenomena.[10-15]

The inverse problem is to calculate the sources within a volume conductor given the empirical potential field on the surface. Without additional information, the inverse problem is mathematically ill-posed and has no unique solution. Because the potential fields of different sources in a volume conductor sum linearly (Helmholtz' principle of superposition), a given surface potential field could be produced by any

*This work was supported by the Veterans Administration and by Grant MH-05286 from the National Institute of Mental Health.

†Wilson and Bayley aptly summarized the importance of Helmholtz's contribution to the problem of electric currents in biological tissue: "Five years hence, a century will have passed since Helmholtz, a young man, 32 years of age, who had already made important contributions to theoretic physics, had been the first to measure the speed of the nervous impulse, and who had invented the ophthalmoscope, became interested in the bioelectric investigations of his friend, Emil du Bois Reymond. The result of this interest was an article, published in 1853, on the distribution of electric currents in volume conductors, which came very close to exhausting the subject" (p. 84–85).

0077–8923/82/0388–0139 $1.75/0 © 1982, NYAS

number of source configurations.[8,9,16,17] However, if the conditions on the inverse problem are restricted by making assumptions about the geometry and conductivity of the medium and the number and form of electrical sources, then an approximate solution can be achieved using iterative numerical minimization techniques. Based on the work of Schneider[4] and Henderson *et al.*,[18] such iterative approaches to the inverse problem have been recently developed and applied to empirical EPs by Darcey and his colleagues,[6,7,19] and by Sidman and his colleagues.[5] Although the Darcey *et al.* and Sidman *et al.* methods will be the main focus of this paper, the issues to be considered apply to other approaches based on similar assumptions. For convenience, the term dipole localization method (DLM) will be used to refer jointly to both methods. I will first review the major assumptions and properties of DLM and then illustrate its application to visual, somatosensory, and auditory EPs. In the final section, I consider the anatomical and physiological validity of the assumptions of DLM and their implications for the interpretation of DLM solutions.

The Darcey et al. and Sidman et al. Dipole Localization Methods

Although the specific implementations of the two methods differ in some respects (e.g., different numerical minimization algorithms), they are based on the same fundamental assumptions regarding the shape and conductivity of the head and the number and form of electrical sources. First, the head is assumed to consist of a sphere corresponding to the brain, surrounded by two concentric shells corresponding to the skull and scalp. Each region is assumed to be a homogeneous, isotropic conductive medium, in which capacitive and reactive effects are assumed to be negligible; in such a medium electrical effects will propagate instantaneously. The brain and scalp are assumed to be equal in resistivity (220 Ω-cm), with the skull 80 times greater.[20] Second, the surface potential field at a given instant in time is assumed to approximate that generated by a dipole source. The dipole assumption is made for two reasons: (a) dipoles are fundamental elements in electrical field theory, and a well-developed body of theory and methods exists to deal with them; and (b) many empirical EP phenomena have potential fields on the scalp that closely approximate fields generated by dipole sources. It should be emphasized that in modeling the electrical sources with dipoles neither Darcey *et al.* nor Sidman *et al.* are asserting that the generators of EPs are physical dipoles. Rather, the dipole is used as an approximation of physiological sources that is mathematically tractable and yet is sufficiently realistic to be useful. The concept of an "equivalent" or "resultant" dipole[16,17] refers to the dipole whose potential field best approximates the summated fields generated by a number of current sources and sinks. For example, the potential field produced by a horizontal layer or sheet of dipoles can be approximated by the field of a single dipole located near the center of the sheet; the latter is said to be the equivalent or resultant dipole of the former. As will be illustrated below, the physiological validity of the solutions given by DLM depends to a considerable extent upon the degree to which the equivalent dipole is a tolerable approximation of the physiological sources and sinks in question.

FIGURE 1 illustrates some of the properties of surface potential fields generated by dipole sources in spherical media of the type assumed by DLM.[7] The two halves of the figure compare surface fields for a homogeneous spherical medium (left) and a three-sphere medium (right) like that assumed by DLM in which the brain, skull, and scalp are represented by concentric spheres of differing conductivity. Within each half of the figure, the left column (a–d) shows fields generated by dipoles oriented radially to the surface of the sphere, and the right column (e–h) shows fields generated by dipoles oriented tangentially to the surface. Dipole locations and orientations are

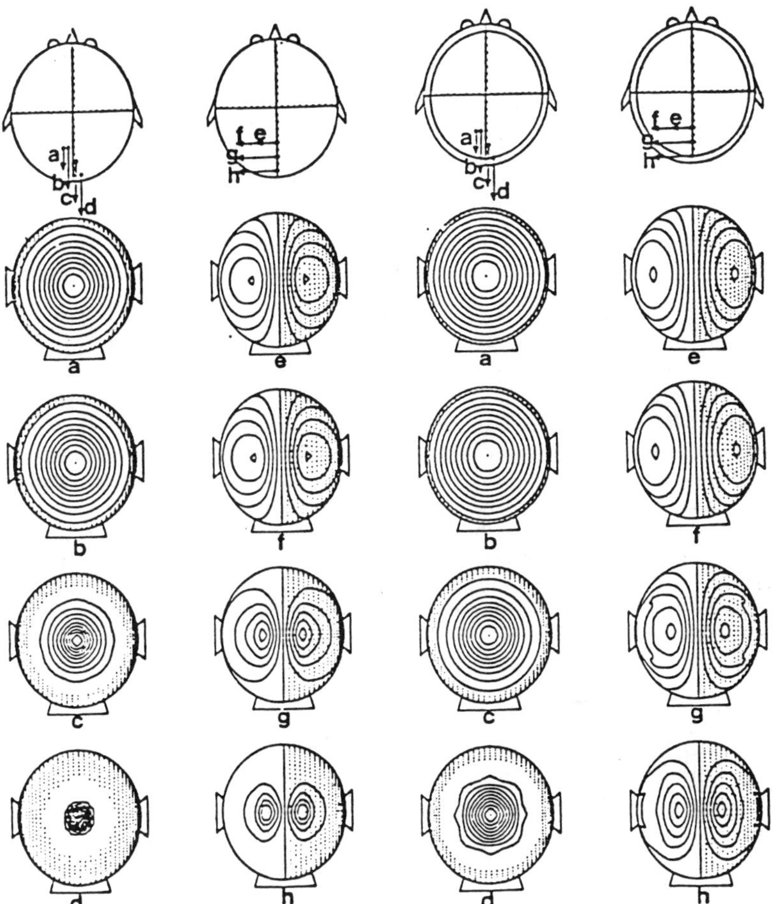

FIGURE 1. Illustration of potential fields on the surface of a one-sphere (homogeneous) model of the head (left) and a three-sphere model of the head (right) like that assumed by the dipole localization method (DLM). The source for each field is a single dipole with the location and orientation indicated at the top of each column. All potential fields are normalized in amplitude to emphasize differences in shape rather than size. (After Darcey.[7])

shown in top views of the model head, and potential field maps are shown on back views. All dipoles are located at different depths on the midline; the locations of the radial dipoles are displaced laterally to aid visualization.

Three aspects of FIGURE 1 should be noted. First, relatively deep sources produce relatively diffuse surface fields with broad amplitude maxima and minima. Second, the three-sphere model produces more diffuse fields than does the one-sphere model, particularly for superficial sources. Third, these fields illustrate that the surface potential fields generated by DLM can capture many of the main features of empirical EP fields.

Based on the assumptions about the conductive medium and electrical source outlined above, DLM estimates the best-fitting dipole for a given instantaneous surface field using numerical minimization techniques. The equation for calculating the surface field generated by a dipole in a spherical medium has six parameters: X, Y, and Z location coordinates, and dipole moment or strength parameters along the X, Y, and Z axes (the vector sum of the latter determines the dipole's orientation). Initial values for these six parameters are chosen arbitrarily (e.g., a dipole at the center of the sphere) and the resulting surface field for these model parameters is compared to the empirical EP field using goodness-of-fit statistics. The model parameters are then altered slightly, and the process is repeated until either a convergence criterion or a maximum iteration count is reached. If the latter is reached without adequate convergence, the results are generally disregarded. Values of the goodness-of-fit measures indicative of adequate solutions have been investigated by applying DLM to simulated data consisting of surface fields generated by dipoles of known location and orientation plus various forms and amounts of random noise.[5,7]

Although detailed parametric comparisons of the Darcey *et al.* and Sidman *et al.* implementations of DLM remain to be performed, we have made preliminary comparisons of the homogeneous (one-sphere) version of the two methods using empirical EP data.[21] For the first few milliseconds after stimulus onset in which no reliable EP activity is present, the solutions obtained by the two methods differ by about the same magnitude as the differences between the solutions given by either one alone at successive time points during this time period. Thus, in absence of reliable EP activity both methods simply reflect the random noise present in the waveforms. However, during time periods in which significant EP activity is present, the two solutions are virtually identical. Such close agreement of the two independently developed versions of DLM supports the mathematical validity of the approach and the robustness of the numerical minimization techniques employed. These results do not address the physiological validity of DLM solutions, which is considered below.

Applications of Dipole Localization Models to Empirical Evoked Potentials

This section illustrates the application of DLM to visual, somatosensory, and auditory EPs, and compares the DLM solutions to the hypothesized sources of such potentials derived from other evidence.

Pattern-Reversal Visual Evoked Potential

Figure 2 (left) illustrates the scalp distribution of the pattern reversal VEP as reported by Barrett *et al.*[22] and Blumhardt and Halliday.[23] For full-field stimulation (left column), the P100 component is largest near the midline, whereas the P100s for left and right half-field stimulation (center and right columns) are maximal over the left and right hemispheres, respectively. Thus, the maxima of the half-field distributions lie over the hemisphere ipsilateral to the stimulated half-field, not over the contralateral hemisphere as might be expected from the crossed projections of the visual pathway. Barrettt *et al.* suggested that this "paradoxical" lateralization of the half-field VEP could be explained in terms of the location of visual cortex on the occipital pole and mesial surface of the hemisphere as shown in Figure 2 (right). According to their interpretation, P100 is indeed generated in the hemisphere contralateral to the visual field stimulated as the anatomy suggests, and maximum

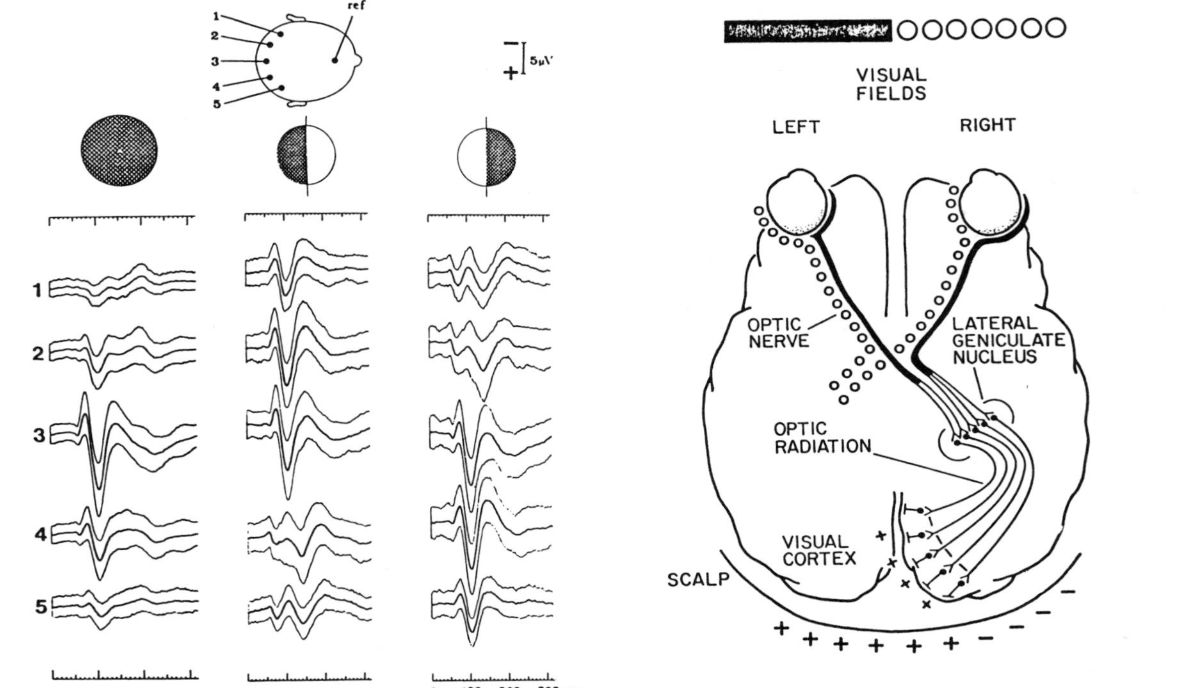

FIGURE 2. Left: Distribution over the occipital scalp of pattern reversal VEPs to full-field (left column), left half-field (center column), and right half-field stimulation (right column). Right: Schematic illustration of the projection of the visual fields onto visual cortex in humans and the sources for the P100 component of the pattern reversal VEP hypothesized by Barrett *et al.* (After Barrett *et al.*[22] and Blumhardt & Halliday.[23])

P100 amplitude on the scalp occurs over the ipsilateral hemisphere because of passive volume conduction. This interpretation is supported by the distribution of half-field VEPs following hemispherectomy[23] and by intracranial VEP recordings.[24,25]

The results of applying DLM to the half-field pattern reversal VEP are shown for three different subjects in FIGURE 3. In this figure, the results are presented for the hemiretina, rather than half-field, stimulated. An isovoltage plot of the P100 amplitude distribution and the location and orientation of the best-fitting single dipole for that distribution are shown for the two hemiretinae for each of the subjects. The dipole solutions are shown in back views (center column) and top views (right column) of the head. The location of the dipole is indicated by the tail of the arrow and its orientation is indicated by the arrow's direction and length in each plane. In all three subjects, the best-fitting dipoles are located near the midline, in the left hemisphere for stimulation of the left hemiretina (right half-field) and in the right hemisphere for stimulation of the right hemiretina (left half-field). In all cases the dipoles are pointed toward the opposite hemisphere. Thus, the DLM solutions for these data provide strong quantitative support for the Barrett *et al.* hypothesis concerning the sources of P100.

Tibial Nerve Somatosensory Evoked Potential

Another example of maximum EP amplitude over the "wrong" hemisphere occurs for the somatosensory EP (SEP) evoked by stimulation of the lower extremities.[26–28] Because the somatosensory pathway is completely crossed, stimulation of the lower extremities should initially activate the hemisphere contralateral to the side stimulated. As shown in FIGURE 4 (top), the initial positivity of the posterior tibial nerve

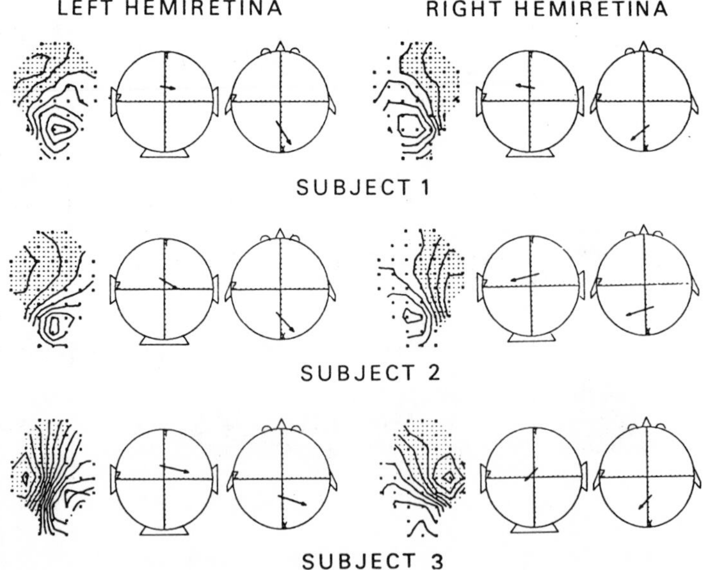

FIGURE 3. Scalp distributions and DLM results for stimulation of left hemiretina (right visual field) and right hemiretina (left visual field). (From Darcey, unpublished observations.)

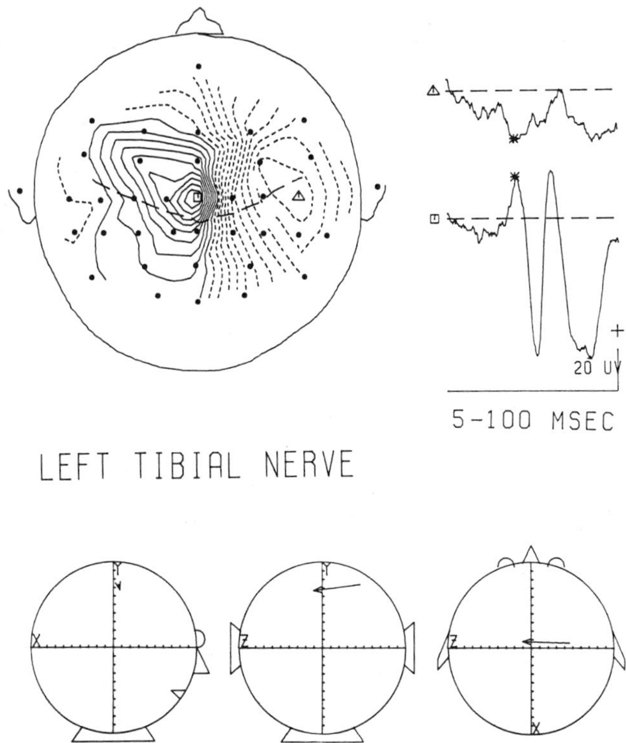

FIGURE 4. Top: Scalp distribution (left) and EP waveforms (right) following stimulation of the left posterior tibial nerve. The scalp distribution is plotted at the peak of the initial positivity at the vertex electrode (square), which occurs simultaneously with a negative peak over the contralateral central scalp (triangle). The distribution was derived from simultaneous recordings at locations indicated by the square, triangle, and solid circles, all referred to a balanced sternovertebral reference. Regions of positive and negative voltage are indicated by solid and dotted isovoltage contours, respectively. Bottom: DLM results at the same latency shown at the top. (From Allison, Wood & McCarthy, unpublished observations.)

SEP (latency 35–40 msec) is largest at the midline and is distributed mainly over the hemisphere ipsilateral to the side stimulated. A negative peak occurs simultaneously over the contralateral hemisphere. These positive and negative peaks could reflect two distinct sources, one in the ipsilateral hemisphere and one in the contralateral hemisphere. However, such a hypothesis would require virtually simultaneous initial activition of cortex in both hemispheres, for which there is no known anatomical basis. An alternative explanation for the ipsilateral positive and contralateral negative peaks is that they reflect a source located on the mesial surface of the contralateral (in this case, right) hemisphere, oriented toward the ipsilateral hemisphere as in the P100 VEP case considered above. Cortical surface stimulation studies in humans[29] and single-unit recordings in monkeys[30] indicate that the foot representation area in somatosensory cortex lies partially on the dorsal surface of the hemisphere and partially on the medial wall. Thus, the general orientation of the foot representation area relative to the overlying scalp is similar to that of the visual cortex, only rotated forward along the midline by roughly 90 degrees (i.e., from the occipital pole to the vertex).

The DLM solution for the initial positivity of the tibial nerve SEP shown at the bottom of FIGURE 4 demonstrates the plausibility of the above interpretation. The best-fitting dipole is located in the hemisphere contralateral to the side of stimulation and is oriented tangentially to the surface of the scalp at the vertex, pointing toward the ipsilateral hemisphere. It is deeper and more lateral than a source located entirely on the mesial surface, suggesting that both the mesial and dorsal surfaces of the hemisphere are involved.

Median Nerve Somatosensory Evoked Potential

A third example of DLM solutions for empirical EPs in the initial cortical response to median nerve stimulation, which consists of a P20-N30 complex at frontal locations and a N20-P30 complex at parietal locations as shown in FIGURE 5. Based on cortical surface recordings, Broughton[31] and Allison[32,33] hypothesized that these potentials originate in a source buried in the posterior bank of the central sulcus (FIGURE 1 of Reference 33, this volume), oriented tangentially to the surface of the scalp. This region corresponds to the somatosensory hand representation area (i.e., cytoarchitechtonic area 3b). FIGURE 5 (top) indicates that the scalp distribution of the peak of parietal P30 is dipolar in shape with a negative frontocentral maximum and a positive parietal maximum. Sidman *et al.*[5] reported DLM results for this P30-N30 field in which the location of the best-fitting equivalent dipole corresponded closely to that predicted by the Broughton-Allison hypothesis. The DLM results shown at the bottom of FIGURE 5 confirm that conclusion. The best-fitting dipole is located just posterior to the approximate location of the central sulcus and is oriented tangentially to the surface of the scalp. An alternative hypothesis for the origins of these potentials is to assume separate precentral and postcentral generators for N30 and P30 instead of a single tangentially oriented source.[34-36] See Allison[33] for discussion of the relative merits of the dual-source and single-source hypotheses.

For the earlier N20-P20 potentials, the Sidman *et al.*[5] DLM results yielded unstable solutions with a poor fit to the obtained data. FIGURE 6 confirms their results and illustrates how DLM's failures as well as its successes can be illuminating. FIGURE 6 presents the scalp distribution of N20-P20 from the same data shown in FIGURE 5 for which DLM reached a numerically acceptable and physiologically reasonable solution. For the N20-P20 field, in contrast, the DLM's fit to the data was poor and the best-fitting dipole was located outside the head (FIGURE 6 bottom).‡ Thus, something is awry; if N20-P20 and P30-N30 are generated by equivalent dipole sources with the same location and opposite orientation as the Broughton-Allison hypothesis suggests, then they should have exactly the same scalp distribution with opposite polarities. One possible explanation for this discrepancy can be seen by comparing the potential fields for N20-P20 and P30-N30 in FIGURES 5 and 6. Unlike the P30-N30 field, which is reasonably symmetrical with the zero potential line roughly half-way between the positive and negative peaks, the polarities of the N20-P20 field are markedly asymmetrical. The latter has the same dipolar shape as the former (even the same relative sizes of frontal and parietal maxima) but the latter is negative in polarity at all locations except for two electrodes nearest the frontal

‡In their more whimsical moments, practioners of DLM (sometimes known as "sourcerers") refer to such dipoles as "transcendental sources."

positive maximum (FIGURE 6). Thus, we need an explanation for the clearly dipolar shape of the N20-P20 field combined with a marked displacement of the zero potential line toward the frontal negative maximum. Others[37-39] have suggested that N20 is generated subcortically rather than cortically, but this hypothesis cannot account for the dipolar shape of the N20-P20 field shown in FIGURE 6. One way to

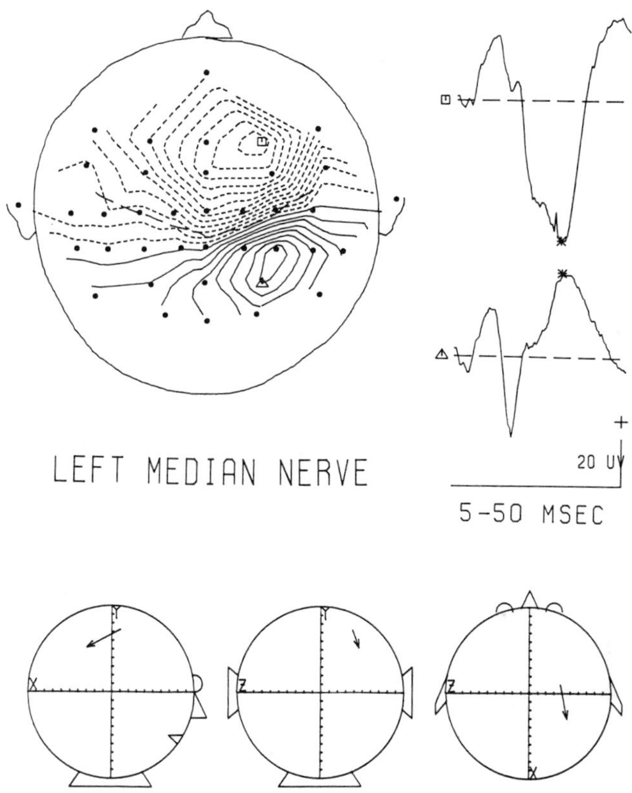

FIGURE 5. Top: Scalp distribution (left) and EP waveforms (right) for the P30-N30 component of the median nerve SEP. Details as in FIGURE 4. Bottom: DLM results at the same latency shown at the top. (From Allison, Wood, and McCarthy, unpublished observations.)

explain the asymmetric dipolar field would be the summation of a somatosensory cortex source[32,33] with a more broadly distributed subcortical negativity.[37-39] This possibility is discussed further by Allison[33] and Kimura.[40]

Auditory "Vertex Potential" and Coincident Activity

This final example has been selected for two reasons. First, all of the previous examples have presented DLM solutions only at major peaks in EP waveforms, with the implication that similar solutions and interpretations apply to the rising and falling

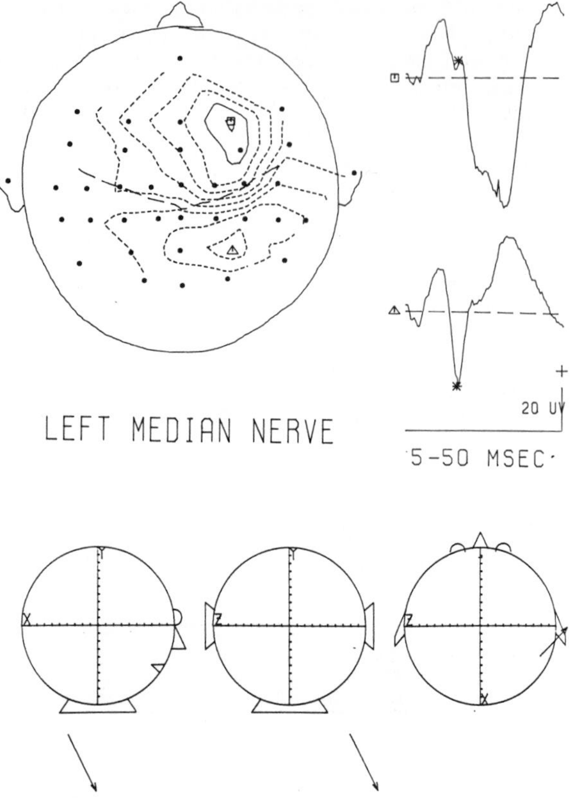

FIGURE 6. Top: Scalp distribution (left) and EP waveforms (right) for the N20-P20 component of the median nerve SEP. Details as in FIGURE 4. Bottom: DLM results at the same latency shown at the top. (From Allison, Wood, and McCarthy, unpublished observations.)

portions of the deflection as well as the peak (i.e., to the entire duration of a "component" as conventionally defined). However, whether or not the scalp potential fields and best-fitting dipoles are stable over time is an important empirical question. For some components (e.g., the P30-N30 median nerve SEP studied by Sidman *et al.*[5] and shown in FIGURE 5), both the scalp distributions and DLM solutions are relatively stable over much of the duration of the deflection. In such cases, it is reasonable to attribute the potential over the entire latency period to the same source or set of sources activated synchronously. However, in cases such as the example to be presented, the scalp distributions and corresponding DLM solutions change systematically over the duration of a major deflection. Such cases imply systematic changes in the location and/or orientation of a single source, or the asynchronous activation of multiple sources.

The second reason for presenting this example is that it illustrates DLM solutions for EPs which other data suggest are not generated by a single relatively focal source as in the preceding examples. In such cases, it is particularly important not to confuse the numerical adequacy of DLM solutions with their physiological validity.

FIGURE 7 presents scalp potential fields (left) and DLM solutions (right) at successive latencies over the duration of the large negative-positive complex commonly termed the auditory "vertex potential" (also N1-P2, N90-P170, etc.). Two alternative explanations for the origins of auditory N90-P170 have been offered, both of which have treated N90 and P170 as unitary components for which a single hypothesized source has been proposed. One hypothesis is that N90-P170 are generated by symmetric bilateral sources in auditory cortex which summate to produce maximal amplitude at the midline.[14,41] The second is that N90-P170 are generated by a more diffuse cortical source, possibly frontal cortex.[42–44]

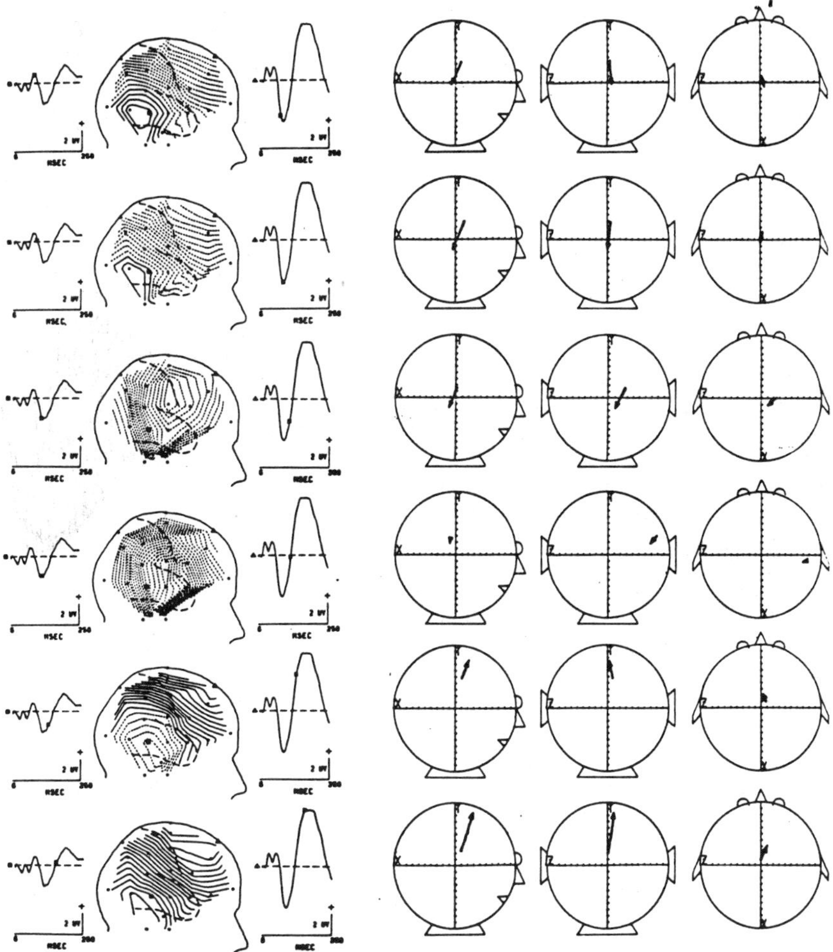

FIGURE 7. Left: Scalp distributions and EP waveforms for AEPs during the latency region of the classical auditory "vertex potential." Inset waveforms are from posterotemporal (left) and frontal midline locations (right). Details as in FIGURE 4. (After Wood & Wolpaw.[45]) Right: DLM results at the same latencies shown on the left. (From Wood, Darcey & Sidman, unpublished observations.)

The latency corresponding to each scalp distribution and DLM solution in FIGURE 7 is indicated by a dot on the inset waveforms, which illustrate EPs at posterior temporal (left) and frontal midline (right) scalp locations. The scalp distribution and DLM solution at the top of FIGURE 7 correspond to a latency on the negative-going slope of N90, followed by those corresponding to latencies at the peak of N90, the positive-going slopes of N90 and P170, finally reaching the peak of P170 at the bottom of the figure. Note that the scalp distributions undergo a change from: (a) negative at frontocentral sites and positive at posterotemporal sites; to (b) a lateral negative peak which moves down the side of the head, to (c) positive at frontocentral sites and negative at posterotemporal sites; and finally to (d) exclusively positive at all locations sampled. The best-fitting DLM solutions at each latency shown on the right of FIGURE 7 also undergo systematic changes over time. Such changes in scalp distribution and best-fitting DLM solution over the duration of N90 and P170 suggest that they reflect the interaction of multiple sources activated asynchronously.[3,45,46] Taken together, all of the available evidence suggests that bilateral sources in or near auditory cortex contribute significantly to auditory N90 and P170. However, additional sources, either in temporal lobe structures or elsewhere, are required to account for the obtained changes in scalp distributions.[3]

Most of the best-fitting single dipole solutions shown in FIGURE 7 are located deep in the brain near the midline over most of N90-P170 (the exceptions are those at more superficial temporal locations in the middle panels of FIGURE 7). Equivalent dipoles located deep near the midline are what might be expected for either the bilateral auditory cortex sources or a diffuse cortical source; thus, in this case the DLM results do not help to discriminate between alternative hypotheses. Although one might expect that a single dipole model would show poor fits to such complex sources, the DLM solutions throughout the latency period shown in FIGURE 7 were characterized by excellent fits. Thus, as discussed in the following section, a numerically adequate solution from DLM means only that the empirical potential field has a shape approximating that of a dipole field; it does not mean that the field was generated by a focal dipolar configuration of sources and sinks. In some cases the ambiguity between dipole fields and more complex fields can be resolved by more extensive recordings over the head. For example, because of limitations on number of simultaneous recording channels, the data shown in FIGURE 7 were obtained only over one hemisphere. Simultaneous recordings over both hemispheres yield distributions at many latencies with more than one pair of maxima and minima which cannot be adequately fit by a dipole field. In other cases, however, (e.g., the exclusively positive scalp field shown at the bottom of FIGURE 7) more extensive sampling will not significantly reduce the goodness-of-fit of the single dipole solution.

THE ANATOMICAL AND PHYSIOLOGICAL VALIDITY OF SOLUTIONS PRODUCED BY DIPOLE LOCALIZATION METHODS

DLM is based on a mathematical model which, like any such model, leads to conclusions that are only as valid as the model's underlying assumptions. The validity of DLM's assumptions about the head as a conductive medium and the consequences of their violation have been discussed elsewhere.[7,47,48] Available evidence suggests that the three-sphere model of the head (each sphere a homogeneous, isotropic conductive medium) is an adequate compromise between mathematical simplicity and anatomical realism. Because of its greater resistivity, variations in shape and thickness of the skull appear to be the most important source of error among the geometrical and electrical assumptions about the head.[47,48]

The validity of the DLM single dipole assumption depends upon the precise manner in which the assumption is formulated. To illustrate, let us consider the following three versions of the dipole assumption.

Literal Dipole Assumption: The instantaneous EP field over the scalp: (a) closely apppoximates that of a dipole field; because (b) it is in fact generated by a current dipole.

Equivalent Dipole Assumption (with Spatial Constraints): The instantaneous EP field over the scalp: (a) closely approximates that of a dipole field; (b) is generated by a spatially restricted region of current sources and sinks; and (c) has an equivalent dipole located in or near the region of sources and sinks.

Equivalent Dipole Assumption (with No Spatial Constraints): The instantaneous EP field over the scalp: (a) closely approximates that of a dipole field; (b) is generated by an arbitrary distribution of current sources and sinks; and (c) has an equivalent dipole that could be located anywhere relative to the distribution of current sources and sinks.

All three versions require that the scalp EP field have a shape approximating that of a dipole field. When the DLM goodness-of-fit statistics indicate that this condition is not met, DLM is inappropriate for the data in question. However, as the results for the SEP N20-P20 field (FIGURE 6) illustrate, even this result can be informative in some cases. What distinguishes the three versions of the dipole assumption is: (a) the extent to which the term "dipole" refers to a literal physical configuration of sources and sinks as opposed to a mathematical description of the potential field; and (b) whether assumptions are made about the configuration of actual sources and sinks and their location relative to the best-fitting equivalent dipole.

As a physical description of sources and sinks, dipole refers to two closely spaced point sources of current having equal and opposite polarity. If this physical definition is what is meant by dipole (as in the Literal Dipole Assumption), then the dipole assumption will never be met by physiological sources in the brain, since the latter consist of current sources and sinks that are always spatially distributed to some degree. As a mathematical description of a potential field, in contrast, dipole refers to the second term in the multipole expansion (monopolar term, dipolar term, quadrupolar term, etc.) that is used to express the potential due to an arbitrary configuration of sources.[49] The reason that the dipole has played such an important role in physics, chemistry, and other physical sciences is that the contribution of the dipolar term to the potential field falls off as the inverse square of the distance from the source, whereas contributions of the higher terms in the expansion fall off as the inverse of increasingly higher powers of distance. Thus, the dipolar term captures much of the variance in many real-world potential fields, becoming an increasingly good approximation with increasing distance from the source.

It is the mathematical sense of the term dipole that is shared by the two versions of the Equivalent Dipole Assumption. In neither case does the use of a mathematical dipole to approximate an empirical potential field imply that the field is generated by a physical dipole. The two different versions of the Equivalent Dipole Assumption are distinguished by whether or not they make additional assumptions regarding the actual distribution of sources and sinks and their location relative to the equivalent dipole. The stronger version (with spatial constraints) assumes that the distribution of sources and sinks is relatively restricted in size and is located near the location of the best-fitting equivalent dipole. In contrast, in weaker version (without spatial constraints) assumes nothing about the configuration of the actual sources and sinks or their location relative to the equivalent dipole. These differences can be illustrated by comparing the DLM results for the SEP P30-N30 field shown in FIGURE 5 with the AEP fields during N90-P170 shown in FIGURE 7, both of which were well fit by dipole fields. For the SEP field, cortical surface and intracranial recordings suggest that the distribution of sources and sinks is restricted to a relatively small area of somatosensory cortex;[32,33] thus, this case would appear to meet the stronger spatially constrained version of the Equivalent Dipole Assumption. For the AEP fields, in contrast, the available data suggest that the sources and sinks are not confined to a single relatively restricted region of tissue;[3] thus, this case would meet the weaker spatially unconstrained version but would not appear to meet the stronger spatially constrained version.

Three important conclusions should be drawn from this discussion and from the examples presented in the preceding section. First, the results of DLM indicate only whether the EP field in question approximates a dipole field; that is, whether the spatially unconstrained version of the Equivalent Dipole Assumption is met. If it is, then DLM provides a good estimate of the location and orientation of the best-fitting equivalent dipole. Whether or not we wish to go further and conclude that the field is generated by a spatially restricted distribution of current sources and sinks in the vicinity of the equivalent dipole (i.e., whether the spatially constrained version is met) *is a matter of interpretation, about which DLM is entirely mute.* Other types of evidence (e.g., lesion data, intracranial recordings, etc.) are required to justify such a conclusion. Second, investigators employing DLM or similar methods should be careful to make explicit which sense of the term dipole they are using in a given case. Much of the controversy regarding DLM and other applications of electric field theory to EP data can be traced to confusion or disagreement over the exact sense of the dipole concept that is at issue. Third, the stability of scalp EP distributions and corresponding DLM solutions over time have important implications for whether a traditionally defined EP "component" (i.e., in terms of peaks and troughs) can be attributed to a single source or synchronous set of sources. Changes in scalp distributions and DLM solutions during a traditionally defined component imply changes in the location and/or orientation of a single source or the interaction of multiple sources activated asynchronously.

In cases where even the spatially unconstrained version of the Equivalent Dipole Assumption is unjustified (e.g., scalp distributions with multiple pairs of maxima and minima), versions of DLM based on multiple dipoles can be employed. Darcey *et al.*[19] successfully used a two-dipole version of DLM for pattern appearance/disappearance VEPs evoked by stimuli activating both hemispheres (whole-field, upper and lower half-fields); such an approach may also be useful for the auditory EPs in FIGURE 7. Finally, distributed sources such as dipolar sheets, spherical caps, and annular sectors have begun to be explored using forward problem techniques.[7,15,50] The value of inverse problem techniques based on such distributed sources remains to be determined.

ACKNOWLEDGMENTS

I am grateful to T. M. Darcey and R. D. Sidman for assistance in applying their methods to the data described herein, and to them, T. Allison, and G. McCarthy for comments on the manuscript.

REFERENCES

1. VAUGHAN, H. G. 1982. The neural origin of sensory evoked potentials. Ann. N. Y. Acad. Sci. This volume.

2. GOFF, W. R., T. ALLISON & H. G. VAUGHAN. 1978. The functional neuroanatomy of event-related potentials. *In* Event–Related Brain Potentials in Man. E. Callaway, P. Tueting, and S. H. Koslow, Eds. Academic Press. New York, N.Y.
3. WOOD, C. C., G. MCCARTHY, N. K. SQUIRES, H. G. VAUGHAN, D. L. WOODS & W. C. MCCALLUM. 1981. Anatomical and physiological substrates of event-related potentials: Two case studies. EPIC VI Panel Report. Lake Forest, Ill.
4. SCHNEIDER, M. R. 1972. A multistage process for computing virtual dipolar sources of EEG discharges from surface information. IEEE Trans. Biomed. Eng. BME-**19:** 1–12.
5. SIDMAN, R. D., V. GIAMBALVO, T. ALLISON & P. BERGEY. 1978. A method for localization of sources of human cerebral potentials evoked by sensory stimuli. Sensory Proc. **2:** 116–129.
6. KAVANAGH, R. N., T. M. DARCEY, D. LEHMANN & D. H. FENDER. 1978. Evaluation of methods for three-dimensional localization of electrical sources in the human brain. IEEE Trans. Biomed. Eng. BME-**25:** 421–429.
7. DARCEY, T. M. 1979. Methods for localization of electrical sources in the human brain and applications to the visual system. Doctoral Thesis. California Institute of Technology, Pasadena, Calif.
8. HELMHOLTZ, H. 1853. Uber einige Gesetze der Vertheilung elektrischer Strome in korperlichen Leitern, mit Anwendung auf die thierischelektrischen Versuche. Ann. Phys. Chem. **29:** 211–233, 353–377.
9. WILSON, F. N. & R. H. BAYLEY. 1950. The electric field of an eccentric dipole in a homogeneous spherical conducting medium. Circulation **1:** 84–92.
10. BRAZIER, M. A. B. 1949. The electrical fields at the surface of the head during sleep. Electroenceph. Clin. Neurophysiol. **1:** 195–204.
11. SHAW, J. C. & M. ROTH. 1955. Potential distribution analysis. I. A new technique for the analysis of electrophysiological phenomena. Electroenceph. Clin. Neurophysiol. **7:** 273–284.
12. SHAW, J. C. & M. ROTH. 1955. Potential distribution analysis. II. A theoretical consideration of its significance in terms of electric field theory. Electroenceph. Clin. Neurophysiol. **7:** 285–292.
13. GEISLER, C. D. & G. L. GERSTEIN. 1961. The surface EEG in relation to its sources. Electroenceph. Clin. Neurophysiol. **13:** 927–934.
14. VAUGHAN, H. G. & W. RITTER. 1970. The sources of auditory evoked responses recorded from the human scalp. Electroenceph. Clin. Neurophysiol. **28:** 360–367.
15. VAUGHAN, H. G. 1974. The analysis of scalp-recorded brain potentials. *In* Bioelectric Recording Techniques: B. Electroencephalography and Human Brain Potentials. R. F. Thompson & M. M. Patterson, Eds. Academic Press, New York, N.Y.
16. GABOR, D. & C. V. NELSON. 1954. Determination of the resultant dipole of the heart from measurements on the body surface. J. Appl. Phys. **25:** 413–416.
17. PLONSEY, R. 1966. Limitations on the equivalent cardiac generator. Biophys. J. **6:** 163–173.
18. HENDERSON, C. J., S. R. BUTLER & A. GLASS. 1975. The localization of equivalent dipoles of EEG sources by the application of electrical field theory. Electroenceph. Clin. Neurophysiol. **39:** 117–130.
19. DARCEY, T. M., J. P. ARY & D. H. FENDER. 1980. Methods for the localization of electrical sources in the human brain. Prog. Brain Res. **54.**
20. RUSH, S. & D. A. DRISCOLL. 1968. Current distribution in the brain from surface electrodes. Anesth. Analg. Curr. Res. **47:** 717–723.
21. DARCEY, T. M., C. C. WOOD & R. D. SIDMAN. Unpublished observations.
22. BARRETT, G., L. D. BLUMHARDT, A. M. HALLIDAY, E. HALLIDAY & A. KRISS. 1976. A paradox in the lateralisation of the visual evoked response. Nature **261:** 253–255.
23. BLUMHARDT, L. D. & A. M. HALLIDAY. 1979. Hemisphere contributions to the composition of the pattern-evoked potential waveform. Exp. Brain Res. **36:** 53–69.
24. LEHMANN, D., T. M. DARCEY & W. SKRANDIES. 1980. Intracerebral and scalp fields evoked by hemiretinal checkerboard reversal and modeling of their dipole generators. Symp. on Clinical Applications of Evoked Potentials in Neurology, Lyon, France.
25. ALLISON, T. Unpublished observations.

26. VAS, G., J. CRACCO & R. CRACCO. 1980. Scalp recorded short latency cortical and subcortical somatosensory evoked potentials to peroneal nerve stimulation. 34th Ann. Meet. American EEG Society, Boston, Mass.
27. TSUMOTO, T., N. HIROSE, S. NONAKA & M. TAKAHASHI. 1972. Analysis of somatosensory evoked potentials to lateral popliteal nerve stimulation in man. Electroenceph. Clin. Neurophysiol. **33:** 379–388.
28. PEROT, P. L. 1981. Scalp recorded somatosensory evoked potentials to stimulation of nerves in the lower extremities and evaluation of patients with spinal cord trauma. Ann. N. Y. Acad. Sci. This volume.
29. PENFIELD, W. & T. RASMUSSEN. 1950. The Cerebral Cortex of Man. Macmillan, New York, N.Y.
30. KAAS, J. H., R. J. NELSON, M. SUR & M. M. MERZENICH. 1981. Organization of somatosensory cortex in primates. *In* The Organization of the Cerebral Cortex. F. O. Schmidt, F. G. Worden, G. Adelman & S. G. Dennis, Eds. MIT Press, Boston, Mass.
31. BROUGHTON, R. 1969. Discussion. *In* Average Evoked Potentials. E. Donchin and D. B. Lindsley, Eds. NASA SP-91. U.S. Government Printing Office, Washington, D.C.
32. ALLISON, T., W. R. GOFF, P. D. WILLIAMSON & J. C. VANGILDER. 1980. On the neural origin of early components of the human somatosensory evoked potential. Prog. Clin. Neurophysiol. **7.**
33. ALLISON, T. 1981. Scalp and cortical recordings of initial somatosensory cortex activity to median nerve stimulation in man. Ann. N. Y. Acad. Sci. This volume.
34. DESMEDT, J. E. & G. CHERON. 1980. Somatosensory evoked potentials to finger stimulation in healthy octogenarians and in young adults: Wave forms, scalp topography and transit times of parietal and frontal components. Electroenceph. Clin. Neurophysiol. **50:** 404–425.
35. DESMEDT, J. E. & E. BRUNKO. 1980. Functional organization of far-field and cortical components of somatosensory evoked potentials in normal adults. Prog. Clin. Neurophysiol. **7.** J. E. Desmedt, Ed. Karger, Basel, Switzerland.
36. PAPAKOSTOPOULOS, D. & H. J. CROW. 1980. Direct recording of the somatosensory evoked potentials from the cerebral cortex of man and the difference between precentral and postcentral potentials. Prog. Clin. Neurophysiol. **7.**
37. CELESIA, G. 1979. Somatosensory evoked potentials recorded directly from human thalamus and Sm I cortical area. Arch. Neurol (Chic.) **36:** 399–405.
38. CHIAPPA, K. H., R. R. YOUNG & W. D. GOLDIE. 1979. Origins of the components of human short-latency somatosensory evoked responses (SER). Neurology **29:** 598.
39. KRITCHEVSKY, M. & W. C. WIEDERHOLT. 1978. Short latency somatosensory evoked responses in man. Arch. Neurol. **35:** 706–711.
40. KIMURA, J. 1981. Short latency somatosensory evoked potentials following median nerve stimulation. Ann. N. Y. Acad. Sci. This volume.
41. SIMSON, R., H. G. VAUGHAN & W. RITTER. 1977. The scalp topography of potentials in auditory and visual discrimination tasks. Electroenceph. Clin. Neurophysiol. **42:** 528–535.
42. KOOI, K. A., A. C. TIPTON & R. E. MARSHALL. 1971. Polarities and field configurations of the vertex components of the human auditory evoked response: A reinterpretation. Electroenceph. Clin. Neurophysiol. **31:** 166–169.
43. PICTON, T., S. A. HILLYARD, H. I. KRAUSZ & R. GALAMBOS. 1974. Human auditory evoked potentials: I. Evaluation of components. Electroenceph. Clin. Neurophysiol. **36:** 179–190.
44. STRELETZ, L. J., L. KATZ, M. HOHENBERGER & R. Q. CRACCO. 1977. Scalp recorded auditory evoked potentials and sonomotor responses: An evaluation of components and recording techniques. Electroenceph. Clin. Neurophysiol. **43:** 192–206.
45. WOOD, C. C. & J. R. WOLPAW. (in press). Scalp distribution of human auditory evoked potentials: II. Evidence for multiple sources and involvement of auditory cortex. Electroenceph. Clin. Neurophysiol.
46. VAUGHAN, H. G., W. RITTER & R. SIMSON. 1980. Topographic analysis of auditory event-related potentials. Progr. Brain Res. **54.**

47. Witwer, J. G., G. J. Trezek & D. L. Jewett. 1972. The effect of media inhomogeneities upon intracranial electrical fields. IEEE Trans. Biomed. Eng. BME-**19:** 352–362.
48. Ary, J. P., S. A. Klein & D. K. Fender. 1981. Location of sources of evoked scalp potentials: Corrections for skull and scalp thicknesses. IEEE Trans. Biomed. Eng. BME-**28:** 447–452.
49. Feynman, R. P. 1963. Lectures on Physics. Addison-Wesley, Palo Alto, Calif.
50. Sidman, R. D., B. Kearfott & V. Giambalvo. 1978. Modeling the sources of evoked cerebral potentials: Single dipole versus dipole layers. Paper presented to the Southern EEG Society.

CHRONOTOPOGRAPHICAL ANALYSIS OF THE HUMAN EVOKED POTENTIAL IN RELATION TO THE VISUAL FIELD (DATA FROM NORMAL INDIVIDUALS AND HEMIANOPIC PATIENTS)

Nicole Lesevre*

Laboratoire d'Electrophysiologie et Neurophysiologie Appliquee (L.E.N.A.)
Hôpital de la Salpêtrière
75651 Paris Cedex 13
France

On theoretical grounds, the exact location and orientation of the intracranial generators of evoked potentials cannot be evaluated from scalp recordings since "a given surface field can be generated by an infinite variety of source configurations."[1] However, interesting and valuable indications concerning the most probable sites of these sources may be obtained through a proper analysis of the spatial distribution of the various components of the evoked response resulting from well-defined experimental variables and taking into account the complexity of the geometry of the sensory cortex.

It is well known that one of the main problems concerning the interpretation of the scalp-recorded evoked potentials and the understanding of their intracranial sources comes chiefly from their being made up of several components overlapping in both time and space. Scalp topography along with the use of experimental variables permitting the isolation of the overlapping components is one means of assessing, in normal humans, the origin of such brain events. Another way is to evaluate the effects of brain lesions on the various components of the response, in particular on their spatial distribution. Both approaches have been used in our laboratory for several years in order to shed some light on the underlying sources of the evoked response obtained for the visual modality.[2–11]

The present paper deals with data obtained by means of spatiotemporal maps[12] concerning: (1) the modifications of the various components of the visual response obtained from "normal" individuals in relation to (a) the part of the visual field stimulated, and (b) the presence or absence of luminance and/or spatial contrast changes; (2) the changes of the pattern-reversal response as seen in patients with lateral homonymous hemianopia of various origins depending on the hemifield stimulated.

Keeping in mind that the visual cortex presents the most complex geometry of all sensory projection areas, the analysis of the modifications of each component as a function of the region of the retina being stimulated will be discussed and interpreted in terms of a simple dipole sheet model of the visual cortex, similar to that proposed by several authors.[1,10,11,13–22]

*Maitre de Recherche, Institut National de la Sante et de la Recherche Medicale (INSERM), Paris.

0077–8923/82/0388–0156 $1.75/0 © 1982, NYAS

Data from Normal People

Methods

Stimulation

Most of our data were obtained by pattern onset stimulation, but in order to isolate some components usually blended together in the pattern onset response—which is made up of both luminance and spatial contrast effects—and to try to understand their respective sources, the same experimental visual field variables were tested in addition to pattern-reversal and blank field onset stimulations. For the three kinds of stimuli, the overall luminance was 60 cd/m^2 and the background luminance 7 cd/m^2.

The full field stimulus consisted of light transmitted through a pattern or a blank transparency of 20° diameter back-projected onto a screen with a central fixation. The pattern was a checkerboard made up of black and white high-contrast squares of 20 minutes of arc each. Onsets and offsets were produced by a shutter whereas pattern reversals were obtained by the displacement of a mirror in such a way that the position of the checkerboard was shifted by one check.[23] For the three kinds of stimuli the rise time was 10 msec and the frequency of occurrence was random with a mean interval of 1500 msec between two successive onsets or pattern reversals.

Various visual field situations were tested: full-field (20° stimulus with central fixation), 4 half-fields and 4 quadrants whose sizes were, respectively, one half and one quarter of the 20° whole field; selective stimulation of the extramacular area by occluding the central 5° or 10° of the full-field or half-field pattern; selective stimulation of the foveal and macular areas (full-field 2° and 5° checkerboards). Two runs per subject were recorded for each experimental condition, at least in the case of the full-field and the four half-field situations. Both eyes were always simultaneously stimulated.

Electrophysiological Recordings

The evoked potentials were recorded with at least 9 electrodes in line forming a "cross montage." The distance between 2 successive electrodes was usually 4 cm but in some cases a smaller interelectrode distance was chosen. The longitudinal part of the cross was on the midline, the last electrode being located on the inion and the most anterior one 16 cm above; the transverse branch crossed the midline 4 cm above the inion and extended 8 cm away from the midline on both hemispheres, in direction of the upper part of the ears and the external border of the eyes. In order to get a better image of the potential field, bipolar and "monopolar" recordings were obtained simultaneously. In the case of monopolar recordings, the reference electrode consisted of linked earlobes; but in several cases the influence of the reference was systematically studied by also using a frontal (FZ) and a noncephalic electrode.[24] The horizontal and vertical eye movements (EOG) were always recorded simultaneously on two other channels in order to control eye fixation.

Data Processing

The electrophysiological data were digitized "on line" (with a sampling period of 2 msec during the whole epoch of stimulation, i.e., approximately 1300 msec; but the figures shown in this work only illustrate the first 350 msec of the averaged response). The digital data were processed by computer (B.G.E. Gamma M40) which calculated off-line the averaged responses. The responses occurring during or just after an eye movement were eliminated from the average. Each average response was thus obtained from 60 to 70 successive stimuli and was presented by computer in the form of chronograms and spatiotemporal maps of gradients (bipolar recordings) or potentials (monopolar recordings). On these maps, amplitude is represented in the form of isogradient or isopotential lines, the values between two successive electrodes being obtained by a mathematical second-order interpolation. All reported results and illustrations (except those on the lower map of FIGURE 4) correspond to potential values ("monopolar" recordings).

Population

To date, the responses to pattern onset and/or to pattern reversal in different visual field conditions have been recorded in our laboratory, with various montages and various stimuli conditions, from over 60 healthy adults (age range: 25–55), all these responses having been analyzed by means of spatiotemporal maps as well as conventional amplitude-versus-time curves.

The pattern-onset and pattern-reversal responses of 20 subjects out of this whole population have been recorded under exactly the same conditions regarding electrode positions, stimuli characteristics, and visual field conditions, i.e., the conditions described above. Part of the statistical analysis of the data concerning the responses of these 20 subjects has been reported in previous papers[10,11] and therefore will not be repeated here. The following results concern chiefly individual data obtained from this population of 20 subjects.

Results

Space and Time Image of the Full-Field Pattern-Onset Response: Its Intraindividual Reliability and Interindividual Variability

The first 350 msec of the pattern-onset full-field response obtained under the experimental conditions described above was made up of four principal components which have been described by many authors, and labeled N60 (wave 0), P100 (wave 1), N140 (wave 2), and P200 (wave 3) (FIGURE 1). The surface of the successive isopotential curves which represent these components, as seen on the maps constructed along both anteroposterior and transversal lines of electrodes, differed from one component to the other, the surface of N60 being the smallest and P200 the largest. Whatever the extension on the scalp of these components, their maximal activity—indicated on the maps by the sign plus or minus, according to their polarity—was always clearly located on the posterior region, between 0_z and the inion for most subjects (longitudinal axis), and on the midline (transversal axis). However, it must be noted that P100, contrary to the other midline components, often appeared (60% of our population) as two distinct symmetrical peaks usually located near the midline; besides, in a few cases (3 out of 20), P100 was so posteriorly located (lower than 0_z, and even lower than the inion) that it did not show up on our montage (FIGURE 7, map 1).

In addition to these posterior-midline components, bilateral activities, peaking 8 cm or more away from the midline, could be clearly seen in most subjects (90%): first

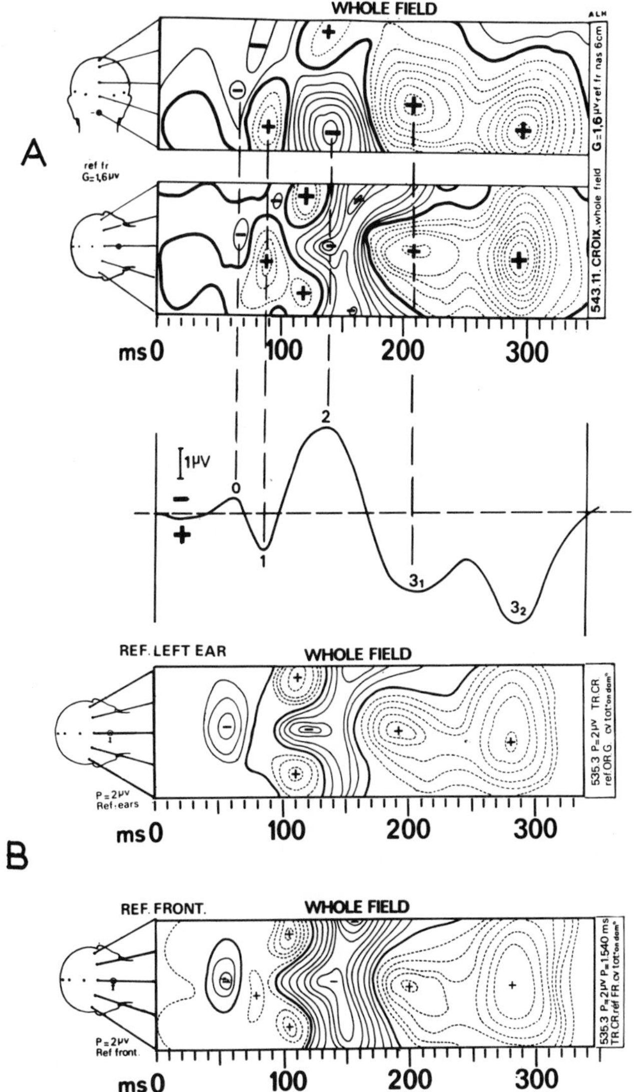

FIGURE 1. Intraindividual reproducibility and influence of the reference. Typical full-field pattern-onset response obtained from a normal subject (ALH): (A) with a frontal reference (longitudinal part of the cross-montage: upper map; transversal part: lower map); (B) 3 months later, with a frontal (lower map) and a linked earlobes (upper map) reference; only the transversal response is shown for (B). On these maps (and those of other figures), the positive potentials are represented by broken lines, the negative ones by plain thin lines, zero potential being marked by a thick line. Each peak of potential is indicated at the exact place where it occurs by a plus sign or a minus sign. Between 2 successive isopotential lines there is an increase of 1.6 μV for (A) and 2 μV for (B). The chronogram placed under the map of (A) represents the response from the midline electrode, 4 cm above the inion.

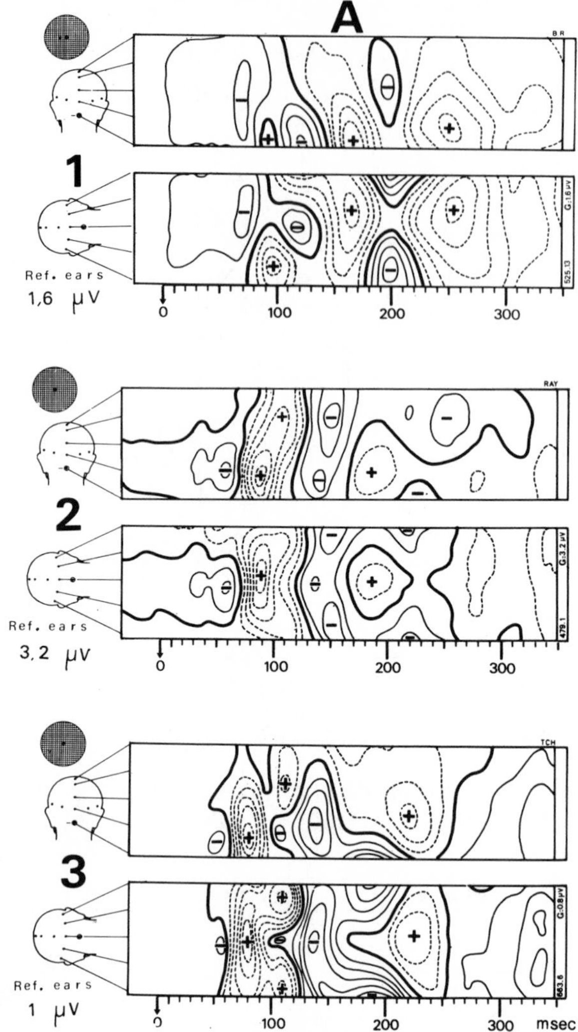

FIGURE 2. Interindividual variability. Full-field (A) and left- and right-half field (B) pattern-onset responses obtained from 3 different subjects: BR (1), RAY (2), and Tch (3). For the 3 subjects, the reference electrode was on the linked earlobes; the difference between 2 successive isopotential lines was 1.6 μV for BR (1), 3.2 μV for RAY (2), and 1 μV for Tch (3). Subject BR (1A) has an asymmetrical LP120, no LN150, and a rather anterior LN200. Right and left field stimulations elicit, for the 3 subjects, a contralateral LP120 and LN150. The N140 peak remains on the midline (1B, lower map), is shifted to the ipsilateral hemisphere (1B upper map, 2B upper map), or disappears (3B: upper map; 2B: lower map).

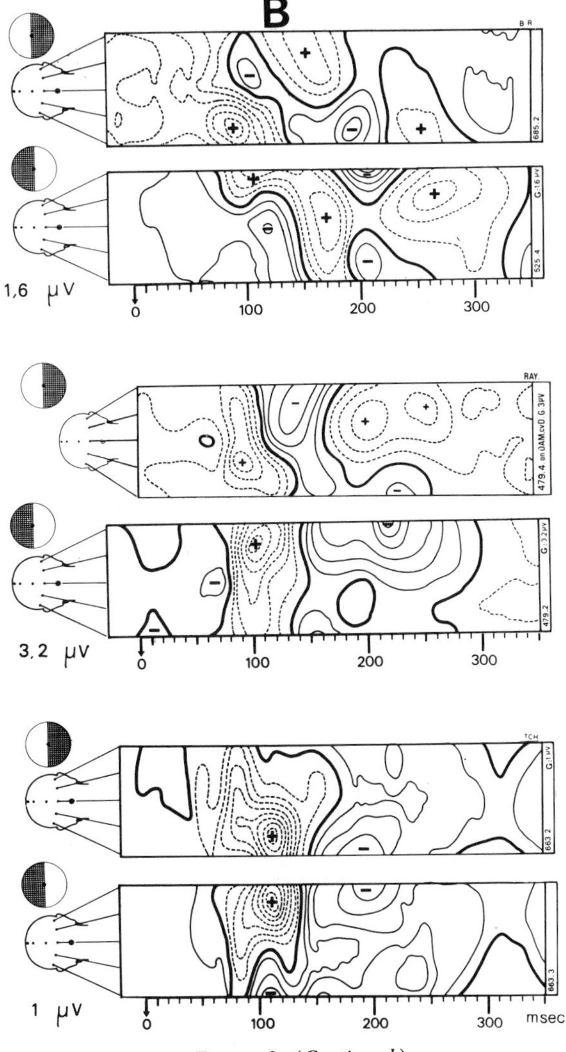

FIGURE 2. (*Continued.*)

of all, two bilateral positive peaks occurred with an average 20 msec delay after P100: these components, well differentiated for 60% of our population, were labeled LP120. The bilateral LP120 occurred at the same time as a positive wave peaking between the vertex and P_z [FIGURES 1, 2A (maps 3), and 7 (map 1)]; they were followed by bilateral negativities, labeled LN150, which also culminated at the same time as an anterior negative wave peaking a little below the vertex (see in particular FIGURE 2A, maps 2 and FIGURE 7). Finally, towards 200–250 msec, a late lateral component, "LN200," appeared as two symmetrical peaks in the response of most subjects (FIGURE 2A), whereas for some others these late negativities were hidden by the presence of a particularly large P200 (FIGURE 1). In contrast to the two other bilateral components, LN200 corresponded on the longitudinal montage to a very posterior negative component, usually peaking lower than the inion (FIGURE 2A, maps 2 & 3).

Some people were characterized by important lateral components and hardly any midline components, others by extended midline components hiding the lateral waves, but the majority, however, showed both types of components. Despite important changes in amplitude, the reproducibility within one subject of this individual time and space image was quite remarkable, even after several months or years insofar as the experimental conditions remained the same (FIGURE 1). In spite of amplitude and latency differences, the overall topographical organization of the response did not change in relation to the position of the reference electrode; in particular, the lateral activities described above remained with the same polarity and same topography (FIGURES 1 & 4).

Changes of the Pattern-Onset Response According to Visual Field Conditions

The changes of the various characteristics (amplitude, peak latency, topography, polarity) of the components of the pattern-onset response according to visual field conditions and their statistical significance are given in TABLE 1.

Right and Left Half-Field Stimulations. When the right or left field was selectively stimulated, the spatiotemporal organization of P100 depended upon the presence or absence of well-differentiated LP120s in the full-field response. For those subjects whose full-field response did not show any clear lateral positivity, the midline P100 peak was shifted to the contralateral hemisphere in response to the 366f-field stimulations. This contralateral P100 peaked approximately 4 cm away from the midline, a little later than the full-field P100 (FIGURE 2, A & B2, FIGURE 4). For these subjects whose whole field response was made up of two distinct lateral positive components following the midline P100, the half-field responses were made up of a high-amplitude contralateral component (always of higher amplitude than the full-field LP120), peaking approximately at 8 cm away from the midline, a little earlier than the full-field LP120, but a little later than the full-field P100 (FIGURE 2, A & B3). Such a spatiotemporal organization suggests that in those cases both midline and lateral components were mingled together. For a few subjects, however, (5 out of 20) the midline P100 was clearly distinguished from LP120 in the right or left field response, and in that case P100 remained on the midline and only LP120 was contralateral (FIGURE 3). This was also observed when only the lower right or left quadrants were stimulated (lower map of FIGURE 3), but then the latency of the second positive peak was a little shorter. Thus the midline P100 could be more or less shifted to the contralateral hemisphere according to subjects, but LP120 was always predominant on the contralateral hemisphere; no exception to this rule was observed among our whole population of 60 healthy subjects, whatever the reference electrode (FIGURE 4). In particular, when bipolar recordings were used, a phase reversal of the gradients, indicating a maximal positive activity (source), was always clearly seen on the contralateral hemisphere (lower map of FIGURE 4). The question whether these midline and lateral positive peaks represent or not two components of different origin and functionnal significance will be discussed later.

In the same time as these contralateral positivities, either a negativity or a positivity of low amplitude could be seen on the ipsilateral hemisphere; in the case of a low-amplitude negative wave occurring on the ipsilateral side, this negativity did not peak at the same latency as the contralateral P100 (only 2 exceptions out of 20

TABLE 1

CHANGES OF THE PATTERN-ONSET RESPONSE ACCORDING TO VISUAL FIELD CONDITIONS*

	Posterior Midline Components				Lateral Components		
	N60	P100	N140	P200	LP120 Ant.	LN150 Ant.	LN200 Post.
Lower vs full 20° field							
Amplitude	↓***	↑***	↓***	I	↓**	↓*	I
Latency	0**	0**	0**	I	↓**	0**	0*
Topo.	0**	0*	0**	0**	0*	0**	0*
Polarity	0***	0***	0***	0***	0***	0**	0***
Upper vs lower field							
Amplitude	↓/D**	↓**	D***	↓D***	?	↓ (NS)	0*
Latency	0	0*	I	I		0	0*
Topo.	0	+P*	A (+)**	I		0	A NS
Polarity	0	−A***	+P***	−NS		0	0***
Right or left vs full field							
Amplitude	↓/D**	I	I	I	↑***	I	I
Latency	0*	0**	0**	0*	↓*	0*	0 NS
Topo.	0*	0/CL	0/IL	0/IL	CL***	CL***	CL***
Polarity	0*	0***	***	***	0***	0***	0***
Foveal versus full field							
Amplitude	0**	↓/0 (NS)	↓/0 (NS)	↓***	↓**	↓**	I
Latency	0**	0 (NS)	0 (NS)	0 (NS)	0*	0*	I
Topo.	0**	P**	P***	0***	0*	0*	0**
Polarity	0***	0***	0***	0***	0*	0*	0***
Extramacular vs full field							
Amplitude	D**	↓**	↓**	↓/0 (NS)	0*	0*	I
Latency		0**	I	0*	0*	0*	I
Topo.		A**	A**	0**	0**	0**	0**
Polarity		0***	0***	0***	0***	0***	0***

*Intraindividual comparisons (20 healthy subjects). Statistical signification (sign test): *** $P < 0.01$; ** $P = 0.01$; * $P = 0.05$; NS, $P > 0.05$. ↑ or ↓, amplitude or latency increase or decrease; 0, no change; I, inconsistent change; D, disappears. Regarding topographical changes, P and A mean that the − (negative) or + (positive) peak becomes more posterior or more anterior; CL and IL mean that the peak is shifted to the contralateral or ipsilateral hemisphere to the stimulated hemi-field.

subjects: FIGURE 7 lower maps 3 & 4, and FIGURE 2 lower map B3). Only 1 subject out of 20 had a negative ipsilateral wave of higher amplitude than the contralateral P100. In the same way, the lateral negativities labeled LN150 and the later LN200s were always of higher amplitude on the contralateral hemisphere (no exception).

In contrast, the location of the midline peaks N60, N140, and P200 either did not change (FIGURE 2, lower map B1), or, regarding N140 and P200, appeared as being shifted to the ipsilateral hemisphere. This was the case when the contralateral LP120 was of particularly high amplitude and long duration (FIGURE 2, upper maps B1 and B2, FIGURE 7). A third possibility was that the midline components disappeared.

Upper Versus Lower Field Stimulations. In spite of significant changes in amplitude (TABLE 1), the overall time and space image of the lower field response was very similar to that of the full-field response; it must be noticed, however, that usually in the lower field response the lateral LP120 waves were not as well differentiated as in the whole-field response. In contrast, the upper field response was characterized by important changes concerning its topography as well as its polarity: P100 became more posterior (FIGURE 5, subjects 1 and 2), to such an extent that for most people (FIGURE 5, subjects 3 and 4) this positive component was too posterior to be seen with our montage. This important topographical shift determined an apparent change in polarity in such a way that the region of the scalp that was occupied by a positive wave in the full-field or lower-field response became negative; this negative wave (N100), of lower amplitude than P100, peaked exactly at the same latency as P100 but slightly more anteriorly. On the transverse montage, the upper field N100 always peaked on the midline even in the case of subjects whose P100 was made up of 2 bilateral peaks located at least 4 cm apart (FIGURE 6). An apparent polarity reversal was also seen for N140, which, indeed, was replaced by a positive wave in the upper field condition. However, this positive wave did not peak at the same latency as N140, but usually with an average 20 msec difference. Concerning the midline P200, it tended to disappear in the upper-field response (FIGURE 5), though, excepted a few cases (FIGURE 5, subject 3), no clear polarity reversal was usually seen.

The symmetrical LP120 waves were not clearly seen in the upper-field response, but it was not possible to tell whether they had disappeared or were hidden by N100 (which could be more widespread than P100), or mingled with the next positive wave of the upper-field response, as was suggested by the comparison of the 4 quadrant

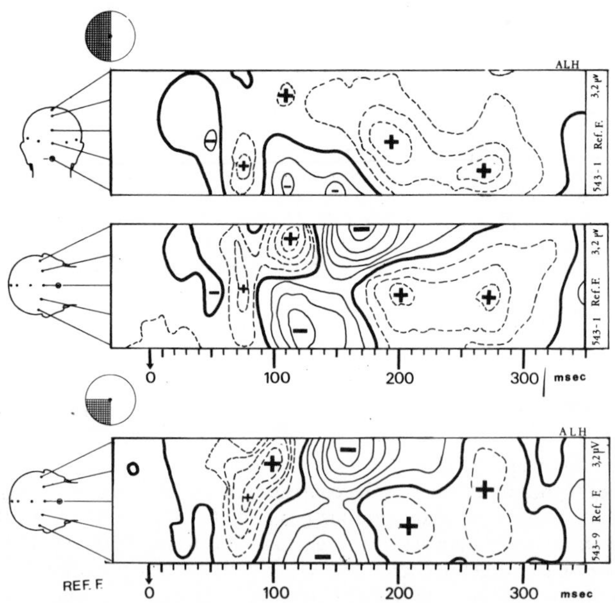

FIGURE 3. Left-field pattern-onset response (longitudinal and transversal upper maps) and left-inferior quadrant response (lower transversal map) obtained from same subject as that of FIGURE 1 (ALH). Reference electrode: linked earlobes; 3.2 μV between two successive isopotential lines. The midline posterior P100 remains on the midline whereas the lateral anterior LP120 is contralateral to the stimulated field (occurring earlier in the quadrant response).

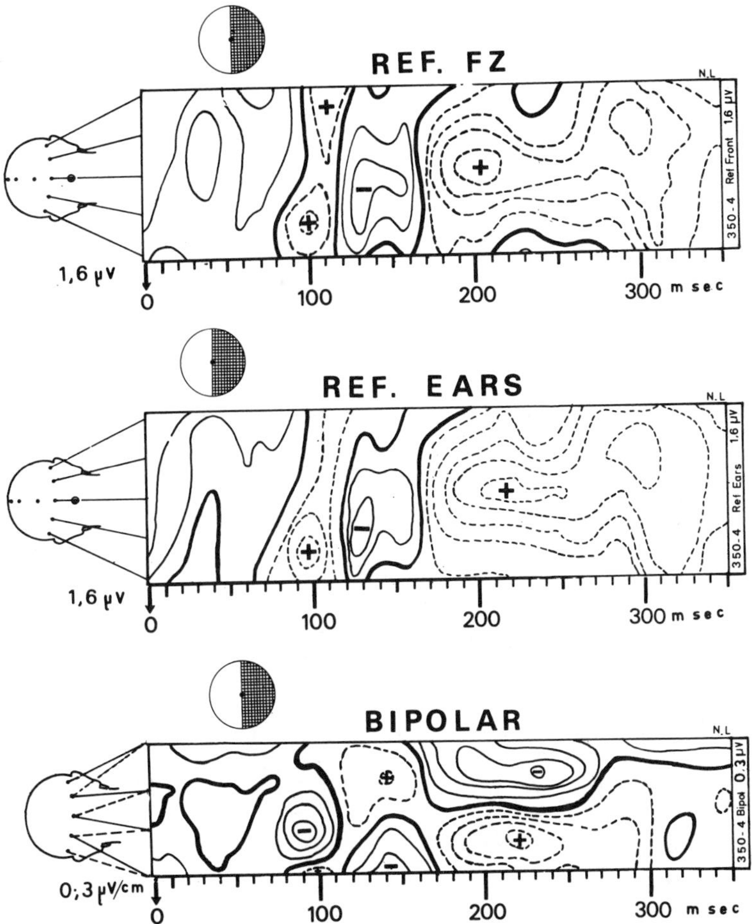

FIGURE 4. Right field pattern-onset response obtained, from one subject (NL), simultaneously, with a frontal reference (upper map), a linked earlobe reference (middle map), and with bipolar leads (lower map). 1.6 μV between each successive isopotential lines (upper and middle maps); 0.30 μV/cm between each isogradient lines (lower map). LP120 and LN150 are not seen for this subject; on the 3 maps, P100 predominates on the contralateral hemisphere. The bipolar recordings show, towards 100 msec, a phase reversal of gradients—indicating a source—occurring on the left hemisphere, whereas the phase reversal for the two other doublets of gradients are located near the midline.

responses. Furthermore no polarity reversal was observed in the upper-field response for the later negative lateral components (LN150 and LN200).

Foveal and Macular Stimulation. N60 was the only component that did not decrease in amplitude and did not change topography when the fovea (2° pattern) was selectively stimulated (TABLE 1), whereas it disappeared when the central 5° of the pattern was occluded (FIGURE 7). When stimulating only the fovea, the other midline components became more posterior (TABLE 1); in particular P100 and N140 became

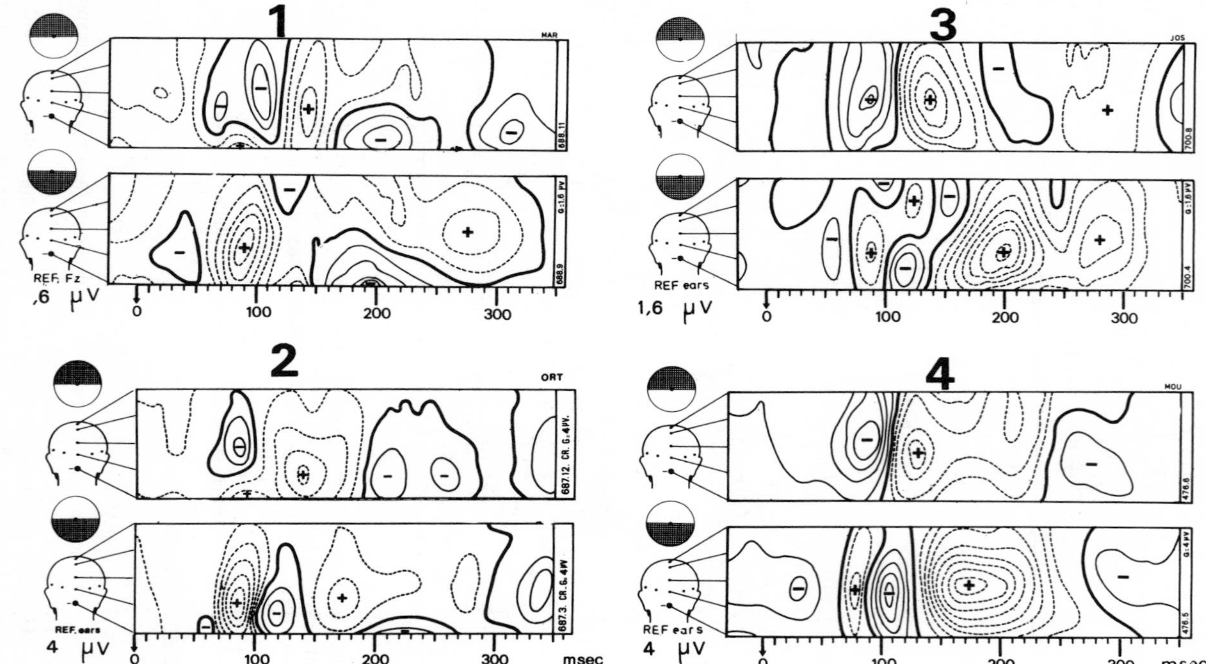

FIGURE 5. Upper- and lower-field pattern-onset responses (longitudinal part of the cross montage) obtained from 4 different subjects MAR (1), ORT (2), JOS (3), and MOU (4). Linked earlobes reference for subjects 2, 3, and 4; frontal reference for subject 1. Between successive isopotential lines: 1.6 μV for subjects 1 and 3; 4 μV for subjects 2 and 4. The upper-field responses of subjects 1 and 2 both show a very posterior P100 peak. This positivity is not visible in the case of subjects 3 and 4.

so posterior for some people that these components could no longer be seen on the maps and therefore only the bilateral LP120 and LN150 were observed but with a much lower amplitude than for the full-field response; regarding P200, its amplitude considerably decreased.

The response obtained when the central 5° was occluded varied from one subject to an other in a rather incoherent way. The only constant changes seen in most people were a significant tendency of N60 to disappear, and an important decrease in amplitude of the midline components P100 and N140 which tended to peak more anteriorly (approximately 4 cm above the full-field response peak). In contrast with the foveal response, the amplitude and topography of P200 remained practically unchanged. Concerning the lateral components, their spatiotemporal image remained unchanged in spite of a decrease in amplitude the importance of which varied according to subjects (a rather marked decrease is seen in FIGURE 7, maps 2).

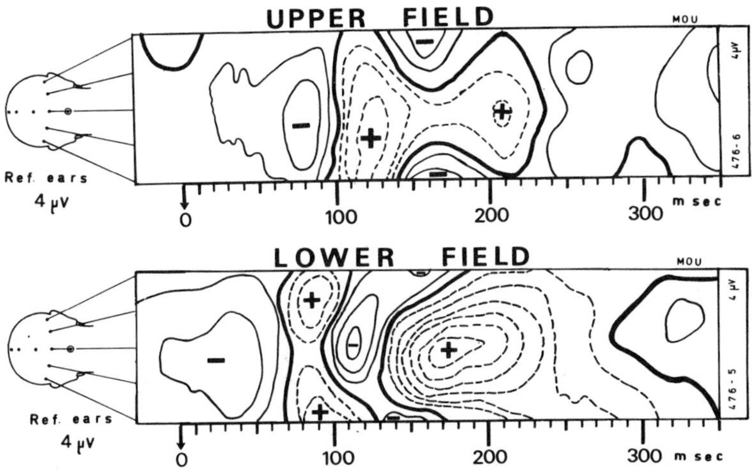

FIGURE 6. Upper- and lower-field pattern-onset responses (transverse montage) obtained from subject MOU (same as FIGURE 5, maps 4). Reference: linked earlobes; 4 μV between successive isopotential lines. The lower field P100 is made up of 2 laterally located peaks (11 μV each) whereas the upper-field N100 peaks on the midline with a lower amplitude (8 μV).

Changes Related to Selective Spatial Contrast and Luminance Effects

In the full-field condition, the pattern-reversal response was made up of the same midline components as the pattern-onset response, i.e., N60, P100, N140, and P200—in spite of slight and inconsistent differences in latency (a nonsignificant trend to shorter latencies for pattern reversal was noted). However, there was in the pattern-reversal response a significant tendency for N140 to increase and for P200 to decrease in amplitude (sign test: $P < 0.01$). In contrast with the midline components—which, on the whole, showed the same topographical organization in both types of responses—the anterior and lateral components LP120 and LN150 tended to disappear (FIGURE 8), whereas the occipital bilateral LN200 increased.

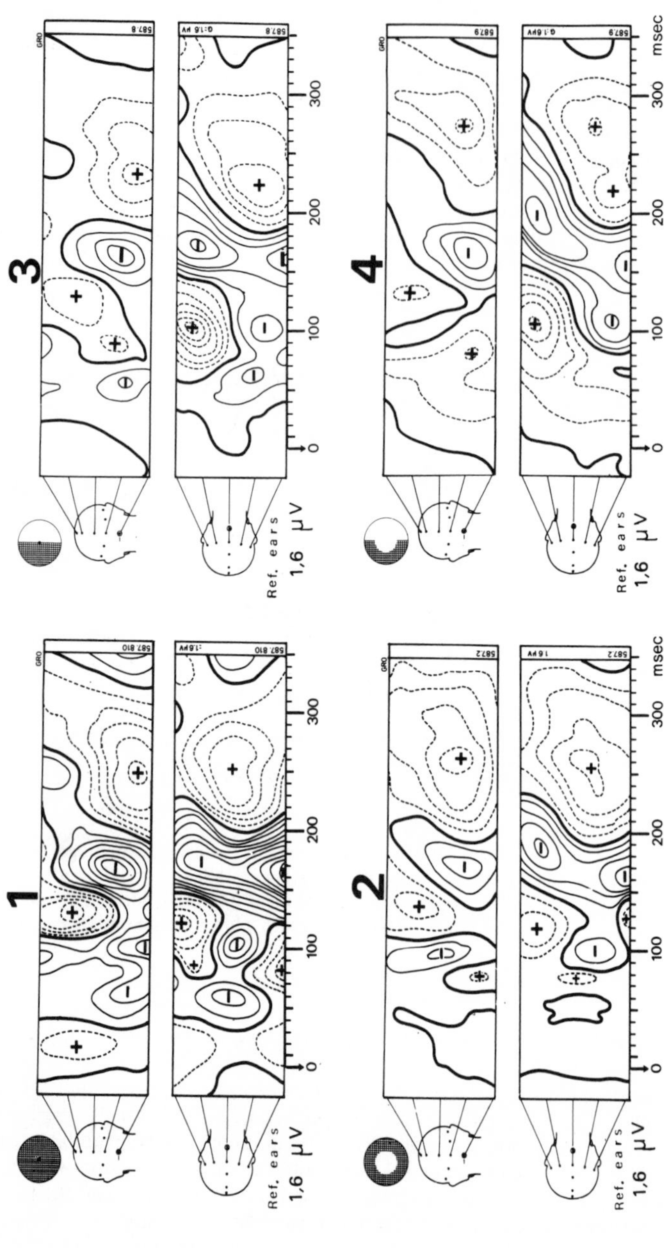

FIGURE 7. Pattern-onset response obtained from the same subject (GRO) in four different visual field conditions: full-field (1); extramacular, (central 5° occluded) (2); left half-field (3); left half-field with the central 5° occluded (4). Reference electrode: linked earlobes; 1.6 μV between each successive isopotential lines. When the central 5° is occluded: N60 disappears but not P100 or N140; the lateral components decrease in amplitude but are still present; P200 does not change significantly.

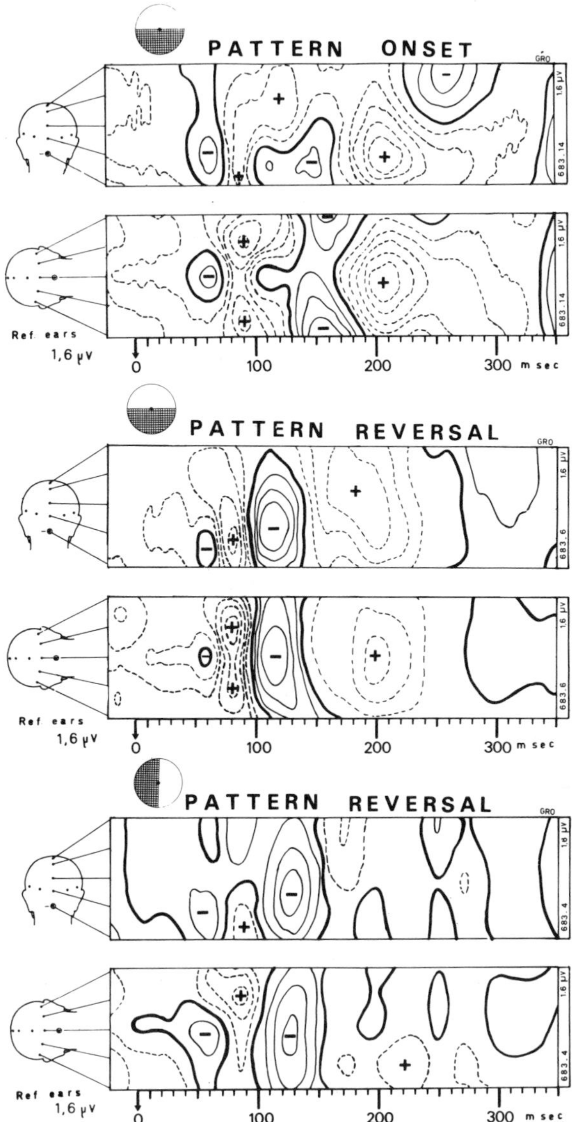

FIGURE 8. Lower-field pattern-onset (upper maps) and pattern-reversal (middle maps) responses and pattern-reversal left half-field response (lower maps). All responses obtained from the same subject (GRO), same as that of FIGURE 7. Reference electrode: linked earlobes; 1.6 μV between successive isopotential lines. LP120 and LN150 disappear in the pattern-reversal response. When pattern reversal stimulates the left visual field, P100 peaks on the right hemisphere whereas N60 and N140 remain on the midline.

Pattern-reversal right and left hemifield stimulations determined a contralateral P100 as did pattern onset (FIGURE 8), but, contrarily to pattern onset, usually no clear contralateral LP120. The other midline components decreased in amplitude for right or left hemifield stimulations but they usually remained peaking on the midline (FIGURE 8). This finding confirms that the ipsilateral topography often seen for the pattern onset N140 was a sort of artifact due to the presence of a very large contralateral LP120.

The response to blank field onset was on the contrary characterized by important lateral and anterior components, by the decrease of the late occipital bilateral LN200, and by the complete disappearance of the midline N140 (no exception) which was replaced by a positive wave of slightly different latency and topography (the other midline components remained very similar to those seen in the pattern onset response).

These findings indicate that whereas the midline N140, and perhaps also the lateral occipital LN200, are chiefly related to spatial contrast, the anterior and lateral waves LP120 and LN150 seem to be, on the contrary, chiefly due to luminance effects.

DATA FROM PATIENTS WITH HOMONYMOUS HEMIANOPIA*

The visual evoked potential of hemianopic patients with lesions well localized by scan should theoretically constitute a particularly useful experimental model for the understanding of the neural generators responsible for the production of each component of the visual scalp-recorded response. However, in practice, the interpretation of the various disturbances observed in these patients' responses is highly complex, insofar as these disturbances depend not only on the nature, the site and extent of the lesion, but also on a great variety of functional factors such as, in particular, the individual recovery capacities which vary according to the time elapsed between the first clinical sign and the VEP recording.

We shall report some results of a study in which pattern-reversal responses were obtained from 23 patients with a lateral homonymous hemianopia. These patients were recorded under experimental conditions similar to those reported in the previous section of this paper except for the following: the checkerboard was seen under a 13° angle; the fixation point was always placed 30′ away from the vertical margin; the overall luminance was 900 cd/m^2. Besides, only the transverse part of the cross-montage was used and only the full-field and right and left half-fields were stimulated. A control group of 20 healthy subjects was recorded under these same conditions.

The hemianopic patients were characterized by a macular sparing of less than 5° (17 out of 23 of them had a macular splitting) and a visual acuity better than 6/10 confirmed by ophthalmologic examination. All lesions were well defined by scan. The lesions were infarctions (9 cases), hematoma (7 cases), and tumors (7 cases); the time elapsed between the VEP recording and the onset of the first clinical sign varied from 2 to 390 weeks.

A detailed clinical and electrophysiological analysis of the data thus obtained is to be published elsewhere;[25] therefore, only some aspects of this study will be emphasized, i.e., those that should help to understand the origin of the VEP components.

*This section has been written with the help of F. Chain, M. LeBlanc, and J. F. Pinel, "Unité de Recherche de Neuropsychologie INSERM U84," Salpêtrière Hospital, Paris, with whom this study was undertaken.

Results

When the normal half-field was stimulated, most patients showed a normal pattern-reversal response made up of components N60, P100, N140, and P200, the time and space characteristics of which did not significantly differ from those of the control group. In particular, as regards topography, P100, and also the lateral activities LP120 and LN150, peaked on the contralateral hemisphere for practically all normal subjects and patients, whatever the reference, (Fz or linked earlobes): only 2 patients out of 23 and 2 controls out of 20 had components peaking on the hemisphere ipsilateral to the stimulated field. It must be noticed that, contrary to the pattern-reversal responses described in earlier in this article, which were characterized by a decrease or a disappearance of the lateral LP120 and LN150 waves seen for the pattern-onset response, these lateral activities were often clearly present in the pattern-reversal response of this control group. This was probably due to the higher luminance of the pattern used for this experiment (900 cd/m^2 instead of the 60 cd/m^2 for the experiment above with normal people).

Regarding the response obtained to the stimulation of the hemianopic half-field, a first observation was that no simple relation was found between the abnormality of this response and the importance of the perceptive defect as evaluated by clinical tests (Goldmann perimetry, Amsler tables). In other terms, patients with a similar macular splitting could show either no response at all, either a high-amplitude abnormal activity or a low-amplitude response normally organized in space and time, whereas some patients with a 5° central sparing could be characterized by an absence of response when stimulating the hemianopic field.

In fact, the type of activity recorded by stimulating the hemianopic field depended chiefly upon the nature of the lesion: patients with vascular lesions (in particular, patients with infarction), showed either no response at all—and in this case N60, P100, and also N140 had entirely disappeared but a late positive wave could sometimes be seen at a longer latency and with a longer duration than P200 (FIGURE 9)—or a response of very low amplitude but not significantly delayed and normally organized in space and time (FIGURE 10). A relation was observed between the degree of hypodensity of the lesion as seen on the scan, and the degree of abnormality of the response (compare FIGURES 9 & 10, both from patients with a hemianopic macular splitting due to infarction of the cerebral posterior artery territory: hypodensity at the scan was much more important for the patient of FIGURE 9, who had no response when his hemianopic field was stimulated, than for the one of FIGURE 10, who showed a low-amplitude but normally organized response).

In contrast to what was seen for infarction cases, the stimulation of the defected field of patients with tumors always elicited a large amplitude activity of delayed latency and unusual organization which, in 5 cases out of 7, was either predominant or selectively recorded on the ipsilateral hemisphere (FIGURE 11). The stimulation of the normal hemifield of these patients usually elicited on the contralateral hemisphere a normally organized response (FIGURE 11).

It must be noted that in spite of hematomas being "occupying space" lesions, most of these vascular cases elicited responses very similar to those seen in patients with infarction rather than to those seen in tumors. This finding can be understood if an other important variable—which is the age of the lesion—is taken into account. Indeed, 5 out of 7 of our hematoma cases were old lesions, hypodense at the scan. Recent hematomas seemed, on the contrary, to elicit high-amplitude abnormal activities similar to those seen with tumors, but more cases would have to be seen before we can draw conclusions.

The site of the lesion is also an important variable which must be taken into account, especially in order to explain the "abnormal" responses seen for some patients when stimulating the normal half-field. Indeed, although the majority of patients showed normal responses when their normal half-field was stimulated, some of them, surprisingly, had in that case delayed and somewhat disorganized responses on their contralateral hemisphere, sometimes accompanied by relatively important ipsilateral activities. A correlative study between these "abnormalities" and the site of the lesion showed that these delayed responses that were obtained when the normal

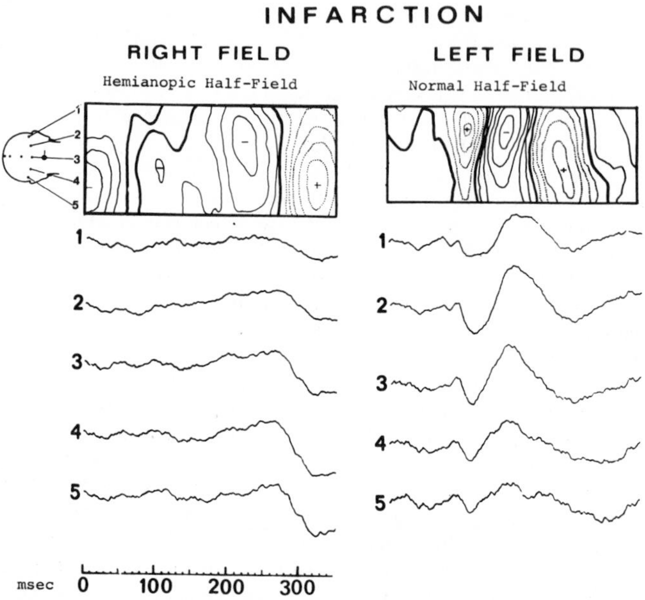

INFARCTION

RIGHT FIELD — Hemianopic Half-Field

LEFT FIELD — Normal Half-Field

msec 0 100 200 300

FIGURE 9. Pattern-reversal responses obtained by stimulation of the right hemianopic hemifield and of the normal left hemifield from a patient with macular splitting hemianopia due to infarction of the cerebral posterior artery, with marked hypodensity at the scan. Reference electrode: linked earlobes; amplitude: 1 μV between successive isopotential lines.

half field was stimulated were found in recent occipital but not in recent temporal lesions (FIGURE 12). This finding suggests that lesions of the occipital inner surface can influence the generators of the other, supposedly normal hemisphere and therefore modify its visual response.

Discussion

The analysis of the space and time image of the pattern evoked response of normal individuals as illustrated by the maps has clearly shown the existence of two different types of components regarding their scalp distribution: posterior-midline components (N60, P100, and N140) and lateral ones, more anteriorly situated than the midline waves (LP120–LN150); in addition, a third type could be differentiated, character-

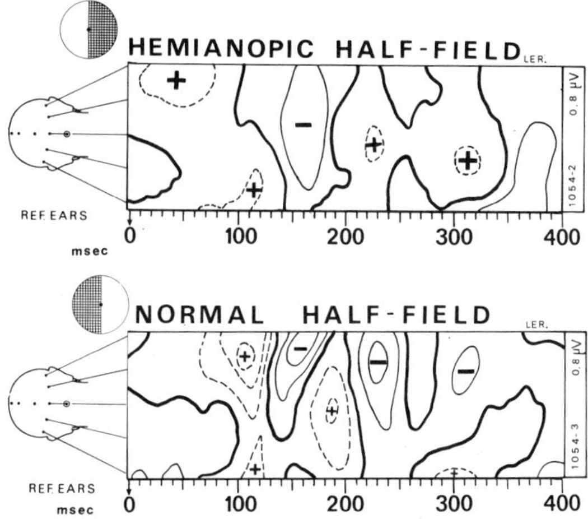

FIGURE 10. Pattern-reversal responses obtained by stimulation of the right hemianopic half-field and left normal half-field, from a patient with macular splitting hemianopia due to infarction of the cerebral posterior artery territory, with a much less accentuated hypodensity at the scan than patient of FIGURE 9. Reference electrode: linked earlobes; amplitude: 1 μV between successive isopotential lines.

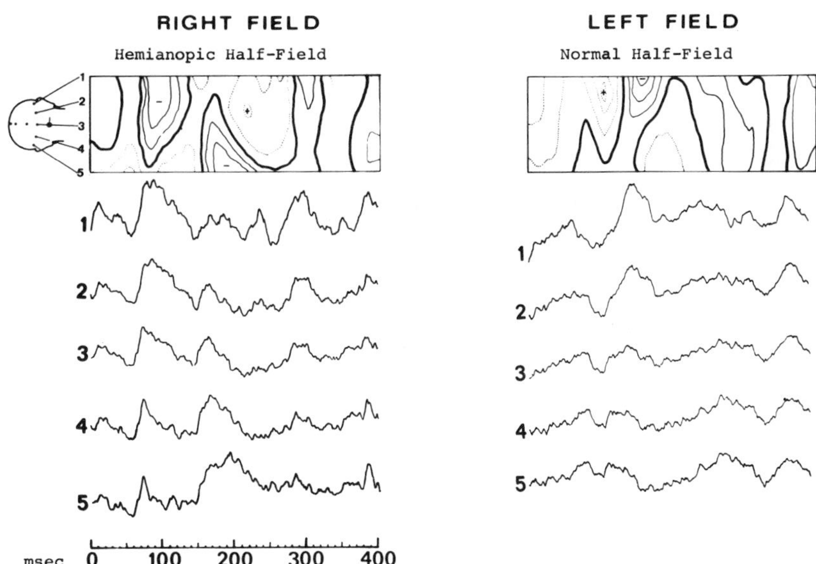

FIGURE 11. Pattern-reversal responses obtained by stimulation of the right hemianopic hemifield and the normal left hemifield from a patient with macular splitting hemianopia due to a left occipital glioma. Reference electrode: linked earlobes; 4 μV between successive isopotential lines.

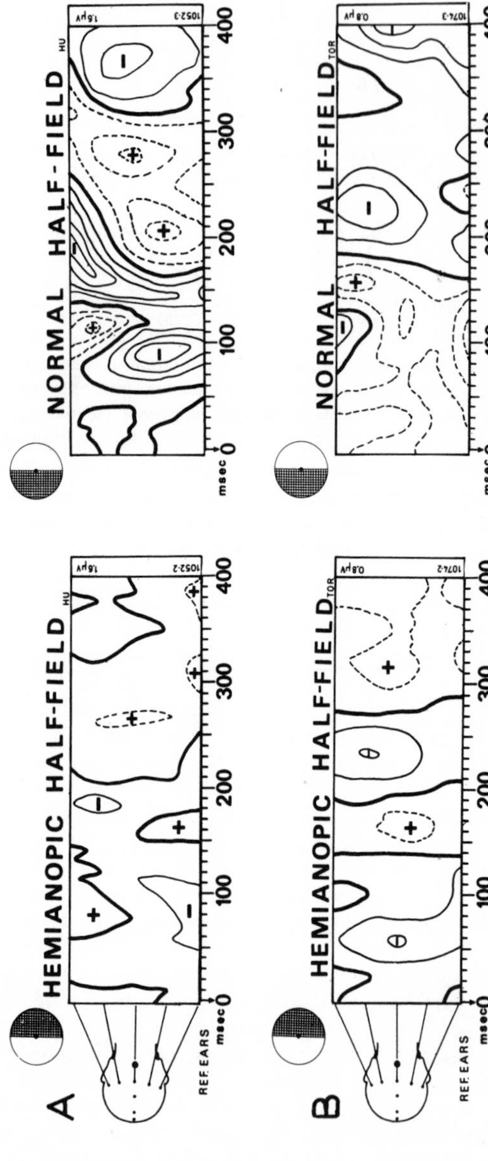

FIGURE 12. Pattern-reversal responses obtained by stimulation of the right hemianopic half-field and the left normal half-field from two patients, both with macular splitting hemianopia due to recent vascular lesions (2 weeks) of different localization: (A) Infarction of the sylvian artery territory. (B) Infarction of the posterior cerebral artery territory. Reference: linked earlobes; difference between successive isopotential lines: 1.6 μV for (A); 0.8 μV for (B).

ized by its lateral peaks located, on the contrary, on regions posterior to the inion (LN200). In spite of marked interindividual differences (in particular concerning the respective extent of the midline and of the lateral components), which is probably due to the well-known variability of the geometry of the various visual areas from one normal person to an other,[26] the changes of this spatiotemporal image according to the part of the visual field stimulated were not only remarkably reproducible within the same subject[10] but also quite coherent across subjects (TABLE 1).

One of the most important topographical changes was the posterior shift of the P100 upper-field response, leading to a polarity reversal of the lower-field P100, since in most cases the upper-field P100 peaks too posteriorly to be seen with our montage. This very posterior location of the P100 upper-field peak compared to that of the lower-field is in agreement with the finding of Lehmann and Skrandes[27] by multichannel EP field mapping. But, contrary to these authors, we cannot give an answer as to the respective delay between the maximum activity occurrences of both upper- and lower-field generators since the extreme positive values of the upper-field P100 were not seen on our maps. However, the negative N100, which replaced more anteriorly the lower-field P100, always peaked at the same time as the latter; thus this negativity could be interpreted, as have done several authors,[14,15,17,21,28] in terms of a dipole sheet orientation effect, as will be discussed further. In contradiction with some authors like Halliday and co-workers,[29-31,40] but in agreement with others,[21,32-34] the second important topographical finding was that P100–LP120 and the late lateral negative waves were never ipsilateral but rather contralateral to the stimulated half-field. We have no ready explanation for these discrepancies; in particular, contrary to Blumhardt *et al.*,[31] neither the size of the field nor the presence of the central 5° area modified the importance of this contralateral predominance when intraindividual comparisons were made. In contrast with Hardline *et al.*,[34] neither did the luminance and contrast levels have an effect on this topographical organization, be it for the onset or for the pattern-reversal response (see our discussion paper for the binocular VEP roundtable session in this volume).

In order to try to interpret these EP changes related to visual field changes in terms of the most probable underlying cortical sources of each component, we shall refer to the classical model, first suggested by Mary Brazier,[35] according to which the cortex is considered as a dipole sheet within a volume-conducting brain.[1,22,36-39] Regarding the visual cortex, as several authors have emphasized,[15-21] the geometry of area 17 suggests that the axis of the equivalent dipole is perpendicular to the overlying surface where are the recording electrodes for the part of the striate area lying within the calcarine fissure along which projects the horizontal meridian; on the contrary, for the part of the striate area lying on the medial surface on which projects the vertical meridian the axis of the dipole is parallel to the surface electrodes. Therefore, in the case of components originating in area 17, the stimulation of right and left half-fields should give rise to bipolar surface activities with a polarity reversal occurring at the midline. On the other hand, the stimulation of the upper and lower field should determine on the midline a unique potential peak, positive or negative depending upon whether the upper or lower field (which project, respectively, on the floor and the roof of the calcarine fissure) is stimulated. Similar dipole sheet effects should also be observed when the different regions of the extra striate areas on which project the upper and lower fields (situated, respectively, on the under and upper surface of the occipital lobe) are selectively activated. With the upper-field generators being further away from the recording electrodes than the lower-field generators, it would be expected that the upper-field response would not only appear with a reverse polarity but also with a lower amplitude and a different scalp distribution than that of the lower field. No such difference should be observed for components due to the upper-

and lower-field generators within area 17. However, this classical model seems to be better applied to area 18 than to area 19 since data concerning the retinotopic organization of area 19 in monkeys[42,43] have shown that this latter area is, contrary to the former, far from representing a functional unity. It is more likely made up of multiple independent areas of various localizations, each of which represent the whole hemiretina projections. If such were the case in man, scalp components coming from area 19 should predominate on the contralateral hemisphere, but, contrary to the propositions we made in a previous paper,[10] these components should not reverse polarity when turning from lower- to upper-field stimulation. On the contrary, components from area 18, (V2 in monkeys) should reverse polarity like those coming from area 17, insofar as only one hemiretina is represented in area 18.

The Midline Posterior N60

Our findings concerning this early negative wave are in favor of its originating in the part of the striate area which emerges on the convexity and on which projects the fovea: indeed, when the fovea was selectively stimulated by a 2° pattern, this component had the same amplitude, latency, and topographical characteristics as when the 20° full-field was stimulated (TABLE 1), whereas it completely disappeared when the central 5° was occluded and only the peripheral retina was stimulated (FIGURE 7). In particular, when turning from the lower to the upper-field, no polarity reversal was observed for N60, and when the right or left half-field was stimulated, this component, when it did not disappear, remained peaking on the midline. Both results rule out the participation of the deep part of area 17 as well as that of area 18. The fact that the importance of this component varied markedly from one subject to an other (for 20% of our population it was not seen at all) is in keeping with the well-known interindividual variability of the size of the exposed foveal area portion of the striate cortex.[26] This would be more difficult to understand if N60 originated in area 18.

The Midline Posterior P100 and the Lateral LP120

Contrary to some of our previous reports,[10,11] we have in the present work differentiated in the pattern-onset full-field response two successive positive components, one peaking on the midline, in regions near the inion, the other one peaking some 10 to 20 msec later, made up of 2 symmetrical peaks located more anteriorly and 4 to 8 cm away from the midline. These two successive components often overlapped in such a way that they were difficult to distinguish from one another, but in some instances they clearly appeared as distinct components [FIGURES 1, 2 (map 3A), 3]. Contrary to the positive lateral waves identified by Blumhardt and Halliday,[40] which could be either earlier or later than the midline positivity, these lateral positivities always peaked at least 10 msec later than the midline P100 (no exception). Before discussing the possible origin of both midline and lateral positivities, the validity of such a description should be questioned. First of all, the fact that these lateral positivities remain with the same spatiotemporal organization, regardless of the location of the reference electrode, rules out an artifactual origin due to the choice of the reference. The decrease of these lateral waves in the lower-field response could suggest that such a lateral location and a time delay as those seen for LP120 in the full-field response are partly an artifact due to the presence in that case of the negative midline component (N100) of the upper-field response; this N100 would push apart and delay the midline P100 of the lower-field response. But in spite of their being less important and peaking slightly earlier—probably indeed because of the absence in that case of the upper-field negativity—the lateral LP120 waves were still seen in the lower-field response. In particular, they clearly appeared as distinct components in the response to the right or left lower quadrant of some subjects where they predominated on the contralateral hemisphere whilst P100 peaked on the midline (FIGURE 3, lower map). This finding rules out the suggestion that this spatiotemporal image could have an artifactual origin in relation to the upper-field N100, and is therefore in favor of the existence of two successive positive components having different scalp distributions, thus originating in different cortical areas and having probably different functional significances (insofar as LP120 was more markedly affected by luminance level or/and luminance changes than was P100).

The first possible source suggested by several authors for the midline P100 was the striate areas inside the calcarine sulcus[16,17,20,21] since this wave, according to the dipole sheet model of area 17, is made up of one unique midline peak which reverses polarity when turning from lower- to upper-field stimulation (radial dipole), whereas the stimulation of right or left hemifields gives rise to two peaks of potentials of different polarity, one on each hemisphere (tangential dipole). In fact, our data showed that the stimulation of the right or left half-field did not always lead to such bipolar distributions (FIGURE 2, maps 2B, upper map 3B, FIGURES 3 & 4). When it did, the waves of opposite polarity seen on both hemispheres were not likely to be due to an orientation dipole sheet effect since they usually did not peak at the same latency. The negative ipsilateral wave seen in these cases is in fact the midline N140 whose peak is shifted to the ipsilateral hemisphere on account of the presence of a particularly important contralateral LP120 with which it overlaps (FIGURE 2, upper map 1B). Another finding that favors such an explanation is that this ipsilateral negativity was never seen in the blank field onset response when the midline N140 is replaced by a positive wave of different topography.[10]

However, the polarity reversal of P100 seen when turning from lower-field to upper-field stimulation can be interpreted as a true dipole sheet orientation effect since both waves of opposite polarity peak at the same time. As suggested by Halliday and co-workers,[14,15] the location of the upper- and lower-field generators in the extrastriate areas, at least in area 18, could also explain such a dipole sheet effect. Our data concerning P100 favors this suggestion: indeed, the significant differences between the scalp distributions of the upper-field N100 and the lower-field P100 are too important (N100 always peaked on the midline and was 4 cm more anterior than P100: FIGURES 5 & 6) to be explained by a striate origin of these components and would be better understood by an origin in area 18 where the opposite dipoles are further from one another.[10] In addition, the finding that the upper-field N100 was usually of lower amplitude than the lower-field P100 better fits such an interpretation. Furthermore, the significant amplitude decrease of P100 when the central 5° was occluded compared to its slight nonsignificant decrease when only the fovea was stimulated (TABLE 1) would be difficult to understand if P100 came from the deep part of the calcarine.

In regard to the lateral positive components LP120, their topographical characteristics as well as their changes according to selective peripheral retina and selective foveal stimulation—which were opposed to those of the midline P100—suggest that they originate in other parts of the extrastriate areas than does P100, more laterally and anteriorly located, thus suggesting area 19 rather than 18. Whereas P100 could originate in area 18 at the boundary of the extrastriate and striate areas located not far from the medial sulcus where projects the vertical meridian, LP120 could on the contrary reflect the activity of parts of area 19 on which projects the peripheral retina.

If this were the case, the absence of polarity reversal of the LP120 waves in the upper-field response could be understood. However, it must be kept in mind that in spite of recent available data on the retinotopic organization of the extrastriate areas in monkeys,[41–43] hardly anything is known concerning this organization across the various parts of area 19 in man.

The Posterior Midline N140 and the Lateral Negativities (LN150 and LN200)

In contrast with the two positive components which were often difficult to isolate from one another, the two negative components N140 and LN150 were easily differentiated in spite of their peaking nearly at the same time: indeed N140 completely disappeared in the pure luminance response whereas LN150 increased with luminance effects. This latter wave is probably the same as the luminance-related negative component described, with the same latency, by Jeffreys.[44] In regard to N140, a component which is known to be selectively related to spatial contrast,[5,7,44] its midline topography and the fact that it appears with a reverse polarity and a different topography in the upper-field response are both in favor, as for P100, of an area 18 origin (in spite of both waves showing slightly different scalp distributions; indeed P100 peaks sometimes more posteriorly and often a little further from the midline than N140).

In contrast, the topographical characteristics of the LN150 waves and their modifications in relation to visual field situations (in particular their contralateral predominance in the right or left hemi-field response and their absence of polarity reversal in the upper-field response) as well as their change with luminance effects are very similar to those observed for the LP120 waves. Therefore a common origin in area 19 could be suggested for these lateral components which both occur some 10 to 15 msec after the midline components of same polarity (in monkeys a similar delay was found between evoked potentials recorded in striate and extrastriate areas for foveal stimulations).[41]

In a previous paper,[10] we had suggested the inferior temporal cortex as a possible source for the late bilateral waves labeled LN200, chiefly on account of their very peculiar topography made up of peaks located posteriorly to the inion, at least 8 cm away from the midline, and of their selective reactivity to pattern stimulation. However, if this were the case, the finding of their peaking always on the contralateral hemisphere would be difficult to explain since it is now well established that, at least in monkeys, neurons through this area almost always include the center of the gaze.[45]

The Midline Posterior P200

This long duration wave, widespread over the posterior region, is probably made up of at least two successive components, the second one peaking usually more posteriorly than the first (FIGURE 1). However, for the 60% of our population for whom P200 was divided into two parts,[10] our experimental conditions did not permit us to differentiate these two components in terms of visual field properties or/and luminance and contrast effects. The suggestion of an area 18 origin, at least for parts of this wave, could be supported by the following findings: the consistency of its midline peaking, whichever half-field is stimulated, and its amplitude decrease or its disappearance in the upper-field response (though with no clear polarity reversal). However, in contrast with P100 and N140, this midline P200 component seems to be related mostly to the activation of the extramacular projections since it did not significantly change when the central 5° was occluded. Therefore this long duration wave is probably the result of various components coming from the various areas related to visual processes, including perhaps the parietal cortex.

Before concluding, a few words must be said regarding the data obtained from hemianopic patients. It must first of all be emphasized that when the stimulation of the hemianopic field elicited either no response or a delayed response of abnormally low amplitude on the contralateral hemisphere, all the components were either absent or delayed. In other words, the "abnormality" was never limited to one component—contrary to what had been reported long ago by Vaughan in a study on flash evoked responses in hemianopic patients,[46] and no relation was observed between the location of the lesion and the absence or abnormality of one particular component. Therefore the data obtained from these patients were of little or no help to interpret each component in terms of their underlying generators. However, an interesting finding was that when the location of the lesion was on the inner surface of the occipital region, the response to the intact half-field could also be abnormal, suggesting an interhemispheric interaction, probably by volume conduction, which was not observed when the lesion was temporal and more superficial.

A last point which must be underlined is the important influence of the nature of the lesion (tumor or ischemia), and also of the presence or not of an edema (recent or old lesions), as shown by our data in agreement with the findings of Holder[47] and of Samson-Dollfus and Pouliquen.[48] This adds to the difficulty of interpreting the hemianopic field response in terms of generators, insofar as these factors probably modify considerably the conductivity of the various tissues. In particular the fact that ipsilateral abnormal activities, which could be of high amplitude, were found in occupying space lesions but not in infarctions, except in cases of very recent lesions showing hyperdense images on the scan, can probably be explained by such changes in the conducting properties of the diseased tissues. The localization on the side of the absent lobe of the preserved field response seen in patients following occipital lobectomy described by Chain *et al.*[6] and by Blumhardt *et al.*[49] is probably also an effect of volume conduction occurring in abnormal conditions.

Conclusion

Has this spatiotemporal analysis of the scalp visual evoked potentials obtained from normal subjects and from patients enabled us to give an answer to the main question this session of this conference had to deal with, that is: "How useful are scalp topographies in the localization of intracranial generators"? First of all, it must be underlined that our study was rather restricted as to its topographical aspect on account of the relatively small number of recording sites and the use of but two arrays of electrodes oriented according to only two perpendicular directions. Indeed, many methodological problems should be discussed concerning the optimal conditions as to the electrodes' position, spacing, and orientation in order to get the most complete description of each EP scalp distribution. A more complete and sophisticated topographical analysis could certainly help to identify better the various overlapping components, and thus permit a better interpretation of some of the latency or topographical shifts we have illustrated. But in fact, even in the case of the most complete topographical analysis possible, the problem of the interpretation of scalp components in terms of underlying generators would still remain unsolved, since no scalp component, even when isolated by a proper spatiotemporal analysis, is ever a "unitary phenomenon" comparable to a single action potential. Given the empirical data, we can make tentative assumptions concerning the possible areas involved in the production of the VEP components, as we have done, based in part on analyses implying the existence of a single dipole generator—which is certainly an over-simplification. Thanks to these assumptions, sensory information-processing models can be proposed which, like all models, will have the advantage of inducing ulterior experiments. But in order to go further, complementary approaches are needed, such as recordings made directly from patients' brains, in spite of the difficulty of inferring normal mechanisms from pathological ones.

Recently, several researchers dealing with the late waves of event-related potentials linked to cognitive processes, such as Donchin in particular, have emphasized that spatial distributions should be, and could be, usefully utilized as variables dependent on time-varying sensory, motor, or cognitive processes rather than as independent variables for localization of sources in terms of anatomical structures.[50,51] Many years of research in the latter domain and a few years in the former have led us to the conclusion that, indeed, such a use of scalp distributions as dependent variables in order to clarify information-processing mechanisms is probably wiser and more immediately profitable than its use for determining where the generators are. However, despite the fact that evoked potentials can be fruitfully utilized without our knowing their sources, we still are persuaded that the understanding of these sources is the most fascinating problem to tackle.

References

1. Vaughan, H. G. 1974. The analysis of scalp-recorded potentials. In R. F. Thompson & M. M. Patterson, Eds. Bio-Electric Recording Techniques, Part B. pp. 157–207. Academic Press, New York, N.Y.
2. Remond, A. 1964. Description des réponses corticale et oculaire à la stimulation lumineuse intermittente chez l'homme. *In* Flicker. H. E. Henkes & L. H. van der Tweels, Eds. pp. 157–193. Junk, The Hague.
3. Remond, A. & N. Lesevre. 1965. Distribution topographique des potentiels évoqués occipitaux chez l'homme normal. Rev. Neurol. **112**(4): 317–330.
4. Lesevre, N & A. Remond. 1968. Etude du champ visuel par les potentials évoqués moyens. Revue neurologique. **118**(6): 419–430.
5. Lesevre, N. & A. Remond. 1972. Potentiels évoqués par l'apparition de patterns: effets de la dimension du pattern et de la densité des contrastes. Electroenceph. Clin. Neurophysiol. **32:** 593–607.
6. Chain, F., N. Lesevre, A. Remond, F. Lhermitte & M. Leblanc. 1972 Etude topographique des réponse visuelles dans un cas de lobectomie occipitale. Rev. Neurol. **126**(5): 372–378.
7. Lesevre, N. 1973. Potentiels évoqués par des patterns chez l'homme: influence de variables caractérisant le stimulus et sa position dans le champ visuel. *In* Activités Évoqués et Leur Conditionnement Chez l'Homme Normal et en Pathologie Mentale. G. Lelord, Ed. pp. 1–22. (public). INSERM, Paris.
8. Lesevre, N. 1976. Topographical Analysis of the pattern Evoked Response: its application to the study of macular and peripheral vision in normal people and in some pathological cases. Doc. Ophthalmol. Proceeding Ser. (XIIth Iscerg Symp.) Alfieri & Sole, Eds. Vol. **10:** 87–102. W. Junk, The Hague.
9. Ragot, R. A. & A. Remond. 1978. EEG field mapping. Electroenceph. Clin. Neurophysiol. **45:** 417–421.
10. Lesevre, N. & J. P. Joseph. 1979. Modifications of the Pattern Evoked Potential related to the part of the visual field stimulated (clues for the most probable origin of its various components recorded on the scalp). Electroenceph. Clin. Neurophysiol. **47:** 183–203.
11. Lesevre, N. & J. P. Joseph. 1980. Hypothesis concerning the most probable sites of origin of the various components of the pattern evoked potential. *In* Evoked Potentials. C. Barber, Ed. pp. 159–166. MTP Press Limited, Lancaster, England.

12. REMOND, A. 1961. Integrative and topological analysis of the EEG. Electroenceph. Clin. Neurophysiol. (Suppl. 20): 64–67.
13. VAUGHAN, H. G. 1969. The relationship of brain activity to scalp recordings of event related potentials. *In* Averaged Evoked Potentials. Methods, Results, Evaluations. E. Donchin & D. B. Lindsley, Eds. SP 191: 45–94. NASA, Washington.
14. HALLIDAY, A. M. & W. F. MICHAEL. 1970. Changes in pattern evoked responses in man associated with the vertical and horizontal meridians of the visual field. J. Physiol. (London) **208:** 499–513.
15. MICHAEL, W. F. & A. M. HALLIDAY. 1971. Differences between the occipital distribution of upper and lower field pattern-evoked responses in man. Brain Res. **32:** 311–324.
16. JEFFREYS, D. A. 1971. Cortical source-locations of patterned-related VEPs (visual evoked potentials) recorded form the human scalp. J. Physiol. (London), **229:** 502–504.
17. JEFFREYS, D. A. & J. C. AXFORD. 1972. Source location of pattern-specific components of human visual evoked potentials. I. Components of striate cortical origin. Exp. Brain Res. **16:** 1–21; and II. Components of extrastriate cortical origin. Exp. Brain Res. **16:** 22–40.
18. RISTANOVIC, D. 1971. The human visual cortex as a system of four generators. Acta Med. Jug. **25:** 379–395.
19. RISTANOVIC, D. & D. DOKIC-RISTANOVIC. 1971. A dipole representation of the human visual cortex. Acta Med. Jug. **25:** 369–378.
20. BIERSDORF, W. R. & L. NAKAMURA. 1973. Localization studies of the human visual evoked response. Doc. Ophthalmol. (Den Haag) **2:** 137–144.
21. BIERSDORF, A. 1974. Cortical evoked responses from stimulation of various regions of the visual field. (XIth Iscerg Symp.) Doc. Ophthalmol. (Den Haag) **4:** 249–259.
22. HENDERSON, C. J., S. R. BUTLER & A. GLASS. 1975. The localization of equivalent dipoles of EEG sources by the application of electrical field theory. Electroenceph. Clin. Neurophysiol. **39:** 117–130.
23. COBB, W. A., H. B. MORTON & G. ETTLINGER. 1967. Cerebral potentials evoked by pattern reversal and their suppression in visual rivalry. Nature (Lond.) **216:** 1124–1125.
24. STEPHENSON, W. A. & F. A. GIBBS. 1951. A balanced non-cephalic reference electrode. Electroenceph. Clin. Neurophysiol. **3:** 237–240.
25. CHAIN, F., N. LESEVRE, J. F. PINEL & M. LEBLANC. (In press). Spatiotemporal study of visual evoked potentials in patients with homonymous hemianopia. J. Courjon, Ed. Raven Press, New York, N.Y.
26. POLYAK, S. 1957. The Vertebrate Visual System. Univ. Chicago Press, Chicago, Ill.
27. LEHMANN, D. D. & W. SKRANDIES. 1979. Multichannel evoked potential fields show different properties of human upper and lower hemiretina systems. Exp. Brain Res. **35:** 151–159.
28. JEFFREYS, D. A. & A. T. SMITH. 1979. The polarity inversion of scalp potentials evoked by upper and lower half-field stimulus patterns: Latency or surface distribution differences. Electroenceph. Clin. Neurophysiol. **46:** 409–416.
29. HALLIDAY, A. M. 1978. New developments in the clinical application of evoked potentials. *In* Contemporary Clinical Neurophysiology. W. A. Cobb & H. Van Duijn, Eds. EEG Suppl. (34): 105–120.
30. HALLIDAY, A. M., G. BARRET, E. HALLIDAY & W. F. MICHAEL. 1977. The topography of the pattern-evoked potential. J. E. Desmedt, Ed. pp. 121–133. Clarendon Press, Oxford.
31. BLUMHARDT, L. D., G. BARRET, M. HALLIDAY & A. KRISS. 1978. The effect of experimental "scotomata" on the ipsilateral and contralateral responses to pattern-reversal in one half-field. Electroenceph. Clin. Neurophysiol. **45:** 376–392.
32. COBB, W. A. & H. MORTON. 1970. Evoked potentials from the human scalp to visual half-field stimulation. J. Physiol. **208:** 39–40.
33. SHAGASS, C., M. AMADEO & A. ROEMER. 1976. Spatial distribution of potentials evoked by half-field pattern reversal and pattern onset stimuli. Electroenceph. Clin. Neurophysiol. **41:** 609–622.
34. HARDING, G. F. A., G. F. SMITH & P. A. SMITH. 1980. The effect of various stimulus parameters on the lateralization of the VEP. *In* Evoked Potentials. C. Barber, Ed. pp. 213–218. MTP Press Limited, Lancaster, England.
35. BRAZIER, M. A. B. 1949. A study of the electrical fields at the surface of the head. Electroenceph. Clin. Neurophysiol. (Suppl. 2): 38–52.

36. ROTH, M. & J. C. SHAW. 1955. Potential distribution analysis. I and II. Electroenceph. Clin. Neurophysiol. **7:** 273–284 and 285–292.
37. GEISLER, C. D. & G. L. GERSTEIN. 1961. The surface EEG in relation to its sources. Electroenceph. Clin. Neurophysiol. **13:** 927–934.
38. SCHNEIDER, M. & P. GERIN. 1970. Une méthode de localisation des dipôles cérébraux. Electroenceph. Clin. Neurophysiol. **28:** 69–78.
39. RYDING, E. 1980. A mathematical model for localization of the source of cortical evoked potentials. Electroenceph. Clin. Neurophysiol. **48:** 312–317.
40. BLUMHARDT, L. D. & A. M. HALLIDAY. 1979. Hemisphere contributions to the composition of the pattern evoked potential wave form. Exp. Brain Research **36:** 53–63.
41. GULD, C. & A. BERTULIS. 1976. Representation of fovea in the striate cortex of vervet monkey. Vis. Res. **16:** 629–631.
42. KAAS, J. S. 1978. The organization of visual cortex in primates. *In* Sensory Systems of Primates. C. R. Noback, Ed. Plenum Press. New York, N.Y.
43. VAN ESSEN, D. C. & S. M. ZEKI. 1978. The topographic organization of rhesus monkey prestriate cortex. J. Physiol. **277:** 193–226.
44. JEFFREYS, D. A. 1977. The physiological significance of pattern visual evoked potentials. *In* Visual Evoked Potentials in Man: New Developments. J. E. Desmedt, Ed. pp. 134–167. Clarendon Press, Oxford.
45. DESIMONE, R. & C. G. GROSS. 1979. Visual areas in the temporal cortex of the macaque. Brain Res. **178** (2–3): 363–380.
46. VAUGHAN, H. G. 1966. The perceptual and physiologic significance of visual evoked responses recorded from the scalp in man. *In* Clinical Electroretinography. pp. 203–223. Pergamon Press, New York, N.Y.
47. HOLDER, G. E. 1980. Abnormalities of the pattern visual evoked potential in patients with homonymous visual field defects. *In* Evoked Potentials. C. Barber Ed. pp. 285–298. MTP Press Limited, Lancaster, England.
48. SAMSON-DOLLFUS, D. & A. POULIQUEN. 1976. Application des PEV à l'étude des hémianopsies latérales homonymes. Rev. D'EEG et Neurophysiol. Clin. **6:** 179–180.
49. BLUMHARDT, L. D., G. BARRETT & A. M. HALLIDAY. 1977. The asymmetrical visual evoked potential to pattern reversal in one half field and its significance for the analysis of visual field defects. Br. J. Ophthalmol. **61:** 454–461.
50. DONCHIN, E. 1978. Use of scalp distribution as a dependent variable in event-related potential studies. *In* Multidisciplinary Perspectives in Event-Related Brain Potentials Research. D. A. Otto, Ed. pp. 501–510. U.S. Govt. Printing Office, Washington, D.C.
51. GOFF, W., T. ALLISON, P. WILLIAMSON & J. VAN GILDER. 1978. Scalp topography in the localization of intracranial evoked potential sources. *In* Multidisciplinary Perspectives in Event-Related Brain Potential Research. D. A. Otto, Ed. pp. 526–532. U.S. Govt. Printing Office, Washington, D.C.

TOPOGRAPHIC DISPLAY OF EVOKED POTENTIALS: CLINICAL APPLICATIONS OF BRAIN ELECTRICAL ACTIVITY MAPPING (BEAM)*

Frank H. Duffy†

Harvard Medical School and
Developmental Neurophysiology Lab
Childrens Hospital Medical Center
Boston, Massachusetts 02115

INTRODUCTION

It is widely agreed that long-latency sensory evoked potentials (EPs) are sensitive not only to stimulus parameters, but to variations of brain state. On this basis, one might predict that such EPs would achieve widespread clinical use paralleling that of the EEG. With a few exceptions, this has not been the case. We do not believe that this results from an inherent insensitivity of EP to cerebral pathophysiology. On the contrary, we propose that such measures of brain electrical activity represent not too little, but too much clinical information to be easily appreciated by unaided inspection or measurement.

To assist clinical appraisal of such data, we have recently developed a system for the topographic mapping and computerized display of scalp-recorded signals referred to as brain electrical activity mapping or BEAM.[1] As routinely used in our laboratory, BEAM images are constructed from data gathered from 20 scalp electrodes placed in the standard 10–20 EEG format (FIGURE 1). Resultant topographic images are displayed on a computer-driven color video monitor using a colored "grey" scale. For EP data, we visualize the dynamic change of electrical activity with time by sequential display of images; the display technique produces an animation effect highlighting the spread of EP activity over the scalp. These methods condense and summarize the spatiotemporal information obtained from multielectrode recordings to facilitate analysis by visual inspection.

BEAM is now a routine clinical neurophysiological test in the Seizure Unit at Childrens Hospital Medical Center, Boston. Clinical referrals over the last 12 months blanket a wide spectrum of neuropathology including epilepsy (21%), learning disability (22%), emotional disturbance and dementia (14%), headache (10%), and unknown neurological disease (33%). BEAM appears to be most applicable to cases where the CT scan is either normal or noncontributory to the clinical question at hand.

Two refinements of the BEAM methodology have improved its applicability as a clinical instrument: (1) Significance probability mapping, a procedure for statistically delineating abnormal regions; and (2) Grid sector analysis, a procedure for quantifying the degree of overall and/or focal abnormality in BEAM images. In the following

*This work was supported in part by Grants R01 HD 13420 from the National Institute of Childrens Health and Diseases and R01 NS 14767 from the National Institute of Neurological, Circulatory and Digestive Diseases of the National Institutes of Health and by a grant from Braintech, Inc. BEAM is a trademark of Braintech, Inc.

†Address for correspondence: Seizure Unit and Developmental Neurophysiology Lab, Childrens Hospital Medical Center, 300 Longwood Ave., Boston, Mass. 02115.

0077-8923/82/0388-0183 $1.75/0 © 1982, NYAS

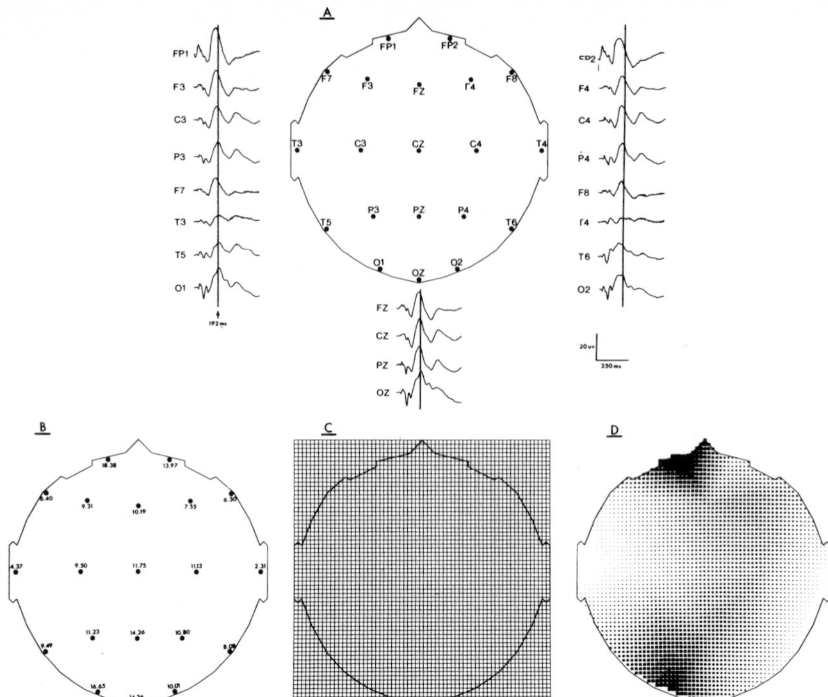

FIGURE 1. Construction of a BEAM Image. Example of the construction of a topographic map for EP data. In this case, mean visual evoked potentials (VEPs) are formed from each of 20 recording sites placed in the standard 10–20 EEG electrode positions. Each VEP spans 512 msec and consists of 128 data points where each point represents 4 msec. The baseline is determined as the average of the 512 msec that preceded the stimulus (not shown). In (A) the individual VEPs are shown for the electrode locations indicated on the head diagram. A vertical line is drawn through all VEPs indicating the data point 192 msec after stimulation. It is this 4-msec epoch that is to be mapped topographically. In (B) the mean voltage values at these locations are shown alongside the electrodes from which they were measured. Note that simple quantification does not provide much assistance in determining the potential distribution across the head. In (C) the head is covered by a 64 by 64 matrix of picture elements (pixels). The 20 pixels that overly original electrode sites are assigned the numerical values obtained from them. The remaining majority of pixels are assigned values based upon the distance from the the three nearest electrodes—a process known as three-point linear interpolation. Three-point linear interpolation is chosen because: first, three points are the smallest number of points that could define an area about a given pixel; second, such triangles can more easily be fitted into irregular geometric outlines of the head or portions thereof; third, there is little evidence from classic EEG to suggest that electrodes more distant than the nearest three known points have additional influence upon intermediate pixel locations; and fourth, comparisons of three-point linear interpolation with four-point interpolation and with polynomial approximation indicate that three-point interpolation more closely approximates the real intermediate values obtained from intervening electrodes. Finally, for display (D) the raw voltage values are fitted to a discrete-level equal-interval scale. This illustration was prepared with a digital plotter in black and white. However, all BEAM images are now displayed on a color TV monitor under computer control. Two display conventions are employed. The first is called "grey scale" where values are assigned different intensities of a single color. For these images positive values are assigned different intensities of red and negative values of blue. The second is called "pseudo-color" where each range of values is assigned a separate color. Although a display image is produced, the underlying numerical pixel matrix is maintained for subsequent statistical manipulations, e.g., significance probability mapping (SPM) and grid sector analysis (GSA).

sections we shall discuss these new methods and show selected examples of the clinical utility of BEAM.

Significance Probability Mapping (SPM)

Symmetry is a primary characteristic of BEAM studies of normal subjects when EP are formed to simple and symmetrical flash, click, or shock stimuli. As one might expect, pathology or lateralized brain state activation paradigms often produce regional asymmetries. However, when such asymmetries are found it may not be obvious by visual inspection which side is the abnormal or changed side. To assist in this determination we have adapted from the field of "image processing,"[2] a technique known as significance probability mapping or SPM. This method was first introduced by Bartels and Subach[3] to extract information from light microscopic and sonar images. We have adapted this technique to the analysis of BEAM data.[4] It consists of the replacement of an image by the results of a statistical manipulation of the numerical data that underly the original image. Results of such data abstraction are presented within the coordinate system or framework of the original data. In general, the goal is to make visible data characteristics that might otherwise remain obscured.

In the analysis of BEAM images two types of SPM are formed. The first is based upon the *Z*-transform statistic[5] and is used to delineate regions in which brain electrical activity from an individual subject differs statistically from that of a reference population (Figure 2A). Z-transform SPMs are primarily used in the delineation of abnormal regions during the clinical evaluation of BEAM images. An example is shown in Figure 3 (also Figures 5A & 5B).

The second is based upon Student's *t*-test statistic[5] or other measures of differences between group (Figure 2B). In general, *t*-statistic SPM are used to statistically delineate regional differences between two groups of subjects. For example, they have been used to locate regions of cortex that differ between dyslexic boys and normals.[6]

It should be emphasized that the *Z* or *t*-statistic SPM are intended to localize regional differences and are not intended to be optimal measures of overall level of group or individual difference. Overall significance is best assessed by multivariate statistical techniques.[7]

Grid Sector Analysis (GSA)

In assessing BEAM images, the clinician is frequently confronted with the difficult task of having to determine whether a given focality signals regional neuropathology or whether it falls within the normal range of variation. SPM assists here by reprocessing BEAM images into maps showing the degree of deviation from normal. But once again, the clinician must judge when statistical abnormality indicates a clinical abnormality. This is a role of grid sector analysis (GSA) which develops, from SPM, numerical measures of the degree of abnormality.[10]

The goal of GSA is the generation of numbers from the numerical matrix that underlies a SPM, which describe or indicate the degree of overall (global) and focal abnormality. Global features are derived to assess general or overall deviation from normal as might be expected, for example, in patients with a diffuse encephalopathy. A global feature might consist of the mean of all values in the SPM matrix or the number of matrix values exceeding a chosen criterion level.

Focal features were derived to detect focal (or multifocal) deviation from normal as might be expected in patients with restricted supratentorial lesions, the location of which, however, might not be known in advance. Focal features are derived from such measures as the difference in the mean values of the right and left hemispheres, the length of time a given region exceeds a criterion level, and so on.

FIGURES 3 and 4 show how GSA could be used to discriminate between a SPM demonstrating a large focal abnormality (illustrated) and one demonstrating a diffuse abnormality simulated by randomly scattering the focal SPM matrix values across the image (not illustrated). The SPM are overlain by grids of varying sector size, with each sector represented by the mean of enclosed SPM matrix values. For the focal lesion, note how the histograms of grid sector mean values remain unchanged as sector size enlarges. In contrast, note the dramatic change of histogram waveshape with enlarging sector size for the diffuse lesion.

We reasoned that GSA could facilitate automated differentiation between focal and diffuse deviations from normal. For example, focal deviations might elevate the integrated value in just one sector; diffuse deviations would slightly elevate all sectors. Moreover, varying the size of the sectors would provide a useful indicator of lesion size independent of lesion location.

To evaluate the clinical utility of BEAM with SPM and GSA, we initiated a trial where the diagnostic success of these methods was compared to more conventional analysis of EEG and EP data. Patients were either normal controls or had diagnosed

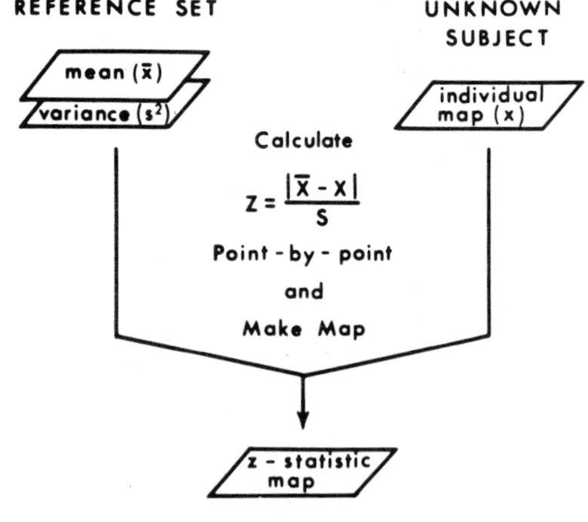

FIGURE 2A. The construction of significance probability maps (SPM): The Z-statistic SPM. The Z-statistic SPM measures the deviation of an individual BEAM matrix from the mean of a set of BEAM matrices. The Z-transform (the number of standard deviations by which an individual observation differs from the mean of a reference set) is calculated for each picture element (pixel) in a BEAM image. These transformed scores compare the individual subject to the mean and variance BEAM images of a reference population. For example, a BEAM image representing the spatial voltage distribution at a given latency for an unknown subject may be compared with the mean BEAM image for the same latency derived from a population of control subjects. The result of this point-by-point Z-transformation is a new matrix of Z values retaining the spatial framework of the original BEAM image. Clusters of high Z values will then define regions in which the individual subject statistically differs from the reference population. For display purposes, an SPM image is formed to illustrate such regions.

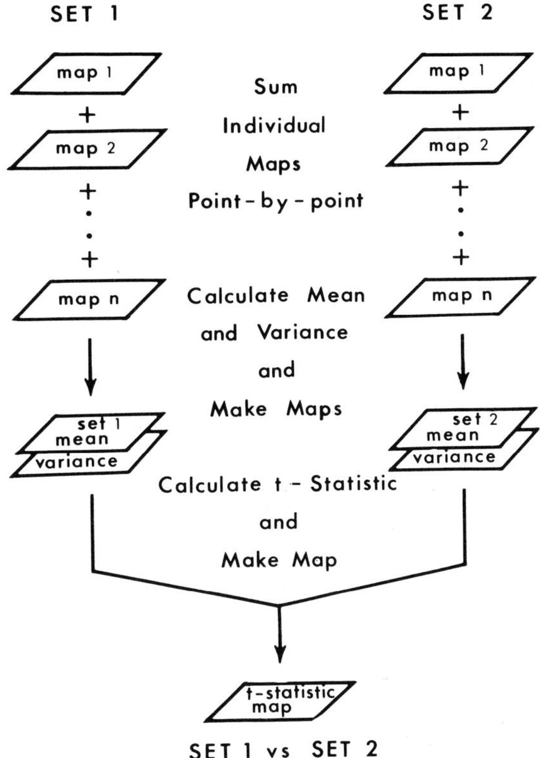

FIGURE 2B. The construction of the *t*-statistic SPM. The *t*-statistic SPM measures the regional differences between two sets of BEAM images derived from two populations of subjects (e.g., sets of maps representing EP activity at a given latency obtained from a group of control subjects and from a group of tumor subjects). The first step in the process is to sum the BEAM images separately for each group. This is done for each of the pixels. Next the mean and variance is calculated for each pixel and two new BEAM matrices are formed representing these respective population statistics. Finally, Student's *t*-statistic is calculated for each pixel based upon the mean and variance matrices of the two groups. In these statistical transformations, the spatial framework of the pixels within the head outline is retained. Then, the *t*-statistic map reveals regions in which the two populations significantly differ from one another. For display purposes, an SPM image is formed representing the distribution of ranges of the value of *t*.

supratentorial brain tumors. BEAM was formed to flash (VEP) and bilateral median nerve shock (BSEP). In our experience, the flash VEP activates the cortex more widely than pattern-reversal VEP does and, thus, is of greater value when screening for supratentorial lesions. Moreover, Yamada *et al.*[8] demonstrated that bilateral somatosensory EPs (BSEP) are more able to demonstrate supratentorial lesions than EPs formed from unilateral stimulation. All data were clinically evaluated on a double-blind basis. Two neurophysiologists read the EEG, the raw EP, the BEAM, and the SPM data. Both were instructed to classify subjects into the normal or tumor category. Only one data type was available for inspection at one time. Automated classification by computer used the *K*-means clustering algorithm[9] and was based upon GSA-derived features. The results are shown in TABLE 1.

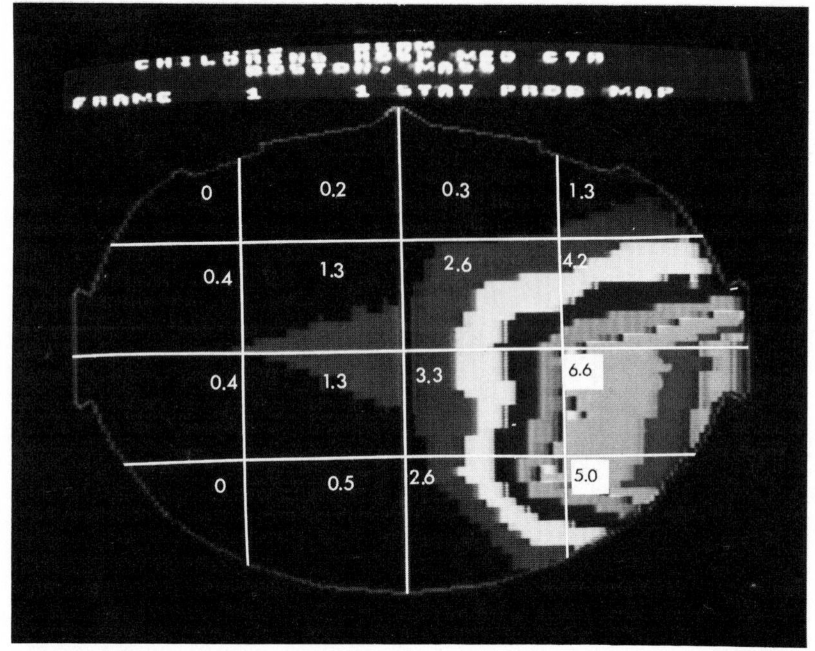
FRAME 1 1 STAT PROB MAP
0 0.2 0.3 1.3
0.4 1.3 2.6 4.2
0.4 1.3 3.3 6.6
0 0.5 2.6 5.0

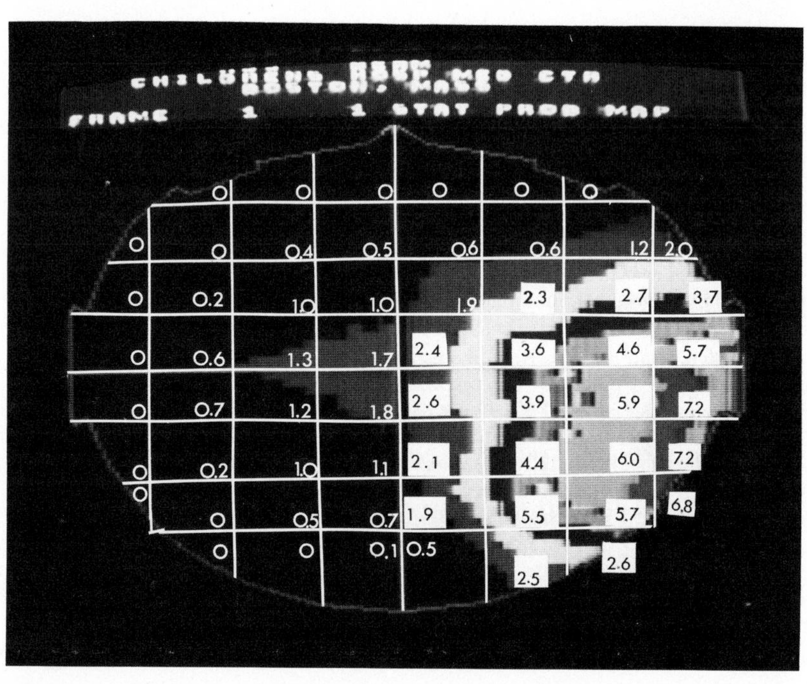
FRAME 1 1 STAT PROB MAP
O O O O O O
O O 0.4 0.5 0.6 0.6 1.2 2.0
O 0.2 1.0 1.0 1.9 2.3 2.7 3.7
O 0.6 1.3 1.7 2.4 3.6 4.6 5.7
O 0.7 1.2 1.8 2.6 3.9 5.9 7.2
O 0.2 1.0 1.1 2.1 4.4 6.0 7.2
O O O 0.5 0.7 1.9 5.5 5.7 6.8
O O 0.1 0.5 2.5 2.6

Note that diagnostic "success" was similar for all comparisons but that, in general:

1. Clinical EEG reading was better than clinical EP reading.
2. Visual inspection of BEAM was better than visual inspection of the raw VEP or BSEP waveforms.
3. Inspection of SPM was more successful than visual reading of the BEAM for both EP modalities and was slightly better than the EEG reading.
4. Automated classification by clustering of GSA-derived measures was surprisingly successful.

Indeed, in our clinical laboratory, GSA has been used to assist in the detection of focalities and asymmetries of all types and especially to assist the clinician in the screening for the focal abnormalities associated with stroke, malformation, abscess, epilepsy, tumor, and so on.

Selected Clinical Examples

Epilepsy. Epilepsy is defined by a history of clinical seizures and is most often confirmed by the presence of spikes or spike-wave discharges in the EEG. In some cases, it becomes necessary to activate discharges when they do not spontaneously occur. Techniques such as hyperventilation, sleep deprivation, and intravenous methohexital are often used to bring out evidence of hyperactive cortex. Nonetheless, despite such activating procedures, some 5%–10% of probable epileptics elude confirmation by EEG. In our experience, BEAM-EP studies of epileptics often demonstrate clinically meaningful evidence of focal hyperactivity in the absence of concurrent spikes in the background EEG. Such an example is shown in FIGURE 5C.

Cerebral Infarction. It has been our experience that, at all ages, cerebral infarcts produce a decrement in EP activity overlying the infarcted cortex. Exceptions to this are infarcts with irritative cortex at their borders. BEAM-EP studies have been useful in gauging the size of infarcts, especially when repeated over time in adults. In newborns and infants, it has been useful in the screening for infarcts that often escape detection by neurological examination and EEG. Such a case of a "silent" infarct is illustrated in FIGURE 5D.

Tumor. Supratentorial mass lesions make themselves evident by abnormalities of overlying brain electrical activity. In general, early EP activity is deficient but later

---

FIGURE 3. Grid sector analysis (GSA): Overlaying SPM with grids of two sizes. This figure illustrates the grid sector analysis (GSA) procedure which is used to detect and measure abnormalities of topographic data. The same Z-statistic SPM is shown twice; once overlaid with a 16-sector grid and once with a 64-sector grid. The SPM represents the deviation from normal of the 290–366 msec latency epoch of a 59-year-old male with a subsequently confirmed glioblastoma multiforme in the right parietal-posterior temporal region. The SPM is shown in black and white with each grey scale level representing one standard deviation. The left hemisphere shows virtually no deviation from normal, i.e., $Z < 2.0$. In contrast, the right hemisphere shows marked deviation from normal reaching a Z value of 8.0 in the mid-temporal region. The focal abnormality of this SPM nicely demarcates the underlying lesion. The average Z value within each grid sector is illustrated. Plots of these values are shown in the top part of FIGURE 4 as "large tumor." Next the individual Z pixel values within the basic SPM were randomly scattered across the entire head map (not illustrated) so as to simulate the SPM of a subject with a diffuse and nonfocal encephalopathic process. The same two grids were once again overlain on the resulting SPM and the grid sectors were once again represented by the average Z value within the sector. Plots of these values are shown in the bottom part of FIGURE 4 as "diffuse abnormality."

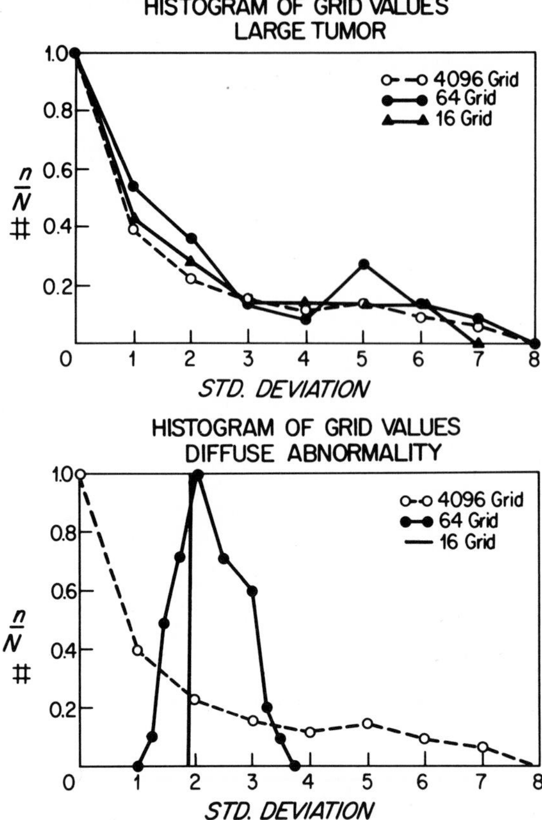

FIGURE 4. Illustration of the use of sector values to discriminate between a focal and a diffuse lesion. These graphs illustrate how a focal lesion can be differentiated from a diffuse encephalopathic process using GSA. For each histogram, the vertical axis represents the number of grid sectors (or pixels) having the average Z value indicated by the horizontal axis. The vertical axis is normalized by the total number of sectors (or pixels) for each plot. There are three plots in each case. The histogram of Z values for the finest grid (original pixels), the medium grid (64 sectors), and the coarse grid (16 sectors) are plotted separately. The three histograms for the large focal lesion of FIGURE 3 are graphed in the upper plot. Note the similarity of waveshape. In general, grid superimposition upon SPM with lesions containing prominent focalities seldom produces changes in histogram waveshape. In contrast, the three histograms for the simulated diffuse lesion (not illustrated) show marked differences in waveshape. In general, grid superimposition causes the histograms of diffuse lesions to converge on a single value as the grid sector size is increased (as the grid becomes more coarse). The average grid sector Z values, the difference in Z values between corresponding sectors of each hemisphere, and the histograms of 3 grid Z values are used to generate numerical measurements sensitive to the degree of focal and/or diffuse abnormality in SPMs. These are not only of immediate use to clinicians but are of use in the automated classification of unknown subjects (see TABLE 1 and text).

activity produces an excessive response that persists for long periods, often over 75 msec.[1] In a few cases, abnormally excessive activity persists for most of the EP duration. An example is shown in FIGURE 5E. In general, atrophic lesions demonstrate a reduced responsivity while irritative lesions produce augmented responses. Tumors represent a class of lesions that may demonstrate both characteristics.

Learning Disabilities. We have found that patients with specific reading disability,[6] attentional deficits, and dyscalculia (work in progress) demonstrate meaningful differences in their brain electrical activity from normal subjects. Dyslexics, for example, demonstrate EEG and EP differences overlying the entire cortical system believed to be active in speech and reading.[6] Whereas many subjects with communication disorders demonstrate reduced BEAM-EP activity overlying left hemisphere speech regions, a few demonstrate regional overactivity. FIGURE 5F demonstrates the latter response in the BEAM-AER of a 4 year old with a history of "developmental aphasia." Such regional overactivity is most often found in subjects with evidence of neurological abnormality such as: history of anoxia, seizures, hemiparesis, and so on.

SUMMARY AND CONCLUSION

The interpretation or evaluation of multichannel EP data requires analysis of large volumes of data across both space and time. We propose that the inherent difficulties involved in making such spatiotemporal correlations by unaided visual inspection place constraints on both the clinical utility and research applicability of EP. The topographic mapping system described herein and elsewhere[1] reduces the dimensionality of data and offers four major advantages:

First, the topographic mapping presents spatial information in a more concise and summarized form, relieving the investigator from having to perform such spatial interpolations in his own mind. A further advantage of this multichannel approach is the constant use of data from all 20 standard electrode positions, which obviates the need to limit analysis for practical reasons to a few sites chosen on an *a priori* basis.

Second, linking together of images (cartooning) enables the visualization of spatiotemporal patterns of brain activity that are not easily perceived otherwise. The temporal sequencing of BEAM-EP images allows the visualization of traveling wavefronts of positive and/or negative activity. In this manner waveform components appearing as differing latencies at different electrode positions may prove to represent the same wavefront slowly progressing across the scalp.

Third, the use of significance probability mapping (SPM) relieves the investigator or clinician from having to maintain normalized standards in his mind.[4] The Z-transform SPM immediately displays regions where a given subject differs from a reference group. As shown in FIGURE 3, this simplifies the identification of the abnormal side when asymmetries are found. The t-statistic SPM demarcates regional between-group differences in population studies of an investigative nature. The investigator chooses the t-statistic threshold level and the locations of group difference, if present, become evident. Three refinements of SPM have proven useful: (1) the cartooning of t and Z images, visually displays changes in location of group or individual difference with time; (2) the use of other statistical hypotheses such as variance ratios and nonparametric tests allows one to adapt to the particular structure of the data under analysis. The d-prime statistic, the Kruskal-Wallis statistic, and the Mahalanobis distance measure[11] have proven quite useful; and (3) the serial application of two or more statistical transforms allows one to form images that delineate regions fulfilling several hypotheses at one time.

Fourth, the development of grid sector analysis (GSA) enables the automated identification and quantification of global and/or focal deviations from normality.

TABLE 1

RELATIVE ACCURACIES OF NINE METHODS FOR DISTINGUISHING PATIENTS FROM CONTROLS*

Subjects	Visual Inspection								Computerized Classification: Clustering
	EEG Reading A	EEG Reading B	Raw VEP	BEAM-VEP	VEP-SPM	Raw BSEP	BEAM-BSEP	BSEP-SPM	
Controls									
C1	–	–	–	–	–	–	–	–	1
C2	–	–	–	–	–	–	–	–	1
C3	(+)	(+)	–	–	–	(+)	–	–	1
C4	–	–	–	–	–	–	–	–	1
C5	–	–	–	–	–	–	–	–	1
C6	–	–	(+)	–	–	–	–	–	1
C7	–	–	–	–	–	–	–	–	1
C8	–	–	–	–	–	–	–	–	1
C9	(+)	–	(+)	(+)	–	(+)	(+)	–	(2)
C10	–	–	–	–	–	–	–	–	1
C11	–	–	–	–	–	–	–	–	1
C12	–	–	–	–	–	–	–	–	1
C13	–	–	(+)	–	–	(+)	–	–	1
C14	–	–	–	–	–	–	–	–	1
C15	–	–	–	–	–	–	–	–	1
C16	–	–	–	–	–	–	–	–	1
C17	–	–	–	–	–	–	–	–	1
C18	–	–	–	–	–	–	–	–	1

Tumors									
T1	+	+	+	+	+	+	+	+	3
T2	+	+	+	+	+	+	+	+	3
T3	+	+	+	+	+	+	+	+	2
T4	(−)	+	(−)	+	+	(−)	(−)	+	3
T5	+	+	+	+	+	+	+	+	3
T6	+	+	+	+	+	+	+	+	3
T7	+	+	+	+	+	+	+	+	2
T8	+	+	+	(−)	+	(−)	+	+	2
T9	(−)	(−)	(−)	(−)	+	+	(−)	+	(1)
T10	(−)	(−)	(−)	(−)	(−)	(−)	(−)	(−)	(1)
T11	+	+	(+)	+	+	+	+	+	3
T12	+	(−)	(−)	(−)	+	+	+	+	2
Correct	83.3%	86.6%	73.3%	83.3%	96.6%	80.0%	86.6%	96.6%	90.0%

*This table compares nine methods of classifying subjects as either controls or as patients with supratentorial brain tumor. A correct classification is indicated by a + or a − sign or a number without brackets. Incorrect classifications are shown in parentheses, e.g., (+), (−), (2), and so on.

EEG readings "A" and "B" refer to the prediction by two electroencephalographers. Raw VEP and raw BSEP refer to classification by inspection of the original set of 20 EP tracings. BEAM-VEP and BEAM-BSEP refer to diagnosis on the basis of visual inspection of the entire set of 128 VEP BEAM images or 128 BSEP images. Finally, VEP-SPM and BSEP-SPM refer to prediction on the basis of the sets of 128 SPM formed from comparison to a normalized population. In every instance the clinical raters were told that each subject was either normal or had a tumor. All evaluations were made without reference to any of the other data types. In other words, the raters could not identify a given BEAM-VEP with its corresponding VEP-SPM because all data were coded so as to prevent this.

The final column represents the automated classification by computerized "clustering" using the standard *K*-means algorithm. The two features used for clustering were derived by GSA. The first represented the value of the most frequently asymmetrical sector of the 64 grid map of the BSEP from 272–368 msec. The second represented maximum difference between corresponding sectors of the 16 grid map (80–464 msec) of the BSEP. In the "clustering," column numbers are used because the algorithm found three separate clusters. The first represented the normals (1 error) and the second and third represented the heterogeneous tumor population (2 errors).

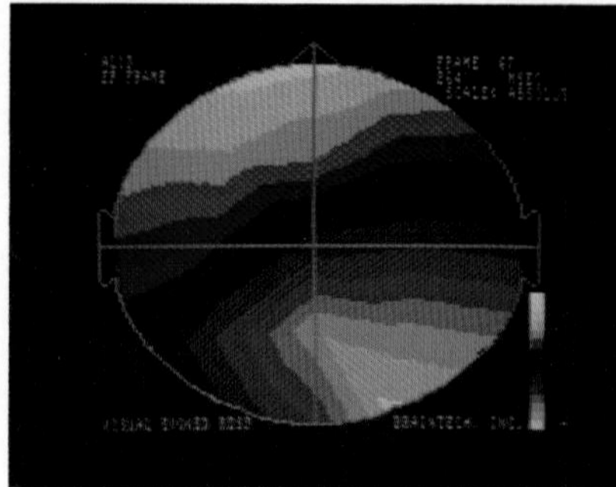

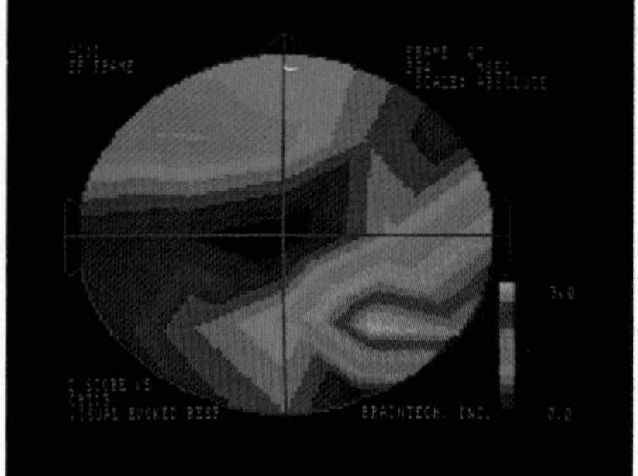

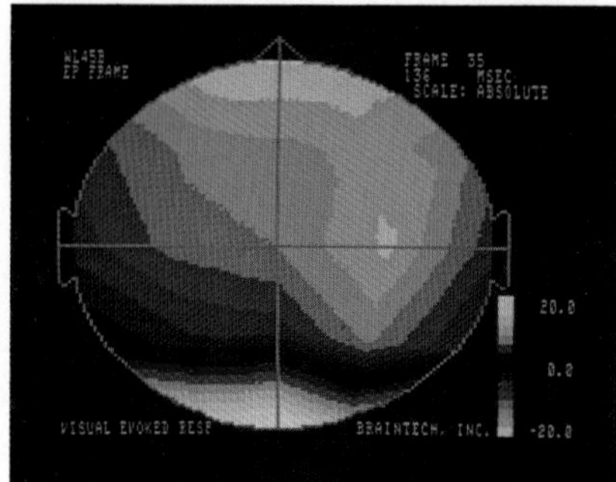

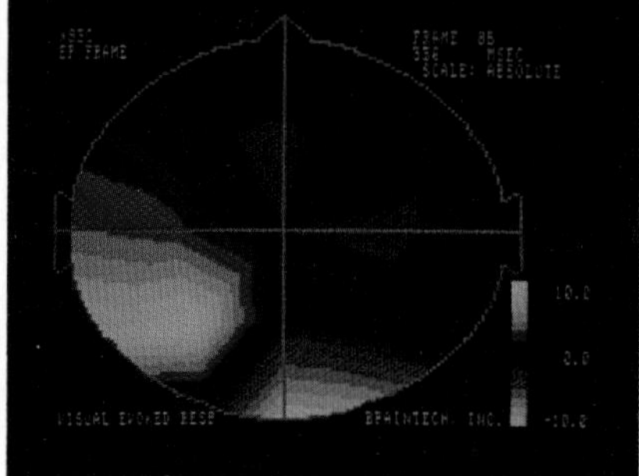

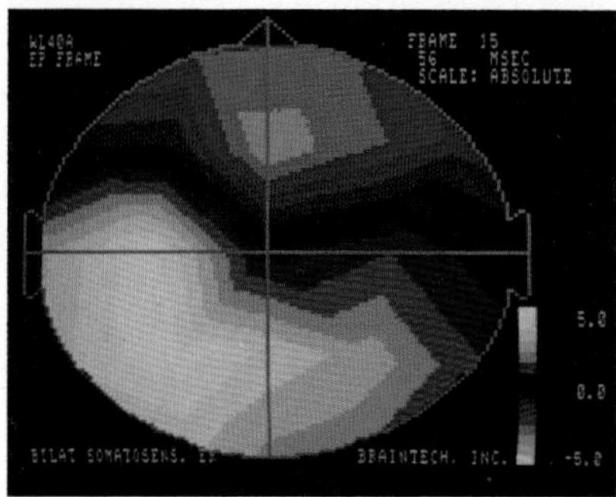

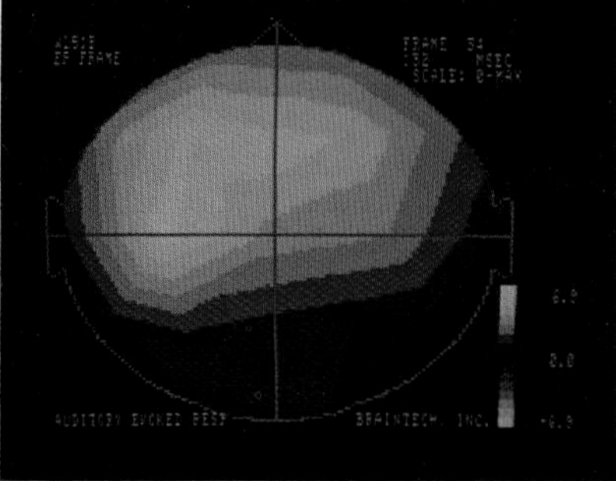

FIGURE 5. (A, upper left) Clinical use of significance probability maps (SPM): BEAM visual evoked response showing an asymmetry. This illustration shows a frame of the BEAM-VEP taken 264 msec following stimulation for a 13-year-old female with a history of speech delay and bilateral midtemporal lobe epileptiform discharges in sleep but no history of clinical epilepsy. Colors ranging from dark red to bright reddish-white represent positive values in microvolts (μV) with reference to linked ear electrodes. Colors ranging from dull blue to bright bluish-white represent negative values. The scale, shown to the lower right, ranges from +10 to −10 μV. Note the striking asymmetry of negative activity (blue) in the posterior quadrants. Although such asymmetries are easily recognized as abnormal, it is not clear whether there is excessive activity on the right or deficient activity on the left.

(B, upper right) SPM of BEAM-VEP containing an asymmetry. This illustration shows the Z-transform SPM derived from (A) by referencing the underlying VEP pixel values to a control

This provides supporting numerical evidence to the clinician as he must decide whether an SPM-delineated asymmetry is clinically significant. Furthermore, GSA-derived features may provide numerical data for further statistical analysis.

Research and clinical experience suggest that the greatest utility of BEAM is in the evaluation of functional lesions where the CT scan is normal or noncontributory. Indeed, brain-imaging procedures such as the CT scan, the PET scan, nuclear magnetic resonance (NMR), and BEAM represent different windows upon brain function. They provide separate but complementary information useful in our continuing search for a fuller understanding of brain function in health and disease.

---

population of similar age. The scale to the lower right shows the Z values starting at zero (black) and progressing through colors of the rainbow to a Z of 3.0 (red-white). Note that by this SPM, the maximal deviation from normal is clearly on the right. Moreover, note that the maximum abnormality is positioned slightly more anteriorly than expected upon inspection of the raw BEAM-VEP (A). Thus, although bilateral discharges were seen on the EEG and the clinical history suggested left-sided dysfunction (speech problems), the SPM demarcated a clear right posterior abnormality. Subsequent EEGS demonstrated a similar lateralization to the right.

(C, middle left) BEAM-VEP abnormalities in an epileptic. This figure shows the BEAM-VEP of an 11-year-old patient with history of a seizure disorder and EEG demonstrating spikes in the right central, mid-temporal, and parietal locations. Display conventions are the same as for (A) but the scale ranges from +20 to −20 μV. The clinical diagnosis was consistent with the sylvian seizure syndrome (Reference 12). The current BEAM-VEP was taken during a period when no clinical seizures were evident and when there were no spikes in the underlying EEG. Note the negative focality in the right central region at 136 msec (illustrated). A positive focality was also noted at 204 msec in the right parietal region (not illustrated). Such regional evidence of cortical hyperactivity is often found in epileptics even in the absence of spikes in the clinical EEG tracing.

(D, middle right) BEAM-VEP in an infant with a covert cerebral infarct. This figure illustrates the BEAM-VEP of a 4-week-old full-term infant. Display conventions are the same as for (A) and the scale ranges from +10 to −10 μV. He was referred for neurological and neurophysiological evaluation with the complaint of deafness. The brainstem auditory evoked response was within normal limits. A special behavioral examination revealed good visual and auditory alerting and good visual tracking. However, auditory tracking was absent. The classic neurological examination and the clinical EEG were within normal limits. The BEAM-VEP demonstrated absent VEP activities over the right posterior quadrant in the parietal and posterior temporal regions, best seen at 336 msec (illustrated). Similar deficiencies of activity were seen on the AEP and EEG BEAM studies (not illustrated). Subsequently, a CT scan demonstrated a large and clinically unsuspected infarct in the right parietal-posterior temporal region. BEAM-EP studies often demonstrate abnormalities in advance of CT scan changes following cerebral infarction.

(E, lower left) BEAM-BSEP of a patient with a supratentorial brain tumor. This figure illustrates the BEAM-EP to bilateral median nerve shock. Reference 8—the bilateral somatosensory EP or BSEP—of a 35-year-old male with history of personality change and subtle right-sided motor findings. A left anterior parietal tumor was shown by CT scan, and a low grade glioma was subsequently proven at surgery. The display conventions are as for (A) but the scale ranges from +5 to −5 μV. Note the prominent focal overactivity which overlays the region of the tumor as delineated by CT scan. Similar focal abnormalities were seen in the BEAM-VEP, BEAM-AEP, and BEAM-EEG studies (not illustrated). BEAM and SPM are useful in following the size of growing and/or changing lesions.

(F, lower right) Focal abnormalities in the BEAM-AEP of a patient with "congenital aphasia." This figure illustrates the BEAM auditory EP (AEP) of a 4-year-old male with history of speech retardation possibly related to perinatal brain damage. Note the overactive response seen in the BEAM-AEP at 132 msec which overlays the left frontal lobe and expressive or anterior speech regions. Whereas many patients with congenital aphasia (and subsequent dyslexia) demonstrate reduced left-sided activity, a few show focal overactivity. This may suggest an irritative process potentially responsive to pharmacotherapy—a possibility under investigation.

ACKNOWLEDGMENTS

The author thanks Dr. Peter Bartels for his statistical expertise, especially the SPM technique, Drs. Gloria McAnulty and James Burchfiel for their technical assistance, and Dr. Cesare Lombroso for his continuing encouragement.

REFERENCES

1. DUFFY, F. H., J. L. BURCHFIEL & C. T. LOMBROSO. 1979. Brain electrical activity mapping (BEAM): A new method for extending the clinical utility of EEG and evoked potential data. Ann. Neurol. **5:** 309–321.
2. GRASSELLI, A., Ed. 1969. Automated Interpretation and Classification of Images. Academic Press, New York, N.Y.
3. BARTELS, P. H. & J. A. SUBACH. 1976. Automated interpretation of complex scenes. *In* Digital Processing of Biomedical Imagery. E. Preston & M. Onoe, Eds. pp. 101–114. Academic Press, New York, N.Y.
4. DUFFY, F. H., P. H. BARTELS & J. L. BURCHFIEL. 1981. Significance probability mapping: An aid in the topographic analyses of brain electrical activity. Electroenceph. Clin. Neurophys. **51:** 455–462.
5. DOWNIE, N. M. & R. W. HEATH. 1974. Basic Statistical Methods. Harper and Row, New York, N.Y.
6. DUFFY, F. H., M. B. DENCKLA, P. H. BARTELS & G. SANDINI. 1980. Dyslexia: Regional differences in brain electrical activity by topographic mapping. Ann. Neurol. **7:** 412–420.
7. DUFFY, F. H., M. B. DENCKLA, P. H. BARTELS, G. SANDINI & L. S. KIESSLING. 1980. Dyslexia: Automated diagnosis by computerized classification of brain electrical activity. Ann. Neurol. **7:** 421–428.
8. YAMADA, T., J. KIMURA, S. YOUNG & M. POWERS. 1978. Somatosensory-evoked potentials elicited by bilateral stimulation of the median nerve and its clinical application. Neurology **28:** 218–223.
9. HARTIGAN, J. A., ed. 1975. Clustering Algorithms. J. Wiley and Sons, New York, N.Y.
10. DUFFY, F. H., P. H. BARTELS & J. L. BURCHFIEL. Manuscript in preparation.
11. BARTELS, P. H. 1979. Numerical evaluation of cytologic data III. Selection of features for discrimination. Analytical Quant. Cytol. J. **1:** 153–159.
12. LOMBROSO, C. T. 1967. Sylvian seizures and midtemporal spike foci in children. Arch Neurol. **17:** 52–59.

MAGNETIC LOCATION OF CORTICAL ACTIVITY*

Lloyd Kaufman† and Samuel J. Williamson‡

New York University
New York, New York 10003

This paper is a review of recently completed studies of neuromagnetic fields preceding motor activity and following sensory stimulation. Since many of the experiments described here have only recently been submitted for publication, this paper may be considered to be a preview of work to be described in considerably expanded detail elsewhere. Our purpose is to bring together an existing body of facts and ideas supporting the notion that evoked fields complement evoked potentials since they permit a high degree of resolution of the locations of active tissue that are the ultimate sources of evoked potentials. This same degree of resolution, in at least some cases, is not achievable by the study of potentials alone.

The problem of determining the location of a source of potentials on the scalp from an analysis of the distribution of potentials is an instance of the inverse problem. As Helmholtz pointed out long ago,[1] the inverse problem has no unique solution. A large number of possible sources could produce precisely the same distribution of potentials. However, by assuming a particular type of source, e.g., the current dipole, and imagining it to be immersed in a conducting medium within a model representing the head, it is possible to compute the potentials that the hypothetical source would produce on the outer surface of the model. If this solution approximates an observed pattern of potentials on the actual scalp, then it might be assumed that the source of the potentials is located at a place in the brain that corresponds to the place of the current dipole within the model. This general approach has been used with a limited degree of success by Sidman *et al.*,[2] for example, who placed the source of components occurring 20 and 30 msec after somatic stimulation near the central sulcus. This conclusion is consistent with the finding of Goff *et al.*[3] that these components are readily detectable near the posterior bank of the central sulcus when recording from the exposed pial surface.

Most workers studying evoked potentials do not go to the trouble of developing sophisticated models in trying to locate sources. One commonly used criterion is the reversal of polarity of a component. For example, the 30-msec component is electrically positive behind the sulcus and negative anterior to it. If we assume that the source can be represented by a current dipole tangential to the scalp and oriented normal to the central sulcus, then its location coincides with the null point, i.e., the place of polarity reversal.

This procedure seems to be both reasonable and convenient, even though we all recognize that conclusions based on it are ultimately model dependent. Unfortunately, it does not always work. Some evoked potential components as conventionally measured do not display abrupt polarity reversals. Some of them display a gradual change in phase as the active electrode is moved across the scalp. The locations of sources of such components are difficult to determine even when explicit model-building techniques are employed.[4]

*This work was supported by the Office of Naval Research (Contract N00014-76-C-0568) and the National Institutes of Health (Grant 1 RO1 EY02059-03).

†Department of Psychology.
‡Department of Physics.

0077–8923/82/0388–0197 $1.75/0 © 1982, NYAS

One possible reason for such difficulties is that more than one source may be active at a given time. It is widely acknowledged that the evoked potential is mediated by volume currents that flow from a source throughout the intracranial space and into the skin. These currents pass through the pia, the cerebral spinal fluids of the subarachnoid space, through the dura, through and around the skull and, ultimately, into the skin where they produce differences in potential. The boundaries that separate these layers of different conductivities strongly affect the distribution of potentials on the scalp, depending upon the dimensions and depth of the primary source. When two or more sources are simultaneously active, their volume currents are superimposable in the passive conducting media of the head. Moreover, anisotropies in conductivity may well distort the distribution of volume currents in ways that existing models do not attempt to handle. In view of all of this, it is little wonder that source location is so difficult a task.

The situation is not entirely hopeless. For example, Donald[5] was able to recover at the scalp what he called the "Rolandic late wave." This wave is evidently the same as the late somatic response discovered by Goff *et al.*[6] in their pial recordings. The reason for singling out Donald's achievement is that the late somatic response is normally masked at the scalp by the vertex potential. This is an instance of how simultaneously active sources may make it difficult to isolate a relatively unitary source. However, the vertex potential is inherently variable and its variability is independent of the variability of the late somatic response. By selective averaging, when the vertex potential was at its minimum, Donald succeeded in recovering the late somatic response. Moreover, owing to the fact that this response reversed polarity across the central sulcus, he was able to conclude that it originated in the somatosensory cortex.

Unfortunately, it is not clear that all sources of responses are independently variable. Difficulties arise even when attempting to localize the 20- and 30-msec components from scalp recordings in some subjects.[6] It may not be unreasonable to suggest that other active sources played a role in masking these components in some subjects.

How can magnetic recording be of help? The answer to this question resides in some unique properties of the neuromagnetic response. One of the most important of these is that normal biological tissue is essentially transparent to low-frequency magnetic fields, e.g., less than 1 kHz. Therefore, a field produced by a source inside the head emerges without distortion. Another major feature of the magnetic response is our finding, to be documented below, that the field is relatively unaffected by the volume currents that underly the evoked potential. Rather, the field is due to the relatively high-density intracellular currents flowing in active neural tissue. These two facts taken together, the transparency of tissue to the fields of interest and the intracellular aspect of the current sources, make it possible to resolve spatially separated sources that cannot be resolved by measuring the potentials that arise from these same sources.

We have quite deliberately taken a strong position in this controversial area. For example, Grynszpan and Geselowitz[7] noted that, in principle, when volume currents flow in media separated by boundaries of substantially different conductivity, the boundaries can perturb the volume currents to effectively set up "secondary sources" of fields. Our observations suggest that this theoretical result is not significant in measurements of brain events although it is certainly important in other domains of biomagnetism, e.g., magnetocardiology. In fact, the important theoretical result of Cuffin and Cohen[30] suggesting that volume currents do not contribute to the normal component of the field outside the scalp seems to be applicable to actual measurements.

This introductory section is not the place in which one should fully justify the claims that neuromagnetic fields are unaffected by intervening media and are not generated by the same volume currents that underly the evoked potential. A full justification depends upon a presentation of the data to be reviewed below. For the present, it is sufficient merely to consider the implications of these claims for source resolution. One implication stems from the fact that the strength of the field of a current dipole varies inversely with the square of the distance between the current dipole and the sensor. Now, if the source of the evoked field can be approximated by a current dipole, then the strength of the detected field will be markedly affected by the position of the sensor on the scalp. If it is far from the source, then the field will be dramatically weaker than the field generated by a relatively nearby source. In view of the widespread nature of volume currents and the anisotropies of the conducting medium in which they flow, such a simple distance dependency could not exist for the magnitude of the evoked potential and the distance between its source and the active electrode.

One caveat worth mentioning here is that the field detected outside the scalp must be generated by current that flows tangential to the surface. In spherical models for the head, a current dipole that is not aligned parallel to the radius at the dipole's position has a tangential component and this, in principle, can generate detectable fields. However, a current dipole that is directed radially will not generate a detectable field outside the sphere. The reason is that secondary sources at the boundary of the sphere create a magnetic field outside that exactly cancels the field from the dipole itself (Ampere's law). This theoretical result indicates that field strength outside the head is not only dependent upon the distance between the sensor and the source, but also upon the orientation of the source. The radially oriented current dipole, however, will produce potential differences on the scalp. A dipole that is tilted so that it does not lie along a radius within a sphere can be thought of as being equivalent to two dipoles, one parallel to the radius and the other tangential to the outer surface. The tangential component alone would generate the external field normal to the surface while both the tangential and radial components would affect the pattern of potentials on the surface. As Williamson and Kaufman point out,[8] in a spherical model the external pattern of the normal field remains constant in shape and position as a current dipole is tilted about its center from the tangential orientation, although the field strength will diminish as the dipole is tilted, with a field strength of zero representing the case where the dipole is tilted so that it is radially oriented. This same dipole will produce a potential pattern on the surface that loses symmetry and shifts in position as the dipole is tilted. In the limit, when the dipole is radially oriented, the original dipolar potential field pattern becomes unipolar in appearance. All of this is illustrated in FIGURE 1 which, though based on a half-space model rather than a spherical model, qualitatively demonstrates the different kinds of behavior to be expected as the current dipole's orientation departs from the tangent to the surface.

Based on such considerations, it seems likely that measured neuromagnetic fields from a single localized source would be distributed in the form of a dipolar pattern, with one member of the symmetrical pair of regions representing the field directed outward from the head and the other field directed inward, as would be produced by an underlying current dipole. The direction of the field, i.e., inward or outward, would depend upon the direction of current flow, in accord with the right-hand rule. Moreover, as illustrated in FIGURE 1, if the current dipole is strictly tangential to the scalp and if it alone is active, then the resulting potential pattern should also be dipolar in shape but rotated 90 deg relative to the magnetic field pattern. If the current dipole has a radial component as well as a tangential component, then the potential pattern need not have a dipolar appearance while the magnetic field pattern would still be dipolar in character. In any event, all of our field measurements entailed seeking signs

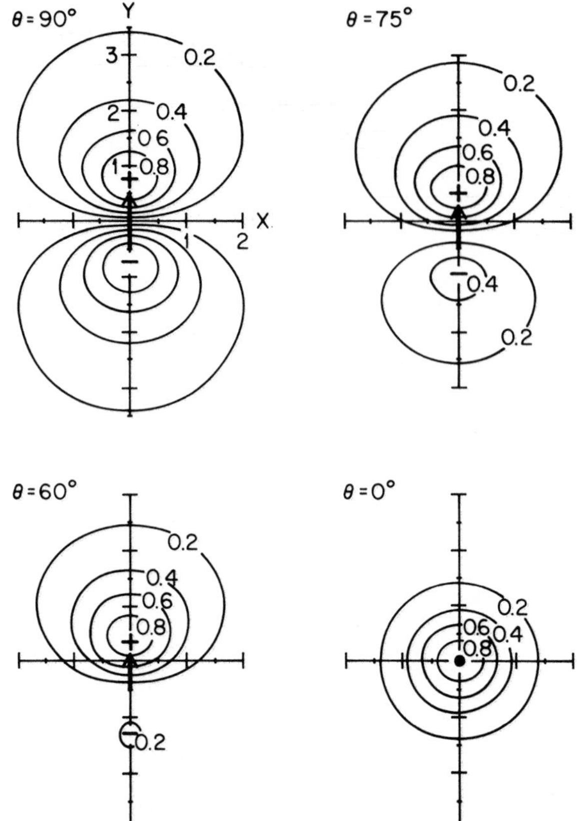

FIGURE 1. Isopotentials on the flat surface (x-y plane) bounding a half space of uniform conducting material in which a current dipole lies. The dipole is tipped by the angle θ from the surface's normal toward the $+y$ axis. The distances along x and y axes are expressed in units of the depth of the dipole below the surface, and isopotentials are expressed relative to the maximum value of the potential. Isofield contours describing the component of the magnetic field lying normal to the surface are identical to the isopotentials for the $\theta = 90°$ case, but the pattern would be rotated by 90° in the x-y plane.

of both inward and outward directed fields inasmuch as these should be present if the underlying source can be modelled as a current dipole. We shall now turn to the actual experiments and then conclude with an overview of the accuracy now attainable in source localization by magnetic means.

METHODOLOGICAL CONSIDERATIONS

Before turning to the results of experiments it is of some importance to review the methods used. The first consideration is that the instruments employed must be sufficiently sensitive to low-frequency fields that are typically one billionth the strength of the earth's steady field. This field sensitivity can only be achieved by using a sensor based on the superconducting quantum interference device (SQUID). The SQUID and a superconducting detection coil shown schematically in FIGURE 2 are kept in a bath of liquid helium within a fiberglass cryogenic Dewar flask. The particular detection coil shown in the figure is known as a second derivative gradiometer since it is wound to be insensitive to uniform fields and to fields with uniform spatial gradient. Consequently, it is relatively immune to fields that arise from distant sources and it is selectively sensitive to fields arising from nearby sources, e.g., those that are close to the bottom-most coil of the gradiometer (the "pick-up coil"). These fields from nearby sources differentially affect the bottom-most coil so that shielding currents are caused to flow in the entire circuit containing the gradiometer to keep the amount of flux trapped within the circuit constant regardless of the external field (the Meisner effect). These shielding currents are magnetically coupled to the SQUID by the input coil shown in the figure. Except for this effect on the SQUID by the input coil, the SQUID itself is isolated from the ambient magnetic field by a superconducting shield placed around it. The output of the SQUID electronics is a voltage that is proportional to the external field linking the pick-up coil. The details of biomagnetic measurements are described in the review article by Williamson and Kaufman.[9]

As shown in FIGURE 3, the Dewar flask is mounted in a holder so that it can be displaced in the X, Y, and Z directions and tilted through an angle of 45 deg, with the tip of the Dewar's tail section as the center of rotation. This makes it possible for the experimenter to easily move and tilt the Dewar from one position to the next and keep the bottom-most pick-up coil tangential to the scalp. Therefore, the pick-up coil senses the field that is normal to the scalp. This coil is kept approximately 0.7 cm away from the scalp by the thickness of the bottom wall of the tail section of the Dewar.

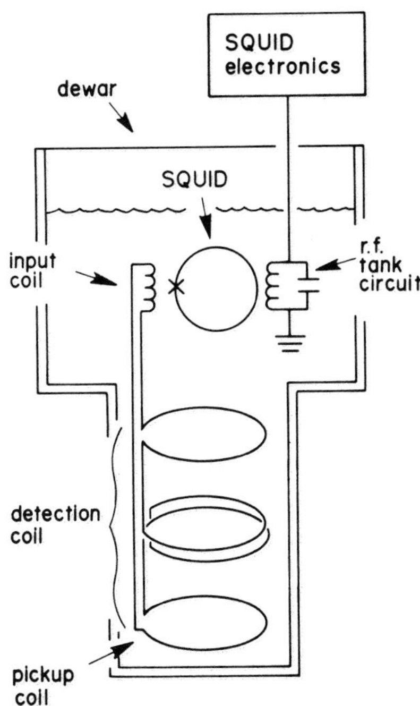

FIGURE 2. Magnetic field sensor for biomagnetic studies. The net magnetic flux linking the superconducting coil causes a current to flow through the input coil, and that in turn applies a magnetic field to the SQUID. The response of the SQUID is detected by the SQUID electronics, which is magnetically coupled to the SQUID by a circuit resonating at 19 MHz.

In the experiments described here, the Dewar is typically positioned so that the center of the pick-up coil is in the vicinity of the skull where evoked fields are likely to be detected, e.g., over one side of the occipital pole in a vision experiment. The stimulus is then periodically presented for a predetermined period of time, usually 1 or 2 minutes, and the SQUID output is averaged after being applied to a bandpass filter. The particular filter settings and the details of stimulation are described below. The position of the Dewar is then varied in steps of 1 cm from trial to trial until a response is detected that is significantly greater in amplitude than the background noise. Once

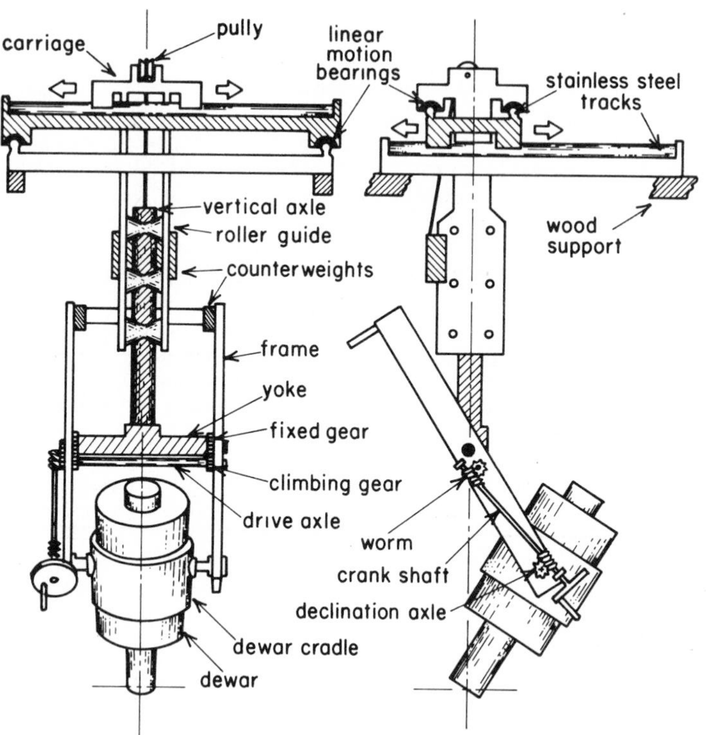

FIGURE 3. Arrangement for supporting the Dewar flask which contains the SQUID magnetic sensor.

such a response is found, the trial is repeated to establish the reliability of the measure. The Dewar is then systematically moved in small positional increments until the entire region of the head from which responses of like sign (directed inward or outward) is determined. Such a procedure may take place in sessions that extend over several days. Once the topography of the response is determined, the Dewar is then moved to discover the region from which a response of opposite sign is found. This region is then mapped in the same way as was the other region. When this mapping is completed we generate an isofield graph in which the contours define the loci of fields of equal strength. These maps include the locations on the scalp of the maximum fields directed outward and inward (the "extrema") and repeated measures are made at these points to insure their reliability. This is particularly important since, as we shall see below, the positions of these extrema permit locating the lateral position of the source as well as its depth.

The procedure just described was used in experiments where auditory, somatic, and visual stimuli were employed. A similar procedure was used to locate the source within the motor cortex that became active prior to the voluntary flexion of a finger. The only difference in the latter case was that the reference signal was the electromyogram recorded from the forearm rather than the time of presentation of a stimulus. Moreover, the procedure of Gilden, Vaughan, and Costa[10] was followed in that the response was averaged backward in time from the time of onset of muscle activity as well as forward in time so that events that precede and follow motor action could be recovered.

We will not go into the details here, but will simply indicate that every effort was made to insure that stimulus artifacts did not influence our results. For example, recordings in visual studies were made as a check with the screen of the stimulating display completely masked by a piece of cardboard. In the case of auditory stimulation the earphones were removed from the subjects' ears and responses recorded to insure that we were not picking up an artifact from the transducer in the headset. Similarly, in the case of somatic stimulation, the Dewar was removed from the scalp by a distance of 1 cm to be sure that the pick-up coil was not sensing the electrical stimulus rather than the activity of the brain.

We conclude this section on methodology with a few comments on the procedures used in locating sources from the field patterns. Williamson and Kaufman[8] demonstrated that it is possible to determine the depth of a current dipole within a conducting half-space merely from knowledge of the distance between the extrema of the field pattern measured across the flat surface. This depth is given by dividing the distance between the extrema by $2^{1/2}$. This same method is approximately correct when the current dipole giving rise to the field is near the outer surface of a sphere. In this case, the curvature of the sphere can be ignored and one can assume that the surface is a plane and the dipole is located a short distance beneath it. The results are unaffected if slabs of uniform but different conductivity are interposed between the current dipole and surface to represent the cerebrospinal fluid, skull, and dermis. However, for deeper sources there are strong differences between the sphere model and the half-space model. In the half-space model there is a linear relationship between the distance separating the extrema and the depth of the current dipole. However, in the spherical model the distance separating the extrema increases more rapidly than the depth as the depth increases. This means that small increments of depth can be more easily resolved for deep sources by measuring the distance separating the extrema on the surface. The actual computation of depth in the spherical model merely requires knowledge of the radius of the sphere and a measure of the distance separating the extrema. The appropriate equations can be found in Williamson and Kaufman.[8] In the case of actual applications it is possible to measure the curvature of the head of the human subject to determine its approximate radius and also to measure the distance along the scalp separating the extrema and, by assuming that the head can be approximated by a sphere, compute the depth of the source of the observed field. The depth values given in the following sections were calculated in this way, and, as we shall see, their values are in reasonable correspondence with the depths as given in stereotaxic atlases of the structures most likely to contain the sources of activity affected by the various stimuli employed.

With this background in mind, we shall now turn to consideration of the results of experiments in four different areas of neuromagnetism. These are the auditory evoked field, the somatic evoked field, the visual evoked field, and the field generated by the motor cortex.

THE AUDITORY EVOKED FIELD

The auditory evoked field (AEF) was detected over the temporal region of the scalp by investigators in several laboratories. Reite and Zimmerman[11] were the first to do so using click stimuli. Farrell *et al.*[12] also used clicks as stimuli for evoking a transient response which had a conspicuous 50-msec component. They noted that a similarly conspicuous component had been reported in the literature for the vertex-detected auditory evoked potential.[13] The relative polarity of the magnetic and potential data is consistent with the predictions of a model in which the source of the auditory evoked potential is a current dipole oriented vertically, and therefore the volume currents arising from the current dipole flow downward 50 msec after stimulation, as indicated by the positivity of the component at this time near the vertex. Hari *et al.*[14] observed the AEF for a 1000-Hz tone presented for 800 msec. Elberling *et al.*[15] used a similar stimulus (100 Hz presented for 500 msec) and verified the finding of Hari *et al.*[14] These responses contained clear 100-msec components as well as a sustained component. The fields evoked by these binaural stimuli were of greatest intensity near the two ends of the Sylvian fissure, such as would be produced by two nearly vertically oriented dipoles, with one in each hemisphere. This result was essentially the same as that obtained by Farrell *et al.*[12]

The most interesting aspect of all of these results, however, is that the direction of the field was predictable from a current dipole whose current orientation was opposed to the direction of the current producing corresponding components of the evoked potential. Therefore, the extracellular currents producing the evoked potential could not be the same as the current underlying the AEF. The latter must flow in the opposite direction. Therefore, Hari *et al.* concluded that the source of the AEF is the net intracellular currents evoked by the stimulus while the corresponding components of the evoked potential are due to the extracellular volume current associated with these intracellular currents. This conclusion is consistent with the finding of Farrell *et al.* concerning the 50-msec component of the transient field and potential.

The foregoing results tell us something about the nature of the source of the AEF and that it is located in the vicinity of the Sylvian fissure. Our own concern, however, is with the degree to which it is possible to further refine measures of the location of the source of the AEF. Toward this end, Romani, Williamson, and Kaufman resorted to a very different kind of stimulus.[16] A pure tone of 200, 600, 2000, or 5000 Hz was amplitude modulated by a 32-Hz sinusoid and presented to the subject binaurally by means of an airline headset. The loudness of each tone was adjusted until the subject judged them to be approximately equal. The 32-Hz modulating signal was used as the trigger (reference signal) for signal averaging. Thus, the responses were at 32 Hz and they were all produced by tones that can be described as composed of three frequencies, i.e., the carrier frequency and the sum and difference of the carrier and modulating frequencies. It was assumed that these spectra are sufficiently narrow relative to all of the central carrier frequencies so that they would each affect a relatively uniform population of cells.

During the experiment, the subject lay on his left side and listened attentively to the stimulus. The pick-up coil of the Dewar was positioned with its center over a point in the temporal region of the skull and all four stimuli were presented in repeated trials and in random order. The Dewar was then moved to other locations in small positional increments and the entire series of measurements was repeated at each location. The position of the Dewar and the amplitudes and phases of the responses at 32 Hz obtained in each 30-sec measurement trial were recorded. The results obtained on repeated trials were averaged and all of the data were then plotted in four graphs, one for each of the stimuli. The coordinate system used in plotting the data employed as the horizontal axis the distance along a line connecting the ear canal and the corner of the eye. Distances (in cm) forward of the ear canal are considered positive and those posterior to the ear canal negative. The ordinate of the graph simply represents the distance of the tail section of the Dewar along a meridian connecting the ear canal and the vertex. These two coordinates permit locating the position of the pick-up coil relative to the scalp.

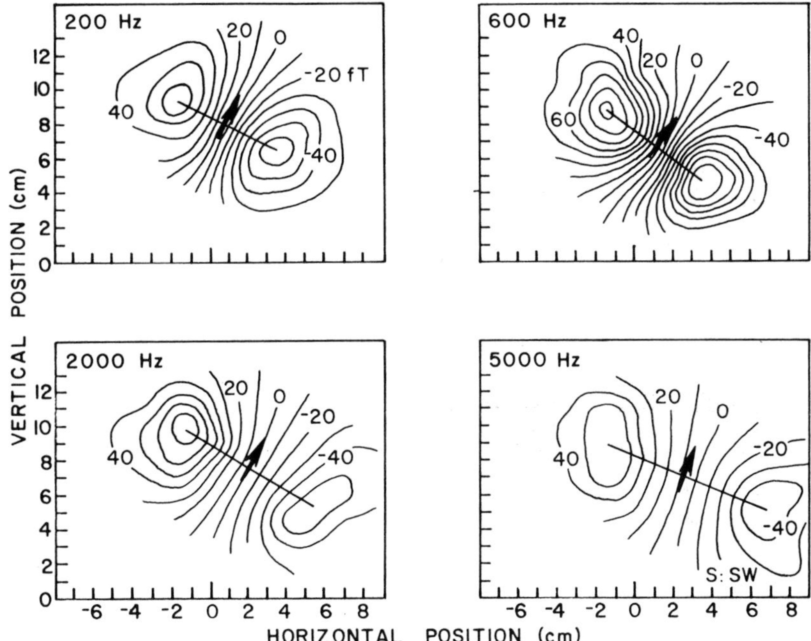

FIGURE 4. Isofield contours for the component of the auditory evoked field detected normal to the scalp over the right hemisphere. The ear canal serves as the origin, and the corner of the eye lies at a horizontal position of +9 cm. Arrows denote the positions of the equivalent current dipole sources for tones at 200, 600, 2000, and 5000 Hz.

Constant field contours were fit to the data for each tone. These are shown in FIGURE 4. These are apparently dipolar in character with one region (+) representing the emerging field and the other region (−) the reentering field. These are so designated because responses in the two regions were 180 deg out of phase. The maximum field strengths in the two regions are indeed comparable. The location of the underlying dipole is midway between the maximum emerging and reentering fields. It is obvious that the current dipole deduced from the field pattern evoked by the 200-Hz stimulus is nearly directly above the ear canal and that the locations of the other stimuli are increasingly shifted toward the anterior portion of the head with frequency. Morever, the separation between the maxima increases with increasing tonal frequency.

With a correction for the dimensions of the gradiometer, it is possible to compute the depth of a current dipole from the distance separating the maximum emerging and reentering fields for a semi-infinite plane model and also for a spherical model of the head. Using the algorithm for the spherical model, which entails estimating the radius of the head from measurements of its curvature and measurements on the scalp of the distances separating the two maxima of each pattern, the following depths were found: 2.3 cm below the scalp for the 200 Hz; 2.6 cm for the 600 Hz tone; 3.0 cm for the 2000 Hz tone; and 3.3 cm for the 5000 Hz tone. This sequence follows a simple logarithmic progression.

These data for one subject were approximately replicated in another. They show that there is indeed a tonotopic map for the human: cell populations selectively sensitive to different frequencies have a spatial ordering correlated with frequency and, moreover, these populations can be resolved by neuromagnetic means. Thus, the main point of this section is that Romani *et al.*[16] were able to resolve sources within auditory cortex that are separated by distances within the cortex of less than 0.5 cm. Moreover, these regions have different functional properties inasmuch as they are selectively affected by stimuli that differ in their acoustic spectral properties.

It may be worthwhile here to briefly mention one supporting reason for this last assertion, i.e., all of the pure tones were modulated by precisely the same modulating frequency which also served as the trigger. Yet the phase lags of the responses to the modulated stimuli differed by as much as 90 deg from each other. This phase difference could only be due to the difference in the carrier frequencies of the stimuli. Such an effect could only occur if neurons tuned to different spectra of sounds were responding.

Somatic Evoked Fields

Using a SQUID device, Brenner, Lipton, Kaufman, and Williamson[17] detected the steady state response of the somatosensory projection area to stimulation of the little finger of one hand. The response could be detected only over the hemisphere contralateral to the stimulated finger and its topography about the scalp was approximately dipolar in character. The pattern of the somatically evoked field (SEF) was consistent with a current dipole source oriented normal to the central sulcus. A similar pattern was observed when the thumb was stimulated rather than the little finger and the data suggested that the source of this response was at a somewhat lower position along the sulcus. This was confirmed in detail by Okada *et al.*[18] who, in a similar experiment, stimulated the thumb, index finger, little finger or ankle of the subject. Each of these stimuli produced patterns of the SEF whose symmetry allowed the lateral localization of sources. Distinct sources of responses to stimulation of the ankle, thumb, index finger, and little finger were resolved. The positions of these sources are consistent with the known positions along the central sulcus of the representations of the various parts of the body.

Kaufman *et al.*[19] recently detected the transient somatic evoked field in response to stimulation of the median nerve. Responses from the left side of the head of one subject during stimulation of the contralateral or ipsilateral wrists are shown in FIGURE 5. These responses were measured with a bandwidth of 1–100 Hz and produced by a stimulus presented at about 1.9 Hz. They are typical of responses observed in five subjects. It will be noticed that sharply defined responses are present only when the stimulus was applied to the right (contralateral) wrist. By contrast, the somatic evoked potential (SEP) recorded under similar experimental conditions is quite different in waveform and many of its components are represented with equal strength on both sides of the head.[6] Similarly strong bilateral representation of the evoked field was not encountered in any of our five subjects. It seems likely that the bilaterality of many of the components of the SEP is due to effects of widespread volume currents which do not have the same effect on the SEF. It is perhaps of greater interest to note that there is a strong resemblance between the waveform of the evoked field and that of the electrical response recorded directly from the pial surface of the exposed brain by Goff *et al.*[3] It is also noteworthy that early components occurring 20 and 30 msec after stimulation are visible in the evoked fields of all of our subjects.

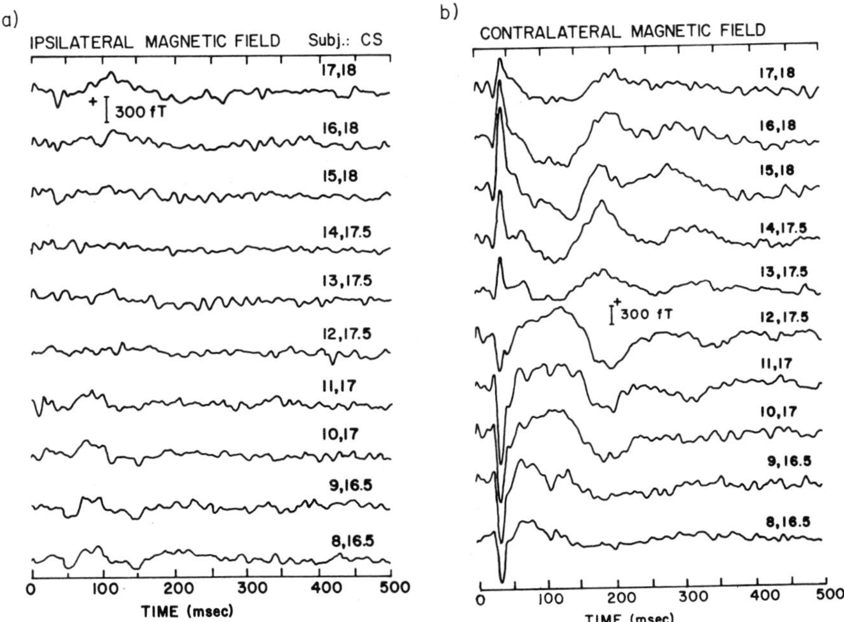

FIGURE 5. Transient somatically evoked responses over the left hemisphere in response to stimulation of (a) the median nerve of the left wrist and (b) the median nerve of the right wrist. A positive field, indicated by upward deflection, designates a field direction emerging from the head. The pair of numbers indicate the position (in cm) of the recorded response above the ear canal and behind the nasion as measured along the midline.

Although this is not true in scalp recordings, the same components are easily seen in electrical recordings from the pial surface.[6] Such evidence implies that the two responses have a common source, but other sources, perhaps from deeper structures, produce effects that contribute to the evoked potential on the scalp. The effects of these other sources seem to be absent in the SEF. This is one of the major reasons for proposing that the evoked field may be capable of resolving sources that cannot normally be resolved in the evoked potential alone.

A particularly important bit of evidence in this connection is that most components of the somatic response on the pial surface recorded posterior to the central sulcus are opposite in polarity to those of the same components recorded anterior to the sulcus, including components with latencies as long as ~150 msec. This suggests that the source of these components is at or near the posterior bank of the sulcus. Now, all components of the somatic evoked field abruptly reverse polarity as the pick-up coil is moved rostrally along the projection of the central sulcus onto the scalp. Those components of the somatic evoked potential that did not reverse polarity as the electrode was moved across the central sulcus in the pial recordings of Goff *et al.*[3] were obviously generated by relatively distant sources, and volume currents from these sources contributed to the response. We did not find any components that did not reverse polarity as the probe was moved at right angles to the direction of movement of the active electrode used in the study by Goff *et al.*[3] Consequently, the somatic evoked field seems to be relatively immune to the volume currents that affect the pial response.

We were able to identify three components in the SEF that obviously had counterparts in the pial SEP. These occurred about 20, 30, and 100 msec after stimulation. By way of example, let us consider the 30-msec component. This component in the pial recordings is positive posterior to the sulcus and negative anterior. By the convention that current flows from positive to negative, this may be taken to mean that the volume current related to this component must be flowing anteriorly 30 msec after stimulation. The same is true when this component is detectable in scalp recordings;[4] i.e., the component is conventionally designated as P30 posterior to the projection of the sulcus onto the scalp and N30 anterior. Now, if the evoked field were generated by the same volume current then, by the right-hand rule, the field from the left side of the head should emerge from the head below the source and reenter the head above the source. FIGURE 5b shows that this is clearly not the case. Upward deflection in the figure indicates an emerging field. The 30-msec component emerges from the head above the source and reenters below. Consequently, the current that gives rise to the evoked field flows in a direction that is *opposite* to the direction of the current that produces this same component in the evoked potential. The same is true of the 20-msec component and the 100-msec component. Now, the only current that could be flowing in opposition to the volume current in the interstitial space is the net intracellular current. It must be concluded, therefore, that the most likely source of the somatic evoked field is the net intracellular current—probably the axial intracellular currents flowing in the apical dendrites of pyramidal cells. More details as to the theoretical reasons for accepting this assertion are provided by Kaufman *et al.*[19]

Our recordings of the transient SEF indicate that the null position—the point at which the external field is tangential to the scalp and, therefore, not sensed by the pick-up coil—defines the lateral position of the source. It is possible to resolve this position with an accuracy of better than 0.5 cm.

The depth of the source of the field evoked by stimulation of the median nerve may be calculated using the same procedure as that used in calculating the depth of the auditory field's source. The distance between the extrema of the 30-msec component of the responses of FIGURE 5 is about 6 cm. Assuming a head radius of 10 cm, and correcting for the effect of the gradiometer on the response, the depth of the source of this component is approximately 3 cm beneath the scalp. This places the source about half-way into the central sulcus.

VISUAL EVOKED FIELDS

Earlier studies of visual evoked fields indicated that they are relatively sharply localized near the occipital pole.[20,21] Other studies have revealed systematic relations between features of the visual stimulus and the visually evoked field.[22–25] We shall not dwell on such subjects here since our purpose is to discuss neuromagnetic source location.

One of the more interesting of our recent findings is illustrated in FIGURE 6. Okada *et al.*[24] employed a contrast-reversal grating pattern to stimulate the right visual field. They measured both the steady-state evoked potential and the steady-state evoked field. When the potential reached its maximum positivity to the right of the midline, current associated with it was flowing from right to left in response to stimulation. If this volume current also produced the evoked field, then the field would emerge lower on the scalp and reenter above. However, as shown in the figure, at this time the field

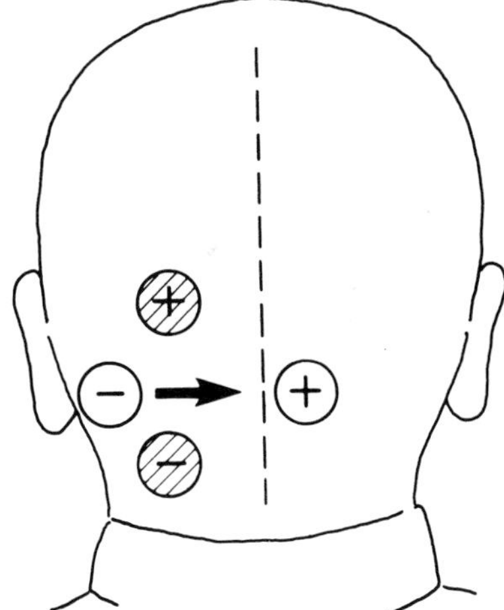

FIGURE 6. Relative polarities of the visual evoked potential (open circles) and field (hatched circles) for a stimulus pattern in the right visual field. The arrow indicates the direction of the current source responsible for the visually evoked magnetic field.

was emerging from *above* the source and reentering below. Therefore, as in the case of the auditory and somatic evoked fields, the current that gives rise to the field flows in the direction shown by the arrow, in opposition to the current underlying the evoked potential. This too is consistent with the theory attributing the evoked field to intracellular currents. Moreover, the source in this case is localizable in the left hemisphere, as would be predicted from the hemifield in which the stimulus is placed.

To gain a clearer idea of the possibilities for source localization inherent in magnetic recording, we present some of the data recently collected by Maclin *et al.*[26] In these experiments a grating of 2 cycles/deg was presented either within a hemicircle with a radius of 1.7 deg or within a semiannulus with an inner radius of 4.4 deg and an outer radius of 6.5 deg. With fixation of the left edge of the hemicircle or of

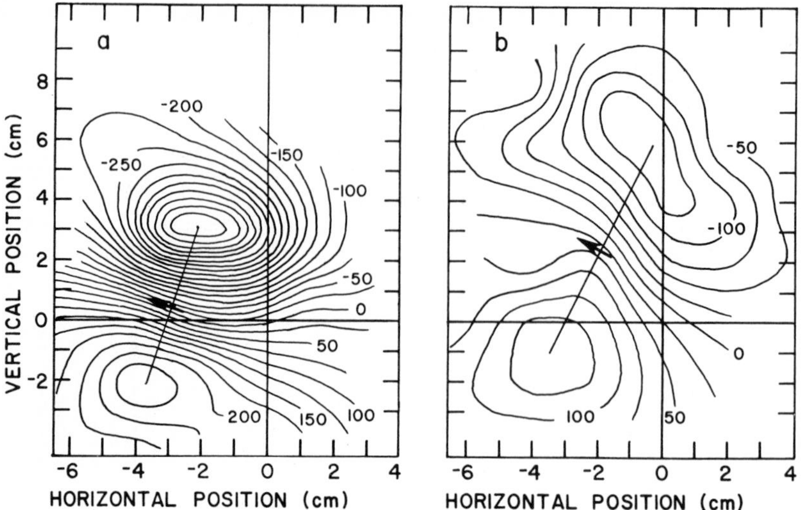

FIGURE 7. Isofield contours for the visually evoked magnetic field measured normal to the scalp for (a) a central stimulus confined to a hemicircle of 1.7 deg radius in the right visual field and (b) a semiannulus of 4.4 deg inner radius and 6.5 deg outer radious. in the right visual field. Vertical position is measured along the midline with the inion serving as the reference. Horizontal position is measured across the scalp toward the right from the midline, in a plane parallel to the plane defined by the inion and ear canals.

the center of the left edge of the semiannulus, the stimuli fell into the right visual field. The hemicircle stimulated the central portion of the hemifield while the semiannulus stimulated a more peripheral region. The contrast-reversal rate of the stimulus was 13 Hz, and this frequency served as the reference signal for measuring the average response at the same frequency. As expected from the basic anatomy of the visual system, the region of the cortex activated by the peripheral stimulus should fall within the longitudinal sulcus while that from the more central stimulus should be at or near the lip of the occipital pole.

The results are shown as isofield contours representing the loci of places that gave equal field amplitudes in response to these stimuli (FIGURE 7). The equivalent current dipole source that best fits the contours lies between the maximum outward and maximum inward fields. Note that with the central stimulus (FIGURE 7a) the dipole is about 3 cm to the left of the midline and for the more peripheral stimulus (FIGURE 7b) it is about 2 cm higher and is closer to the midline. More important is the fact that the distance separating the maxima in the field patterns is greater for the peripheral stimulus than for the central. This separation is not accurately portrayed in FIGURE 7 since the graphs are plotted into rectangular coordinates with the inion represented as the origin. Actual measurements on the scalp show the separation of the maxima for the central stimulus to be 6.5 cm and for the peripheral stimulus 10.5 cm. These values correspond to depths of 3.3 cm and 5.0 cm, thus confirming that the more peripheral stimulus leads to a field generated by a deeper source than does the more central stimulus. This too illustrates one of the more important features of magnetic studies: a directly measured quantity—the separation between field extrema—is approximately twice as large as the depth of the source. Consequently, small changes in depth can be detected with correspondingly greater sensitivity.

VOLUNTARY MOTOR FIELDS

We conclude with a brief account of work by Okada, Williamson, and Kaufman,[27] which is only now being readied for publication. This work is directly related to the study of Gilden, Vaughan, and Costa[10] and Kornhuber and Deecke[28] who succeeded in detecting cortical activity prior to voluntary flexion of the wrist. In our experiment, the subject flexed his forefinger in response to the reversal in contrast of a grating pattern. The EMG recorded during finger flexion served as a reference signal. The pick-up coil of the Dewar was moved along the projection of the central sulcus on the scalp and responses to 100 finger flexions were averaged at each of several positions. The top pair of tracings in FIGURE 8 show the average time variation of the magnetic field normal to the scalp prior to, during, and after motion of the index finger on the contralateral side. Trace A was obtained 21.5 cm in back of the nasion measured along the midline and 16.0 cm above the line joining the ear canal and the eye. Trace B, which was recorded at a lower position, i.e., 10 cm up and 19.5 cm back, displays the polarity inversion typical of those that occur when recordings are made on different sides of an underlying source. The lowermost traces C and D show the electromyogram obtained simultaneously for the flexor extensor muscles. It is apparent that activity in the motor cortex begins about 40 msec prior to the onset of the myogram. Moreover, when the same finger is moved passively by the subject's other hand (on the ipsilateral side), then the early components of the activity of the motor cortex disappear. This preliminary report suggests the possibility that the source of activity of motor cortex prior to finger flexion can be resolved with an accuracy similar to that obtainable with the somatic evoked response.

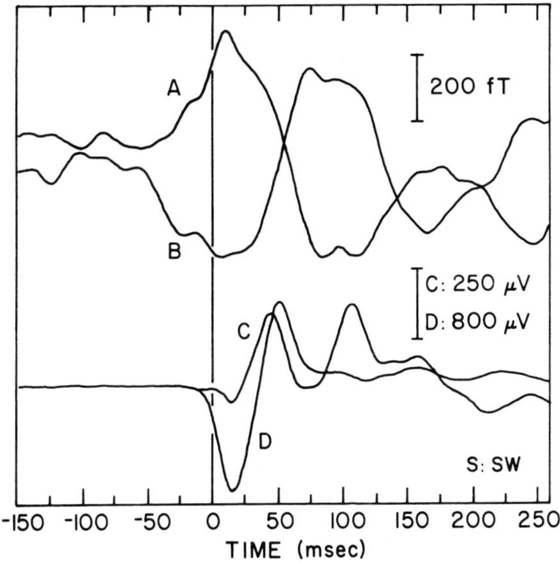

FIGURE 8. Magnetic field over the right hemisphere preceding and during voluntary flexture of the left index finger, with traces A and B obtained at different positions along the central sulcus showing a reversal in field direction. The lower two traces show the simultaneously recorded electromyograms obtained for the C extensor muscle and D flexure muscle.

Conclusions

The main conclusion to be drawn from this review is that the magnetic technique makes it possible to localize sources of evoked fields with an accuracy of a few millimeters. Thus far, this has been true only of sources of cortical responses since subcortical sources have yet to be detected. However, in view of the fact that cortical sources deep inside the longitudinal fissures have been resolved, it is well within the realm of technical feasibility to study the activity of subcortical sources. In any event, it is apparent that the magnetic technique is a useful complement to conventional potential measures since it provides a noninvasive means for identifying relatively unitary sources that contribute to the complex wave known as the evoked potential.

[NOTE ADDED IN PROOF: Recently, Okada, Kaufman, and Williamson[29] observed magnetic activity that is correlated with the P300 complex detected in the event-related potential at the vertex. The field pattern can be interpreted as arising from an equivalent current dipole source in each hemisphere lying deep within the brain in the hippocampal formation. This finding is evidence that the magnetic technique has the capability of determining the location of subcortical activity, as well as the cortical activity described above.]

References

1. Helmholtz, H. 1853. Uber einige Gesetz der Verteilung elektrischer Strome in korperlichen Leitern, mit Anwendung auf die thierischelektuschen Versuche. Ann. Phys. Chem. (Ser. 3) **89:** 211–233, 353–377.
2. Sidman, R. D., V. Giambalvo, T. Allison & P. Bergey. 1977. A dipole localization method for determination of sources of human cerebral evoked potentials. Am. EEG Soc. (Miami).
3. Goff, W. R., P. D. Williamson, J. C. Van Gilder, T. Allison & T. C. Fisher. 1980. Neural origins of long latency evoked potentials recorded from the depth and cortical surface of the brain in man. Prog. Clin. Neurophysiol. **7:** 126–145.
4. Goff, W. R., T. Allison & H. E. Vaughan, Jr. 1978. The functional neuroanatomy of event related potentials. *In* Event Related Brain Potentials in Man, E. Callaway, P. Tueting & S. Koslow, Eds. pp. 1–79. Academic Press, New York, N.Y.
5. Donald, M. W. 1976. Topography of evoked potential amplitude fluctuations. *In* The Responsive Brain. W. C. McCallum & J. R. Knott, Eds. pp. 10–14. J. Wright and Sons, Ltd., Bristol, England.
6. Goff, G. D., Y. Matsumiya, T. Allison & W. R. Goff. 1977. The scalp topography of human somatosensory and auditory evoked potentials. Electroenceph. Clin. Neurophysiol. **42:** 57–76.
7. Grynszpan, F. & D. B. Geselowitz. 1973. Model studies of the magnetocardiogram. Biophys. J. **13:** 911–925.
8. Williamson, S. J. & L. Kaufman. 1981. Evoked cortical magnetic fields. *In* Biomagnetism. S. N. Erné, H. D. Hahlbohm & H. Lübbig, Eds. pp. 353–402. Walter de Gruyter, Berlin.
9. Williamson, S. J. & L. Kaufman. 1981. Biomagnetism. J. Magn. Magn. Mat. **22:** 129–202.
10. Gilden, L., H. G. Vaughan, Jr. & L. D. Costa. 1966. Summated human EEG potentials associated with voluntary movement. Electroenceph. Clin. Neurophysiol. **20:** 433–438.
11. Reite, M., & J. E. Zimmerman. 1978. Magnetic phenomena of the central nervous system. Ann. Rev. Biophys. Bioeng. **7:** 167–188.
12. Farrell, D. E., J. H. Tripp, R. Norgren, & T. J. Teyler. 1980. A study of the auditory evoked field of the human brain. Electroenceph. Clin. Neurophysiol. **49:** 31–37.
13. Picton, T. W., S. A. Hillyard, H. I. Krausz, & R. Galambos. 1974. Human auditory evoked potentials. I: Evaluation of components. Electroenceph. Clin. Neurophysiol. **36:** 179–190.

14. HARI, R., K. AITTONIEMI, M. L. JÄRVINEN, T. KATILA & T. VARPULA. 1980. Auditory evoked transient and sustained magnetic fields of the human brain. Exp. Brain Res. **40:** 237–240.
15. ELBERLING, C., C. BAK, B. KOFOED, J. LEBECH & K. SAERMARK. 1980. Magnetic auditory responses from the human brain. A preliminary report. Scand. Audiol. **9:** 185–190.
16. ROMANI, G.-L., S. J. WILLIAMSON & L. KAUFMAN. 1982. Tonotopic organization of the human auditory cortex. Science (in press).
17. BRENNER, D., J. LIPTON, L. KAUFMAN & S. J. WILLIAMSON. 1978. Somatically evoked fields of the human brain. Science **199:** 81–83.
18. OKADA, Y., R. TANENBAUM, L. KAUFMAN, & S. J. WILLIAMSON. 1982. Projection areas of human primary somatosensory cortex determined by neuromagnetic techniques. (In preparation).
19. KAUFMAN, L., Y. OKADA, D. BRENNER & S. J. WILLIAMSON. 1982. On the relation between somatic evoked potentials and fields. Intl. J. Neurophysiol. (in press).
20. ZIMMERMAN, J. T., N. J. EDRICH, J. E. ZIMMERMAN & M. L. REITE. 1978. The human magnetoencephalographic averaged visual evoked field. *In* Proc. San Diego Biomed. Symp. J. I. Martin & E. A. Calvert, Eds. Vol. 17: 217–221. Academic Press, New York, N.Y.
21. BRENNER, D., S. J. WILLIAMSON & L. KAUFMAN. 1975. Visually evoked magnetic fields of the human brain. Science **190:** 480–482.
22. TEYLER, T. J., B. N. CUFFIN & D. COHEN. 1975. The visual evoked magnetoencephalogram. Life Sci. **17:** 683–692.
23. WILLIAMSON, S. J., D. BRENNER & L. KAUFMAN. 1978. Biomedical applications of SQUIDs. A.I.P. Conf. Proc. **44:** 106–116.
24. OKADA, Y., L. KAUFMAN, D. BRENNER & S. J. WILLIAMSON. 1982. Spatial and temporal modulation transfer functions of the human visual system revealed by visually evoked magnetic field. Vision Res. (in press).
25. BRENNER, D., Y. OKADA, E. MACLIN, S. J. WILLIAMSON & L. KAUFMAN. 1981. Evoked magnetic fields reveal different visual areas in human cortex. *In* Biomagnetism. S. N. Erné, H. D. Hahlbohm, and H. Lübbig, Eds. pp. 431–444. Walter de Gruyter, Berlin.
26. MACLIN, E., Y. OKADA, S. J. WILLIAMSON & L. KAUFMAN. Topography of human visual response revealed by neuromagnetic measurements. (In preparation).
27. OKADA, Y., S. J. WILLIAMSON & L. KAUFMAN. 1982. Magnetic field of the human sensorimotor cortex. (Submitted for publication).
28. KORNHUBER, H. & L. DEECKE. 1965. Hernpotentialanderungen bei Willkeurbewegungen und passiven Bewegungen des Menschen; Bereitschaftspotential und Potentiale. Pflugers Arch. Ges. Physiol. **284:** 1–17.
29. OKADA, Y. C., L. KAUFMAN, & S. J. WILLIAMSON. 1982. Hippocampal formation as a source of endogenous slow potentials. (Submitted for publication).
30. CUFFIN, B. N. & D. COHEN. 1977. Magnetic fields of a dipole in special volume conductor shapes. IEEE Trans. Biomed. Eng. BME-**24:** 372–381.

SIMULTANEOUS PATTERN-REVERSAL ELECTRORETINOGRAMS AND VISUAL EVOKED POTENTIALS IN DISEASES OF THE MACULA AND OPTIC NERVE

Jerome Sherman

State University of New York
State College of Optometry
University Optometric Center
New York, New York 10010

INTRODUCTION

Nearly two decades ago, Vaughan and Katzman[1] reported to this Academy that electroretinograms (ERGs) from the right and left eye and visual evoked potentials (VEPs) from the right and left occipital cortex could be obtained simultaneously to help localize the site of involvement in patients with various visual disorders. Since the stimulus employed was an unstructured bright flash, the resultant ERGs and VEPs were rather easily recordable but could not be expected to reflect subtle deficits in visual system functioning. Nevertheless, the results of their four-channel electrophysiological recordings were uniformly in agreement with the results from other neurological and radiological procedures.

Although the VEP elicited by flash still has clinical utility, numerous more recent studies have demonstrated that the VEP elicited by a pattern is a more sensitive indicator of subtle visual system dysfunction. Both types of VEPs arise from the central retina (primarily the macula) and related pathways, but the VEP elicited by an unstructured flash is generated from a somewhat larger retinal area. In contrast, the ERG to pattern and the ERG to flash originate from vastly different retinal loci as discussed below.

To the clinician, the macula corresponds to the central 5° of the posterior pole and if it is assumed that the functional retina extends about 200° in both the horizontal and vertical meridians, then the macula comprises only 5/200 or 1/40 of the linear dimension in each meridian. The area of the macula as a percentage of the total area of the retina is, therefore, approximately $(1/40)^2$ or less than one-tenth of one percent. The standard full-field flash ERG, typically recorded with a scleral lens which degrades optical imagery, reflects overall retinal activity. The macula, comprising such a small area of the retina, contributes little to the total ERG response. It is not surprising, therefore, that patients with advanced macular disease and very reduced visual acuity still generate normal or near normal flash ERGs.

Instead of using a bright flash, local ERGs (from the macula, for example) can be recorded. If a TV monitor is used to produce a pattern-reversal checkerboard, stimulation can be limited to the central retina. It seems reasonable to assume that, in patients with macular degeneration, such a pattern-reversal ERG (PERG) will be grossly abnormal while the standard flash ERG will remain unaffected. With the recent introduction of the gold foil electrode (by Arden *et al.*[2]) and the fiber electrode (by Dawson *et al.*[3]), both of which do not disturb the optical properties of the eye, pattern-reversal ERGs can now be recorded in the clinical environment.

Recalling Vaughan and Katzman's simultaneous ERGs and VEPs to flash, one

0077–8923/82/0388–0214 $1.75/0

might be tempted to utilize the gold foil or the fiber electrode and record simultaneous ERGs and VEPs to pattern reversal. Potentially, simultaneous PERGs and PVEPs might be an additional procedure of great value for the differential diagnosis of visual disorders in patients presenting with marked symptoms but with no observable ocular abnormality.[4]

For the last two years, we have performed simultaneous PERGs and PVEPs utilizing a clinical averager (a Nicolet CA-1000, see below, Methods) as a clinical tool in cases where traditional testing failed to reveal the etiology of the patients' complaint. Subsequent examinations (including neurological, radiological, and ophthalmological procedures) have usually revealed the site of involvement in these cases. Our experience over this two-year period has been that both the PERG and PVEP were abnormal in patients who were eventually diagnosed as having macular disease but that *only* the PVEP was affected in optic nerve disorders.[5,6] Since it is well established that the flash ERG reflects activity of the photoreceptor layer and the inner nuclear layer (probably from the Muller cells[7] in response to extra cellular potassium[8]) but not of the ganglion cells layer, it was not surprising that the PERG was normal in patients with optic nerve disease. (In carefully controlled studies, the ERG does appear to be reduced in amplitude in some cases of optic nerve disease, but these decrements are minor in degree.[9])

Recently, however, Maffei and Fiorentini[10] have argued that the PERG, unlike the flash ERG, might show some sign of activity of ganglion cells, since the pattern-reversal checkerboard is a much more effective stimulus for ganglion cells than is unpatterned light. After unilateral section of the optic nerve in cat, Maffei and Fiorentini[10] found that a previously recordable PERG completely disappeared about four months after the section whereas the flash ERG remained unaltered. Based upon these findings, which were repeated in seven cats, they suggested that ganglion cell activity is the main source of pattern-reversal ERG. Furthermore, they "expect that in patients with ganglion cell degeneration resulting either from compression of the optic nerve or retina, or from chronic optic nerve neuritis, the pattern reversal ERG should be absent or abnormal." If Maffei and Fiorentini are correct, both the pattern-reversal VEP and ERG will be abnormal in patients with *either* optic nerve disease *or* macular disease. Hence, simultaneous pattern-reversal ERGs and VEPs would be of little value in the differential diagnosis of retinal versus optic nerve disorders.

Spekreijse, Estevez, and Van der Tweel[11] have argued that the PERG is the result of the addition of focal ERGs, i.e., ERG components directly related to local luminance change. In an unpublished manuscript, Vaegan[12] compared the PERG to focal ERGs under a wide variety of parameter variations and concluded that the PERG cannot be a straightforward sum of focal ERG components.

The PERG has been investigated in patients with amblyopia by Sokol *et al.*[13] and Vaegan *et al.*[14] These studies suggest that the PERG is often abnormal in amblyopia although the site of the defect was not established.

This report was initially intended to summarize the results, which have been accumulated over the past two years, of simultaneous PERGs and PVEPs in patients with visual loss and no obvious ocular or visual abnormality. However, as a consequence of Maffei and Fiorentini's recent findings in cat, it was decided to report the results obtained from a small group of patients with either macular or optic nerve disease where the diagnosis was unquestionable. If patients with well-documented optic nerve disease generate normal PERGs, then the main source of the PERG must be preganglionic. Furthermore, such findings would validate the use of simultaneous PERGs and PVEPs for the differential diagnosis of macular versus optic nerve disease.

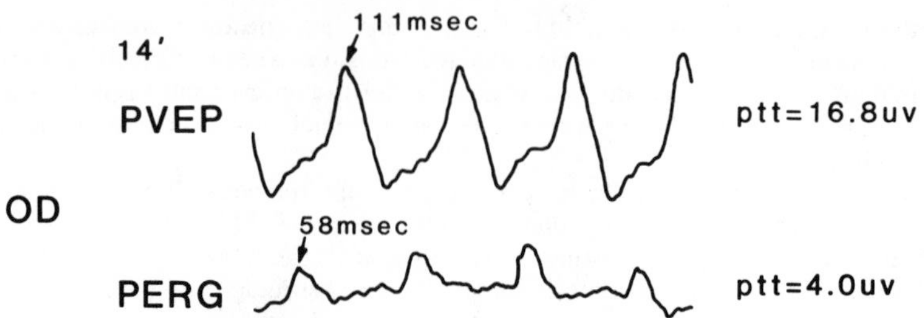

FIGURE 1. Simultaneous pattern-reversal electroretinograms (PERGs) and pattern-reversal visual evoked potentials (PVEPs) recorded from a normal subject to a checkerboard stimulus pattern (with square sizes of 14 minutes) reversing at a rate of 7.5/sec. Positive polarity is up and the analysis time is 500 msec in this and all other traces. The peak times and peak-to-trough (ptt) amplitudes are within normal limits as discussed in the text. Ptt values represent the mean of the four responses shown.

METHODS

Pattern-reversal ERGs and VEPs were recorded simultaneously using a clinical averager (Nicolet CA-1000) and a visual stimulator (Nicolet 1005). Either a gold foil electrode[2] or a fiber electrode[3] was used to record the ERGs and corneal touch was insured throughout the trial by careful monitoring. A gold cup electrode (Grass) was used as the ERG reference (affixed above the eyebrow of the eye being tested) and a second gold cup electrode was used 1 cm above the inion along the midline to record the (monopolar) VEP. An ear clip electrode was used for the VEP reference and a second ear clip for the common ground. Electrode impedence was maintained below 6000 kΩ. In addition to the averaging function, the CA-1000 incorporates an artifact rejection buffer which was utilized to reject any artifacts greater than about 50 μV which might be created by blinks or large eye movements.

Squares of 14, 18, 28, 36, 56, or 112 minutes of arc in a checkerboard pattern were pattern-reversed at 7.5/sec. The stimulus distance was either 0.5, 0.75, or 1 meter. The overall field size at 1 m was 12° vertical × 15° horizontal and proportionately larger at the shorter viewing distances. Mean luminance was maintained at 130 cd/m^2 and contrast $(L_{max} - L_{min})/(L_{max} + L_{min})$ at 92%. Bandpass filters were set at 1–30 Hz. All patients were optically corrected for the stimulus distance, and mydriatics and cycloplegics were not utilized. The analysis time was set at 500 msec and between 100 and 400 responses were averaged for each trial.

RESULTS

Normals

FIGURE 1 depicts typical simultaneous PERGs and PVEPs from a patient with normal vision to a 14-min pattern-reversed checkerboard (see Methods for details). Positive polarity is up in this and all other traces. The peak time (the time from the onset of the pattern reversal until the first maximum response) of the PERG was shorter than the peak time of the PVEP, but the amplitude of the PVEP was considerably larger. Normative values to the 14-min checkerboard were calculated: PERG peak time, 62.1 ± 4.0 msec; PERG peak-to-trough (ptt) amplitude, 3.0 ± 0.5

μV; PVEP peak time, 113.2 ± 6.7 msec; and PVEP peak-to-trough amplitude, 16.9 ± 4.5 μV. Although normative data have not been compiled for larger squares, it appeared that the PERG and PVEP peak time occurred slightly earlier and the amplitude of the PERG was slightly larger whereas the PVEP amplitude was somewhat reduced.

Macular Disease

As expected, both the PERG and the PVEP were abnormal in patients with macular involvement. In mild macular disease with nearly normal visual acuity, the PERG and PVEP were somewhat reduced in amplitude and occasionally slightly delayed. In gross macular pathology, the PERG and PVEP were extinguished to small squares (i.e., 14 and 28 min) and very reduced or extinguished with large squares. The following two cases of patients with monocular macular disease exemplify the typical PERG and PVEP findings in normal and affected eyes.

Case 1

An 18-year-old black female presented with a history of poor vision in the right eye since age 4. Best corrected visual acuity was 5/120 in the right eye and 20/20 in the left eye. In the right eye a central macular lesion, approximately 10° × 14°, was observed (FIGURE 2).

The PERG from the right eye was flat to 28-min squares and reduced in amplitude and poorly formed to 56-min squares. The PVEP was flat to 28 and 56-min squares. FIGURE 2 contrasts the abnormal PERGs and PVEPs from the right eye to the normal responses from the left eye. The flash ERG was normal in the right eye, as expected in macula disease, and the VEP to flash was nearly extinguished.

Case 2

A 30-year-old hispanic female presented with a history of reduced vision in the left eye since birth. Best corrected visual acuity was measured at the 20/20 level in the right eye and 20/70 in the left. The patient had natural black hair and green eyes. Both irides transilluminated equally. Although the right fundus was normal, no foveal or macular landmarks were observed in the left fundus (FIGURE 3).

Testing of the left eye revealed a flat PERG to 28- and 56-min squares. The PVEP was flat to 7- and 14-min squares and reduced in amplitude and delayed to 28- and 56-min squares. FIGURE 3 contrasts the normal simultaneous PERGs and PVEPs from the right eye to the abnormal responses from the left eye in this patient with presumed monocular albinism.[15]

Optic Nerve Disease

Simultaneous PERGs and PVEPs were recorded from a group of patients with a wide variety of optic nerve disease. Hereditary, congenital, traumatic, inflammatory, and ischemic etiologies were encountered. Patients with macular or other retinal disease, in addition to optic nerve disease, will not be reported here since the

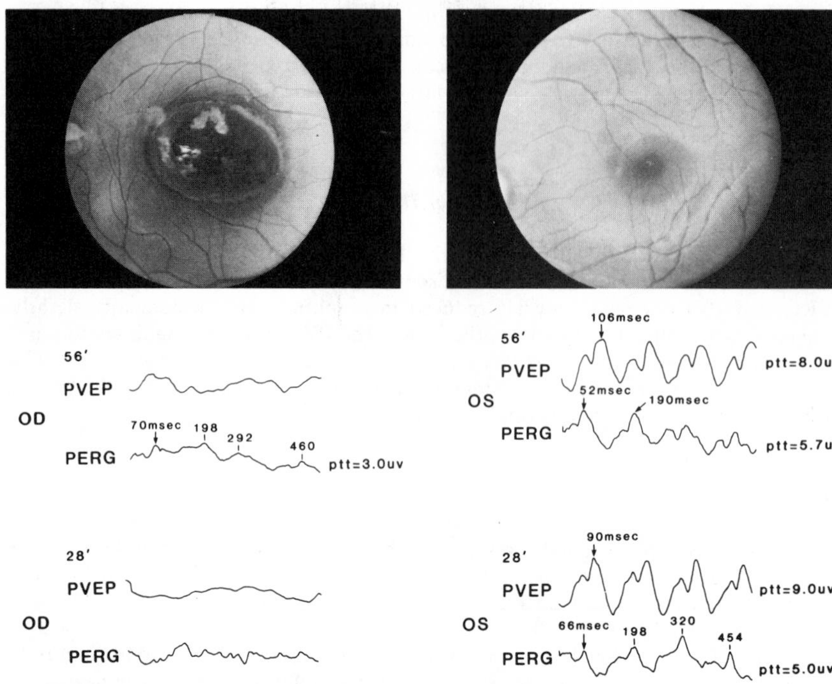

FIGURE 2. The macular lesion in the right eye as depicted above measured about 10° × 14° (the white spot in both fundus photos is a photographic artifact). Note the abnormal PERGs and PVEPs from the right eye as contrasted to the normal responses from the left eye. Except for the square sizes of 28 min and 56 min, the conditions are the same as in FIGURE 1. Since the pattern-reversal rate is 7.5/sec, the positive peaks should be about 133 msec apart.

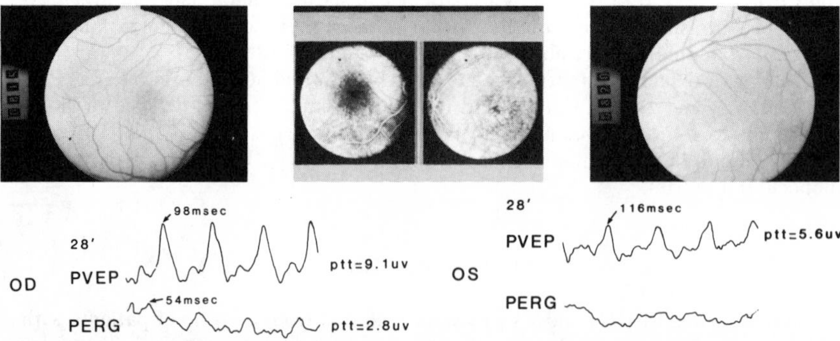

FIGURE 3. Although the right fundus appeared normal, no foveal or macular landmarks were observed in the left. Fluorescein angiography revealed abnormal macular hyperfluorescence in the left eye. Note the flat PERG and reduced and delayed PVEP from the left eye to the 28-min checkerboard stimulus. See text for more details.

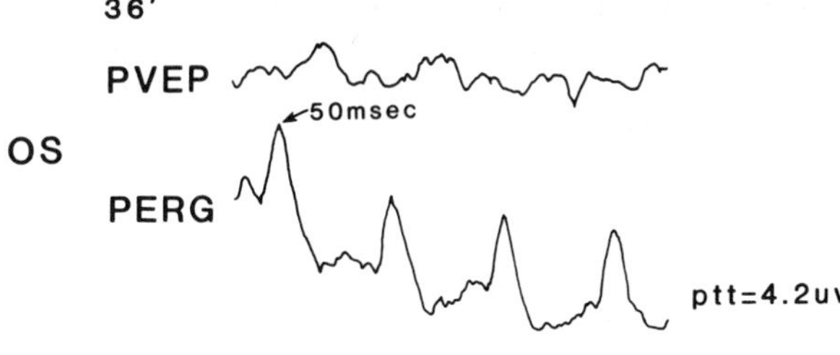

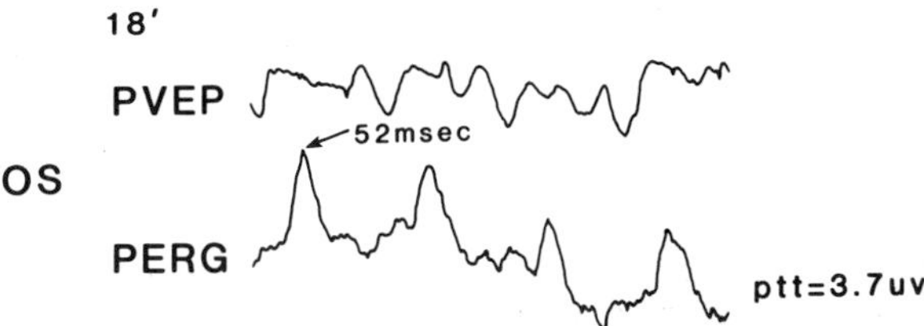

FIGURE 4. Patient has bilateral optic nerve disease. Note the flat PVEPs but the normal PERGs to squares of 18 min and 36 min. Visual acuity in the left eye was 20/200.

interpretation of the results in such cases becomes questionable. In the seven patients studied with obvious optic nerve disease without macular involvement, the PERGs were normal or near normal while the PVEPs were extinguished (or in one case, quite delayed).

Case 1

A 34-year-old white female presented with a history of poor vision, probably since birth. No family history of visual problems was reported. The best corrected visual acuities have remained at the 20/70 level in the right eye and 20/200 in the left eye during the past 10 years. Slight nystagmus, temporal pallor, a minor color vision defect, and visual fields reduced to within ten degrees, were noted in both eyes on routine examination. The flash ERG was normal, thereby ruling out any overall retinal dysfunction. The specific etiology of this optic nerve disease is uncertain.

Simultaneous PERGs and PVEPs were performed to the left eye only. The PERGs to 18- and 36-min checks appear normal whereas the PVEPs to the same check sizes are extinguished (FIGURE 4.)

Case 2

A 54-year-old white male presented with a history of reduced vision since age 4 or 5. A detailed history revealed an extensive pedigree of optic nerve disease consistent with a diagnosis of autosomal dominant optic atrophy.[16] Best corrected visual acuity was 10/180 in the right eye and 10/140 in the left eye. Temporal pallor was observed in both eyes.

Simultaneous PERGs and PVEPs were recorded through the left eye to squares of 14, 28, 56, and 112 min of arc. The PERGs appeared normal to all 4 square sizes although the PVEPs were extinguished. When a +38 diopter lens was placed in front of the left eye, the PERG was not recordable (FIGURE 5).

Case 3

A 25-year-old white female presented with a history of retrobulbar neuritis in her left eye which occurred 2 years earlier. Visual acuity was measured at the 20/20 level although the previous record indicated a reduction to 20/50 during the attack. Temporal pallor of the left optic disc was observed.

Simultaneous PERGs and PVEPs were obtained through the left eye to 14-min squares. The PERG was normal in amplitude and peak time. Although the PVEP amplitude was normal, the response was delayed by about 45 msec (FIGURE 6).

Case 4

A 19-year-old black male presented with a history of trauma to the right side of his head 4½ years earlier. Vision began to decrease in the right eye shortly thereafter. Visual acuity was measured to be no light perception (NLP) in the right eye and

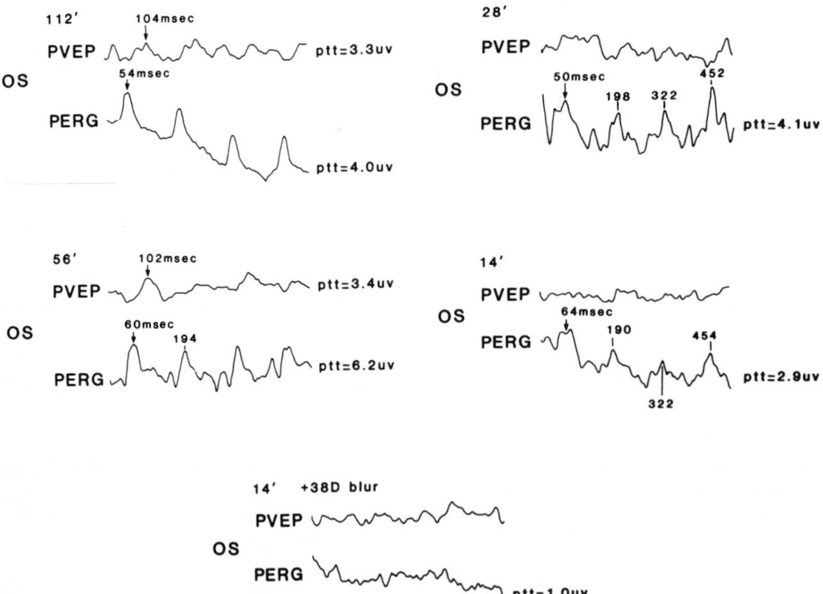

FIGURE 5. Patient has a history of bilateral optic nerve disease for 50 years. Simultaneous PERGs and PVEPs from the left eye (visual acuity = 10/140) are shown to 112-, 56-, 28-, and 14-min squares. Note the normal (or near normal) PERGs and the flat PVEPs to the four square sizes tested. Below, the PERG was flat when a +38 diopter lens was introduced. See Discussion in text concerning the condition with induced blur.

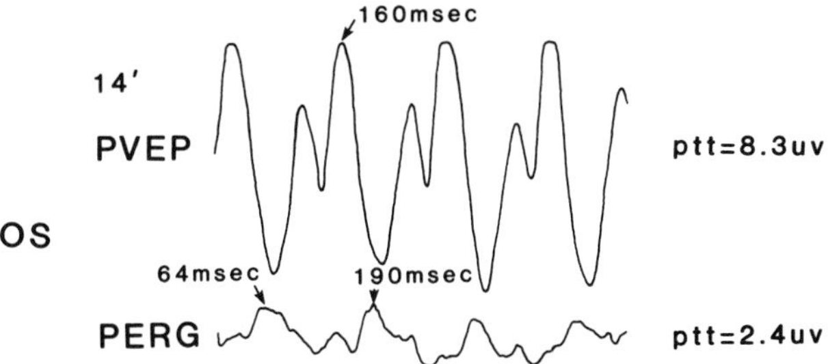

FIGURE 6. Patient had retrobulbar neuritis in the left eye two years ago, but now the VA is 20/20. Note the normal PERG and the delayed PVEP. The PVEP delay was essentially the same with a 1.88 reversal/sec stimulus (not shown).

20/20 in the left eye. Light perception was not demonstrable through the right eye even with the highest intensity focused light of the binocular indirect ophthalmoscope. A blink reflex was absent when the left eye was occluded. On observation, the right eye was noted to be exotropic although the degree of turn would vary from moment to moment. No direct pupillary reflex was noted in the right eye, even with the bright light and 10× magnification of the biomicroscope. The indirect pupillary reflex to the right eye was intact. Ophthalmoscopy of the right eye revealed a totally white optic disc (FIGURE 7), no small vessels traversing the disc border, and no nerve fiber layer with red-free light. The rest of the fundus was unremarkable. The fundus of the left eye was well within normal limits.

Simultaneous PERGs and PVEPs were obtained through the right eye to 28- and 56-min squares and through the left eye to 28-min squares. Since the patient could not fixate with the right eye, one of two methods was used. First, PERGs and PVEPs to stimulation of the right eye were obtained with left eye fixation at such a point in space to the left of the TV monitor so as to center the reflection of the checks in the pupil of the right eye. Alternatively, a mirror was introduced in front of the left eye to insure fixation through the left eye at a distant point. With either of the two methods, appropriate stimulation of the right eye was accomplished, although with some difficulty because the angle of strabismus would vary somewhat unpredictably.

The PERGs from the right eye to both 28- and 56-min squares appeared normal although the PVEPs were extinguished. Both PERGs and PVEPs to 28-min squares from the left eye were normal (FIGURE 7). When PERGs and PVEPs were repeated from the right eye to 28-min squares with a +40 diopter lens, both the PERG and PVEP were extinguished. As anticipated in a blind eye, the VEP to flash of the right eye was extinguished.

The results from three other patients with optic nerve disease (ischemic optic neuropathy,[17] Lebers Optical Atrophy,[18] and autosomal dominant optic atrophy) were similar to the findings in the four cases presented above. Grossly abnormal PERGs have not been found in any patient with optic nerve disease as the sole pathology, although mild reductions in amplitude were sometimes encountered.

DISCUSSION

In order to support the contention that simultaneous PERGs and PVEPs can be utilized for the differential diagnosis of macular versus optic nerve disorders, PERGs must be normal (or near normal) in patients with confirmed optic nerve involvement. Of the seven cases of optic nerve disease thus far studied, normal or near normal PERGs were found in all seven. Probably the most convincing clinical evidence that the PERG is preganglionic in origin is the recording of normal or near normal PERGs in a patient with long-standing, undeniable optic nerve disease. For example, in case 4, the diagnosis of optic nerve disease in the right eye was unquestionable, based upon no

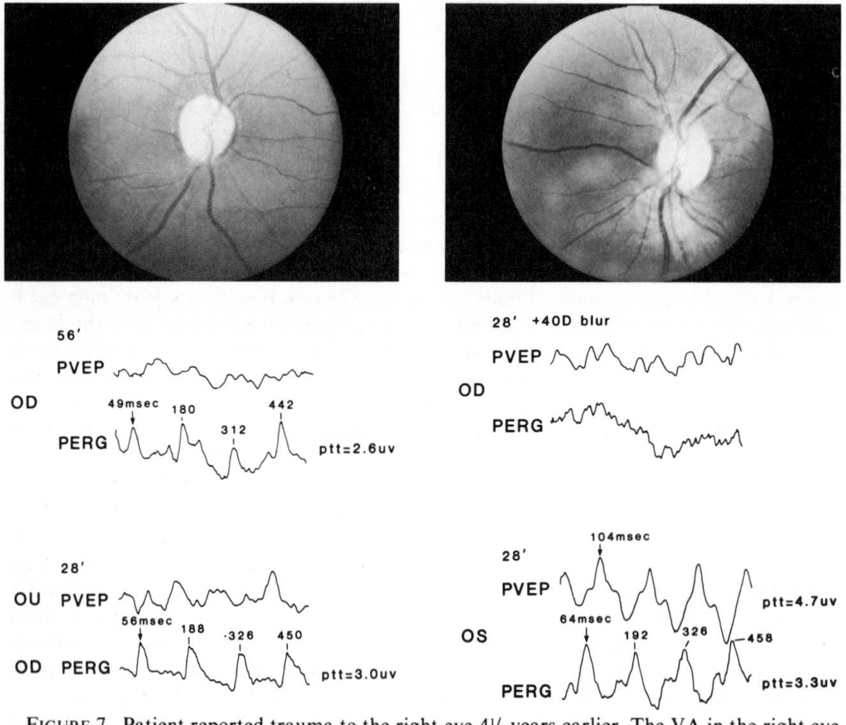

FIGURE 7. Patient reported trauma to the right eye 4½ years earlier. The VA in the right eye was NLP and the optic disc was white. Note the difference in optic nerve head color and the absence of nerve fiber layer reflexes from the right fundus. The PERGs from the right eye to 28- and 56-min squares were normal or near normal. (When the 28-min PERG from the right eye was recorded, the left eye was used to insure fixation as described in text. Hence, the "OU PVEP" notation.) As in FIGURE 5, the PERG to blur was flat. The normal PERGs and PVEPs from the left eye to 28-min squares are shown for comparison. The peak-to-trough (ptt) amplitudes shown represent the mean of the four responses in this and the previous figures. The trough immediately preceding the peak was chosen for the ptt amplitude measures.

subjective awareness of even the brightest light, no direct pupillary reflex, a chalk white optic disc, no observable nerve fiber layer and a flat VEP to pattern and to flash. Since the PERG was normal, the electrical source responsible for its generation must be preganglionic.

An argument can be made that the ERGs recorded in this study were the result of mean luminance flicker which is sometimes encountered in inexpensive TV monitors.

If the ERGs presented here were due to flicker and not pattern, then the introduction of a high plus lens, which greatly degrades the pattern but does not affect any flicker, should result in normal ERGs. However, when a high plus lens was used, the ERGs were extinguished, demonstrating that the response must be to the pattern reversal.

Upon casual observation, the normal PERG depicted in FIGURE 1 might be considered to be a far field potential which represents an attenuated and inverted PVEP. However, the PERG persisted even when the PVEP was extinguished (FIGURES 4, 5 & 7). The possibility that a normal PERG recorded from an eye with optic nerve disease was generated by stimulation of the normal fellow eye was eliminated by simple occlusion experiments. Moreover, four of the seven patients with optic nerve disease and normal PERGs had bilateral involvement. Another point of contention might be that the PERG represents a photic-induced electrode potential, but of course the extinguished PERG in macular disease (FIGURES 2 & 3) is not supportive of this notion.

When performing simultaneous PERGs and PVEPs in normals, the PVEP is usually recognizable from noise after about 10 averages whereas the PERG typically requires about 50. The relatively small amplitude of the PERG and its relatively unfavorable signal-to-noise ratio are probably responsible for a PERG which is typically extinguished early in macular disease in spite of the persistence of the PVEP. The responses from the left eye with a congenital macular abnormality depicted in FIGURE 3 illustrate an extinguished PERG and a recordable (although reduced and delayed) PVEP.

As discussed by Maffei and Fiorentini,[10] ganglion cell degeneration does not occur immediately after optic nerve compromise. They sacrificed their cats after the last recording session and later observed that the whole-mounted retinas revealed degeneration of the ganglion cell layer four months after the optic nerve section. (Recall that the PERGs also disappeared completely four months after the section.) In this present clinical study, since normal or near normal PERGs were recorded from patients with long histories of optic nerve disease (approximately 50 years in case 2), one is hard pressed to argue that ganglion cell degeneration has not yet occurred.

The results reported by Maffei and Fiorentini in cats with surgically severed optic nerves are apparently not generalizable to humans with less traumatic optic nerve disorders. Perhaps severing the optic nerve in a cat leads to retrograde transneuronal degeneration of preganglionic structures and such degeneration does not occur in the types of human optic nerve pathology reported in this present study. If we assume for a moment that the results reported by Maffei and Fiorentini and the results reported here are replicated by others, than at least two explanations must be entertained: First, retrograde transneuronal degeneration may have occurred in the cats but was limited to the area centralis. This might explain the extinguished PERG. The normal ERG to flash and to flicker might have resulted from stray light, since a bright surround typically used to minimize stray light effects was apparently not utilized by Maffei and Fiorentini. Second, the main source of the PERG is neither Muller cells (because the flash ERG, generally attributable primarily to the Muller cells, was normal) nor the ganglion cells. Further clinical and experimental evidence suggesting that the Muller cell is not the main source of the PERG will be presented at this meeting by Arden and Vaegan.[19] Although they have argued that the PERG is generated at or near the ganglion cell layer, this present clinical study certainly does not implicate the ganglion cell layer. Another possible source mentioned by Arden and Vaegan are the amacrine cells although no direct evidence has yet been furnished.

Although the results of this preliminary research clearly indicate a preganglionic source of the PERG, further, more detailed study is necessary to confirm if mild to moderate abnormalities of the PERG are typically encountered in optic nerve disease as reported by Arden and Vaegan.[19] Abnormality of the PERG would not be surprising since even the flash ERG is mildly abnormal in some eyes with optic nerve disease and might reflect retrograde transneuronal degeneration.[9]

SUMMARY

Although it is well established that the flash electroretinogram (ERG) is preganglionic in origin, conflicting evidence exists concerning the main source of the pattern-reversal ERG (PERG).

In this study, PERGs were recorded simultaneously with visual evoked potentials (PVEPs) in a group of patients and controls. In those patients with obvious macular disease, both the PERG and PVEP were reduced, whereas the standard flash ERG was normal. This result is consistent with either a preganglionic or ganglion cell origin of the PERG.

In those patients with obvious optic nerve disease ($n = 7$), normal or near-normal PERGs were recordable even when PVEPs were completely absent. If the PERG reflects ganglion cell activity, one would anticipate *both* abnormal PVEPs *and* PERGs in these patients.

These results suggest a preganglionic origin of the PERG in humans.

ACKNOWLEDGMENT

Victor Richardson has been of great technical assistance in the construction of the gold foil electrodes and in performing simultaneous PERGs and PVEPs.

REFERENCES

1. VAUGHAN, H. G. & R. KATZMAN. 1964. Evoked responses in visual disorders. Ann. N.Y. Acad. Sci. **112:** 305–319.
2. ARDEN, G. B., R. M. CARTER, C. HOGG, I. M. SIEGEL & S. MARGOLIS. 1979. A gold foil electrode: Extending the horizons for clinical electroretinography. Invest. Ophthalmol. Vis. Sci. **18**(4): 421–426.
3. DAWSON, W. W., G. L. TRICK & C. A. LITZKOW. 1979. Improved electrode for electroretinography. Invest. Ophthalmol. Vis. Sci. **18**(9): 988–991.
4. SHERMAN, J., S. J. RICHTER & A. EPSTEIN. 1980. The differential diagnosis of visual disorders in patients presenting with marked symptoms but with no observable ocular abnormality. Am. J. Optom. Physiol. Opt. **57**(8): 516–522.
5. SHERMAN, J. 1980. Electrophysiological assessment of macular integrity in suspected macular disease. ARVO abstracts. Invest. Ophthalmol. Vis. Sci. (Suppl.): 93.
6. SHERMAN, J., S. J. BASS & V. RICHARDSON. 1981. The differential diagnosis of macular vs. optic nerve disease. J. Am. Optom. Assoc. **52**(7): 933–939.
7. MILLER, R. F. & J. E. DOWLING. 1970. J. Neurophysiol. **33:** 323–341.
8. DICK, E. & R. F. MILLER. 1978. Brain Res. **154:** 388–394.
9. IKEDA, H., K. E. TREMAIN & M. D. SANDERS. 1978. Neurophysiological investigation in optic nerve disease: Combined assessment of the visual evoked response and electroretinogram. Br. J. Ophthalmol. **62:** 227–239.
10. MAFFEI, L. & A. FIORENTINI. 1981. Electroretinographic responses to alternating gratings before and after section of the optic nerve. Science **211:** 953–955.
11. SPEKREIJSE, H., O. ESTEVEZ & L. M. VAN DER TWEEL. 1972. Luminance responses to pattern reversal. Doc. Ophthalmol. (Proc. Series) **2:** 205–211.

12. VAEGAN. Pigeon pattern electroretinograms differ from focal electroretinograms. Unpublished manuscript.
13. SOKOL, S. & D. NADLER. 1979. Simultaneous electroretinograms and visually evoked potentials from adult amblyopes in response to a pattern stimulus. Invest. Ophthalmol. Vis. Sci. **18**(8): 848–855.
14. VAEGAN, G. B. ARDEN, C. R. HOGG & L. HOLDEN. Neural basis of pattern electroretinograms: Reduction in amblyopia and optic nerve disease. Unpublished manuscript.
15. NATH, S. & J. SHERMAN. 1981. Isolated foveal hypoplasia is a variant of ocular albinism. ARVO Abstract. Invest. Ophthalmol. Vis. Sci. (Suppl.): 237.
16. SMITH, V. C., J. POKORNY & J. T. ERNEST. 1977. Primary hereditary optic atrophies. *In* Hereditary Retinal and Choroidal Diseases Vol. II. Clinical Characteristics. A. E. Krill, Ed. p. 1117. Harper and Row Publishers, Hagerstown, Md.
17. SHERMAN, J. & S. J. BASS. 1980. An atypical case of optic nerve disease. Rev. Optom. **117**(12): 31–37.
18. SHERMAN, J. & I. BODIS-WOLLNER. VEPs In Hereditary Optic Atrophy. In preparation.
19. ARDEN, G. B., VAEGAN & C. R. HOGG. 1982. Clinical and experimental evidence that the pattern ERG (PERG) is generated by the innermost retinal layers. Ann. N.Y. Acad. Sci. This volume.

DISCUSSION OF THE PAPER

H. SPEKREIJSE: When the pattern ERG can be generated, that does not tell you whether you are dealing with a pattern-generated, contrast-generated response or a luminance-generated one.

J. SHERMAN: Agreed.

SPEKREIJSE: Of course you did the control for the overall luminance variation and that is fine. However, I observed that the amplitude of the response for the 28- and 56-minute checks remained the same. That is exactly what you would expect if it is just a local luminance variation.

SHERMAN: There is a difference in the size of the responses to the various size checks. In our hands, using the five check sizes noted, the largest responses are usually found with either the 56-minute check or the 28-minute check. We certainly do not get the largest PERG with the 14-minute checks as we typically do with the pattern VEP, but we do get a definite curve when we plot check size versus amplitude.

L. MAFFEI: I think that one must be very cautious to go from clinical experiments to some physiological deduction. I mean one should do all experiments that are needed before a conclusion is reached, and I think controls about local luminance and spatial frequencies probably are needed. I would like to say that the speaker agrees and we are going to do some experiments to make sure that such an important point is valid or not.

SHERMAN: Drs. Maffei and Bodis-Wollner and I are planning to run a few additional experiments (with this patient with no vision in one eye and normal vision in the other eye). We will utilize even higher spatial frequencies and substitute sign-wave gratings for the checkerboard.

VAEGAN: The pattern ERG recorded in our hands is approximately 7 msec faster than that normally recorded by Dr. Sherman. Also, we always record the responses from both eyes simultaneously and calculations of loss are usually based on the ratio difference between the two eyes. We have never seen a total loss except in one case of base of skull fracture where the fixation was very poor, and so we are not confident of the tracings.

Now I think there may be differences between the gratings and the checkerboard, but the difference of 7 msec we have here between the two laboratories is not a trivial one. Seven milliseconds is a shifting peak time which I cannot produce under any manipulation of the stimulus I have so far tried except for two. One of these is a rather trivial one of luminance and a shift of 7 msec would require something like half a log unit of shift of luminance. The other thing is an increase in spatial frequency at below an eighth of a degree check. These are quite substantial differences, and I think it will remain for other workers to find out which sort of responses they get and how this difference relates to the difference between the clinical result of the two laboratories.

THE EFFECTS OF VISUAL FIELD CHANGES AND OCULAR HYPERTENSION ON THE VISUAL EVOKED POTENTIAL

G. Bartl

University Eye Hospital
A-8036 Graz
Austria

INTRODUCTION

A diminution of the sensitivity of the visual field caused by damage to nerve structures in the visual pathway may lead to changes in the visual evoked potentials (EP) such as a decrease in amplitude and/or an increase in latency. Both the topographical position and the dimension and degree of the diminished sensitivity of the visual field are important for changes in the EP. The nearer to the center the visual field defect is localized, the larger the changes of the EPs are expected to be.[1] Thus, for example, a small relative scotoma located near the center may effect significant changes on the EP while a large absolute scotoma in the periphery may cause only minor changes in the EP. The visual field defect in chronic glaucoma occurs mostly paracentrally as an insignificant relative scotoma.[2] Acute glaucoma, going hand in hand with a diminution of the sensitivity in the peripheral visual field, is usually associated with a central scotoma.[2] According to current opinion, the glaucoma represents a sickness in which damage of nerve structures and defects in the visual field are caused by a diminished blood supply.[3-6] This is based on a disproportion between the intraocular pressure and the blood pressure of the ophthalmic artery, especially the peripapillary choroid, supplying the prelaminary part of the optic nerve (FIGURE 1). In consequence, depending on the atrophy of optic nerve fibers, a quantitative restriction in the conduction of signals to the visual cortex arises together with a decrease of the EP. Therefore both types of glaucoma cause one to expect changes in the EP whereby, from the electro-ophthalmological view, the following questions may be of interest: Firstly, which nerve structures are affected in their functions and how do visual field defects caused by glaucoma influence the EP? Secondly, which preoperative prognosis for the visual acuity produced by the EP can be given to patients who have a dense cataract in addition to the glaucoma? Thirdly, does a decrease of intraocular pressure in chronic and acute glaucoma influence the EP? And fourthly, what is the behavior of the EP in normal and glaucomatous eyes under experimentally elevated intraocular pressure?

METHOD

Photopic electroretinograms (ERG), luminance evoked potentials (L-EP), and pattern-reversal EP (PR-EP) were recorded.[7] In all examinations, when one eye was recorded, the other was closed. On the ERG, the latencies and amplitudes of the *a* and *b* waves were measured. For recording the ERG, a Henkes-electrode in reference to an earlobe electrode was used. On the EP, the amplitude between a negative peak (N) of about 70 msec and a positive peak (P) of about 110 msec was measured as well as the latencies themselves of a bipolar leading. The positions of the scalp electrodes were on the midline, 5% and 25% above the inion. For stimulation a xenon-flash (Strobotest II

0077-8923/82/0388-0227 $1.75/0 © 1982, NYAS

Typ BN 940) with a light energy of 0.375 J at a distance of 30 cm was used in front of the eye. The luminance could be lowered by density filters (D). At the same distance, a blue adaption field (cinemoid No. 19 dark blue) was fastened. The luminance amounted to 2100 lux without the blue filter. The registration of the ERG and the L-EP was performed simultaneously. After amplification and filtering, 128 signals were averaged with a Nicolet 1072 computer and recorded by a *xy*-plotter. The stimulus frequency was 3.7 Hz. TABLE 1 represents the normal values of the ERGs and TABLE 2 those of the L-EPs at three stimulus intensities obtained from 44 normal subjects.

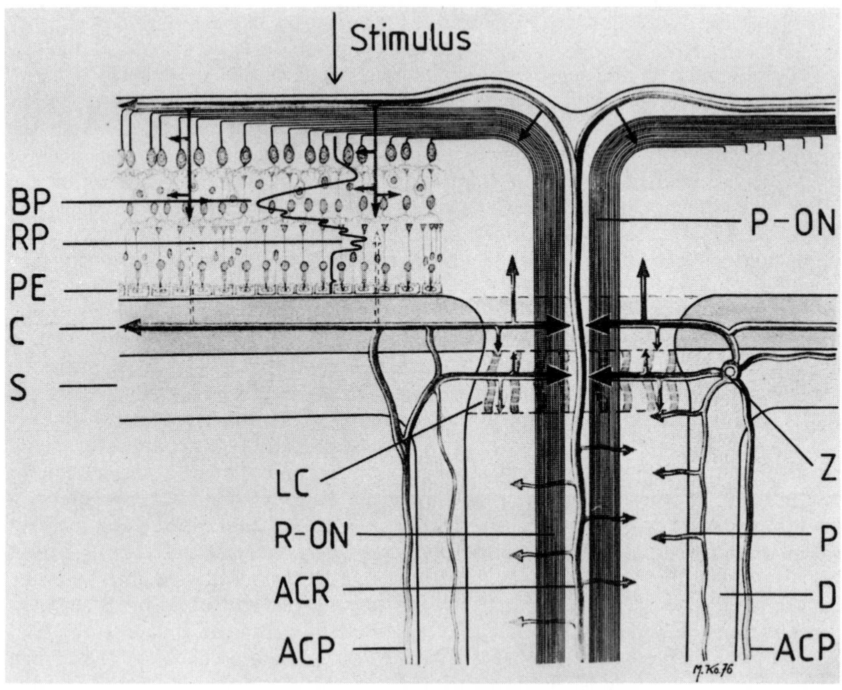

FIGURE 1. Blood supply of the retina and the pre- and retrolaminary part of the optic nerve. After stimulation of the retina, the signals are conducted into the optic nerve to evoke the visual potential in the cortex. ACP, arteria ciliaris posterioris; ACR, arteria centralis retinae; BP, bipolar potential; C, choroid; D, dura; EP, evoked potential; LC, lamina cribrosa; P-ON, R-ON, pre- and retrolaminary optic nerve; P, pia; PE, pigmentepithelium; RP, receptor potential; S, sclera; Z, circle of Zinn.

For stimulating the PR-EPs, a black-and-white checkerboard pattern in reversal stimulation, presented by a monitor, was used. The subject sat 2 m in front of the monitor and had to fixate on a point in the center of the screen during the examination. The diameter of the screen represented a visual angle of 18°. The check size was 30 minutes of arc; the modulation depth, 20%; the average brightness, 400 lux; and the stimulus frequency 5, cycles/sec. After amplification, PR-EPs were filtered by an active filter tuned to 10.0 Hz. The position of the electrodes and the

TABLE 1

PHOTOPIC ELECTRORETINOGRAPHS: NORMAL VALUES, *a* AND *b* WAVES

	Stimulus Intensity 1 Phot. (D 2.0)				Stimulus Intensity 2 Phot. (D 1.5)				Stimulus Intensity 3 Phot. (D 1.0)			
	Amplitude (μV)		Peak Time		Amplitude (μV)		Peak Time		Amplitude (μV)		Peak Time	
	a	*b*	*a*	*b*	*a*	*b*	*a*	*b*	*a*	*b*	*a*	*b*
Mean	33.4	114.8	21.6	41.9	50.4	164.3	20.0	41.2	79.1	185.3	18.4	40.5
Standard deviation	10.8	32.7	2.0	3.5	14.3	42.1	1.3	2.5	21.1	50.7	1.3	2.5

TABLE 2

LUMINANCE EVOKED POTENTIALS: NORMAL VALUES

	Stimulus Intensity 1 (D 2.0)			Stimulus Intensity 2 (D 1.5)			Stimulus Intensity 3 (D 1.0)		
	Amplitude (μV)	Peak Time (msec)		Amplitude (μV)	Peak Time (msec)		Amplitude (μV)	Peak Time (msec)	
		Neg	Pos		Neg	Pos		Neg	Pos
Mean	7.4	70.6	112.6	7.5	67.3	110.5	7.1	66.4	111.0
Standard deviation	2.95	13.2	13.4	3.0	12.4	12.0	2.8	12.4	15.0

technique for recording were the same as for recording the L-EPs. The mean value of 40 normal subjects was 4.5 μV (standard deviation = 2.0).

The visual acuity was examined by a Rodavist and the visual field by an automatic perimeter (Octopus; stimulus, 3; 30° eccentricity).

During all psychophysical and electro-ophthalmological examinations the pupils were dilated by mydriaticum Roche, except for patients with acute glaucoma.

The elevation of the intraocular pressure was performed by a trained person with a Müller spring-dynamometer, a number of millimeters lateral of the corneo-scleral edge. The photopic ERGs and L-EPs were recorded at five different levels of the intraocular pressure depending on the blood pressure of the ophthalmic artery (P_o = normal intraocular pressure, P_d = intraocular pressure at the diastolic –, P_s = systolic blood pressure of the ophthalmic artery, $P_m = (P_s + P_d)/2$, $P_{dm} = (P_d + P_m)/2$; and $P_{ms} = (P_m + P_s)/2$. The beginning of the measurements of the photopic ERGs and L-EPs at elevated intraocular pressure were delayed 20 sec having more constant blood circulation as was found out in previous experiments. Between the recordings of one session and another, there was always a rest of 5 min given to the subjects.

EP in Visual Field Defects Caused by Glaucoma

Patients

For investigation into the questions of which nerve structures are affected in their functions by glaucoma, and to what grade of exactness can visual field defects be determined by means of the EP, 70 pressure-regulated glaucomatous eyes with different visual field losses were examined. All subjects showed visual field defects either on one or on both eyes, and the visual acuity ranged from 0.3 to 1.0. In all cases, visual field defects were caused by glaucoma only and the refractive media have been clear.

Results

Figures 2 and 3a represent the mean amplitudes and the mean latencies of the photopic ERGs and the L-EPs of normal and glaucomatous eyes at three stimulus intensities. All changes of the photopic ERG and L-EP of the glaucomatous eyes are within the normal standard deviation. At all stimulus intensities the mean values of both the *b* and the *a* waves of the photopic ERG of the glaucomatous eyes are diminished and the latencies are increased. At the lowest stimulus intensity the mean value of the L-EPs of the glaucomatous eyes is diminished by about 30% and decreases with increasing stimulus intensity. A sample is given in Figure 3b. The mean latencies of the N and P peaks are slightly increased. The field defects mainly consist of relative and absolute scotomas in the Bjerrum area. In general, if visual acuity were normal and the visual field defects did not extend into the 10° boundary, the EPs would be mostly diminished but within the standard deviation of the control group. An almost significant decrease of the EPs could be noticed when visual field defects are within the 10° boundary or a diminution of the visual acuity were present. A sample is given in Figure 4.

If we compared the visual field defects and the EPs of both eyes in each subject, the smaller response was always obtained from the more damaged eye (Figure 5).

After a period of 2 years, 28 patients with visual field defects were reexamined. In 21 cases the visual fields remained almost constant and no significant changes were

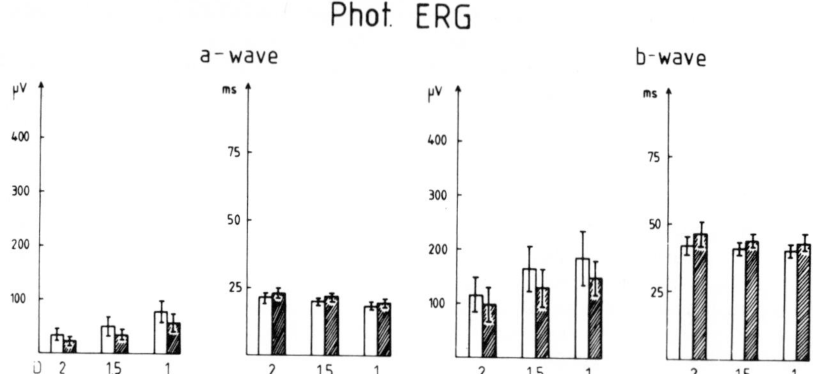

FIGURE 2. The histogram shows the mean values and the standard deviation of the amplitudes and of the latencies of the *a* and *b* waves of the photopic ERG of 70 pressure-regulated glaucomatous eyes (shaded bars) compared to the values of normal eyes (open bars). The mean values of the *a* wave as well as the *b* wave of the glaucomatous eyes are diminished; their latencies are increased compared to the values of the normal eyes.

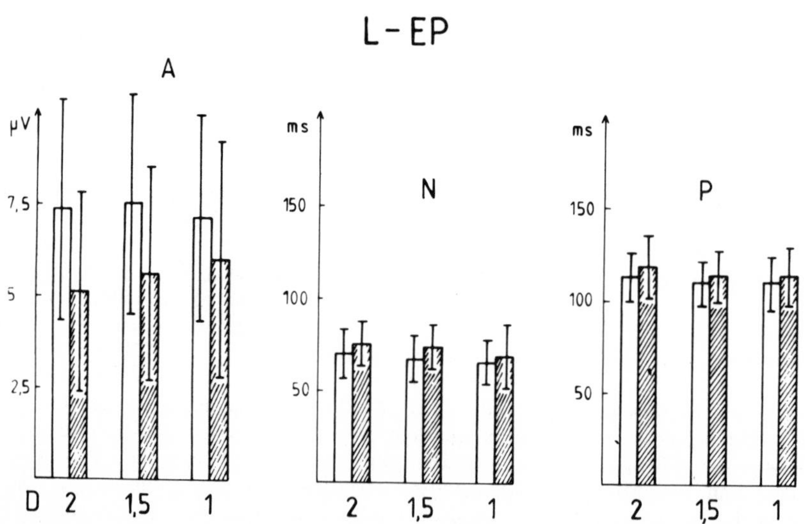

FIGURE 3a. The histogram shows the mean values and the standard deviation of the amplitudes and of the latencies of the luminance evoked potentials (L-EPs) of 70 pressure-regulated glaucomatous eyes (shaded bars) compared to the values of the normal eyes (open bars). At the stimulus intensity D 2 the mean amplitude is diminished by about 30% and becomes less with increasing stimulus intensity. The mean latencies of the N and P peaks of the glaucomatous eyes show a little increase.

seen on the EPs. In 7 cases relative scotomas changed into absolute scotomas, defects progressed to the center, and EPs were diminished when compared to the results of first examination. In general, changes of the sensitivity of the visual field are better reflected by the PR-EPs than by the L-EPs.

Discussion

While latencies show little differences, the mean amplitude of the L-EPs decreases by about 30% at the lowest stimulus intensity and becomes smaller with increasing stimulus intensity (FIGURE 3b). This decrease of the L-EP may first of all be caused by the ophthalmoscopically evident loss of optic nerve fibers. In addition to this damage

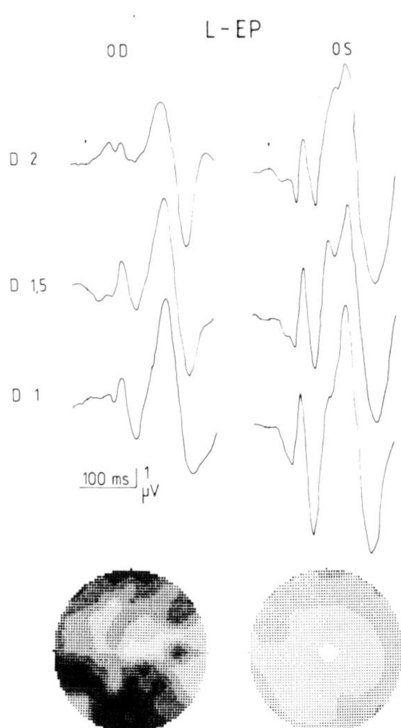

FIGURE 3b. L-EPs of a patient with glaucoma capsulare of the right eye. The visual acuity on both eyes is 0.8. On the bottom the visual field (30° eccentricity) obtained from computer shows relative as well as absolute scotomas. At the stimulus intensity D 2, the amplitude of the L-EP is significantly diminished compared to the normal left eye (OS). Note that the difference of amplitudes of both eyes becomes smaller with increasing stimulus intensity.

of the optic nerve fibers, a functional diminution of the first and second neuron may be of consequence, as is reflected by the decrease of amplitude and the increase of latencies on the photopic *a* and *b* waves of the ERG. Summarizing, we can state that the decrease of the EP of glaucomatous eyes results from a functional diminution of all intraocular nerve structures in which the prelaminary part of the optic nerve is affected most. For the investigation of visual field defects by means of the EP, two technical innovations of the last few years, namely the computer perimeter and the pattern-stimulator, produced results that allow a better comparison of defects in the visual field with changes in EP. Compared to the usual kinetic perimetry, computer perimetry has the advantage of obtaining static results from only one examination and the data is checked by double stimulation.[8] The pattern EPs, especially the pattern-

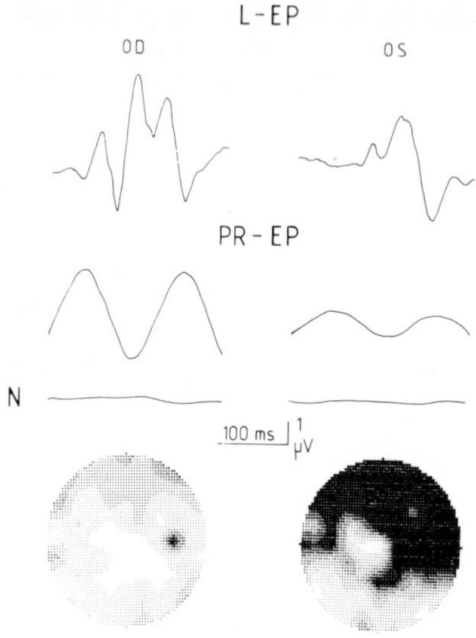

FIGURE 4. The L-EPs, the pattern-reversal evoked potentials (PR-EPs), and the visual fields (30° eccentricity) of a patient with pigmentary glaucoma on both eyes and normal visual acuity. Note the well-developed EPs of the right eye (OD) with only little changes in the visual field and the distinct decrease of the EPs of the left eye (OS) with marked paracentral defect of the visual field (N = noise).

reversal EPs, show high contrast sensitivity.[9] If the refractive media are clear, the advantage of pattern-reversal EP compared with luminance EP is that with pattern-reversal EP there is little diminution in the sensitivity of visual field effect changes such as decreases in amplitude and increases in latency.[7] If there exist absolute as well as relative visual field defects on glaucomatous eyes with normal visual acuity, which do not extend inside the 10° boundary, then the responses will be mostly below the mean value of the control group but within the standard deviation. If the defects markedly extend inside the 10° bondary, or additionally, if the visual acuity is diminished, then a decrease without the normal standard deviation can be expected (FIGURE 4). Comparing the responses of both eyes having one- or both-sided glaucoma with different visual field losses, the smaller response was always obtained from the more damaged eye (FIGURE 5). The observation of the course of the glaucoma by

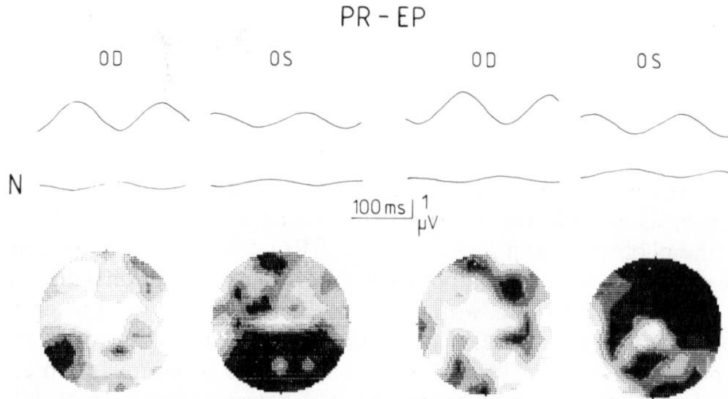

FIGURE 5. The PR-EPs and the visual fields of two patients with both-sided open-angle glaucoma and different visual field defects but normal visual acuity. The minor response is always obtained from the more damaged eye (N: noise).

means of the EP scarcely showed changes in the EP in cases with stationary visual field defects. However, if the visual field defects progress, or turn from a paracentral relative scotoma into an absolute one or if a relative scotoma proceeds towards the center, a decrease in the EP will occur.

Evoked Potentials in Glaucomatous Eyes with Dense Cataracts

Patients

For estimation of the visual acuity in glaucomatous eyes with dense cataracts, the L-EPs were recorded on 14 glaucomatous eyes before cataract surgery. In all cases the visual acuity was below 0.07 and ophthalmological judgment was impossible.

Results

After cataract surgery the visual acuity ranged with correction in 6 cases from 0.07 to 0.3, in 3 cases from 0.3 to 0.5 and in 5 cases from 0.5. to 1.0. Before the operation the L-EPs of all cases were distinctly diminished at the lowest stimulus intensity, and while the first and second group showed no or scarcely any increase with increasing stimulus intensity, the third group showed a significant increase. Two typical samples are given in Figure 6.

Discussion

An important topic in the clinical treatment of glaucoma is the preoperative estimation of the function of the visual pathway by means of the EP, in case that, together with the glaucoma, the refractive media might be turbid, which might, for example, be caused by cataracts. A postoperative increase of visual acuity, or, of the sensitivity of the visual field, can be expected if the L-EPs are within the standard deviation, taking into account that the saturation of the L-EPs will be obtained at a higher level of stimulus intensity because of the turbid refractive media.

However, sometimes there is no improvement of visual acuity after an operation although well-developed L-EPs were recorded before the operation. In these cases there are often pathological changes in the retina, especially changes of the macula. For example, there may be a central venous thrombosis with a macular edema in the early stage.[10] If the EP is significantly diminished and does not increase to higher levels of stimulus intensity, or if the L-EP is not recordable, then there will be no hope for an improvement of the visual acuity after the operation (Figure 6).

EP in Chronic and Acute Glaucoma Before and After Pressure Regulation

Patients and Results

Intraocular pressure values of 15 glaucomatous eyes with chronic pressure increases ranged from 28 to 44 mmHg. In all cases visual defects caused by glaucoma existed. After pressure regulation by Timolol, Pilocarpine, and trabeculectomy, the values of intraocular pressure was below 25 mmHg and systemic blood pressures was almost unaltered. On reexamination 4 weeks after treatment, slight improvements of the sensitivity of the visual field and increases of the amplitude of the EPs could be noticed in 4 cases (with initial pressure levels of 28, 32, 40, and 44 mmHg), whereas the visual fields and the EPs from the others remained almost unchanged. A sample for an improvement of the visual field and the EPs is given in FIGURE 7.

Before treatment, the intraocular pressure of 10 acute glaucomatous eyes ranged from 50 to 74 mmHg. After treatment with Diamox and Pilocarpine, and in 1 case by tapping the anterior chamber, the intraocular pressure was decreased below 25 mm Hg in 9 cases while systemic blood pressure remained constant. Simultaneously with the decreasing intraocular pressure, a significant increase of the EP occurred (FIGURE 8). In one case with acute angle closure glaucoma treated by ½% Isoglaucon, the intraocular pressure was reduced from 62 to 30 mmHg and the systemic blood pressure from 190/100 to 130/100. The L-EP and the photopic ERG showed a significant decrease in amplitude (FIGURE 9).

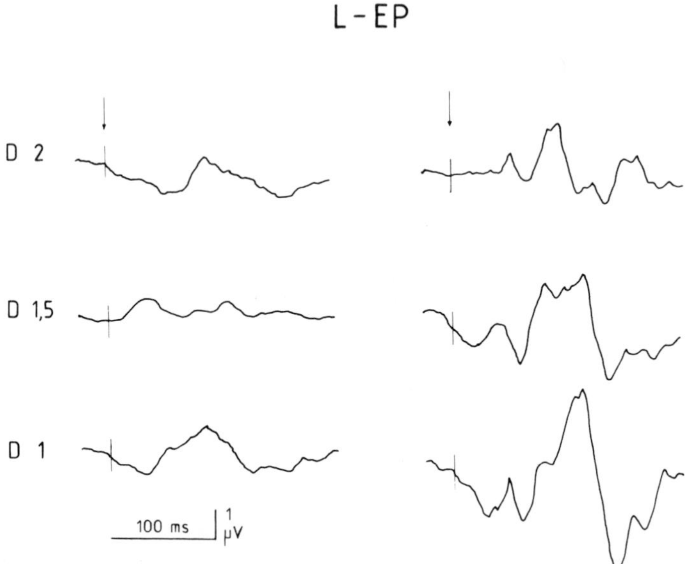

FIGURE 6. The left and right side represents the L-EPs at three stimulus intensities of two patients with dense cataract and visual acuity of only light perception. After the operation of the left case no useful visual acuity was obtained, whereas the right case showed an improvement to 0.8. Note that the responses of the left case are almost extinguished and do not increase with increasing stimulus intensity. The responses of the right case are well developed and show an increase in amplitude with increasing stimulus intensity.

Discussion

When elevated intraocular pressure induces damage of nerve structures together with visual field defects, regulation of the intraocular pressure may prevent the progression of these defects if the systemic blood pressure remains constant. In some

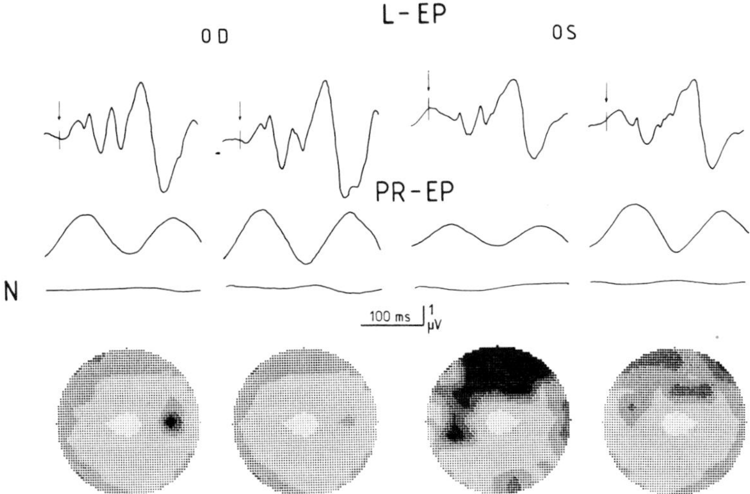

FIGURE 7. The L-EPs and PR-EPs of a patient with both-sided open-angle glaucoma before (OD = 27 mmHg, OS = 32 mmHg) and after pressure regulation (OD = 16 mmHg, OS = 14 mmHg). After pressure regulation the sensitivity of the visual fields (30° eccentricity) of both eyes shows an improvement and at the same time an increase of the EPs.

cases, an improvement may occur depending on the duration and the amplitude of the elevated intraocular pressure. The reason for maintaining or even improving the function lies in the increase of blood circulation caused by the decrease of the intraocular pressure. The damage caused by the diminished blood flow may first of all be localized in the prelaminary part of the optic nerve.

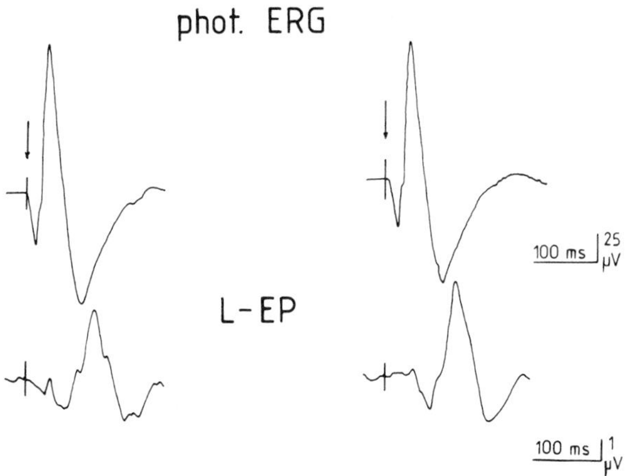

FIGURE 8. The left side shows the photopic ERG and the L-EP of a patient with secondary glaucoma caused by a melanoma of the iris at an intraocular pressure of 55 mmHg. The right side represents the recordings after regulation of the intraocular pressure to 10 mmHg. The decrease of the intraocular pressure leads to a significant increase of the L-EP whereas the photopic ERG remains almost unchanged.

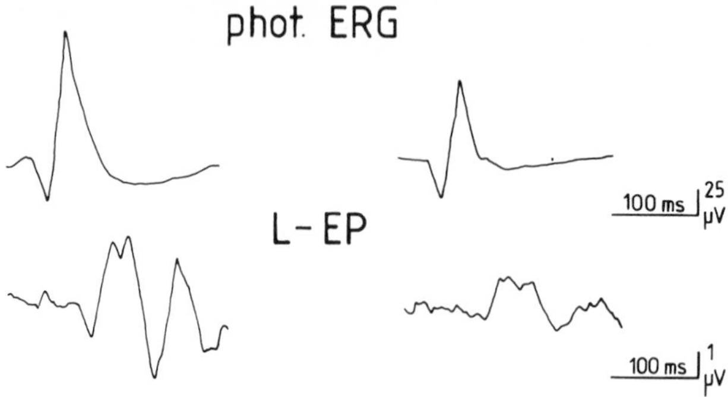

FIGURE 9. Significant diminution of the photopic ERG and the L-EP after a decrease of the intraocular pressure from 62 to 32 mmHg and simultaneous decrease of the blood pressure from 190/100 to 130/100 of a patient with acute angle-closure glaucoma after treatment with 0.5% Isoglaucon eye drops.

Fluorescent angiographical studies showed that the peripapillary choroid is most sensitive to pressure elevation[11–14] which may be caused by the fact that the choroid has no autoregulation. In the beginning of our studies we examined the visual field, determined by kinetic perimetry and the EP, on patients with chronic glaucoma at pressure levels from 25 to 40 mmHg before and after pressure regulation. It was remarkable that the pattern-reversal EPs showed phase shifts and little changes in amplitude.[7] The small observed changes in the visual field and the insufficiently precise method of kinetic perimetry made interpretations of the results impossible. Later examinations performed with the computer perimeter method showed that an increase of amplitude on the pattern-reversal EPs and on the L-EPs may occur in patients with chronic glaucoma after pressure regulation at an initial pressure level of about 30 mmHg.[15] Simultaneously an improvement of the visual field could be noticed (FIGURE 7). In glaucomatous eyes with an acute increase of the intraocular pressure by values of about 50 mmHg and more, a marked increase of the EP and a recovery of the sensitivity of the visual field could always be noticed (FIGURE 8). An important point in the therapy of acute glaucoma with eye drops is that besides a decrease of the intraocular pressure, there also occurs a decrease of the systemic blood pressure. In such cases, the perfusion pressure within the eyes may be reduced in spite of a decrease in the intraocular pressure, and may lead to a functional diminution of the intraocular nerve structures, which can be seen by a decrease of the EP (FIGURE 9).

EP IN NORMAL AND GLAUCOMATOUS EYES WITH AN EXPERIMENTALLY ELEVATED INTRAOCULAR PRESSURE

Patients and Results

Elevation of intraocular pressure was performed on 7 normal and on 10 pressure-regulated glaucomatous eyes. All subjects showed the same changes in the photopic ERG and L-EP at elevated intraocular pressure. FIGURE 10 represents a normal subject with the typical behavior of the ERG and L-EP upon stepwise elevation of the intraocular pressure dependent on the blood pressure of the ophthalmic artery. With increasing intraocular pressure the *a* and *b* waves of the photopic ERG show a gradual decrease in amplitude, with the *b* wave being more altered than the *a* wave. At the highest pressure level P_s, the *a* and *b* waves are distinctly diminished but well developed. The L-EPs show no significant changes up to the pressure P_m. Above P_m, a significant decrease of the amplitude of the EP can be noticed, and at P_s the response is almost extinguished. The behavior of the photopic ERGs and L-EPs of all 10 pressure-regulated glaucomatous eyes upon stepwise pressure elevation is almost the same. The *a* and *b* waves of the ERG show a gradual decrease with increasing intraocular pressure and are well developed at P_s. The L-EPs show a significant decrease when intraocular pressure exceeded P_{dm} and are more or less extinguished up to the pressure level P_s. FIGURE 11 represents the characteristic changes of the L-EPs at elevated intraocular pressure of a patient with glaucoma.

Discussion

In experimental investigations, we elevated the intraocular pressure on normal eyes, dependent on the blood pressure of the ophthalmic artery, is a stepwise manner by means of a spring dynamometer, and the photopic ERG and the L-EP were recorded. Consequently, when intraocular pressure was elevated and exceeded the mean blood pressure of the ophthalmic artery, the *a* and *b* waves of the ERG remained almost unchanged, whereas the EP came to a significant decrease (FIGURE 10). This phenomenon can be explained by a disturbed conductivity in the third neuron effected by a pressure-caused vascular insufficiency in the peripapillary choroid. This behavior of the EP at elevated intraocular pressure in normal eyes may be a criterion for the pressure sensitivity of the peripapillary choroid. A comparison of normal eyes with glaucomatous eyes that have relatively low pressure levels, namely, below 25 mmHg, showed little differences in the *a* and *b* waves of the ERG at elevated intraocular pressure, but the EPs of the glaucomatous eyes were almost extinguished at a pressure level equal to the mean blood pressure of the ophthalmic artery (FIGURE 11). Therefore the conclusion seems to be right that the increased pressure sensitivity of the EPs on glaucomatous eyes is caused by an increased pressure sensitivity of the peripapillary choroid which supplies the prelaminary part of the optic nerve.[16] If an

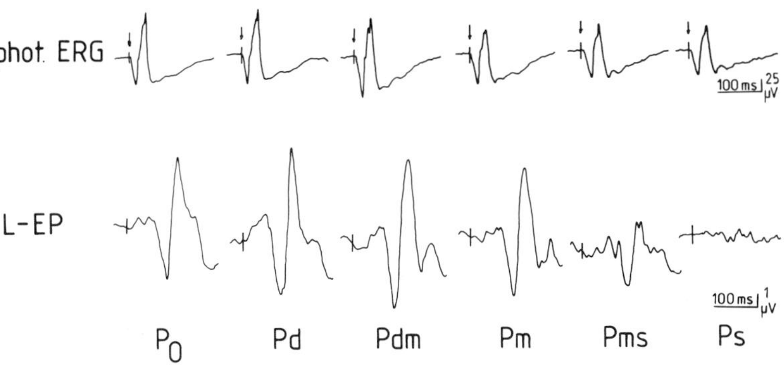

FIGURE 10. The behavior of the photopic ERG and L-EP of a normal eye at stepwise elevation of the intraocular pressure (P_o = normal intraocular pressure; P_d = intraocular pressure at the diastolic –, P_s = systolic blood pressure of the ophthalmic artery; $P_m = (P_s + P_d)/2$; $P_{dm} = (P_d + P_m)/2$; and $P_{ms} = (P_m + PS)/2$ with a spring dynamometer. Note when the intraocular pressure exceeds P_m, the ERG remains almost unchanged, whereas the L-EP shows a significant decrease.

eye with limited pressure values, with a suspicious optic disc but normal visual field, shows a significant decrease in the EP at an intraocular pressure level equal to the mean ophthalmic artery pressure, then we may suspect that this eye is exposed to danger of glaucoma.

Summary

The effects of visual field changes and ocular hypertension on visual evoked potentials were investigated by photopic ERG and by luminance and pattern-reversal EPs on 116 glaucomatous and on 7 normal eyes. The problem was approached by way of four investigations:

Firstly, which nerve structures are affected by glaucoma and how do visual field

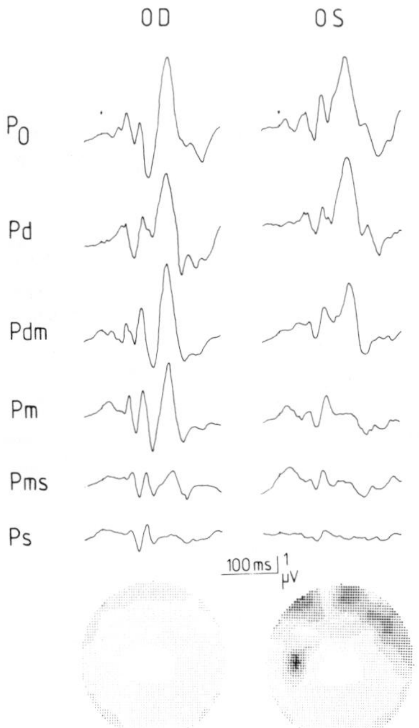

FIGURE 11. The L-EPs at stepwise elevation of the intraocular pressure of a patient with a pressure-regulated glaucoma and visual field defects (visual field, 30° eccentricity) of the left eye. The visual acuity of both eyes is 1.0. At the pressure level P_m the L-EP of the normal right eye is well developed whereas the L-EP of the glaucomatous eye is distinctly diminished.

defects caused by glaucoma influence the EP? The results show a functional diminution of all intraocular nerve structures in which the prelaminary part of the optic nerve is most affected. The EPs, especially the pattern-reversal EPs, are markedly diminished if the visual field defects extend inside the 10° boundary. Differences in the visual field defects of both eyes and the course of the sickness can be well observed by the EPs.

Secondly, which preoperative prognosis for visual acuity produced by the EP can be given to patients who have a dense cataract in addition to glaucoma? A postoperative improvement of the visual acuity can be expected if the L-EPs are within the standard deviation. If the EP is distinctly diminished and does not increase with increasing stimulus intensity, then there is no hope for an improvement of the visual acuity after the operation.

Thirdly, does a decrease of intraocular pressure in chronic and acute glaucoma influence the EP? In acute glaucoma with pressure levels of 50 mmHg or more, and sometimes in chronic glaucoma with pressure levels of about 30 mmHg, an increase of the amplitude of the EP and an improvement of the visual field could be noticed after pressure regulation.

Fourthly, what is the behavior of the EP in normal and glaucomatous eyes at experimentally elevated intraocular pressure? The amplitudes of the ERG components show a gradual decrease in normal as well as in glaucomatous eyes when intraocular pressure is increased and are maintained when intraocular pressure reaches systolic ophthalmic blood pressure. On the other hand, the EPs show a strong decrease in amplitude when intraocular pressure exceeds the mean ophthalmic blood pressure, particularly in the case of glaucomatous eyes.

References

1. Simschitz, E., G. Bartl, H. Gruber, H. Hanselmayer & H. Hiti. 1977. Elektroophthalmologische Untersuchungen bei Läsionen im Sehbahnbereich mittels Halbfeld- und Quadranten-Fernsehmusterstimulation. Vortrag zur 18. Tagung der Österr. Ophthalmolog. Ges. **18:** 9.0–11.6.
2. Kolker, A. E. & Hetherington, Eds. 1970. Becker-Shaffer's Diagnosis and Therapy of the Glaucomas. 3rd edit. Section 4. Mosby Company, Saint Louis, Mo.
3. Drance, S. M., V. P. Sweeny, R. W. Morgan & F. Feldmann. 1973. Studies of factors involved in the production of low tension glaucoma. Arch. Ophthalmol. (Chicago) **89:** 457–465.
4. Hayreh, S. S. & E. S. Perkins, Eds. 1969. Effects of the raised intraocular pressure on the retinal and optic disc circulation in rhesus monkeys. Symp. Int. sur l'angiographic Fluoresceinique. I. Albi, Ed. P. Amalric. Karger, Basel, Switzerland. pp. 323–328.
5. Gafner, F. & H. Goldmann. 1955. Experimentelle Untersuchungen über den Zusammenhang von Augendrucksteigerungen und Gesichtsfeldschädigung. Ophthalmologica (Basel) **130:** 357–377.
6. Goldmann, H. 1973. Über die Pathophysiologie des Glaukoms. Klin. Mbl. Augenheilk. **162:** 427–436.
7. Bartl, G. 1978. Das Elektroretinogramm und das evozierte Sehrindenpotential bei normalen und an Glaukom erkrankten Augen. Albrecht v. Graefes Arch. Klin. Exp. Ophthalmol. **207:** 243–269.
8. Fankhauser, F., H. Bebie & J. Sphar, Eds. 1977. Three years of experience with the Octopus automatic perimeter. Proc. 2nd Int. Visual Field Symp., Tübingen, 1976. Doc. Ophthalmol. **14:** 7.
9. Spekreijse, H., Ed. 1966. Analysis of EEG responses in man. Thesis. Junk Publishers, The Hague.
10. Bartl, G., H. Hofmann, C. Faschinger, B. Vidic & E. Simschitz. 1980. Elektroophthalmologisches Verhalten von vaskulären Papillenschwellungen. Ber. Dtsch. Ophthalmol. Ges. **77:** 429–433.
11. Ernest, J. T. & A. M. Potts. 1968. Pathophysiology of the distal portion of the optic nerve. II. Vascular relationship. Am. J. Ophthal. **66:** 380–387.
12. Blumenthal, M., K. A. Gitter, M. Best & M. A. Galin. 1970. Fluorescein angiography during induced ocular hypertension in man. Am. J. Ophthalmol. **69:** 39–43.
13. Blumenthal, M., K. A. Gitter, M. Best & M. A. Galin. 1971. Ocular Circulation: Analysis of the effect of induced ocular hypertension on retinal and choroidal blood flow in man. Am. J. Ophthalmol. **71:** 819–825.
14. Blumenthal, M., M. Best, M. A. Galin & H. Toyofuku. 1971. Peripapillary choroidal circulation in glaucoma. Arch. Ophthalmol. (Chicago) **86:** 31–38.

15. Faschinger, C., G. Bartl, W. Wutz & E. Wocheslãnder. 1980. Computerperimetrie bei Glaukomaugen vor und nach Druckregulation. Ber. Dtsch. Ophthalmol. Ges. **77:** 527–530.

16. Bartl, G., O. Benedikt & H. Hiti. 1975. Das elektrophysiologische Verhalten gesunder und glaukomkranker menschlicher Augen bei kurzzeitiger intraokularer Druckbelastung. Albrecht v. Graefes Arch. Klin. Exp. Ophthalmol. **195:** 201–206.

Discussion of the Paper

S. Sokol: It was not clear to me if you are saying that you find the flash VEP more sensitive in glaucomatous damage than the pattern-reversal VEP.

G. Bartl: The pattern-reversal VEP is more sensitive.

Question: You did not mention anything about the medications that these patients were on or about the pupil size. Some of these medications are neuroactive, like *pilocarpine* and epitrate. Do you think that plays a role in the results?

Bartl: I think so. We examine all patients with dilated pupils but if pupil size is not controlled, changes in latency might occur.

H. Spekrejse: Why have you chosen amplitude instead of phase for a criterion? Was the phase not as good a criterion as amplitude?

Bartl: Yes.

Spekreijse: Yes, but you have recognized phase shifts now.

Bartl: When you elevate the intraocular pressure we have seen different phase shifts.

Spekreijse: But not consistently in one direction?

Bartl: Yes.

I. Bodis-Wollner: What was the size of the total stimulus field?

Bartl: 18 degrees in visual angle.

Bodis-Wollner: Based on this answer, I ask Dr. Spekreijse, if the total visual field size of the stimulus is 18 degrees, do you think that it would be easy or would it be possible to measure phase shift of the steady-state evoked potential as a criterion for visual field defects?

Bartl: Visual field defects in glaucoma may be significantly smaller. In the pattern EP, the macula region is dominated so strongly in the response that when you have field defects at 10–50 degrees you will not see too much abnormality in the response.

J. Sherman: Cappin and Nissim about a half a dozen years ago occluded the central five degrees in an attempt to record from one quadrant in one eye and compare to the corresponding quadrant of the other eye. This approach appeared to be successful in unmasking field defects outside of the macula, which is, of course, typical in glaucoma. Have you ever attempted anything like that?

Bartl: No, I have not.

THE PATHOPHYSIOLOGY OF AMBLYOPIA: ELECTROPHYSIOLOGICAL STUDIES*

Dennis M. Levi and Ruth E. Manny

University of Houston
College of Optometry
Houston, Texas 77004

Functional amblyopia may be defined as a loss of visual acuity caused by form deprivation and/or abnormal binocular interaction, for which no organic cause can be detected by the physical examination of the eye.[1] Amblyopia is usually associated with strabismus (turned eye), anisometropia (unequal refractive error), or form deprivation early in life, and thus represents a developmental disorder. There appears to be a critical period for the development of amblyopia in humans and it may be preventable if the causative factors are discovered and eliminated within this critical period.[1] Interestingly, amblyopia may also be reversible much later in life.[2]

Over the past 15 to 20 years, there has been a renewed interest in the study of amblyopia, which is in large part attributable to the single-unit studies in cats and monkeys reared with experimentally induced amblyopia. These studies have shown that normal visual experience is essential for the development and maintenance of the physiological characteristics of cells in the visual cortex, and that disruption of the normal visual process during an early period of susceptibility by light or form deprivation, strabismus, or anisometropia, may result in a marked disturbance of the physiological organization of the visual cortex, cell shrinkage in the LGN, and severe amblyopia.[3–9] However, it is not clear whether strabismus and form deprivation (due to lid suture or occlusion) affect similar cell populations in the visual pathway. There is evidence that the effects of monocular lid suture are selective for the Y or transient cells,[10] while in kittens reared with unilateral strabismus, the X or sustained cells appear to be most affected.[11,12] Although such methods of investigation are valuable in dealing with experimental animals, studies of human amblyopia have, for the most part been restricted to psychophysics and visual evoked potentials (VEP). In humans, with naturally occurring amblyopia, the VEP is generally the only direct method available for studying cortical responses to visual stimuli and may provide a valuable tool for the assessment of visual function in amblyopia for several reasons: (1) it provides an objective measure of function which may be useful in infants and nonverbal patients; (2) it may aid in localizing function and dysfunction; (3) it may be of value in assessing prognosis and monitoring therapy; and (4) it may be helpful in bridging the gap between psychophysics and physiology.

Over the past few years we have been studying the VEP of observers with naturally occurring amblyopia using luminance and pattern stimuli. The present paper reviews some of the results.

LUMINANCE EVOKED POTENTIALS

There have been a large number of studies of the VEP in amblyopia using unpatterned stimuli; however, the results are quite equivocal. Some investigators

*This work was supported by a grant (R01 EY 01728) from the National Eye Institute.

0077–8923/82/0388–0243 $1.75/0 © 1982, NYAS

reported irregularities in amplitudes and latencies of the amblyopic eye,[13-15] while others did not.[16-18] One possible reason for this ambiguity is that amblyopes of varying etiology have often been presented as a homogenous group, with little specificity provided as to the nature of the condition. A second important aspect, which has received little attention in the study of amblyopia, is the effect of the temporal frequency of stimulation. Spekreijse, Khoe, and van der Tweel[17] have studied the VEP of a single anisometropic amblyope to low- and high-frequency luminance modulation; however, the length of time necessary to acquire the data for each eye of an amblyopic observer over a large range of temporal frequencies has discouraged this undertaking using conventional procedures.

Recently, the application of pseudorandom binary sequence stimulation to VEP recording has provided an alternative technique which avoids the difficulties of numerous averages at each separate temporal frequency.[19-22] This stimulus, when analyzed in the frequency domain, has a continuous power spectrum with equal power at each temporal frequency over a large range of frequencies. This stimulus has several other advantages, since analysis may be accomplished both in the frequency and in the time domain. Srebro and Wright[20] have shown that this technique is extremely sensitive to disease of the macular and optic nerve.

FIGURE 1 shows the stimulus schematically in the time (A) and frequency domains (B). A microprocessor was programmed to produce a pseudorandom sequence consisting of ones and zeros based on the output of an 8-bit shift register. Depending upon the output of the shift register, the luminance of the monitor (Tektronix 608 with P45 Phosphor) was set to one of two values by modulation of a reference D.C. voltage

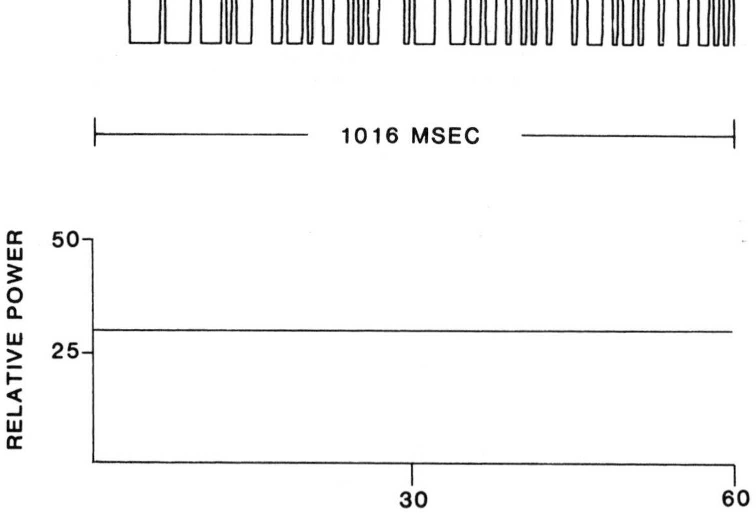

FIGURE 1. (A) Schematic representation of the modulation of the stimulus display by the pseudorandom binary sequence. (B) The power spectrum of the stimulus shown above (A).

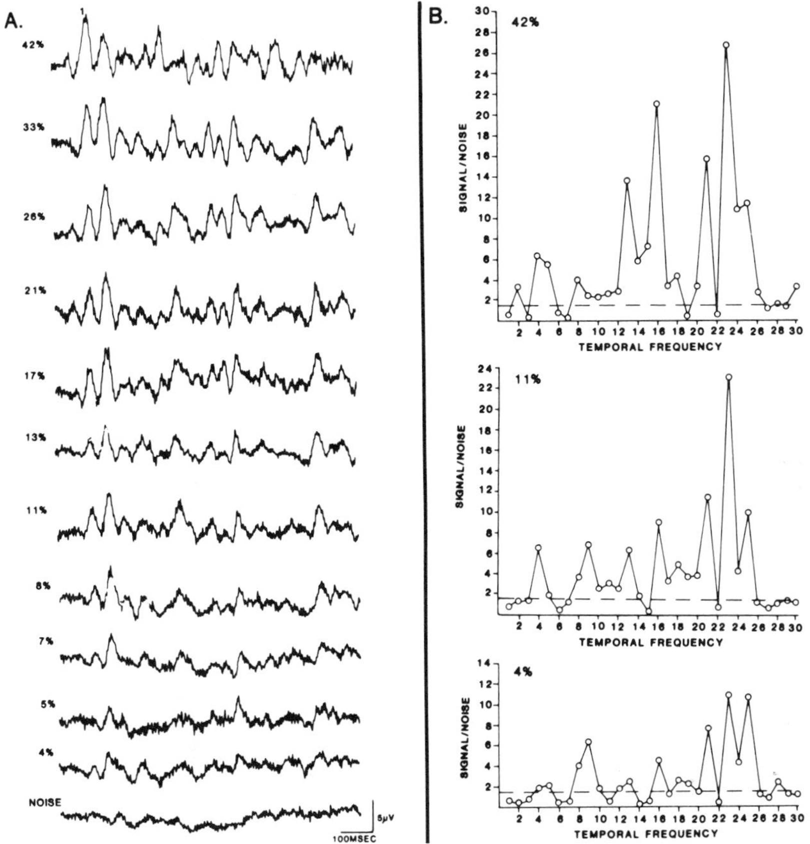

FIGURE 2. (A) Averaged evoked potentials recorded binocularly in control observer D.L. for modulation at contrast levels ranging from 42% to 4%. The bottom trace was recorded in response to a steady field, i.e., the noise condition. (B) The signal-to-noise ratio as a function of temporal frequency at 42, 11 and 4% contrast derived from the corresponding records shown in (A).

applied to a digital/analog converter. These two luminance values were selected to produce a time averaged luminance of 32 cd/m^2. The duration of the sequence was slightly greater than 1 sec (1016 msec) with a 0.5 probability that the luminance would change every 8 msec. Following each modulation sequence, the luminance of the monitor returned to the average value for a random interval (between 256 and 4095 msec) after which the modulation sequence was repeated.

The monitor was masked by a circular diffusing cone which provided a light surround approximately the color and luminance of the test stimulus, which subtended 9° at the 57 cm viewing distance. The stimulus appeared to the observer as a somewhat rhythmic, irregular flicker.[20] Specific details of the stimulus and recording of VEPs are provided elsewhere.[23]

FIGURE 2A shows the averaged evoked response (recorded with a monopolar referential electrode configuration) for a normal observer D.L. over a range of stimulus modulation depths from 42% to 4%. The bottom trace represents the noise level (i.e., the averaged EEG obtained when the observer viewed a homogenous screen of the same average luminance as the test stimulus (with 0 modulation depth).

It is clear that the early peaks of the averaged responses are quite dependent upon the stimulus modulation depth; in particular, the component labeled 1 appears to change systematically as a function of contrast. The power spectra on the right (plotted in terms of the signal-to-noise ratio) (FIGURE 2B) show the temporal tuning of the VEP at three different modulation depths (42%, 11%, and 4%). It is of interest to note that the major peaks in the temporal tuning functions are evident at each contrast level, and that certain components such as the 16 Hz component, show very systematic variations in amplitude as a function of modulation depth.[19]

FIGURE 3A shows the average evoked responses and power spectra obtained with a modulation depth of 42% and binocular viewing for D.L. on two separate days. It is clear that the results are repeatable over time, and for normal observers there is little difference in amplitude or latencies of the responses of the two eyes, as shown in FIGURE 3B.

FIGURE 4 shows the averaged evoked potentials and power spectra in response to 42% modulation for each eye of amblyopic observer R.G. There are several points of interest in this figure. (1) There appears to be little difference in amplitude or latency of the major components of the averaged responses of the two eyes. (2) The temporal tuning appears to be quite similar in the two eyes; however, there are small, but significant differences in the signal-to-noise ratios obtained for the two eyes, particularly in the low- and medium-frequency range. This is most easily seen in the inset which shows the log of the ratio of the power (nonamblyopic eye to amblyopic eye). While some strabismic/anisometropic amblyopes show no difference between the two eyes, the results of R.G. are typical of many strabismic and anisometropic amblyopes in showing small reductions in the signal-to-noise ratio of the amblyopic eye at low and medium temporal frequencies. In contrast, the results obtained for a severe amblyope due to stimulus deprivation (unilateral congenital cataract) are shown in FIGURE 5. The averaged evoked potentials of the two eyes of this observer are clearly different in both amplitude and waveform. The temporal tuning specificity shown in the signal-to-noise ratio plot is similar in the two eyes, with peaks occurring at or near the same temporal frequencies; however, the amblyopic eye shows large reductions in the signal-to-noise ratio in the midtemporal frequency range (12–22 Hz). This is more easily seen in the inset. These results are consistent with recent psychophysical data[24] suggesting that some, but by no means all, strabismic and anisometropic amblyopes show small but significant losses in sensitivity to temporal modulation of a uniform field at low and medium temporal frequencies, while amblyopia resulting from stimulus deprivation results in much more substantial losses in temporal sensitivity over a wide range of temporal frequencies.

PATTERN STIMULATION

By comparison to the small losses in luminance processing of the amblyopic eye, much larger losses are evident in the processing of spatial information.

FIGURE 6A shows the signal-to-noise ratio of the steady-state evoked potential for each eye of amblyopic observer T.T. in response to the appearance/disappearance of sinusoidal gratings at 8 Hz. The contrast of the gratings was 44%. The data of the nonamblyopic eye are shown by circles, those of the amblyopic eye, by squares. The

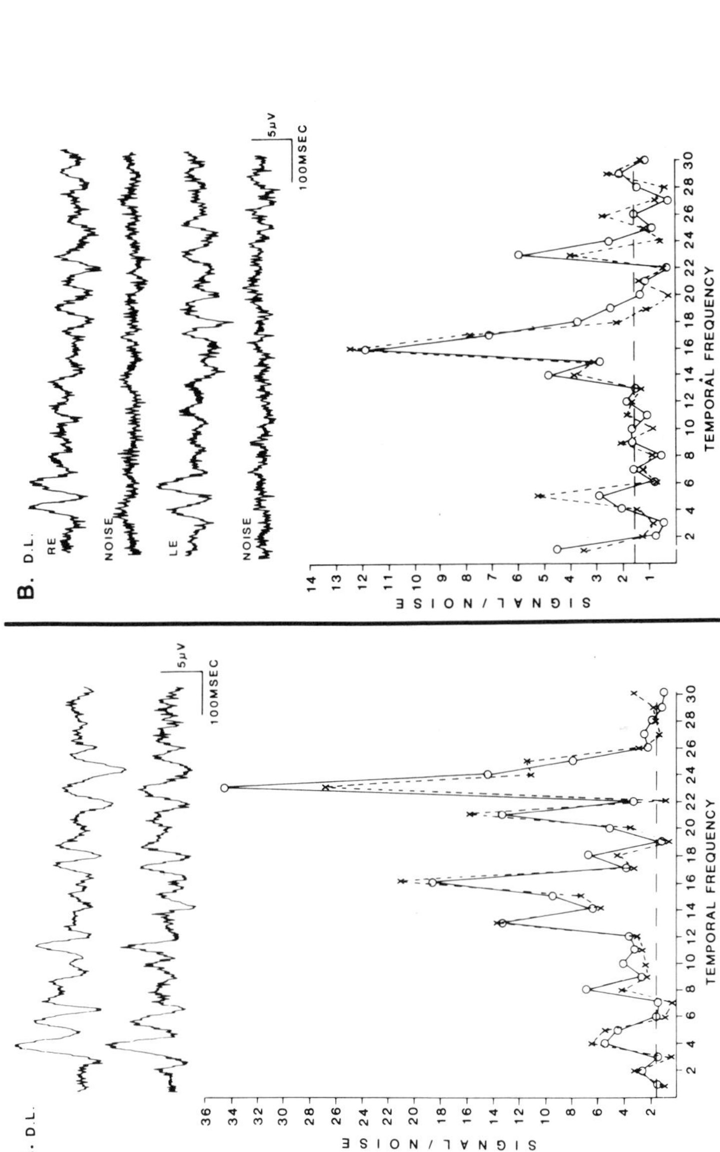

FIGURE 3. (A) Binocular visual evoked potentials recorded on two different days for control observer, D.L. (above). The signal-to-noise ratio as a function of temporal frequency is shown below for the upper trace (circles) and lower trace (crosses). (B) Monocular visual evoked potentials recorded in response to stimulation of the right eye (first trace) and left eye (third trace) of normal observer D.L. The second (right eye) and fourth (left eye) trace were recorded in response to a steady field (noise condition). The signal-to-noise ratio is plotted below as a function of temporal frequency for the right eye (circles) and left eye (crosses).

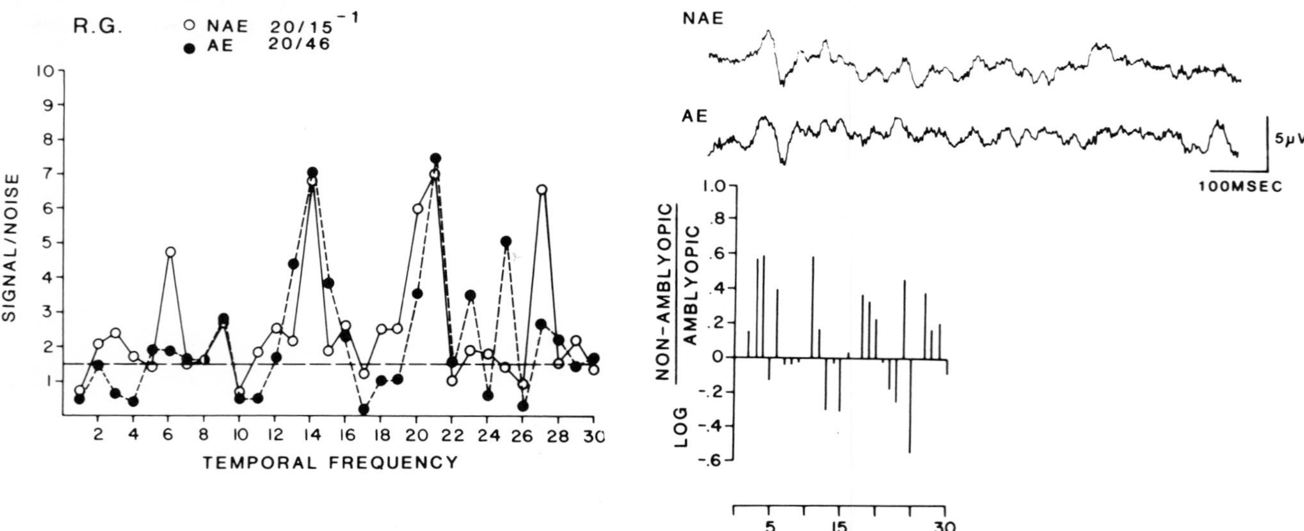

FIGURE 4. Averaged visual evoked potential (right) for the nonamblyopic eye (upper trace) and amblyopic eye (lower trace) of strabismic/anisometropic amblyopic observer R.G. The signal-to-noise ratio as a function of temporal frequency (nonamblyopic eye, open symbols; amblyopic eye, filled symbols) is to the left. The logarithm of the ratio (nonamblyopic eye/amblyopic eye) for those temporal frequencies where the signal-to-noise ratio is greater than 1.5 for either eye is shown in the lower right.

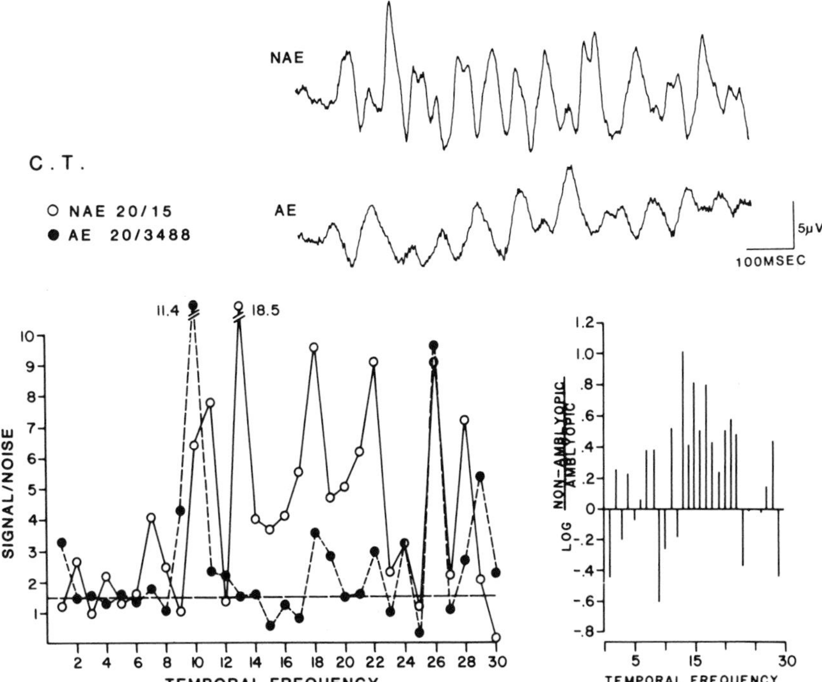

FIGURE 5. The averaged evoked potential for the nonamblyopic eye (above) and amblyopic eye (below) of stimulus deprivation amblyope, C.T., is shown on the right. The signal-to-noise ratio as a function of temporal frequency is shown on the left (nonamblyopic eye, open symbols; amblyopic eye, filled symbols). The logarithm of the ratio (nonamblyopic eye/amblyopic eye) at those temporal frequencies where the signal-to-noise ratio was greater than 1.5 in either eye is shown below (lower left).

spatial frequency tuning curves for the amblyopic and nonamblyopic eye show a peak between 2 and 4 cycles/degree (c/deg); however, the data of the amblyopic eye shows reduced signal-to-noise ratios across the spatial frequency spectrum, with the largest difference between the two eyes occurring around the peak of the spatial frequency tuning function (perhaps as a result of "floor effects" at the highest spatial frequencies). The finding that amblyopes show reduced VEP amplitudes for high spatial frequency stimuli appears to be quite general, and has been reported under many different stimulus and recording conditions.[17,18,25–28] At low spatial frequencies there appears to be considerably more individual variation, with some amblyopic observers showing reduced amplitudes for low spatial frequency stimuli while others do not.[28] Similar variation in the psychophysical contrast sensitivity functions of amblyopes has been reported.[29,30]

The lower part of FIGURE 6B shows the phase lag of the VEP as a function of spatial frequency for each eye of observer T.T. Open diamonds are the nonamblyopic eye, filled diamonds, the amblyopic eye. For both the nonamblyopic and amblyopic eye, the phase lag of the fundamental component of the VEP increased with increasing spatial frequency. This finding has been previously reported in the VEP of normal observers[31] and is consistent with the longer reaction time for high spatial frequencies observed psychophysically[30,32–34] and shown by open and filled circles for the nonam-

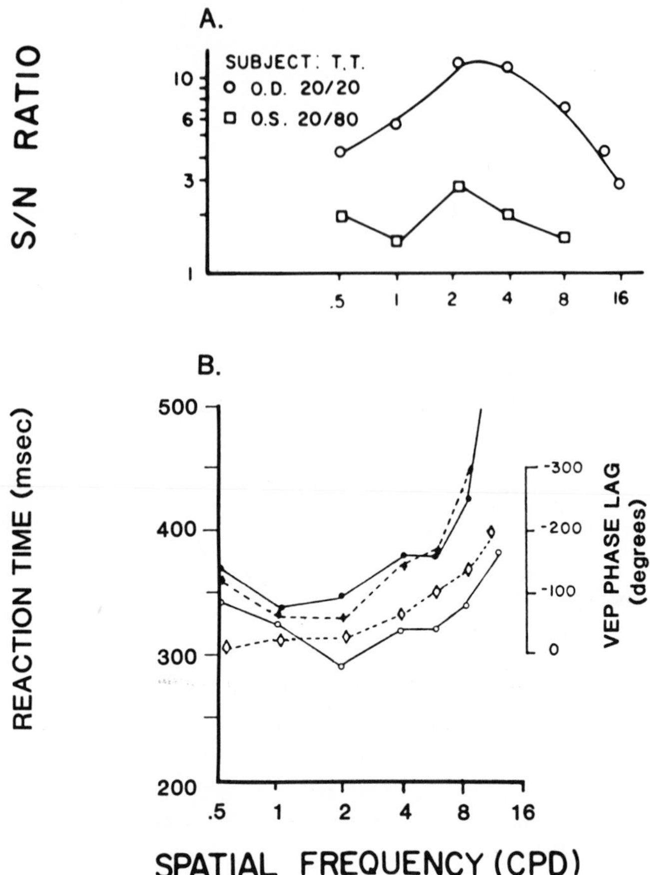

FIGURE 6. (A) VEP spatial frequency tuning function for the nonamblyopic (circles) and amblyopic (squares) eyes of anisometropic amblyope T.T. (OD 20/20; OS 20/80). Stimulus contrast was 0.44. (B) VEP phase lag (diamonds) as a function of spatial frequency for the same observer; open diamonds are the nonamblyopic eye; solid diamonds are the amblyopic eye. The open and filled circles show reaction time to the onset of gratings as a function of spatial frequency for the nonamblyopic and amblyopic eyes, respectively. (After Levi, Harwerth & Manny.[34])

blyopic and amblyopic eyes of observer T.T. In addition for a given spatial frequency, the phase lag of the amblyopic eye (and the reaction time) was generally longer than that of the nonamblyopic eye, particularly at higher spatial frequencies, suggesting processing delays in the amblyopic visual nervous system.

EFFECTS OF CONTRAST

The response of the amblyopic eye to spatial contrast is abnormal over a wide range of contrast values.[28] FIGURE 7A and B show examples of VEP versus contrast functions for two amblyopes (M.M. and S.H.) obtained with gratings of 4 c/deg. The abscissa is the contrast of the stimulus (plotted on a logarithmic scale), and the ordinate represents the signal-to-noise ratio of the VEP. For both the nonamblyopic and amblyopic eyes, the relationship between the logarithm of the stimulus contrast and the VEP amplitude is approximately linear over some range of contrast values.[28,34] Characteristically, the data for the amblyopic eyes showed a lower slope than those of the nonamblyopic eyes; this was especially true in the high spatial frequency range.[28]

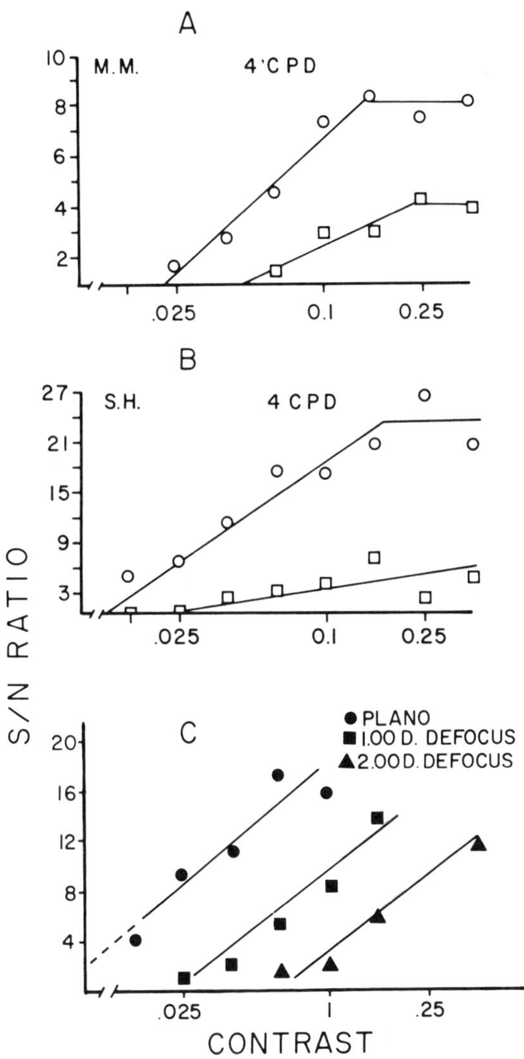

FIGURE 7. VEP signal-to-noise ratio versus stimulus contrast for nonamblyopic (circles) and amblyopic (squares) eyes of strabismic/anisometropic amblyope M.M. (A); strabismic amblyope S.H. (B); and for normal observer R.S.H. with 0.0, 1.00, and 2.00 diopter of induced defocus (C). The stimulus was the appearance/disappearance of a 4.0 c/deg sine-wave grating (From Levi & Harwerth.[28] By permission of *Investigative Ophthalmology and Visual Science.*)

The lower slope of the contrast function of the amblyopic eye is not explained on the basis of optical blur, since the effect of blur is to simply lower the effective contrast of the stimulus. Thus an optical blur should result in a shift of the line to the right, without a change in its slope as shown in FIGURE 7C.

EFFECTS OF RETINAL LOCUS

The reduced function of the amblyopic eye evident in the VEP to spatial contrast is greater for high than for low spatial frequencies, and probably reflects abnormalities of the central portion of the visual field. FIGURE 8 shows the amplitude of the VEP at the rate of pattern reversal (F1) and at twice the reversal rate (F2). The open histograms show the results of the nonamblyopic eye, the solid histograms, the amblyopic eye. The results shown in FIGURES 8A and B were obtained to checks with a fundamental spatial frequency of 0.5 (A) and 2.2 (B) c/deg reversed in counterphase at a pattern reversal rate of 8 Hz (i.e., twice the rate of local luminance variation). The stimuli were presented in a 3.5° disc. As has been previously reported in normal observers most of the power is at the frequency of pattern reversal (twice the frequency of local luminance variation),[26,27,35] however, it is of interest to note that there is some activity present at twice the reversal frequency. More interestingly, the amplitude of the responses of the amblyopic eye are markedly lower than those of the nonamblyopic eye, particularly with the smaller checks.

The results shown in FIGURE 8C and D were obtained with stimuli presented in an annulus with an inner diameter of 6 deg and an outer diameter of 14 deg. It was found that this type of stimulus (with an area about 14 times greater than that of the 3.5° disc) was quite effective in eliciting VEPs from the periphery in both nonamblyopic and amblyopic eyes.[27] Under these conditions, the lower spatial frequency stimuli were clearly better stimuli for peripheral VEPs than the higher spatial frequency stimuli. Interestingly, under these conditions, there was also a substantial response at twice the pattern reversal rate, and for both the fundamental and second harmonic there was little difference between the amplitude of the responses of the two eyes. These results are in agreement with those of other investigators.[19,36]

Spatiotemporal Interactions

The electrical responses of the brain evident in the VEP to pattern depend critically on both the spatial and the temporal parameters of the stimuli.[37,38] In order to investigate spatiotemporal *interactions,* VEPs were measured in response to sine-wave gratings alternated in counterphase by the same pseudorandom binary sequence that was used to generate the uniform field modulation described above (FIGURE 1). Srebro, Sokol, and Wright[22] have used a similar stimulus to record the VEP in normal observers.

FIGURE 9 shows the results for each eye of amblyopic observer R.G. for gratings of various spatial frequencies. The response of the nonamblyopic eye is shown in the top trace; that of the amblyopic eye in the lower trace. Shown below are the responses obtained for a steady uniform field; i.e., the "noise" condition. At each spatial frequency the amplitude of the early components of the averaged VEP is reduced in the amblyopic eye compared to the nonamblyopic eye. This is in agreement with the results of many previous studies[18,26–28,39] showing a reduced VEP amplitude of the amblyopic eye to patterned stimuli. In fact, at higher spatial frequencies (4 c/deg and above), the responses of the amblyopic eye are indistinguishable from noise. The effect of spatial frequency on the temporal tuning of each eye is shown to the left of each pair of records in the form of a signal-to-noise ratio plot (nonamblyopic eye, open symbols; amblyopic eye, filled symbols). As noted for the uniform field results, the temporal tuning of the nonamblyopic and amblyopic eyes appears to be quite similar; however,

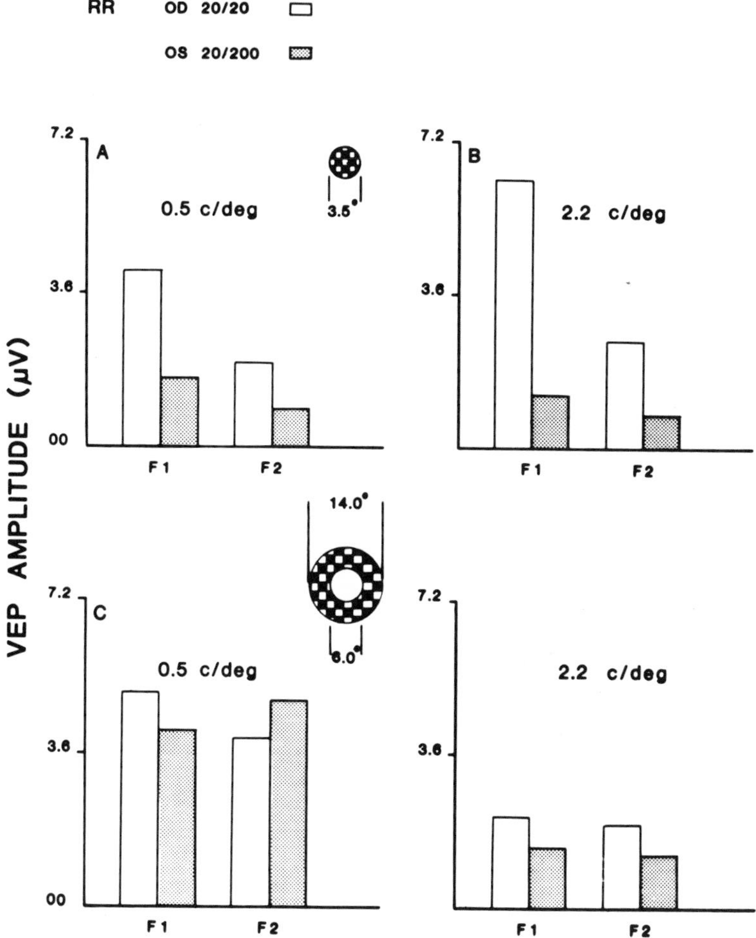

FIGURE 8. The amplitude of the VEP to an 8 Hz counterphase alternated checkerboard of 0.5 and 2.2 c/deg in a 3.50 disc (A & B); and in an annulus with a 6° center diameter and a 14° outer diameter (C & D). The results are shown for the amplitude of the VEP recorded at the rate of pattern reversal (F1) and at twice the pattern reversal frequency (F2).

at spatial frequencies greater than 0.5 c/deg, the signal-to-noise ratio of the VEP of the amblyopic eye is reduced at most temporal frequencies. This is more easily seen when the logarithm of the ratio of the power (nonamblyopic eye/amblyopic eye) is plotted for all temporal frequencies at which the signal-to-noise ratio was 1.5 or more

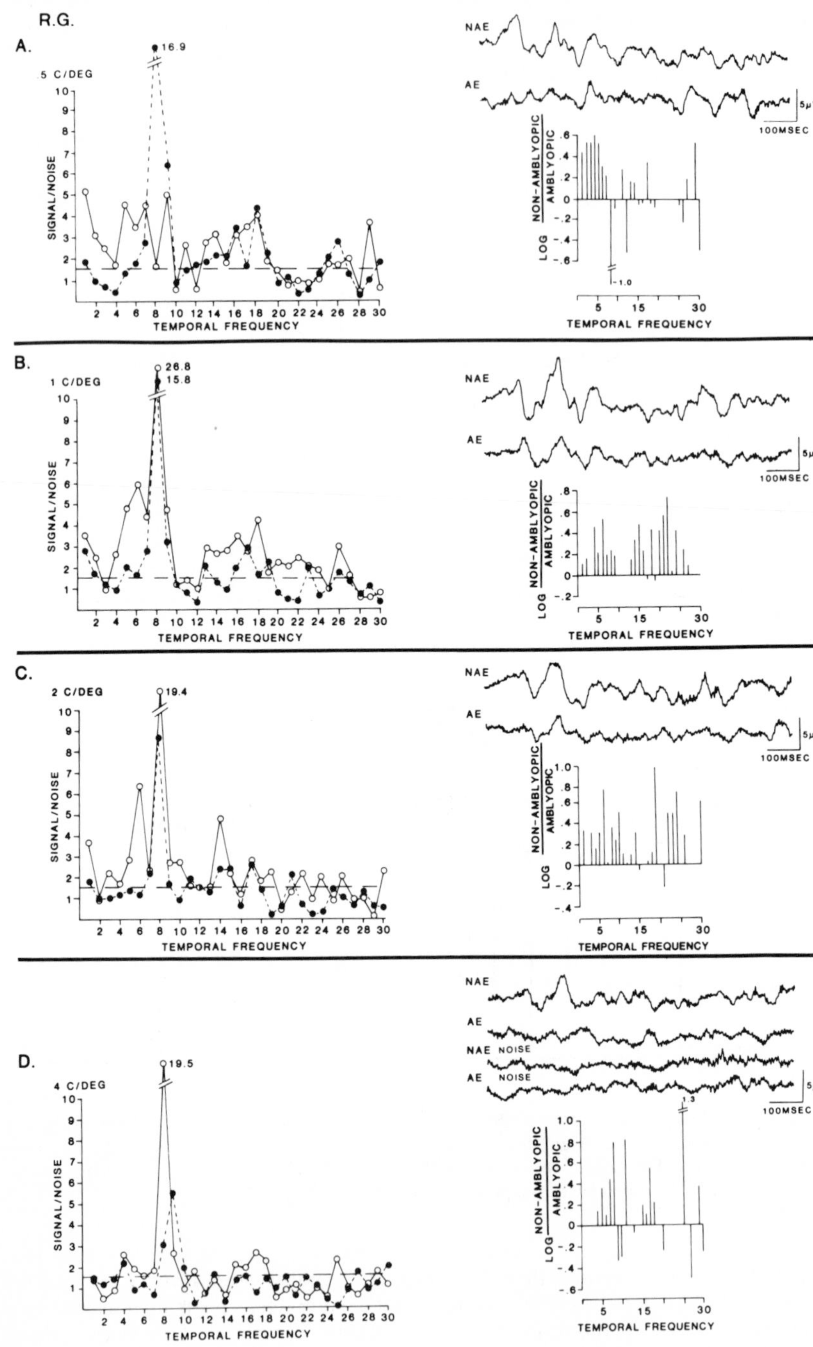
R.G.
A.
.5 C/DEG
16.9
SIGNAL/NOISE
TEMPORAL FREQUENCY
NAE
AE
5μV
100MSEC
LOG NON-AMBLYOPIC / AMBLYOPIC
-1.0
TEMPORAL FREQUENCY
B.
1 C/DEG
26.8
15.8
C.
2 C/DEG
19.4
D.
4 C/DEG
19.5
NAE NOISE
AE NOISE
1.3

in either eye (shown below the averaged VEPs). Similar results have been obtained on amblyopes due to strabismus, anisometropia and stimulus deprivation.[23]

The Locus of the Abnormality in Amblyopia

The VEP results suggest that in amblyopia due to strabismus/anisometropia pattern processing is markedly abnormal at or before the site of generation of the VEP, while luminance processing appears to be either not affected, or only slightly affected at low and medium temporal frequencies. Interestingly, while amblyopia due to stimulus deprivation appears to result in similar anomalies in spatial vision, luminance evoked potentials (as well as psychophysical temporal modulation sensitivity functions) are also abnormal.

Most early investigations of retinal responses in amblyopes have used unpatterned flashing lights to elicit electrical responses from the retina. Some authors reported abnormalities in the ERG of amblyopes, while others have not.[15,17,39,40]

We have measured retinal responses of amblyopes evoked by modulation of a uniform field using the pseudorandom binary sequence described above for eliciting VEPs. A gold foil electrode similar to that described by Arden *et al.*[41] was used in order to avoid interference with the optics of the eye, and fixation was controlled by the normally fixing eye in order to avoid eye movement artifacts. When recording with the amblyopic eye, a fixation point was positioned so as to ensure that the fovea of the amblyopic eye was directed at the center of the 9° field. An artifact rejection program was used during data acquisition to eliminate artifacts due to blinks, eye movements, and so on. FIGURE 10A shows 3 ERGs (above) obtained in this manner for each eye of strabismic amblyope C.W. (nonamblyopic eye, left; amblyopic eye, right). Each record is an average of 50 presentations of the pseudorandom binary sequence. The average of the three records is shown in the center trace, while the impulse response is shown below. None of the strabismic and/or anisometropic amblyopes tested showed any significant difference in the amplitude or latency of the *a* and *b* waves in the impulse response of the luminance ERG of the two eyes.

The results of the stimulus deprivation amblyope, C.T., are shown below. The impulse response of the amblyopic eye of this observer appears to be abnormal, with a reduced amplitude and prolonged latency of the *b* wave of the amblyopic eye.

These results suggest that the retinal responses to luminance stimuli are normal in the more common forms of amblyopia, i.e., those due to strabismus and/or anisometropia; however, the increased latency and decreased amplitude of the *b* wave of the amblyopic eye of an observer with profound amblyopia due to stimulus deprivation indicates that an anomaly may be present, at least in part, early in the visual pathway.

The ERG results with uniform field stimulation in strabismic/anisometropic amblyopes are not surprising, since the major anomaly in amblyopia appears to be in the processing of spatial information. We are presently studying the retinal responses

---

FIGURE 9. Averaged visual evoked potentials (right column) for the nonamblyopic eye (upper trace) and amblyopic eye (lower trace) obtained in response to pseudorandom counterphase modulation of sine-wave gratings of 0.5 c/deg (A), 1 c/deg (B), 2 c/deg (C), and 4 c/deg (D). The averaged response to a steady field (noise condition) is also shown in A (nonamblyopic eye-third trace; amblyopic eye-fourth trace). The left column contains the signal to noise ratio as a function of temporal frequency for the nonamblyopic (open symbols) and amblyopic eye (filled symbols). The logarithm of the ratio (nonamblyopic eye/amblyopic eye) at those temporal frequencies where the signal-to-noise ratio is greater than 1.5 in either eye is shown below the averaged responses on the right.

of amblyopes elicited by pseudorandom counterphase alteration of sine-wave grating patterns. FIGURE 11 shows the electrical responses of the retina of the nonamblyopic eye (left) and amblyopic eye (right), of R.G., evoked by pseudorandom counterphase alternation of a 2 c/deg sine-wave grating. The two upper traces are each the average of 125 responses, the center trace being the average of the two records shown above.

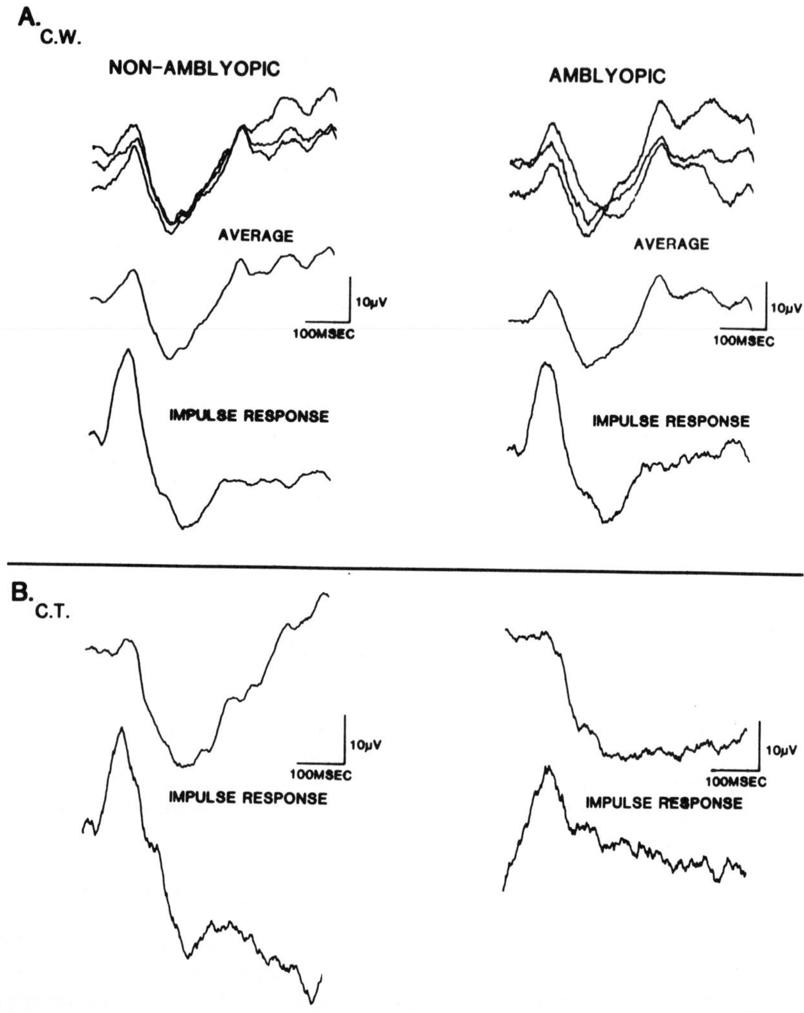

FIGURE 10. (A) Three electroretinograms (above) obtained in response to pseudorandom modulation of a uniform field in strabismic/anisometropic amblyope C.W. (nonamblyopic eye, left; amblyopic eye, right). Each record is an average of 50 presentations of the pseudorandom sequence. The average of these three records is shown in the middle trace while the lower trace is the impulse response. (B) The electroretinogram (nonamblyopic eye, left; amblyopic eye, right) recorded in response to pseudorandom modulation of a uniform field in stimulus deprivation amblyope, C.T. Each record is an average of 50 presentations of the pseudorandom sequence. The impulse response is shown below each record.

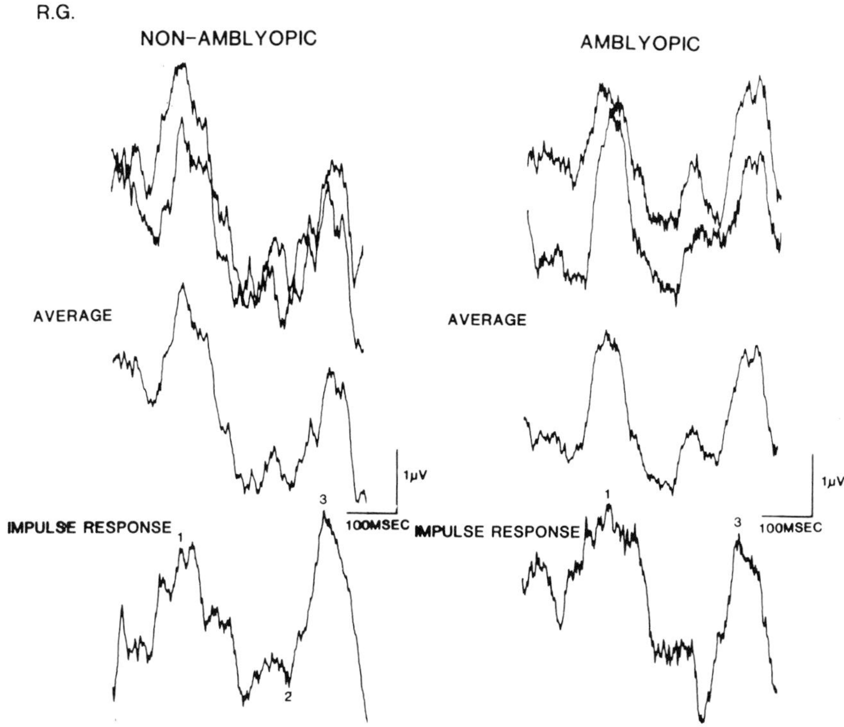

FIGURE 11. (Top) Two ERG traces recorded in response to pseudorandom counterphase alternation of a 2 c/deg sine-wave grating. The nonamblyopic eye is shown on the left; the amblyopic eye on the right for mild amblyope R.G. (20/40). Each trace represents the average of 125 responses. (Middle) The mean of the two traces shown above. (Bottom) The impulse response of the average for each eye.

The lowest trace is the impulse response of the averaged data. For this observer with a moderate degree of amblyopia (20/46) there is little difference, in latency or amplitude of the major components, between the two eyes. The results for a more severe amblyope (J.V.) are shown in FIGURE 12. The results at the top of FIGURE 12 are the grand mean of three separate sets of data (each representing 125 averages) for the nonamblyopic (left) and amblyopic (right) eyes. Below are the impulse responses of the averaged data. For this observer, the latency of the response of the amblyopic eye to the three major components (labeled 1, 2, and 3) appear to be delayed with respect to the nonamblyopic eye, and the amplitude of the component labeled 3 appears to be depressed.

These results are quite preliminary; taken in conjunction with the recent report of Sokol and Nadler,[39] however, it appears that deep amblyopes may show anomalies in retinal responses elicited by patterned stimuli. It remains to be tested whether milder amblyopes, such as R.G. (FIGURE 11) and the subjects of Tuttle[42] might also show abnormalities with higher spatial frequency stimuli. These results are also consistent with the recent report that in kittens reared with experimental unilateral strabismus, ganglion cells of the strabismic eye show reduced spatial resolution.[43]

FIGURE 12. (Top) The mean of three separate ERG traces recorded in response to pseudorandom counterphase alternation of a 2 c/deg sine-wave grating. The nonamblyopic eye is shown on the left; the amblyopic eye on the right, for moderate amblyope J.V. (20/80). (Bottom) The impulse response of the average ERG for each eye.

CONCLUSIONS

The results of these studies are consistent with many previous studies in showing that amblyopia due to strabismus/anisometropia results in marked abnormalities in the cortical responses of the amblyopic eye to patterned stimuli. In amblyopia due to stimulus deprivation, marked abnormalities occur in response to luminance as well as patterned stimuli. A comparison of the cortical evoked responses of amblyopes with responses evoked from the retina and psychophysical results may provide valuable information as to the mechanism and site(s) of amblyopia.

ACKNOWLEDGMENTS

Some of the work reported here was done in collaboration with Dr. Ronald S. Harwerth, and with Dr. James Walters.

REFERENCES

1. VON NOORDEN, G. K. 1977. Mechanisms of amblyopia. Adv. Ophthalmol. **34:** 93–115.
2. BIRNBAUM, M. A., K. KOSLOWE & R. SANET. 1977. Success in amblyopia therapy as a function of age: A literature survey. Am. J. Optom. Physiol. Opt. **54:** 269–275.
3. HUBEL, D. H. & T. N. WIESEL. 1963. Receptive fields of cells in striate cortex of very young, visually inexperienced kittens. J. Neurophysiol. **26**(6): 994–1002.

4. Hubel, D. H. & T. N. Wiesel. 1965. Binocular interaction in striate cortex of kittens reared with artificial squint. J. Neurophysiol. **28**(b): 1041–1059.

5. Wiesel, T. N. & D. H. Hubel. 1963. Single-cell responses in striate cortex of kittens deprived of vision in one eye. J. Neurophysiol. **26**(b): 1003–1017.

6. Wiesel, T. N. & D. H. Hubel. 1965. Comparison of the effects of unilateral and bilateral eye closure on cortical unit responses in kittens. J. Neurophysiol. **28**(b): 1029–1040.

7. Crawford, M. L. J., R. Blake, S. J. Cool & G. K. von Noorden. 1975. Physiological consequences of unilateral and bilateral eye closure in macaque monkeys: Some further observations. Brain Res. **84**(1): 150–154.

8. Crawford, M. L. J. & G. K. von Noorden. 1979. The effects of short-term experimental strabismus on the visual system of *Macaca Mulatta*. Invest. Ophthalmol. Vis. Sci. **18:** 496–505.

9. Eggers, H. M. & C. Blakemore. 1978. Physiological basis of anisometropic amblyopia. Science. **201:** 264–267.

10. Sherman, S. M., K. P. Hoffman & J. Stone. 1972. Loss of a specific cell type from dorsal lateral geniculate nucleus in visually deprived cats. J. Neurophysiol. **35**(4): 532–541.

11. Ikeda, H. & M. J. Wright. 1976. Properties of LGN cells in kittens reared with convergent squint: A neurophysiological demonstration of amblyopia. Exp. Brain Res. **25**(1): 63–77.

12. Ikeda, H., K. E. Tremain & G. Einon. 1978. Loss of spatial resolution of lateral geniculate neurone in kittens reared with convergent squint produced at stages in development. Exp. Brain Res. **31:** 207–220.

13. Potts, A. M. & T. Nagaya. 1969. Studies on the visual evoked response: III. Strabismus amblyopia and hysterical amblyopia. Doc. Ophthalmol. **26:** 394–402.

14. Shipley, T. 1969. The visually evoked occipitogram in strabismic amblyopia under direct view ophthalmoscopy. J. Pediatr. Ophthalmol. **6**(2): 97–112.

15. Nawratski, I., E. Auerbach & H. Rowe. 1966. Amblyopia ex anopsia; the electrical response in retina and occipital cortex following photic stimulation of normal and amblyopic eyes. Am. J. Ophthalmol. **61**(3): 430–435.

16. Fishman, R. S. & R. M. Copenhaver. 1967. Macular Disease and Amblyopia; the visual evoked response. Arch. Ophthalmol. **77**(6): 718–725.

17. Spekreijse, H., L. H. Khoe & L. H. van der Tweel. 1972. A case of amblyopia; Electrophysiology and psychophysics of luminance and contrast. *In* The Visual System; Proc. 9th ISCERG Symp. G. B. Arden, Ed. Plenum Publishing Co., New York, N.Y.

18. Levi, D. M. 1975. Patterned and unpatterned visual evoked responses in strabismic and anisometropic amblyopia. Am. J. Optom. Physiol. Optics **52**(7): 455–464.

19. Spekreijse, H., O. Estevez & D. Reits. 1977. Visual evoked potentials and the physiological analysis of visual processes in man. *In* Visual Evoked Potentials in Man: New Developments. J. E. Desmedt, Ed. pp 16–89. Clarendon Press, Oxford, England.

20. Fricker, S. J. & M. Kuperwaser. 1979. Use of pseudo randomly timed stimuli and cross-correlation techniques for visual evoked response measurements. Computers in ophthalmology, April, 1978, St. Louis, Mo. pp. 242–248. IEEE, New York, N.Y.

21. Srebro, R. & W. W. Wright. 1980. Visually evoked potentials to pseudorandom binary sequence stimulation: preliminary clinical trials. Arch. Ophthalmol. **98:**296–298.

22. Srebro, R., B. Sokol & W. W. Wright. 1981. The power spectra of visually evoked potentials to pseudorandom contrast reversal of gratings. Electroencephalogr. Clin. Neurophysiol. **51:**63–68.

23. Manny, R. E. 1981. Psychophysical and electrophysiological investigations in amblyopia: uniform field and spatio-temporal interactions. Doctoral dissertation, Univ. of Houston, Houston, Tex.

24. Manny, R. E. & D. M. Levi. 1981. Psychophysical investigations of the temporal modulation sensitivity function in amblyopia: uniform field flicker. Invest. Ophthalmol. Vis. Sci. (in press).

25. Lombroso, C. T., F. H. Duffy & R. M. Robb. 1969. Selective suppression of cerebral evoked potentials to patterned light in amblyopia exanopsia. Electroencephalogr. Clin. Neurophysiol. **27**(3): 238–247.

26. SOKOL, S. & B. BLOOM. 1973. Visual evoked cortical responses of amblyopes to a spatially alternating stimulus. Invest. Ophthalmol. **12**(12): 938–939.
27. LEVI, D. M. & J. W. WALTERS. 1977. Visual evoked responses in strabismic and anisometropic amblyopia: effects of check size and retinal locus. Am. J. Optom. Physiol. Optics **54**(10): 691–698.
28. LEVI, D. M. & R. S. HARWERTH. 1978. Contrast evoked potentials in strabismic and anisometropic amblyopia. Invest. Ophthalmol. Vis. Sci. **17**(6): 571–575.
29. HESS, R. F. & E. R. HOWELL. 1977. The threshold contrast sensitivity function in strabismic amblyopia: Evidence for a two type classification. Vis. Res. **17:** 1049–1055.
30. HARWERTH, R. S. & D. M. LEVI. 1978. A sensory mechanism for amblyopia: Psychophysical studies Am. J. Optom. Physiol. Optics **55**(3): 151–162.
31. PARKER, D. M. & E. A. SALZEN. 1977. Latency changes in the human visual evoked response to sinusoidal gratings. Vis. Res. **17**(10): 1201–1204.
32. BREITMEYER, B. G. 1975. Simple reaction time as a measure of the temporal response properties of transient and sustained channels. Vis. Res. **15**(12): 1411–1412.
33. HARWERTH, R. S. & D. M. LEVI. 1978. Reaction time as a measure of suprathreshold grating detection. Vis. Res. **18:** 1579–1586.
34. LEVI, D. M., R. S. HARWERTH & R. E. MANNY. 1979. Suprathreshold spatial frequency detection and binocular interaction in strabismic and anisometropic amblyopia. Invest. Ophthalmol. Vis. Sci. **18:** 714–725.
35. CAMPBELL, F. W. & L. MAFFEI. 1970. Electrophysiological evidence for the existence of orientation and size detectors in the human visual system. J. Physiol. **207:** 635–652.
36. SOKOL, S. 1977. Visual evoked potentials stimuli in strabismic amblyopia. *In* Visual Evoked Potentials in Man: New Developments. J. E. Desmedt, Ed. pp. 410–417. Clarendon Press, Oxford, England.
37. REGAN, D. 1978. Assessment of visual acuity by evoked potential recording: ambiguity caused by temporal dependence of spatial frequency selectivity. Vis. Res. **18:** 439–443.
38. TYLER, C. W., P. APKARIAN & K. NAKAYAMA. 1978. Multiple spatial-frequency tuning of electrical responses from human visual cortex. Exp. Brain Res. **33:** 535–550.
39. SOKOL, S. & D. NADLER. 1979. Simultaneous electroretinograms and visually evoked potentials from adult amblyopes in response to pattern stimulus. Invest. Ophthalmol. Vis. Sci. **18:** 848–855.
40. BURIAN, H. & T. LAWWILL. 1966. Electroretinographic studies in strabismic amblyopia. Am. J. Ophthalmol. **61:** 422–430.
41. ARDEN, G. B., R. M. CARTER, C. HOGG, J. M. SIEGEL & S. MARGOLIS. 1979. A gold foil electrode: Extending the horizons for clinical electroretinography. Invest. Ophthalmol. Vis. Sci. **18:**421–426.
42. TUTTLE, D. R. 1973. Electrophysiological studies of functional amblyopia utilizing pattern reversal techniques. Thesis, Univ. of Louisville, Louisville, Ky.
43. IKEDA, H. & K. E. TREMAIN. 1979. Amblyopia occurs in retinal ganglion cells in cats reared with convergent squint without alternating fixation. Exp. Brain Res. **35:** 559–582.

DISCUSSION OF THE PAPER

QUESTION: I find your temporal frequency differences between the amblyopic and nonamblyopic eye very interesting. I just wondered if you could comment on any differences you might see between the nonamblyopic eye of the amblyope and a normal eye of the nonamblyope.

D. LEVI: That is a tough question. In terms of our psychophysical results, we see the same sort of variation in nonamblyopic eyes as we see in normal eyes in terms of their temporal characteristics. The VEP temporal specificity seems to be quite different from one observer to another. It seems to be quite repeatable for a given observer but changes from one to another. This is well documented by Tyler and others as well (Tyler, Apkarian, and Nakayama, 1978). And so it is difficult to answer that question. It is true that there are abnormalities that have been reported in nonamblyopic eyes; however, these seem to be mostly in the motor system; that is, the eye movements of strabismic amblyopes in the nonamblyopic eye are not quite normal.

R. SREBRO: I have a question and a comment on the binary pseudorandom ERG. I have tried that, and the problem I ran into was that the subjects vary enormously in how well they can keep their eyes steady. And if you try to get some sort of crude baseline by having them observe a constant field, for example, it is disconcerting how much difference there is. If you look at the data, record by record, the signal-to-noise is really rather good for the ERG; one does not have to average. But there are obvious artifacts, so the data needs some scrubbing. With amblyopes we know that their fixation is not very good. Could it be possible that the differences you are finding really represent difficulties in fixation?

LEVI: That is an excellent question and I would like to answer it in two ways. We ran into the same problems with that, with ERGs both to luminance and pattern, and we attempted to deal with them in several ways. Firstly, we record with an Arden gold foil electrode to preserve the optics of the eye. Secondly, we maintain fixation using the nonamblyopic eye. When we test the amblyopic eye, the nonamblyopic eye is given a fixation point and the screen is positioned so that the nine degree field is essentially in the foveal area. Thirdly, we have an artifact rejection program which eliminates traces from buffer memory that include blinks, eye movements, or any kind of sudden change in voltage. These are the ways in which we have attempted to deal with a lot of the technical problems associated with measuring ERGs.

We also try to train our observers over a period to time to be able to keep their eye open and not blink until we tell them to blink, which we do in between averages.

SREBRO: Could you comment about the luminance modulation VEPs using the binary psueudorandom stimuli? Your data showed some losses around the 6 Hz region. We found that a number of people with macula degeneration had similar findings. I would point out two things. One is that we did report in the *EEG Journal* (Srebro, Sokol, and Wright. 1981) that if you do pseudorandom pattern inversion for sine-wave gratings, and you push the sine-wave spatial frequency up to between 7 and 15 Hz, that the most consistent finding amongst a lot of the subjects is that your power to localize to that 6 Hz region may have some kind of speciality.

There is another thing that I did not show yesterday and which has cropped up. If you recall, the ternary pseudorandoms do not have any even power components, any even harmonics, but that 4 to 7 Hz region is almost all made up of the even power harmonics. In other words, it is a selectively highly nonlinear region, so it has some special properties.

I have looked at three amblyopes in great detail using pattern sinusoidal reversal of gratings with binary pseudorandoms and I could not convince myself that they had any spatiotemporal interaction in their responses. Their results look more or less like yours, but the VEP power was down; it tended to be down a little more at the high spatial frequencies than the low. But there seemed to be no clear spatiotemporal interaction in that surface.

LEVI: That is 100% compatible with our results; that is, for pattern stimuli the temporal tuning seems to be essentially the same in the two eyes with pretty much overall losses in amplitude, which seem to get somewhat bigger at medium and higher spatial frequencies.

We have not yet performed any nonlinear analysis on the recorded data, but we are hoping to do that.

I. SIEGEL (*NYU*): I suppose I believed along with many of us that optical blur selectively reduces high spatial frequencies, yet there is a recent paper by Hess and Woo in which they looked at cataracts of various kinds, and I must tell you that most of the graphs that they showed indicated that there was almost as much attenuation at the low and middle frequencies as there was in the high spatial frequency domain. On the basis of that, can we completely eliminate subtle optical effects from being included in amblyopia such as Enoch's photoreceptor to tilt?

LEVI: We probably cannot altogether eliminate that. On the other hand, with Hess and Woo's cataract data, it is quite possible that many of the cataract patients have tiny areas of the lens that allow more or less clear imagery allowing high spatial frequencies through while more diffuse regions of opacities affect low spatial frequencies much more.

Harold Bedell essentially repeated the receptor tilt experiments and was unable to demonstrate a receptor tilt that could account for the acuity losses of amblyopes.

I. BODIS-WOLLNER: The Hess and Woo experiments demonstrated a 6 dB or twofold loss, and it is very difficult with EPs to be so refined as to demonstrate a twofold loss, so I am not sure that it would enter tremendously into this consideration.

VAEGAN: I am very glad to see you have been able to get pattern losses in the ERG because we also have seen losses in a substantial series of about 27 amblyopes. Also, I have gotten much more substantial pattern losses in deprivation amblyopia than in any of our other amblyopes that we have so far seen.

We have amongst this series only 4 of the 27 for whom the response in the amblyopic eye through pattern stimulation is larger than or equal to the nonamblyopic eye. When we carefully looked at these 4 patients there was only one feature which was different from the other 23 and that is that these patients had responded to treatment. At this stage, have you looked at therapy?

LEVI: We have not looked at therapy directly and we really have not measured pattern ERGs on enough amblyopes with different levels of acuity to make a real comment. I will say that the amblyope that I showed who essentially did not show a difference had 20/46 acuity in his amblyopic eye which is not very deep amblyopia. He had previously been treated and was also a subject in many different experiments in which he was being tested with his amblyopic eye, so we cannot rule out that that was essentially a treatment effect.

VAEGAN: Some of the ratio losses we see are very small and we do not find any direct correlation between the magnitude of the loss in most cases, and the level of amblyopia. However, if we take the ratio between the eyes, there is a rough correlation between that ratio and the ratio of the acuities, but it is extremely rough.

S. SOKOL: In the mild amblyope where you found no problem with the pattern ERG, did you talk about his VEP?

LEVI: Yes, his VEP was markedly abnormal at all spatial frequencies. The results that I showed for the spatiotemporal pseudorandom binary VEPs were for the same amblyope.

TYLER: Dr. Levi, I was very interested in the high temporal frequency responses that you were getting. In general, it seemed to me that you had in several observers two peaks at around 40 and 70 Hz. Were these peaks going up and down with the physiological state in such a way as to indicate that they were unitary peaks? And do you have any comment on the relationship between the presence of those two peaks and the Regan–Spekreijse picture, i.e., that there is one high temporal frequency mechanism?

LEVI: In the Fourier domain we have only really looked at the first 30 Hz, we did not extend our analysis beyond that. Now that is not because there may not be responses at higher frequencies but we know that the stimulus was flat to at least 30 Hz, so I cannot really make a direct comment.

TYLER: Was I misreading the axis then?

LEVI: Maybe. It only went out to 30 Hz.

TYLER: I was obviously projecting my own data on to yours. My data do in fact show what I just asked.

LEVI: Initially we looked at higher frequencies than the first 30 Hz. We were looking at the amplitudes, and the amplitudes get very low by 30 Hz. However, as you well know, and have previously reported, so does the noise level. In terms of the signal-to-noise level, we have looked up to 30 Hz, where some observers very clearly have signals significantly higher than the noise, but I cannot really say much beyond that.

S. FRICKER: What were the ages of the amblyopes you were testing?

LEVI: We have tested amblyopes from between 12 and 30.

FRICKER: That is obviously a bit of a problem since the sort of effective period perhaps for dealing with them is when they are younger and something may have happened when they are older. But even with the younger ones we have always found that there is a possibility of a wide variation in responses depending on how the patient feels. I would like to say, as Dr. Bodis-Wollner has said, if you are dealing with signals with a signal-to-noise ratio of 3 to 6 dB, I think we are all wasting our time looking for fine differences because fine differences will occur statistically. You must take care that you are not just interpreting the noise and thinking it is a signal.

LEVI: Yes. We have looked at these differences more quantitatively than I have presented today. And in terms of comparing the two eyes, we look at data where the signal-to-noise ratio is above the noise level; that is, we only look at data where the signal-to-noise ratio is at least 1.5, but what you say is absolutely right. Where there are small differences, for instance, in the luminance responses, those differences are small and should be taken in the light of the VEPs repeatability and reliability.

The other thing that I might point out is that we have also repeated these experiments on all of these observers several times and essentially have obtained the same results.

THE PATTERN-EVOKED POTENTIAL IN LESIONS OF THE POSTERIOR VISUAL PATHWAYS

L. D. Blumhardt, G. Barrett, A. Kriss, and A. M. Halliday

Medical Research Council
Institute of Neurology
National Hospital
Queen Square
London WC1N 3BG
England

Following the introduction of averaging techniques,[1] attempts were made to use the visual evoked response (VER) to enable lesions affecting the visual pathways to be accurately localized. Early studies of the highly variable occipital potentials evoked by flash stimuli produced conflicting results and encountered considerable difficulty in separating normal and pathological responses, even in the presence of gross hemisphere lesions.[2,40] The greater consistency of the waveform of the potentials evoked by structured stimuli, together with their high sensitivity to pathological processes, offered considerable advantages over the responses evoked by luminance-change stimuli. Since the early 1970s the occipital potential evoked by pattern-reversal stimulation has been widely utilized in clinical neurophysiology for the detection of pathological lesions in the anterior visual pathways. Published accounts of their use for the detection of lesions of the cerebral hemispheres are mainly limited to single case studies or small series of patients.[3–9] These reports often lack comparative data on the healthy population as well as the clinical and radiological evidence required to precisely locate pathological lesions although two recent studies of occipital lobe and retrochiasmatic lesions have included clinical and neuroradiological data.[10,36] It has been suggested that the pattern response obtained by recording the full-field potentials at one or two active electrodes may be insensitive even to lesions resulting in dense hemianopic visual field defects[7,9] and is incapable of separating them from healthy controls.[10,36] Abnormalities may be found in such patients when recordings are made from the lateral occipital scalp.[8–11,36] However, such interpretations must allow for the high degree of asymmetry and waveform variance that is encountered in healthy subjects when recordings are made at some distance lateral to the midline.[12] These must be distinguished from the abnormal distribution of potentials ("uncrossed asymmetry") that is characteristic of the responses in patients with homonymous visual field defects.[11–15] The high variance of the waveform over the occipital scalp in the healthy can be reduced significantly by comparing the individually recorded half-field responses.[12]

In this paper we present the results of multichannel recordings of full- and half-field responses of 32 patients with a variety of hemisphere lesions which have been clearly demarcated by clinical, neuro-ophthalmological and neuroradiological techniques. Preliminary reports of this investigation have appeared elsewhere.[13–15]

Controls and Clinical Material

Fifty healthy medical students and paramedical hospital personnel with normal vision and negative visual histories were recorded. The pathological lesions and field

0077–8923/82/0388–0264 $1.75/0 © 1982, NYAS

defects in the patients recorded in the study are outlined in TABLE 1. Patients with well-demarcated lesions which followed a single pathological event were selected. All cases of trauma or infarction were recorded remote in time from the onset of the lesion to avoid possible nonspecific effects on the visual pathways from oedema or displacement of intracranial structures. The CAT scans showed the extent and localization of the acute lesion soon after presentation, which would thus, in most cases, be expected to be more extensive than at the time of the later VER recording. The residual deficit in these cases is reflected by the extent of the visual field defect plotted at the time of

TABLE 1
CLINICAL MATERIAL*

Case	Age	Sex	Visual Field Defect	Pathological Lesion
1. IP	50	F	R ho. he.	L hemispherectomy
2. ET	47	F	L ho. he.	R hemispherectomy
3. AM	28	F	L ho. scot.	R subtotal occipital lobectomy
4. FB	35	M	L ho. he.	R subtotal occipital lobectomy
5. FM	52	M	L ho. he.	R subtotal occipital lobectomy
6. AS	73	M	R ho. he.	L Calc. A infarction
7. DF	42	M	L ho. he.	R. occipital infarction
8. GS	66	F	R ho. he.	L parietal/occipital infarction
9. LP	59	M	L ho. he.	R occipital meningioma (ex.)
10. AR	43	M	R ho. he.	L posterior CA thrombosis‡
11. SK	26	M	R ho. he.	?L posterior CA infarction‡
12. RP	47	M	R ho. scot.	L Calc. A infarction
13. MB	31	F	R ho. he.	L occipital horn plaque (MS)
14. AL	30	F	L ho. scot.	?R optic tract plaque (MS)
15. EM	28	F	L ho. he.	R cerebral infarction‡
16. PH	32	M	R u. quad.	L temporal lobe abscess (ex.)
17. PM	59	F	R u. quad.	L posterior parietal infarction
18. WW	56	M	L u. quad.	R posterior parietal infarction
19. MA	76	F	L u. quad.	R occipital infarction
20. SH	20	M	L l. quad.	R occipito/parietal abscess (ex.)
21. LC	50	M	L l. quad.	R occipital infarction
22. RW	68	M	R l. quad.†	Bilateral occipital infarction
23. WS	62	M	R l. quad.	L. posterior cerebral infarction‡
24. LA	27	F	L con. ho. he.	R fronto/parietal AVM (ex.)
25. TC	27	M	L con. ho. he.	Bilateral occipital trauma
26. LA	55	M	L con. ho. he.	Bilateral occipital metastases
27. BL	52	M	Sym. con. fields	Bilateral occipital infarctions
28. KC	42	M	Nil	L parietal AVM
29. EH	34	F	Nil	L fronto/parietal AVM
30. IR	42	F	Nil	R temporal glioma
31. CB	47	M	Nil	L middle CA infarction
32. LT	64	F	Nil	R thalamic haemorrhage

*Key: ho. he. = homonymous hemianopia; scot. = scotoma; u. quad. = upper quadrantanopia; l. quad. = lower quadrantanopia; con. = constricted; sym. = symmetrical; MS = multiple sclerosis; ex. = excised; AVM = arterio-venous malformation; CA = cerebral artery; Calc. A = calcarine artery; R = right; L = left.

†Field defect crosses vertical meridian.

‡No abnormality detected on CAT scan.

the recording. A few cases of particular interest with tumors or plaques of demyelination were included who were recorded soon after their initial presentation.

Fifteen patients (cases 6–8, 10–12, 15, 17–19, 21–23, 27, and 31) were recorded following well-documented episodes of cerebral infarction. Eleven of these had clear evidence on their CAT scans of the site of the lesion. In four cases the CAT scan was normal but the clinical history, signs, and angiographic evidence indicated a unilateral retrochiasmatic lesion. Two vascular cases had clinical and CAT scan evidence of bilateral hemisphere involvement. One additional patient was recorded after recovery from a thalamic hemorrhage.

Nine patients (cases 1–5, 9, 16, 20, and 24) were recorded after surgical trauma in which there was damage to the posterior visual pathways. These included two cases (16 and 20) recorded after surgical excision of cerebral abscesses. Two additional cases with bilateral cerebral pathology were included, one with trauma to both occipital lobes (case 25) and the other with bilateral occipital metastases (case 26). Three cases with extensive space-occupying lesions sparing the primary geniculocalcarine pathways were recorded prior to surgery (cases 28–30). Two patients (cases 13 and 14) were recorded within days of an acute postchiasmal plaque of demyelination. When classified by visual field defect there were 18 cases with either complete or incomplete homonymous hemianopic visual field defects (cases 1–15). Three patients with bilateral pathology had markedly constricted visual fields (cases 24–26). There were eight patients with homonymous quadrantanopias (cases 16–23), four with upper and four with lower field defects. Five patients with extensive hemisphere pathology on the CAT scan had no detectable defects of their visual fields (cases 28–32).

Methods

The techniques are identical to those described in previous studies of the pattern VER in healthy subjects[12,16] or in patients.[11,13–15]

The patient sat facing a circular translucent screen whose radius subtended 16° at the eye, on to which was back-projected a slide of a black and white checkerboard. A small spot of light (40′ diameter) was placed at the center of the screen to provide a fixation point. The screen could be masked to provide half-field or quadrantic stimuli. Individual checkerboard squares subtended 50′ and the brightness levels were 227 cd/m^2 and 8.2 cd/m^2, respectively, for the white and black squares.

Pattern reversal was produced once every 605 msec by a rapid lateral displacement of the pattern through one square width, the transition taking approximately 10 msec. The average response to 200 pattern reversals was recorded at least twice for each stimulus condition. Fixation was monitored throughout the recording by closed circuit television.

The occipital potentials were recorded from an array of 16 electrodes each referred to a common reference placed 12 cm posterior to the nasion in the midline. A standard transverse chain of five electrodes was sited 5 cm above the inion; one electrode was placed on the midline with two electrodes to either side at an interelectrode distance of 5 cm. The additional eleven electrodes were disposed symmetrically in subsidiary chains above and below this level.

At the time of the recording all patients had their visual fields plotted on a Goldmann perimeter. Both patients and controls had their visual acuity measured on a Snellen card and those requiring refraction wore their glasses during the recording.

The responses recorded at each of the 16 electrode sites were visually inspected. Detailed analyses of component peak latencies and amplitudes (measured from the preceding peak of opposite polarity) were carried out on the recordings from the standard transverse chain of five electrodes. Measurements were also made at other electrode sites where the responses showed clear differences from the potentials recorded in the standard chain.

RESULTS

Physiological Asymmetries and Their Analysis by Half-Field Stimulation

Detailed analysis of the healthy control data has been previously published.[12] The results to be presented here are limited to examples of healthy waveform variation which may be misinterpreted as pathological.

A wide spectrum of waveform variations are encountered in healthy subjects, which, for descriptive purposes, can be grouped into six broad categories (FIGURE 1). A major factor in the variability is the transverse occipital voltage gradient. The amplitude of the major full-field positivity, for example, at the electrode 10 cm lateral to the midline, varied from 0%–40% of the midline value. In about 6% of healthy subjects there was difficulty identifying a clear major positive wave at the lateral scalp electrodes (subject 2, FIGURE 1). In other subjects the major positive wave was widespread and clearly identifiable at all electrodes (subject 3, FIGURE 1). This normal variation may create difficulties in the recognition and measurement of the major potentials in some full-field responses.

Asymmetries of the monocular full-field responses are frequently encountered in healthy subjects and are invariably homonymous or "uncrossed." Eleven of the 50 healthy controls had "minor asymmetries" with amplitude ratios for the P100 waves (i.e., the positive peaks at about 100 msec) at the 5 cm electrodes of less than 4:3, but in 14 of the 50 the ratios exceeded 2:1. In the other 25 individuals the ratios lay between these two limits. Asymmetries of latency may also occur. Apparent interchannel latency differences, particularly between the midline and lateral responses, were common. In about 6% of healthy subjects the midline major positive deflection was bifid or "W-shaped," creating problems for interindividual and interchannel comparisons (subject 6, FIGURE 1).

The difficulties caused by such variable waveforms could be largely eliminated by recording the half-field responses. The above "physiological variants" were caused by the variable amplitude ratios of ipsilateral and contralateral components from each visual half-field, as the full-field waveform arises from a simple algebraic summation of these potentials.[11,12] Where the ipsilateral and contralateral components are of similar amplitude and opposite polarity, the "cancellation" accounts for the apparent lack of activity at lateral scalp electrodes. Symmetrical cancellation may occur on both sides of the scalp (FIGURE 2). In other subjects an asymmetrical full-field response distribution could be attributed to the differences between the *contralateral* components recorded from each visual half-field (FIGURE 3). In an occasional subject variations in form of *both* ipsilateral and contralateral potentials from each hemisphere were responsible for the asymmetry of the full-field response (FIGURE 4). In a minority of subjects different *relative distributions* of ipsilateral and contralateral potentials of similar form arising from each hemisphere resulted in waveform variations near, or at, the midline (FIGURE 5).

Limits of Normal Asymmetry

Normal limits were established with this data obtained from 50 healthy subjects, details of which have been previously reported.[12] The normal amplitude and latency

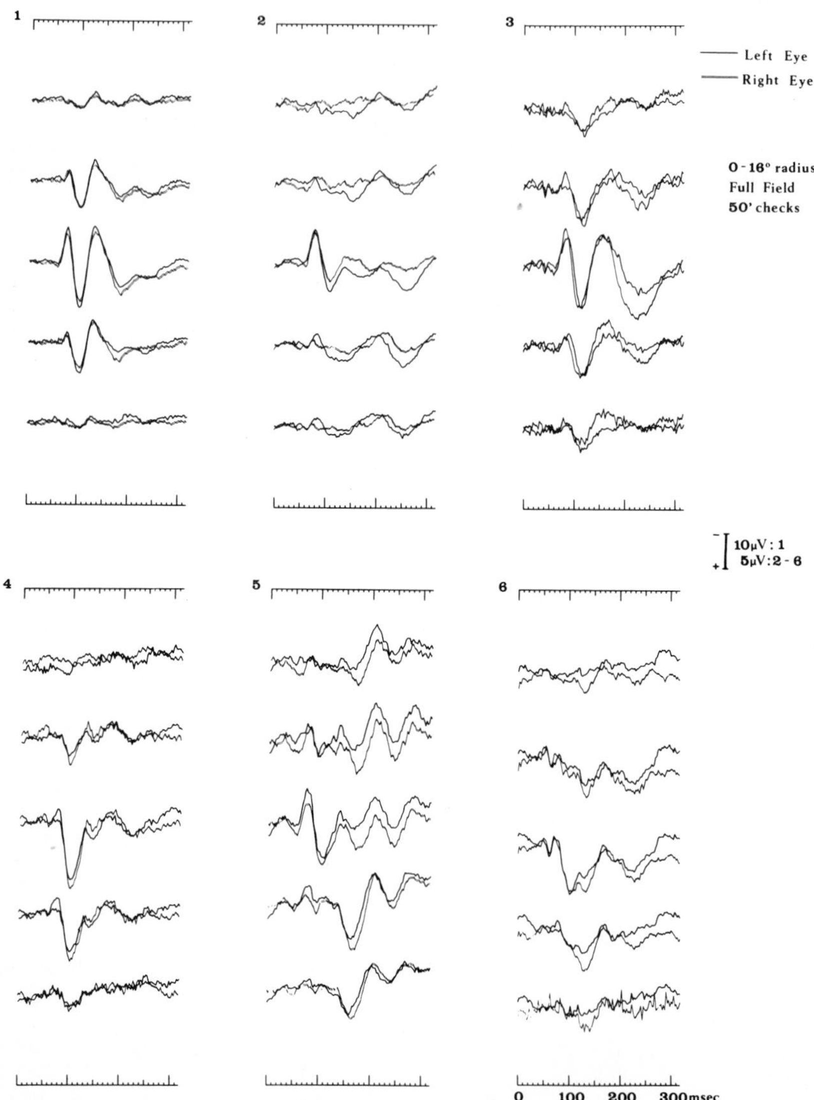

FIGURE 1. Examples of full-field response waveform variation Blumhardt 18/636ts. Left and right eye responses superimposed. Subjects 1–3 show variable lateral spread of components across occipital scalp. Asymmetric "uncrossed" distribution of P100 components is shown in subject 4 and for the positive wave at 165 msec by subject 5. Subject 6 shows complex "W-shaped" midline response. For montage, see FIGURES 2–11.

criteria were established for the most consistent potentials at electrode sites that provided the least variable intersubject results; for full-field responses, measurements of the major positivity were made in the midline and at electrodes 5 cm to either side (TABLES 2 & 3). The ratio of the larger to the smaller amplitude on either side of the

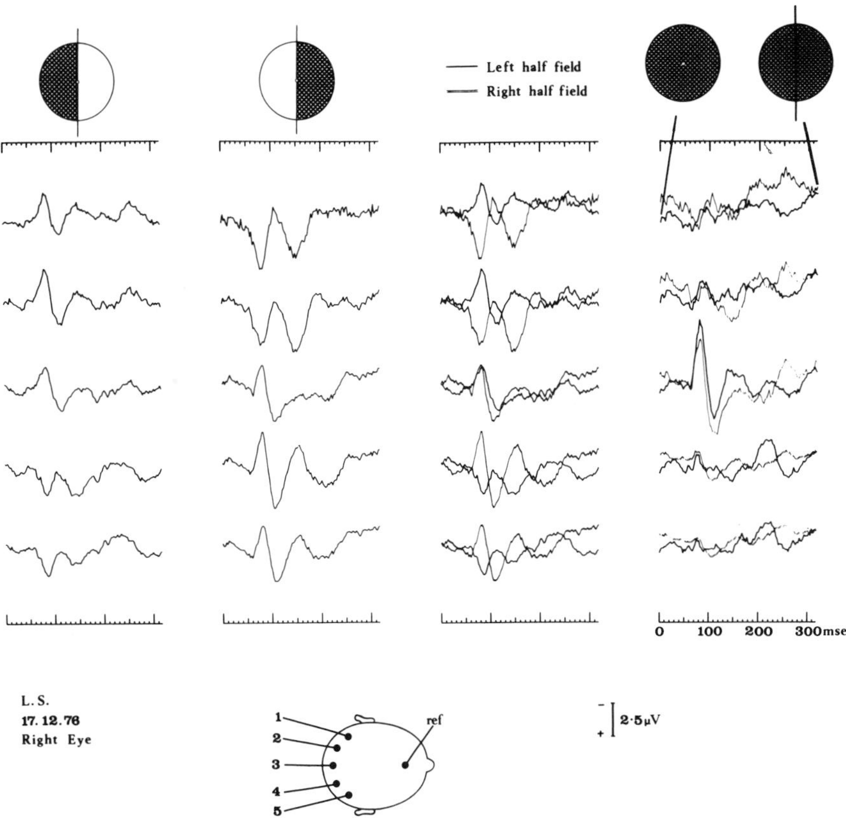

FIGURE 2. Component cancellations in the full-field response: (a) An example of the "midline response." Half-field, summed half-field and full-field responses of subject 2, FIGURE 1. Components from each half-field are well-defined and widespread over at least 10 cm of lateral scalp to either side, but full-field response appears largely restricted to the midline electrode. Superimposition of half-field responses in column 3 demonstrates how components of opposite polarity from each half-field "cancel" at lateral electrodes while similar polarity responses at the midline summate. The right-hand column demonstrates the similarity between the algebraically summed half-field responses and the separately recorded full-field response as for FIGURES 3–5.

scalp was used as an estimate of asymmetry. From this data an upper limit for full-field amplitude asymmetry could be set at 1:3 (TABLE 3).

For the half-field responses the intersubject and interhemisphere variability of the *contralateral* half-field components was found to be too high to allow clear separation of the normal and pathological, for these particular potentials.[12,16] The most reliable basis for half-field comparisons was found to be the P100 waves at ipsilateral electrodes. This wave had a narrow latency and amplitude range (TABLES 2 & 3). From this data the upper limit of amplitude asymmetry between half-field responses could be set for this component at 1:2.2 for the left eye and 1:2.1 for the right eye (TABLE 3).

Pathological Asymmetries of the Full-Field Response

The dominant feature of the monocular *full-field* VER in patients with homonymous visual defects is an asymmetric distribution of the response components which is

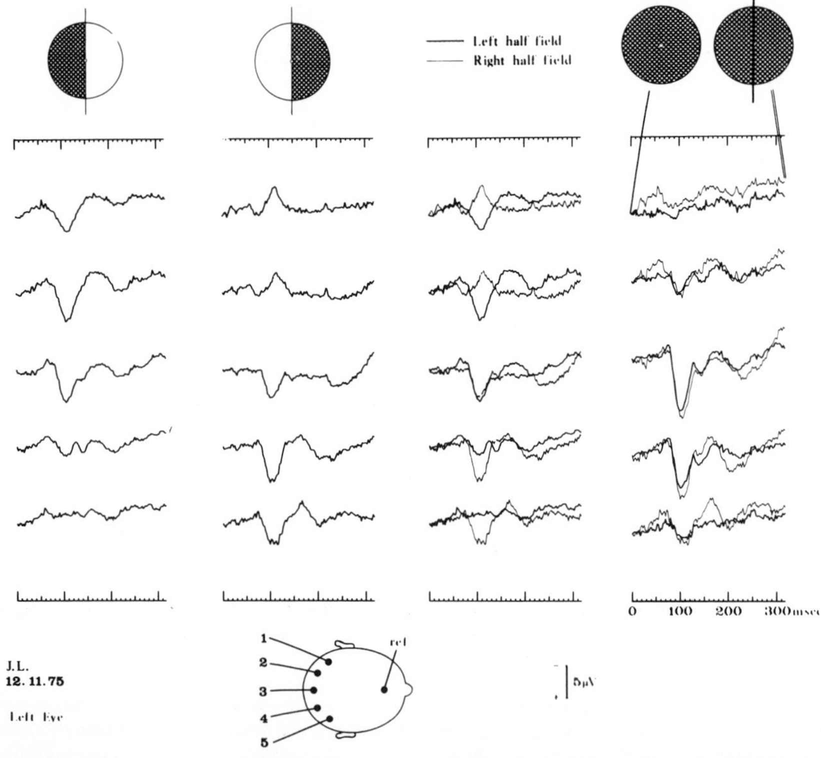

FIGURE 3. Component cancellations in the full-field response: (b) The "asymmetrical response." Half-field, summed half-field and full-field responses of subject 4, FIGURE 1. Full-field positivity is larger over the right side (compare channels 2 and 4). Compared with similarity of ipsilateral response waveforms from each half-field there are marked differences in the contralateral recordings (compare the most lateral channels in first two columns). Summation results in partial cancellation at left-sided electrodes and slight enhancement to the right of the midline because of the more widely spreading P100 component of the left half-field response and the presence of a contralateral negativity at 100 msec from the right half-field response.

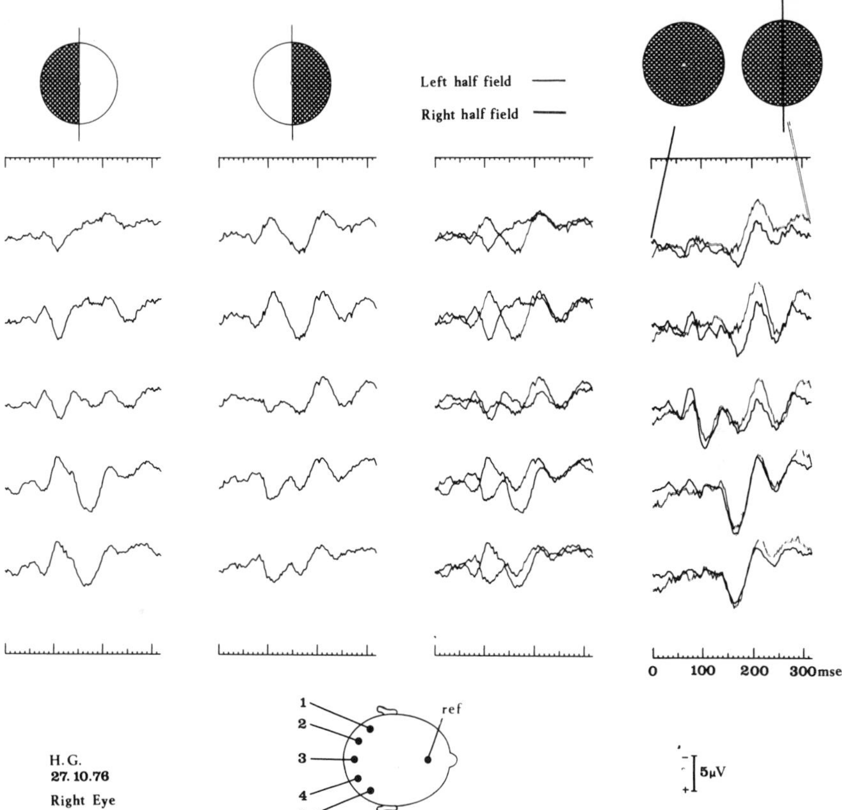

FIGURE 4. Component cancellations in the full-field response: (c) The "asymmetrical response." Half-field, summed half-field and full-field responses of subject 5, FIGURE 1. At about 100 msec cancellation occurs laterally because of unusually prominent contralateral complexes. The large positive wave in left-sided channels at 165 msec (P135) is largely balanced by broad surface-negative wave from the left half-field response. The equivalent negativity on the right is small and relatively "early" and followed by a positivity so that the left half-field P135 is not attenuated but actually reinforced at the right-sided electrodes in full-field response.

similar from the left and right eyes, as we have previously described.[11,13–15] Although the asymmetry may be more marked than those seen in healthy subjects, some do not exceed normal amplitude ratio limits (TABLE 4). The asymmetry is caused by attenuation, abolition, or replacement of the normally recorded activity by potentials of opposite polarity, at the electrodes situated over the scalp ipsilateral to the visual field defect. In contrast, the responses recorded from electrodes contralateral to the visual field defect are invariably of normal waveform.

With a complete or dense hemianopic defect the asymmetry of the full-field responses results from the loss of all components generated from the affected half-field or hemisphere. The corresponding potentials are abolished on *both* sides of the scalp.

The resultant response to full-field stimulation is then identical to the intact half-field response (e.g., FIGURE 7). Thus the "pathological asymmetry" of the recorded activity in such cases depends entirely on the distribution of the potentials from the *unaffected* hemisphere. While most half-field responses are markedly asymmetric, the responses recorded at contralateral electrodes are highly variable and may show either widely spreading ipsilateral potentials, a contralateral positive-negative-positive complex

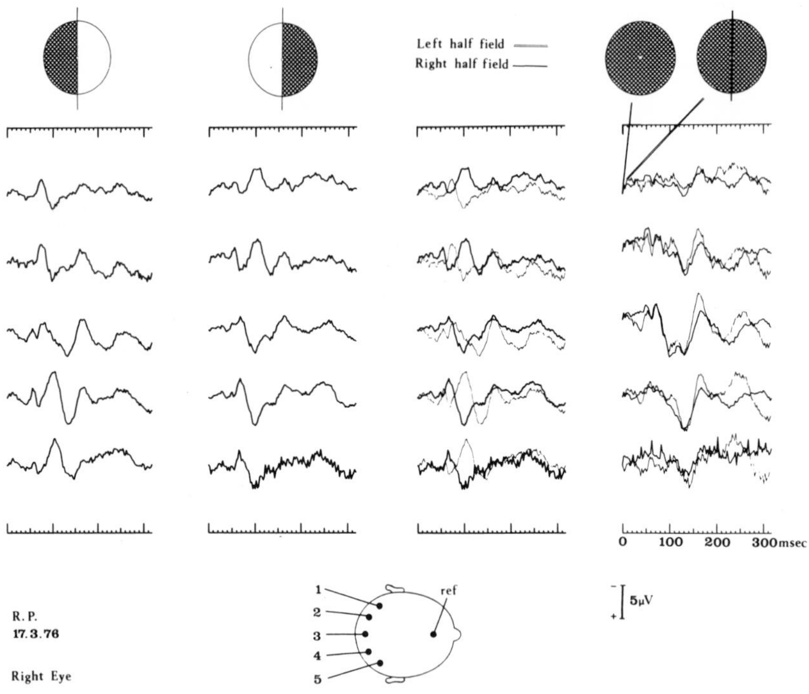

FIGURE 5. Component cancellations in the full-field response: (d) The "W-waveform." Half-field, summed half-field and full-field responses of subject 6, FIGURE 1. Component distributions vary for each half-field response. A more usual distribution of the P100 component in ipsilateral and midline channels is seen for the right half-field response. For the left half-field response the P100 is lateralized to channels 1 and 2 and the positive peak in the midline coincides with large contralateral P135. Midline response for left half-field is thus "transitional."[16] Superimposition in column 3 shows components cancelling at about 75 and 100 msec leaving only positive waves unopposed by negativity. First and second positive peaks in midline full-field response are predominantly derived from the ipsilateral P100 of the right half-field and the contralateral P135 of the left half-field, respectively. In lateral channels of full-field response the P100 component from each half-field is "cancelled" and P135 waves dominate. This accounts for the apparent "latency difference" between midline and lateral full-field response in this subject.

(PNP) or hybrid waveforms.[16] Thus it could be predicted that variable waveforms would also be obtained from electrodes situated contralateral to the *intact* half-fields of hemianopic patients. In 12 of the 26 patients with homonymous visual field defects, the activity recorded ipsilateral to the abnormal field on full-field stimulation was a

TABLE 2

MEAN LATENCIES AND UPPER LIMITS (n = 50)

	Eye	Mean Latency and SD	Upper Limit (2.5 SD)
(a) Full-field major positive wave			
Electrode			
Midline	Left	105.0 ± 4.0	115
	Right	105.7 ± 4.6	117
5 cm left of midline	Left	105.5 ± 10.3	131
	Right	107.6 ± 9.4	131
5 cm right of midline	Left	107.2 ± 7.6	126
	Right	107.3 ± 8.9	130
(b) Half-field P100 at electrodes 5 cm ipsilateral to stimulated field			
Half-field			
Left	Left	105.3 ± 4.9	118
Left	Right	106.0 ± 4.7	118
Right	Left	106.0 ± 3.9	116
Right	Right	104.8 ± 4.0	115

clearly defined contralateral PNP complex (e.g., FIGURE 7). The asymmetries of the full-field response in such cases are described as "polarity reversals" in TABLE 4.

In other patients the activity recorded at the same latency on either side of the scalp was similar in polarity but differed in amplitude (FIGURE 6). The asymmetry in these cases is described as an amplitude reduction (either ipsilateral or contralateral with respect to the field defect) in TABLE 4. For example, in this group, the major positivity was smaller on the side of the scalp ipsilateral to the visual field defect in 12 of 14 patients (e.g., FIGURE 6). In the majority of this group the asymmetry did not exceed the normal full-field ratio limits (TABLE 4). Two patients (cases 21 and 23) with lower quadrantanopias had unexpected asymmetries in which the P100 was slightly smaller contralateral to the affected half-fields. Both these cases had sparing of macular vision (e.g., FIGURE 9).

Thus 24 of 26 patients (92%) with various homonymous visual field defects had asymmetries which could be predicted from the side of the hemisphere lesion. However, only 14 of the 26 (54%) had full-field responses with abnormal monocular

TABLE 3

MEAN AMPLITUDE RATIOS FOR ELECTRODES 5 cm LATERAL TO MIDLINE

Eye	Ratio	Upper Limit (2.5 SD)
*(a) Full-field major positive wave**		
Left	1.58 ± 0.559	2.98
Right	1.60 ± 0.533	2.93
(b) Half-field P100†		
Left	1.36 ± 0.331	2.19
Right	1.34 ± 0.296	2.08

*Ratio of larger wave over smaller wave.

†Ratio of larger wave over smaller wave from left and right half-fields in the same eye.

TABLE 4

P100 AMPLITUDE ASYMMETRIES IN POSTERIOR VISUAL PATHWAY AND HEMISPHERE LESIONS

Case No.*	Full-Field Asymmetry†	Full-Field Ratio‡ Left Eye	Full-Field Ratio‡ Right Eye	Half-Field Ratio‡ Left Eye	Half-Field Ratio‡ Right Eye
1. IP	Polarity reversal	∝	∝	∝	∝
2. ET	Polarity reversal	∝	∝	∝	∝
5. FM	Polarity reversal	∝	∝	∝	∝
6. AS	Polarity reversal	∝	∝	∝	∝
20. SH	Polarity reversal	∝	∝	∝	∝
24. LA	Polarity reversal	∝	∝	∝	∝
7. DF	Polarity reversal	∝	∝	3.3§	2.1§
8. GS	Polarity reversal	∝	∝	∝	2.7§
11. SK	Polarity reversal	∝	∝	5.0§	3.5§
16. PH	Polarity reversal	∝	∝	3.7§	2.1§
19. MA	Polarity reversal	∝	∝	∝	5.4§
13. MB	Polarity reversal	∝	∝	NR	NR
3. AM	Ipsilateral reduction	5.5§	3.6§	∝	∝
4. FB	Ipsilateral reduction	2.0	2.1	∝	∝
10. AR	Ipsilateral reduction	1.9	3.0	2.4§	∝
9. LP	Ipsilateral reduction	1.5	1.8	∝	∝
14. AL	Ipsilateral reduction	2.0	1.7	2.5§	2.2§
25. TC¶	Ipsilateral reduction	1.8	1.7	∝	∝
26. LA¶	Ipsilateral reduction	2.2	2.4	∝	∝
15. EM	Ipsilateral reduction	1.9	3.1§	1.4	2.5§
22. RW¶	Ipsilateral reduction	4.0§	2.5	3.9§	1.7
12. RP	Ipsilateral reduction	2.1	2.3	1.7	1.1
18. WW	Ipsilateral reduction	1.6	1.5	1.0	1.1
17. PM	Ipsilateral reduction	3.2§	4.6§	1.3	1.2
23. WS	Contralateral reduction	1.2	1.2	1.3	4.5§
21. LC	Contralateral reduction	1.0	1.2	1.5	1.1
29. EH**	Contralateral reduction	1.3	2.0	1.5	1.4
28. KC**	Contralateral reduction	1.5	1.1	1.1	1.3
30. IR**	Contralateral reduction	1.2	1.7	1.1	1.1
31. CB**	Contralateral reduction	1.2	1.2	1.4	1.5
32. LT**	Contralateral reduction	4.3§	4.5§	1.6	1.1

*Note that case 27 has been excluded from the analysis because of symmetric constriction of visual fields.

†Laterality of reduced amplitudes are with respect to visual field defect *or* field contralateral to hemisphere pathology if no field defect is present.

‡The symbol ∝ denotes abnormal ratios due to absence of component from one side of the scalp. NR, Not recorded.

§Abnormal ratios.

¶Cases with bilateral hemisphere pathology.

**Cases with hemisphere pathology but no detectable visual field defect.

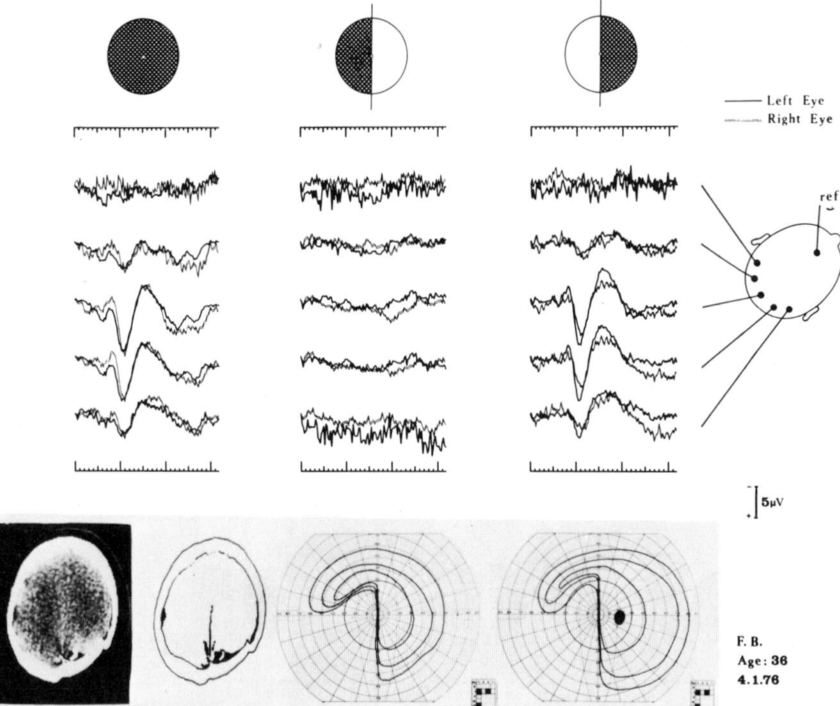

FIGURE 6. Full- and half-field responses after subtotal right occipital lobectomy. Note that the preserved periphery of left upper field is outside the stimulus screen and thus no response is detectable from the homonymous left half-fields. The full- and right half-field responses thus approximate. Note that the ipsilateral activity from the right half-field spreads over the midline with no clearly defined contralateral PNP complex. The severe field defect is thus not apparent in the full-field response (case 4).

amplitude ratios for both eyes. If the two patients with abnormal uniocular ratios (cases 15 and 21) are included, the detection rate for such lesions is increased to 62%. When the three cases with bilateral hemisphere damage are excluded the detection rate for homonymous visual field defects on full-field stimulation was 15/23 (65%).*

In five patients whose visual fields were clinically intact despite gross hemisphere pathology, there were homonymous asymmetries which were the reverse of those that would have been predicted from the knowledge of the side of the lesion. They were within normal limits in four of the five cases (case 32 exceeded the normal full-field ratio limit). Overall, 14 of 31 patients (45%) with gross but asymmetric hemisphere damage, with or without visual field defects, had amplitude asymmetries that were indistinguishable from those seen in healthy subjects.†

*Exclusions on these grounds may not be justified as the asymmetries appear to reflect only the visual field defects (see below).

†One patient with bilateral cortical lesions (case 27) had small, symmetrically-constricted fields.

Analysis of the Results of Half-Field Stimulation in Patients with Hemisphere Lesions

The results of half-field stimulation confirmed that the asymmetries associated with dense hemianopic field defects were frequently due to the loss of all activity from the affected fields. This was a common finding in the twelve patients whose full-field asymmetries were characterised by "polarity reversals" (TABLE 4). For example, when stimulation was restricted to the affected half-fields in either eye, no consistent P100 component could be recognized above the background noise in 6 of the 11 in this group who had half-field recordings (e.g., FIGURE 7). The remaining 5 patients with this type of full-field asymmetry had residual, abnormally attenuated components ipsilateral to the visual field defect in one or both eyes on half-field stimulation (TABLE 4). These small waves were obscured in the full-field response by the contralateral components from the intact half-fields (FIGURE 8). Nevertheless, the "unmasking" of half-field PNP complexes at electrodes on one side of the scalp in the full-field response appears to be a strong pointer to the presence of hemianopic field defects in the ipsilateral half-field. In the remaining group of 14 patients with homonymous field defects, in whom the P100 was still present on both sides in the full-field response, the

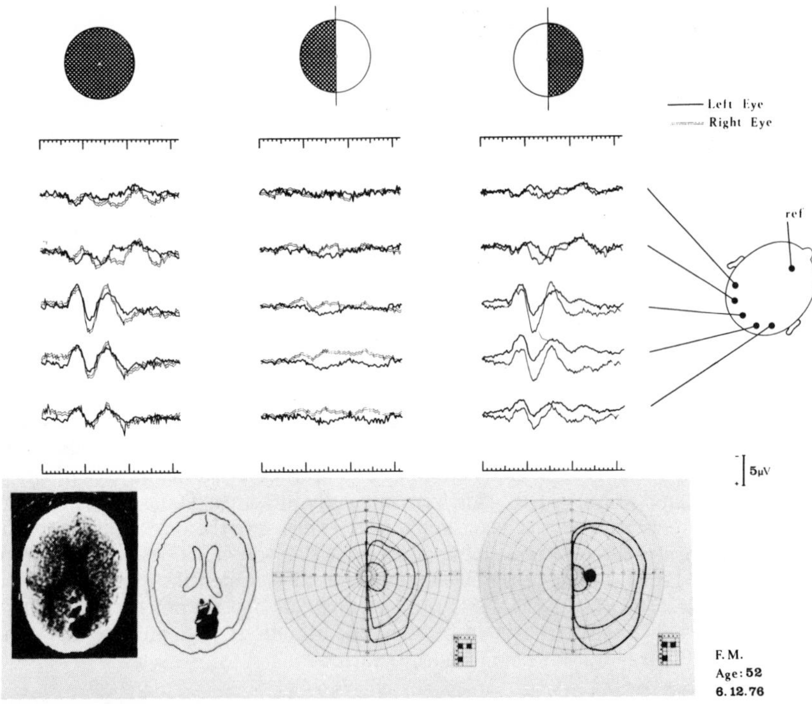

FIGURE 7. Full- and half-field responses after extensive right occipital lobectomy. Fields show macular-splitting complete homonymous hemianopia. There are no consistent responses from the left half-field. P100 components and accompanying ipsilateral components are clearly lateralized on scalp ipsilateral to intact half-fields, while a typical contralateral waveform (PNP) is recorded over the left scalp on right half-field stimulation (case 5).

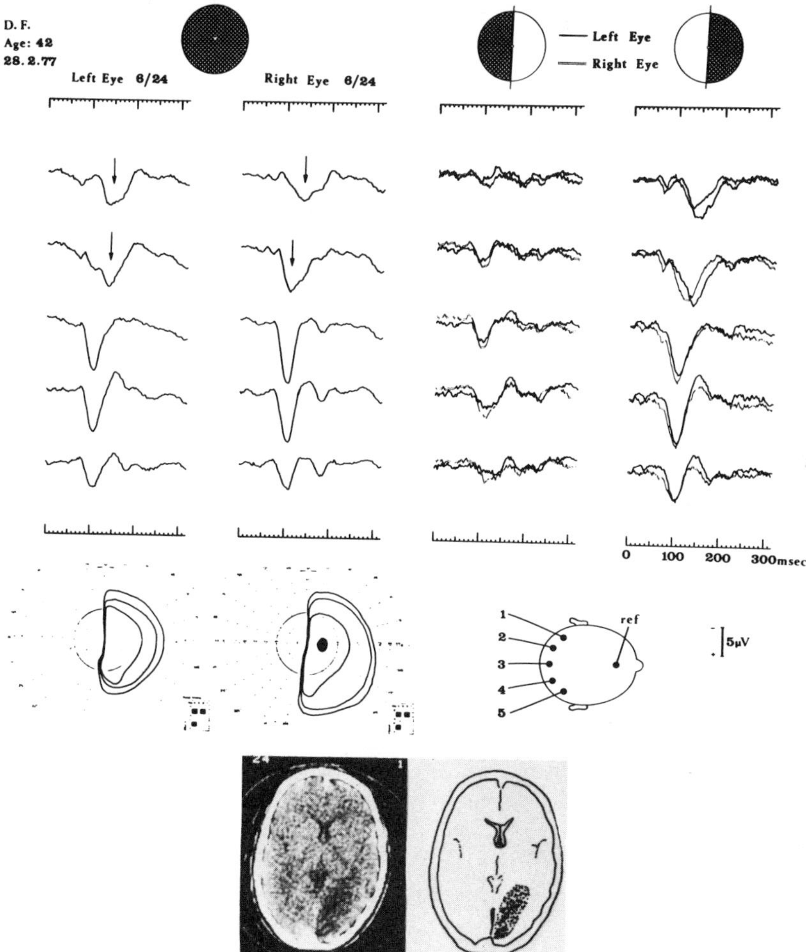

FIGURE 8. "Uncrossed" asymmetry with residual activity from affected hemianopic field. Right occipital infarction. Full-field responses show apparent "delay" of large positive wave contralateral to intact fields (arrows). This wave and the preceding negativity which is prominent in some channels (e.g., channel 1 left eye response) are largely composed of large P135 and N105 waves, respectively, arising from intact right half-fields. Residual left half-field activity arises from spared strip parallel to vertical meridian in lower field and is widespread and undelayed. Distribution of N75 and P100 waves in left half-field response clearly indicate an origin from affected field rather than inadvertent stimulation of intact right half-field near the vertical meridian (case 7).

full-field asymmetry was within normal limits for both eyes in ten cases (TABLE 4). The discrepancy between the activity arising from each hemisphere was revealed in seven of these *only on half-field stimulation.* In four of these cases (4, 9, 25, 26) *all* the recorded activity on both sides of the scalp was seen to be arising from the intact half-fields (e.g., FIGURE 6). The failure of the full-field response to demonstrate the pathological asymmetry in these cases is due to the spread of the ipsilateral NPN complex across the midline from the unaffected hemisphere, presumably because of a more posterior orientation of the macular generator areas.

There were two cases (17 and 32) with abnormal full-field response asymmetries but normal half-field amplitude ratios. Unexpected asymmetries of the full-field

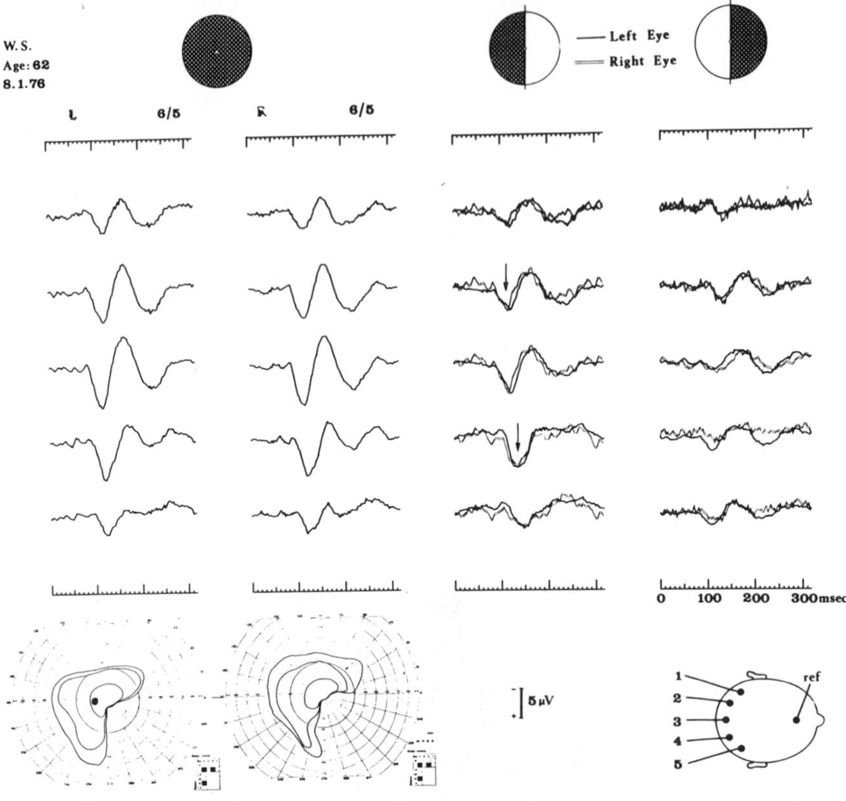

FIGURE 9. Symmetrical full-field responses in lower quadrantanopia with macular sparing. Left posterior cerebral artery infarction (CT scan normal). Full-field responses are slightly smaller over left scalp and are within normal limits. Left half-field responses show P100 and P135 waves at normal latency (arrows, channels 2 and 4, respectively). Responses from right half-fields are of reduced amplitude with less well-defined components. The latter have normal latencies and distributions but the responses from the right eye have an abnormal amplitude ratio. Summation of P100 with P135 waves obscures the asymmetry in the full-field responses (case 23).

response in which the major positive wave was reduced contralateral to the visual field defect, occurred in two cases (22 and 23). These misleading results were caused by cancellation in the full-field response while the half-field asymmetries were appropriate to the visual field defect, although only one case (23, FIGURE 9) exceeded the normal limits.

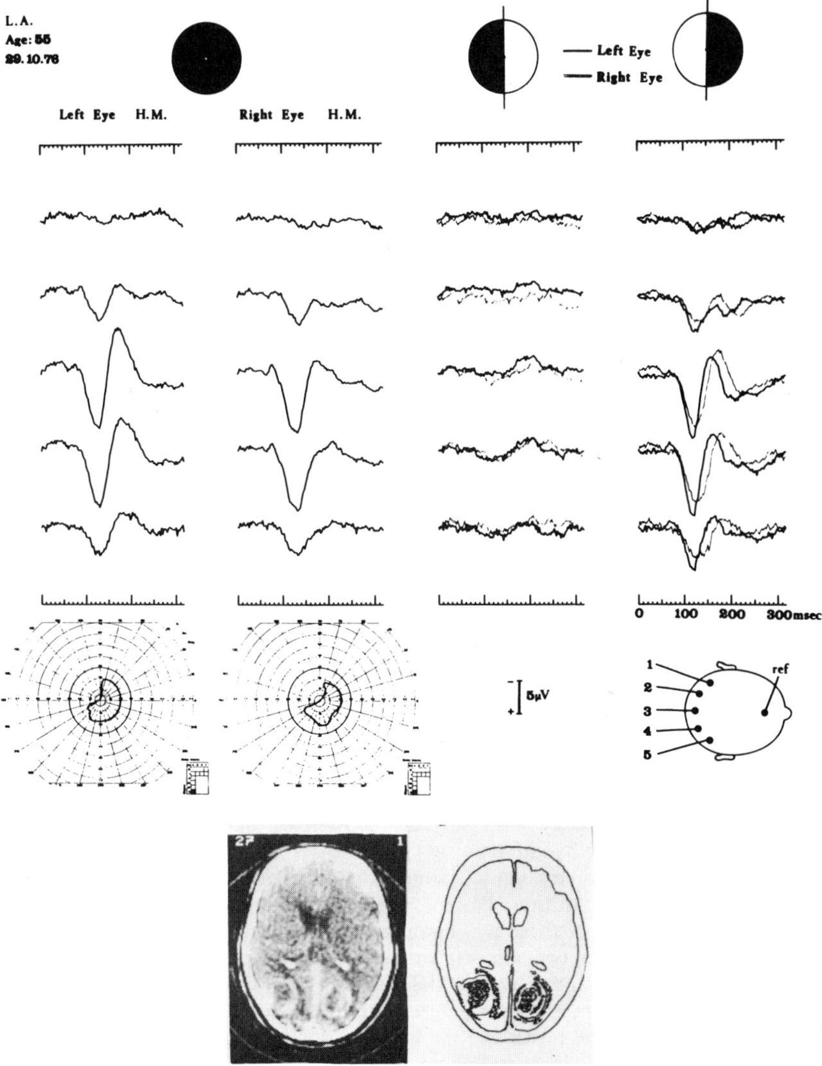

FIGURE 10. Relative indifference of the pattern VER to hemisphere pathology sparing the primary visual cortex. Similar metastatic melanomatous deposits in each occipital lobe produce severe effects on fields and vision. Right occipital lesion more mesially situated in the lobe, causes relatively more severe disruption of the left half-field. Full-field responses show an "uncrossed asymmetry," which does not exceed normal limits. However, almost all the recorded activity arises from the right half-fields. The grossly asymmetric half-field amplitudes parallel the more severe involvement of the left half-field despite the extensive bilateral pathology (case 26).

The abnormality detection rate of the half-field technique in patients with various homonymous visual field defects was thus 18 out of 25 (72%), or 21 out of 25 (84%) if monocular response abnormalities were included. Note that this analysis excludes case 13 whose abnormal full-field ratios were not examined by half-field recordings (TABLE 4). The four patients whose half-field responses remained within normal limits (cases 12, 17, 18, and 21) all had incomplete quadrantic or scotomatous field defects (TABLE 1). While such defects could be demonstrated by restricting the stimulus to the affected area the responses at this level were very variable and normal limits have not been determined. The responses in three patients with homonymous hemianopias and bilateral hemisphere damage exhibited abnormal uncrossed asymmetries. Despite reasonably symmetrical pathological lesions the response asymmetries paralleled the findings on perimetry rather than the CAT scan appearances (e.g., FIGURE 10).

Patients with cerebral lesions sparing the visual fields proved to have remarkably symmetrical half-field amplitude ratios which were all within normal limits (e.g., TABLE 4 and FIGURE 11). The "abnormalities" suggested by the full-field responses in one of these patients, as well as the minor asymmetries in the remainder, were due to the asymmetrical summation of normal half-field components.

A comparison of the amplitudes of the P100 components from the affected and unaffected heteronymous fields in the 25 patients with homonymous field defects gave mean values of 1.46 ± 1.78 μV and 5.52 ± 2.12 μV, respectively ($p < 0.01$). The P100 waves from the unaffected half-fields in these patients (5.52 ± 2.12 μV) were not significantly different from the equivalent waves in the same half-field of the healthy subjects (5.76 ± 2.14 μV). In the most severe hemianopic defects all components were equally attenuated, but in some patients where the half-field defect was incomplete there appeared to be relative sparing of the N145 waves compared with earlier components. However, in most cases there appeared to be no particular differential effect on the various components of the ipsilateral or contralateral complexes.

The Effect of Posterior Lesions on Latency

On full-field stimulation only 2 of the 32 patients had P100 latencies that exceeded the upper limit in the midline channel *for both eyes* (TABLE 5). One patient had bilateral occipital pathology with severely constricted fields and rather variable responses while the other had intact visual fields (cases 25 and 32, respectively). Thus none of the patients with one completely intact field had delayed midline responses for both eyes. However, four of these patients had delayed responses from one eye (cases 6, 14, 16, and 26). One of these had bilateral pathology and constricted fields (case 26). At lateral electrodes two subjects with constricted fields and bilateral occipital pathology exceeded the upper limit in the response from one eye (cases 25 and 26). However, another patient with severely constricted fields had responses of normal latency (FIGURE 12). The mean midline latencies for all patients with lesions of the posterior visual pathways or hemispheres are shown in TABLE 6. Despite a wider variance in the patient group and mean values for the left and right eyes which were, respectively, 3.5 and 3.6 msec longer than for the healthy control values, these group "delays" were not significant. In those patients with homonymous asymmetries due to polarity reversals on one side of the scalp, the major negativity (N105) which replaced the P100 ipsilateral to the visual field defect, had mean peak latencies of 103.4 ± 8.36 msec and 105.1 ± 10.12 msec for the left and right eyes, respectively. These values are not significantly different from the N105 latencies in healthy controls. Similarly, there were no significant delays for any other major component.

As a consequence of the severe attenuation of the affected half-field responses,

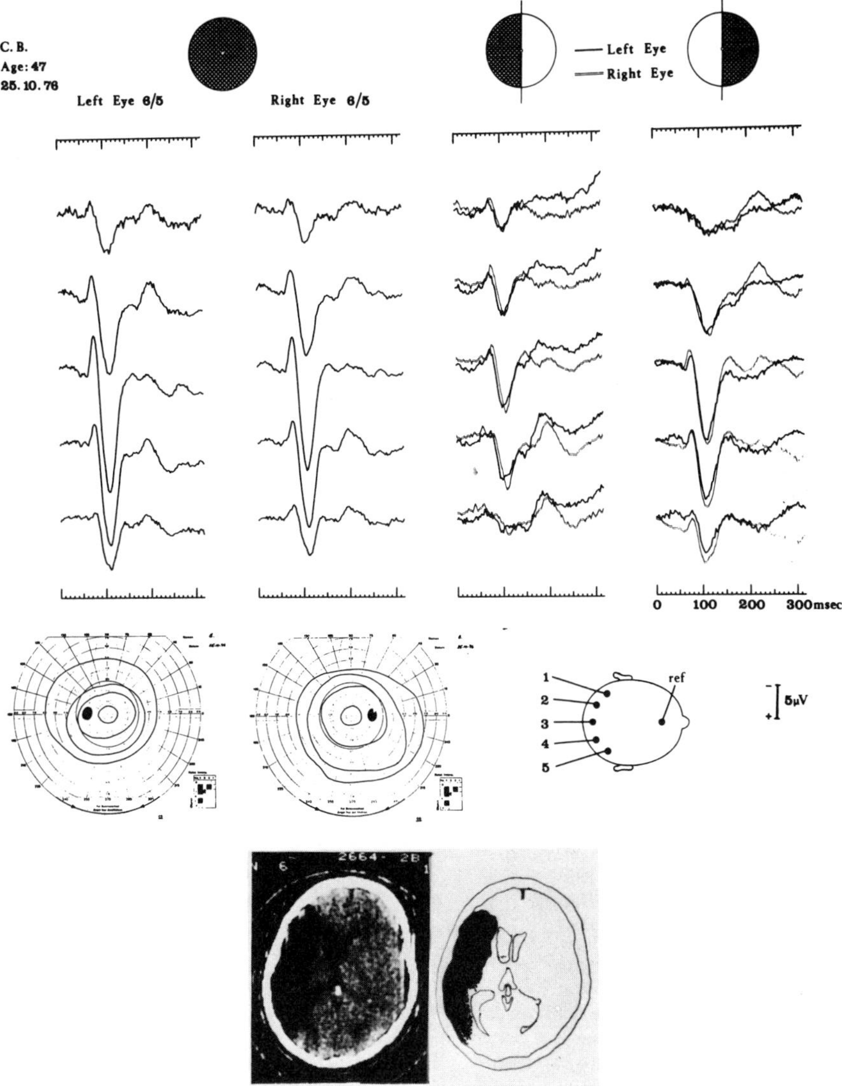

FIGURE 11. Symmetrical responses with massive hemisphere damage sparing the visual pathways. Left middle cerebral artery occlusion. Note the slight uncrossed asymmetry of the full-field responses which are smaller over the left scalp but within normal limits. Half-field responses evoked from the damaged left hemisphere are actually larger than those from the healthy right hemisphere but the degree of asymmetry is well within normal limits. Note the widely spreading full-field potentials due to the lack of significant contralateral PNP activity from each half-field as seen in some healthy subjects (e.g., FIGURE 1 subject 3) (case 31).

TABLE 5

P100 COMPONENT LATENCIES IN POSTERIOR VISUAL PATHWAY AND HEMISPHERE LESIONS*

Case	Full-Field Response						Half-Field Response			
	Ipsi. Int.		Midline		Ipsi. VFD		Ipsi. Int.		Ipsi. VFD	
No.	LE	RE	LE	RE	LE	RE	LE	RE	LE	RE
1	109	111	107	111	(118)	(123)	108	109	Abs	Abs
2	101	99	100	96	(103)	(99)	100	98	Abs	Abs
3	104	102	102	100	99	101	102	101	Abs	Abs
4	106	106	107	106	108	105	109	106	Abs	Abs
5	113	113	113	112	(103)	(102)	116	116†	Abs	Abs
6	117	123	115	125†	(101)	(112)	117	115	Abs	Abs
7	108	105	108	105	(91)	(90)	106	106	102	98
8	108	101	108	110	(99)	(95)	106	108	113	117†
9	114	114	114	114	(114)	(122)	118†	118†	Abs	Abs
10	98	100	98	101	99	102	100	106	103	Abs
11	105	104	106	106	(100)	(102)	105	105	108	103
12	110	113	108	109	111	113	114	110	113	112
13	95	97	100	100	(96)	(95)	—	—	—	—
14	111	118	115	118†	(96)	(96)	115	116†	113	114
15	96	108	104	95	108	104	96	96	108	105
16	110	105	106	124†	(95)	(102)	103	103	108	90
17	100	103	99	101	100	104	100	102	103	102
18	108	111	108	109	108	110	108	113	114	115
19	111	103	115	107	(108)	(105)	112	112	Abs	118
20	108	106	108	106	(107)	(115)	110	103	Abs	115
21	115	113	113	110	121	113	111	111	110	103
22	116	115	115	107	(101)	(103)	113	108	111	107
23	120	118	115	115	121	122	118	115	110	115
24	112	108	103	110	(119)	(115)	116	108	Abs	Abs
25	116	116	119†	120†	121	130†	109	118†	Abs	Abs
26	116	128	114	125†	115	138†	117†	125†	Abs	Abs
27	115	112	112	112	111	112	—	—	—	—
28	120	125	114	113	114	118	118	118	118†	115
29	100	115	100	103	98	103	96	93	107	106
30	105	111	105	108	118	118	109	115	113	106
31	110	109	109	108	110	112	101	108	108	111
32	101	111	116†	119†	107	101	115	122†	122†	113

*Values in parentheses are for the N105 component when it replaces the P100 in the full-field response. Ipsi. Int.: Component ipsilateral to the intact or relatively intact field. Ipsi. VFD: Component ipsilateral to the field defect or the suspect half-field if no field defect present (cases 28–32). Abs.: Absent component. LE: Left eye. RE: Right eye.

†Values exceed upper limits of normal.

only 15 of the 25 patients with homonymous visual field defects had measurable P100 components arising from stimulation of the affected half-fields in one or both eyes (TABLE 5). P100 waves were only recordable in 27 of the 30 monocular records from these 15 patients and were within normal limits in 26 of them. One case had a minor delay of 2 msec. When the intact, or relatively intact, half-field responses were included, 69 of 77 recorded P100 components were within normal latency limits. For the 8 slightly prolonged responses the average delay was less than 3 msec. The mean group values were not significantly different from those for the healthy controls (TABLE 6).

Discussion

The human visual cortex is a complex structure in which the orientation and topography of the primary and association areas show a high degree of intersubject and interhemispheric variation.[17–19] The pattern VER is also complex with contributions of varying polarity, latency and surface distribution from different areas of the visual field.[16,20–23]

Recent multichannel studies have shown that the occipital response to pattern-reversal stimulation is a composite of at least six major potentials from each half-field which can be separated by their distinctive form and spatial distribution.[12] Although the precise origin of these different potentials remains unknown, the half-field pattern-reversal stimulus appears to evoke a true hemisphere response, as all the major components recorded on half-field stimulation in healthy subjects are present following total hemispherectomy and have characteristics indistinguishable from those seen in the normal response.[12,15] All the activity that is normally recorded on both sides of the scalp to half-field stimulation must thus arise predominantly, if not entirely, from the hemisphere contralateral to the stimulated field. Similar conclusions were drawn

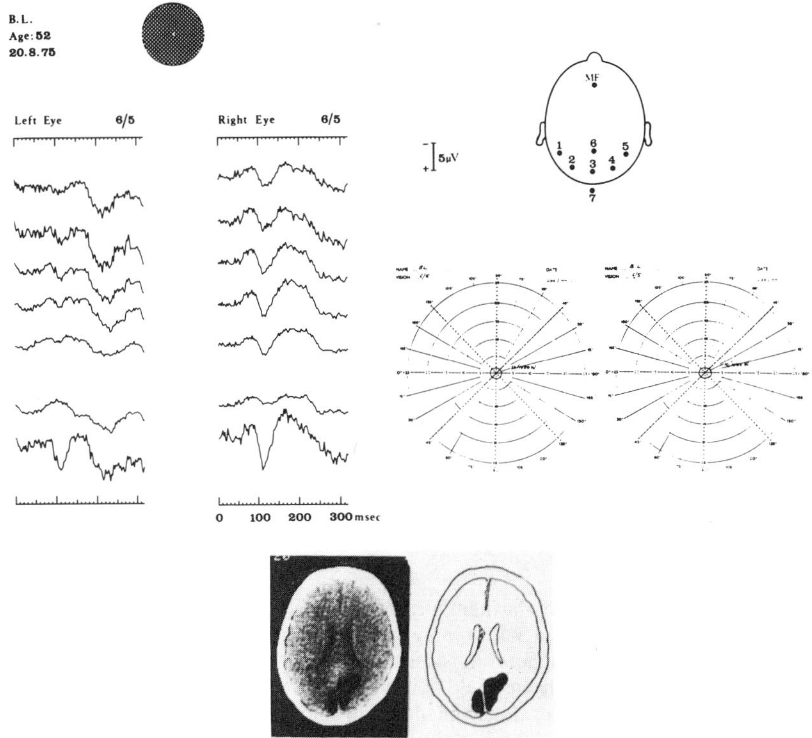

FIGURE 12. Localization and character of foveal responses in bilateral occipital lobe infarction with macular sparing. Fields restricted to 3° diameter. Potentials above the inion are small with shallow lateral voltage gradients. The P100 wave is clearly maximal at the electrode on the inion overlying the foveal cortical representation. Despite extensive bilateral occipital damage, the responses are at normal latency (case 27).

from a study of the flash VER in a patient with a split optic chiasm.[24] Common reference recordings have also shown that the simplistic, but widely held, view that the visual evoked potentials to half-field stimulation, or at least their amplitude maxima, should be restricted to scalp overlying the hemisphere of origin, does not apply either to flash[24,25] or pattern responses.[22,26–28]

The new understanding of the composite nature of the pattern response has not been considered in previous studies of the effects of hemisphere lesions. It is hardly surprising that in the majority of published accounts, where recordings are restricted to single electrode sites over each hemisphere, the results should be variable and often conflicting.

In the present study the value of separating and comparing half-field responses has been demonstrated. The clearest results were obtained in patients recorded after various surgical resections, either occipital lobectomy or hemispherectomy. When the

TABLE 6

GROUP MEAN P100 LATENCIES (msec) IN PATIENTS WITH POSTERIOR VISUAL PATHWAY OR HEMISPHERE LESIONS

	Ipsi. Intact*	Midline	Ipsi. VFD†	*n*
(a) Hemianopic patients: Full-field responses				
LE	108.7 ± 6.67	108.5 ± 5.83	110.1 ± 8.62	
	(*n* = 26)	(*n* = 26)	(*n* = 11)	
RE	109.2 ± 7.67	109.3 ± 8.29	112.9 ± 12.23	
	(*n* = 26)	(*n* = 26)	(*n* = 11)	
(b) Hemianopic patients: Half-field responses				
LE	109.2 ± 6.49		108.9 ± 4.11	
	(*n* = 25)		(*n* = 13)	
RE	109.1 ± 6.89		108.1 ± 8.32	
	(*n* = 25)		(*n* = 14)	
(c) Hemisphere lesions without visual field defects: Full-field responses				
LE	107.2 ± 8.17	108.5 ± 6.53	109.4 ± 7.60	5
RE	114.2 ± 6.42	110.2 ± 6.06	110.4 ± 8.08	5
(d) Hemisphere lesions without visual field defects: Half-field responses				
LE	107.8 ± 9.26		113.6 ± 6.43	5
RE	111.2 ± 11.39		110.2 ± 4.09	5

*Values obtained at 5 cm electrode ipsilateral to the intact, or relatively intact, visual half-field.

†Values obtained at 5 cm electrode ipsilateral to the visual field defect, or "suspect" field in (c) and (d).

visual stimulus is limited to the hemianopic fields of such patients, all potentials generated within the affected hemisphere are lost on both sides of the scalp. The complete abolition of hemisphere responses following occipital lobectomy has been previously reported.[6,11] With full-field stimulation the asymmetry of the residual activity in such patients is thus determined by the characteristics of the response from the *intact* half-field or hemisphere. Stimulation of the latter via the contralateral half-fields produces responses indistinguishable from those recorded in healthy subjects. The P100 and its accompanying negative waves which make up the ipsilateral complex are maximally recorded from the scalp overlying the affected or absent hemisphere, while the contralateral activity over the clinically and radiologically unaffected cerebral hemisphere is small, absent or shows phase-reversed complexes. Other pathologies resulting in dense hemianopias produce similar results.

The findings in patients recorded following total hemispherectomy provide no evidence, either of significant transcallosal conduction of any major half-field potential in the first 150 msec[8] or of the "enhanced transmission" of electrical activity through the empty cranial vault as proposed by Chain and colleagues.[6] The latter suggestion takes no account of the variation in the distribution of hemisphere components between individuals. The present findings are consistent with previous authors who concluded that the site and orientation of the cortical generator areas are major factors in determining the distribution of potentials on the scalp.[20–23,27] The gross asymmetries of the pattern potentials which resulted in all cases after partial or total resection of an occipital lobe contrasts with the relative insensitivity of the flash evoked potential in early studies of similar patients.[29] The lack of any differential effect on the various half-field pattern potentials also contrasts with the reported sparing of early and late flash EPs by similar lesions[30] and provides no clues to possibly distinct origins in specific or association cortex. The total obliteration of all pattern-generated half-field activity on *both* sides of the scalp by a discrete lesion in the striate cortex suggests either generators in common or in series.

The effects of cortical anatomy on the scalp distribution of pattern potentials also appears to account for the limitations of full-field stimulation in the detection of field defects. The usefulness of this technique is limited by the occurrence of asymmetries in healthy subjects, as others have also reported.[10,36] Such asymmetries arise from unequal cancellation of hemisphere components on opposite sides of the scalp. Furthermore, ipsilateral half-field potentials may spread so widely across the midline that they may mimic a healthy asymmetric full-field response and "mask" a severe hemianopia. It is only with careful restriction of the stimulus to each half-field that normals and hemianopics can be more reliably distinguished. The use of the half-field technique significantly increased the detection of field defects. However, similar limitations apply to the half-field technique itself when field defects are quadrantic or smaller in extent. This increasing variability of evoked activity from smaller subunits of the visual field is consistent with the marked increase in variability of the surface exposure and orientation of the cortical projection areas at this level. Although the potentials obtained from residual areas of an affected visual field may be of small amplitude, the alteration may not exceed the wide normal limits at this level. The amplitude ratios of the responses from some patients with even dense quadrantanopias may be within the range of the normal half-field response, whereas in other cases relatively minor field defects may result in an abnormal amplitude ratio. The variability of the activity recorded from the upper quadrants is so great that normal and pathological responses cannot be reliably separated.

Numerous publications have reported delayed flash evoked potentials with hemisphere lesions though the delays did not correlate well with the pathology and were frequently bilateral or even on the opposite side from the lesion.[31–34,40] Latency has been considered to be the most reliable indicator of a posterior visual pathway lesion.[34] The results obtained with EPs to structured stimuli in small series of patients with homonymous hemianopias have been inconsistent. Some authors have found no evidence of delays with full-field stimulation and midline recording in homonymous hemianopia[7,9] but there have been other reports of delays with unilateral or bilateral involvement of the posterior visual pathways.[10,35,36] Phase shifts of the steady-state pattern-reversal response in patients with homonymous hemianopia have also been described.[8]

In the present study the few component delays found were of minor degree compared with those reported to occur, for example, in multiple sclerosis. Furthermore their significance is in doubt due to the lack of correlation with either the presence or laterality of the visual field defect. In some cases, even where the visual fields were severely affected, the residual small potentials obtained were at normal latency. The slightly increased group average latency and variance for the patients in this study compared with the healthy controls may reflect the practical problem of accurately determining peak latencies in attenuated responses. In addition, the delays appeared to be independent of the pathology. For example, the effects of posterior pathway lesions in two patients with multiple sclerosis were undoubtedly more marked on the amplitude of the responses than on their latency. Hence components were reduced and at normal latency. This is to be expected in cases with acute plaques of demyelination when the major responsible pathological mechanism is likely to be conduction block rather than decreased conduction velocity.[37] The recorded response comes largely from the unaffected fibers subserving the preserved field.

A relatively high proportion of patients in this study had lesions involving the generator areas of the occipital lobes. On first principles it might be expected that delays would be more likely to result from lesions of the optic tract or radiation. However, the responses of patients whose lesions affected the optic radiations generally showed reduced amplitudes and undelayed potentials. Of the four patients with slightly delayed P100 components from the affected fields, there were two with more anteriorly placed lesions (one thalamic and one optic tract lesion) and four with lesions involving the occipital cortex. It is possible that even these "delays" could be attributed to waveform distortions associated with amplitude changes. However, a further comparison between larger numbers of patients with selected optic tract and radiation lesions and others with damage limited to the occipital lobes would be required to confirm these preliminary conclusions.

The reason for the high incidence of delays in earlier studies may be due to a fundamental difference in the flash and pattern-evoked potential or more probably a failure to identify clearly the various potentials in the pathological response with the consequent comparison of nonequivalent components. It has been demonstrated in this study that apparent delays on one side of the scalp to full-field stimulation may be due to the unmasking of contralateral components at normal latency from the intact half-field (or hemisphere). Furthermore, occasional patients or even healthy controls with transitional midline responses on half-field stimulation[16] may be considered to have delays if information from laterally placed electrodes is not available. Two recent publications which reported delays in 11% and 25% of cases with retrochiasmatic lesions[10,36] did not clearly distinguish between the distinct ipsilateral and contralateral half-field positivities. Without accurate identification of components and determination of their origins from the intact or affected fields by scalp-lateralization of the half-field potentials, the interpretation of the results will be in doubt. The highly variable waveform of the flash response may make such comparisons very difficult, if not impossible. Similarly, phase shifts of the "steady state" pattern-reversal response may come about by the loss or addition of certain components and cannot be reliably attributed to prolonged component latencies.[38]

In contrast with previous reports for the flash-evoked potentials, abnormalities could not be detected in the absence of lesions of the primary geniculo-calcarine pathways despite the presence of gross hemisphere damage. Although these experiments were limited to a small number of such patients, the lesions selected were gross, as determined clinically and by CAT scanning. It seems unlikely that abnormalities would be revealed in patients with more subtle hemisphere lesions sparing the visual fields even if large numbers of cases were examined. The retention of normal responses in a patient with gross tunnel vision due to bilateral infarctions of the striate cortical representation of the peripheral visual field (FIGURE 12) and the marked asymmetry in cases with bilaterally similar pathology but asymmetrical field defects (FIGURE 10) strongly suggests that the pattern EP is insensitive to lesions outside the specific visual pathway, and its alteration by pathology reflects primarily the extent, location, and density of the central visual field defect.

These results provide clear evidence to support the suggestions in previous reports that multichannel recording of half-field responses is essential for the detection of lesions of the posterior visual pathways.[11–15,36] While 16-channel records were made in all patients, the study has shown that quantitative analysis of the transverse distribution of the responses at five electrodes across the occipital scalp will enable the objective detection of all severe hemianopias as well as a proportion of incomplete visual field defects. The usefulness of the half-field technique is, however, limited by the high variance imposed by the inequality of the hemisphere responses. This variance increases as the half-field is fractionated in agreement with other studies which have demonstrated very large quadrant variability in the normal population.[39] The data from the present study suggests that normal intersubject and interhemisphere variation in half-field potentials is so marked that quantitative analysis is unlikely to be reliable for the detection of smaller than quadrantic defects. The profound waveform alterations associated with homonymous hemianopia have characteristic features, but their recognition and the avoidance of erroneous pathophysiological interpretations requires the careful identification of the various hemisphere components.

ACKNOWLEDGMENTS

Our thanks are due to the physicians and surgeons of the National Hospitals for allowing us to study their patients. We should also like to thank J. R. Pitman for invaluable technical assistance and Mrs. E. Halliday for help in preparing the illustrations.

REFERENCES

1. DAWSON, G. D. 1954. A summation technique for the detection of small evoked potentials. Electroencephalogr. Clin. Neurophysiol. **6:** 65–84.
2. HALLIDAY, A. M. 1975. The effect of lesions of the visual pathway and cerebrum on the visual evoked response. *In* Evoked Potentials: Handbook of Electroencephalography and Clinical Neurophysiology. W. S. Van Leeuwen, F. H. Lopes da Silva & A. Kamp, Eds., Vol. 8A: 119–129. Elsevier, Amsterdam.
3. REGAN, D. & J. R. HERON. 1969. Clinical investigation of lesions of the visual pathway: A new objective technique. J. Neurol. Neurosurg. Psychiatry **32:** 479–483.
4. REGAN, D. & J. R. HERON. 1970. Simultaneous recording of visual evoked potentials from the left and right hemispheres in migraine. *In* Background to Migraine. A. L. Cochrane, Ed. pp. 66–77. Heinemann, London.
5. REGAN, D. & R. F. CARTWRIGHT. 1970. A method of measuring the potentials evoked by simultaneous stimulation of different retinal regions. Electroencephalogr. Clin. Neurophysiol. **28:** 314–319.
6. CHAIN, F., N. LESÈVRE, M. LEBLANC, A. RÉMOND & F. LHERMITTE. 1972. Étude topographique des réponses évoqués visuelles dans un cas de lobectomie occipitale. Rev. Neurol. **126:** 372–378.

7. ASSELMAN, P., D. W. CHADWICK & C. D. MARSDEN. 1975. Visual evoked responses in the diagnosis and management of patients suspected of multiple sclerosis. Brain **98:** 261–282.
8. WILDBERGER, H. G. H., G. H. M. VAN LITH, R. WIJNGAARDE & G. T. M. MAK. 1976. Visually evoked cortical potentials in the evaluation of homonymous and bitemporal visual field defects. Br. J. Ophthalmol. **60:** 273–278.
9. CELESIA, G. G. & R. DALY. 1977. Visual electroencephalographic computer analysis (VECA). Neurology **27:** 637–641.
10. STRELETZ, L. J., S. H. BAE, R. M. ROESHMAN, N. J. SCHATZ & P. J. SAVINO. 1981. Visual evoked potentials in occipital lobe lesions. Arch. Neurol. (Chicago) **38:** 80–85.
11. BLUMHARDT, L. D., G. BARRETT & A. M. HALLIDAY. 1977. The asymmetrical visual evoked potential to pattern reversal in one half field and its significance for the analysis of visual field defects. Br. J. Ophthalmol. **61:** 454–461.
12. BLUMHARDT, L. D. & A. M. HALLIDAY. 1979. Hemisphere contributions to the composition of the pattern-evoked potential waveform. Exp. Brain. Res. **36:** 53–69.
13. BLUMHARDT, L. D. & A. M. HALLIDAY. 1978. The pattern-reversal response in lesions of the posterior visual pathways. Neurosci. Lett. Suppl. 1: 369.
14. BLUMHARDT, L. D. & A. M. HALLIDAY. 1980. The effect of lesions of the posterior visual pathways on the pattern-reversal response. Electroencephalogr. Clin. Neurophysiol. **49:** 95P.
15. BLUMHARDT, L. D. & A. M. HALLIDAY. 1981. Cortical abnormalities and the visual evoked response. *In* Visual Pathways Electrophysiology and Pathology. Proc. 18th ISCEV Symp., Amsterdam, May 18–22, 1980. H. Spekreijse & P. A. Apkarian, Eds. Doc. Ophthalmol. Proc. Ser. **27:** 347–365. Junk, The Hague.
16. BLUMHARDT, L. D., G. BARRETT, A. M. HALLIDAY & A. KRISS. 1978. The effect of experimental 'scotomata' on the ipsilateral and contralateral responses to pattern-reversal in one half-field. Electroencephalogr. Clin. Neurophysiol. **45:** 376–392.
17. POLYAK, S. 1957. The Vertebrate Visual System. H. Klüver, Ed. Univ. of Chicago Press, Chicago, Ill.
18. BRINDLEY, G. S. 1972. The variability of the human striate cortex. J. Physiol. (London) **225:** 1–3P.
19. STENSAAS, S. S., D. K. EDDINGTON & W. H. DOBELLE. 1974. The topography and variability of the primary visual cortex in man. J. Neurosurg. **40:** 747–755.
20. HALLIDAY, A. M. & W. F. MICHAEL. 1970. Changes in pattern-evoked responses in man associated with the vertical and horizontal meridians of the visual field. J. Physiol. (London) **208:** 499–513.
21. MICHAEL, W. F. & A. M. HALLIDAY. 1971. Differences between the occipital distribution of upper and lower field pattern-evoked responses in man. Brain Res. **32:** 311–324.
22. JEFFREYS, D. A. & J. G. AXFORD. 1972. Source locations of pattern-specific components of human visual evoked potentials. I. Component of striate cortical origin. Exp. Brain Res. **16:** 1–21.
23. JEFFREYS, D. A. & J. G. AXFORD. 1972. Source locations of pattern-specific components of human visual evoked potentials. II. Component of extrastriate cortical origin. Exp. Brain Res. **16:** 22–40.
24. LEHMANN, D., R. N. KAVANAGH & D. H. FENDER. 1969. Field studies of averaged visually evoked EEG potentials in a patient with a split chiasm. Electroencephalogr. Clin. Neurophysiol. **26:** 193–199.
25. NAKAMURA, Z. & W. R. BIERSDORF. 1971. Localization of the human visual evoked response. Early components specific to visual stimulation. Am. J. Ophthalmol. **72:** 988–997.
26. BIERSDORF, W. R. 1974. Cortical evoked responses from stimulation of various regions of the visual field. *In* Proc. XIth ISCERG Symp., Bad Neuheim, May 1973. E. Dodt & J. T. Pearlman, Eds. Doc. Ophthalmol. Proc. Ser. **4:** 249–260. Junk, The Hague.
27. BARRETT, G., L. D. BLUMHARDT, A. M. HALLIDAY, E. HALLIDAY & A. KRISS. 1976. A paradox in the lateralisation of the visual evoked response. Nature (London) **261:** 253–255.
28. SHAGASS, C., M. AMADEO & R. A. ROEMER. 1976. Spatial distribution of potentials evoked by half-field pattern-reversal and pattern-onset stimuli. Electroencephalogr. Clin. Neurophysiol. **41:** 609–622.

29. CRIGHEL, E. & M. BOTEZ. 1966. Photic-evoked potentials in man in lesions of the occipital lobe. Brain **89:** 311–316.
30. CORLETTO, F., A. GENTILOMO, G. ROSADINI, G. F. ROSSI & J. ZATTONI. 1967. Visual evoked potentials as recorded from the scalp and from the visual cortex before and after surgical removal of the occipital pole in man. Electroencephalogr. Clin. Neurophysiol. **22:** 378–380.
31. EBE, M., T. MIKAMI, H. ITO, M. AKI & M. MIYAZAKI. 1963. Photically evoked potentials (PEPs) in brain disorders. Tohoku J. Exp. Med. **80:** 323–372.
32. COHN, R. 1963. Evoked visual cortical response in homonymous hemianopic defects in man. Electroencephalogr. Clin. Neurophysiol. **15:** 922P.
33. VAUGHAN, H. G. & R. KATZMAN. 1964. Evoked responses in visual disorders. Ann. N.Y. Acad. Sci. **112:** 315–319.
34. KOOI, K. A., A. M. GUEVENER & B. K. BAGCHI. 1965. Visual evoked responses in lesions of the higher optic pathways. Neurology **15:** 841–854.
35. ASHWORTH, B., A. F. J. MALONEY & H. R. A. TOWNSEND. 1978. Delayed visual evoked potentials with bilateral disease of the posterior visual pathway. J. Neurol. Neurosurg. Psychiatry **41:** 449–451.
36. KUROIWA, Y. & G. G. CELESIA. 1981. Visual evoked potentials with hemifield pattern stimulation. Arch. Neurol. (Chicago) **38:** 86–90.
37. HALLIDAY, A. M. & W. I. MCDONALD. 1977. Pathophysiology of demyelinating disease. Br. Med. Bull. **33:** 21–27.
38. DUWAER, A. L. & H. SPEKREIJSE. 1978. Latency of luminance and contrast evoked potentials in multiple sclerosis patients. Electroencephalogr. Clin. Neurophysiol. **45:** 244–258.
39. REGAN, D. & B. A. MILNER. 1978. Objective perimetry by evoked potential recording: Limitations. Electroencephalogr. Clin. Neurophysiol. **44:** 393–397.
40. VAUGHAN, H. G., R. KATZMAN & J. TAYLOR. 1963. Alterations of visual evoked response in the presence of homonymous visual defects. Electroencephalogr. Clin. Neurophysiol. **15:** 737–746.

STEADY-STATE AND TRANSIENT VISUAL EVOKED POTENTIALS IN CLINICAL PRACTICE

Gastone G. Celesia

Department of Neurology
William S. Middleton Memorial Veterans Hospital
University of Wisconsin
Madison, Wisconsin 53705

Neurophysiological studies in cats and primates have shown that the most effective stimulus for cortical neurons is contrast rather than absolute intensity of light.[1,2] Barlow[3] was the first to suggest that visual neurons may require a particular pattern of stimulation in order to produce a maximal response. Hubel and Wiesel[1,2] found that neurons respond selectively to visual patterns of progressively greater complexity at ascending levels in the hierarchy of the cortex. Cortical neurons practically ignore uniform illumination of the retina but are selectively activated by specific forms and shapes. It thus appears that at different levels of the visual system there are different optimal stimuli to activate the neuronal receptive fields (FIGURE 1). These findings in experimental animals have been extrapolated to humans and have led researchers to utilize complex visual stimuli to activate consistent response waveforms.[4,5] Among the various stimuli tried, pattern-reversal checkerboard stimulation has emerged as a relatively simple method to obtain stable and reliable visual evoked potentials.[6,7] Thus, pattern-reversal stimulation has received wide acceptance as a tool for the assessment of visual dysfunctions. This success, however, should not deter us from the use of other visual stimuli and must be kept in perspective to avoid the temptation to select only one stimulus as the panacea for every visual diagnostic problem. The need of utilizing different visual stimuli and different recording methods to improve our diagnostic yield will be discussed in this presentation.

METHODS

Silver–silver chloride electrodes were applied to the scalp with collodion. Electrode impedance was always below 5000 ohms. The electrode placement of Halliday's group[8] was adopted in most of the recordings because of the extensive data base existing with these montages. A mid-occipital electrode (MO) was placed 5 cm above the external occipital protuberance. The lateral occipital electrodes were placed 5 cm to the left or to the right of the MO electrode. These electrodes were referred to a common midfrontal reference electrode 12 cm above the nasion.

Electroretinogram (ERG) was monitored from silver–silver chloride electrodes placed with collodion over the periorbital region. The active electrode was positioned at the middle of the infraorbital ridge just below the eye, this electrode was then connected to a second electrode positioned 2.5 cm laterally to the outer canthus of the same eye.

Input from the electrodes was led to eight preamplifiers adjusted to a bandwidth ranging from 0.3 Hz to 1 kHz. Preamplifier output was fed simultaneously to an oscilloscope and a minicomputer. The computer sampling rate was 3,700 samples/sec.

Three visual stimulation paradigms were tested in each subject.

0077–8923/82/0388–0290 $1.75/0 © 1982, NYAS

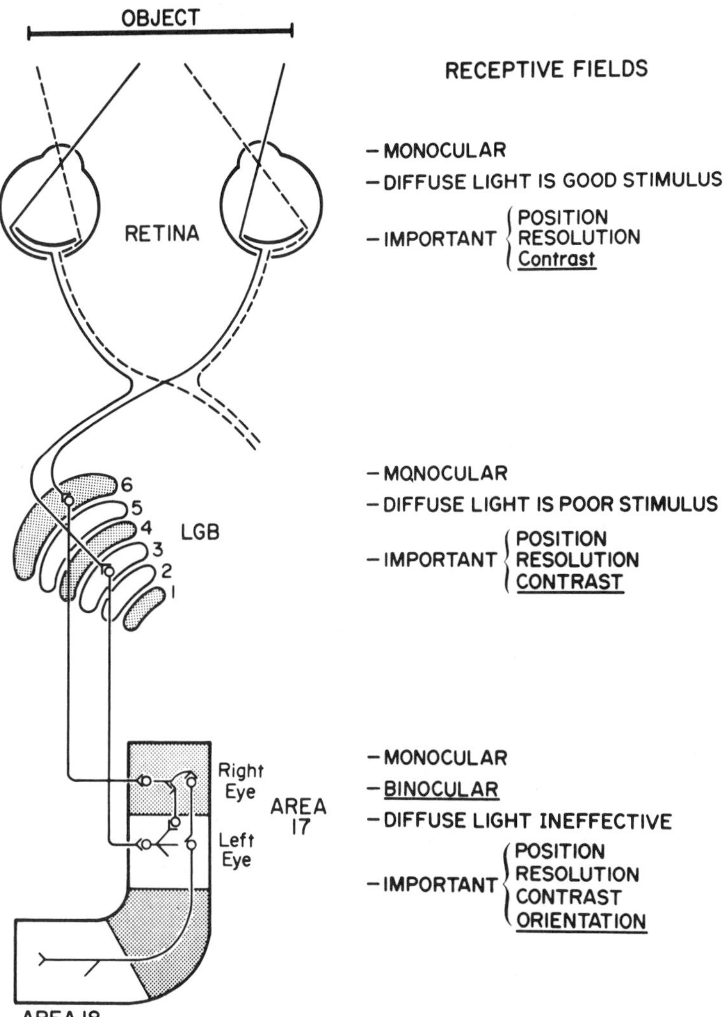

FIGURE 1. Diagram of the retino-geniculostriate pathways of primates showing at each level the optimal pattern of stimuli required to activate the receptive fields of the neurons. Note that neurons in the various layers of the lateral geniculate body (LGB) are predominantly innervated from one eye or the other. Interaction between the two eyes occurs at the cortical level.

Paradigm A consisted of a pattern-reversal checkerboard. The pattern was projected from the rear onto a translucent screen. Pattern reversals were produced by a rapid lateral displacement of the checkerboard through one square. Monocular full-field stimulation was carried out with two sizes of squares (or checks), subtending 31′ 05″ and 1° 2′ of arc to the subject's eye. For both sizes, the contrast between alternating checks was 57%. The rate of reversal was every 600 msec. The full-field luminance was 34 foot-lamberts (ft-L) with a background luminance of 0.06 ft-L.

Monocular half-field stimulation was carried out only with a check size of 1° 2′. During half-field stimulation the subject fixated on a white dot placed 1° of arc laterally from the stimulated hemi-field.

Paradigm B was used to determine the critical frequency of photic driving (CFPD) and consisted of flashes generated by a photic stimulator (Grass, Model PS-2), set at intensity 4 and placed 45 cm from the subject's eye.

Stimulation began at low frequency of flashes and was gradually increased until no photic driving could be obtained. The following frequencies were used: 1, 20, 32, 44, 53, 62, and 73 Hz.

Paradigm C was used to produce the visual spectrum array and consisted of six trains of flashes of 8-sec duration. The frequency of flashes varied in each train. The following frequencies were applied: 7, 10, 13, 16, 19, and 22 flashes per second.

Results and Discussion

Transient Visual Evoked Potentials (T-VEPs)

T-VEPs are electrical potentials resulting from the *transient* change of brain waves following an intermittent photic stimulus. Conventionally, the term "transient" has been omitted but will be retained here for clarity.

Pattern T-VEPs are critically dependent on many parameters. Arden *et al.*[9] have recommended the following parameters be controlled and specified: (1) stimulus luminance or brightness; (2) type of pattern (checkerboard, grating, etc.); (3) size of pattern specified in terms of visual angle; (4) total field size and shape and its relation to the fixation point, to be specified in terms of visual angle and retinal eccentricity; (5) method of presentation of the pattern (pattern reversal, brief pattern onset or off-set patterned light, etc.). The presentation rate of the pattern should also be specified. The effect of the rate of pattern reversal is shown in Figure 2. At reversal intervals less than 600 msec, the amplitude of the response is decreased. The amplitude becomes smaller proportionally to the decreased interval. At intervals equal or shorter than every 100 msec, the responses overlap and no individual response cycle can be related to any particular stimulus cycle. Intervals of 600 msec between each reversal is a satisfactory compromise to record undistorted T-VEPs in most subjects.

Researchers in the field have arbitrarily selected squares (or checks) of one size at one contrast level, although changes in these parameters are known to affect T-VEPs. Different check sizes produce responses of different amplitude. Potentials of maximum amplitude are evoked by check sizes ranging from 11 min to 18 min subtense.[10] Asselman *et al.*[11] compared pattern-reversal checks of 57 min to small checks of 30 min with a constant field of 18 deg and found that the T-VEPs had identical latencies; however, the amplitude was slightly smaller for the larger checks. T-VEPs obtained with checks of less than 15 min are mostly due to stimulation of the macular region while the T-VEPs produced with checks larger than 15 min are the result of foveal and extrafoveal stimulation.[4,5,10] Refractive errors may affect both the latency and amplitude of evoked responses.[12,13] Van Lith[13] has shown that the effect of refractive anomalies and medial eye opacities upon T-VEPs varies according to various check sizes and contrasts. Relatively large check patterns and contrasts greater than 50% circumvent this potential problem. Retinal eccentricity is another important variable. The effect of stimulation of discrete areas of the retina was studied by Celesia and Meredith.[14] T-VEPs could be obtained at the fixation point with stimuli as small as 6 min and 54 sec subtense. No T-VEPs could be obtained with these small fields outside the 4-deg horizontal and vertical meridians. T-VEPs could, however, be elicited by increasing the size of the stimulus. The size required to evoke a response increased in relation to the distance from the fixation point. These data support the invariance principle[15] stating that photopic stimuli presented anywhere in the visual field are equally effective if the stimuli are equivalent in terms of numbers of neurons activated.

The choice of the square size may have to be changed according to the region of the visual pathways to be studied. To detect small demyelinating lesions affecting optic nerve fibers originating from ganglion cells in the foveal region (the papillo-macular bundle), smaller checks with a small total field should be used. Hennerici *et al.*[16] have shown that a stimulus subtending 45 min of arc was more sensitive in detecting abnormalities of T-VEPs in multiple sclerosis patients than a stimulus field subtending 20 deg of arc at the subject's eye. On the other hand, if the aim is to detect retrochiasmatic lesions, checks larger than 1 deg may be a better choice. Every subject

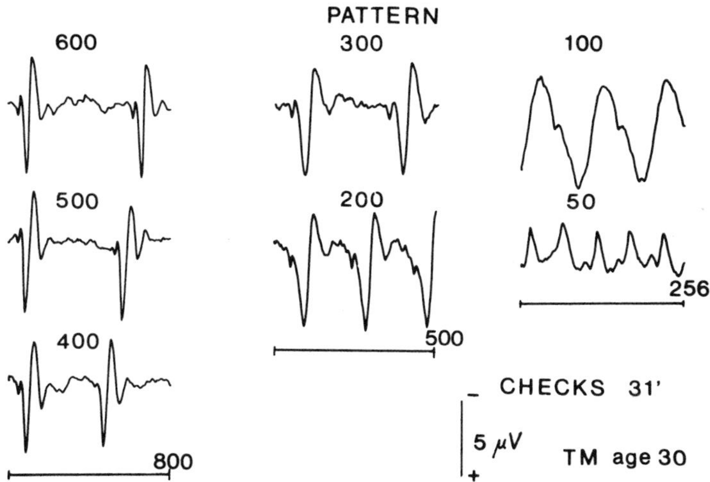

FIGURE 2. Effect of the rate of pattern reversal on the amplitude and morphology of evoked potentials. The number above the response refers to the interval in milliseconds between reversals. Note that the analysis time changes in each column.

undergoing a pattern-reversal stimulation should be tested for his or her ability to see the pattern, and should wear corrective glasses whenever appropriate.

Pattern reversal of squares can be produced with commercially available televisions, projectors, or LEDs. These instruments produce visual images with different characteristics and, evoke T-VEPs with different values. FIGURE 3 illustrates response obtained in normals with checks of roughly the same size but produced by three different stimulators. The smallest potentials were evoked by LEDs and the largest and most stable potentials by the projector. The mean amplitude and latency of these responses is tabulated in TABLE 1. The variation of T-VEPs was most likely related to the difference in contrast among the three stimuli and by the relative blur of the television and LED images compared to the sharp edges of back-projected checks. Furthermore, LED checks were red-white. Colored checks cannot be considered a stimulus equivalent to black-white checks. These data are presented to emphasize the lack of standardization of visual stimulators and the importance of monitoring the luminance, contrast, and sharpness of the visual images employed.

T-VEPs evoked by pattern-reversal as described in paradigm A of methods are influenced by age. With advancing age, there is an increase in peak latency of both the N_1 (N_{70} of International Classification) and P_1 (P_{100} of International Classification) waves.[17]

This increase is most prominent after age 40. It further varies with the luminance of the pattern. Shaw and Cant[19] did not detect any latency change with high luminance pattern. They did demonstrate changes in latencies with the same patterns at lower levels of luminance. This increase in latency probably reflects a slowing of conduction velocity in the optic nerve and/or optic pathways. Age is an important variable that must be taken into account when establishing the boundary of normality. It is suggested that the boundaries of normal be placed 2.5 times above the standard

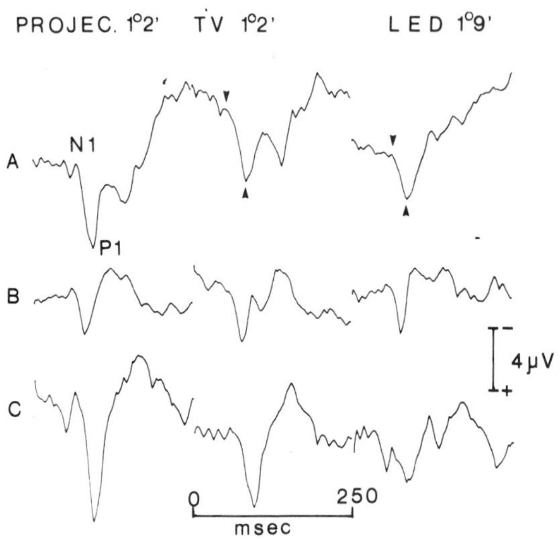

FIGURE 3. Comparison of T-VEPs recorded from the midoccipital region and evoked by projector, television, and LEDs in three normal subjects A, B, and C.

deviation of the regression line; 99.5% of normal subjects will have scores below this line.[7,17] The effects of aging on the latency differences between the right and left eyes are too small to be significant. The limit of normal for latency differences for N_1 is 5 msec (mean + 2.5 SD) and for latency differences of P_1 is 6 msec. Assuming a normal distribution for these parameters, the use of the mean + 2.5 SD reduces the overlap between normal and abnormal values to 0.5%. Similarly, a strict criterion of defining abnormal values as those responses with latencies falling beyond two and a half times the standard deviation of the mean was used by Halliday *et al.*[20]

Recently controversy has risen concerning the topography of pattern T-VEPs to full- and half-field stimulation and the reliability of measuring amplitude asymmetries to detect retrochiasmatic lesions.[21] Most of the discrepancies result from the utilization of different electrode montages and stimulation parameters. Pattern stimulation with checks as described in paradigm A results in symmetrical T-VEP distribution over both occipital regions. In contrast to this symmetry, half-field

TABLE 1

NORMATIVE DATA OF T-VEPS PRODUCED BY COMMERCIALLY AVAILABLE STIMULATORS IN 10 VOLUNTEERS

Stimulator	Check Size	Field Size	Contrast %	Luminance*	N_1† Amplitude	N_1† Latency	P_1† Amplitude	P_1† Latency
Pattern-reversal projector	1° 2′	14° 8′	57	38.2	2.1 ± 1.1	67.2 ± 3.2	8.7 ± 2.6	92.6 ± 5.1
Television	1° 2′	16° 44′ × 12° 4′	38	62	1.2 ± 0.5	65.2 ± 4.3	4.2 ± 1.8	93.7 ± 7.2
LEDs	1° 9′	7°	18	17.2	0.8 ± 0.3	63.4 ± 3.2	3.8 ± 1.6	87.7 ± 4.7

*Luminance is expressed in foot-lamberts

†Mean amplitude is expressed in μV and mean latency in msec.

stimulation (a field of 9 deg) results in T-VEPs with the largest amplitude over the lateral occipital scalp ipsilateral to the half-field stimulated.[22] Barrett *et al.*[23] suggested that this paradoxical lateralization of the major positive wave of T-VEP is related to the mesial location of the potential generators on the hemisphere contralateral to the field stimulated. The location of ipsilateral electrodes is optimal to record a potential from the posteromedial aspect of the contralateral occipital lobe. Analysis of our normative data indicates that the utilization of amplitude ratios will normalize interpersonal variations and permit statistical quantification of the data. The lateral occipital ratio was found to be useful. Lateral occipital ratio is defined as the amplitude in microvolts of T-VEP at occipital scalp contralateral to hemifield stimulated divided by the amplitude of T-VEP at occipital scalp ipsilateral to hemifield stimulated. The mean normal value of this ratio is 0.61 ± 0.30. The boundary of normal for the lateral occipital ratio is 1.36 (mean ± 2.5 SD).

Steady-State Visual Evoked Potentials

Steady-state potentials are electrical events evoked by rapid repetitive sensory stimulation. Rapid continuous stimulation produces evoked responses of constant amplitude and frequency; each potential overlaps another so that no individual response can be related to any particular stimulus cycle.[15] It is presumed that the brain has achieved a "steady-state" of excitability.

Steady-state visual evoked potentials (S-VEPs) are utilized for the determination of *critical frequency of photic driving* (CFPD). CFPD is defined as the highest frequency of photic driving response in flashes per second. Flash stimulation begins at low frequencies of flashes and gradually increases until no photic driving can be obtained.[7,17] For each frequency tested, 200 to 300 samples are summated. CFPD is recorded simultaneously with ERGs at the retinal level (*retinal CFPD*) and with occipital scalp electrodes (cortical CFPD). The responses to flashes of high frequency consist of sinusoidal waves following the stimulus (FIGURE 4). Subharmonic waves are often seen, particularly at lower frequencies of flashes, both at the retinal and cortical levels. Retinal and cortical CFPD values are related to the intensity of the flash as well as to the brightness of the background illumination and the pupillary size. They are also influenced by age and decrease as age progresses.[17] In normals, retinal and cortical CFPDs have similar values. The following mean values were obtained: 72 flashes per second at ages 20 to 30; 69 flashes per second at ages 30 to 60; and 62 flashes per second above age 60. Under this stimulation condition, differences between retinal and cortical CFPD had to be higher than 10 flashes per second to be considered abnormal. Cortical CFPD was never greater than retinal CFPD.

Regan[5] was the first to apply Fourier analysis to obtain a precise description of S-VEPs to harmonically simple light. He studied both phase and amplitude of S-VEPs. The method was effective in detecting retrobulbar neuritis.[5,24] S-VEPs were delayed and had low amplitude with monocular stimulation of the affected eye. In our laboratory we combine steady-state flash stimulation (paradigm C as described in methods) with Fourier analysis of the recorded EEG to obtain the *visual evoked spectrum array*.[25] Visual stimulation consists of 8-sec trains of flashes. Average compressed arrays are computed using the fast Fourier transform. Each array consists of the summation of two 4-sec epochs of EEG during a specific steady frequency of stimulation.

Quantification is achieved by calculating the ratios of spectral energy at each frequency for homologous regions of the right and left hemispheres. If the two hemispheres contain equal energy, the ratio would be one. Normal spectral ratios were calculated for 19 normal subjects with recordings from 02-CZ and 01-CZ (International nomenclature). Spectral ratios were less than or equal to 2.0 in all normal subjects.

Clinical Application of Visual Evoked Potentials

Visual disturbances can now be studied objectively with neurophysiological methods.[7,11,20,26] Abnormalities in visual evoked potentials indicate dysfunction somewhere along the visual pathways. Different disease processes affecting the same region will produce similar disturbances.

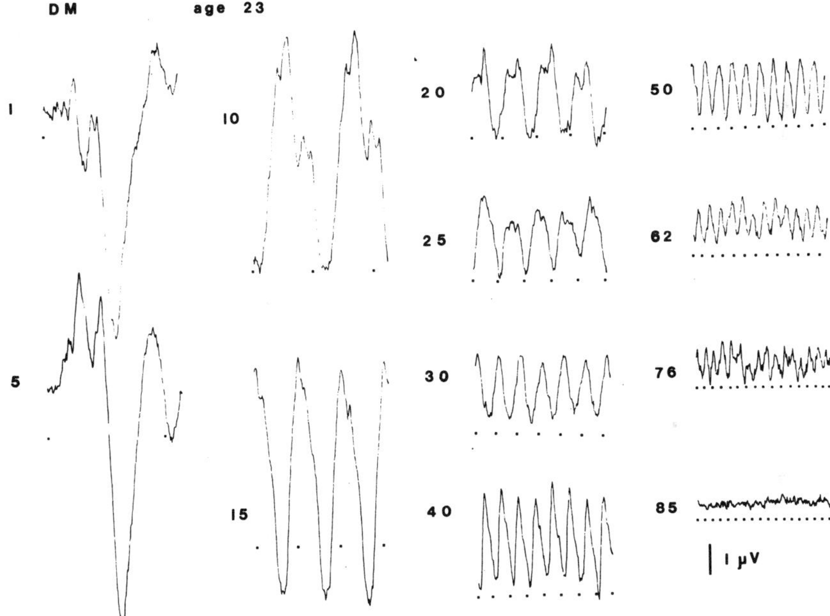

FIGURE 4. Evoked responses at different flash frequencies. Monocular stimulation of the left eye in a 23-year-old subject. Each response is a summation of 250 samples averaged for 204 msec. Numbers at the left of the response indicate flash frequency per second. Dots represent stimulus artifact. Note sinusoidal responses up to frequency of 76 flashes per second and absence of response at 85 flashes per second.

Utilizing the three major types of visual stimuli and data analysis described in the methods and after screening more than 500 patients with visual dysfunction, we have identified three abnormal profiles. These profiles permit the differentiation of involvement of the eye proper, the optic nerve, and the retrochiasmatic pathways.

In most instances, diseases of the vitreous, lens, anterior chamber, and cornea produce T-VEP amplitude attenuation without affecting their latencies, provided that checks of high contrast and of a size larger than 20′ of arc are employed.[12,13] However, these lesions are best diagnosed by clinical examination. Diseases of the retina and optic nerve may result in pattern-reversal T-VEPs of low amplitude and delayed latencies. In most cases, the differentiation between the two can be made by identification of retinal lesions on fundoscopic examination. The simultaneous determination of retinal and cortical CFPD is useful in these cases. A retinopathy will equally affect retinal and cortical CFPD while a lesion of the optic nerve will produce a dissociation of CFPD with normal retinal and decreased cortical CFPD (FIGURE 5). In more difficult cases, the differential diagnosis between retinal and optic nerve involvement may require the utilization of electroretinography with contact lens electrodes.

Unilateral retrochiasmatic lesions do not alter CFPD. On the other hand, bilateral destruction of the visual cortices will affect cortical CFPD while retinal CFPD remains normal (FIGURE 6). This CFPD dissociation is present to monocular stimulation of either eye as well as to binocular stimulation. CFPD dissociation related to optic nerve lesions is monocular. Three adult patients with cortical blindness related to bilateral infarction of the occipital cortex were studied with T-VEPs and CFPD. T-VEPs to pattern-reversal checks were present in every patient. The morphology of

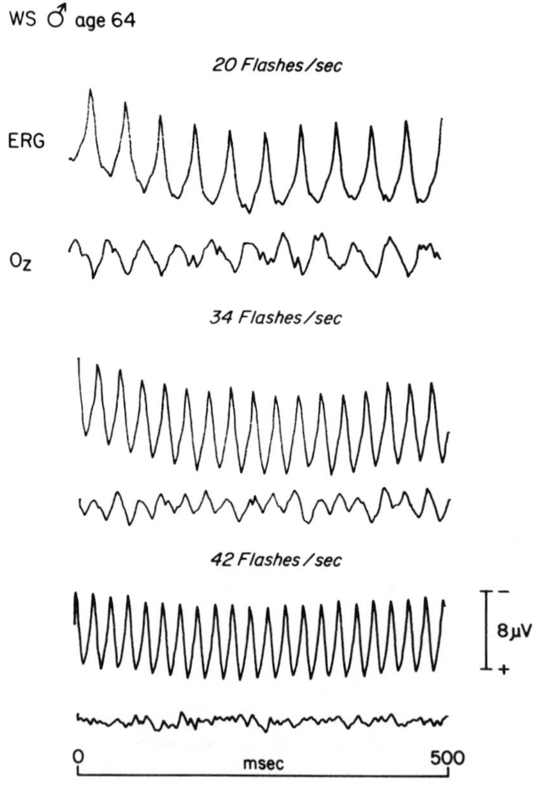

FIGURE 5. CFPD from stimulation of the right eye in a 64-year-old paraplegic with a 10-year history of progressive chronic myelopathy. Note normal ERG responses while cortical evoked responses were absent after the frequency of flashes was increased above 34 flashes per second. This dissociation between retinal and cortical CFPD suggests a dysfunction of the right optic nerve.

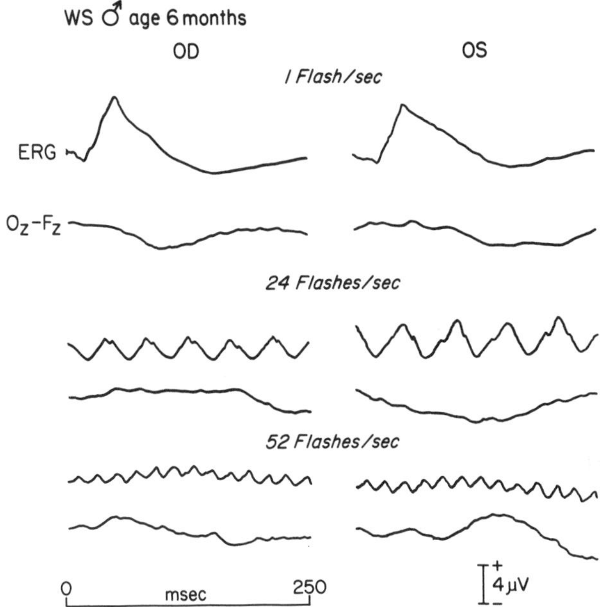

FIGURE 6. T-VEPs and S-VEPs in an infant suspected to be blind. Fundoscopic examination and pupillary reactions were normal. Note the excellent retinal responses with a retinal CFPD of 52 but the total absence of cortical responses from stimulation of either eye. This pattern is highly suggestive of bilateral retrochiasmatic dysfunction.

the responses was normal. In one subject, the latencies of both N_1 and P_1 were delayed whereas they were normal on the other two subjects. In every patient cortical CFPD was abnormally low with stimulation of either eye, but retinal CFPD was normal. CFPD has also been found useful in our laboratory to assess infants with suspected blindness. Although a normal response does not exclude impairment of visual perception[27] and/or moderate visual deficit, an abnormal cortical CFPD usually confirms the clinical suspicion. Severely impaired cortical CFPD to either eye or to binocular stimulation with preserved retinal CFPD were found in four of six newborns with suspected cortical blindness (FIGURE 6).

Delayed or absent T-VEPs have now been reported in optic and retrobulbar neuritis, optic atrophy, ischemic optic neuropathy, and compression of the optic nerve (FIGURES 7 & 8). A T-VEP is considered to have a delayed latency when the peak latency for wave N_1 and P_1 falls outside the boundary of normality (mean ± 2.5 SD according to age) or when the peak latency difference for both waves N_1 and P_1 between the stimulation of the right and left eyes is greater than 6 msec. Celesia[28] studied 74 multiple sclerosis patients and found delayed or absent T-VEPs in 55 (74%) of the cases. Every patient with a central or a paracentral scotoma had an abnormal T-VEP. Of 865 patients suffering from multiple sclerosis, 556 (65%) have so far been reported to have abnormal T-VEPs (TABLE 2). More important is the great sensitivity of the test and its ability to show abnormal pattern evoked responses in early optic nerve lesions when other clinical signs of visual impairment are lacking.

Celesia[28] demonstrated delayed T-VEPs in 16 (55%) of 29 multiple sclerosis patients who were without signs or symptoms of visual dysfunction. Delays were also found in 6 out of 7 cases of multiple sclerosis who had normal visual acuity, visual fields, and no subjective visual complaints, but who had a past history of optic neuritis. Similarly, Halliday and co-workers[6,20] found delayed T-VEPs in 12 of 14 patients with normal optic discs and no history of optic neuritis. Asselman *et al.*[11] reported delayed VEPs in 28% of eyes assessed as normal by other criteria. Seven of 10 patients with a diagnosis of multiple sclerosis but no visual symptoms were similarly found to have abnormal VEPs to a sinusoidal grating pattern.[26] Not only are T-VEPs useful in determining subclinical and/or early lesions of the optic nerves, but they can also be

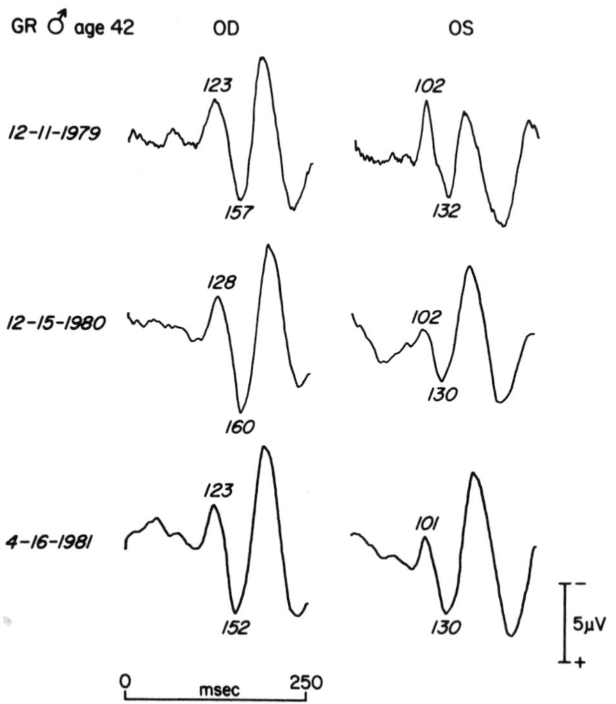

FIGURE 7. 42-year-old man with a diagnosis of definite multiple sclerosis. The patient had a history of right optic neuritis 10 years previously. At the time of these recordings, his visual acuity was 20/30 in both eyes, and perimetry was normal. T-VEPs have normal amplitude but prolonged latencies suggestive of optic nerve dysfunction. Note the remarkable reliability of the responses, waveforms, and latencies during a period of 16 months.

used to monitor early compression of the optic nerve and chiasma by sellar and parasellar tumors. T-VEPs can also be utilized to quantify the effect of surgery and/or irradiation. Halliday *et al.*[29] showed marked improvement of VEP in the postoperative recordings of 9 parasellar neoplasms. The improvement of VEPs was associated with improved visual function. Craniopharyngiomas and pituitary tumors compressing the chiasma resulted in abnormal VEPs in both eyes in 9 out of 10 patients.

Critical frequency of photic driving was studied in 74 multiple sclerosis patients. It was abnormal in 44% of the cases. When compared to T-VEPs, CFPD was a less

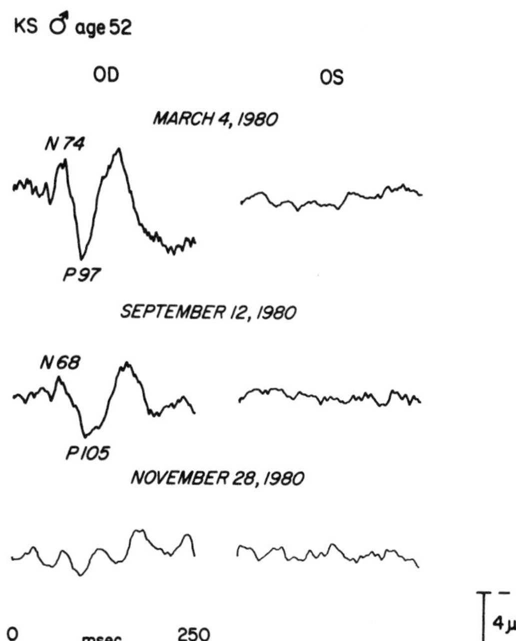

FIGURE 8. T-VEPs to pattern reversal in a patient with ischemic optic neuropathy. The patient developed a sudden onset of visual blurring in February, 1980. Visual fields showed a central scotoma in the left eye (OS). T-VEPs in March showed absence of response to stimulation of OS and normal response to stimulation of the right eye (OD). A follow-up 6 months later showed a change in the potential evoked from OD with broadened P_1 and a P_1 latency 8 msec longer on March 7, although still within the normal range of values for his age. At this time the patient was complaining of transient obscurations but Goldman's perimetry of OD was normal. Worsening of vision was noted by the patient in November of the same year; at that time he had bilateral cortical scotoma and absent T-VEPs.

TABLE 2

PATTERN-REVERSAL VISUAL EVOKED RESPONSES IN MULTIPLE SCLEROSIS

Reference	Year	Check Size (min of arc)	Number of Patients Studied	Abnormal VER	
				Number	%
Halliday *et al.*[20]	1973	50	51	49	96
Asselman *et al.*[11]	1975	30	51	34	67
Lowitzch *et al.*[35]	1976	50	135	98	73
Zeese[36]	1977	45	30	20	67
Hennerici *et al.*[16]	1977	45	57	50	88
Celesia[28]	1978	15	74	55	74
Mastaglia *et al.*[37]	1977	*	102	52	51
Hoeppner & Lolas[38]	1978	50	104	49	47
Shahrokhi *et al.*[39]	1978	26	149	85	57
Bodis-Wollner *et al.*[26]	1979	13†	103	64	62
Totals			856	556	65

*Information not given.
†Gratings.

sensitive indicator of optic nerve pathology than T-VEPs but the two tests were not mutually exclusive because CFPD was the sole abnormality in 2 of the 74 patients.[28] The effectiveness of CFPD for the diagnosis of optic nerve involvement in multiple sclerosis was confirmed by Cohen *et al.*[30] Similarly, Regan *et al.*[31] identified a distinct group of multiple sclerosis patients having a defect to medium frequency flicker independently of the delay in pattern-reversal VEPs. The utilization of more than one test will enhance the reliability and the yield of the procedure.

There is considerable controversy about the effect of retrochiasmatic lesions on the amplitude of T-VEPs.[8,11,21,23,28,29] Halliday's group, using full- and half-field pattern stimulation, were able to demonstrate clear asymmetries of amplitude distribution in hemianopic patients. These findings were confirmed by Kuroiwa and Celesia.[22] In a study of 14 hemianopic patients, the following amplitude distribution abnormalities were noted: (1) absent T-VEPs to stimulation of the affected half-field with normal amplitude distribution to stimulation of the normal half-field; (2) reversal of normal amplitude pattern to the half-field stimulated; and (3) lateral occipital ratios above 1.36. One or two of these abnormalities were present in 12 patients (FIGURE 9 & 10).

Another method that has shown promise for the detection of hemianopic fields is the visual evoked spectrum array (VESA). The following abnormalities were noted in 16 hemianopic patients: (1) Small spectral amplitude over the occipital region contralateral to the affected field; (2) visual evoked spectral ratio plot with high peaks and valleys; (3) spectral ratios above 2.5 (FIGURE 11). The major strength of VESA is

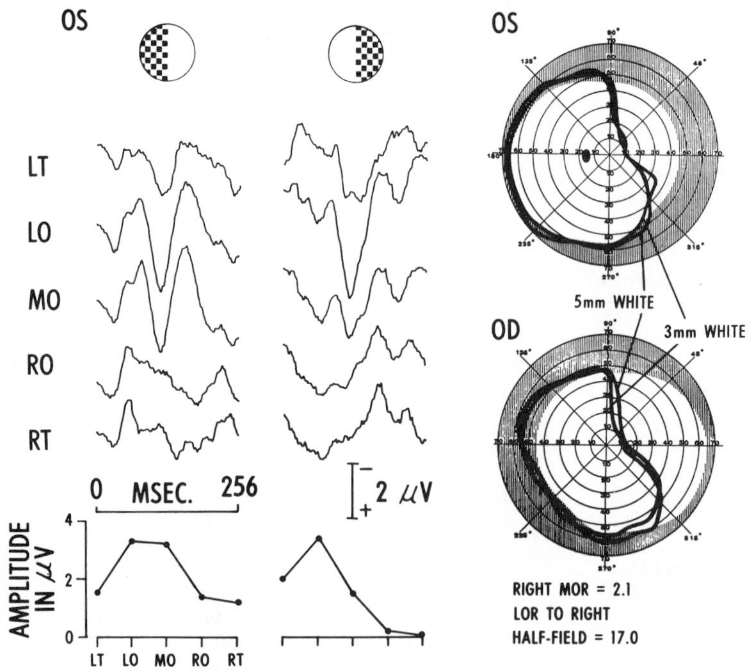

FIGURE 9. Patient with right homonymous hemianopsia from an infarct of the left occipital lobe. Note the pathological reversal of amplitude asymmetry with contralateral preponderance to stimulation of affected right hemifield. Amplitude of large positive wave is plotted in the lower part of the figure. LOR, lateral occipital ratio; LT, left temporal; LO, left occipital; MO, midoccipital; RO, right occipital; RT, right temporal.

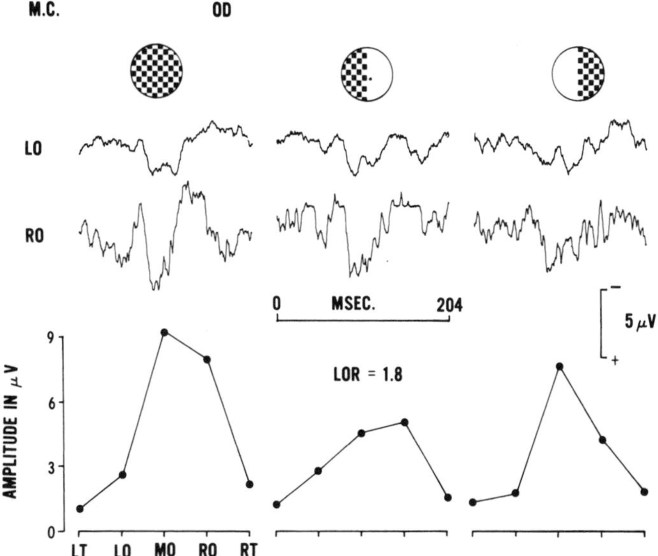

FIGURE 10. Patient with left homonymous hemianopsia. Selective T-VEP to full field, right and left hemifield, are shown in the upper half of the figure. Note the lack of ipsilateral amplitude preponderance to the left hemifield (affected field) stimulation, that is a reversal of normal amplitude asymmetry.

its ability to be carried out in uncooperative patients. Visual stimulation at each frequency lasts 8 sec and the test is over in less than 2 min. As promising as these tests appear for the objective determination of retrochiasmatic lesions, a note of caution is in order. While either T-VEP amplitude distribution of VESA is reliable, neither test has yet proved as sensitive as the visual field perimetry. Further testing of these methods is needed.

Transient visual evoked potentials have been used successfully to demonstrate disorganization of retinogeniculo striate projections in human albinos.[32,33] Albinism is not only characterized by hypomelanosis but also by aberrant retinal projections. Guillery *et al.*[34] have verified these aberrant optic projections in several brains of human oculocutaneous albinos. In contrast to the normal pattern of decussation, albinos have crossing fibers arising from the temporal retina near the zero vertical meridian while the more peripheral temporal retina produces nondecussating projections. This disorganization alters the orderly representation within the lateral geniculate and the striate cortex. Coleman *et al.*[33] found that their albino subjects had abnormal amplitude distribution of visual evoked responses suggesting aberrant visual pathways. The albinos showing abnormal VEP had defective stereopsis. These authors suggest that human albinos have separate representations of both the ipsilateral and abnormal contralateral visual fields in each hemisphere.

SUMMARY

The electrophysiological analysis of visual evoked responses is a powerful tool for the study of visual function. The combined application of pattern-reversal transient

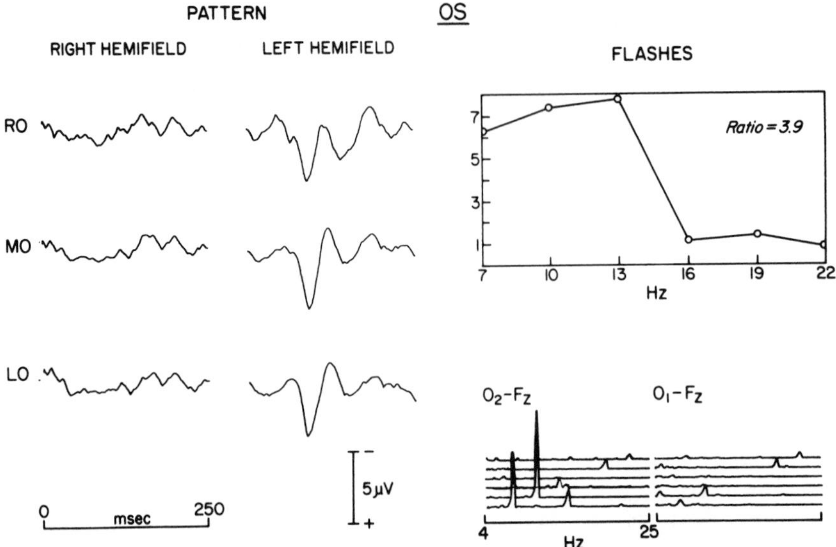

FIGURE 11. Transient visual evoked responses to pattern reversal are shown in the right half of the illustration. Note the absence of T-VEPs with stimulation of the affected field. The visual evoked spectrum array to steady-state flash stimuli is displayed in the lower part of the right half and shows very small energy peaks over the left occipital region (O_1–F_z). The ratio of the peak energy is a 0.5 Hz band for each of the six frequencies of stimulation, and is plotted in the upper part of the right half of the figure. The dominant energy side was the right occipital region with a mean ratio of 3.9.

visual evoked potentials, critical frequency of photic driving, and visual evoked spectrum array has enhanced the reliability and the yield of these tests for the diagnosis of visual dysfunctions. Prechiasmatic and retrochiasmatic lesions are characterized by different abnormalities. Prechiasmatic lesions often can be further differentiated into retinal and optic nerve lesions by the simultaneous recording of retinal and cortical potentials.

REFERENCES

1. HUBEL, D. M. & T. N. WIESEL. 1959. J. Physiol. (London) **148:** 574–591.
2. HUBEL, D. H. & T. N. WIESEL. 1974. J. Comp. Neurol. **158:** 295–306.
3. BARLOW, H. D. 1953. J. Physiol. (London) **119:** 69–88.
4. HARTER, M. R. & C. T. WHITE. 1970. Electroencephalogr. Clin. Neurophysiol. **28:** 48–54.
5. REGAN, D. 1972. Evoked Potentials in Psychology, Sensory Physiology, and Clinical Medicine. Chapman and Hall, London.
6. HALLIDAY, A. M. & W. I. MCDONALD & J. MUSHIN. 1972. Lancet (1): 982–985.
7. CELESIA, G. G. & R. F. DALY. 1977b. Neurology **27:** 637–641.
8. HALLIDAY, A. M., G. BARRETT, E. HALLIDAY & W. F. MICHAEL. 1977. The topography of the pattern evoked potential. *In* Visual Evoked Potentials in Man: New Developments. J. E. Desmedt, Ed. pp. 121–133. Oxford Univ. Press, Oxford, England.
9. ARDEN, G. B., I. BODIS-WOLLNER, À. M. HALLIDAY, A. JEFFREYS, J. J. KULIKOWSKI, H.

SPEKREIJSE & D. REGAN. 1977. Methodology of patterned visual stimulation. *In* Visual Evoked Potentials in Man: New Developments. J. E. DESMEDT, Ed. pp. 3–15. Oxford Univ. Press, Oxford, England.

10. REGAN, D. & W. RICHARDS. 1971. Vis. Res. **11:** 679–684.
11. ASSELMAN, P., T. W. CHADWICK & C. D. MARSDEN. 1975. Brain **98:** 261–282.
12. COLLINS, D. W., W. M. CARROLL, J. L. BLACK & M. WALSH. 1979. Br. Med. J. (1): 231–232.
13. VAN LITH, G. H. M. 1980. Clin. Neurol. Neurosurg. **82**(2): 85.
14. CELESIA, G. G. & J. T. MEREDITH. 1982. Ann. N.Y. Acad. Sci. This volume.
15. ROVAMO, J. & V. VIRSU. 1979. Exp. Brain Res. **37:** 495–510.
16. HENNERICI, M., D. WENZEL & H. J. FREUND. 1977. Brain **100:** 119–126.
17. CELESIA, G. G. & R. F. DALY. 1977. Arch. Neurol. (Chicago) **34:** 403–407.
18. ALLISON, T., W. R. GODD & C. S. WOOD. 1979. Auditory, somatosensory and visual evoked potentials in the diagnosis of neuropathology: Recording consideration and normative data. In Human Evoked Potentials: Application and Problems/NATO Conference Series of Human Factors. D. Lehmann & E. Callaway, Eds. Vol. 9: 1–16. Plenum Press, New York, N.Y.
19. SHAW, N. A. & B. R. CANT. 1980. Electroencephalogr. Clin. Neurophysiol. **48:** 237–241.
20. HALLIDAY, A. M., W. I. MCDONALD & J. MUSHIN, 1973b. Br. Med. J. (4): 661–664.
21. STARR, A., H. SOHMER & G. G. CELESIA. 1978. Some applications of evoked potentials to patients with neurological and sensory impairment. *In* Event Related Brain Potentials in Man. E. CALLAWAY, P. TUETING & S. H. KOSLOW, Eds. pp. 155–196. Academic Press, New York, N.Y.
22. KUROIWA, Y. & G. G. CELESIA. 1981. Arch. Neurol. (Chicago) **38:** 86–90.
23. BARRETT, G., L. D. BLUMHARDT, A. M. HALLIDAY, E. HALLIDAY & A. KRISS. 1976. Nature (London) **261:** 253–255.
24. REGAN, D. 1975. Nature (London) **253:** 401–407.
25. CELESIA, G. G., V. K. SONI & W. S. RHODE. 1978. Arch. Neurol. (Chicago) **35:** 678–682.
26. BODIS-WOLLNER, I., C. D. HENDLEY, L. H. MYLIN & J. THORNTON. 1979. Ann. Neurol. **5:** 40–47.
27. CELESIA, G. G., C. R. ARCHER, Y. KUROIWA & P. R. GOLDFADER. 1980. Arch. Neurol. (Chicago) **37:** 704–706.
28. CELESIA, G. G. 1978. Am. J. EEG Technol. **18:** 47–59.
29. HALLIDAY, A. M., E. HALLIDAY, A. KRISS, E. I. MCDONALD & J. MUSHIN. 1976. Brain **99:** 357–374.
30. COHEN, S. N., K. SYNDULKO, W. W. TOURTELLOTTE & A. R. POTVIN. 1980. Arch. Neurol. (Chicago) **37:** 80–83.
31. REGAN, D., B. A. MILNER & J. R. HERON. 1976. Brain **99:** 43–66.
32. CREEL, D., F. E. O'DONNELL JR., & C. J. WITKOP JR. 1978. Science **201:** 931–933.
33. COLEMAN, J., C. E. SYDNOR, M. L. WOLBARSHT & M. BESSLER. 1979. Exp. Neurol. **65:** 667–679.
34. GUILLERY, R. W., A. N. OKORO & C. J. WITKOP JR. 1975. Brain Res. **96:** 373–377.
35. LOWITZSCH, K., U. KUHNT, C. H. SAKMANN, K. MAURER, H. C. HOPF, D. SCHOTT & K. THATER. 1976. J. Neurol. **213:** 17–32.
36. ZEESE, J. A. 1977. Arch. Neurol. (Chicago) **34:** 314–316.
37. MASTAGLIA, F. L., J. L. BLACK, L. A. CALA & D. W. K. COLLINS. 1977. Br. Med. J. (1): 1315–1317.
38. HOEPPNER, T. & F. LOLAS. 1978. J. Neurol. Neurosurg. Psychiatry **41:** 493–498.
39. SHAHROKHI, F., K. H. CHIAPPA & R. R. YOUNG. 1978. Arch. Neurol. (Chicago) **35:** 65–71.

GENERAL DISCUSSION

Ivan Bodis-Wollner, *Moderator*

Mount Sinai School of Medicine
of the City University of New York
New York, New York 10029

S. SOKOL: I have a question for Dr. Celesia about his normals. Did the flicker ERGs and the flicker VEPs break down, so to speak, at the same flicker rate, or were there differences between the ERG and VEP?

G. CELESIA: No, usually they broke down at approximately the same frequency.

SOKOL: Years ago, van der Tweel and Riggs both found that the ERG will continue to follow flicker at much higher rates than the VEP does. I noticed the same thing in some of your clinical examples and was wondering whether this was also happening in normals.

CELESIA: We consider it abnormal if the differential is greater than ten.

G. H. M. VAN LITH: Dr. Celesia, you said that in optic nerve disease or in disturbances of the conduction system, the ERG is normal. In our opinion, the conventional ERG, the flicker ERG, is not always normal. There are optic atrophies and inherited optic neuropathies where we see absolutely abnormal ERGs. We see them also in cases of alcohol abuse. You cannot say that when the ERG is abnormal that it is primarily a retinal disturbance. It can be secondary to the optic nerve disease or parallel with it.

R. J. SCLABASSI: The relationship of the EEG power spectral analyses to the evoked potential is not entirely clear to me in your data.

CELESIA: I do not claim that there is any direct relationship between evoked potentials and the power spectrum.

K. CHIAPPA: Dr. Blumhardt mentioned the fixation difficulty, together with the difficulty of assessing whether or not the patient is concentrating. It was shown by Dr. Epstein at the last EEG meeting that there are amplitude differences which occur as a function of attention. These problems, to a large extent, can be gotten around by the sequential stimulation technique that was presented first by Dr. Rowe. With his method the whole field is presented to the patient first, then just the left field checks, and then just the right field checks are shifted only 300 milliseconds apart, and the response to both is obtained in one sweep. The question to Dr. Blumhardt is: Do you see any clinical utility in stimulating not only half-fields but only the central part of one half-field, and only the peripheral part of one half-field as a method of increasing sensitivity and revealing abnormalities that might be masked by the central components?

L. D. BLUMHARDT: First of all, I do not think I did mention the fixation problem. But it is in fact a very small problem, provided we have a responsive and cooperative subject. Nevertheless, our technique is only useful when fixation is well maintained as one of the effects of eccentric stimulation is to reduce the ipsilateral potentials and either to enhance or to leave the contralateral components relatively unaffected. Whether or not stimulating parts of the half-field is useful depends on what you want to do. Using quadrants and sub-units of the visual half-fields you are up against the topographic problem. It is impossible to define normal limits for small parts of the field; while annuli and small sectors may sum in some patients, they will produce no response at all in others. So I do not think that will be very practical. It is important, however, to identify the component you are measuring, because so many of the

0077-8923/82/0388-0306 $1.75/0 © 1982, NYAS

pathologies we have been hearing about, particularly those affecting the anterior pathways, will create central scotomas. As a consequence, measuring and comparing the pathological response with the normal VEP is, in essence, comparing different waves: those coming from the paracentral areas of the retina in the scotomatous response with potentials arising from the central field, in the normal. That is a very important consideration from the point of view of interpreting responses, but I do not believe it is going to increase the accuracy of detecting small field defects.

N. LESEVRE: Dr. Blumhardt, I wanted to know whether you found a relation between the type of abnormality of the response and the nature of the lesions on the one hand, and with the age of the lesion on the other. The reason I am asking this is that we have done a study on hemianopic patients who all had macular splitting, and the people with tumors had high activity and abnormal activity, whereas the infarction cases had either no response whatever, or a low amplitude abnormal response. We even found a relationship between the pathology and the degree of high potency of the scalp, and I was wondering if you could comment on this. Did you find anything of this sort?

BLUMHARDT: I wonder if you are looking at rather longer latency VEP components with chronotopograms than we are. We find no correlation with the type of pathology at all. A plaque of demyelination which produces a field defect will, as I said, produce exactly the same type of abnormality as a tumor or infarct which has been there for six weeks or six years. The abnormality of the VEP is dependent on the field defect and not the pathology. It is very difficult, of course, to find cases with well-identified single episodes of demyelination. We have been lucky enough to get a series of patients presenting *de novo* with single plaques in the posterior visual pathways and no known problems beforehand and, as you would expect, these show, in the acute situation, asymmetries and waveform distortions paralleling the visual field defect, and not delays.

H. SPEKREIJSE: First a comment. Dr. Chiappa stated that attention has an influence on the amplitude of the pattern EPs. I want to deny that. I mean that attention *per se* does not influence the amplitude of the EP at all—we have studied that extensively. What you have to be certain about is that the subject accommodates on the screen. It is not attention. If attention were a factor, all baby and infant work would be useless. For children, cartoons are being used to enhance the child's accommodation at the screen. Attention does not really influence the VEP, certainly not the first 120 milliseconds or so.

I. BODIS-WOLLNER: Thank you, Dr. Spekreijse, I agree with you.

INTRAOPERATIVE MONITORING OF EVOKED POTENTIALS

Peter A. Raudzens

Department of Neurology
Barrow Neurological Institute
St. Joseph's Hospital
Phoenix, Arizona 85013

Significant developments in evoked potential (EP) studies, many of them detailed at this conference, have made it easier to test EPs for clinical applications. Essential features of stimulus presentation and signal acquisition have been characterized and accepted as conventional recording techniques. The availability of microprocessor-based signal averagers has increased while the cost and size of the equipment has decreased. Pathological correlates have been established between changes in EP waveform parameters and specific diseases.[1,2] It is accepted that changes in latency, amplitude, or phase relation values of the EP waveform components are an objective measure of sensory function and have been applied clinically for the detection and diagnosis of nervous system disorders.

This is particularly useful in the assessment of unresponsive patients such as infants and comatose or anesthetized patients. There is a compelling need to monitor the central nervous system (CNS) during surgery as reliably as the cardiovascular system is monitored. Sensory EP testing may satisfy this need.

The features that make EPs attractive for intraoperative monitoring are the following:

1. The waveforms are recorded by noninvasive techniques from surface scalp electrodes.
2. The waveforms are relatively stable and reproducible and the short latency components are independent of the subject's level of arousal.
3. The short latency EPs, such as the brainstem auditory evoked potential (BAEP), are resistant to the therapeutic levels of most drugs,[3] including anesthetics, and even the longer latency cortical events have a predictable dose-dependent relationship to volatile anesthetics.[4]
4. The signal averager is portable and compact, suitable for intraoperative or Intensive Care monitoring.
5. Many of the nonneurological factors that can influence EP latency or amplitude values such as temperature,[5,6] age, gender, stimulus features, stimulus phase, stimulus intensity, stimulus rate, filter settings, and signal-to-noise characteristics can be controlled or eliminated by using the contralateral sensory pathway as the control for the tested side.

This paper summerizes our initial experience with intraoperative monitoring of sensory EPs in 173 neurological procedures from 1979 to 1981. It was obvious early in the study that successful EP recording required a co-operative surgeon, an informed anesthesiologist, and an expert technician. Many problems had to be solved to get a "clean" signal quickly and to reduce data analysis. Interpretation of EP changes was limited to a binary "on-off" demonstration of waveform reproducibility.

But do intraoperative EP changes detect neurological deficits? Can this early

0077–8923/82/0388–0308 $1.75/0 © 1982, NYAS

recognition prevent postoperative sensory deficits by timely corrective surgical intervention?

Three questions had to be satisfied:

1. Can EPs be reliably recorded in the Operating Room (OR) without added risk or inconvenience?
2. Do intraoperative changes in EPs correlate with postoperative sensory deficits?
3. Can intraoperative EP changes be used to prevent postoperative sensory deficits?

Unlike other clinical applications that evaluate the utility of EP testing as a diagnostic technique, we were interested in EPs as a predictive index of neurological outcome.

TABLE 1

SURGICAL PROCEDURES PERFORMED WHILE MONITORING EVOKED POTENTIALS

Sensory Evoked Potentials	Surgical Procedures	Number of Patients
Visual evoked potentials	Pituitary tumors	36
	Suprasellar tumors or cysts	20
	Craniopharyngiomas	3
	Optic nerve decompression (Arachnoiditis, Abcess)	4
	Aneurysms	8
Brainstem auditory evoked potentials	Posterior fossa explorations for neoplastic or vascular lesions	35
	Microvascular decompression of CN V or VII	19
	Vestibular nerve section for intractable vertigo	8
	Acoustic Neuroma	4
Somatosensory evoked potentials	Cervical decompression (herniated nucleus pulposis or cervical spondylosis)	14
	Spinal cord tumor	3
	Harrington rod fixations	9
	Thoracic spine fracture	3
	Brachial plexus exploration	2

METHODS AND MATERIALS

Intraoperative sensory EPs were recorded in 173 surgical procedures consisting of 71 visual evoked potentials (VEP) during pituitary or aneurysm surgery, 66 brainstem auditory evoked potentials (BAEP) during posterior or middle fossa explorations, and 31 somatosensory evoked potentials (SSEP) during spinal cord surgery. The other 5 cases had multimodality EP testing. Each sensory modality will be considered separately.

Informed consent was obtained preoperatively for EP testing. TABLE 1 lists the operative procedures.

Visual Evoked Potentials

Flash stimuli were presented monocularly through the closed lid by an array of 15 light-emitting diodes (LED) mounted in an opaque eye patch. Stimulus rates of 1.8 to 2.4/sec with 3 msec duration were used. Red LEDs were flashed at maximum intensity. The Active A_g cup electrode was secured at the occiput (O_z) referenced to the forehead (F_z) with ground electrode on the earlobe (A_1). Interelectrode impedances were kept at 3 kΩ or less. Suitable VEPs required 64 to 128 averages with amplification times 50,000 and filter bandpass settings of 1 and 30 Hz. Sample times of 250 and 500 msec, following the flash stimulus, were recorded.

In the majority of cases, the intraoperative VEP recordings were compared to pre- and postoperative VEPs generated by both the LED flash and pattern-reversal stimuli using an alterating checkerboard TV image with 2 cm check size. Visual testing of Snellen acuity, color vision, contrast sensitivity functions and visual field analysis was done in most cases.

Brainstem Auditory Evoked Potentials

All patients were tested preoperatively to obtain a baseline BAEP. Hearing threshold was measured. The acoustic stimulus was presented monaurally. A click was generated by passing a 0.1-msec square-wave pulse through shielded earphones. The test stimulus intensity was 60 dBHL at a rate of 11.3/sec ipsilateral with 40 dB masking noise contralateral. In the OR, molded ear inserts were fitted to an acoustically matched transformer and the stimulus intensity was increased to 85 dBHL. The bipolar EEG activity was recorded from vertex (C_z) to earlobe (A_1) ipsilateral to the acoustic stimulation. Amplification times 100,000 and a bandpass filter at 150 and 1500 Hz were used. The signal following the stimulus for 10.24 msec was averaged 1000 to 2000 times.

Postoperative BAEPs and auditory testing of speech discrimination and pure-tone audiometry were done to document hearing deficits.

Somatosensory Evoked Potentials

Somatosensory signals were constant current stimuli applied by disc or needle electrodes through the skin overlying a mixed peripheral nerve. The median, ulnar, or peroneal nerves were frequently used. Stimulus intensity was gradually increased until an appropriate muscle twitch could be driven at the stimulating frequency. Intraoperatively this stimulus was increased by 1.5 ×. A large ground electrode was secured proximal to the stimulating electrodes.

For upper extremity recording, the median or ulnar nerves were stimulated. Four channels of simultaneous recording were acquired from these locations:

Channel 1, Erb's point to FP_z (+*ve*);
2, C_7 to FP_z (+*ve*);
3, C_2 to FP_z (+*ve*);
4, Contralateral cortex to FP_z (+*ve*).

For lower extremity recording, the peroneal or tibial nerves were stimulated. The following four channels of simultaneous recording were acquired from these locations:

TABLE 2

CHANGES IN EVOKED POTENTIALS IN THE OPERATING ROOM (OR)

Sensory Evoked Potentials	Number of Patients	Technically Satisfactory EPs	Artifacts	No OR EP Change	Transient OR EP Change	Permanent EP Change with Sensory Change
VEP	71	62 (87%)	9 (13%)	0	58 (81%)	4 (6%)
BAEP	66	66 (100%)	0	50 (75%)	10 (15%)	6 (10%)
SSEP	31	25 (81%)	6 (18%)	23 (75%)	2 (6%)	0

Channel 1, L_3 to L_1 (+*ve*);
2, T_{10} to T_6 (+*ve*);
3, C_3 or C_4 to C_z (+*ve*);
4, Contralateral cortex to C_z (+*ve*).

The stimuli were presented at a rate of 4.1/sec and 300 msec duration. Good waveforms required 1048 averages at 50,000 × amplification and bandpass filter settings of 30–1500 Hz. The initial 50 msec following stimulus presentation was sampled.

Pre- and postoperative SSEP and sensory testing was done in most of the patients. A few emergency cases were excluded from the complete study.

EPs were recorded continuously in the OR throughout the surgical procedure. Averaged responses were superimposed to demonstrate reproducibility. Latency and amplitude values were measured with a cursor calibrated to the display CRT. The patients were exposed to a variety of anesthetic agents. Each patient's normal side, when possible, was used as a control for the operated side to aid EP interpretation.

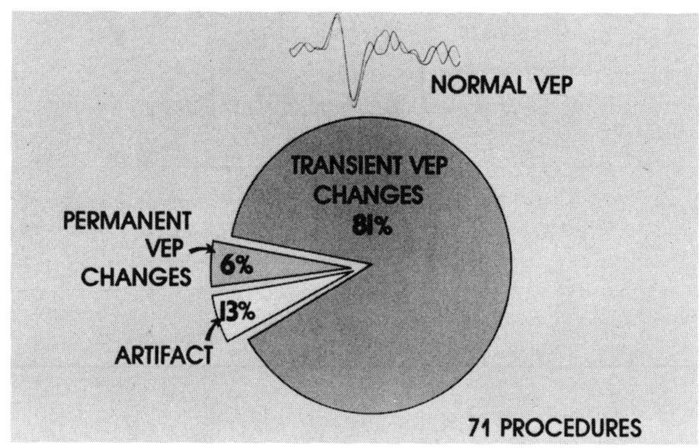

FIGURE 1. Changes in visual evoked potentials (VEPs) during surgery.

RESULTS

Early in this study it was obvious that the short-latency BAEP were technically easier to record from anesthetized patients in the OR than the longer-latency cortical VEPs and SSEPs. It was also evident that the yield of positive correlations between postoperative sensory deficits and predictive EP changes was low. This is summarized in TABLE 2 and FIGURES 1–3.

Visual Evoked Potentials

The interpretation of VEPs generated by an LED flash through closed eyelids was confounded by an 81% variability and both false positive and negative results.

This problem is a function of stimulus presentation. Luminance could not be controlled because the diffuse red flash had to pass through the eyelid before illuminating the retina. The pupil size and direction of gaze and attention to the

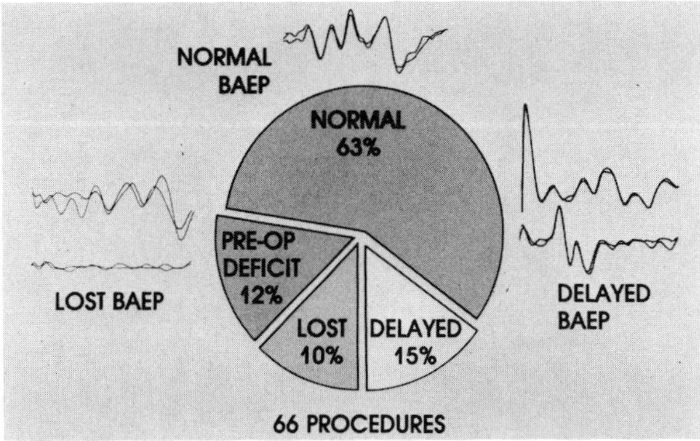

FIGURE 2. Changes in brainstem auditory evoked potentials (BAEPs) during surgery.

stimulus also could not be controlled in an anesthetized patient. The anesthetic agents can also influence the VEP by either an inhibiting effect on the metabolic activity of the pigment epithelium of the retina[7] or on the neural discharge of cells in area 17. All these factors contribute to distort the flash VEP.

Instead of the usual VEP peak positivities at about 60 and 120 msec when a constant luminance flash is presented to the central 6–12° of the visual field in awake subjects,[8] latency values of the positive peaks varied from 90 to 200 msec. Although significant reproducibility of these responses was recorded, the latency and amplitude values shifted randomly.

We tried several different patterns of stimuli presentation to improve signal stability but without success. For instance, green or yellow LEDs, to select the peak VEP sensitivity, failed to improve the variability. Different interstimulus intervals

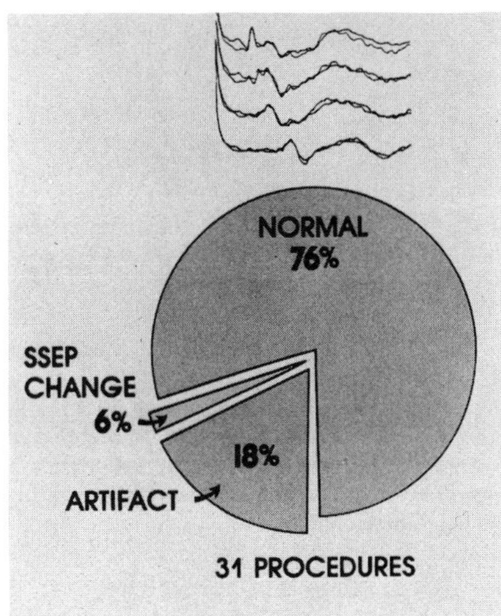

FIGURE 3. Changes in somatosensory evoked potentials (SSEPs) during surgery. Normal SSEP recordings are at the top of the figure.

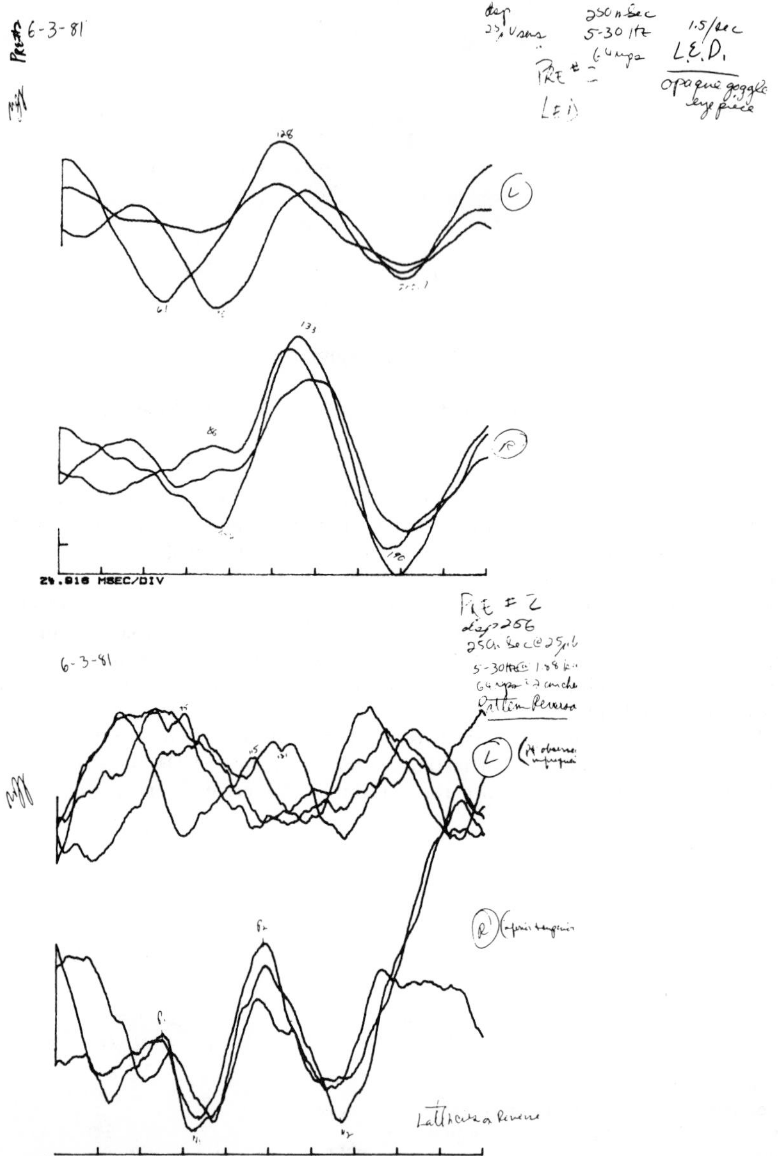

FIGURE 4. VEPs to both LED flash stimuli and pattern-reversal stimuli are shown prior to drainage of a pituitary abscess. Note that preoperatively the patient had no reproducible pattern-reversal VEP and no functional vision in the left eye (O.S.) but had a normal flash VEP. FIGURES 4–6 are actual recordings made during the operation and include the original markings made by the technician.

between 200 to 400 msec did not help. Altering the rate of presentation from 1.5 to 10 Hz had little effect except for amplitude attentuation at higher frequencies. Binocular stimulation did not improve the variability and only marginally increased waveform amplitude.

But a more disturbing result than VEP variability was the presence of both false negative and positive correlations between VEP changes and visual acuity. In five cases reproducible flash VEPs were recorded in the absence of pattern-reversal VEPs and visual acuity. This is illustrated in FIGURE 4 and 5.

False negative results can be understood easier because of the possibility of recording artifact. Electrodes can slip, stimulators can fail, and a patient with functionally normal vision could present with an abnormal VEP. But if a patient without visual function can generate a flash VEP, then interpretation of the visual pathway integrity is impossible. Clearly the pattern-reversal VEP, which is a function of visual acuity, is processed by different neural channels than the flash VEP. To be meaningful, a more specific visual stimulus, such as pattern-reversal, or more sensitive data analysis of VEPs, such as steady-state parameters will have to be used for OR monitoring.

With these reservations, we recorded several convincing changes in VEP amplitude and latency during mechanical retraction of the optic chiasm and induced hypotension. Prompt recovery of the VEP with return of normal perfusion pressures and removal of retraction is evident in FIGURE 6. This suggests that the VEP reflected a transient disturbance in the visual system.

In this series only two patients had visual deficits postoperatively. More frequent morbidity was aphasia, third nerve palsy and hemiparesis secondary to edema of associated structures such as the temporal lobe. Two patients developed intractable intracranial hypertension after aneurysm clipping and never recovered.

The most satisfying cases were drainage of a pituitary abscess. The patient presented with failing vision of the left eye and a computer tomography scan revealed a pituitary fossa lesion with suprasellar extension. This was explored through a transphenoidal approach, identified as an abscess and drained. Intraoperatively the amplitude of the VEP improved although latencies were prolonged to 190 msec. Postoperatively both the pattern and flash VEPs were only slightly delayed and visual acuity was remarkably improved.

This case is interesting for two reasons. It seems as if the improvement in the flash VEP intraoperatively predicted improved visual function following decompression of the optic chiasm. But the fact that preoperatively the patient had a normal flash VEP on the side with reduced visual acuity and absent pattern-reversal VEPs confuses interpretation of the VEP changes recorded.

Brainstem Auditory Evoked Potentials

The most convincing results were with this group of patients. Technically satisfactory BAEPs were recorded in all 66 cases (100%). Analysis was limited to latency values because of the variability of the amplitude changes. No intraoperative changes in BAEP latencies were found in 50/66 (75%) of these patients. A typical case is illustrated in FIGURE 7. The mean latency values are listed in TABLE 3. This group includes 9 (13%) patients with preoperative hearing and corresponding BAEP abnormalities.

These responses persisted throughout the surgical procedure. Our longest continu-

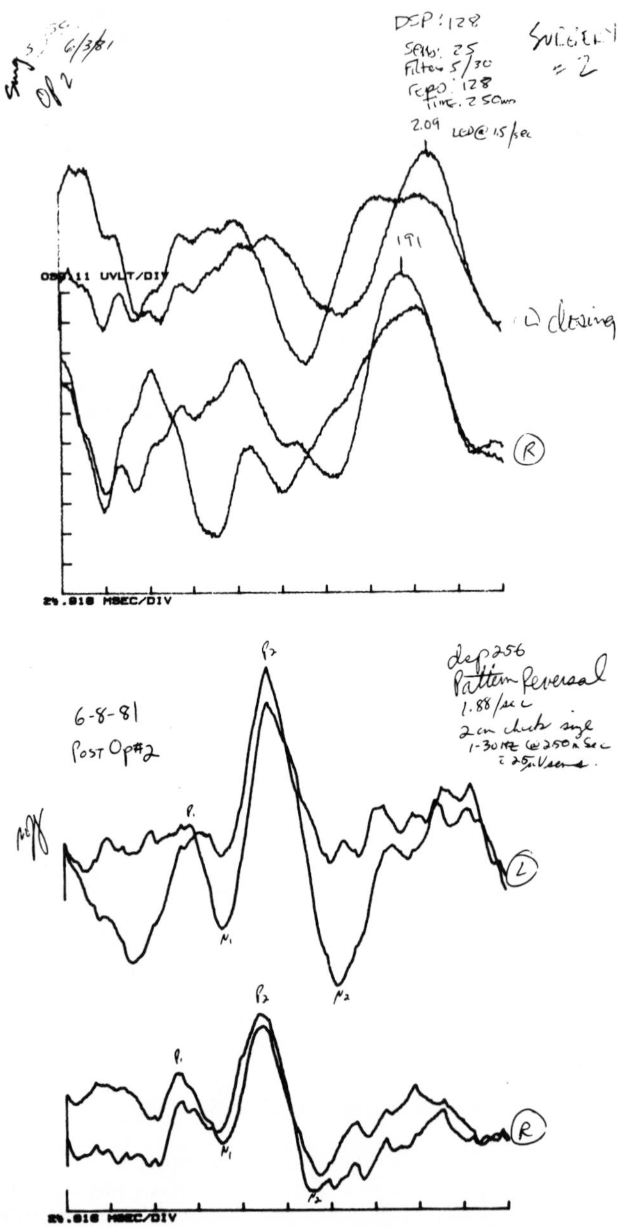

FIGURE 5. VEPs to both LED flash stimuli and pattern-reversal stimuli after drainage and decompression of the optic chiasm in the same patient as in FIGURE 4. Note that postoperatively vision and pattern-reversal VEP had recovered in the left eye. Recovery of the VEP at the end of surgery suggested improved visual function.

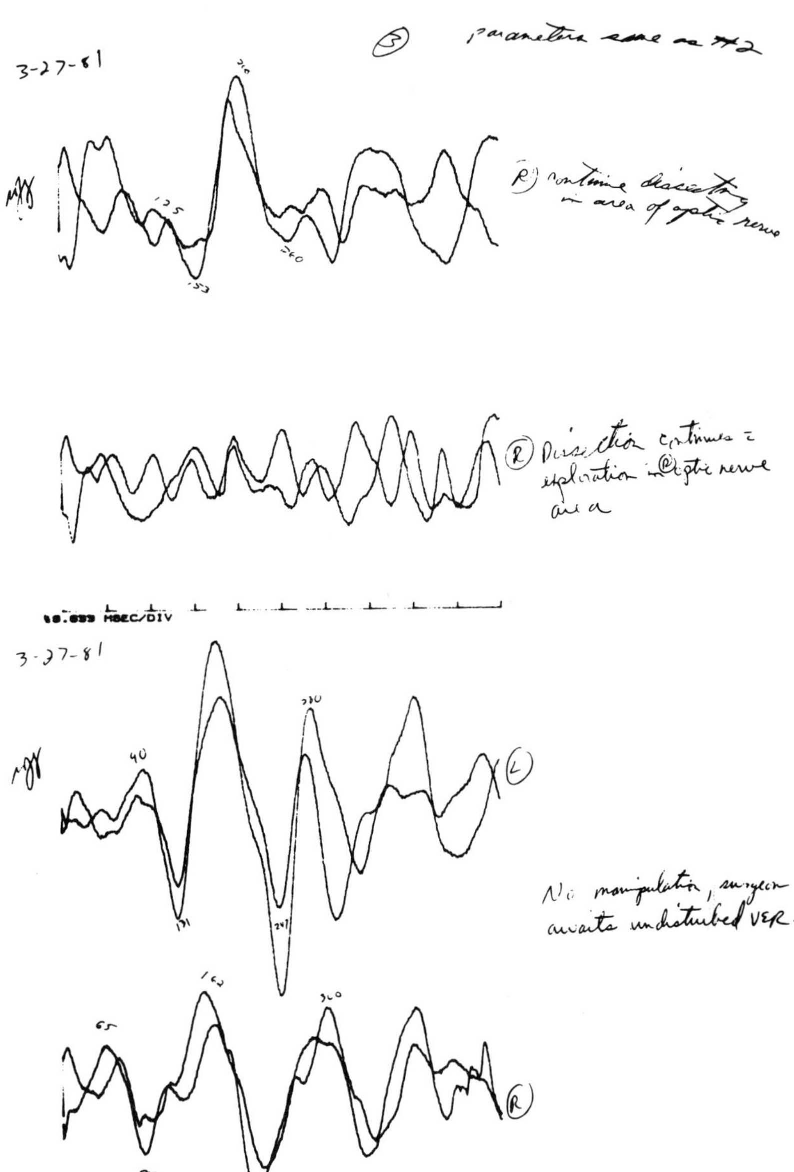

FIGURE 6. VEPs recorded during optic nerve decompression demonstrate the transient loss of a reproducible positive peak around 200 msec with retraction of the optic chiasm. The surgeon was alerted to the VEP change and removed his instruments. The VEP recovered within 3 minutes. Note also the fluctuating latency and amplitude changes seen in more than 80% of our cases.

ous recording time was 16 hours for acoustic neuroma excision. Auditory or brainstem function was unaffected in this group postoperatively.

Two unique intraoperative influences were recorded. A transient increase in latency and reduction in amplitude resulted from drilling around the ear. Also a "jitter" of latency values was often recorded. This transient variability of up to 1.5 msec may be due to temperature effects or mechanical distortion or ischemic effects on CN VIII, or the brainstem. We considered this transient degree of variability normal.

However, two different types of BAEP changes were identified in the remaining cases. The first change was a delay of BAEP latencies. This delay intraoperatively was recorded in 10/66 (15%) patients. Seven patients had microvascular decompression of CN VII and CN V and three patients had postfossa explorations. Postoperatively speech discrimination was reduced up to 52% but pure-tone audiometry was unaffected. Recovery of both the auditory function and normal BAEP latency values occurred over 30 days. This is illustrated in FIGURE 8.

The second change was an abrupt loss of the BAEP waveform beyond wave I despite a grossly intact CN VIII. No recovery of the EP was recorded. This change was invariably associated with profound sensory-neural hearing loss on the operated side. This was recorded in 6/66 (10%) of the cases.

Only one case of brainstem dysfunction was recorded in this group of patients. This patient developed uncontrollable cerebellar edema during acoustic neuroma

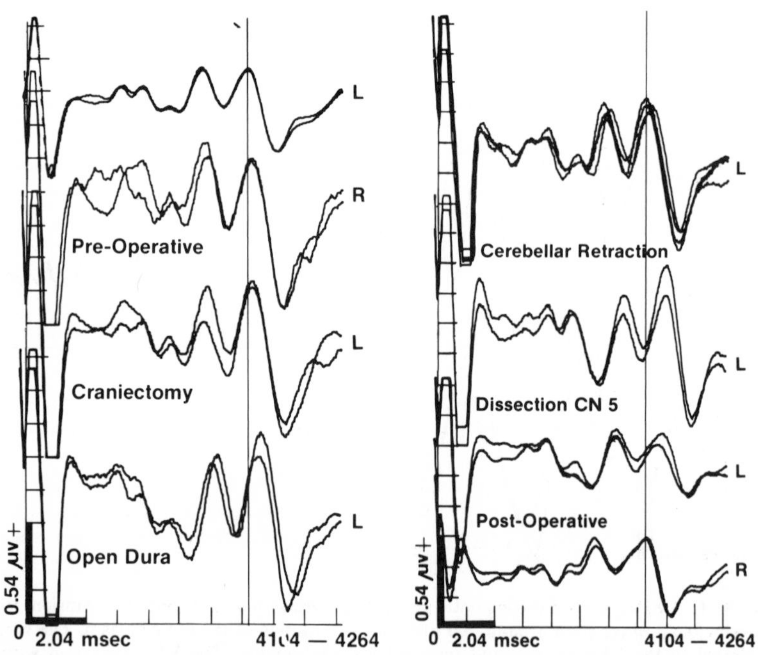

FIGURE 7. BAEPs recorded during microvascular decompression of the left CN V for trigeminal neuralgia demonstrate the remarkable stability characteristic of this EP. Small amplitude and latency increases are evident but not significant. Postoperatively auditory and brainstem function was normal.

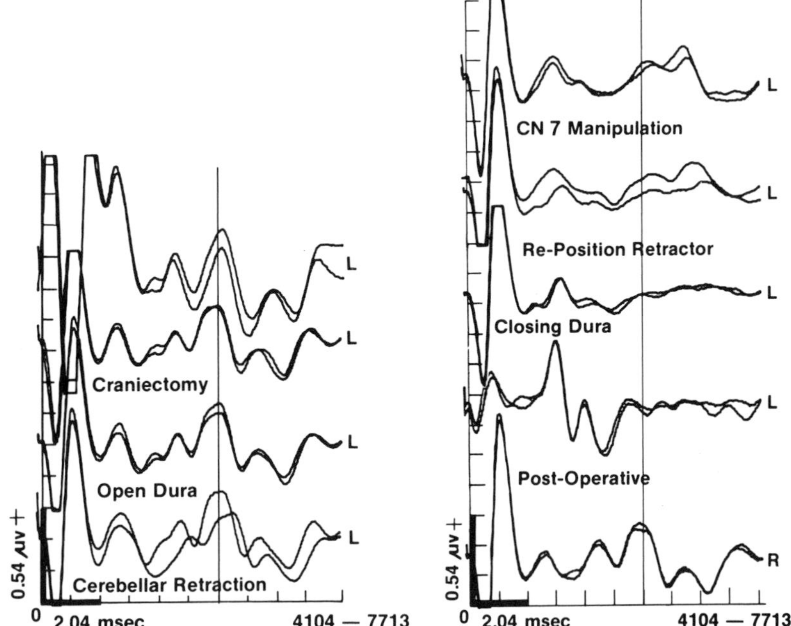

FIGURE 8. BAEPs recorded during microvascular decompression of the Left CN VII for hemifacial spasm are reproducible until the cerebellum is retracted to expose the CN VII, CN VIII root entry zones. Immediately two averages fail to superimpose. The response deteriorates until only waves I and II are obvious on the operated side. The BAEP on the opposite side is normal. Postoperatively this patient had a 52% reduction in speech discrimination on the left side.

excision and required extensive removal of the cerebellar hemisphere to facilitate closure. Vital signs and the BAEP brainstem deteriorated. The patient never regained consciousness and died one month postoperatively. Only waves I and II were present. Another patient failed to awaken following removal of a lateral ventricle meningioma but her BAEPs intraoperatively and postoperatively were unchanged.

BAEP changes correlated well with auditory function. We were fortunate to have no false negative results. Care with recording technique allowed us to obtain BAEPs in all these patients. We also had no false positive results. Persistence of the BAEP waveform was associated with intact auditory acuity, deterioration or loss of the BAEP waveform was associated with a minimal or profound postoperative hearing deficit. One patient with extensive midbrain edema remained comatose postoperatively.

Somatosensory Evoked Potentials

This is the most promising sensory modality to record intraoperatively but unfortunately it is also technically the most difficult. SSEP recording was totally contaminated by artifact in 6/31 (19%) cases. Different electrode locations, stimulus

TABLE 3
ABSOLUTE AND INTERWAVE LATENCIES

Sensory Evoked Potentials	Absolute Latencies of Waves (msec)			Interwave Latencies (msec)	
BAEP	I	III	V	I–V	III–V
Induction	2.47 ± 0.62	4.92 ± 0.55	6.99 ± 0.86	4.48 ± 0.64	2.03 ± 0.45
Closing	2.24 ± 0.47	4.68 ± 0.49	6.57 ± 0.54	4.42 ± 0.51	1.88 ± 0.30
Normal (BNI)	1.90 ± 0.40	3.96 ± 0.40	5.91 ± 0.63	4.02 ± 0.55	1.95 ± 0.35
SSEP					
Lower	L_3–L_1	T_{10}–T_6		C_z–A_1	L_1–C_z
extremity	12.04 ± 2.15	15.04 ± 2.62		28.00 ± 4.48	15.95 ± 3.30
Upper	Erbs	C_7	C_2	C_3–A_1	C_7–C'_3
extremity	10.31 ± 2.03	13.66 ± 2.04	13.90 ± 2.18	19.52 ± 2.63	5.88 ± 1.16

intensities or recording parameters failed to improve signal resolution. All six cases involved peroneal nerve stimulation during spinal cord manipulation. Our normal values for SSEPs are listed in TABLE 3. Only 2 (6%) of the remaining 25 cases demonstrated SSEP latency or amplitude changes related to the surgical procedure. One patient had a significant loss of the cortical EP amplitude during cervical cord decompression. This peak recovered with removal of methylmethacrylate that presumably was pressing against the cervical cord. Postoperatively, SSEPs were within normal limits and no sensory or motor deficit was evident. Another patient had consistent but insignificant reduction of the cervical-cortical interwave latency after successful micro anastomosis of the superficial temporal artery to the middle cerebral artery. Postoperatively, SSEPs and sensory motor function were unchanged.

The remaining 23/31 (75%) cases did not have significant SSEP changes related to the surgical procedure.

Most anesthetic agents, either the volatile gasses like ethrane and halothane or intravenous agents like pentothal and innovar, have a dose-dependent effect on the later cortical components. At "light" levels of anesthesia the shorter spinal EPs were well preserved.

Transient attenuations of amplitude or increase in latency was recorded but good reproducibility of SSEPs persisted in this group of patients. Postoperatively, both the SSEPs and the sensory function were unchanged.

DISCUSSION

The utility of intraoperative monitoring of sensory EPs have to be evaluated separately for each modality.

Visual Evoked Potentials

Our experience with 71 cases of intraoperative monitoring of VEPs generated by an LED flash emphasizes the need for a more specific visual stimulus. Interpretation of intraoperative changes and correlation with postoperative visual function is distorted by 13% artifact and 81% variability. This compares to a study by Allen[9] who reported a 68% variability and 12% artifact. Clearly it is possible to record VEPs intraoperatively. But, is interpretation of VEP changes meaningful in terms of postoperative visual function and do these changes reliably predict visual acuity postoperatively? Probably not, because this data is contaminated by both false positive and false negative results. Of our patients, 13% had VEPs lost in "noise" of artifact but retained unchanged visual function. These false negative results reflect technical difficulties with electrodes, line transients, stimulus intensity, and electrode placement. The patient's pre-existing visual disturbance and the combined stresses of changing anesthetic levels and surgical manipulation all contribute to increase the "noise" in this system.

A more compelling reason to question the validity of the VEP as an index of visual function is the presence of false positive results. We recorded normal VEPs to LED flash stimuli in 5/71 (7%) patients who had significant decrease in visual acuity and no pattern-reversal VEPs. Allen[9] also reported 2/29 (8%) patients with flash VEPs intact but visual deficits.

The variability of the diffuse flash VEP has been described[10,11] and Bodis-Wollner[12] has suggested that the changes in VEP are not specific for discrete visual deficits. It is unlikely that this variability can be reduced by simply increasing the flash luminance, changing the rate of presentation, or selecting a different spectral sensitivity. Our results also suggest that the flash VEP is processed by a different neural mechanism from the one responsible for visual acuity and pattern-reversal VEPs. The primary source for the VEP recorded over area 17 arises from retinal receptors which reach the occipital lobe via the lateral geniculate nucleus. A parallel channel transmits retinal activity to the occipital vortex via the reticular formation and diffuse thalamic projection system.[13] Supporting evidence for these separate channels comes from binocular interaction. When patterned light is presented to one eye and diffuse light to the other, both monocular and binocular VEPs indicate that the response to diffuse light is suppressed.[14]

In order to increase the specificity of VEPs either the visual stimulus or data analysis will have to be improved. A better presentation could be a pattern-reversal stimulus. Unfortunately the constraints of the operating room impose severe limitations on the size of any proposed stimulator and its proximity to the eye and the visual field. Bagolini[15] reported on the effects of pattern-reversal VEPs under general anesthesia in four patients but suggested that the VEPs produced by phase-alternating square-wave gratings could not be correlated with acuity. Not only should more specific visual stimuli, such as pattern-reversal presentation, be tested, but also VEPs should be analyzed differently.

Stimuli can be presented in a pseudorandom sequence to improve averaging of transient VEPs. Also, steady-state VEPs can be recorded and the signals analyzed in terms of their Fourier components. At flicker frequencies greater than 7/sec, the steady-state VEP can be adequately described by 4 harmonics. A Fourier analyzer splits up the VEP waveform into its constituent harmonic components and describes each component by specifying its amplitude and phase. Regan[16,17] has described some advantages of this technique such as speed and improved signal resolution.

Improvements in either stimulus specificity or data analysis are necessary before VEPs can be reliably interpreted.

Brainstem Auditory Evoked Potentials

BAEPs can be reliably recorded in the OR without added risk or inconvenience. In our experience about 10% of the patients had an abrupt loss of the BAEP waveform components beyond wave I following manipulation of CN VIII, although the nerve was grossly intact. The lesion was undetermined but vascular insufficiency was suggested. All these patients had a unilateral hearing deficit postoperatively. The surgeon could accurately predict the outcome of the procedure in terms of the CN VIII function based on intraoperative BAEP changes.

A more interesting group of patients is represented by the 15% who had a delay of BAEP component latencies intraoperatively. The early detection of these EP changes alerted the surgeon to a potential CN VIII problem. The surgical technique could be altered to correct the BAEP abnormality by repositioning retractors or changing the area of surgical dissection. In the majority of cases the latency delay persisted but did not deteriorate. In two cases the latency delay returned to normal. In all these patients the hearing deficit was minimal and reversible. It is possible that this group of patients had hearing preserved because of early detection of intraoperative BAEP changes and timely surgical intervention.

Only one patient had BAEP changes associated with brainstem dysfunction. This patient died without recovery of consciousness or the BAEP. This compares to the study by Hashimoto.[18]

Intraoperative BAEP monitoring is a more specific test of auditory than brainstem function. It is particularly useful during microvascular decompression of CN V and CN VII. Early recongition of BAEP changes can be used to alert the surgeon to potentially reversible hearing deficits. This is our most reliable intraoperative sensory EP.

Somatosensory Evoked Potentials

Intraoperative SSEPs cannot yet be confidently recorded or interpreted in the OR. Approximately 20% of our recordings were contaminated by artifact. Only 6% of the cases studied demonstrated consistent changes in SSEP amplitude or latency related to manipulation of neural structures. No correlations were established between SSEP changes and postoperative sensory function. More important, no postoperative motor deficits occurred in this group. SSEP detection of a potential motor deficit would be extremely valuable.

SSEPs are sensitive to all the influences that can alter EPs, such as temperature, anesthetic level, blood pressure, acid–base balance, and surgical retraction of associated neural structures. The early components are more resistant to these effects than the later cortical components. This is evident by comparing the stable BAEP waveform to the highly variable VEP. By limiting our analysis of SSEP components to 50 msec, we improved the stability of the resulting EP. But the sensitivity of the test was reduced.

The lack of intraoperative SSEP changes reflects patient selection and surgical technique as well as recording limitations. This initial group of patients did not have significant postoperative deficits. The report by Engler[20] described similar results.

One promising area for future study is central conduction time (CCT)[21,22] as a function of cortical blood flow during cerebral vascular surgery. Another is the determination of motor influences on SSEPs.

Summary

Sensory EPs were recorded intraoperatively in 173 neurosurgical procedures (71 VEPs, 66 BAEPs, and 31 SSEPs) to evaluate the utility of this technique. EPs could be safely recorded in all cases, but the yield of useful results varied with each sensory modality. BAEPs were recorded reliably in 100% of the cases and intraoperative latency changes accurately predicted postoperative hearing deficits in 10%. Potential hearing deficits were detected in another 15%. BAEP changes were associated with brainstem dysfunction in only one case.

VEP changes were difficult to interpret intraoperatively because of contamination by a high degree of variability and both false negative and false positive results. Changes in VEP amplitudes related to surgical manipulation of the optic chiasm were only suggested.

SSEP changes were recorded reliably in only 75% of the cases and no correlations between SSEP changes and postoperative sensory function were established. Again, intraoperative amplitude attenuation of the SSEP waveform with surgical manipulation only suggested a potential sensory deficit.

Intraoperative EP monitoring is a valuable technique that provides a functional analysis of the sensory nervous system during surgical procedures. Specific sensory stimuli and improved data analysis will increase the utility of this CNS monitor.

Acknowledgments

I am very grateful for the considerate contribution of resources and time given to me by my colleagues at the Barrow Neurological Institute. Without their direct help this project would not have been possible.

I wish to single out Dr. Elizabeth Wilkinson for giving me the time to work on this study, Drs. John Green and Robert Crowell for providing the necessary surgical co-operation and support, Drs. Volker Sonntag, William White, & Andrew Shetter and many others who generously let me use their case material, and Martin Nunemacher whose technical expertise was the critical element in the study.

I am also grateful for the equipment support provided by the Nicolet Instrument Company.

References

1. Starr, A. & A. C. Hamilton. 1976. Correlation between confirmed sites of neurological lesions and abnormalities of far-field auditory brainstem responses. Electroenceph. Clin. Neurophysiol. **41:** 595–608.
2. Stockard & J. J., J. E. Stockard & F. W. Sharbrough. 1977. Detection and localization of occult lesions with brainstem auditory responses. Mayo Clin. Proc. **52:** 761–769.
3. Stockard & J. J., J. E. Stockard & F. W. Sharbrough. 1980. Brainstem auditory evoked potentials in neurology. Methodology interpretation and clinical application. *In* Electrophysiologic Approaches to Neurologic Diagnosis. M. J. Aminoff, Ed. pp. 370–443. Churchill Livingstone, New York, N.Y.
4. Uhl, R. R., K. C. Squires, D. L. Bruce & A. Starr. 1980. Effect of halothane anesthesia on the human cortical visual evoked response. Anesthesiology **53:** 273–276.
5. Stockard, J. J., J. F. Hughes & F. W. Sharbrough. 1979. Visually evoked potentials to electronic pattern-reveral latency variations with gender, age and technical Factors. Am. J. EEG Technol. **19:** 171–204.
6. Stockard, J. J., J. E. Stockard & F. W. Sharbrough. 1978. Nonpathologic factors influencing brainstem auditory evoked potentials. Am. J. EEG Technol. **18:** 177–209.
7. Raitta, C., V. Karhunen, A. M. Seppalainen & M. Nankkarinen. 1979. Changes in the electroretinogram and visual evoked potentials during general anesthesia. Arch. Klin. Exp. Ophthal. **211:** 139-144.
8. Sokol, S. 1976. Visually evoked potentials. Theory, techniques and clinical applications, 1976. Survey Ophthalmol. **21**(1): 18–44.
9. Allen, A., K. Nudleman & A. Starr. 1980. Assessment of sensory function in the operating room utilizing cerebral evoked potentials. A study of 56 surgically anesthetized patients. Clin. Neurosurg. (in press).
10. Borda, R. P. 1977. Visual evoked potentials to flash in the clinical evaluation of the optic pathways. *In* Visual Evoked Potentials in Man. New Developments. J. E. Desmedt, Ed. pp. 481–489. Clarendon Press, Oxford, England.
11. Harding, G. F. A. 1977. The use of visual evoked potentials to flash stimuli in the diagnosis of visual defects. *In* Visual Evoked Potentials in Man. New Developments. J. E. Desmedt, Ed. pp. 500–508. Clarendon Press, Oxford, England.
12. Bodis-Wollner, I. & C. D. Hendley. 1977. Relation of evoked potentials to pattern and local luminance detectors in the human visual system. *In* Visual Evoked Potentials in Man. New Developments. J. E. Desmedt, Ed. pp. 197–207. Clarendon Press, Oxford, England.
13. Jeffreys, D. 1977. The physiological significance of pattern visual evoked potentials. *In* Visual Evoked Potentials in Man. New Developments. J. E. Desmedt, Ed. pp. 134–167. Clarendon Press, Oxford, England.
14. Harter, M. R. 1977. Binocular interaction: Evoked potentials to dichoptic stimulation. *In* Visual Evoked Potentials in Man. New Developments. J. E. Desmedt, Ed. pp. 208–233. Clarendon Press, Oxford, England.

15. BAGOLINI, B., A. PENNE, S. FONDA & A. MAZZETTI. 1979. pattern reversal visually evoked potentials in general anesthesia. Arch. Klin. Exp. Ophthalmol. **209:** 231–238.
16. REGAN D. 1977. Fourier analysis of evoked potentials: Some methods based on Fourier analysis. *In* Visual Evoked Potentials in Man. New Developments. J. E. Desmedt, Ed. pp. 110–117. Clarendon Press, Oxford, England.
17. REGAN, D. 1976. Latencies of evoked potentials to flicker and to pattern speedily estimated by simultaneous simulation methods. Electroenceph. Clin. Neurophysiol. **40:** 654–660.
18. HASHIMOTO, I., Y. ISHIYAMA, G. TOTSUKA & H. MIZUTANI. 1980. Monitoring brainstem function during posterior fossa surgery with brainstem evoked potentials. *In* Evoked Potentials. C. Barker, Ed. University Press.
19. GRUNDY, B. L., C. L. NASH & R. H. BROWN. 1981. Arterial pressure manipulation alters spinal cord function during correction of scoliosis. Anesthesiology **54:**249–253.
20. ENGLER, G. L., N. I. SPIELHOLZ, W. W. BERNHARD, F. DANSIGER, H. MERKIN & T. WOLFE. 1978. Somatosensory evoked potentials during Harrington instrumentation in scoliosis. J. Bone Joint Surg. **60:**-A (4): 528–532.
21. SYMON, L., J. HARGADINE, M. ZAWIRSKI & N. BRANSTON. 1979. Central conduction time as an index of ischemia in subarachnoid hemmorhage. J. Neurol. Sci. **44:** 95–103.
22. HUME, A. L. & B. R. CANT. 1978. Conduction time in central somatosensory pathways in man. Electroenceph. Clin. Neurophysiol. **45:** 361-375.

DISCUSSION OF THE PAPER

I. BODIS-WOLLNER: Twenty years ago, I believe, Dr. Galambos sectioned the optic tract of the cat and was able to demonstrate that it is possible to get a normal flash evoked response even when you have less than 5% of the fibers intact.

G. CELESIA: For your VEP studies you selected the worst stimulus possible. We have used the identical system of LED goggles. We recorded 15 normals, all either pretty nurses, students, or medical students, and gave up trying because the variation, even in normals, is so enormous. I do not think you could possibly get a reliable response. So, first of all, you should use a different stimulation technique. Second, the visual evoked potential and the brainstem auditory evoked potential are not comparable. In the VEP one is recording near field evoked potentials in contrast to the far field of the brain stem evoked potentials. We have shown in human, recording from the auditory cortex A1, that nitrous oxide statistically delays the response and that Halothane essentially wipes it out. There are many complicating factors, and I am afraid you have got to find a better stimulus for the visual evoked potential.

P. RAUDZENS: Right—I was making the same point in my talk. There has to be a better stimulus for the visual system.

M. WOLBARSHT: One thing that might be very helpful would be to roll your averaging along so that you drop out the first one, and add a new one on so that you get a continuous recording. That is very simple to manage and, of course, involves a much shorter integration time. You might be able to accomplish that by using pattern reversal, which would be a real test to the system. I do not think it would be too difficult to keep the patient's eyes open and to use anesthesia while you have a pattern in focus on the retina.

RAUDZENS: We have developments along those lines right now. With respect to anesthesia, there is a paper by Uhl and his colleagues in one of the anesthesia journals in which they described a kind of dose-dependent relationship of the flash evoked response. Quite clearly, though, they also show a high degree of variability. I think we had better get away from this stimulus altogether, and look at either a more specific stimulus, or more sensitive data analysis.

H. SPEKREIJSE: I disagree completely with you. We have been monitoring EPs for a completely different reason. We used flash evoked and auditory brainstem potentials during open heart surgery. We observed that the flash EP was superior to the brainstem potential for monitoring the state of the subject, the reason being that since the subject is not moving, we can consider that a variation in amplitude of the flash EP by 5% or 10% was significant. For instance, if something went wrong with the patient's blood pressure, the brainstem potential dropped out instantly. But, just as you observed, the flash EP went down gradually. So it was a more sensitive measure than the potential which simply is or isn't there.

RAUDZENS: I would like to point out that the confounding part of this whole study is that we are looking at multi-varied analysis, not merely at blood pressure or perfusion effects. Quite clearly, however, the most profound effect is that of the surgeon actually manipulating the neural structures themselves.

SPEKREIJSE: You can exclude the variability introduced by the LEDs by using a flash EP.

K. HECOX: A variability may be quite useful in fact, if you look at a systematic relationship between the duration of drilling and the amount of depression of the response and the degree of latency shift, and so on. We found this in otolarnygology patients in whom we monitored the procedures which were being attempted to save the auditory nerve. We found that there was a tight relationship between the amount of preoperative hearing loss and the intraoperative sensitivity to all of these manipulations. I wonder if Dr. Raudzens had preoperative audiograms performed to find out whether or not those people were the ones who showed more variability.

RAUDZENS: In fact, that was included in my summary. Thirteen percent of our patients had hearing deficits related to a lesion or whatever problem they presented with. Clearly their brainstem responses are quite abnormal, while the patients with normal thresholds and normal auditory acuity had fairly normal brainstem responses.

SYSTEM DISEASES AND VISUAL EVOKED POTENTIAL DIAGNOSIS IN NEUROLOGY: CHANGES DUE TO SYNAPTIC MALFUNCTION*

Ivan Bodis-Wollner†‡ and Marco Onofrj†

Departments of †Neurology and ‡Ophthalmology
Mount Sinai School of Medicine
New York, New York 10029

INTRODUCTION

In this chapter we will review recent visual evoked potential (VEP) studies in degenerative and system diseases in neurology, especially in Parkinson's disease, and compare VEP abnormalities that occur in these conditions with those that occur in demyelinating diseases and in a nondemyelinating optic neuropathy: glaucoma. Some of these studies are described in detail elsewhere. Here a few selected issues will be discussed. One of them is whether or not temporal dispersion of the VEP is a specific concept with respect to the presumed pathophysiology of these diseases. Another is the mechanism of recovery of VEP latency under treatment in Parkinson's disease and pernicious anemia. In conjunction with these human studies we will review VEP changes that occur in the rat under dopaminergic manipulations. Based on these considerations we will propose synaptic malfunction as an additional mechanism of VEP delays, besides the well-known explanation based on conduction defects.

We will primarily consider VEP data obtained with patterned stimuli. Although flash VEPs can provide gross electrophysiological correlates of abnormalities in neurologic afflictions such as multiple sclerosis (MS), hydrocephalus, storage and other metabolic diseases, glaucoma, and degenerative or malformative diseases,[1-12] the evaluation of a flash VEP is difficult and depends on the expertise of each laboratory. While several of the flash VEP components can be discerned in the normal,[2,7,13-15] in most clinical instances only a criterion of presence, absence, or general abnormality can be reliably used.[7,8,10] Thus it is not surprising that following the introduction of patterns as stimuli for visual evoked potential studies of human visual physiology[16-19] and the first demonstration of the clinical validity of pattern VEP measurements in MS,[20] these have played an increasingly more important role in clinical neurophysiology. While some of the original controversies concerning pattern and flash[1,2,20] emerge on occasion, so far pattern VEPs have better withstood the test of time. In a discussion involving the diagnostic use of VEPs in MS, Halliday[13] aptly remarked that:

> the flash response is, on the average, much less delayed than the pattern response. Fewer of the individual flash responses from the affected eyes exceeded the limits of the normal latency range. . . . The smaller latency shift in the flash response was compounded with its much greater variability in latency in the normal population to render the flash response particularly insensitive as a test of the normal-abnormal dichotomy."

Although this view is commonly accepted,[21-23] there is a need for standardization as recognized by the American Electroencephalography Society.[24] Nevertheless, even

*This work was supported in part by Grant EY01708 from the National Eye Institute; Grant NS11631 from the Clinical Center for Research in Parkinson's and Allied Diseases; Core Center Grant EY01867 from the National Eye Institute; and National Institutes of Health Grant RR-00071, Division of Research Resources, General Clinical Research Center Branch.

0077-8923/82/0388-0327 $1.75/0 © 1982, NYAS

when somewhat different methods of presentation are followed,[13,21,22,25–30] pattern VEP measurements have acceptable interindividual statistics.

Besides the reliability of clinical data, another compelling reason to use pattern VEPs is that the processing of patterns can be related to knowledge gained from studies of single cell responses at different levels of the afferent visual pathway of mammalians.[31–33] Furthermore, even though prudence is required in extrapolating from psychophysical to VEP studies,[34,35] pattern VEPs obtained by manipulating either the stimulus contrast or frequency show correspondence between psychophysical and VEP thresholds in the same observer.[31,36,37]

For diagnostic applications, the latency of the major positive component, called P1 or P100,[13,38] is considered to be statistically the most reliable of the peaks and troughs of the triphasic complex of the initial 200 msec of the 1-Hz pattern-reversal EP. The time of occurrence of P1 can be unambiguously measured as "latency" up to 4 reversals per second.[39] Following the original studies by Halliday,[20] delayed VEPs have been widely reported in multiple sclerosis and optic neuritis, and this work undoubtedly has provided a major impetus for the widespread use of VEP measurements in clinical neurology. As a result of these initial studies, it has been accepted that VEP delays in multiple sclerosis relate to its pathognomic feature, i.e., demyelination. Over the last decade, however, delays have been noted in other diseases such as open angle glaucoma,[12,23,40,41] sector ischemic neuropathy,[23] compression of the anterior visual pathway,[42] adrenoleucodystrophy,[43] chronic uremia,[44] hydrocephalus,[45] B_{12} deficiency,[46–48] Friedreich's ataxia and other forms of progressive ataxia,[49–52] certain forms of sporadic progressive spastic paraparesis,[53] and Parkinson's disease.[54,55] What is the mechanism of VEP delays in these diseases?

As mentioned above, based on MS as an experimental model, VEP delays in general have been attributed to slowed conduction in the optic nerve or tract. Indeed the first VEP studies in MS were initiated following the studies of McDonald[56] on the effects of demyelination on central conduction velocity. By injecting diphtheria toxin, he experimentally induced demyelination of the posterior columns of the spinal cord and, in the presence of large lesions, found complete conduction block. When the demyelinating lesions were small, slowed conduction could be measured in single fibers. Individual fibers were also unable to transmit fast trains of impulses.[56] Although McDonald concluded that "slowing of conduction contributes directly to the delays seen in multiple sclerosis and optic neuritis," he did pose the question of to what extent slowed conduction alone accounts for increased EP latencies. He pointed out that internodal conduction is likely to fail completely in large demyelinated plaques,[57] and introduced the concept that continuous conduction along the optic nerve instead of internodal conduction could explain the marked delays (up to 100 msec) found in MS.[58,112] McDonald himself did consider other possible sites and modes for VEP delays, however.

Following an anatomical sequence, one may consider the retina as the first possible site of VEP delays. Indeed, McDonald considered retinal pathology or, more precisely, nerve "fiber bundle defects" in diagnosing VEP abnormalities.[59] When we examined the relationship between evoked response latencies to gratings of 2.3 cycles/degree in a 4° foveal patch and the detection of the same pattern(s) by multiple sclerosis patients,[30] we found little correspondence and concluded that retinal pathology is not likely to be the only factor in explaining VEP delays in most MS patients.

McDonald also discussed the possibility that VEP delays could be due to selective loss of the faster retino-geniculate fibers, leaving only the slower fibers to mediate the response. However, he pointed out that "the predominant affect in optic neuritis is of central vision and the fibers subserving this are the slower, not the faster ones."[57,60] There is strong suggestive evidence that in the mammalian visual system a difference between individual neurons that respond preferentially to coarse versus fine patterns correlates with a division of fast and slow conducting fibers. When we evaluated the psychophysical data in 98 MS patients concerning their ability to detect gratings ranging from 1 to 30 cycles/degree (the "visuogram"), it was apparent that the contrast sensitivity losses in this group of patients were not consistent with the concept of a selective loss of fibers responding to either low or high spatial frequencies.[30] However, these data, revealing a dissociation of psychophysical and VEP measures of grating pattern responses in MS patients, did not exclude at least partial pathology at the cortex where the discrepant results are easier to visualize. Indeed, McDonald also considered possible mechanisms of cortical delays, and felt that the EP could be delayed as a result of conduction block in many fibers, or of dispersion of the afferent volley due to unequal slowing in different fibers.[57] This explanation is consistent with abnormal conduction as the major explanation of VEP delays. However, it was also pointed out by Halliday and McDonald[61] that VEP delays could occur as a result of changes in dendritic potentials, possibly due to humoral factors present in MS patients.[58,62] This explanation is clearly a departure from the "classical" notion of conduction in demyelinated axons. In our recent studies, when we compared checkerboard and grating pattern responses in 56 MS patients,[63] we also studied VEP latency as a function of pattern orientation. These studies surprisingly revealed orientation-dependent VEP delays in MS. Surprising though this may be, there is evidence from another laboratory concerning this issue. Using psychophysical methods, Regan[64] also found orientation-dependent losses in MS patients. Since precortical neurons of the visual pathways of primates respond equally well to all orientations of bars or gratings, our findings are not easy to explain by subcortical pathology in MS. As pointed out recently,[65] asymmetrical "ovoid" scotomas may possibly cause such meridional changes; however, most of our patients had no visual field defects. Thus we entertain the possibility that cortical pathology may contribute to VEP delays in MS. Demyelination at the cortex may occur, but it is difficult to see how such a process could be so selective as to cause disruption of some but not all orientation-selective neurons or columns at the cortex.

Thus, all in all, several aspects of VEP findings in MS itself lead one to consider additional and/or alternate explanations of VEP delays besides conduction defects. Indeed, the purpose of this chapter in reviewing VEP findings in diseases other than MS and describing the effects of central dopamine manipulation in humans and in rats is to call attention to synaptic malfunction as another general concept of VEP delays.

Method

Stimulus Definitions

Spatial frequency of a grating pattern is expressed as the number of adjacent pairs of dark and bright bands (a cycle) subtended in one degree of visual angle at the observer's eye.

Bar width of a grating refers to one-half of a full cycle (which consists of a dark and a light bar).

Check spatial frequency. The diagonal of an individual checkerboard square (or check) corresponds to one complete cycle of the fundamental sine wave component of that checkerboard. Therefore the spatial frequency of a checkerboard is the number of checks measured along the diagonal subtended in one degree of visual angle (see Fourier synthesis, below).

Check width refers to the dimension of one side of an individual check square.

Contrast of a pattern is defined as the difference of maximum luminance and minimum luminance over their sum.

Mean luminance is one-half the sum of the maximum and minimum luminance.

Fourier Synthesis of Checkerboards

In terms of Fourier analysis, a pattern with a one-dimensional sinusoidal luminance profile is the simplest visual pattern. Analytically, any other two-dimensional visual pattern can be decomposed into a series of sine waves. A checkerboard could be synthesized from sinusoidal gratings by properly arranging a range of sinusoidal gratings of specific spatial frequencies and orientations. The contrast (amplitude) at various spatial frequencies and orientations needed for this synthesis would depend on the size and orientation of the check stimulus one wished to produce. FIGURE 1(a) illustrates a vertical sine wave grating and to the right its amplitude spectrum. Notice that all the amplitude is in one spatial frequency and one orientation. This is the simplest spectrum possible for a two-dimensional pattern. The diagram below in FIGURE 1B represents the amplitude spectrum of a checkerboard pattern with its edges aligned normally (horizontal and vertical). The fundamental spatial frequency of this checkerboard pattern is represented by the two large dots, and higher spatial frequency components by smaller and smaller dots representing a decreasing contribution of higher spatial frequencies to the pattern.

Consideration of the power spectrum analysis of checkerboard patterns is relevant for two reasons. One is that the appropriate measure to use in order to equate the fundamental spatial frequency of the check pattern and with that of a grating is the grating cycle length, which equals the check diagonal. The other is that when the edges of a checkerboard are aligned vertically and horizontally, *all* the power is concentrated in the oblique orientation. One can see from the spectrum that the *major power* of a checkerboard pattern is concentrated at 135° and 225°, with power dropping to zero as one moves towards 0° or 90°. A checkerboard with its edges aligned along the major axes has neither vertical nor horizontal components.[30]

Displays. Vertical gratings with a sinusoidal luminance profile were generated on the cathode ray tube (CRT) of an oscilloscope (Tektronix 561)[66] or of a large screen special display unit (Joyce Electronics). The screen subtended 4° or 9° at the observer's eye. The smaller CRT screen was surrounded by another screen which was back-illuminated at about the same hue and luminance as the CRT. The space-time average luminance of the screen was constant either at 0.7 log footlambert (ft-L) or at 1.7 log ft-L. The stimulus for VEP measurement was a grating pattern of 2.3 cycles/degree (c/deg) at 50% contrast. This was reversed twice per second, i.e, dark bands turned abruptly into bright bands and bright into dark. The frequency of modulation for steady-state responses was 8 Hz. The signal effecting the pattern reversal triggered a signal averager (Nicolet, model 1074). Luminance of the CRT had been calibrated for AC and DC *Z*-axis input. Contrast was calibrated at different spatial frequencies.[67]

Checkerboard stimuli were presented by means of either a TV generator or a rear projection system (Digitimer). With this apparatus, a checkerboard pattern is focused onto a small mirror, which is pivoted electromagnetically. The mirror projects the image of the checkerboard through a lens onto a translucent screen. The movement of the mirror takes 5 msec (abrupt square wave movement), and is adjusted so that the squares are displayed exactly 1 square width. The circular check display subtended 4° at the eye. It was surrounded with a diffusing mask so that the total field size was 6° square. The mean luminance was 1.7 log ft-L and the contrast was approximately 0.9.

Grating Orientation. In initial experiments comparing only the effects of vertical and horizontal gratings, the oscilloscope was simply rotated so that it rested on its side. For subsequent measurements, using horizontal and oblique as well as vertical gratings, observers viewed the grating display through a dove prism which was held in front of the eye by a rigidly fixed stand. The orientation of the grating was changed by rotation of the prism. In recent studies the electromagnetic coils of the display unit were rotated.

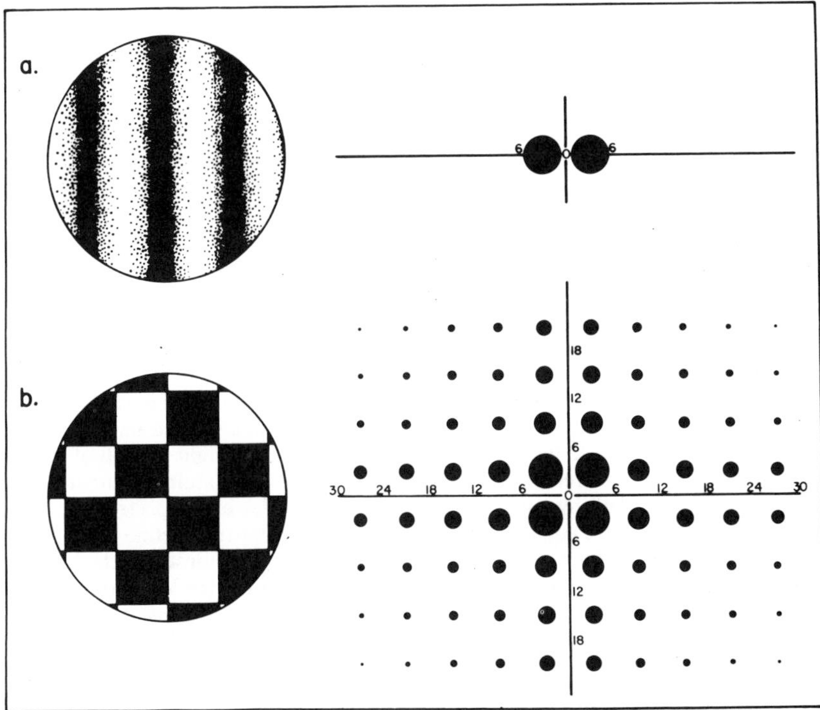

FIGURE 1. An illustration of (a) the sinusoidal grating pattern and (b) the checkerboard pattern with the Fourier spectrum of each. The numbers indicate spatial frequency and the radial dimensions and orientation of the spatial frequency components of each pattern. Note that the sine grating contains energy *only* at the fundamental frequency (3 cycles/degree), while the checkerboard contains energy at a number of frequencies. The relative contribution of each harmonic to the total amplitude of the pattern is indicated by the size of the dot at each spatial frequency. The orientation of the sinusoidal components of the check pattern are all oblique, with no energy at either the vertical or horizontal meridians. (From Camisa *et al.*[63] By permission of the *Annals of Neurology*.)

Evoked Potential Recordings were obtained by placing gold disc electrodes 2.5 cm above the patient's inion, referenced to the midfrontal position. The ground electrode was placed on the forehead. Occasionally lateral electrodes were also used, and a second reference site (posterior temporal) was added. Potentials were differentially amplified with preamplifiers (Grass, model 5PS) and filtered with corner frequencies of 0.15 and 50 Hz.

At a stimulation rate of 1 Hz, the evoked potential waveform is labeled for successive positive and negative deflection according to the nomenclature of Ciganek.[68] The critical measurement in our studies was the latency of the major positive wave. For steady-state measurements the amplitude of the second harmonic component was measured using a spectrum analyzer (Hewlett-Packard 3580A).

Ophthalmologic examination of each patient included: visual acuity measurement with the Snellen chart, measurement of pupil reaction to a light flash, clinical evaluation of pursuit and saccadic eye movement, and optokinetic nystagmus, tested with a regular tape measure. Patients with visual acuity of worse than 20/80 were excluded from this study. In addition, in patients who had EP latency differences to gratings of specific orientations, contrast sensitivity measurements to high spatial frequencies were also obtained. This was done because it is known that even if corrected later on in life, astigmatism early in life can cause reduced high-spatial-frequency contrast sensitivity.[69] Thus, using psychophysical methods[66] to establish high-spatial-frequency contrast sensitivity, we wished to ascertain if any of these patients had a hidden form of meridional amblyopia. However, none of the patients reported here had such losses. In addition, we explored the effect of artificial or pathological astigmatism on EP latency to 2.3-c/deg gratings in patients not affected by MS. We found no effect of moderate astigmatism (less than 1 diopter) on the latency to our standard 2.3 c/deg sinusoidal grating stimulus.

Statistical Analysis of the Data

As described elsewhere in detail,[30] normality or abnormality of the VEP latency is considered with reference to the data of a control population studied under identical conditions. The covariance matrix and the mean of all of the right and all of the left eye latencies were compared between controls and patients. A multivariate analysis[70] was used to test the null hypothesis that the mean response of the pretreatment and posttreatment groups are equal. The normal population statistics define an ellipse on the scattergram of right versus left eye latencies. Within this ellipse fall the latencies of 97% of the control population. Based on the measurement of the VEP latency in the right and left eye, a "C^2 score" was derived for each patient, representing the distance of an observed data point from the center of the normal probability ellipse. Five parameters are required to completely satisfy this distribution function. The detailed methodology of converting absolute latencies into the statistical parameter "C^2" is described below.[71] This method is especially useful in evaluating the normality or abnormality of VEP latency in statistical terms rather than as actual time, since, as the RESULTS section below shows, in some patients checks and in others grating stimuli were used, and the absolute VEP latency is stimulus dependent.

The C^2 Score

The bivariate normal distribution is used to approximate the joint probability distribution of the right and left eye latencies found in a normal population. Five parameters are required to completely specify this distribution function. The parameters are the expected (mean) value, μ_R, and standard deviation, σ_R, of a normal individual's right eye latency; the expected (mean) value, μ_L, and standard deviation, σ_L, of a normal individual's left eye latency, and the correlation coefficient, ρ, between the right and left eye latencies. These parameters can be estimated from latencies measured from a random sample of the normal population.

For the bivariate normal distribution, ellipsoidal contours can be constructed which contain specified proportions, α, of the distribution. The equation of the ellipse which contains $100\alpha\%$ of the distribution is

$$\left[\frac{R-\mu_R}{\sigma_R}\right]^2 - 2\rho\left[\frac{R-\mu_R}{\sigma_R}\right]\left[\frac{L-\mu_L}{\sigma_L}\right] + \left[\frac{L-\mu_L}{\sigma_L}\right]^2 = -2\,(1-\rho^2)\log\,(1-\alpha) = C^2.$$

Using estimates of the five parameters obtained from a study of the normal population and the measured latencies of an individual, we can use the equation to calculate a C^2 score. The resulting score can be used to assess the likelihood that the individual is a member of the normal population. The C^2 score is directly related to the probability of belonging to the control population. The smaller the probability, the further away the observed value is from the center of the ellipse, the higher the C^2 score (TABLE 1). The ellipse, containing 97.5% of the distribution (the "normal" ellipse) is illustrated in the left lower corner of FIGURE 2.

RESULTS

Multiple Sclerosis

The VEP diagnostic yield is around 80% of all examined MS patients [for a recent review of the results obtained by different researchers, see Halliday].[13] The impor-

TABLE 1

MEAN NORMAL VALUES OF VEP LATENCY (in msec) FOR THE RIGHT (μ_R) AND LEFT (μ_L) EYE WITH THEIR STANDARD DEVIATIONS (σ_R, σ_L)*

	Gratings	Checks
μ_R	115.86	106.97
μ_L	115.96	107.47
σ_R	8.20	7.89
σ_L	7.71	8.43
ρ	0.81	0.93

*ρ represents the correlation coefficient for normals for their right and left eye latency. The normal population consisted of 50 volunteers ranging in age from 18 to 82 years.

tance of VEP measurements in the diagnostic evaluation of suspected MS patients, and in complementing the clinical assessment of MS using the criteria of McAlpine,[72] is well documented. The precise diagnostic yield depends on several stimulus-specific factors. Here we wish to point out some observations which lead one to question whether or not all of the delay in MS can be explained by delayed central conduction.

Normal VEP Latency in MS. It is not uncommon to find normal latency recorded over a midline electrode in patients with unilateral retrochiasmal plaques.[27,73] One of our patients suffering from acute MS with a demyelinating plaque in the occipital lobe identified by CT scan had normal visual fields, determined by campimetry. His pattern-reversal VEP recorded with a horizontal array of electrodes revealed an "increased lateralization" of the response (indicative of a homonymous field defect). The normal VEP compatible with a unilateral defect may be explained by assuming that intact conduction in half of the fibers (one optic radiation) is sufficient to produce normal latency. However, this interpretation makes one question whether or not a

VEP delay could be caused by a dispersion of the afferent volley since in this patient there was impaired conduction presumably in half of the fibers.

Stimulus-Specific VEP Abnormalities. The VEP delay in MS is crucially dependent on the spatial and/or temporal modalities of stimulation, which cannot be explained on the basis of reduced conduction velocity alone. Luminance is another important parameter to be considered.[74] We compared the VEP latency in patients and controls at two luminance levels each (1.7 and 0.7 log ft–L), and found that grating and check stimuli at the lower luminance level elicit delayed VEPs in a significantly larger number of MS patients.[63] Thus the VEP diagnostic yield is dependent on overall luminance, which is likely to be related to either retinal or lateral geniculate nucleus network properties.[75] As discussed earlier, the latency of the VEP

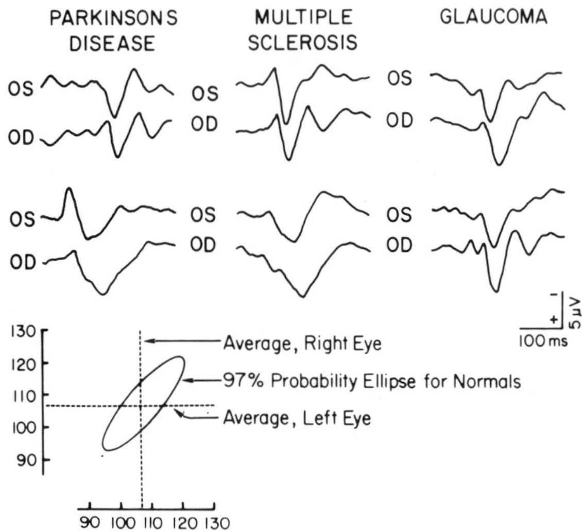

FIGURE 2. Delayed visual evoked potential traces (top part of figure) in the left (OS) and right eyes (OD) of six patients, two suffering from Parkinson's disease, two from multiple sclerosis, and two from glaucoma. Apparently a widened response shape ("temporal dispersion") can occur in PD and MS alike. The lower part of the figure shows the normative scattergram upon which the classification of an "abnormal" or "normal" VEP is based on our laboratory. All the VEPs illustrated in this figure are delayed compared to the normal ellipse.

in MS is dependent on the orientation of the stimulus,[63] a finding supported by psychophysical studies.[64] Besides luminance and orientation, one must also consider pattern element size, as first suggested by Hennerici[76] as another stimulus specific aspect of VEP latency in MS. Observations by Sherman[77] and by us confirm that spatial frequency of the stimulus may be critical for the diagnostic yield and differential diagnosis in MS and other diseases. More precisely, in evaluating MS versus maculopathy, glaucoma, and nondemyelinating retrobulbar pathologies, the consideration of precise stimulus parameters is crucial. Based on these facts, it is difficult to maintain that VEP delays in MS are only caused by unselective demyelination of the optic nerve. Rather, one must consider that lateral neuronal connections, which are essential for establishing stimulus specificity of individual neurons from retina to cortex may suffer, perhaps selectively, in the disease process.

Recovery of the VEP. It is generally believed that once the VEP is delayed it remains so, and therefore the amount of delay depends on stable anatomical features, such as the length of demyelinated fibers. However, recovery of the VEP latency and/or amplitude has been documented in several patients[27,61,78,79] and our own data in patients who recovered from optic neuritis have also convinced us that this may occur. It would be possible to argue that early amplitude recovery of the VEP reflects the restoration of conduction in fibers initially blocked by edema, and recovery of latency represents remyelination.[58] However, while the occurrence of remyelination in central nerve fibers has been demonstrated in MS,[110] and experimental demyelination in the cat is followed by restored conduction due to remyelination,[111] in MS in the human remyelination does not seem to be a significant mechanism in restoring delayed conduction.[112]

Temporal Dispersion of the Abnormal VEP in MS and Other Diseases

In the evaluation of VEP latencies, temporal dispersion of the response was introduced as a supplementary measurement by several authors.[26,50,61] "Temporal dispersion" is commonly evaluated by measuring the difference in latency of the first negative (N_1) and second negative (N_2) component of the VEP.[26,50] The term is borrowed from studies of peripheral nerve neurophysiology,[80–82] and qualifies the abnormal duration of an electrical potential. It is a useful concept when applied to a response that is the result of the activity of a known number of fibers with identical individual response latencies although it is believed that ambiguities arise when these are unknown. In the case of pattern reversal VEPs, temporal dispersion was described concomitantly with delayed VEPs in some but not all patients suffering from Friedreich's and other forms of ataxia,[50,51] compression of the anterior optical pathways,[42] and ischemic optic neuropathy.[23,61] According to several authors, while the VEP is delayed in MS patients, it does not show significant temporal dispersion[26,50,61,83] and this lack of significant temporal dispersion of the VEP in MS patients is used to support the argument that a simple reduction of conduction velocity underlies VEP delays. Furthermore, it is thought that the presence or absence of temporal dispersion can differentiate two mechanisms of delayed VEPs. One nonspecific mechanism is relevant for compressive, ischemic, and other pathological processes and implies temporal dispersion of the response. The second, without temporal dispersion, is common in MS.[27,50,61,83] The optic nerve is composed of fibers with different conduction velocities. The implication of a specific VEP delay, therefore, is that in MS or any other disease, a delayed but not dispersed VEP would be caused by generally decreased conduction velocity of all types of fibers.

We compared the temporal dispersion of VEPs in a group of normal subjects and patients suffering from definite MS[72] and other afflictions including Parkinson's disease and glaucoma, in which delayed VEPs are thought to be caused by different mechanisms than in MS. FIGURE 2 illustrates VEPs with and without temporal dispersion in different patients in these categories. The purpose of this illustration is to show that temporal dispersion is not a specific index with regard to pathophysiology. Statistical comparison of patients with Parkinson's disease and MS using two-tailed Student's t-tests showed a significant difference in temporal dispersion *vis-à-vis* control subjects and glaucoma patients. Temporal dispersion was ±2 SD (standard deviation) outside the normal mean in 44% of MS, and 42% of Parkinson's disease patients, and was normal in patients suffering from glaucoma. There was no statistically significant difference between Parkinson's disease (PD) and MS (TABLE 2).

In other disease categories, we tested only a few patients and found that temporal dispersion was outside mean values +2 SD in acute hydrocephalus, ischemic optic neuropathy, and radiation treatment (FIGURE 3), while it was normal in both eyes of one patient (FIGURE 4), and in one eye of another patient, both suffering from pernicious anemia. We conclude that in a pure optic nerve disease—glaucoma—the measurement of temporal dispersion reveals no statistically significant abnormality, whereas both MS and PD patients may have abnormal temporal dispersion. These data do not lend support to the notion that temporal dispersion is a simple criterion to rule in or rule out a conduction defect of the optic nerve caused by presumed demyelination.

TABLE 2

MEAN LATENCY OF THE MAJOR POSITIVE COMPONENT AND THE TEMPORAL DISPERSION (MEASURED AS THE TIME BETWEEN THE N_2 AND P COMPONENTS) OF VEPs IN MULTIPLE SCLEROSIS (MS), GLAUCOMA, PARKINSON'S DISEASE (PD), AND CONTROL PATIENTS*

	n	P_1 (in msec)	Standard Deviation
Normal	96	106	6
MS	47	145.1	12.8
PD	24	149	18.9
Glaucoma	10	140.2	12.8
	n	Temporal Dispersion $N_1 - N_2$ (in msec)	Standard Deviation
Normal	44	84.9	22.29
MS	47	111.8	34.77
PD	24	109.6	29.53
Glaucoma	10	82.6	17.44

*For this table it can be deduced that among patients with abnormal VEPs the mean latency of (P_1) is almost the same in MS, PD, and glaucoma, and that "temporal dispersion" is abnormal in patients with MS and PD but indistinguishable from controls in patients with glaucoma. The stimulus was a 2.3-cycles/degree sinusoidal grating pattern at 50 contrast. The field subtended 9° at the eye and had an average luminance of 1.7 log footlamberts.

VEPs in Parkinson's Disease and in Other Conditions Associated With Dopamine Deficiency

In Friedreich's ataxia and other forms of degenerative ataxias, ischemic optic neuropathy, hydrocephalus, compressions of the optic nerve, and B_{12} deficiency, VEP delays may be explained by slowed conduction of the optic nerve. However, over 50% of Parkinsonian patients have delayed VEPs,[53] and their VEP delay cannot be explained by reduced conduction velocity since neither obvious involvement of visual pathways nor demyelination is a characteristic feature of this disease. VEP latency correlates with the clinical severity of the disease: the more severe the disease, the higher the C^2 score. This score represents the statistical transformation of the VEP latency of an individual (see METHODS). Thus, VEP latency changes in PD are not an epiphenomenon, but rather reflect an essential aspect of the progression and severity of this disease.[84] An important feature of the C^2 score is that it expresses interocular as well as absolute latency abnormality. The average interocular latency difference was 13 ± 17 msec in Parkinsonian patients, compared to the control population with an interocular difference of 3 ± 3.5 msec. In almost half of the Parkinsonian patients the

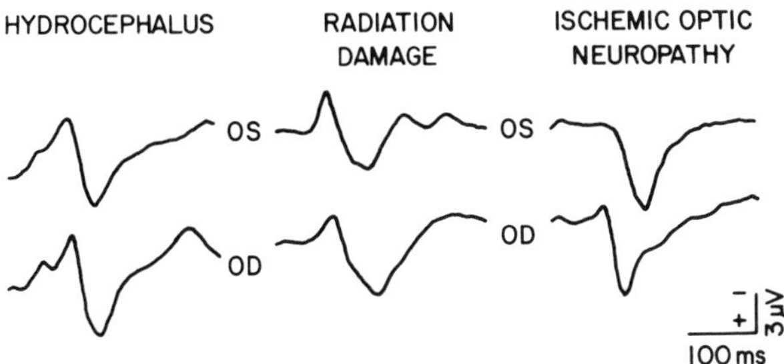

FIGURE 3. Delayed and "temporally dispersed" VEPs in three patients in three different diagnostic categories.

interocular difference is above 14 msec. In over two-thirds of the patients with interocular differences, the greater delay is in the VEP recorded from the eye ipsilateral to the affected side of the body.[85] While the anatomy of the VEP abnormality is not clear, its pharmacology is most noteworthy. It has been demonstrated in two laboratories that the VEP delay is reversible after dopamine precursor therapy[54,55,84] in Parkinsonian patients whose pretreatment EP is abnormal.

The influence of monoamines on VEPs was first pointed out by Schafer.[86] There is indirect evidence that dopamine blockade could increase the latency of the VEP. Saletu[87] reported that antipsychotic drugs increased the latency of flash VEPs in schizophrenic patients. Later, a "slowing of retinogeniculate transmission" was attributed to haloperidol.[88] Delayed evoked responses to flash VEPs were also recorded after administration of chlorpromazine in the monkey.[89] Fluphenazine and sulpiride markedly increase the latency of flash VEPs in schizophrenic patients.[90] There is further evidence that dopamine plays a role in modifying not only the amplitude, but also the latency[86,91] of the VEP. Our finding of the reversibility of VEP delays through dopamine treatment[92] led us to consider how much of the VEP change is due to dopamine itself or possibly (through its conversion) to norepinephrine. Several studies have reported that besides dopaminergic deficiency, norepinephrine is also deficient in PD. In order to assess whether dopamine or norepinephrine causes VEP changes, we studied 14 schizophrenic patients whose pretreatment VEP latency was normal. Following at least 2 weeks of therapy with known dopaminergic blockers, 7 of the 14

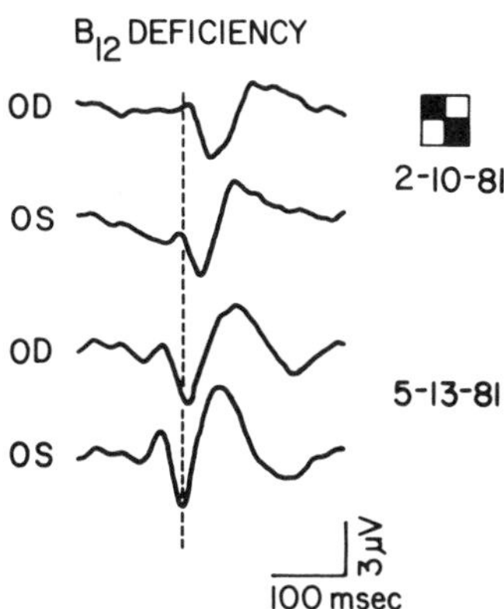

FIGURE 4. Delayed VEPs without temporal dispersion in both eyes of a patient suffering from pernicious anemia prior to treatment (date: 2/10/81) and a return toward normal latency during intensive therapy (5/13/81). The VEP trace in the right eye (OD) is somewhat broadened (but within normal limits of the "temporal dispersion" measure), while the latency of the major positive wave shows considerable improvement in both eyes.

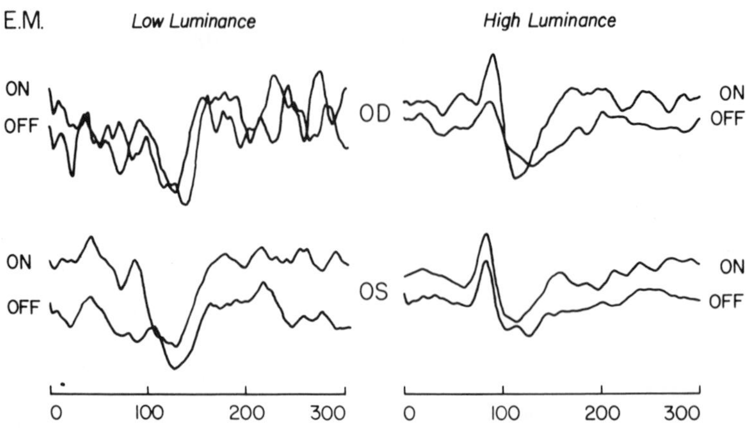

FIGURE 5. Visual evoked potential traces obtained in a patient suffering from PD. Traces labeled ON and OFF represent VEPs obtained 30 minutes apart while his clinical condition changed from "on" to "off," representing, respectively, good and poor drug control. VEPs were recorded at 2 luminance levels in each state and for each eye separately. Spatial frequency 2.3 c/deg, 1 Hz modulation rate, 9° screen. Note, especially in the high luminance condition, improved latency in the "on" condition. Besides latency there are other waveshape changes, more difficult to quantify.

developed significantly delayed VEPs.[92] Furthermore, our preliminary data in three patients suggest that VEP latency in Parkinsonian patients treated with dopamine and dopa-decarboxylase inhibitors is related to the "on-off" feature of their clinical symptoms[93,94] in that the delay in patients with severe Parkinson's disease who have been treated for many years with dopamine is reduced if the patient is recorded in the "on" rather than in the "off" phase of the disease (FIGURE 5). This finding can be related to the concept that the "off" phases in Parkinsonian patients is due to the fact that dopaminergic receptors become temporarily unresponsive, or else to the fact that these patients fail to maintain a uniform level of levodopa.[94] Thus these human data suggest a major, albeit not exclusive role for dopamine in maintaining normal VEP latency, and a place for VEP measurements in clinical neuropharmacology and therapeutics.

An Animal Model for Studying Dopaminergic Mechanisms of the Pattern VEP

Recent animal studies provide further evidence that depletion or blockade of dopamine can cause VEP delays. Dyer[95] was the first to describe delayed flash VEPs in hooded rats in which a dopamine deficiency had been induced by neuropharmacological means. Haloperidol (a dopamine receptor blocker) and alphamethylparatyrosine (an inhibitor of tyrosine hydroxylase) caused consistent delays of the P_{40}–N_{60} components of the rat VEP, which are stable even under sedation.[96–100] The VEP remained normal following adrenergic blockade. Furthermore, the cortically recorded EP to direct optic nerve stimulation was not delayed in the dopamine-depleted rats. This latter result suggests that the site of dopamine depletion causing VEP delays is at a level preceding the optic nerve, i.e., the retina. Indeed, flash evoked potentials recorded directly from the optic nerve were delayed in the same rats.

Our rat studies confirmed the major findings of Dyer.[97] Furthermore, we were able to demonstrate the effects of haloperidol and alphamethylparatyrosine on contrast-reversal pattern VEPs to a range of temporal frequencies, from transient to so-called steady-state responses. The VEPs obtained using high temporal frequency stimulation were more delayed than the major early components of the transient VEP. As a further step, we demonstrated that haloperidol-induced VEP delays could be reversed with the administration of apomorphine, a pharmacologic dopamine agonist. In other words, a dopamine agonist can reverse the slowing effect of dopamine blockade on the visual system. These experiments affirm that dopamine depletion may be one cause of delayed pattern VEPs. Most importantly, these data emphasize that neurotransmitter deficiency may be causally linked to EP delays[84,92] and suggest (as we shall discuss) that synaptic malfunction may be another cause of VEP latency changes.

Chronic Versus Acute Recovery of VEP Delay in Parkinson's Disease, Optic Neuritis, and Pernicious Anemia

As mentioned earlier, VEP latency normalized following L-Dopa with carbidopa therapy in Parkinsonian patients whose pretreatment latency was abnormal. This became apparent as early as 20 to 40 days after the beginning of treatment. The mean latency reduction was 15 msec.[84] Our initial experiments in patients with severe Parkinsonian symptoms, i.e., who are rated above 30 according to scoring methods of Hoehn and Yahr,[101] reveal that in 15 minutes after the parenteral administration of apomorphine (2 mg), the VEP latency starts to decrease. In 30 minutes the latency is reduced by 8–16 msec. At the same time, apomorphine causes clinical improvement in these patients, reducing their clinical score to 20 on the average.[102] FIGURE 6 illustrates the effect of apomorphine in two patients.

An analogous effect of apomorphine occurs in rats. Haloperidol (3 mg/kg) increased the VEP latency to both flash and patterned stimuli in 30–50 minutes. The magnitude of the delay was stimulus-dependent (FIGURE 7). Intraperitoneal administration of 0.25 mg/kg of apomorphine corrected the VEP delay by 65%–70% in 15–30 minutes. Such a rapid change of VEP latency as occurs using dopamine antagonist and agonist therapy both in humans and rats has never been reported in other conditions. Frequent VEP testing in patients with optic neuritis reveals that the

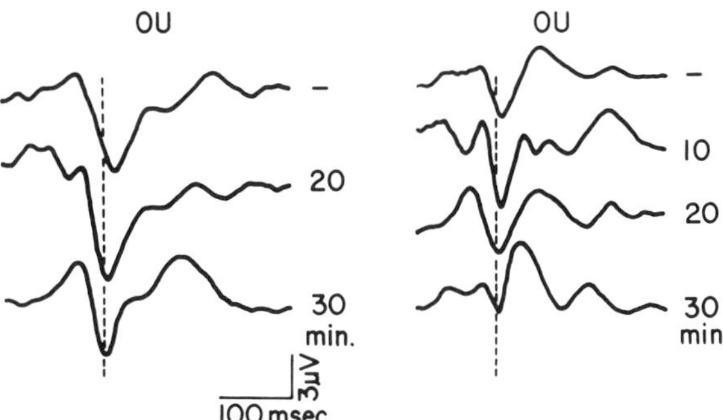

FIGURE 6. The effect of intravenous apomorphine injection on the VEP latency in two patients suffering from PD and having abnormal VEPs (left and right columns). A progressive return to normal VEP latency can be seen within 30 min following injection.

latency decreases by more than 20 msec one to two months after the onset of optic neuritis.[79] In hydrocephalus, VEP latency returns to normal 3 weeks after a shunting procedure.[45] As mentioned earlier, the VEP is delayed in pernicious anemia[46,48] and responds to treatment.[84] We studied the effect of treatment in two patients without optic atrophy and found that in one patient the VEP delay normalized in 20 days,[92] whereas in the other the latency gradually decreased by more than 30 msec over three months. Recently the effect of B_{12} treatment on VEP latency was affirmed in two patients, but the authors did not report whether or not they attempted VEP testing within one year of therapy.[47] Based on our findings, as already stated,[48] we feel that a relatively rapid recovery (in 20 days) should make one doubt that remyelination is the mechanism of the restoration of normal VEP latency in pernicious anemia. Thus, it seems that while chronic VEP changes in MS may be compatible with demyelination

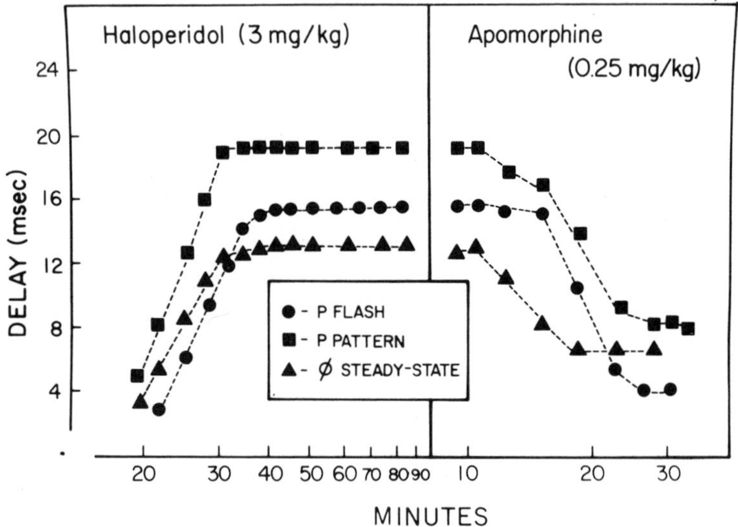

FIGURE 7. A graph showing the effect of haloperidol and apomorphine on the VEP latency of the same rat. Note the progressive delay introduced following an intraperitoneal injection of haloperidol and progressive restoration of the pretreatment VEP latency. Both pattern and flash VEPs were measured, and delays calculated for steady-state VEPs. Apparently dopaminergic mechanisms are important for all, but especially for high-contrast patterned stimuli.

(although other mechanisms may also play a role), and acute changes in PD may represent restoration of synaptic transmission, the mechanism of the delay and the effect of treatment in pernicious anemia remains obscure.

DISCUSSION

Delayed VEPs occur in several diseases, some of which predominantly affect the myelin of the visual pathway (multiple sclerosis, adrenoleucodystrophy), some of which show sporadically demyelinated optic fibers (Friedreich's ataxia and progressive congenital ataxias, B_{12} deficiency, compression of the optic pathway), and some of which supposedly do not interfere with myelin integrity (Parkinson's disease, glaucoma).

VEP latency changes were never claimed to be specific for MS by Halliday and his colleagues,[61] and we would add that the rough shape of a delayed transient VEP is also unspecific for any particular disease. While it is true, on the whole, that compression or ischemia of the optic nerve is likely to produce a higher degree of temporal dispersion of the VEP than MS,[61] the same degree an incidence of temporal dispersion are found both in MS and Parkinson's disease. It has been generally assumed that temporal dispersion is characteristic of delayed VEPs when axons are also damaged, while a pure demyelinating disease causes a simple delay.[50] Quite surprisingly, only in glaucoma patients did we not find temporal dispersion. However, too few patients have been studied up to this time (References 12, 23, and 41 and the first part of this paper), and further studies are required in order to draw any conclusion regarding waveshape changes in glaucoma. Nevertheless, even in MS it is clear that the VEP delay cannot be attributed solely to reduced central conduction velocity.

Mustering our arguments, we pointed out that stimulus-specific aspects of the VEP diagnostic yield in MS challenges demyelination as the only tenable mechanism of VEP delays. Another questionable aspect concerning mechanisms of VEP delays in MS is the magnitude of recovery which may take place in optic neuritis[27,61,79] and in B_{12} deficiency.[47] Originally it was proposed by McDonald[57] that delayed VEPs are caused by an alteration of the internodal conduction in demyelinated plaques. Based on the reasonable calculation of a delay of 25 msec for every centimeter of demyelination, one would assume that a recovery of the VEP latency by 15 msec or more[27,47,61,79,92] is caused by remyelination of more than 6 mm of the optic pathway. Remyelination in the encephalon may occur in short internodes;[113] however, it is thought that an extension of the myelin sheath is only supplied by single oligodendrocytes.[103] The recovery of latency is indeed difficult to attribute to remyelination alone.[112]

Both in clinical studies in Parkinson's disease, and in studies in rats, we found that EP latency changes may be due to dopamine depletion or blockade. Dyer explicitly attributed such delays to a malfunction of amacrine cells of the retina. The maximum concentration of dopamine in the visual system is in the inner nuclear layer of the retina, where amacrine cells are located.[104,105] Ehinger[104] demonstrated that the dopamine-containing neurons correspond to a subpopulation of amacrine cells. The functional properties of certain amacrine cells in the Necturus[106] could make them crucial for the timing of retinal ganglion cell responses, since Werblin's studies[106] show that these cells have a major function in the conversion of graded potentials of bipolar cells into the action potentials of the ganglion cells. Amacrine cells undergo rapid depolarization several milliseconds before the slow hyperpolarization of the preceding neurons is completed, and several milliseconds before the ganglion cells start their firing. Amacrine cells could therefore serve as "triggers" for the ganglion cells, enabling them to fire before bipolar cells have reached their peak response. This function of the amacrine cells is especially significant in view of the fact that "Y" cat retinal ganglion cells speed up their response as a function of contrast.[107,108] Shapley[107] described a "contrast gain control born in the inner plexiform layer [affecting] latency changes and changes in dynamics in the responses of X and Y cells to square wave contrast reversal of sine gratings." As a site of this contrast gain control, which affects temporal properties and latencies of ganglion cell responses to grating stimuli, a common retinal locus among the amacrine cells was suggested.[108] Thus, partial impairment of amacrine cells (such as by retinal or generalized dopamine depletion) should provide a desynchronizing influence on ganglion cell responsiveness. If one assumes therefore that dopaminergic amacrine cells have properties as described above, one may be tempted to propose, on the basis of the aforesaid facts, the effect of generalized CNS dopamine depletion on the latency of the VEP.

This hypothesis of VEP delays, though described primarily for a single synaptic site and with reference to Parkinson's disease, is suggested as a model to emphasize synaptic malfunction as the major mechanism of VEP delays in man. Neurotransmitter deficiency as a concept for explaining EP delays could be generalized to suggest involvement of other than retinal synapses. Although our recent studies in maculopathies are consistent with the notion that retinal pathology may cause clinically important EP delays,[109] we also feel that it may be fruitful to explore central synaptic function in animals using EP recordings in conjunction with appropriate neuropharmacological manipulations.

In summary, this chapter has emphasized the importance of considering synaptic mechanisms as vital links in the generation of VEPs, and has postulated that VEP delays may result from neurotransmitter deficiency. We propose that VEP delays in diseases other than PD may not be uniquely related to decreased central conduction, and suggest that a combination of VEP measurements and neuropharmacological assays could prove a useful tool in the *in vivo* assessment of neurotransmitter deficiencies in man.

Acknowledgments

We thank Prof. W. I. McDonald for a critical review of our manuscript and for several helpful comments. We are grateful to Leland H. Mylin, A.T., for many hours spent in patient testing, to Dr. G. Cohen for suggestions concerning methods of neuropharmacological experiments, and to many physicians and colleagues for patient referrals. We thank Miss Caroline Leake, who prepared the illustrations for this article and typed the manuscript.

References

1. Namerow, N. S. & N. Enns. 1972. Visual evoked responses in patients with multiple sclerosis. J. Neurol. Neurosurg. Psychiatry **35:** 829–833.
2. Feinsod, M., O. Abrasky & E. Auerbach. 1973. Electrophysiological examinations of the visual system in multiple sclerosis. J. Neurol. Sci. **20:** 161–175.
3. Hamel, B., J. R. Bourne & J. W. Ward. 1978. Visually evoked cortical potentials in renal failure: Transient potentials. Electroencephalogr. Clin. Neurophysiol. **44:** 606–614.
4. Visser, S. L., F. C. Stam & W. van Tilberg. 1976. Visual evoked response in senile and presenile dementia. Electroencephalogr. Clin. Neurophysiol. **40:** 385–392.
5. Ellenberger, C., D. J. Petro & S. B. Ziegler. 1978. The visually evoked potential in Huntington's disease. Neurology **28:** 95–97.
6. Sklar, F. H., A. L. Ehle & W. K. Clark. 1979. Visual evoked potentials: A noninvasive technique to monitor patients with shunted hydrocephalus. Neurosurgery **4:** 529–534.
7. Pampiglione, G. & A. Harden. 1977. So-called neuronal ceroid lipofuscinosis. Neurophysiological studies in 60 children. J. Neurol. Neurosurg. Psychiatry **40:** 323–330.
8. Aicardi, J., P. Plouin & F. Goutieres. 1978. Les ceroide-lipofuscinoses. Rev. EEG Neurophysiol. Clin. **8:** 149–155.
9. Godel, V., M. Blumenthal & B. Goldman. 1977. Visual functions in Tay Sachs diseased patients following enzyme replacement therapy. Metab. Ophthalmol. **2:** 27–29.
10. Jones, R. J. & T. D. France. 1978. Electrophysiological responses in hydranencephaly. Am. J. Ophthalmol. **85:** 478–480.
11. Borda, R. P. 1977. Visual evoked potentials to flash in the clinical evaluation of the optic

pathways. *In* Visual Evoked Potentials in Man. J. E. Desmedt, Ed. pp. 481–489. Clarendon Press, Oxford, England.
12. CAPPIN, J. & S. NISSIM. 1975. Visual evoked responses in the assessment of field defects in glaucoma. Arch. Ophthalmol. **93:** 9–18.
13. HALLIDAY, A. M. & J. MUSHIN. 1980. The visual evoked potential in neuro-ophthalmology. *In* International Ophthalmology Clinics. S. Sokol, Ed. Vol. **20:** 155–183. Little, Brown & Co., Boston, Mass.
14. TAGUCHI, K., H. AKAI, M. WATANABE, R. KURODA & M. IOKU. 1980. Subjective visual symptoms and electroencephalographic analysis before and after removal of occipital falx meningioma. Electroencephalogr. Clin. Neurophysiol. **49:** 162–167.
15. PERRY, N. W. & D. G. CHILDESS. 1969. The Human Visual Evoked Response. p. 187. Charles C Thomas, Publisher. Springfield, Mass.
16. KULIKOWSKI, J. J. & W. KOZAK. 1967. ERG, visual evoked responses and pattern detection in man. *In* Advances in Electrophysiology and Pathology of the Visual System (Proc. 6th Int. Symp. Electroretinography). pp. 187–192. George Thieme, Leipzig.
17. RIGGS, L. A. & P. WHITTLE. 1967. Human occipital and retinal potentials evoked by subjectively faded visual stimuli. Vision Res. **7:** 441–451.
18. ARMINGTON, J. C., K. GAARDNER & A. M. L. SCHICK. 1967. Variation of spontaneous ocular and occipital responses with stimulus patterns. J. Opt. Soc. Am. **57:** 1534–1539.
19. COBB, W. A., H. B. MORTON & G. ETTLINGER. 1967. Cerebral potentials evoked by pattern reversal and their suppression in visual rivalry. Nature **216:** 1123–1125.
20. HALLIDAY, A. M., W. I. MCDONALD & J. MUSHIN. 1972. Delayed visual evoked response in optic neuritis. Lancet (1): 982–985.
21. DUWAER, A. L. & H. SPEKREIJSE. 1978. Latency of luminance and contrast evoked potentials in multiple sclerosis patients. Electroencephalogr. Clin. Neurophysiol. **45:** 244–252.
22. LOWITZSCH, K. (In press.) Pattern evoked visual potentials in 251 MS patients in relation to ophthalmological findings and diagnostic classification. *In* Proc. Int. MS Symp., Gottingen, September 1978.
23. NEETENS, A., Y. HENDRATA & J. VAN ROMPAEY. 1979. Pattern and flash visual evoked responses in multiple sclerosis. J. Neurol. **280:** 113–124.
24. CHATRIAN, G. E. 1981. Personal communication.
25. CELESIA, G. G. & R. F. DALY. 1977. VECA—a new electrophysiological test for the diagnosis of optic nerve lesions. Neurology (Minn.) **27:** 637–642.
26. COLLINS, D. W. K., M. L. BLACK & F. L. MASTAGLIA. 1978. Pattern reversal visual evoked potential: Method of analysis and results in multiple sclerosis. J. Neurol. Sci. **36:** 83–95.
27. ASSELMAN, P., D. W. CHADWICK & C. D. MARSDEN. 1975. Visual evoked responses in the diagnosis and management of patients suspected of multiple sclerosis. Brain **98:** 201–282.
28. GAMBI, D., P. M. ROSSINI & M. ONOFRJ. 1980. Visual evoked cortical potentials (V.E.C.P.) by television presentation of different patterned stimuli to patients with multiple sclerosis. Ital. J. Neurol. Sci. **1:** 45–50.
29. NILSSON, B. Y. 1978. Visual evoked responses in multiple sclerosis: Comparison of two methods for pattern reversal. J. Neurol. Neurosurg. Psychiatry **41:** 499–506.
30. BODIS-WOLLNER, I., C. D. HENDLEY, L. H. MYLIN & J. THORNTON. 1979. Visual evoked potentials and the visuogram in multiple sclerosis. Ann. Neurol. **5:** 40–47.
31. CAMPBELL, F. W. & L. MAFFEI. 1970. Electrophysiological evidence for the existence of orientation and size detectors in the human visual system. J. Physiol. **207:** 635–652.
32. CAMPBELL, F. W. & J. J. KULIKOWSKI. 1972. The visual evoked potential as a function of contrast of a grating pattern. J. Physiol. **222:** 345–356.
33. DE VALOIS, K. K., R. L. DE VALOIS & E. W. YUND. 1979. Responses of striate cortex cells to grating and checkerboard patterns. J. Physiol. **291:** 483–505.
34. TYLER, C. W., P. APKARIAN & K. NAKAYAMA. 1978. Multiple spatial frequency tuning of electrical responses from human visual cortex. Exp. Brain. Res. **33:** 535–550.
35. REGAN, D. 1978. Assessment of visual acuity by evoked potential recording: Ambiguity caused by temporal dependence of spatial frequency selectivity. Vision Res. **18:** 439–443.

36. BODIS-WOLLNER, I., C. D. HENDLEY & J. J. KULIKOWSKI. 1972. Electrophysiological and psychophysical responses to modulation of contrast of a grating pattern. Perception **1:** 341–349.
37. SPEKREIJSE, H., L. H. VAN DER TWEEL & T. ZUIDEMA. 1973. Contrast evoked responses in man. Vision Res. **13:** 1577–1601.
38. CHAIN, H., J. MALLECOURT & M. LE BLANC. 1977. Apport de l'enregistrement des potentiels evoques visuels au diagnostic de la sclerose en plaques. Rev. Neurol. (Paris) **133:** 81–88.
39. SOKOL, S. 1976. Visually evoked potentials: Theory, techniques, and clinical applications. Surv. Ophthalmol. **21:** 18–44.
40. BARTL, G. 1978. The electroretinogram and the visual evoked potential in normal and glaucomatous eyes. Albrecht v. Graefs. Arch. Klin. Exp. Ophthalmol. **207:** 243–269.
41. TOWLE, V. L., S. SOKOL, A. MOSKOWITZ & B. SCHWARTZ. 1981. The visually evoked potential in glaucoma: Effects of check size, field size, and alternation rate. Invest. Ophthalmol. Visual Sci. (Suppl.) **20:** 22.
42. HALLIDAY, A. M., E. HALLIDAY & A. KRISS. 1976. The pattern evoked potential in compression of the anterior visual pathways. Brain **99:** 357–370.
43. MALMOLI, B., M. GRAF & K. TOIFL. 1979. EEG, pattern-evoked potentials and nerve conduction velocity in a family with adrenoleucodystrophy. Electroencephalogr. Clin. Neurophysiol. **47:** 411–416.
44. ONOFRJ, M., D. GAMBI & P. M. ROSSINI. 1979. New electrophysiological assessment in chronic uremia. Abstracts of the A.S.A.I.O. Congress, New York, N.Y.
45. ONOFRJ, M., I. BODIS-WOLLNER & L. H. MYLIN. 1981. Visual evoked potential latencies in papilledema and hydrocephalus. Neuro-Ophthalmology **2:** 85–92.
46. TRONCOSO, J., E. L. MANCALL & N. J. SCHATZ. 1979. Visual evoked responses in pernicious anemia. Arch. Neurol. **36:** 168–171.
47. KRUMHOLZ, A., H. G. WEISS, P. J. GOLDSTEIN & K. C. HARRIS. 1981. Evoked responses in vitamin B_{12} deficiency. Ann. Neurol. **9:** 407–409.
48. BODIS-WOLLNER, I. & A. KORCZYN. 1980. Dissociated sensory loss and visual evoked potentials in a patient with pernicious anemia. Mt. Sinai J. Med. **47:** 579–582.
49. BIRD, T. D. & W. B. CRILL. 1981. Pattern-reversal visual evoked potentials in the hereditary ataxias and spinal degeneration. Ann. Neurol. **9:** 243–250.
50. LIVINGSTONE, I. R., F. L. MASTAGLIA, R. EDIS & J. W. HOWE. 1981. Visual involvement in Friedreich's ataxia and hereditary spastic ataxia. Arch. Neurol. **38:** 75–79.
51. CARROLL, W. M., A. KRISS & M. BARAITSER. 1980. The incidence and nature of visual pathway involvement in Friedreich's ataxia: A clinical and visual evoked potential study of 22 patients. Brain **103:** 413–434.
52. HAPPEL, L. T., H. ROTHSCHILD & C. GARCIA. 1980. Visual evoked potentials in two forms of hereditary spastic paraplegia. Electroencephalogr. Clin. Neurophysiol. **48:** 233–236.
53. LIVINGSTONE, I. R., F. L. MASTAGLIA, R. EDIS & J. W. HOWE. 1981. Pattern visual evoked responses in hereditary spastic paraplegia. J. Neurol. Neurosurg. Psychiatry **44:** 176–178.
54. BODIS-WOLLNER, I. & M. D. YAHR. 1978. Measurement of visual evoked potentials in Parkinson's disease. Brain **101:** 661–671.
55. GAWEL, M. J., P. DAS & S. VINCENT. 1981. Visual and auditory evoked responses in Parkinson's disease. *In* Research Progress in Parkinson's Disease. F. C. Rose and R. Capildeo, Eds. pp. 138–146. Pitman Medical, Bath, England.
56. MCDONALD, W. I. & T. A. SEARS. 1970. The effects of experimental demyelination on conduction in the central nervous system. Brain **93:** 583–598.
57. MCDONALD, W. I. 1977. Pathophysiology of conduction in central nerve fibres. *In* Visual Evoked Potentials in Man. J. E. Desmedt, Ed. pp. 427–437. Clarendon Press, Oxford, England.
58. MCDONALD, W. I. 1974. Remyelination in relation to clinical lesions of the central nervous system. Br. Med. Bull. **30:** 186–189.
59. FRISEN, L. & W. F. HOYT. 1974. Insidious atrophy of retinal nerve fibers in multiple sclerosis. Arch. Ophthalmol. **92:** 91–97.
60. FREUND, H. J. 1973. Neuronal mechanisms of the lateral geniculate body. *In* Handbook

of Sensory Physiology. R. Jung, Ed. "Central Processing of Visual Information" Vol. 8/3B: 178–246. Springer Publishing, Berlin.

61. HALLIDAY, A. M. & W. I. MCDONALD. 1977. Pathophysiology of demyelinating disease. Br. Med. Bull. **33:** 21–29.
62. BORNSTEIN, M. B. & S. M. CRAIN. 1965. Functional studies of cultured brain tissues as related to "demyelinative" disorders. Science **148:** 1242–1244.
63. CAMISA, J., I. BODIS-WOLLNER & L. H. MYLIN. 1981. The effect of stimulus orientation on the visual evoked potential in multiple sclerosis. Ann. Neurol. **10:** 532–539.
64. REGAN, D., J. WHITLOCK, T. J. MURRAY & K. I. BEVERLEY. 1980. Orientation-specific losses of contrast sensitivity in multiple sclerosis. Invest. Ophthalmol. Visual Sci. **19:** 324–328.
65. SREBRO, R. 1981. Personal communication.
66. BODIS-WOLLNER, I. 1972. Visual acuity and contrast sensitivity in patients with cerebral lesions. Science **178:** 769–771.
67. BODIS-WOLLNER, I. & S. P. DIAMOND. 1976. The measurement of spatial contrast sensitivity in cases of blurred vision associated with cerebral lesions. Brain **99:** 695–710.
68. CIGANEK, L. 1961. The EEG response (evoked potential) to light stimulus in man. Electroen. Clin. Neurophysiol. **13:** 165–172.
69. MITCHELL, D. E., R. FREEMAN, M. MILLODOT & G. HAEGERSTROM. 1973. Meridional amblyopia. Evidence for modification of the human visual system by early visual experience. Vision Res. **13:** 535–557.
70. TIMM, N. H. 1975. Multivariate Analyses. Wadsworth Publishing, Belmont, Calif.
71. JOHNSON, N. L. & S. ROTZ. 1972. Distributions in Statistics: Continuous Multivariate Distributions. John Wiley & Sons, New York, N.Y.
72. MCALPINE, D., C. E. LUMSDEN & E. D. ACHESON. 1972. Multiple Sclerosis: a Reappraisal. Churchill-Livingstone, Edinburgh, Scotland.
73. HALLIDAY, A. M. 1978. Commentary: Evoked potentials in neurological disorders. *In* Event-Related Brain Potentials in Man. E. Callaway, P. Tueting & S. H. Koslow, Eds. pp. 197–218. Academic Press, New York, N.Y.
74. CANT, B. R., A. L. HUME & N. A. SHAW. 1978. Effects of luminance on the pattern visual evoked potential in multiple sclerosis. Electroencephalogr. Clin. Neurophysiol. **45:** 496–504.
75. BROWN, J. L. & J. E. BLACK. 1976. Critical duration for resolution of acuity targets. Vision Res. **16:** 309–315.
76. HENNERICI, M., D. WENZEL & H.-J. FREUND. 1977. The comparison of small-size rectangle and checkerboard stimulation for the evaluation of delayed visual evoked potentials. Brain **100:** 119–136.
77. SHERMAN, J. 1982. Ann. N.Y. Acad. Sci. (this volume).
78. SPEKREIJSE, H., O. ESTEVEZ & D. REITS. 1977. Visual evoked potentials and the physiological analysis of visual processes in man. *In* Visual Evoked Potentials in Man. J. E. Desmedt, Ed. pp. 16–89. Clarendon Press, Oxford, England.
79. ROSEN, I., H. BYNKE & M. SANDBERG. 1980. Pattern-reversal VEP after unilateral optic neuritis. *In* Evoked Potentials. Colin Barber, Ed. pp. 567–574. University Park Press, Baltimore, Md.
80. CERRA, D. & E. W. JOHNSON. 1961. Motor nerve conduction velocity in "idiopathic" polyneuritis. Arch. Phys. Med. Rehab. **42:** 159–168.
81. SMORTO, M. P. & J. V. BASMAJIAN. 1979. Basic clinical considerations. *In* Clinical Electroneurography. M. P. Smorto & J. V. Basmajian, Eds. pp. 57–91. William & Wilkins, Baltimore, Md.
82. GILLIAT, R. W. & R. J. HJORTH. 1972. Nerve conduction during Wallerian degeneration in the baboon. J. Neurol. Neurosurg. Psychiatry **35:** 335–341.
83. HALLIDAY, A. M., W. I. MCDONALD & J. MUSHIN. 1977. Visual evoked potentials in patients with demyelinating disease. *In* Visual Evoked Potentials in Man. J. E. Desmedt, Ed. pp. 438–449. Clarendon Press, Oxford, England.
84. BODIS-WOLLNER, I., M. D. YAHR & J. THORNTON. 1981. Visual evoked potentials and the severity of Parkinson's disease. *In* Research Progress in Parkinson's Disease. C. Rose & R. Capildeo, Eds. pp. 126–137. Pitman Medical, Bath, England.

85. CLOUGH, C. & I. BODIS-WOLLNER. 1981. Unpublished observations.
86. SCHAFER, E. W. P. & C. MCKEAN. 1975. Evidence that monoamines influence human evoked potentials. Brain Res. **99:** 49–58.
87. SALETU, B. 1974. Classification of psychotropic drugs based on human evoked potentials. *In* Psychotropic Drugs and the Human EEG: Modern Problems of Pharmacopsychiatry. T. M. Itil, Ed. Vol. 8: 258–285. Karger, Basel, Switzerland.
88. MOORE, R. H., K. KATADA & E. F. DOMINO. 1976. Effects of N_1N-dimethyltryptamine on electrically evoked responses in the cat visual system and modification by neuroleptic agents. Neuropharmacology **15:** 535–539.
89. BAKAY, E. P. & A. F. MIRSKY. 1977. Effect of secobarbital and chlorpromazine on cortical and subcortical visual-evoked potentials in the monkey. Psychopharmacol. Bull. **13:** 61–64.
90. BROSTEANU, E. R. & L. FLORU. 1980. Correlation between visual evoked potentials and psychopathological findings in schizophrenic patients under treatment with various psychopharmacological drugs. *In* Evoked Potentials. C. Barber, Ed. pp. 549–556. University Park Press, Baltimore, Md.
91. HOLDER, G. E., J. R. BARTLETT, P. K. BRIDGES, B. D. KANTAMANEI & G. CURZON. 1980. Correlation between transmitter metabolite concentration in human ventricular cerebrospinal fluid and pattern visual evoked potentials. Brain Res. **188:** 582–586.
92. BODIS-WOLLNER, I., M. D. YAHR & L. H. MYLIN. 1980. Visual evoked potential latency changes due to neurotransmitter deficiency in humans. Trans. Am. Neurol. Assoc. **105:** 215–218.
93. YAHR, M. D. 1974. Variation in the "on-off" effect in Parkinson's disease. Adv. Neurol. **5:** 397–404.
94. YAHR, M. D. 1978. Overview of present day treatment of Parkinson's disease. J. Neural Trans. **43:** 227–238.
95. DYER, R. S., W. E. HOWELL & R. C. MACPHAIL. 1981. Dopamine depletion slows retinal transmission. Exp. Neurol. **71:** 326–340.
96. DYER, R. S. & H. S. SWARTZEBDER. 1978. Sex and strain differences in the visual evoked potentials of albino and hooded rats. Pharmacol. Biochem. Behav. **9:** 301–306.
97. ONOFRJ, M. & I. BODIS-WOLLNER. 1982. Dopaminergic deficiency causes delayed VEPs in rats. Ann. Neurol. (in press).
98. ROGER, G. & L. GALAND. 1980. Cortical visual evoked potential conditioning in rats. Exp. Neurol. **30:** 497–596.
99. BIGLER, E. D. & D. E. FLEMING. 1976. Pharmacological suppression of photically evoked after-discharges in rats: Incremental dose, hippocampal EEG and behavioral activity correlates. Psychopharmacologia **46:** 73–82.
100. EBE, M., I. HOMMA & Y. ISHIYAMA. 1972. Laminal analysis of VEP and unit discharge on visual cortex of albino rat. Tohoku J. Exp. Med. **108:** 39–54.
101. HOEHN, M. M. & M. D. YAHR. 1967. Parkinsonism: Onset, progression and mortality. Neurology (Minn.) **17:** 427–442.
102. CLOUGH, C. 1981. Investigation of the "on-off" syndrome in parkinsonism. (Doctoral dissertation in preparation.)
103. MORREL, P. & W. T. NORTON. 1980. Myelin. Sci. Am. **242:** 88–118.
104. EHINGER, B. 1976. Biogenic amines as transmitters in the retina. *In* Transmitters in the Visual Process. S. L. Boting, Ed. Pergamon Press, Oxford, England.
105. LINDVALL, O. & A. BJORKLUND. 1978. Anatomy of the dopaminergic neuron systems in the rat brain. *In* Advances in Biochemical Psychopharmacology. P. J. Roberts, Ed. pp. 1–23. Raven Press, New York, N.Y.
106. WERBLIN, F. S. 1977. Regenerative amacrine cells depolarization and formation of on-off ganglion cell response. J. Physiol. **264:** 767–785.
107. SHAPLEY, R. M. & J. D. VICTOR. 1978. The effect of contrast on the transfer properties of cat retinal ganglion cells. J. Physiol. **285:** 275–298.
108. SHAPLEY, R. M. & J. D. VICTOR. 1979. Nonlinear spatial summation and contrast gain control of cat retinal ganglion cells. J. Physiol. **290:** 141–161.
109. BODIS-WOLLNER, I. & R. FELDMAN. 1982. Old perimacular pathology causes VEP delays in man. Electroencephalogr. Clin. Neurophysiol. (in press).

110. PRINEAS, J. & F. CONNELL. 1979. Remyelination in multiple sclerosis. Ann. Neurol. **5:** 22–31.
111. SMITH, K. J., W. F. BLAKEMORE & W. I. MCDONALD. 1981. The restoration of conduction by central remyelination. Brain **104:** 383–404.
112. BOSTOCK, H. & W. I. MCDONALD. 1982. Recovery of function after demyelination. *In* Neuronal-glial Cell Interrelationships: Ontogeny, Maintenance, Injury and Repair. Dahlem Konferenzen, Berlin (in press).
113. GLEDHILL, R. F., B. M. HARRISON & W. I. MCDONALD. 1973. Pattern of remyelination in the central nervous system. Nature (Lond.) **244:** 443–444.

DISCUSSION OF THE PAPER

G. CELESIA: Well, I am happy you dropped this little apple of discord concerning synaptic mechanisms. I think your interpretation may be correct in schizophrenia. It becomes rather difficult to believe it in multiple sclerosis, when we know that the major lesion is the demyelination, and we know that autopsy cases show 99% of the optic nerves of multiple sclerosis patients to have demyelinated lesions. However, besides synaptic mechanisms and demyelination, there is a third possible explanation. We have had some data in normals that if you stimulate the macular region with checks (and I am sure it would be the same with gratings), the major positive wave will have a specific latency—let's say at 110 milliseconds. If you then stimulate the periphery of the retina selectively with the same luminance but with a different check size (let's say at 14° eccentricity), the major positive wave is now at 120 or 130 milliseconds. So it is possible that some of the abnormality which we are observing is simply the disappearance of the response to the pattern of certain regions of the retina (say, the macula), and then one records a response which is normal for the peripheral retina, except we have no way to differentiate, and we believe that it is simply delayed due to demyelination.

I. BODIS-WOLLNER: To correct a misunderstanding, I do not suggest that in multiple sclerosis the pathology is at the retina but in fact that precisely because we found orientation selective changes, one must consider the cortex. I certainly do not suggest that in multiple sclerosis or in any disease, all delays could be necessarily explained by the action of a single mechanism. I was offering synaptic malfunction as an alternate mechanism of VEP delays based on Parkinson's disease and rat experiments, as a model.

As far as the third mechanism you mentioned is concerned, I cannot completely agree with you. For instance, some of the data of Hennerici could be explained when he used a single small foveal check and looked only at the center electrode latency. However, when one looks at the response at central and lateralized electrodes, one can usually eliminate the "field" theory you mentioned, although in certain instances (rarely MS) you would be right.

M. KUPERSMITH; Dr. Bodis-Wollner, we have duplicated the work you have done in the parkinsonian patients and found VEP delays, but in addition we have attempted to look at the retina to see whether that is a source of dopaminergic dysfunction, perhaps at the inner plexiform cells or at the amacrine cells. We have recorded oscillatory potentials on a whole series of patients. There were no flash ERG abnormalities at all in any of these patients, not even in stage 5 of Parkinson's disease. There was a group in Greece who reported some abnormalities in RPE layer function using EOG which were corrected using L-Dopa several years ago, but I have not seen any duplication of that.

BODIS-WOLLNER: Well, thank you. I cannot comment on it because we did not do flash ERG studies. The pattern visual evoked potential studies were also duplicated in England by Gawel and his colleagues at Charing Cross. As far as the retina being the site of dopaminergic deficiency is concerned, we were staying away from this explanation. However, Dyer and his colleagues have recorded VEPs at the cortex and directly from behind a globe in the rat following Haloperidol therapy and suggested that the major delay is in the retina. From our human VEP data, of course, we cannot directly infer whether or not the retina or extrageniculostriate connections are responsible for the delay, but Dyer's data are suggestive.

COMMENT: I think that age is a very important factor both in latency of VEPs and stage of Parkinson's disease. I can only insist that pathological cases should not be extrapolated into physiopathological considerations and hypotheses.

BODIS-WOLLNER: First of all, we used age-matched controls. Second, whether or not you can extrapolate depends on the particular care you take, and I think that our understanding of most neurological diseases, including Parkinson's disease, has been fostered by scientists who could formulate testable laboratory hypotheses based on astute clinical observations.

COMMENT: I just wanted to reiterate what Dr. Bodis-Wollner said. Several papers have been presented where the clinical aspects of the cases were worked out less than optimally, and someone commented that one should not extrapolate from these cases. I think it is true that if these cases were worked out in detail clinically, there would be less ambiguity in the literature.

L. MAFFEI: I would like to emphasize the importance of animal experiments in reinforcing clinical applications and suggesting clinical applications. And in this regard I like very much the experiments on rats. I think it is a nice experiment to record from the optic nerve and use the same electrode to stimulate and record from the cortex in such a way that one can disentangle retinal from cortical components. This will be useful because it can give us knowledge about the transmitters that play a role in the retina and behind the retina.

STIMULATING AND RECORDING METHODS USED IN OBTAINING SHORT-LATENCY SOMATOSENSORY EVOKED POTENTIALS (SEPs) IN PATIENTS WITH CENTRAL AND PERIPHERAL NEUROLOGIC DISORDERS*

W. C. Wiederholt, E. Meyer-Hardting,† B. Budnick, and K. L. McKeown

Department of Neurosciences
School of Medicine
University of California, San Diego
La Jolla, California 92093

Veterans Administration Medical Center
San Diego, California 92161

BACKGROUND

Repetitive electrical stimulation of both mixed motor and sensory and sensory nerves in the upper and lower extremities, or natural stimuli such as movement, pressure, or touch, will produce, after analysis with averaging methods, a number of potentials wich are locked in time to the stimulus and which can be recorded from the scalp of animals and man. Surface and depth recordings have shown that such somatosensory evoked potentials (SEPs), which occur between 20 and 70 msec following stimulation of the median nerve at the wrist, are localized to the contralateral somatosensory cortex.[1-5] These intermediate latency potentials are followed by potentials with longer latencies, which have a diffuse scalp distribution and are sensitive to tension, sleep, and sedative medication.[6-10] Lieberson and Kim[11] observed an early component of the SEP with a latency of approximately 12 msec evoked by stimulation of the median nerve at the wrist. Since then, several investigators have recorded several potentials in the 10 msec to 14 msec latency range following median nerve stimulation.[12-20] The definition and understanding of early components of the SEP has been greatly advanced since investigators began using long interelectrode distances and particularly since they began using noncephalic reference electrodes.[16,18,20] Discussion in this paper will be limited to potentials evoked by peripheral nerve stimulation at the wrist with latencies of approximately 10 to 25 msec.

All potentials recorded from the scalp are generated at a distance, with some being closer (near-field potentials) and some further away (far-field potentials). Electromagnetic field changes produced by generators close to the recording electrodes are likely to be relatively restricted, with steep slopes, while generators at greater distances will produce rather widespread electromagnetic field changes with only subtle gradients on the surface of the head. Consequently, closely spaced electrode pairs are more likely to record activity generated not far from their location, but such an arrangement will be inadequate for recording potentials with a rather wide distribution and with small electromagnetic field gradients because the electrode pair is likely to be on isopotential field lines, which would result in cancellation if a

*This work was supported by the Medical Research Service of the Veterans Administration.
†Supported by Deutscher Akademischer Austauschdienst No. 430-402-556-0.

0077-8923/82/0388-0349 $1.75/0 © 1982, NYAS

differential amplifier is used. Since it is presumed that the earliest components of the SEP reflect activity in the spinal cord, the brain stem, and other subcortical structures, these components cannot be adequately recorded with closely spaced scalp electrodes. The distribution of these early SEPs is rather widespread,[16–21] and they can be recorded from the earlobes when a noncephalic reference is used.[19–21] Recently Desmedt and Cheron[21] have shown that the earliest two components of the SEP are larger when the G 1 electrode is on the ipsileratеral ear but that subsequent potentials are larger when the G 1 electrode is on the contralateral ear.

Because of the wide distribution on the scalp, including the ears, and the rather short latencies, it has been assumed that the earliest components of the SEP are generated not in the cortex but in the peripheral nerve, spinal cord, brain stem, and sensory radiation. Because of their short latencies, the generators must be in relatively fast conducting pathways, including large myelinated peripheral nerves and medial lemniscus and sensory radiation fibers. It is not likely that electrical activity generated in "closed-field" anatomical structures such as the posterior column nuclei or the ventro-basal thalamic nuclei is recorded at a distance because electromagnetic field changes, at even a small distance from these structures, are negligible.[22] As shown by Arezzo,[23] synchronized volleys of action potentials traveling upstream in sensory pathways generate a coherent electromagnetic current field change, which may be recorded at a distance. Therefore, early components of the SEP, unlike later components, appear to reflect activity in fast-conducting sensory fibers of peripheral nerves, posterior column, medial lemniscus, and sensory radiation. Following unilateral or bilateral median nerve stimulation, several potentials can be recognized, which are usually labeled by their average latency and their polarity as recorded with a differential amplifier. These potentials are: P10 (9.9 ± 0.2 msec), P12 (11.9 ± 0.2 msec), P14 (14.2 ± 0.2 msec), N19 (18.6 ± 0.3 msec), P20 (19.7 ± 0.2 msec), and P23 (22.7 ± 0.3 msec) [values are mean ± SE].[19] The absolute latencies to onset and peak of the potentials directly correlated with the length of the peripheral conduction pathway.[15,20] Very often subpotentials can be recognized as part of P10, P14, and N19. P12 can only be clearly identified in approximately one-half of normal subjects.

Studies in rats, cats, and monkeys[19,23,24–27] have shown that likely generators responsible for the early components of the SEP are located in brachial plexus, posterior columns, medial lemniscus, and sensory radiation. While it is tenuous to extrapolate from animal data to human data, detailed studies of normal subjects and of patients with well-documented lesions and their effect on short-latency SEPs allow certain generalizations.[28–30] Patients with avulsion of cervical roots show preservation of P10 with disappearance of all subsequent components. This evidence clearly suggests that at least part of P10 is generated in the brachial plexus or cervical roots. The occasionally observed subpotentials of P10 suggest that activity both in the brachial plexus and cervical roots may contribute to this potential. In addition, it appears possible that early activity in the posterior columns also contributes to P10. Lesions in the lateral medullary field have no effect on SEPs. A medial medullary lesion, on the other hand, may abolish all potentials beyond P10 or produce prolongation of subsequent components and/or decrease of amplitudes of these components. Lateral lesions in the upper pons and mesencephalon abolish components beyond P12 or produce a prolonged interpeak latency from P10 to N19 with or without a change in amplitude. Purely cortical lesions abolish P23. Lesions in the thalamus and/or the thalamo-cortical sensory radiation abolish N19 or produce an increase in latency of this component, with or without amplitude changes. In some subjects, potentials can be identified on the slope from P14 to N19 which suggests that several generators may contribute to this potential. P14 probably reflects activity in the medial lemniscus. The potential is frequently bilobed, which may reflect activity in anatomically different structures or activities with different latencies in the same structure. Several investigators have raised the question of contribution of electrical activity in the cerebellum to early components of the SEP. Schieppati and Ducati[31] recently reported that, in the cat, total cerebellar ablation by suction did not change the somatosensory evoked potential recorded from an electrode placed on the somatosensory cortex with another in the muscles of the neck. Whether or not this evidence clearly excludes the cerebellum as a potential contributor to early components of the SEP remains to be seen. Some of the likely generators contributing to early components of the SEP are diagramatically shown in FIGURE 1.

Unfortunately, a bewildering multitude of labeling systems for evoked potentials exists.[16,21,29,32,33] We have adopted the nomenclature as recommended by the Committee on Publication Criteria.[32] Each potential is identified by its average peak latency and polarity. Latencies of early components are rounded to the nearest msec. Most investiagtors prefer to display SEPs with an upward deflection when activity at grid 1 in a differential amplifier is relatively more positive than activity at grid 2. Others still prefer to display an upward deflection when grid 1 becomes relatively more negative than grid 2.

STIMULATION METHODS

In most studies of SEPs, the peripheral stimulus is an electrical current applied either transcutaneously or subcutaneously to a peripheral nerve which may be a mixed nerve or a pure sensory nerve. When mechanical stimuli are used,[34–36] potentials recorded from the scalp are very similar to those recorded following electrical stimulation except that their latencies are proportionately prolonged, and, since fewer peripheral nerve fibers are activated, their amplitude is lower. Most investigators apply peripheral stimuli through silver chloride electrodes applied to one or several fingers or through electrodes placed directly over the median, radial, or ulnar nerves at the wrist. Surface electrode separation is usually 2 to 3 cm with the cathode in the proximal location. The common ground electrode is best located proximal to the stimulating electrodes on the same limb. It is essential that skin resistance be reduced as much as possible by slightly abrating the superficial layers of the skin and rubbing electrode paste into this area. In some subjects, in spite of proper technique, a large shock artifact will obscure early components of the SEP. In this situation, subcutaneously placed bare needle electrodes will give very satisfactory results. Standard EEG needle electrodes can be used for this purpose. The needle should be inserted subcutaneously because it minimizes insertional pain and should be perpendicular to the nerve to be stimulated. Interelectrode distance should be at least 1 cm. When using subcutaneous stimulating electrodes, voltage or current can usually be reduced by a factor of ten, which accounts for the much reduced stimulus artifact. Except for the initial insertional pain, which with good technique can be minimized, subcutaneous stimulating electrodes are extremely well tolerated and are advantageous when prolonged recording sessions are required. Stimulus duration is usually from 0.1 msec to 0.3 msec. When stimulating a mixed peripheral nerve, the intensity of the stimulus should be adjusted at a level which produces a moderate twitch of the distal muscle. This usually produces maximal SEPs. When stimulating pure sensory nerves such as the radial nerve at the wrist or digital nerves, optimal SEPs are usually obtained when current or voltage is adjusted to approximately three times sensory threshold. Amplitude of the evoked potential is directly related to the stimulus intensity.[20] In

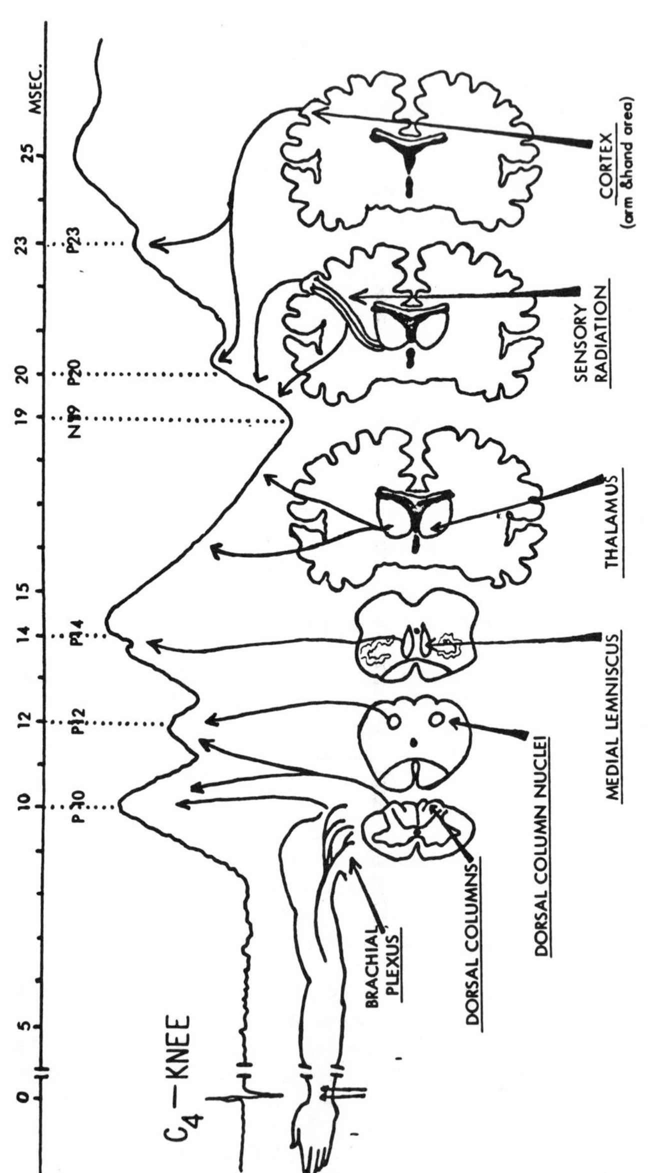

FIGURE 1. Tentative sites of origin of short-latency somatosensory evoked potentials (SEPs).

most instances, the median nerve is used for stimulation either at the wrist or through ring electrodes around fingers. Similar evoked potentials may be obtained with stimulation of the ulnar and radial nerves at the wrist.[37,38] There appear to be no differences between early components of the SEP elicited by median, ulnar, or radial nerve stimulation except P10 is of smallest amplitude following radial nerve stimulation, intermediate with ulnar nerve stimulation, and largest with median nerve stimulation.[37] This can probably be accounted for by the fact that median and ulnar nerve stimulation at the wrist activates antidromic impulses in motor fibers, which is not the case with radial nerve stimulation, and both carry more sensory fibers. In the above study, P10 following radial nerve stimulation was frequently bilobed, suggesting two generators with slightly different latencies. This bilobed character of P10 is not as apparent with median and ulnar nerve stimulation, probably because of the larger amplitude of this potential and the fusion of the two subpotentials. P12 was often easier to recognize following radial nerve stimulation, which is probably related to the observation that the amplitude of P10 was lower with radial nerve stimulation than with stimulation of the other nerves and, therefore, fusion of P10 with P12 was less likely. Stimulation of all three peripheral nerves (median, ulnar, radial) at the wrist may allow a more precise localization of lesions involving the brachial plexus and/or cervical roots for reasons given below. Median nerve stimulation at the wrist activates sensory fibers that enter the spinal cord through upper and middle trunks of the brachial plexus and posterior roots C6 and C7. Antidromic median motor potentials travel through the medial cord and lower trunk of the plexus to enter the spinal cord through C8 and T1 anterior roots. Sensory action potentials following radial nerve stimulation at the wrist travel through the posterior cord and upper trunk of the brachial plexus to reach C6. Ulnar sensory potentials traverse through the medial cord and lower trunk of the brachial plexus and C8. Antidromic ulnar motor potentials reach the spinal cord through the medial cord and lower trunk of the plexus and T1 root.

Components of the SEP beyond 30 msec following nerve stimulation at the wrist decline in amplitude when stimulus rates faster than 1 per second are used. With stimulus rates up to 10 per second, there is little or no change in the amplitude of SEP components up to approximately P23. Stimulus rates between 4 and 6 per second are comfortably tolerated by most subjects while higher rates of stimulation are unpleasant. As shown in FIGURE 2, when using paired stimuli with different interstimulus intervals, little change in early components of the SEP is seen with interstimulus intervals of 100 msec or longer. The effect of paired stimuli is not identical to that of continuous stimuli at rates comparable to the interstimulus interval of paired stimuli. Therefore, continuous stimulation at 10 stimuli per second may produce some decrease in amplitude of the P23 potential. The shape of the recovery curves in FIGURE 2 suggests that P10 is generated in a pathway without an interposed synapse, because the amplitude of the second potential of a pair of potentials shows no attenuation even with interstimulus intervals as short as 5 msec. Slopes of P14, N19, and P23 clearly indicate that their generators are at least one synapse removed from the peripheral nerve stimulated. The progressive increase in steepness of the slope from P14 to N19 to P23 suggests that for each potential the generator(s) is at least one additional synapse removed from the nerve stimulated. These findings are in agreement with other evidence alluded to above and to the generator sites for these potentials.

A minimum number of stimuli per nerve stimulated is 1,000, but, more often than not, 2,000 stimuli or more are required to produce adequate averages. Fewer stimuli in the range of 500 to 1,000 are satisfactory when primary cortical components of the SEP are recorded with closely spaced electrodes.

Recording Methods

Standard EEG surface electrodes properly applied with impedances below 3,000 ohms are satisfactory. Subcutaneously placed needle electrodes for recording purposes are equally acceptable. They are advantageous in situations where electrodes are difficult to apply and when prolonged recording is necessary.

In many instances, an evoked potential study will be carried out after an EEG has been done. Therefore, it appears to be logical to use the already applied electrodes to record SEPs. When the number of channels for recording is limited, C3 and C4 electrodes are most appropriate for recording SEPs following upper extremity stimulation. P3 and P4 electrodes are almost as satisfactory as the C electrodes. When more channels are available, the choice of electrode locations depends on the specific investigative or clinical situation. Some investigators, notably Chiappa *et al.*,[29] prefer to use three channels, each having a specific electrode arrangement, which allows better recognition of certain individual potentials or groups of potentials. Specifically, these investigators use in one channel an electrode pair consisting of F_Z-EP (Erb's point) to enhance activity generated in the brachial plexus, in another channel F_Z-C2 (spinous process of C2) to record their waves A and B which they feel are similar to P12 and P14, and in the last channel F_Z-C4 to record their N2 and P2 which resemble N19 and P23. Such an arrangement may be useful for the practical reasons for which it was designed. It is unwarranted, though, to reason that potentials that may have similar latencies but are recorded with different electrode arrangements are identical.

The choice of the grid 2 electrode largely determines which potentials can be recorded and what their configuration and latencies will be. As shown in FIGURE 3, recording from C3 or C4 referenced to Erb's point produces an initial negative deflection followed by three positive deflections, followed by a large negative deflec-

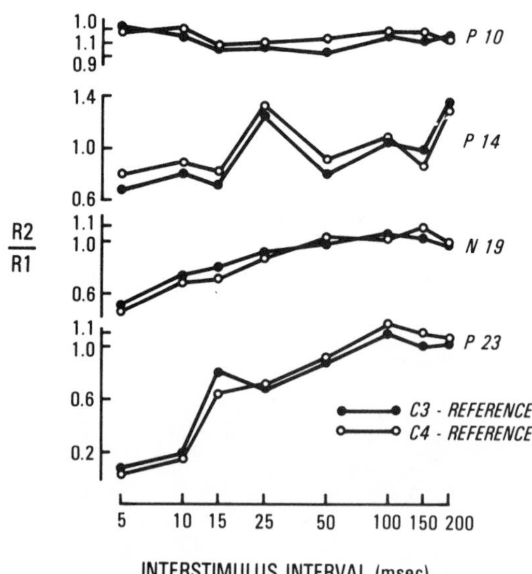

FIGURE 2. Far-field recorded somatosensory evoked potentials: Latencies and proposed origins. Recovery curves of short-latency components of the SEP expressed as the ratio of R2 to R1. Cumulative data from 14 normal subjects.

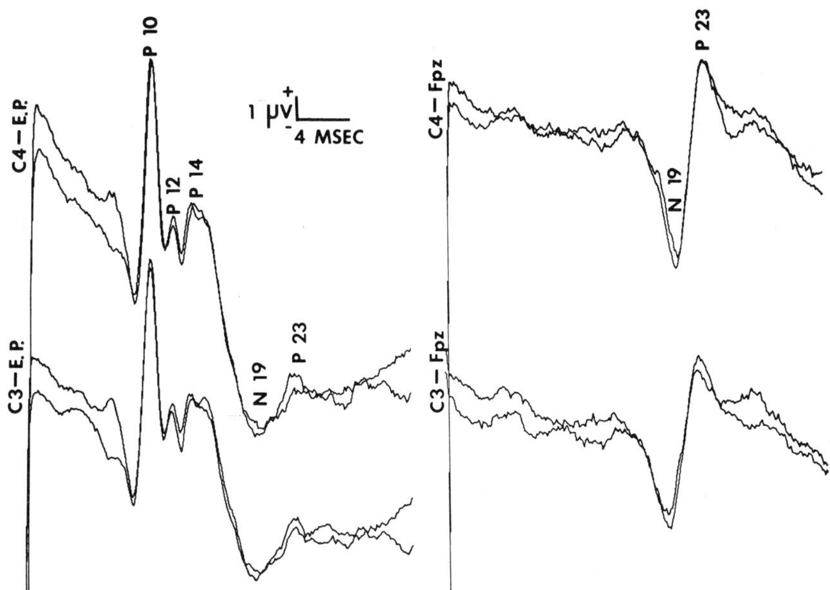

FIGURE 3. Short-latency SEPs following median nerve stimulation at the wrist recorded from C3 and C4 with reference electrodes placed at Erb's point (E.P.) and FPz.

tion (N19) and a small positive deflection (P23). P10 in this illustration is so large that subpotentials, if they exist, cannot be recognized. If one wishes to study such subpotentials that may be buried in the large P10 when recording with the reference placed at Erb's point, a reference on an extremity should be used as well as radial nerve stimulation. When C3 or C4 are referenced to FP_Z (FIGURE 3), the previously described three early positive potentials cannot be recognized, but N19 is equally well defined, and P23 is more prominent than when recorded with longer interelectrode distances. It appears that, when several recording channels are available, at least one should be used for recording from an electrode pair having a long interelectrode distance, which will show potentials prior to N19 rather well, and another channel should record activity from C3 or C4 or P3 or P4 referenced to the forehead, which will allow ready recognition of waves N19 and P23. It would be most helpful if all investigators could agree on using, in one recording channel, an identical electrode montage which would help immensely when comparing data from different laboratories.

Analog filters and, to a lesser degree, digital filters may produce distortion of latencies and voltages depending on filter bandwidth.[39–42] Desmedt[39] found that filtering below 1 kHz produced severe distortion of latencies and voltages of short-latency SEPs and recommended that low-pass filtering should not be done below 3 kHz. Chiappa *et al.*[29] recommended 3-Hz high-pass filters. These investigators noted that even though there is little low frequency activity in the early components of the SEP, the use of 10 Hz or 30 Hz high-pass filters increased the likelihood of stimulus-induced baseline shifts. These baseline shifts were accentuated when using a 60 Hz notch filter. King and Green[43] studied short-latency SEPs using amplifiers with a frequency limit of 100 to 3,000 Hz. These authors suggest that high-pass filtering produces better definition of early components of the SEP and that these components are similar to those recorded with a greater bandwidth, taking into account some latency and amplitude distortions. It appears that at this time the question of filter bandwidth for optimal recording of early components of the SEP cannot be definitively answered.

The recording of early components of the SEP with long interelectode distances is technically not easy. Electrical interference from a variety of sources is rather common, and a relaxed subject—at times using a mild sedative—is extremely helpful. An unavoidable artifact with long interelectrode distance recording is the EKG potential. It is certainly possible to trigger the stimulator and the averaging process from any of the components of the EKG complex and, with appropriate delays, place the evoked potential between the T wave and the next P wave. The drawbacks of this procedure are that an apparent straight baseline when recording in the millivolt range for the EKG may have an unacceptably steep slope when recorded in the microvolt range required for recording SEPs. Furthermore, the heart rate—which in many resting subjects is below 70 per minute—seriously limits the frequency with which the subject can be stimulated, unnecessarily prolonging the procedure. A much more practical approach to minimize the EKG artifact is to use a sweep rejection technique. Most commercially available averagers have built in rejection features. Averaging programs may incorporate variable levels of rejection with the ability to change both the amplitude and the duration at which rejection of sweeps occurs. Using such rejection techniques will produce a rejection of approximately 5% to 15% of sweeps when using stimulus rates of 4 to 5 per second.

It is desirable for each and every averaging run to do duplicate or triplicate averages because optimal superimposition of sequential averages is the best indicator that potentials observed are not due to some artifact. It is helpful to use at least one averaging run without actually stimulating the subject, which allows one to assess ongoing background activity to some degree. Sampling rates of up to 0.14 msec are adequate for averages of early components of the SEP recorded with long interelectrode distances. Most available "hardwired" averagers will allow analog-to-digital conversion at rates of 0.03 msec to 0.08 msec. As previously indicated, investigators display polarity of potentials in different ways. Some will show an upward deflection when grid 1 is relatively more positive than grid 2, while others will show an upward deflection when grid 1 is relatively more negative than grid 2. Beyond a certain inconvenience to readers of papers on SEPs, there appears to be no compelling reason to use one or the other method preferentially.

Conclusions

Improved stimulation, recording, and averaging techniques permit more reliable recording of both short- and long-latency SEPs. Many different nerves, both in the upper and lower extremities, and possibly also in the face, can be stimulated electrically or mechanically. While the precise origin of generators responsible for short-latency SEP components is not as yet settled, nevertheless, solid evidence is accumulating that these potentials reflect activity in the peripheral nerves, spinal cord, brain stem, and forebrain. Thus, the investigator both of basic and clinical problems has a ready tool to assess, with noninvasive techniques, fast-conducting components of the peripheral and central somatosensory system.

References

1. Goff, W. R., B. S. Rosner & T. Allison. 1962. Distribution of somatosensory evoked responses in normal man. Electroencephalogr. Clin. Neurophysiol. **14:** 697–713.

2. DOMINO, E. F., S. MATSUOAKA, J. WALTZ & I. COOPER. 1964. Simultaneous recordings of scalp and epidural somatosensory evoked responses in man. Science **145:** 1199–1200.
3. DOMINO, E. F., S. MATSUOAKA, J. WALTZ & I. COOPER. 1965. Effects of cryogenic thalamic lesions on the somesthetic evoked response in man. Electroencephalogr. Clin. Neurophysiol. **19:** 127–138.
4. GIBLIN, D. R. 1964. Somatosensory evoked potentials in healthy subjects and in patients with lesions of the nervous system. Ann. N.Y. Acad. Sci. **112:** 93–142.
5. ALLISON, T., W. R. GOFF, P. D. WILLIAMSON & J. C. VANGILDER. 1980. On the neural origin of early components of the human somatosensory evoked potential. *In* Clinical Uses of Cerebral, Brainstem and Spinal Somatosensory Evoked Potentials. Prog. Clin. Neurophysiol. J. E. Desmedt, Ed. Vol. **7:** 51–68. Karger, Basel, Switzerland.
6. GOFF, W. R., P. D. WILLIAMSON, J. C. VANGILDER, T. ALLISON & T. C. FISHER. 1980. Neural origins of long latency evoked potentials recorded from the depth and from the cortical surface of the brain in man. *In* Clinical Uses of Cerebral, Brainstem and Spinal Somatosensory Evoked Potentials. Prog. Clin. Neurophysiol. J. E. Desmedt, Ed. Vol. **7:** 126–145. Karger, Basel, Switzerland.
7. ALLISON, T. 1962. Recovery functions of somatosensory evoked responses in man. Electroencepalogr. Clin. Neurophysiol. **14:** 331–343.
8. ALLISON, T., W. R. GOFF, H. A. ABRAHAMIAN & B. S. ROSNER. 1963. The effects of barbiturate anesthesia upon human somatosensory evoked responses. Electroencephalogr. Clin. Neurophysiol. Suppl. **24:** 68–75.
9. GOFF, W. R., T. ALLISON, A. SHAPIRO & B. S. ROSNER. 1966. Cerebral somatosensory responses evoked during sleep in man. Electroencephalogr. Clin. Neurophysiol. **21:** 1–9.
10. DESMEDT, J. E. & D. ROBERTSON. 1977. Differential enhancement of early and late components of the cerebral somatosensory evoked potentials during forced-paced cognitive tasks in man. J. Physiol. (London) **271:** 761–782.
11. LIEBERSON, W. T. & K. C. KIM. 1963. The mapping out of evoked potentials elicited by stimulation of the median and peroneal nerves. Electroencephalogr. Clin. Neurophysiol. **15:** 721.
12. CRACCO, R. W. & R. G. BICKFORD. 1968. Somatomotor and somatosensory evoked response. Arch. Neurol. (Chicago) **18:** 52–68.
13. CRACCO, R. Q. 1972. The initial positive potential of the human scalp-recorded somatosensory evoked response. Electroencephalogr. Clin. Neurophysiol. **32:** 623–629.
14. CRACCO, R. Q. 1972. Travelling waves of the human scalp-recorded somatosensory evoked response. Electroencephalogr. Clin. Neurophysiol. **33:** 557–566.
15. MATTHEWS, W. B., M. BEAUCHAMP & D. G. SMALL. 1974. Cervical somatosensory evoked responses in man. Nature (London) **52:** 230–232.
16. CRACCO, R. Q. & J. B. CRACCO. 1976. Somatosensory evoked potentials in man: Farfield potentials. Electroencephalogr. Clin. Neurophysiol. **41:** 460–466.
17. SMALL, D. G. & S. J. JONES. 1977. Subcortical somatosensory evoked potentials following stimulation at the wrist or the ankle in normal subjects. Electroencephalogr. Clin. Neurophysiol. *43:* 536.
18. JONES, S. J. & D. G. SMALL. 1977. The dipolar distribution of subcortical somatosensory evoked potentials in man. Electroencephalogr. Clin. Neurophysiol. *43:* 537–538.
19. WIEDERHOLT, W. C. 1980. Early components of the somatosensory evoked potential in man, cat, and rat. *In* Clinical Uses of Cerebral, Brainstem and Spinal Somatosensory Evoked Potentials. Prog. Clin. Neurophysiol. J. E. Desmedt, Ed. Vol. **7:** 105–117.
20. KRITCHEVSKY, M. & W. C. WIEDERHOLT. 1978. Short latency somatosensory evoked potentials. Arch. Neurol. **35:** 706–711.
21. DESMEDT, J. E. & G. CHERON. 1980. Central somatosensory conduction in man: Neural generators and interpeak latencies of the far-field components recorded from neck and right or left scalp and earlobes. Electroencepalogr. Clin. Neurophysiol. **50:** 382–403.
22. KLEE, M. & W. RALL. 1977. Computed potentials of cortically arranged populations of neurons. J. Neurophysiol. **40:** 647–666.
23. AREZZO, J., A. D. LEGATT & H. G. VAUGHAN. 1979. Topography and intracranial sources of somatosensory evoked potentials in the monkey: Early components. Electroencephalogr. Clin. Neurophysiol. **46:** 155–172.
24. WIEDERHOLT, W. C. 1975. Far-field reflections of brain stem somatosensory evoked responses. Trans. Am. Neurol. Assoc. **100:** 259–261.

25. WIEDERHOLT, W. C. & V. J. IRAGUI-MADOZ. 1977. Far-field somatosensory evoked potentials in the rat. Electroencephalogr. Clin. Neurophysiol. **42:** 456–465.
26. IRAGUI-MADOZ, V. J. & W. C. WIEDERHOLT. 1977. Far-field somatosensory evoked potentials in the cat. Electroencephalogr. Clin. Neurophysiol. **43:** 646–657.
27. BUDNICK, B., K. L. MCKEOWN & W. C. WIEDERHOLT. 1981. Hypothermia-induced changes in rat short latency somatosensory evoked potentials. Electroencephalogr. Clin. Neurophysiol. **51:** 19–31.
28. NOEL, P. & J. E. DESMEDT. 1980. Cerebral and far-field somatosensory evoked potentials in neurological disorders involving the cervical spinal cord, brain stem, thalamus, and cortex. *In* Clinical Uses of Cerebral, Brainstem and Spinal Somatosensory Evoked Potentials. Prog. Clin. Neurophysiol. J. E. Desmedt, Ed. Vol. **7:** 205–230. Karger, Basel, Switzerland.
29. CHIAPPA, K. H., S. K. CHOI & R. R. YOUNG. 1980. Short-latency somatosensory evoked potentials following median nerve stimulation in patients with neurological lesions. *In* Clinical Uses of Cerebral, Brainstem and Spinal Somatosensory Evoked Potentials. Prog. Clin. Neurophysiol. J. E. Dlesmedt, Ed. Vol. **7:** 264–281. Karger, Basel, Switzerland.
30. ANZISKA, B. & R. Q. CRACCO. 1980. Short latency somatosensory evoked potentials: Studies in patients with focal neurological disease. Electroencephalogr. Clin. Neurophysiol. **49:** 227–239.
31. SCHIEPPATI, M. & A. DUCATI. 1981. Effects of stimulus intensity, cervical cord tractotomies and cerebellectomy on somatosensory evoked potentials from skin and muscle afference of cat hind limb. Electroencephalogr. Clin. Neurophysiol. **51:** 363–372.
32. DONCHIN, E., E. CALLOWAY, R. COOPER, *et al.* 1977. Publication criteria for studies of evoked potentials in man: Report of a committee. *In* Attention, Voluntary Contraction and Event-Related Cerebral Potentials in Man. Prog. Clin. Neurophysiol. J. E. Desmedt, Ed. Vol. **1:** 1–11. Karger, Basel, Switzerland.
33. SMALL, D. G., M. BEAUCHAMP & W. B. MATTHEWS. 1980. Subcortical somatosensory evoked potentials in normal man and in patients with central nervous system lesions. *In* Clinical Uses of Cerebral, Brainstem and Spinal Somatosensory Evoked Potentials. Prog. Clin. Neurophysiol. J. E. Desmedt, Ed. Vol. **7:** 190–204. Karger, Basel, Switzerland.
34. PRATT, H., R. N. AMLIE & A. STARR. 1979. Short latency mechanically evoked somatosensory potentials in humans. Electroencephalogr. Clin. Neurophysiol. **47:** 524–531.
35. PRATT, H., D. POLITOSKE & A. STARR. 1980. Mechanically and electrically evoked somatosensory potentials in humans: Effects of stimulus presentation rate. Electroencephalogr. Clin. Neurophysiol. **49:** 240–249.
36. PRATT, H. & A. STARR. 1981. Mechanically and electrically evoked somatosensory potentials in humans: Scalp and neck distribution of short latency components. Electroencephalogr. Clin. Neurophysiol. **51:** 138–147.
37. GRISOLIA, J. S. & W. C. WIEDERHOLT. 1980. Short latency somatosensory evoked potentials from radial, median, and ulnar nerve stimulation in man. Electroencephalogr. Clin. Neurophysiol. **50:** 375–381.
38. EISEN, A. & G. ELLEKER. 1980. Sensory nerve stimulation and evoked cerebral potentials. Neurology **30:** 1097–1105.
39. DESMEDT, J. E., J. BRUNKO, J. DEBECKER & J. CARMELIET. 1974. The system bandpass required to avoid distortion of early components when averaging somatosensory evoked potentials. Electroencephalogr. Clin. Neurophysiol. **37:** 407–410.
40. WASTELL, D. G. 1979. The application of low-pass linear filters to evoked potential data: Filtering without phase distortion. Electroencephalogr. Clin Neurophysiol. **46:** 355–356.
41. BOSTON, J. R. & P. J. AINSLIE. 1980. Effects of analog and digital filtering on brain stem auditory evoked potentials. Electroencephalogr. Clin. Neurophysiol. **48:** 361–364.
42. DOYLE, D. J. & M. L. HYDE. 1981. Bassel filtering of brain stem auditory evoked potentials. Electroencephalogr. Clin. Neurophysiol. **51:** 446–448.
43. KING, D. W. & J. B. GREEN. 1979. Short latency somatosensory potentials in humans. Electroencephalogr. Clin. Neurophysiol. **46:** 702–708.

SCALP-RECORDED SOMATOSENSORY EVOKED POTENTIALS TO STIMULATION OF NERVES IN THE LOWER EXTREMITIES AND EVALUATION OF PATIENTS WITH SPINAL CORD TRAUMA*

Phanor L. Perot, Jr. and Cristian L. Vera

Department of Neurosurgery
Medical University of South Carolina
Charleston, South Carolina 29425

INTRODUCTION

The use of evoked response techniques in the evaluation of patients with nervous system trauma is now an accepted procedure and an area of intense investigative activity. In comatose head injuries, studies with a combination of visual evoked potentials, somatosensory evoked potentials (SEPs), auditory evoked potentials, and auditory brainstem responses have shown that there is a close correlation between the degree of evoked response abnormality and the clinical outcome of the patient.[1] This was particularly evident with regard to the SEP since it was found that the presence of a relatively undistorted response early after the injury was the single most important prognostic indicator of a good recovery.

There has been increasing interest lately in the application of the SEP technique to the study of patients with spinal cord injury.[2–10] The technique would seem well suited to the noninvasive study of spinal cord conduction through an injury site. Potential limitations include the fact that the response is mediated mainly through the posterior columns and the fact that the response may be affected by factors unrelated to the extent of spinal injury, e.g., the patient's age, level of arousal, drugs, and the presence of any associated brain injury. Despite these caveats, serial recordings of SEPs in patients with spinal cord injury can provide valuable information concerning the extent of the lesion (diagnosis), an index of changes in cord function (monitoring), and predictive data regarding outcome (prognosis). Evoked responses recorded over the spine,[11–14] and short-latency or far-field SEPs[15–18] also show promise as electrodiagnostic tools for the analysis of pathologic lesions of the spinal cord. At this time there is also great interest in the use of SEP recording intraoperatively as a monitor of cord function during corrective surgery for scoliosis and the removal of intraspinal neoplastic, vascular, or other lesions.[19–25]

METHODS

A total of 80 control subjects and 92 patients with cervical spinal cord injury were studied. Of the patients, 37 had a complete spinal cord lesion (no detectable motor or sensory function preserved below the level of the lesion), and 55 had an incomplete cord lesion (some motor or sensory function preserved below the level of the lesion). As soon as possible after admission and initial stabilization of the patient, percutaneous or needle electrodes were applied over the right and left median nerves at the wrist and

*This investigation was supported by Research Grant NS 11066 from the National Institutes of Health, U.S. Public Health Service.

0077–8923/82/0388–0359 $1.75/0 © 1982, NYAS

over the right and left posterior tibial nerves at the level of the medial maleolus of the ankle. Cup-type or needle electrodes were applied to the scalp over the post-central gyrus area for hand (approximately C3, C4 of the 10–20 system) for recording median nerve SEPs. Posterior tibial SEPs were recorded from the midline vertex at a point 55% of the nasion-inion distance behind the nasion. Active or G1 electrodes were referenced to the contralateral ear lobe. A ground scalp electrode was applied to the midline forehead and ground electrodes were placed on the limbs proximal to the stimulating electrodes as needed for stimulus artifact control. Each pair of stimulating electrodes was connected to a constant current, computer-controlled stimulator. Stimulus parameters were the following: current: just sufficient to produce a muscle twitch distally in the muscle innervated; duration: 0.1 msec; frequency: 2/sec. SEPs were recorded from each nerve in sequence, right median, left median, right posterior tibial, left posterior tibial, and bilateral simultaneous posterior tibials. Each patient's digitized averaged responses were stored on an individual floppy disk and displayed by an *x–y* plotter. The frequency response of the recording apparatus was 1–500 Hz. Analysis times of 160 msec were used and 256 responses were summated. The computer horizontal resolution was 320 μsec per point with 512 points/channel. In all the recordings, relative negativity at G1 resulted in an upward deflection. Serial SEPs were monitored on a frequent basis for the first few days after injury (3–4 per day for the first week), then daily for 1–2 weeks and at weekly intervals thereafter until the patient's condition stabilized.

RESULTS

Posterior Tibial SEPs

Most neck injuries involve spinal vertebrae and cord segments C5, C6, and C7, and the main afferent input from the median nerve enters the cord at the C6 and C7 segments. Therefore, the degree of abnormality of median nerve SEPs in cervical cord

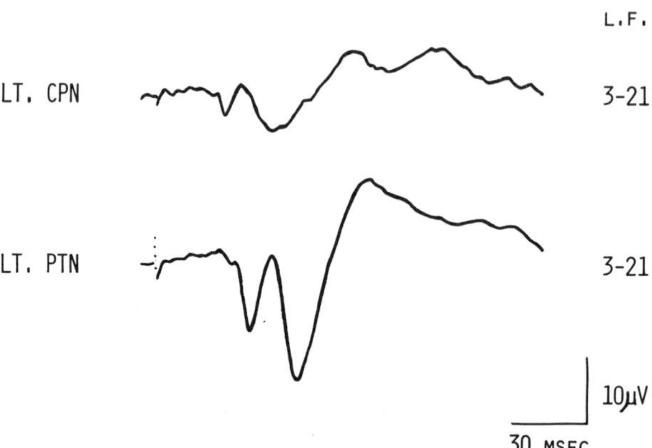

FIGURE 1. Comparison of common peroneal and posterior tibial somatosensory evoked potentials (SEPs). The top trace is an SEP recorded from the vertex referred to the left ear with stimulation of the left common peroneal nerve. The bottom trace is an SEP recorded from the same electrodes in the same control subject with left posterior tibial nerve stimulation. Stimulus parameters, amplification, and averaging processes were identical for both responses.

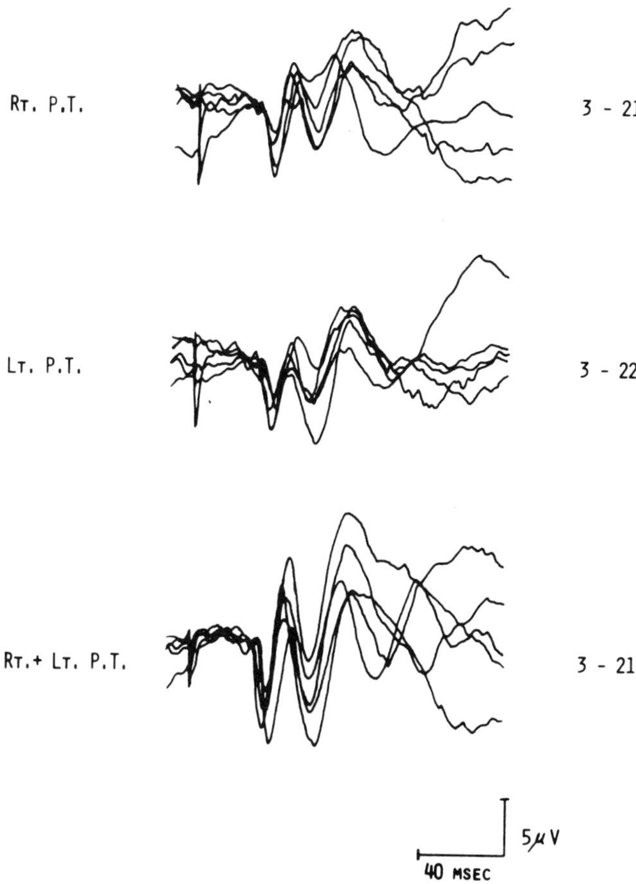

FIGURE 2. SEPs of the posterior tibial nerve. SEPs recorded at the vertex (electrode position 3) and referred to the ear (21, left ear; 22, right ear) in response to stimulation of right posterior tibial (Rt. P.T.), left posterior tibial (Lt. P.T.), and both posterior tibial nerves. Traces were superimposed from five normal subjects. Note the enhancement with bilateral stimulation.

injury depends not only on the severity (completeness or incompleteness) of the lesion, but also on its anatomic level. For this reason it might be expected that SEPs from leg nerves would give a clearer picture of conduction through the injured cervical segment. At first we utilized the common peroneal nerve as the stimulus site because of its easy accessibility in the popliteal fossa, but unilateral common peroneal stimulation failed to produce a well-developed response in approximately 10%–15% of our control subjects. Sural nerve stimulation proved no better and had the added disadvantage of no motor response by which the stimulus intensity could be regulated. We then found that consistent, well-developed responses could be obtained regularly in all control subjects by stimulation of the posterior tibial nerve just posterior to the medial maleolus of the ankle, FIGURE 1. In addition, as reported earlier with common peroneal SEPs,[2] bilateral simultaneous posterior tibial stimulation produced well-defined enhancement of the evoked response, FIGURE 2.

A typical response to bilateral simultaneous posterior tibial stimulation consists of seven peaks with mean latencies of P30, N32, P39, N49, P58, N76, and P117, FIGURE 3. A topographic analysis of the scalp distribution of these components indicates that peak P30 is widely distributed and stable in its polarity and amplitude, and therefore most likely subcortical in origin. The P30 component does not invert in coronal recordings made on either side of the midline, but there is inversion of the P39 component, FIGURE 4. Thus, it seems highly probable that all the components of the response except P30 are cortical in origin.

Diagnosis of the Extent of Cord Injury

TABLE 1 summarizes our initial findings in 92 patients with acute cervical spinal cord injury. Out of 30 patients with a complete cervical cord lesion, none had a

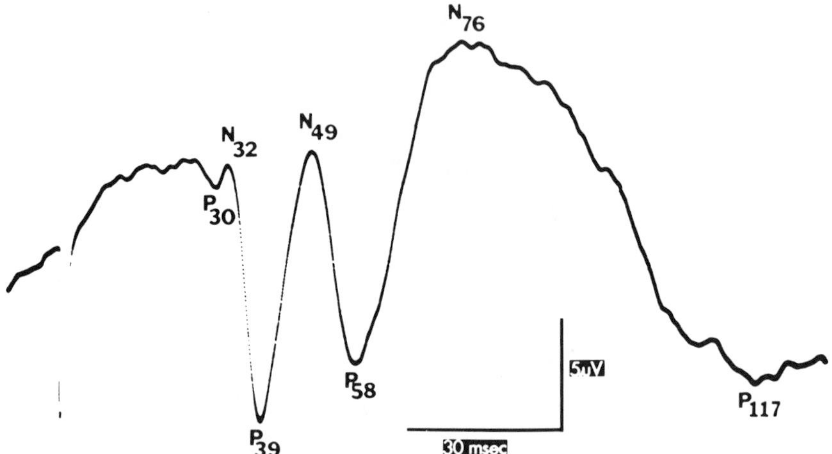

FIGURE 3. A typical bilateral posterior tibial SEP recorded from the vertex referred to one ear. Letters and numbers indicate polarity and mean latency of each peak. Data obtained from 54 normal subjects.

response to bilateral tibial posterior stimulation on the initial examination. Median nerve SEPs were present or absent depending upon the level and severity of the lesion. By contrast, 51 of 54 patients with incomplete cord lesions had posterior tibial SEPs. One incomplete case could not be evaluated because of EMG interference; three incomplete cases failed to show responses. Two may have had a complicating peripheral neuropathy. There is thus a high correlation between the absence or presence of a bilateral posterior tibial SEP and the completeness or incompleteness of the spinal lesion. It is extremely rare for a patient with a complete cord lesion to later regain any useful sensory or motor function. When this happens, it is usually the return of some posterior column sensation. Only one patient in our complete group developed some appreciation of pressure, light touch, and joint position sense in his lower limbs some weeks after the injury. He also showed some return of a definite but very abnormal SEP to bilateral posterior tibial stimulation.

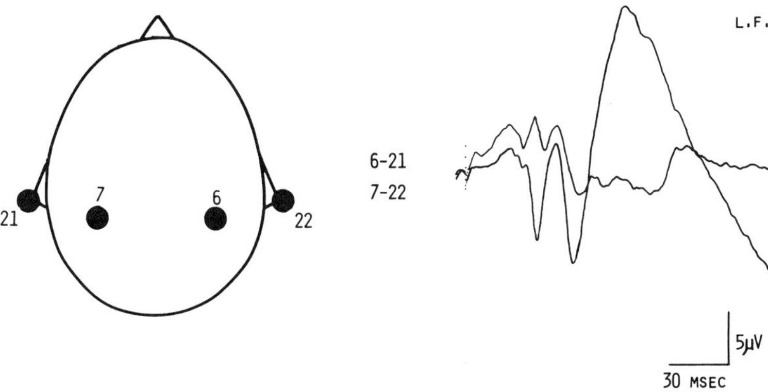

FIGURE 4. SEPs to left posterior tibial nerve stimulation recorded 7 cm on either side of the midline. Note inversion of the P39 component. The larger response is recorded ipsilateral to the side of stimulation.

Depending on the anatomic cross-sectional extent of the lesion, certain clinical cord syndromes can be identified in some incomplete injuries. Patients with a "central cord syndrome" due to hemorrhagic necrosis of the center of the cervical cord exhibit a characteristic clinical picture of marked weakness in their hands and arms, but less or no paresis in the trunk and legs. In this group, variable changes are seen in posterior tibial SEPs, but there is a positive correlation between abnormality of median nerve SEPs and the severity of the neurologic deficit. Patients with "Brown-Sequard syndrome" due to a partial hemisection of the cord show pronounced abnormalities of the posterior tibial SEP and/or the median nerve SEP on the same side as the lesion. The relative involvement of the two depends upon the cervical level and extent of the lesion. "Anterior cord syndrome" patients present a striking picture of complete paralysis and loss of pain and temperature sense below the level of the lesion with varying degrees of preservation of posterior column function (touch, toe position, and vibration sense). Bilateral posterior tibial responses were often present and, when well developed, indicated a favorable prognosis. Median responses were present depending again upon the level of the lesion.

TABLE 1

INITIAL BILATERAL POSTERIOR TIBIAL SOMATOSENSORY EVOKED POTENTIAL FINDINGS IN 92 CASES OF ACUTE CERVICAL SPINAL CORD INJURY*

	Complete Injury	Incomplete Injury
Total patients	37	55
Patients not tested	2	
EMG interference	5	1
Ratio: patients with SEPs/patients studied	0/30	51/54

*Fractions indicate proportion of patients who had SEPs, e.g., 0/30 means no patient among the 30 studied had an evoked response. The table indicates that the study was not done in 2 patients. EMG interference indicates the number of patients for whom the study could not be completed because of myogenic potential interference.

Monitoring Cord Function with SEPs

Since patients with complete spinal injuries do not have posterior tibial SEPs, monitoring is of most value in incomplete injuries.

FIGURE 5 indicates serial posterior tibial SEPs recorded over a 12-day period in a patient with an anterior cord syndrome due to a unilateral rotatory fracture-dislocation at C4–C5. On admission, the patient was quadriplegic with complete paralysis of all muscle function in the hands, trunk, and legs. There was loss of pain and temperature sensation below the C7 spinal level except for some sacral sparing. Touch sensation was only slightly diminished below C7. Position sense was impaired but present to very large angular displacements of the great toes; it was more severely impaired in the left foot than the right. Vibration sense in the feet was present but reduced. Over the first 15 days there was steady improvement in posterior column function in the right foot and some improvement but less in the left foot. This is reflected very well in the changes in configuration of the right and left posterior tibial SEPs. Note that on days 1–4 there was no definitely identifiable left posterior tibial SEP. By the 22nd post-injury day, posterior column function became normal in both feet, and on the 23rd day the patient began to move the toes of her right foot. Over the next two years sensation returned to normal and she regained enough motor power to walk with assistance and drive a standard car.

FIGURE 6 illustrates SEPs recorded over a 13-day interval in a patient with a mainly right-sided Brown-Sequard type lesion. On admission the patient had complete paralysis of the right leg, marked weakness of the left leg, impairment of pain and

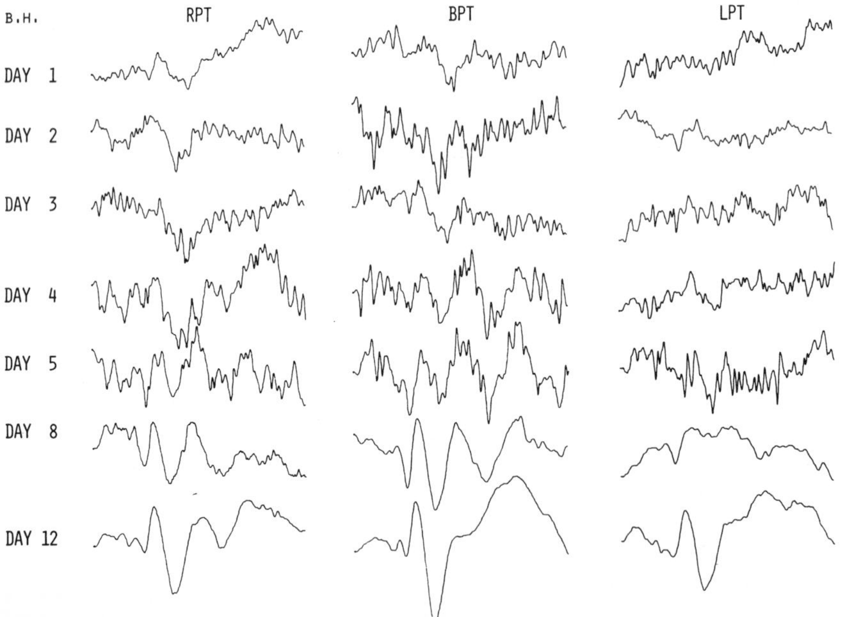

FIGURE 5. Serial SEPs recorded in a patient with an anterior cord syndrome due to a fracture dislocation at C4–C5. See text for details. RPT, right posterior tibial; BPT, bilateral posterior tibial; LPT, left posterior tibial. Day numbers indicate time after injury.

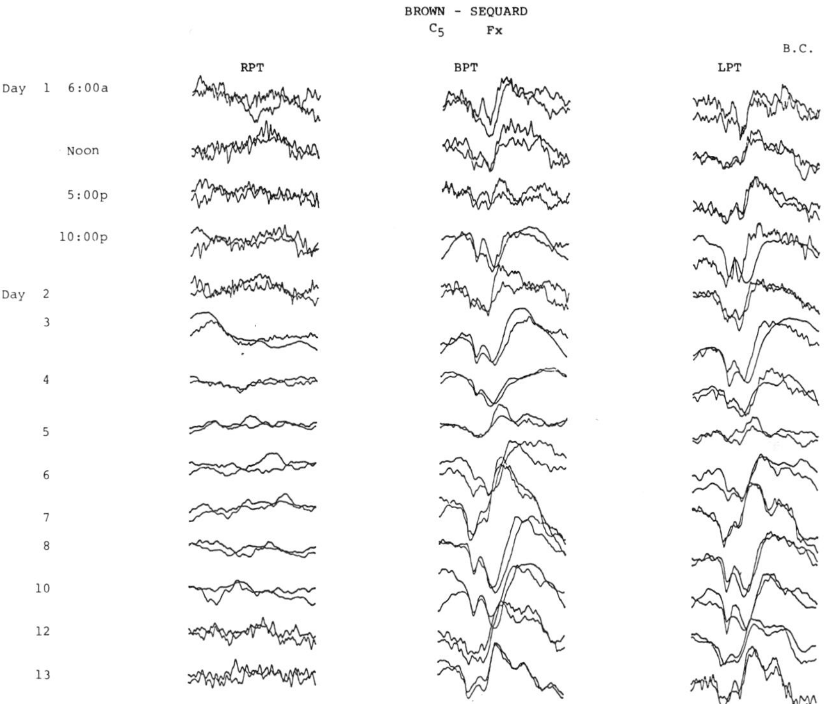

FIGURE 6. Serial SEPs recorded during a 13-day period in a patient with a right-sided Brown-Sequard type lesion due to a C5 vertebra fracture. Abbreviations are the same as in FIGURE 5. See text for details.

temperature sensation, mainly on the left side of the body and loss of posterior column sensation in the right leg. The initial SEP studies from the posterior tibial nerves performed about seven hours after injury showed no definite response from the right posterior tibial nerve (ipsilateral to the main extent of the spinal lesion) but a fairly well-developed response from the left posterior tibial nerve. Over the next 13 days there was rapid improvement in the patient's condition so that some muscular movement returned in the right leg by the third day after injury; however, position and vibration sense in the right foot remained absent until they returned three weeks after injury. Posterior column function remained normal in the left leg and motor power improved to normal within two weeks. In this patient, there was considerable involvement of the right posterior column so that the right posterior tibial response remained absent throughout the 13-day period of observation. There was no indication from the right posterior tibial SEP of the improving motor status of the right limb. In other patients with better posterior column function and with posterior tibial SEPs, we have seen improvement or deterioration of the SEP paralleling improvement or deterioration of muscle function in the lower limb, and yet at the same time there was no corresponding clinical detectable change in posterior column function. As is illustrated in FIGURE 6, there was decreased amplitude of the left posterior tibial SEP on the fifth day after injury, yet there was no definite change in posterior column function in the left foot. We have frequently seen transient abnormalities in the evoked responses usually on the third to sixth day after injury that are presumably related to edema of the spinal cord. They may or may not be associated with clinically detectable changes in the patient's status.

We have been surprised at the steady, progressive improvement that can occur in posterior tibial SEPs over weeks or months after spinal cord injury. As patients with incomplete injuries slowly improve, there is progressive improvement in the amplitude and latency of their evoked responses. Absent or grossly abnormal responses may return to a normal configuration several weeks or months after injury in patients who make a good recovery.

We would agree with other investigators that the posterior tibial SEP is likely to be absent in the presence of severe posterior column loss; however, when there is posterior column function in an incomplete injury (as is usually the case), then, in some patients, one may see changes in posterior tibial SEPs that can be correlated with improvement or deterioration in motor function in the limb. Monitoring of spinal injuries with SEPs is therefore of the most value in incompletely injured patients with some posterior column function spared. In this situation, the SEP can be a sensitive index of spinal cord conduction and it may reveal subtle changes in function not ordinarily detected by the bedside clinical examination.

Prognosis

As confirmed by others,[5] our studies have shown that the SEPs can provide information of considerable prognostic importance in patients with spinal cord injuries. A detailed analysis of the follow-up status of all our patients monitored with SEPs has not been completed at this time but is in progress. Our preliminary findings indicate that, in general, patients with well-preserved median and posterior tibial SEPs early in the post-injury period have the best prognosis for useful recovery of motor and sensory function. Of particular interest are the initial findings in the incomplete patients with central cord and anterior cord syndromes. Patients with central cord syndromes and little or no abnormality in median nerve SEPs usually have an excellent prognosis. The presence of well-developed median and/or posterior tibial SEPs in anterior cord syndrome patients indicates a favorable prognosis for the return of useful motor function.

Summary and Conclusions

Our studies in control subjects have provided normative data regarding unilateral and bilateral posterior tibial SEPs. Well-developed, stable responses could be obtained in all subjects, and bilateral stimulation increased the amplitude of the response. Topographic distribution studies indicate that all the components of the response except P30 are probably cortically generated.

Patients with complete functional transection of the spinal cord in the neck do not have posterior tibial SEPs. Median nerve SEPs will be present or absent depending on the segmental level of the lesion. With rare exceptions, posterior tibial SEPs are present in patients with incomplete cervical cord lesions. As recorded with our stimulus parameters, posterior tibial SEPs depend largely upon the integrity of the posterior column. Monitoring cord function with posterior tibial SEPs is therefore most informative in those incomplete patients who have some posterior column function preserved. In these patients, SEP changes not only reflect improvement in posterior column function, but they may also indicate improvement or deterioration of motor function. The early preservation after injury of median and posterior tibial SEPs argues for a good prognosis. This seems particularly true in patients with either central cord or anterior cord syndromes.

REFERENCES

1. GREENBERG, R. P., D. P. BECKER, J. D. MILLER & D. J. MAYER. 1977. Evaluation of brain function in severe human head trauma with multimodality evoked potentials. Part 2: Localization of brain dysfunction and correlation with posttraumatic neurological conditions. J. Neurosurg. **47:** 163–177.
2. PEROT, P. L., JR. 1973. The clinical use of somatosensory evoked potentials in spinal cord injury. Clin. Neurosurg. **20:** 367–381.
3. PEROT, P. L., JR. 1976. Evoked potentials assessment of patients with neural trauma. *In* Head Injuries. R. McLaurin, Ed. pp. 77–79. Grune & Stratton, Inc., New York, N.Y.
4. PEROT, P. L., JR. 1976. Somatosensory evoked potential in the evaluation of patients with spinal cord injury. *In* Current Controversies in Neurosurgery. T. Morley, Ed. pp. 159–168. W. B. Saunders Company, Philadelphia, Pa.
5. ROWED, D. W., J. A. G. MCLEAN & C. H. TATOR. 1978. Somatosensory evoked potentials in acute spinal cord injury: Prognostic value. Surg. Neurol. **9:** 203–210.
6. KATZ, S., J. G. BLACKBURN, P. L. PEROT, JR. & C. F. LAM. 1978. The effects of low spinal injury on somatosensory evoked potentials from fore-limb stimulation. Electroencephalogr. Clin. Neurophysiol. **44:** 236–238.
7. VLAHOVITCH, B., J. M. FUENTES, Y. SCHOUCAIR, G. ORST, J. G. ASENCIO, B. MAILLE & A. C. VERGER. 1975. Elements de pronostic a la phase precoce des traumatismes vertebro-medullaires graves. Neuro-Chirugie. Paris **21:** 447–468.
8. CADILHAC, J., M. GEORGESCO, J. BENEZECH, H. DUDAY & G. DAPRES. 1977. Potentiel evoque cerebral somesthesique et reflexe d'hoffmann dans les lesions medullaires aigues interet physiopathologique et pronostic. Electroencephalogr. Clin. Neurophysiol. **43:** 160–167.
9. SPIELHOLZ, N. I., M. V. BENJAMIN, G. ENGLER & J. RANSOHOFF. 1979. Somatosensory evoked potentials and clinical outcome in spinal cord injury. *In* Neural Trauma. A. J. Popp *et al.*, Eds. pp. 217–222. Raven Press, New York, N.Y.
10. DORFMAN, L. J., I. PERKASH, T. M. BOSLEY & K. L. CUMMINS. 1980. Use of cerebral evoked potentials to evaluate spinal somatosensory function in patients with traumatic and surgical myelopathies. J. Neurosurg. **52:** 654–660.
11. CRACCO, J. B., R. Q. CRACCO & L. J. GRAZIANI. 1975. The spinal evoked response in infants and children. Neurology **25:** 31–36.
12. CRACCO, R. Q. 1973. Spinal evoked response: Peripheral nerve stimulation in man. Electroencephalogr. Clin. Neurophysiol. **35:** 379–386.
13. CRACCO, J. B., R. Q. CRACCO & R. STOLOVE. 1979. Spinal evoked potential in man: A maturational study. Electroencephalogr. Clin. Neurophysiol. **46:** 58–64.
14. CRACCO, J. B., V. V. BOSCH & R. Q. CRACCO. 1980. Cerebral and spinal evoked potentials in children with CNS degenerative disease. Electroencephalogr. Clin. Neurophysiol. **49:** 437–445.
15. CRACCO, R. Q. & J. B. CRACCO. 1976. Somatosensory evoked potentials in man: Far-field potentials. Electroencephalogr. Clin. Neurophysiol. **41:** 460–466.
16. ANZISKA, B., R. Q. CRACCO, A. W. COOK & B. W. FELD. 1978. Somatosensory far-field potentials: Studies in normal subjects and patients with multiple sclerosis. Electroencephalogr. Clin. Neurophysiol. **45:** 602–610.
17. ANZISKA, B. & R. Q. CRACCO. 1980. Short latency somatosensory evoked potentials: Studies in patients with focal neurological disease. Electroencephalogr. Clin. Neurophysiol. **41:** 227–239.
18. STOLOVE, R., J. B. CRACCO, G. VAS & R. Q. CRACCO. 1980. Scalp-recorded subcortical evoked potentials to peroneal nerve stimulation in man. Electroencephalogr. Clin. Neurophysiol. **49:** 26P.

19. SPIELHOLZ, N. I., M. V. BENJAMIN, G. L. ENGLER & J. RANSOHOFF. 1979. Somatosensory evoked potentials during decompression and stabilization of the spine. Spine **4:** 500–505.
20. OWEN, M. P., R. H. BROWN, R. F. SPETZLER, C. L. NASH, JR., J. F. BRODKEY & F. E. NULSEN. 1979. Excision of intramedullary arteriovenous malformation using intraoperative spinal cord monitoring. Surg. Neurol. **12:** 271–276.
21. MCCALLUM, J. E. & M. H. BENNETT. 1975. Electrophysiologic monitoring of spinal cord function during intraspinal surgery. Surg. Forum **26:** 469–471.
22. NASH, C. L., JR., R. A. LORIG, L. A. SCHATZINGER & R. H. BROWN. 1977. Spinal cord monitoring during operative treatment of the spine. Clin. Orthoped. Related Res. **126:** 100–105.
23. ENGLER, G. L., SPIELHOLZ, N. I., W. N. BERNHARD, F. DANZIGER, H. MERKIN & T. WOLFF. 1978. Somatosensory evoked potentials during Harrington instrumentation for scoliosis. J. Bone Joint Surg. **60-A:** 528–532.
24. BROWN, R. H. & C. L. NASH, JR. 1979. Current status of spinal cord monitoring. Spine **4:** 466–470.
25. SPETZLER, R. F., W. R. SELMAN, C. L. NASH, JR. & R. H. BROWN. 1979. Transoral microsurgical odontoid resection and spinal cord monitoring. Spine **4:** 506–510.

DISCUSSION OF THE PAPER

S. J. JONES: With respect to the possibility of monitoring the motor side of the cord, direct stimulation of motor areas of the cortex through the skull, which elicited motor responses from the extremities, has recently been reported.

P. L. PEROT, JR.: Obviously if one could focus enough current density over the motor cortex and evoke EMG activity in the lower extremities, this would be an ideal approach.

CLINICAL APPLICATIONS OF SHORT-LATENCY SOMATOSENSORY EVOKED POTENTIALS

S. J. Jones

Medical Research Council
National Hospital for Nervous Diseases
Queen Square, London WC1N 3BG
England

Before the development of techniques for recording peripheral, spinal, and subcortical components of the human somatosensory evoked potential (SEP), the number of distinguishable ways in which a recording might deviate from the normal pattern were relatively few, and the prospects for clinical application were correspondingly somewhat limited. Now that it is possible, broadly speaking, to trace the course of a sensory volley from the periphery through to the cortex via the spinal cord and brainstem, a new tool is available which can be used in diagnosis, pathognosis (research into the nature of nerve conduction defects), prognosis, and also as a passive monitoring technique. This paper is intended to illustrate the use of SEPs in each of these four areas.

METHODS

The recordings were obtained from inpatients and outpatients at the National Hospital for Nervous Diseases, London. The stimulus (50 μsec capacitor discharge or 200 μsec constant current square wave) was delivered at 2 impulses/sec to the median nerve at the wrist, with an intensity sufficient to produce a moderate motor twitch. Eight "active" recording electrodes (silver/silver chloride discs) were attached to the skin over the midpoint of either clavicle, the spinal vertebrae at T3, C7, and C2, the mastoid processes and the hand area of the sensory cortex contralateral to the stimulated arm (2 cm posterior to the vertex and 7 cm towards the tragus of the ear). The principal "reference" electrode was at Fz (international 10–20 system), but the mastoid sites were referred to the hand area in order to enhance the subcortical N14 component.[1] The amplifier high-frequency response extended to 5 kHz (−3dB) and the time constant was 1 sec. Usually 400 responses were averaged with a window of 32 msec (8 sample points/msec) starting 3 msec after the stimulus, but a longer window of 128 msec (2 sample points/msec) was sometimes required. Two or more runs could subsequently be combined off-line into "grand average" recordings.

An epidural recording technique was developed for patients undergoing surgical treatment (Harrington instrumentation) for spinal scoliosis at the Royal National Orthopaedic Hospital, London. A 200 μsec square-wave stimulus was delivered at 10/sec to the posterior tibial nerve in the popliteal fossa, with an intensity sufficient to induce flexion of the toes. The recording electrode (an autoclavable lead of 0.9 mm diameter with 3 mm of the insulation at the tip replaced by a stainless steel sleeve of similar diameter) was introduced into the epidural space by means of an intravenous cannula inserted between the exposed vertebrae at upper thoracic level, and threaded upwards by 3–5 cm. The reference electrode was a stainless steel needle inserted into the musculature of the back. Usually 200 responses were averaged (Medelec MS6, high-frequency response approximately 1.6 kHz) with a window of 10 msec (10

0077–8923/82/0388–0369 $1.75/0 © 1982, NYAS

sample points/msec) starting 8–14 msec after the stimulus. In 8 cases, recordings were also obtained from the lower thoracic and lumbar regions.

RESULTS

The Definition of Normal Limits

Normative data was obtained from 31 healthy volunteers aged 19 to 33 years with a mean age of 25 years. The upper limit of normal latency was defined as the mean plus 2.5 standard deviations (TABLE 1), and this was derived for interpeak latencies and left–right differences also. Amplitude values were not normally distributed, being more tightly clustered below the mean than above, and it was considered of greater validity to adopt the extreme values encountered (left–right differences in percentage terms) as the limits of normality.

The amplitude of the clavicular N9 component was of interest in the context of peripheral neuropathies and brachial plexus lesions, and in these and other cases the latency was useful as a marker of conduction time in the arm. N9 usually measures at least 3 μV with a mean of 8–9 μV in the normal group, but in one subject the amplitude was only 1–1.2 μV. There was a great majority of females in the normal group (25 compared with 6 males), and this would have influenced mean latency values which are known to be related to body height.[1] No such correlation has been shown to exist for interpeak separations N9–N13, N13–N20, and N9–N20, however, and so, in assessing patient recordings, greater importance was attached to these than to absolute latency values.

Since the cervical N11 component is sometimes obscured, even in healthy subjects, by the larger N13, the former was only regarded as abnormal if delayed or absent when N13 was also delayed or greatly attenuated. N14, likewise, is often obscured by N13, and its absence was only regarded as significant if N13 was greatly attenuated or N20 delayed. Norms were also derived for wrist–elbow conduction time, from the shift in N20 latency when the stimulus was delivered to the median nerve more proximally.

Diagnosis

Multiple Sclerosis in Cases of "Isolated" Optic Neuritis

A great deal of recent work in the SEP field has been devoted to multiple sclerosis (MS),[2-4] with the principal object of establishing the diagnosis with greater certainty by demonstrating the existence of "clinically silent" lesions. However, since MS presents with an immense variety of early symptoms, it is not sufficient to note the incidence of abnormal SEPs in "definite," "probable," and "possible" groups, without also taking into account the location of clinically established lesions. Clearly an abnormal SEP is of much greater diagnostic significance in patients presenting with an isolated episode of optic neuritis (ON) than in those for whom the clinical symptoms are suggestive of a spinal cord lesion. This section, therefore, concerns an attempt to demonstrate in what proportion of ON cases truly "subclinical" SEP abnormalities may be found.

A group of 43 patients (10 males, 33 females; mean age, 39 years; range, 21–64) was carefully selected for an initial presentation of unilateral ON (32 patients) or bilateral ON (11), the diagnosis supported by abnormal pattern visual evoked potentials in 37 cases. The mean time elapsed since the attack was 5 years (range 1

TABLE 1

SHORT-LATENCY SOMATOSENSORY EVOKED POTENTIAL (SEP) COMPONENTS FOLLOWING MEDIAN NERVE STIMULATION AT THE WRIST, LIMITS OF NORMALITY. MAXIMUM 31 SUBJECTS; 6 MALES, 25 FEMALES; MEAN AGE, 25 YEARS

	Amplitude (μV) (Extremes of Normal Group)		Latency (msec) (Mean + 2.5 Standard Deviations)			
	Lower Limit	L-R Upper Limit (% of larger side)	Upper Limit*	Left-Right Upper Limit	After N9 Upper Limit	After N13 Upper Limit
N9 (clavicle–F_z)	1.0†	40	10.8	0.7	—	—
N11 (C7–F_z)	Present	—	12.8	0.8	2.4	—
N13 (C7 or C2–F_z)	1.2	55	14.5	0.8	4.1	—
N14 (mastoid–hand area)	Present	—	15.8	0.9	5.6	2.3
N20 (hand area–F_z)	0.5	75	20.8	1.0	10.9	6.7
Wrist–elbow shift in N20 latency			4.6			

*Latency limits are somewhat low on account of the preponderance of female subjects.

†The N9 amplitude limit may also be low on account of one anomalous subject. The remainder had amplitudes of at least 3 μV. N11 and N14 abnormalities were only recognized when N13 and/or N20 were also abnormal.

TABLE 2

SHORT-LATENCY SEPs IN ISOLATED OPTIC NEURITIS (ON); 43 SUBJECTS, 10 MALES, 33 FEMALES; MEAN AGE, 39 YEARS

SEPs	Unilateral ON	Bilateral ON
Normal	30	10
Borderline	1	0
Abnormal	1	1
Total	32	11

month to 19 years) and at no time before or since had there been any symptoms or signs that might relate to a lesion of the central nervous system outside the optic pathways. The mean time elapsed since the last clinical examination was 6 months (range, 0–2 years).

SEP recordings from either arm were within normal limits in all but 3 cases (TABLE 2). In one of these (female, age 43, unilateral ON 12 years previously) the left–right latency difference of N20 was at the borderline value of 1.0 msec. In another (female, age 63, unilateral ON 11 years previously) the left–right latency difference of N20 was 1.3 msec, in keeping with a slightly prolonged N13–N20 separation of 7.2 msec (FIGURE 1). In the third case (female, age 56, bilateral ON 6 years previously) a left–right N20 latency difference of 0.8 msec was doubtfully significant, particularly since N9 occurred 0.4 msec later on this side, but N13 and N14 were also slightly delayed and N13 measured only 0.9 μV at C2.

In view of this very low incidence of mild SEP abnormalities, it was concluded that

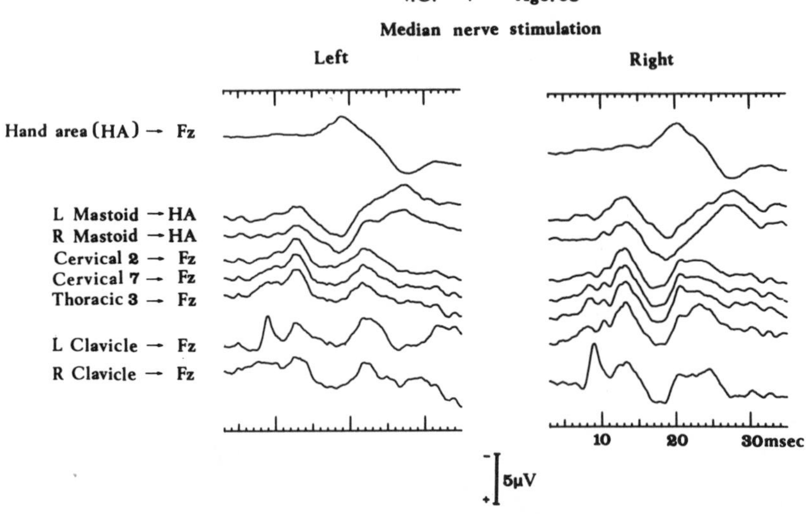

FIGURE 1. Somatosensory evoked potentials (SEPs) in a patient presenting with unilateral optic neuritis (ON) 11 years previously, with no other neurological symptoms or signs. The N20 component is relatively delayed by 1.2 msec for the right arm, and the N13–N20 separation is slightly prolonged on this side. Abnormalities were present in only 3 out of 43 ON cases, and were all of this magnitude or even less marked.

routine screening of all patients presenting with apparently isolated optic neuritis is likely to add little to a thorough neurological examination. Certain other studies (notably Reference 3) have interpreted SEP findings in the light of the locus of clinically established lesions, but more detailed information is still required on the incidence of "subclinical" lesions in the early stages of MS.

Multiple Sclerosis in Other Undiagnosed Cases

Although perhaps not disclosing the existence of any "clinically silent" lesions, the SEP abnormalities encountered in definite MS cases are to some degree characteristic of the disease, and may therefore be of use in diagnosing doubtful cases. In order to define the characteristic features, SEPs were recorded from both arms in 75 clinically definite cases (34 males, 41 females; mean age, 34 years; range, 18–35). The mean duration of the disease was 8 years (range, 0.5–33 years) and 70 patients (93%) had abnormal pattern visual evoked potentials.

TABLE 3

PATTERNS OF SEP ABNORMALITY IN CLINICALLY DEFINITE MULTIPLE SCLEROSIS. TOTAL OF 150 ARMS (75 PATIENTS; 34 MALES, 41 FEMALES; MEAN AGE, 34 YEARS)

Category		N9	N11	N13	N14	N20
1	Normal	48	48	48	48	48
	Att/Abs	—	—	—	—	—
	Delayed	—	—	—	—	—
2	Normal	16	16	—	14	16
	Att/Abs	—	—	14	1	—
	Delayed	—	—	2	1	—
3	Normal	18	18	18	—	—
	Att/Abs	—	—	—	2	0
	Delayed	—	—	—	16	18
4	Normal	57	57	—	6	—
	Att/Abs	—	—	54	15	7
	Delayed	—	—	17	36	53
5	Normal	11	—	—	—	—
	Att/Abs	—	5	11	0	0
	Delayed	—	6	2	11	11

SEPs were judged to be abnormal in 58 cases (77%) and 102 arms (68%). The total of 150 arms was partitioned into 5 categories according to the nature of the SEP abnormality, if any (TABLE 3, FIGURE 2). Category 1 contained 48 arms with SEPs within normal limits. In category 2 (16 arms) N9, N11, N20, and usually N14 were all preserved while N13 was of abnormally low amplitude, absent, or delayed. In category 3 (18 arms) N9, N11, and N13 were preserved (although N13 was frequently low in amplitude) while N20 was always, and N14 usually, delayed. In category 4 (57 arms) N13 and N20 were abnormal, usually with delayed N14 also. Category 5 (not illustrated) contained 11 arms in which N11 was abnormal in addition to all later components. The overall abnormality rates were: N9, 0%; N11, 7%; N13, 56%; N14, 51%; N20, 57%. Where N13 was abnormal it was usually attenuated or absent rather than delayed, while the converse was true of N14 and

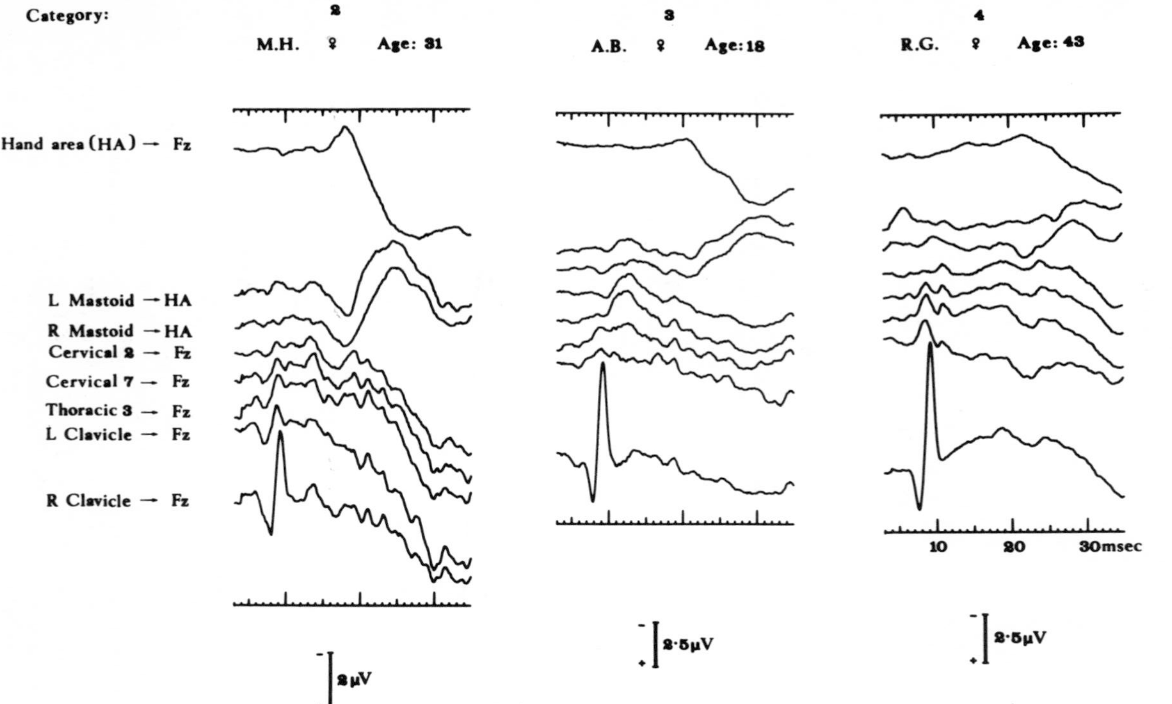

FIGURE 2. The three most common patterns of SEP abnormality in clinically definite multiple sclerosis (MS): In category 2, the N13 component is of abnormally low amplitude with all other components preserved. In category 3, N9, N11, and N13 are within normal limits, but N20 is slightly delayed when measured from N9 or N13 (in this record, N14 is perhaps present as a slightly delayed, low amplitude peak in the right mastoid trace). In category 4, N9 and N11 are preserved, N13 and N14 are of abnormally low amplitude or absent, and N20 is delayed. Categories 1 (normal SEP) and 5 (abnormal N11–N20) are not illustrated.

N20. In 16 arms of categories 4 and 5, N13 was abnormal in terms of both amplitude and latency, reflecting the tendency for this component to be fragmented with some constituents delayed. Where N14 was identifiable in the presence of a delayed N20, it was often delayed to a very similar degree (FIGURE 3).

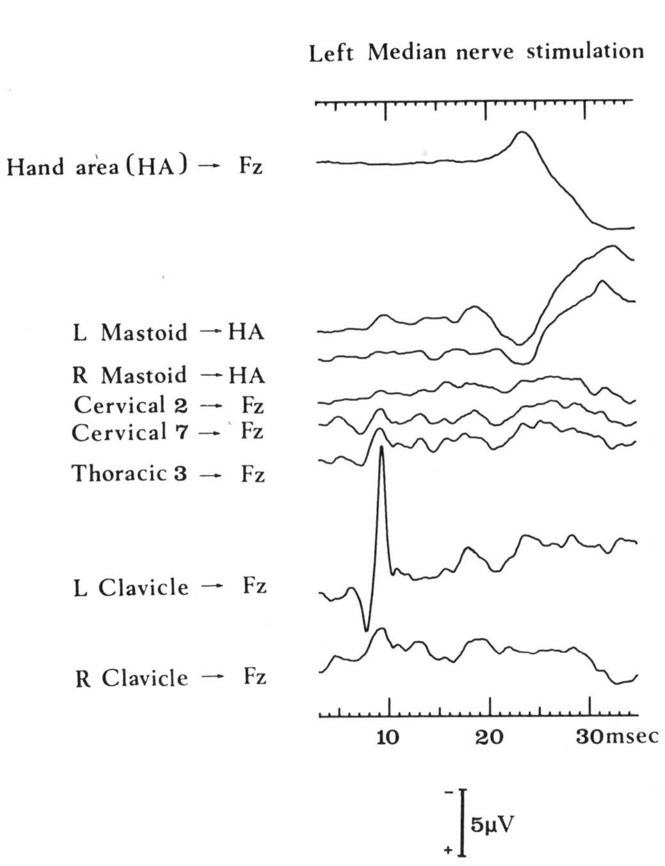

FIGURE 3. Abnormal SEPs (category 4 or 5) in a patient with clinically definite MS, showing a similar delay of N14 (recorded at the left mastoid process) and N20.

There were three main features, therefore, which were very often seen in definite MS and which, if encountered in doubtful cases, might suggest this diagnosis in preference to other possibilities:

1. Preservation of N9 and probably N11 also.
2. Attenuation and fragmentation of N13.
3. Delay of N14 and N20.

It was also apparent that features 2 and 3 might occur independently, particularly in less severely affected cases.

FIGURES 4–6 illustrate contrasting patterns seen in other conditions:

1. Friedreich's ataxia (FIGURE 4). Delayed cortical potentials (top channel) and low-amplitude cervical waveform (channels 3 & 4) might be thought reminiscent of MS, but N13 is not fragmented and N9 is of abnormally low amplitude. In more severe cases N9 and N13 might both be absent. Many cases also exhibit a prolongation of N20 onset-to-peak.

2. Charcot-Marie-Tooth disease, type 1 (FIGURE 5). N9 is not identifiable but N13 and N20 are both fairly well preserved, albeit with prolonged latency. A normal N13–N20 separation, in contrast to the example of Friedreich's ataxia, suggests the absence of central pathway involvement. In more severe cases with N13 absent, the pattern can be distinguished from that of Friedreich's ataxia by a prolonged wrist–elbow conduction time (latency shift of N20).

3. Cortical infarction due to occlusion of right internal carotid artery (FIGURE 6). N20 and subsequent waves are attenuated and only slightly delayed on the affected side, whereas N9 to N14 are normal and symmetrical. This pattern might also conceivably be encountered in MS, due to a lesion of the diencephalon or thalamocortical radiation.

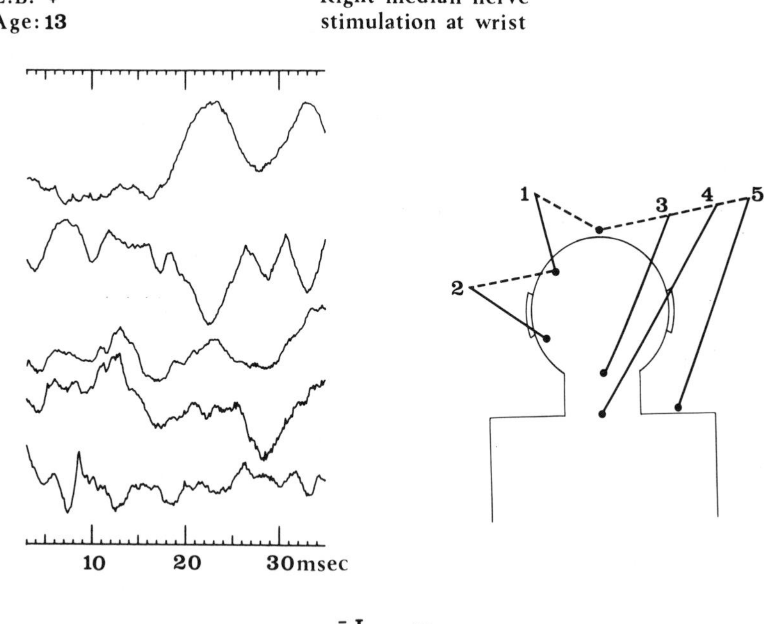

FIGURE 4. Abnormal SEP pattern in Friedreich's ataxia. Clavicular and cervical components are of low amplitude but preserved waveform and latency, while the scalp N20 is slightly delayed.

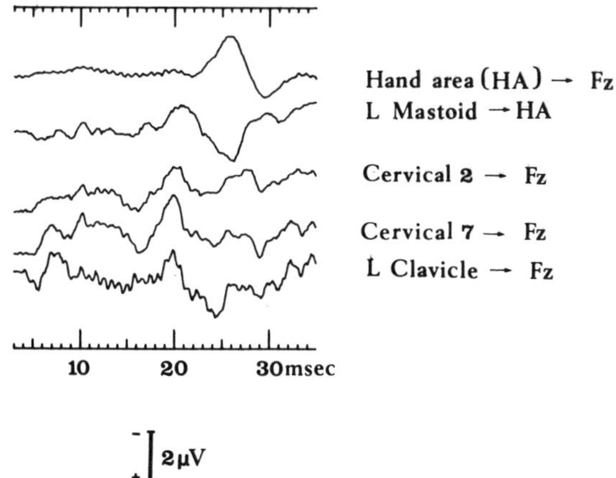

FIGURE 5. Abnormal SEP pattern in Charcot-Marie-Tooth disease, type 1. N13 and N20 are the only short-latency components identifiable, both delayed but with a normal interpeak separation.

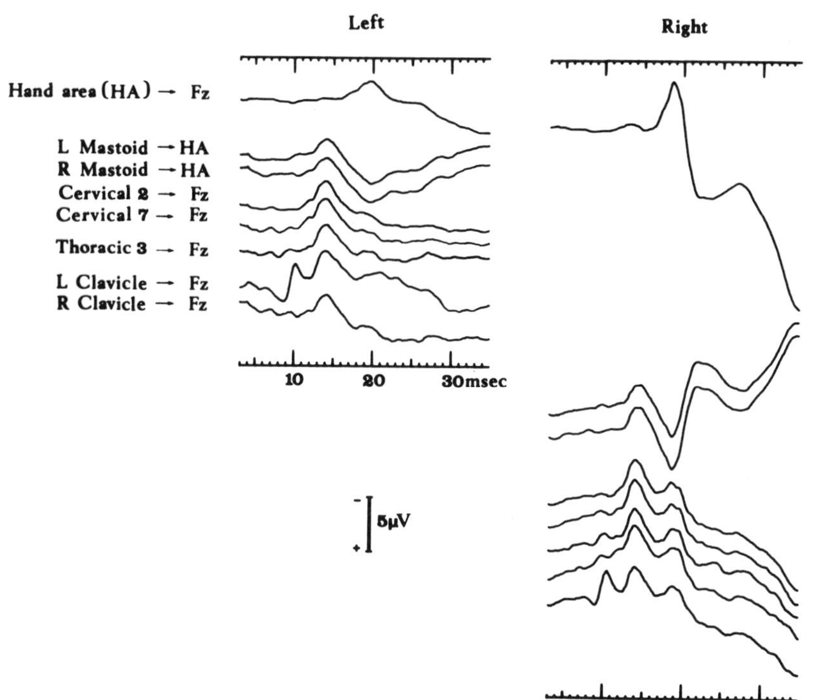

FIGURE 6. Cortical infarction (occlusion of right internal carotid artery) causing relative attenuation of N20 and succeeding potentials following stimulation of the left median nerve.

Pathognosis

Multiple Sclerosis

Of the 150 arms of clinically definite MS cases summarized in TABLE 3, 113 had a past or present history of sensory disturbance and 74% of the latter had abnormal SEPs. However, a fairly high incidence (49%) of abnormal SEPs was also discovered for the 37 arms with no sensory symptoms or signs. The breakdown of SEP abnormalities into several patterns suggested the possibility of correlating these with particular clinical symptoms and signs, and the results of this analysis are presented in TABLE 4.

All symptoms and signs were less frequent in category 1 (normal SEPs), although there was a fairly high incidence of cerebellar ataxia. The highest incidence of Lhermitte's sign and weakness occurred in category 2 (N13 abnormalities), while increased tone and brisk reflexes were most associated with category 3 (N14 and N20 abnormal). Categories 4 and 5 had the highest incidence of symptomatic numbness, light touch impairment on sensory testing, loss of joint-position sense, and cerebellar ataxia. Paraesthesiae, however, were more often associated with categories 2 and 3. None of these correlations reached a high level of statistical significance.

The associations of mild sensory symptoms (paraesthesiae and Lhermitte's sign) with less severe SEP abnormalities (categories 2 and 3) and of more pronounced symptoms with category 4 and 5 SEPs were not surprising, and it is difficult to conceive of topographically distinct lesions which might account for the more particular correlations. The distinction between categories 4 and 5 was sometimes doubtful on electrophysiological grounds, and was not made clearer by any particular pattern of clinical differences.

Friedreich's Ataxia

A number of laboratories have now published detailed SEP findings in this condition,[5–7] directed not so much at the problem of diagnosis as at achieving greater

TABLE 4

PERCENTAGE TABLE SHOWING THE CORRELATION OF CLINICAL SYMPTOMS AND SIGNS WITH THE SEP CATEGORY IN CLINICALLY DEFINITE MS CASES (150 ARMS AS IN TABLE 3)

	Category 1	2	3	4	5
Components abnormal:	—	N13	N14, N20	N13–N20	N11–N20
Total arms:	48	16	18	57	11
Numbness	29%	31%	17%	56%	18%
Paraesthesiae	31	63	67	32	9
Lhermitte's sign	19	50	17	32	27
JPS impairment	2	13	6	18	27
LT impairment	17	25	17	51	55
Pain/Temp impairment	19	6	11	16	18
Weakness	23	63	39	53	18
Cerebellar ataxia	63	63	83	93	91
Increased tone, reflexes	38	38	61	51	36
Reduced tone, reflexes	2	0	11	11	0

TABLE 5

MAJOR SHORT-LATENCY SEP COMPONENTS IN CHARCOT-MARIE-TOOTH DISEASE. 14 PATIENTS (7 MALES, 7 FEMALES; MEAN AGE, 33 YEARS), LATENCIES COMPARED WITH 10 CONTROLS (3 MALES, 7 FEMALES, MEAN AGE, 26 YEARS), USING THE MANN-WHITNEY U TEST

	Mean Latencies (msec) Components					
	N9	N13	N20	N9–N13	N13–N20	Wrist-Elbow
Controls	9.2	12.5	18.0	3.3	5.5	3.3
5 patients with N9 and N13 present	13.7	19.9	—	6.2*	—	—
8 patients with N13 and N20 present	—	20.7	26.5	—	5.8†	—
14 patients with N20 present	—	—	30.7	—	—	7.9*

*$p < 0.002$.
†Not significant.

understanding of the disease process itself. The case illustrated in FIGURE 4 is fairly typical of the early stages, the very small amplitude but normal latency of N9 suggesting an axonal form of peripheral neuropathy without slowing of conduction in the surviving fibers. When peripheral potentials, including sensory nerve action potentials (SNAPs), are absent, it is still possible to demonstrate a normal sensory conduction velocity between the wrist and the elbow by stimulating at either locus and noting the shift in the latency of N20 (a more robust potential, presumably on account of trans-synaptic amplification).[7] N20 itself, however, is usually markedly delayed with a broadening of onset-to-peak[7] or a break-up into multiple peaks,[5] which cannot be due to the peripheral neuropathy, and must therefore be attributable to temporal dispersion of the central volley or reverberation of a cortico-thalamo-cortical loop.[5] It is, perhaps, most likely that the pathological process causing axonal degeneration in the peripheral nerve may give rise to dispersion of the central volley by impairment of synaptic mechanisms at the junction between dorsal column axons and second-order neurons in the dorsal column nuclei, rather than by a slowing of axonal conduction velocity, which has been shown not to occur peripherally.

Charcot-Marie-Tooth Disease (CMTD)

In the more common type 1 of CMTD, a major diagnostic criterion is an abnormally slow distal motor conduction velocity, and where SNAPs are present these are also delayed. Very frequently, however, SNAPs cannot be recorded from the fingers or the median nerve, and so wrist–elbow sensory conduction velocity must be derived from the shift in cortical SEP latency. The N13–N20 separation can be used as a measure of conduction time in the central nervous system (better N20 latency from the onset of N11), but, where neither N11 nor N13 can be recorded, central conduction time (CCT) must be estimated by subtracting an estimate of peripheral conduction time (PCT) from the latency of N20. In healthy subjects the PCT from the wrist to the spinal cord is usually 10–11 msec (or approximately three times the wrist–elbow conduction time) and CCT therefore 8–10 msec. In patients with CMTD, three times the wrist–elbow time should be a *maximal* estimate of PCT, since a lesser degree of slowing occurs in segments proximal to the elbow.[5] Assuming, then, a normal CCT of around 9 msec from cord entry to the peak of N20, we should be able to account for the total latency of N20 by the estimate of PCT plus 9 msec. If N20 occurs later than this, there must be additional slowing proximal to the elbow, probably within the central nervous system.

Fourteen cases of CMTD (7 males, 7 females) were recorded, aged 15–59 years with a mean age of 33 years. Motor conduction velocity in the ulnar nerve ranged from normal (2 cases) to 10 m/sec. With median nerve stimulation at the wrist, N9 was identifiable in 6 cases, N11 in 6, N13 in 8, N14 in 4, and N20 in all 14, although in 2 cases the amplitude of N20 was effectively zero. TABLE 5 gives mean latencies of the 3 major components (N9, N13, N20) plus mean interpeak latencies and the mean wrist–elbow conduction time, compared with 10 normal controls (3 males, 7 females; mean age, 26 years).

Although each component was markedly delayed in the group as a whole, with a significant prolongation of the N9–N13 separation and wrist–elbow conduction time, the N13–N20 interval was not significantly increased. When individual cases were compared with the normal range, however, the N13–N20 separation was slightly enhanced for 2 arms in 2 patients (illustrated by FIGURE 7, although this record was obtained with stimulation at the elbow). There were also 3 arms (3 patients) for which the estimate of PCT plus 9 msec did not account for the total latency of N20. In all there was some evidence of central involvement in 5 out of 14 cases.

These findings are of some interest with regard to the pathognosis of CMTD, and have not been reported in earlier accounts.[5,6] It may also be of importance with regard to the management of individual cases to establish whether particular symptoms suggestive of a central lesion might be attributed to the established diagnosis of CMTD, or whether an additional cause should be sought.

Prognosis

Brachial Plexus Traction Lesions

Several groups have used the SEP, with considerable success, as an indicator of prognosis following spinal or cerebral trauma.[8–10] Trauma to the brachial plexus is also a frequent cause of major disability, and SEPs may play a part in establishing the prognosis by revealing the precise locus of the lesion. With a lesion in continuity on the distal side of the dorsal root ganglia, spontaneous axonal regeneration can occur, while a rupture in this area may be repaired by nerve graft. On the other hand, an avulsed root (torn from the intervertebral foramen and ruptured on the proximal side of its ganglion) has no prospect of recovery.

It has long been recognized[11] that the preservation of SNAPs in an anesthetic part of the arm is indicative of an avulsed root, with continuity preserved between the sensory axons and their cell bodies in the dorsal root ganglia. SEPs, however, can be of more quantitative use, since the attenuation of N13 compared with the intact arm is indicative of the total proportion of damaged fibers, while the attenuation of N9 gives a measure of the proportion damaged on the distal side of their ganglia.[12] Responses to stimulation of the median nerve at the wrist appear to rely heavily on the C6 and C7 roots, while the ulnar nerve can be stimulated to evoke similar, although smaller, SEPs via the C8 and T1 roots. It is therefore possible to estimate separately for the C6/C7 and C8/T1 roots whether the lesion is entirely on the proximal side of the dorsal root ganglia, entirely distal, or a mixture of proximal and distal. For example (FIGURE 8), a 50% attenuation of the median nerve N13 compared with the intact arm, but no reduction of N9, would suggest a proximal avulsion of either the C6 or the C7 root, with the other remaining intact. If there were, however, a 50% reduction of N9 with

N13 attenuated by 80%, this would suggest a distal lesion of one root in addition to a proximal avulsion of the other.

The results of SEP and SNAP studies have now been collected for 42 cases, 16 of whom were surgically explored.[13] The estimated locus of the lesion at C6/C7 and

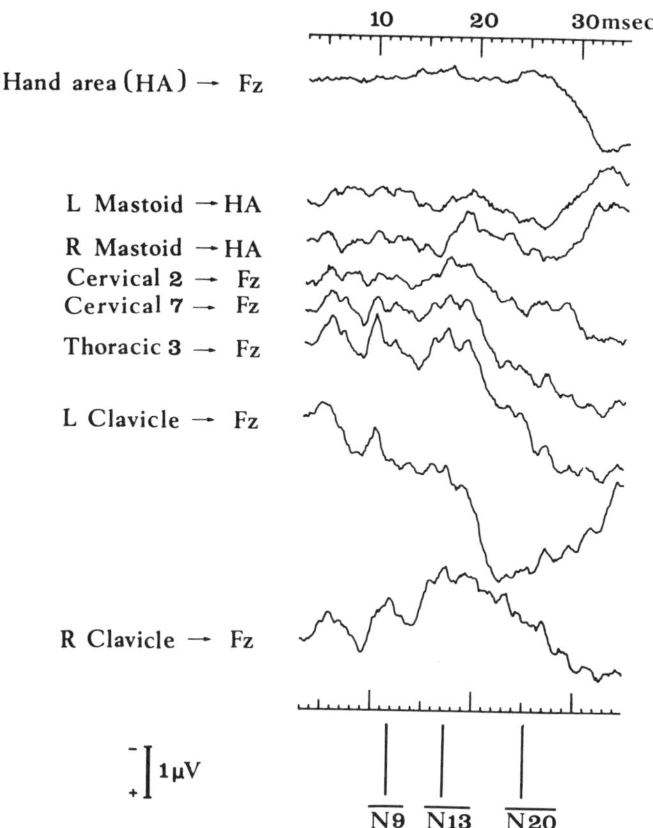

FIGURE 7. Delayed SEPs following median nerve stimulation at the elbow in Charcot-Marie-Tooth disease, type 1. In contrast to FIGURE 5, the N13–N20 separation is prolonged, suggesting a degree of slowing in the central somatosensory pathway.

C8/T1 was found to be accurate in 10/16 cases and substantially correct in another 3. The principal weaknesses of the technique, however, were found to lie in its inability to detect a root avulsion in the presence of distal damage to the same fibers, and in the inaccessibility of the C5 root, which has no representation in the median or ulnar nerve.

In the majority of cases undergoing surgery it was possible to determine whether

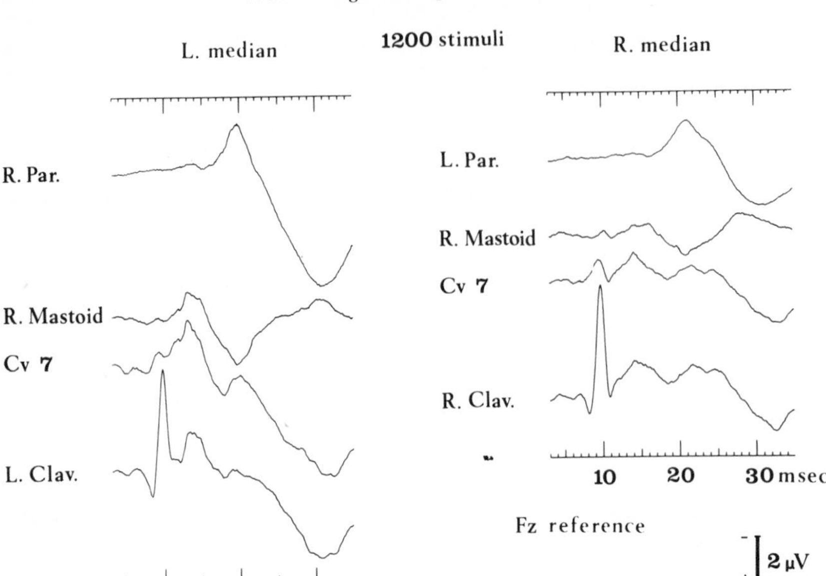

FIGURE 8. Attenuation particularly of cervical SEP components following a mild traction injury of the brachial plexus, with full preservation of N9, indicating a lesion proximal to the dorsal root ganglia. The mastoid electrode (channel 2) is referred to the parietal scalp (hand area).

or not a given root was in continuity with the central nervous system by applying a conventional stimulus to the exposed nerve trunk and recording SEPs over the contralateral hand area (FIGURE 9). The responses were often of very small amplitude (less than 1 μV), but the waveform and latency (10–14 msec for the initial negativity) were consistent with those of a neurogenic cortical response.

Patient Monitoring

Spinal Cord Stimulation in the Treatment of MS

In any disease causing abnormalities of the SEP, there is clearly the possibility of performing serial recordings to assess the efficacy of treatment—for example, decompression of the spinal cord for cervical myelopathy and hemodialysis or transplantation for renal failure. In MS, the majority of regimes that have been tried (dietary control, immunosuppression, and so on) aim, at best, to arrest the course of the disease by preventing further exacerbations. Under these circumstances, serial SEP recordings might not be expected to show any improvement, but might detect evidence of new lesions. Illis and colleagues,[14] however, reported that cervical responses to median nerve stimulation might show marked improvement of amplitude and waveform during and for a short while after continuous epidural stimulation of the spinal cord (SCS), in association with improvement of lower limb symptoms and sphincter control.

In an attempt to reproduce these findings SEPs were recorded before and after approximately 6 days of SCS in 16 clinically definite MS cases. The patients were selected for a relatively stable condition of the disease, the majority with marked locomotor impairment and loss of sphincter control. The epidural electrodes were implanted at mid-thoracic level and delivered brief impulses at 33/sec, the intensity adjusted by the patient so as to produce a warm or tingling sensation in the lower limbs. Although the majority of patients noticed a mild but consistent improvement of bladder symptoms, and a few found gratifying pain relief, other clinical benefits were minimal. Four patients showed a change towards a more normal SEP waveform, although in 3 of these the increase in the amplitude of N13 may have been within normal test–retest variability. There was one patient, however, in whom an initially bifid N20 became fused, with an increase in the amplitude and reduction in the latency of the main peak (FIGURE 10). No convincing explanation for the mechanism of this effect can yet be offered.

Monitoring of Spinal Cord Function During Surgery

The correction of spinal scoliosis by applied traction using a Harrington rod has become a popular and widely practiced procedure. There is, however, a significant risk of postoperative paraplegia, and it was in order to detect evidence of insult to the

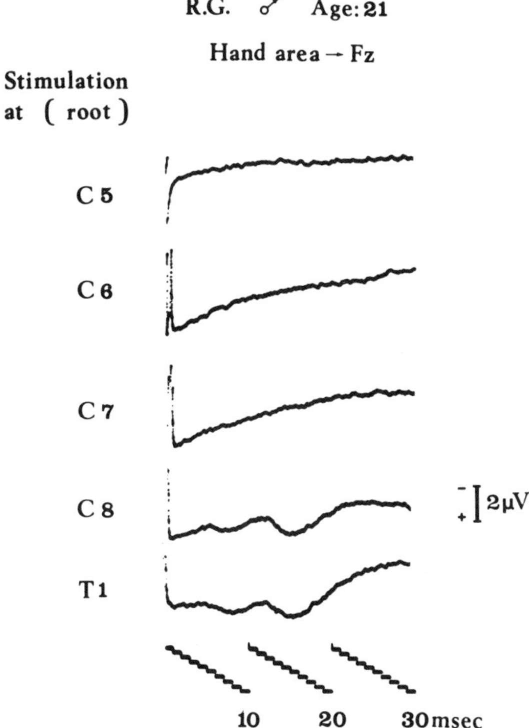

FIGURE 9. Cortical responses to stimulation of individual roots during surgical exploration of the brachial plexus. There were no apparent ruptures, but only the C8 and T1 roots were shown to be in central continuity.

spinal cord at an early, and perhaps reversible, stage that many centers have developed techniques for continuously monitoring sensory conduction through the segments at risk.

In a pilot study performed at the Royal National Orthopaedic Hospital, London, the cortical responses to posterior tibial nerve stimulation at the knee were found to fluctuate widely with changes in anesthetic state and blood pressure. This did not apply to sensory potentials recorded epidurally from the cervical spinal cord, and these also provided faster feedback since they were found to be stable at much higher stimulus repetition rates of up to 20/sec. Results using the latter technique have now been collected for 67 cases (22 males, 45 females; mean age, 15 years; range 11–25), the majority with a diagnosis of adolescent idiopathic scoliosis, who were undergoing treatment by application of traction using a Harrington rod.

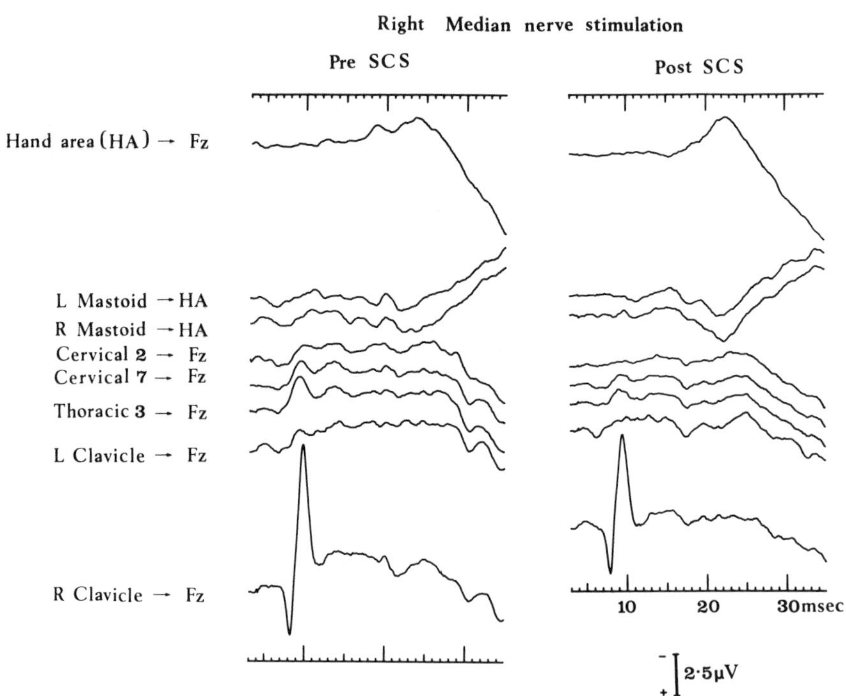

FIGURE 10. SEPs recorded before and after 6 days of spinal cord stimulation for the treatment of MS symptoms. The N20 component was initially bifid but became fused into a single component of larger amplitude and shorter latency.

The epidural cervical responses had an amplitude between 0.5 and 6 μV (mean 2.8 μV) and could be resolved into 3 or more negative components with mean latencies of 14.9, 16.7, and 18.9 msec. The first peak was usually preceded and followed by positive troughs, and resembled a conventional SNAP, while the later waves were sometimes fused into a broad negativity (FIGURE 11). The waveform recorded from the lower thoracic or lumbar region (8 cases) usually consisted of a single major negativity, but where 2 or more components could be resolved it was possible to deduce

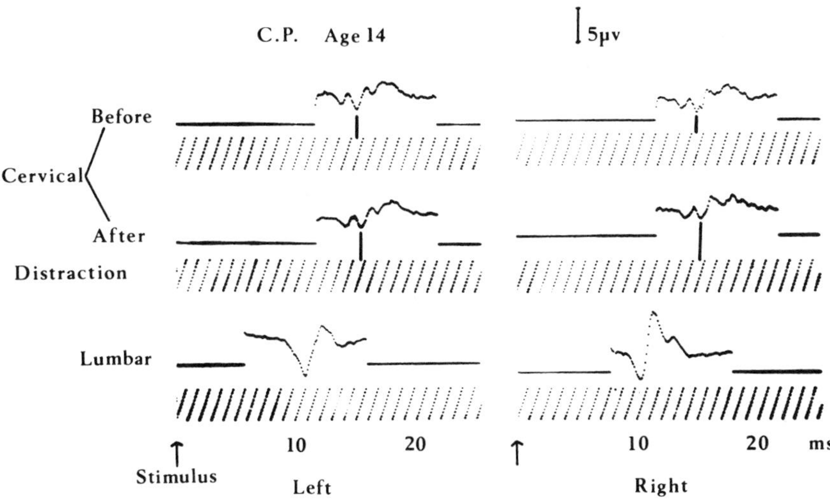

FIGURE 11. Cervical and lumbar potentials evoked by posterior tibial nerve stimulation in a patient undergoing surgical correction of spinal scoliosis. No change was detected in the amplitude or waveform of the cervical response, but there was a slight latency increase of 0.3–0.5 msec.

conduction velocities between lumbar and cervical electrodes, ranging from about 75 m/sec for the fastest to 50 m/sec for the slowest component. The latter was sometimes selectively abolished by reducing the stimulus intensity, suggesting that the break-up of the cervical wave into distinct components was likely to be due to the recruitment of separate fiber groups in the periphery as well as the spinal cord. Increasing the stimulation frequency from 20 to 50 per sec resulted in slight attenuation of the second component, which might therefore be due to postsynaptic activation of, for example, the spinocerebellar tracts.

TABLE 6

CERVICAL EPIDURAL POTENTIALS FOLLOWING POSTERIOR TIBIAL NERVE STIMULATION IN 67 PATIENTS UNDERGOING SURGERY FOR THE CORRECTION OF SPINAL SCOLIOSIS (22 MALES, 45 FEMALES; MEAN AGE, 15 YEARS; 102 LIMBS MONITORED)

Before Spinal Distraction	
Mean amplitude (maximal peak–peak)	2.8 μV
Mean latencies	
Peak 1	14.9 msec
Peak 2	16.7 msec
Peak 3	18.9 msec
After Spinal Distraction	
Number of limbs with	
No change	37
Loss of amplitude	21
Increase of amplitude	3
Waveform alteration	6
Increase in latency	58
Flat response	1

The changes observed after distraction of the spine were mostly very slight (TABLE 6). Out of 102 limbs monitored, 21 showed reduced SEP amplitude (by less than 30% in 15 limbs, 30%–50% in the remainder), 3 showed an increase in amplitude, and 6 showed a waveform alteration with no overall change in amplitude. In as many as 58 recordings, however, there was a latency increase of between 0.3 and 1.0 msec. The latter was not always directly associated with the application of the tractive force, and sometimes reverted to the predistraction value. In one case, the waveform became flat after distraction but recovered after reduction of the force and was intermittently present until closure of the wound and removal of the electrode. This was the only patient to develop any marked postoperative neurological signs (spasticity in both lower limbs), which happily resolved after 24 hours. The overall value of the technique has yet to be convincingly demonstrated, but a general view amongst orthopedic surgeons is that, in this context, any reasonably straightforward safety measure is worthwhile.

ACKNOWLEDGMENTS

I am greatly indebted to a large number of collaborators, including Dr. A. M. Halliday for advice and discussion and Mrs. E. Halliday for technical assistance. Patients were supplied by Prof. W. I. McDonald, Dr. A. Compston, Dr. M. Baraitser, Dr. W. M. Carroll, Dr. C. B. Wynn Parry, Mr. A. Landi, Dr. C. Hawkes, Mr. M. A. Edgar, and Mr. A. O. Ransford.

REFERENCES

1. MATTHEWS, W. B., M. BEAUCHAMP & D. G. SMALL. 1974. Nature **252:** 230–232.
2. MASTAGLIA, F. L., J. L. BLACK & D. W. K. COLLINS. 1976. Br. Med. J. (3): 732.
3. SMALL, D. G., W. B. MATTHEWS & M. SMALL. 1978. J. Neurol. Sci. **35:** 211–224.
4. EISEN, A., J. STEWART, K. NUDLEMAN & J. B. R. COSGROVE. 1979. Neurology (Minn.) **29:** 827–834.
5. NOËL, P. & J. E. DESMEDT. 1980. *In* Clinical Uses of Cerebral, Brainstem and Spinal Somatosensory Evoked Potentials. J. E. Desmedt, Ed. Prog. Clin. Neurophysiol. Vol. 7: 205–230. S. A. Karger, Basel, Switzerland.
6. SAUER, M. 1980. Arch. Psychiatr. Nervenkr. **228:** 223–242.
7. JONES, S. J., M. BARAITSER & A. M. HALLIDAY. 1980. J. Neurol. Neurosurg. Psychiatry **43:** 495–503.
8. PEROT, P. L. 1973. Clin. Neurosurg. **20:** 367–381.
9. GREENBERG, R. P., D. P. BECKER, J. D. MILLER & D. J. MAYER. 1977. J. Neurosurg. **47:** 163–177.
10. HUME, A. L., B. R. CANT & N. A. SHAW. 1979. Ann. Neurol. **5:** 379–384.
11. BONNEY, G. & R. W. GILLIATT. 1958. Proc. R. Soc. Med. **51:** 365–367.
12. JONES, S. J. 1979. J. Neurol. Neurosurg. Psychiatry **42:** 107–116.
13. JONES, S. J., C. B. WYNN PARRY & A. LANDI. 1981. Injury **12:** 376–382.
14. ILLIS, L. S., E. EL-NEGAMY, E. M. SEDGWICK & A. R. D. THORNTON. 1978. Electroencephalog. Clin. Neurophysiol. **44:** 131P.

DISCUSSION OF THE PAPER

J. E. DESMEDT: The fact that one can get an isolated abnormality of N13 strongly suggests that this is not a potential which is generated in the course of the volley up through the medial lemniscus because, if that were the case, one would expect all succeeding potentials to be abnormal. So we do seem to be picking up a different system which probably breaks off at the spinal level.

B. ANZISKA: We recently studied several patients with diffuse neurological disease. There were three interesting findings. First, several patients with diseases like ALS, Huntington's disease, and Wilson's disease had extensive abnormalities involving both rostral and caudal parts of the neuraxis in spite of no clinical evidence of involvement of these areas. Second, the degree of SEP abnormality had little correlation with the degree of clinical abnormality. For example, the patient with Wilson's disease had only a mild tremor, yet she had marked SEP abnormalities. Third, going along with your findings in peripheral neuropathy, we had a patient with a progressive sensory neuropathy of unknown etiology who had a delayed N13–N20, transit time. These were all unexpected findings.

W. WIEDERHOLT: It should not be unexpected to find sensory abnormalities in patients who have nonsensory hereditary degenerative disease. In familial ALS, for example, rather striking pathological abnormalities have been demonstrated in these patients' central sensory pathways and these patients do not have sensory clinical findings.

J. KIMURA: With respect to the patients with Charcot-Marie-Tooth disease, I think it is very difficult to talk about the central conduction in the presence of peripheral nerve disease. We have found that there is a differential effect on shorter and longer latency components of the somatosensory evoked potentials following forearm ischemia. If there is prolongation between, let's say, N13 and N20, that does not necessarily mean that there is a central delay because it may be on the basis of peripheral nerve disease.

S. J. JONES: That is a good point because we have already established that N13 and N20 are unlikely to be generated in the same afferent system. It is quite conceivable that N20 might be mediated by different peripheral fibers than N13.

SOMATOSENSORY EVOKED POTENTIALS IN MAN: SUBCORTICAL AND CORTICAL COMPONENTS AND THEIR NEURAL BASIS*

John E. Desmedt and Guy Cheron

Brain Research Unit
University of Brussels
Brussels 1000
Belgium

Electronic averaging of somatosensory evoked potentials (SEP) in man is receiving increased attention, and SEP studies are developing vigorously. The cerebral somatosensory projection was the first to be mapped out in detail in experimental studies on primates,[62] and it was also the first to be recorded from an intact man.[12] Detailed SEP studies have, however, somewhat lagged behind those in the visual[15] and the auditory[56] modalities, but they are now rapidly gaining ground.[14,17,33] The possibility of recording noninvasively from the first peripheral sensory neuron and of studying the subcortical,[11] primary cortical,[20,23] and late cortical[18,24,29] responses is rather unique to the somatosensory modality. Another feature of the somatosensory pathway is its remarkable length from peripheral skin to cortex, which makes it more vulnerable to a wide variety of pathological conditions: the wide possibilities offered by accurate SEP studies for critical contributions to neurological diagnosis have barely been charted as yet, but progress is coming along fast.

The present paper considers SEPs only in normal adult man and attempts to present the latest data on the functional organization of neural generators involved. Such data are essential for the steady upgrading of pathophysiological applications and of diagnostically relevant uses.[14,17,25,52]

The data selected for this paper deal with SEPs elicited by delivering brief electric stimuli to the fingers or to the median nerve. The potentials are recorded along the peripheral nerve and neck and over the scalp. Blurring or distortion of early SEP components must be avoided by using an adequate bandpass, with a fidelity of the amplifier-recorder system extending from 2 Hz to 2 kHz and with a bin width of 200 μsec or shorter for averaging.[16,19] Since SEP components have a rather small amplitude, it is also essential to minimize or avoid interference from unwanted activities of muscles close to the recording electrodes. The subjects should be fully relaxed, and some device for on-line or off-line rejection of contaminated samples from the current averages should be used. Details of methods can be found in recent publications.[16,20,21,22] The SEP components are labeled by their positive (P) or negative polarity (N) and their modal peak latency, as recommended by an international committee.[30] For simplicity, the modal latencies considered are those of SEP components elicited by stimulation of the median nerve at the wrist, namely, P_9, N_{11}, and N_{13} at the posterior neck, the P_9, P_{11}, P_{13}, and P_{14} far-fields at the scalp, and the components N_{18}, N_{20}, P_{22}, P_{27}, N_{30}, and P_{45}. Later components, which are strongly influenced by cognitive processing activities of the subject,[18,24] are not considered here. In all traces presented, negativity of the active recording electrode produces an

*This work was supported by grants from the Fonds de la Recherche Scientifique Médicale and the Fonds National de la Recherche Scientifique, Belgium, and by the Muscular Dystrophy Association of America.

0077-8923/82/0388-0388 $1.75/0 © 1982, NYAS

upward deflexion. The noncephalic reference electrode used is placed on the dorsum of the hand opposite to the one stimulated.

ESTIMATION OF THE SPINAL ENTRY TIME OF THE AFFERENT SENSORY VOLLEY

It is essential to obtain accurate timing of the subcortical SEP components, and a critical element is the time of entry into the spinal cord at levels C_6–C_7 for an afferent volley generated in the first three fingers of the hand (median nerve).[35,41] Current uncertainties about time relationships of early SEP components have been resolved by comparing direct recordings of the sensory nerve potentials along the peripheral nerve.[14,28] For consistent results it was necessary to stimulate fingers rather than the mixed nerve so as to exclude the group I muscle afferents and the antidromic motor axons potentials. The sensory nerve potentials were recorded with fine uncoated stainless steel needles inserted close to the median nerve (with a reference electrode 3 cm at right angle to the nerve course). These fast potentials were averaged with a bin width of 20 μsec. They are displayed with a vertical separation of traces that is proportional to actual conduction distances along the nerve in FIGURE 1 A–E. The onset of the negative phase of the triphasic nerve potentials indicates the arrival of the nerve volley under the recording electrode.[13,34] The segment from the fingers to the wrist usually presents a slightly slower conduction velocity (CV) due to distal tapering of the nerve fibers. The latency data for sites between the wrist and Erb's point (over the brachial plexus, just above the clavicle) consistently fit a linear regression with a calculated CV of 72 m/sec in this example (FIGURE 1). The mean afferent CV over this stretch of the nerve was 71.1 ± 4.0 (SD) m/sec for 25 healthy adults with a mean age of 22 years.[20] Extrapolation of the regression to the level of the C_6 vertebra provides a fairly quantitative measure of spinal entry time, and this fits in remarkably with the onset of the negative SEP component N_{11} which is recorded over the C_6 spinous process at the posterior aspect of the neck (FIGURE 1 F,G).[20,22]

Two traces are shown for this C_6 neck recording: the first is recorded with a noncephalic reference on the opposite hand and shows a P_9 far-field potential that precedes the N_{11} spinal response. The latter is followed by another negativity N_{13} to be considered below (FIGURE 1 F). The second C_6 trace is recorded with an earlobe reference whereby the P_9 far-field potential (which also appears at the earlobe) thus cancels out while the N_{11} and N_{13} are enhanced (FIGURE 1 G); the latter effect is because the earlobe reference is far from indifferent and actually "sees" concomitant P_{11} and P_{13-14} far-field potentials (compare FIGURES 6 B–C, 7, 9 C). The time of onset of the spinal N_{11} is quite clear is both traces.

A noncephalic reference recording from the contralateral parietal scalp (primary somatosensory projection) also shows a P_9 far-field potential that is followed by far-fields P_{11} and P_{14}. The spinal entry time defined from peripheral nerve potentials also fits in remarkably with the onset of the positive P_{11} far-field, which we interpret as being generated by the spinal dorsal column volley indexed by the neck N_{11} SEP component.

THE VOLUME-CONDUCTED PERIPHERAL NERVE FAR-FIELD P_9 POTENTIAL

The first SEP event recorded at neck or scalp after stimulation of the fingers or median nerve stimulation is a widespread P_9 positivity that is recorded with an identical latency over the entire scalp (FIGURE 9), earlobes and neck, provided a noncephalic reference is used.[11,20,31,42,60] The P_9 phenomenon is annihilated when the

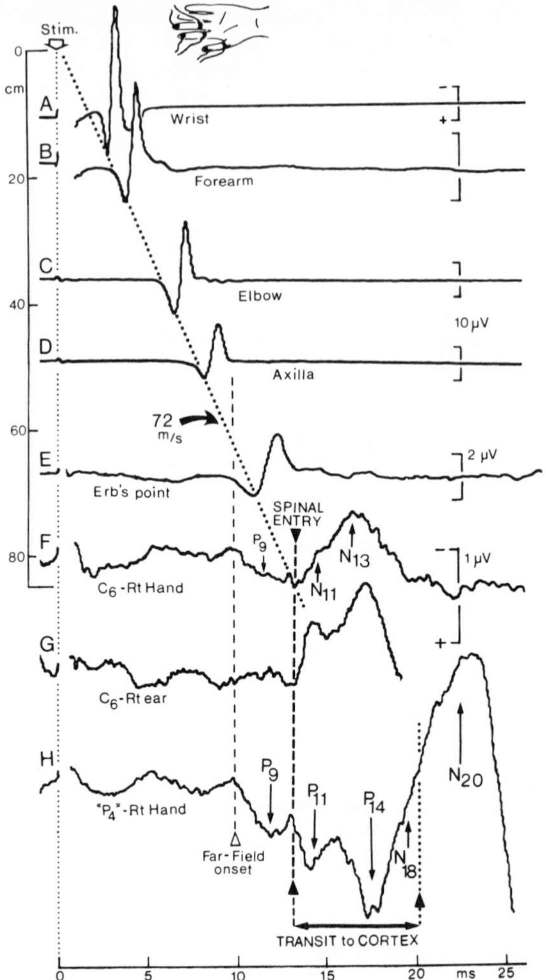

FIGURE 1. Sensory nerve potentials recorded from the median nerve at the wrist, forearm, elbow, axilla, and Erb's point (A–E) with different amplifications. Stimulation was at 3 times the subjective threshold of fingers I-II-III of the left hand. The vertical separation of the averaged traces is proportional to the conduction distances between the recording sites along the median nerve. The calculated linear regression gives a conduction velocity (CV) of 72 m/sec. Extrapolation of CV to the lower neck C_6 recording site fits the onset of the negative spinal component N_{11} (F,G), and the onset of the positive scalp far-field potential P_{11} (H). The onset of the cortical N_{20} recorded from the contralateral parietal scalp with noncephalic reference (H) indicates a spino-cortical transit time of 7.0 msec in this subject. The onset of the first scalp far-field P_9 (white arrowhead and vertical interrupted line) is earlier than Erb's point potential, but later than the nerve potential at the axilla. In (H), the first inflection on the ascending limb of N_{20} corresponds to the N_{18} component, while the second inflection corresponds to the genuine N_{20} onset that takes off from the underlying N_{18}. (From Desmedt & Cheron.[20] By permission of *Electroencephalography and Clinical Neurophysiology*.)

reference electrode is placed on the head (FIGURE 1 G), and it can even be converted into a spurious "N_9" when the reference electrode is placed on the front scalp and records a larger P_9 than the one seen by the active electrode (FIGURE 2 E–F).[20] This and other evidence points to the complications that can arise when using cephalic reference electrodes.

It is now agreed that P_9 represents a volume-conducted potential generated in the peripheral nerve in the distal part of the brachial plexus. One argument is that its onset (about 6 msec for median nerve stimulation at the wrist and 9 msec for finger stimulation) precedes the arrival of the peripheral volley at Erb's point, but is later than the onset of the median nerve action potential recorded directly from the axilla (FIGURE 1 D–F, H).[20] The second argument is that P_9 persists while all subsequent SEP components are lost in the patients with traction injuries of the brachial plexus when the spinal roots have been avulsed.[3] The reason why P_9 reflects the peripheral nerve volley at the distal part of the brachial plexus preferentially is not yet clear, but may have to do with the geometry of the generator orientation: cadaver dissections indicate that the more proximal brachial plexus is horizontally oriented while the more distal part is more vertical.[22] Another factor may be the nature and electrical resistance of tissue surrounding the nerve trunks at different levels.

FACTORS INFLUENCING THE ARRIVAL TIME AT THE SPINAL CORD

Any lowering of the tissue temperature around the peripheral nerve increases the SEP latency since the afferent CV drops with a Q_{10} of 1.5 to 2.0 with temperature. It is important to check local temperature with a thermistor and maintain it within the physiological range by local heating whenever necessary.[14,16] Another factor is the body size and arm length, which influences the distance actually traveled from the stimulation point to the spinal cord.[47,55] In normal adults the onset latencies of the cortical N_{20} to finger stimulation can vary from 15.5 to 22 msec in persons with body sizes ranging from 1.5 to 2.0 m. In patients with neuropathies or nerve entrapment, the peripheral conduction slowing results in SEP delays that can actually be used for estimating the abnormal CV in pathological nerves.[14,25]

THE EARLY SPINAL NEGATIVITY N_{11} AND ITS GENERATOR

In all subjects, the recording from the posterior neck C_6 site with a noncephalic reference discloses a N_{11} SEP component whose onset coincides with spinal entry time. Electrodes placed along the neck at higher levels up to C_2 record a similar negative component, but with a clear latency shift (FIGURE 2 C–D, 3 C–E). None of these potentials is "myogenic" since they are recorded from completely relaxed subjects with no residual electromyographic (EMG) activity in the recorded channels. The latency shift of N_{11} along the neck was missed by previous researchers[47,55] because of their use of a scalp (instead of noncephalic) reference whereby the spinal potentials were spuriously distorted by the scalp far-field P_{11} that is injected into grid No. 2 of the amplifiers in that montage. Since the same P_{11} (with onset corresponding to that of N_{11} at the neck C_6 level) was injected into all recordings from C_6 to C_2 under such conditions, this blurred the onset time of the negativity and made the record from neck C_2 appear with a abnormally early latency (FIGURE 2 E–F).[20,22]

The mean transit time for the spinal N_{11} from levels C_6 to C_2 at the neck is 0.94 ±

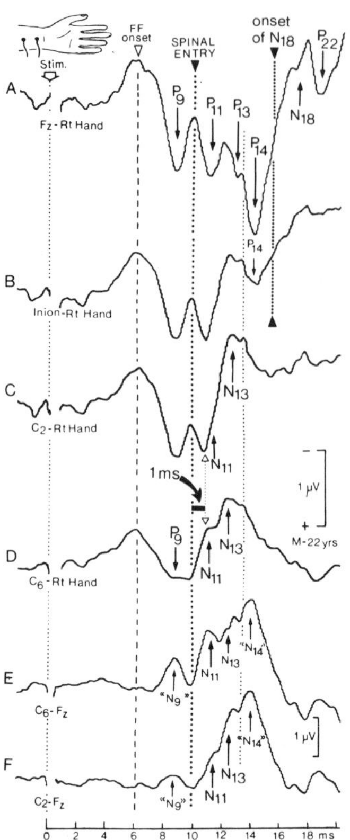

FIGURE 2. Far-field SEP components recorded with noncephalic reference (right hand) in a male subject of 22 years. Stimulation of the left median nerve at the time indicated by the vertical dotted line on the left side. Calibration is the same for all traces. Active electrode at the frontal midline F_z (A) and 2 cm above the inion (B). The neck recording sites are 2 cm below the inion (C, F) or over the spinous process of the sixth vertebra (D, E). White arrowhead and vertical interrupted line indicate onset of first far-field P_9 which is congruent in all the traces. Black arrowheads and thicker dotted lines indicate spinal entry time coinciding with onset of spinal negativity N_{11} at the lower neck (D) or onset of scalp P_{11} far-field (A, B). The transit time from spinal entry to cortex is 6.1 msec. Other vertical lines correspond to the onset of N_{13} (D), of P_{13} and of P_{14} (A–C). The latency shift of the negative spinal component is 1.0 msec. from lower to upper neck recordings. The spinal N_{11} is followed by another negativity N_{13}. The large P_{14} far-field potential observed at the scalp (A, B) does not appear at the neck electrodes (C, D). The onset of the scalp N_{18} is indicated as well as the prerolandic P_{22} component (A). When a scalp F_z reference is connected to grid 2 of the amplifier (instead of the noncephalic hand reference), the posterior neck recordings display spurious "N_{14}" components, which are generated above the foramen magnum and are picked up by the frontal reference (E, F). This addition of the positive far-fields at the reference results in distortions of the spinal waveform and in increased amplitude (see reduced amplification used in E, F).

The latency shift from lower to upper posterior neck is blurred by the frontal reference (E, F), but is clearly recorded with a noncephalic reference (C, D): a latency shift of 1.0 msec. is recorded in this subject. (From Desmedt & Cheron.[20] By permission of *Electroencephalography and Clinical Neurophysiology*.)

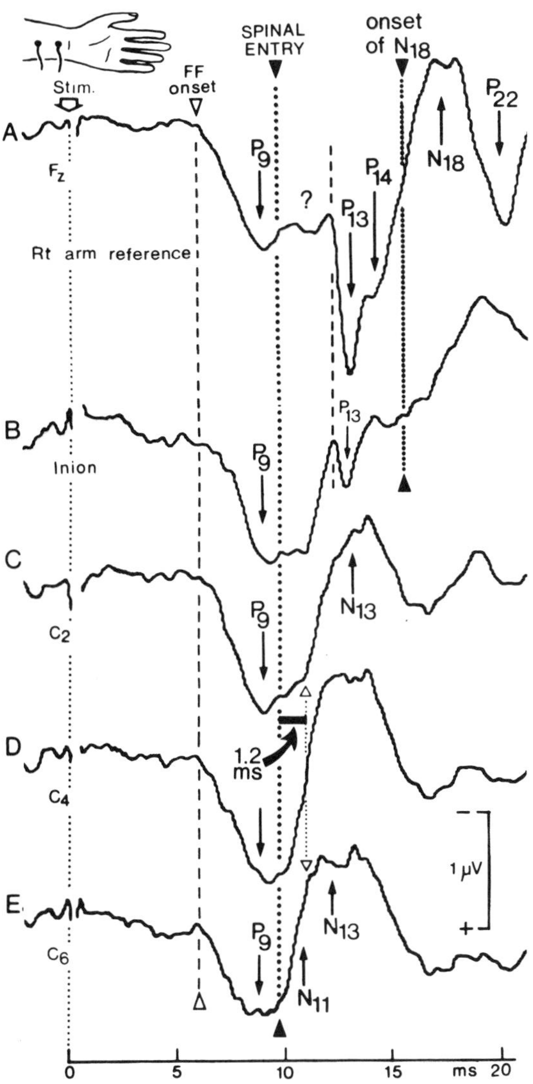

FIGURE 3. Far-field SEP components in a female subject of 20 years. Stimulation of left median nerve. Same presentation as in FIGURE 1. Notice the virtual lack of P_{11} and the preponderance of P_{13} over P_{14} in the scalp recordings. The latency shift of the negative spinal component is 1.2 msec from lower to upper neck recordings. The spinal N_{11} is followed by an N_{13} component (C–E), which is of clearly longer duration that the scalp P_{13} far-field (A, B). The onset of the component N_{18} is indicated as well as the prerolandic P_{22} component at the frontal electrode (A). (From Desmedt & Cheron.[20] By permission of *Electroencephalography and Clinical Neurophysiology*.)

0.24 (SD) msec in 18 normal adult subjects.[22] Anatomical dissections of the spinal cord and spinal roots provided evidence about the actual conduction distance involved, which turns out to be about 55 mm instead of the 100 mm or so that separate the recording electrodes at the skin. Detailed discussion and illustration of this point are provided by Desmedt and Cheron.[20] Thus the mean maximum conduction velocity for N_{11} is 55/0.94 = 58.5 m/sec. This velocity corresponds to that in the centrally directed branch of the primary afferent neuron, making up the dorsal column pathway. It is about 12 m/sec slower than the corresponding peripheral CV measured in the peripheral part of the same primary sensory neurons (FIGUE 1) and this data agrees with animal evidence.[20] Interpretations of N_{11} had been confused by the use of a cephalic reference (see above) and a fixed generator had been hypothesized.[47,55] With the recent evidence of a consistent upwards conduction of N_{11} from spinal entry, its interpretation as reflecting ascending conduction in the dorsal column appears inescapable.[20,22]

THE SPINAL N_{13}–P_{13} COMPONENT AND ITS GENERATOR

Recordings from the posterior neck with a noncephalic reference disclose an N_{13} component starting with an inflection upon the N_{11} (FIGURE 1 F, 2 C–D, 3 C–E). The nature of N_{13} has only recently been clarified through the use of esophageal recording electrodes which offer a strategic access to the anterior aspect of the vertebral bodies. The method used is fully described elsewhere[22] and X-rays were used to identify the exact levels of each recording electrode with respect to the cervical spine (FIGURE 4). The features of prevertebral recording at levels C_7 to Th_3 relate to spinal roots and can be found in the original paper.[22] At prevertebral levels above C_7 (or spinal entry of the afferent volley) the P_9 far-field is recorded as well as at the posterior neck, but it is followed by a different profile (FIGURE 4 A) corresponding to a mixture of negative spinal roots potential and dorsal column N_{11} (see star indicating presumed peak of that small N_{11}) (details in Reference 22). Then the anterior and posterior records diverge dramatically and disclose a clear phase reversal between a posterior N_{13} and a prevertebral P_{13}. The peaks of these components were 13.1 msec and did not differ significantly in 12 subjects. The trace in FIGURE 4 B results from an electronic subtraction of these two records whereby the P_9 far-field potential virtually cancels out, the root spike shows up (with onset slightly *before* spinal entry, as to be expected), and component P_{13} gets quite large through the addition of N_{13} injected into grid No. 2 of the amplifier in this montage. The spinal N_{13}–P_{13} phenomenon is quite distinct from any scalp-recorded far-field potential and its onset is clearly earlier than that of the scalp P_{13-14} far-field (FIGURE 4 C). The duration of N_{13}–P_{13} is longer than that of the scalp far-fields (FIGURE 10).[23]

An important finding is that P_{13} is recorded prevertebrally between levels C_7 to C_3, but sharply drops in amplitude below or above. Moreover, the onset latency of P_{13} is stable in any subject all along the spinal cord, no matter how its amplitude changes with the level of the recording electrode.[22] Therefore the N_{13}–P_{13} must be related to a fixed generator, located in the cervical spinal cord, that does not produce any electrical sign at the scalp. In fact, the vertical extent of P_{13} corresponds with the anatomical distribution of collateral branches given by the central part of the primary sensory neuron to the dorsal horn. A detailed discussion of these and other data provides a strong suggestion for ascribing the N_{13}–P_{13} phenomenon to a fixed generator in the dorsal horn.[22]

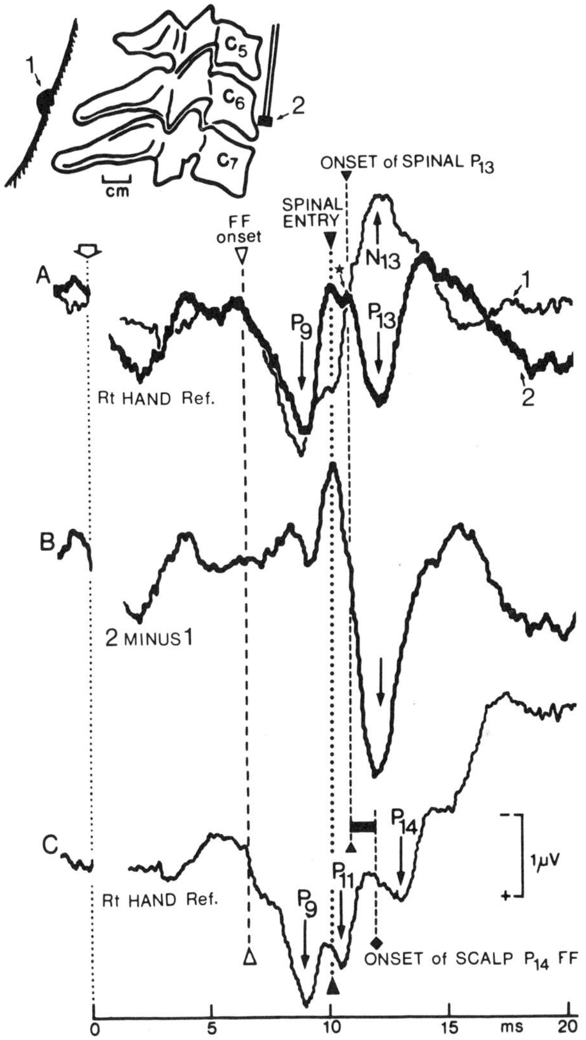

FIGURE 4. Comparison of posterior neck recording with prevertebral (esophageal) recording at the same level near root entry. A noncephalic reference on the right hand is used. (A) Superimposed averaged SEPs to median nerve stimulation, recorded in front (electrode 2, thicker trace) or behind (electrode 1) vertebra C_6. (B) Subtraction of record 1 from record 2 which virtually eliminates the P_9 far-field, isolates the root spike, and artificially increases the P_{13} phenomenon (since the posterior neck N_{13} is fed into grid 2 of the amplifier). There is a clear phase reverseal between P_{13} and N_{13} in (A). (C) Recording of cephalic far-fields P_9, P_{11}, and P_{14} from the right earlobe. The vertical interrupted line indicates the onset of P_9 at 7 msec. The next vertical dotted line to the right corresponds to the spinal entry time and to the onset of spinal N_{11} and scalp far-field P_{11}. The onset of spinal N_{13}–P_{13} precedes that of the P_{14} far-field. (From Desmedt & Cheron.[22] By permission of *Electroencephalography and Clinical Neurophysiology*.)

The Scalp P_{14} Far-Field Potential and Its Generator

The P_{14} positive component is a constant feature of all scalp recordings when a noncephalic reference is used (FIGURE 1 H). In quite a few subjects, a subcomponent P_{13} is seen at the onset of P_{14} (FIGURE 2 A), and this scalp P_{13} can even be a major peak in the far-field complex (FIGURE 3 A). In about a third of the subjects, P_{13} cannot be identified and P_{14} appears as a single component (FIGURE 10 A). The P_{13}–P_{14} complex will be considered as a unit for a while because there appears to be no strong data to differentiate them as yet. It is interesting that the scalp P_{11} far-field potential can be missing in certain subjects (FIGURE 3 A) as if the spinal N_{11} generator had failed to provide a sufficient dipole for volume conduction to the scalp. The P_{14} is recorded even as far back as the inion (FIGURE 2 B), and, when a P_{13} is preponderant, the inion electrode shows a rather clear P_{13} effect as well (FIGURE 3 B). Neither P_{13} nor P_{14} extend to the posterior neck at C_2 where the wider spinal N_{13} is recorded, and there is an obvious difference in the duration of the two phenomena (FIGURE 10). We have suggested that P_{14} is a volume-conducted potential related to a generator located above the foramen magnum and, more precisely, near or in the medial lemniscus.[20,22]

This hypothesis is supported by several data. For example, a calculation of transit time from spinal entry to the cuneate nucleus, considering a mean conduction distance of 70 mm (estimated by dissections on cadavers; compare Reference 20), is 70/58.5 = 1.2 msec. The value of CV of 58.5 m/sec has been calculated above on the basis of latency shift along the posterior neck of the spinal N_{11} component (FIGURES 2 & 3). Before involving the medial lemniscus, the afferent action potentials must be transmitted across the cuneate synapses which present large synaptic boutons around clusters of neurons and dendrites; the synaptic delay is anyone's guess, but it can be taken as 0.3 to 0.5 msec, considering the high safety factor for transmission documented in animal experiments (compare Reference 44). Thus the medial lemniscus fibers would be activated 1.5 to 1.7 msec after spinal entry, and these figures fit in very well with the measured transit time between the onset of P_{11} (or spinal entry time) and the onset of the P_{13}–P_{14} far-fields, that is, in our series of normal adults, 1.75 ± 0.30 (SD) msec.

The conduction time in the fastest lemniscal axons up to the thalamus can also be evaluated. Dissection data suggest a mean conduction distance of 70 mm while the largest lemniscal axons related to the hand input in man appear to have a diameter of 9 μm.[20] Since few of the axons are that big, it may be safer to use a diameter value of 8 μm, which would correspond to a CV of 36 m/sec, if one is willing to apply a conversion factor of 4.5 m/sec per μm (see the detailed discussion in Desmedt & Cheron).[20] Thus the conduction time in medial lemniscus would be roughly estimated as 70/36 = 1.94 msec. This is very close indeed to the mean time measured between the onset of the P_{13}–P_{14} far-field at the scalp and the peak of P_{14}, namely, 1.96 ± 0.29 msec.[23] Thus we interpret the P_{13}–P_{14} far-field as a volume-conducted potential related to the summation of the traveling activity in medial lemniscus, the far-field onset corresponding to the activation of the caudal lemniscal fibers and the far-field peak reflecting the arrival of the action potentials via the fastest lemniscal axons at the thalamus. These results exclude the possibility that P_{13}–P_{14} could be generated at or above the thalamus.[20,22]

This interpretation is supported by clinical evidence that the P_{14} far-field is preserved in patients with a thalamic vascular lesion that eliminates all subsequent SEP activity.[48–50]

If the P_{13}–P_{14} far-fields are generated above foramen magnum and after the somatosensory decussation, they should present some asymmetry between the two sides of the head because the medial lemniscus runs contralaterally, a few mm from

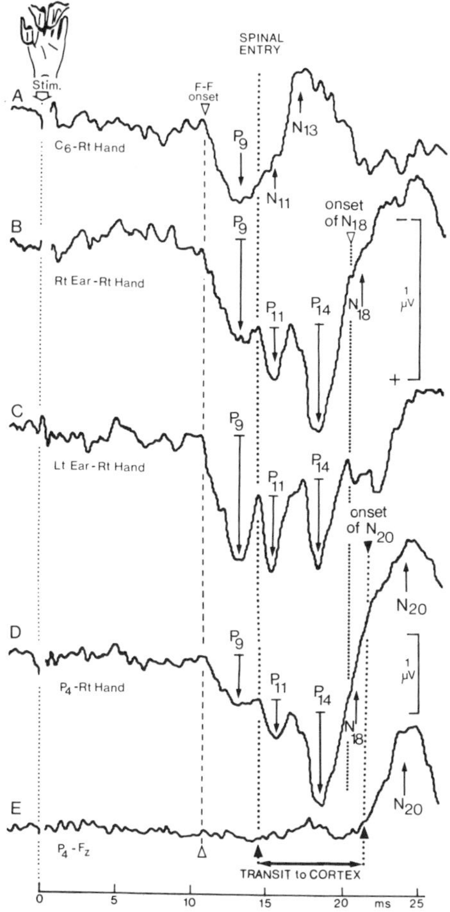

FIGURE 5. Asymmetries of far-field potentials at the two earlobes. The SEPs are evoked by electric stimuli to fingers I-II-III of the left hand. The stimulus to finger I is delayed by 0.5 msec in order to make up for the somewhat shorter conduction distance to the spinal cord. Noncephalic reference electrode in (A) to (D) and frontal scalp reference in (E). (A) Derivation from the posterior neck at the level of C_6. After the P_9 far-field, the N_{11} onset indicates spinal entry time. The N_{11} component is followed by an N_{13}. (B) At the right (contralateral) earlobe, the P_{11} far-field is rather small while the P_{14} far-field is quite large. The onset of the N_{18} component is indicated by vertical dotted line. (C) At the left (ipsilateral) earlobe conversely, the P_{11} far-field is larger while the P_{14} far-field is smaller than contralaterally. (D) Contralateral parietal scalp electrode recorded with half amplification; the P_9 far-field has the same amplitude as at the earlobes while the P_{14} far-field is larger. (E) the same active parietal electrode is recorded with a scalp frontal reference whereby the far-fields cancel out and the parietal N_{20} is prominent. In this montage, the N_{20} can be preceded by a negative bump because the N_{14} far-field at the front is larger than at the parietal region (see FIGURE 7): the larger positivity picked up by grid 2 of the amplifier results in a negative deflexion. Furthermore the P_{22} SEP component picked up by the frontal electrode connected to grid 2 of the amplifier results in a spurious enhancement of the N_{20}. (From Desmedt & Cheron.[20] By permission of *Electroencephalography and Clinical Neurophysiology*.)

the midline. Recent data indicate that the far-fields exhibit striking differences in amplitude (but not in latency) at the left or right earlobe, when using a noncephalic reference.[20] While P_9 is fairly symmetrical, P_{11} which relates to the dorsal column activity ipsilateral to the stimulated hand (see above) is larger at the ipsilateral ear (FIGURE 5 B–C). By contrast, P_{13}–P_{14} is larger at the contralateral ear. These differences present a high degree of significance when considering the regression lines for pooled data in young adult subjects (FIGURE 6). These reverse differentials for P_{11} and for P_{13}–P_{14} appear rather remarkable when one considers that the corresponding activities are ascending along the axis of the brainstem–spinal cord only very few millimeters from either sides of the midline.[20]

The SEP far-fields are currently said to be widespread over the scalp, and our

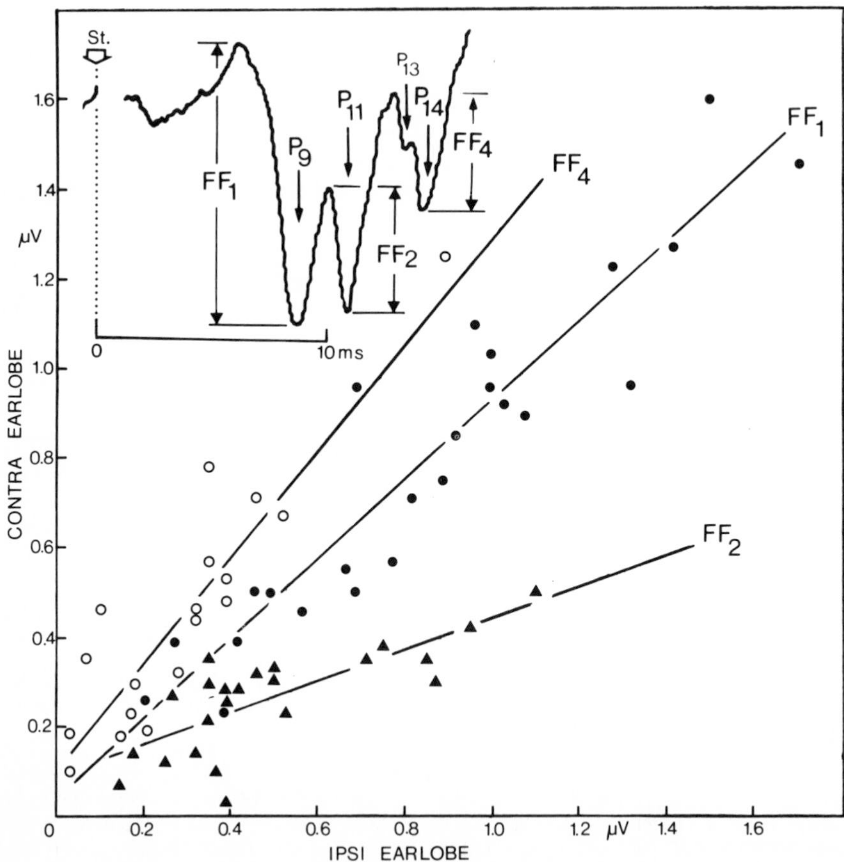

FIGURE 6. Peak amplitudes of early SEP components simultaneously recorded from the ipsilateral (μV, abscissa) and contralateral (μV, ordinate) earlobes with noncephalic reference on the right hand. Stimulation of fingers I-II-III or of median nerve at the wrist on the left side. Amplitudes were measured as depicted on the trace shown (insert). The calculated linear regressions were for P_9 (dots): $y = 0.89x + 0.04$ ($r_{xy} = 0.94$); for P_{11} (triangles): $y = 0.35x + 0.09$ ($r_{xy} = 0.74$); and for the combined P_{13}–P_{14} (circles): $y = 1.17x + 0.10$ ($r_{xy} = 0.88$). (From Desmedt & Cheron.[20] By permission of *Electroencephalography and Clinical Neurophysiology*.)

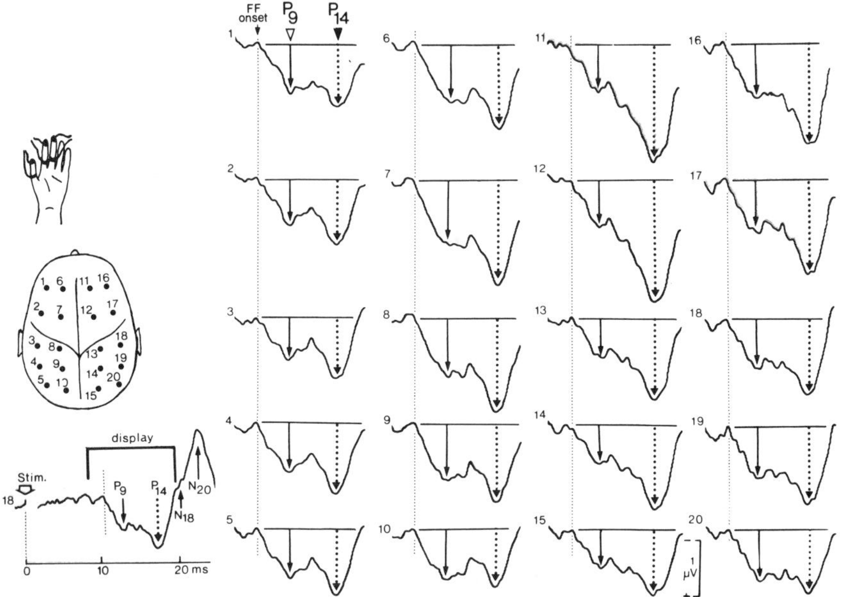

FIGURE 7. Scalp topography of early SEP components in a male subject of 22 years. Stimulation of fingers I-II-III of the left hand. Noncephalic reference on the right hand. The 20 electrode sites are indicated on the figurine. A complete trace recorded from electrode 18 is shown on the left to indicate the portion that is enlarged and displayed (from about 8 to 19 msec after stimulation) in the other traces. The peak amplitude of P_9 (line) and P_{14} (dots) are indicated. The P_{11} is not well delineated in this experiment. (From Desmedt & Cheron.[20] By permission of *Electroencephalography and Clinical Neurophysiology*.)

findings at the earlobes prompted a detailed mapping of scalp far-fields in several subjects. In the experiment illustrated by FIGURE 7, which involved stimulation of the fingers, the P_{11} far-field was not consistent (as occasionally happens) while the P_9 and P_{13}–P_{14} were well delineated. P_9 presented a roughly similar amplitude at all scalp sites and its variations did not appear to fit any consistent pattern. By contrast, the P_{13}–P_{14} far-field was much larger contralaterally, and more so at 35 mm from the midline (traces 11–12) than at 70 mm from midline (traces 16–17). The side differences were most obvious in the frontal region where the largest P_{14} were recorded. These findings must be related to geometrical relationships with the oblique caudo-rostral axis of the brainstem generators and their location on one side of the midline.[20]

The internal consistency of all these data support the suggestion of a lemniscal generator for P_{13}–P_{14}. It should also be pointed out that such analysis should indeed be performed in man because the profiles of far-fields differ between animal species and extrapolations between data may be difficult.[4,5,60]

THE N_{18} AND N_{20} SEP COMPONENTS

It has generally been held that the earliest cortical response revealed in the scalp-recorded SEP was the negative N_{20} component at the parietal scalp opposite to the stimulated hand.[11,14,20,33,37] The N_{20} is larger and longer in normal newborn babies than in adults,[27] and it shows a remarkable increase to about twice adult size during

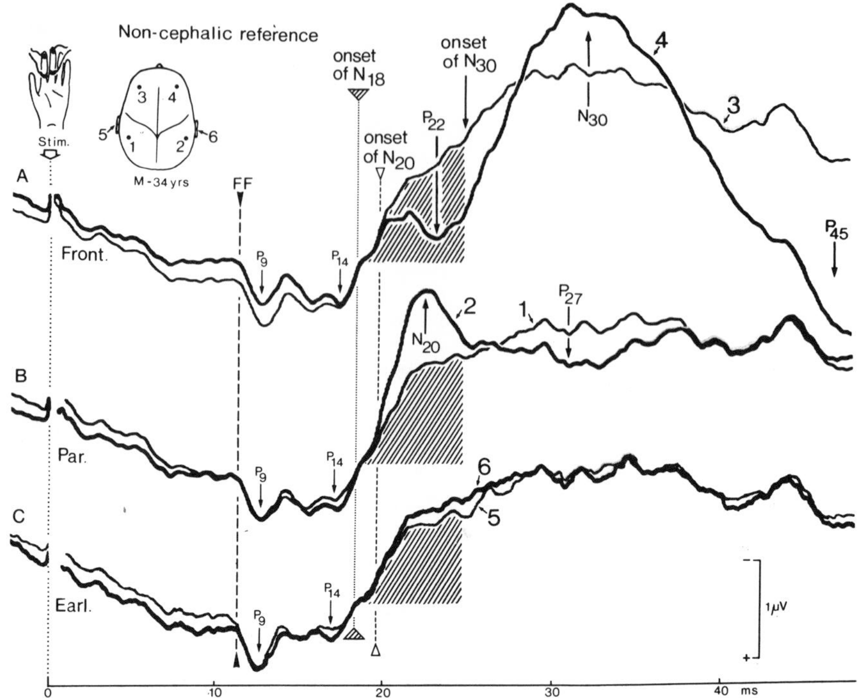

FIGURE 8. Noncephalic reference recording of SEP components recorded at the frontal scalp (A), at the parietal scalp (B), and at the earlobes (C). The traces from the contralateral side are presented by thicker lines (drawn by several writings of the computer) and are superimposed on the traces from the ipsilateral side. Electrode positions and traces are numbered from 1 to 4. The electric stimulus is delivered to the left fingers II and III. The far-field onset and the P_9 and P_{14} far-fields are well delineated. After P_{14}, a first inflection on the negative-going traces indicates the onset of component N_{18} (vertical fine-dotted line) while the onset of the parietal N_{20} is indicated as corresponding to the divergence of the contralateral and ipsilateral parietal traces in (B). The N_{18} phenomenon is represented by hatched area, which is arbitrarily terminated at the onset of the frontal N_{30} component (A). The N_{18} is present at all electrodes ipsilaterally, and is supposed to underlie the focal SEP components N_{20} (B) or P_{22} (A) on the contralateral side. (From Desmedt & Cheron.[23] *Electroencephalography and Clinical Neurophysiology.*)

normal aging in healthy octogenarians.[21] However, scalp recordings with a non-cephalic reference electrode recently appeared to suggest a wider bilateral distribution of the SEP negativity than hitherto supposed,[43] and thus raised the question of the significance of "N_{20}" as an index of the primary contralateral response. This question has now been looked into by extensive scalp mapping with a noncephalic reference recording.[9,23]

It is indeed true that the noncephalic electrode actually provides a more neutral reference whereby the true response features can better be delineated, and that such a reference tends to lift up the traces towards more negativity after the positive far-fields (FIGURE 8). However, several negative phenomena must be involved. At the parietal scalp contralateral to the stimulated hand, several inflections can be identified on the negative-rising limb of the N_{20}. The first inflection seen after the P_{14} far-field indeed appears also at the ipsilateral parietal scalp (FIGURE 8 B), at the front (A) and at the earlobes (C) on both sides. This phenomenon with widespread distribution is designated as a "N_{18}" component that must be distinguished from the true N_{20} generated in the contralateral parietal region. In fact, the actual onset of N_{20} is taken as the point in time when the ipsilateral and contralateral traces diverge from each other, on the assumption that the N_{18} (hatched area in FIGURE 8) is probably of equal amplitude on both sides. This divergence corresponds roughly to the second inflection on the rising limb of N_{20}.[23]

It is interesting that at the contralateral frontal electrode, another divergence towards positivity occurs at a latency generally slightly longer than that of N_{20} (FIGURE 8 A). Indeed this subject shows a prerolandic P_{22} component that is virtually absent at the ipsilateral front, which allows a clear differentiation of N_{18} and P_{22}. The actual duration of N_{18} is difficult to determine because the traces generally linger at a somewhat negative level for some time, and we proposed, without any specific reason to date, that N_{18} terminates or decreases at the time of onset of the frontal N_{30}, which generally develops at the front on both sides but is larger contralaterally (FIGURE 8 A).

When now displaying recordings from the same scalp electrodes with an earlobe reference, it is obvious that the early positive far-fields and the N_{18} must virtually cancel out since they are also picked up at the earlobe reference (FIGURE 9 B–C). The

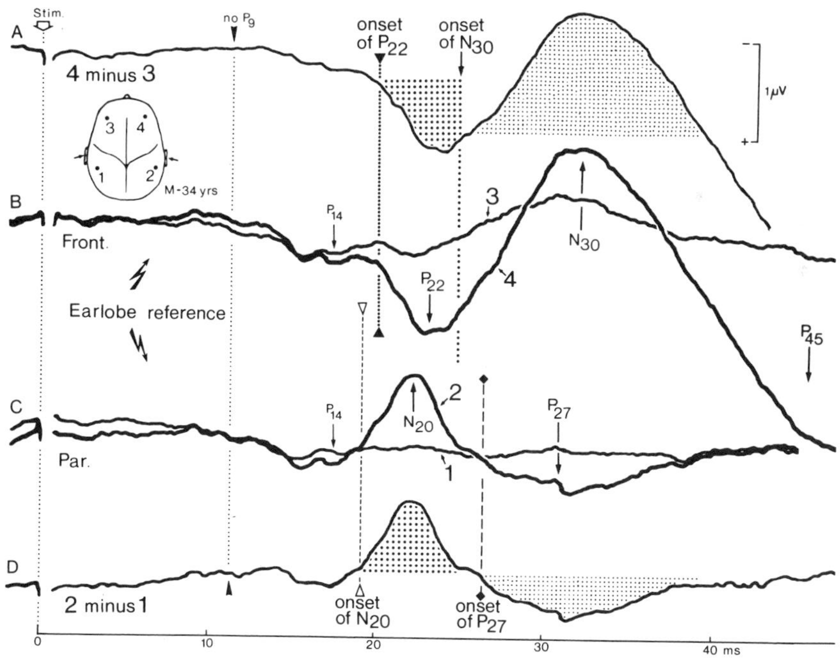

FIGURE 9. Same experiment as in FIGURE 8. The same averaged data are here presented with earlobe reference (B & C) or after subtraction of the ipsilateral front (A) or parietal (D) trace from the symmetrical contralateral trace. These traces present little, if any, far-field and N_{18} components and the positivities are better delineated (such as the P_{22} at the front and the P_{27} at the parietal electrodes). There is virtually no early negativity at the ipsilateral parietal electrode. (From Desmedt & Cheron.[23] *Electroencephalography and Clinical Neurophysiology*.)

negativity upon which the various components appeared to be riding is now considerably reduced, and the cortical positive components P_{22} (prerolandic) and P_{27} (parietal) are better delineated in earlobe reference recordings. Furthermore, the ipsilateral scalp records can be electronically subtracted from the contralateral traces to provide another display of the N_{20}, P_{27} (FIGURE 9 D), P_{22}, and N_{30} (FIGURE 9 A).

DURATION OF THE SCALP FAR-FIELD POTENTIALS AND OF THE NECK N_{13}–P_{13} COMPONENT

The positive far-fields at the scalp are usually quite brief components. They are thought to reflect volume-conduction of synchronized volleys of action potentials in bundles of nerve fibers[4,20,57] whereby an "open field" system[46] generates a coherent current field at a distance. We proposed that the relevant segments of the somatosensory afferent volley are the peripheral brachial plexus (for P_9), the dorsal column (for P_{11}), and the medial lemniscus (for P_{13}–P_{14}).[20] The oblique rostro-caudal orientation of these ascending depolarizations in nerve fibers bundles are thus seen as rather large positive deflexions over the head, thus beyond the termination of the tracts that actually generate the corresponding action potentials (FIGURE 1 H, 2 A–B, 3 A–B, 7, 8).

Our definition of the onset of the N_{18} component (FIGURE 8) now allows an accurate estimation of the duration of the P_{13}–P_{14} far-field (FIGURE 10 A). The duration of the P_{11} far-field raises no problem when this component is clearly delineated (FIGURE 1 A, 10 A). The spinal N_{13}–P_{13} (discussed above) is best measured from esophageal recordings and its duration appears definitely longer than that of the scalp far-fields (FIGURE 10 B–C).[22] The results are in line with the proposed relationship of the spinal N_{13}–P_{13} to a fixed generator in the dorsal horn, rather than to any volume-conducted spike volley.

THE SIGNIFICANCE OF THE N_{18} COMPONENT

It appears indeed surprising that the negative N_{18} presents such a widespread distribution over the scalp and earlobes. The diffuse scalp potentials that reflect volume conduction of subcortical activities are generally thought of as positive deflexions. On the other hand, the N_{18} cannot be considered as a cortically generated potential for two reasons: (a) it is too diffuse to represent any specific regional or areal generator that would be consistent with the known anatomy of the brain somatosensory system, and it is too early to be viewed as an association cortex potential; (b) the N_{18} onset is in fact too early to be compatible with a conduction time from the thalamus to the cortex.

The afferent volley was considered to arrive at the thalamus about 3.5 to 3.7 msec after spinal entry time (see above, and Reference 20) and the onset of N_{18} occurs at a mean delay of 4.74 msec (TABLE 1). Taking into account a synaptic delay of 0.3 msec between the lemniscal axons and the ventrobasal neurons in the thalamus, this would leave only 0.74 to 0.94 msec of radiation conduction time. With a thalamo-cortical conduction distance of about 60 mm, this would result in a quite unreasonably high estimate of 81 to 64 m/sec for the CV in thalamo-cortical axons.

The N_{18} must therefore reflect subcortical activities. These could be viewed either as delayed potentials in thalamus or brainstem, or as action potentials in the lower part of the thalamo-cortical axons. The observation that N_{18} presents a sizeable amplitude on the ipsilateral side invites further questions since the thalamo-cortical radiation potentials are elicited from the contralateral thalamus only.[2,53]

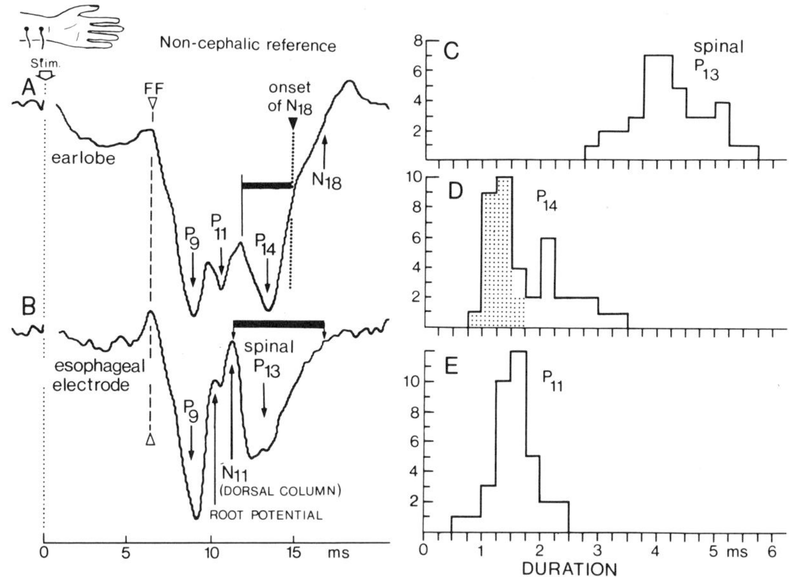

FIGURE 10. Duration of far-field potentials and of the spinal P_{13}–N_{13} component. (A, B) Noncephalic reference recordings of SEP to median nerve stimulation in a young adult subject. (A) Earlobe derivation showing far-fields P_9, P_{11}, and P_{14} (P_{13} can be clearly identified in this experiment). The duration of P_{14} is taken from its onset to the onset of the N_{18} component. (B) Prevertebral (esophageal) recording from the level C_5 showing, after the P_9 nerve far-field, a spinal root action potential, then a small negativity N_{11} from the dorsal column (which is better seen at the posterior neck, FIGURE 4A), and the P_{13} wave whose duration is indicated. The graphs (C) to (E) present pooled data in different adult subjects for the duration of the spinal P_{13} (C), of the scalp or earlobe recorded P_{14} (dotted area correspond to P_{14} when preceded by clear P_{13} while the white area corresponds to P_{14} when P_{13} is apparently lacking) (D), and of the scalp P_{11} (E). (From Desmedt & Cheron.[23] By permission of *Electroencephalography and Clinical Neurophysiology.*)

MAPPING THE CORTICAL SEP COMPONENTS ON THE SCALP

Mapping studies are important to throw light on the organization of neural generators (FIGURE 7) and have been documented and discussed in detail in recent papers.[9,23] One unexpected feature of studies of normal aging of the somatosensory system has been that SEP components are better delineated and parietal components

TABLE 1

MEAN TRANSIT TIMES FROM SPINAL ENTRY (ONSET OF NECK N_{11} OR SCALP P_{11}) (in msec)*

To onset of P_{13}–P_{14} scalp far-fields	1.75 ± 0.30
To peak of P_{14} scalp far-field	3.71 ± 0.55
To onset of scalp N_{18}	4.74 ± 0.27
To onset of parietal N_{20}	6.4 ± 0.45
To onset of prerolandic P_{22}	7.02 ± 0.51

*Pooled data from 40 normal adult subjects.[20, 22, 23]

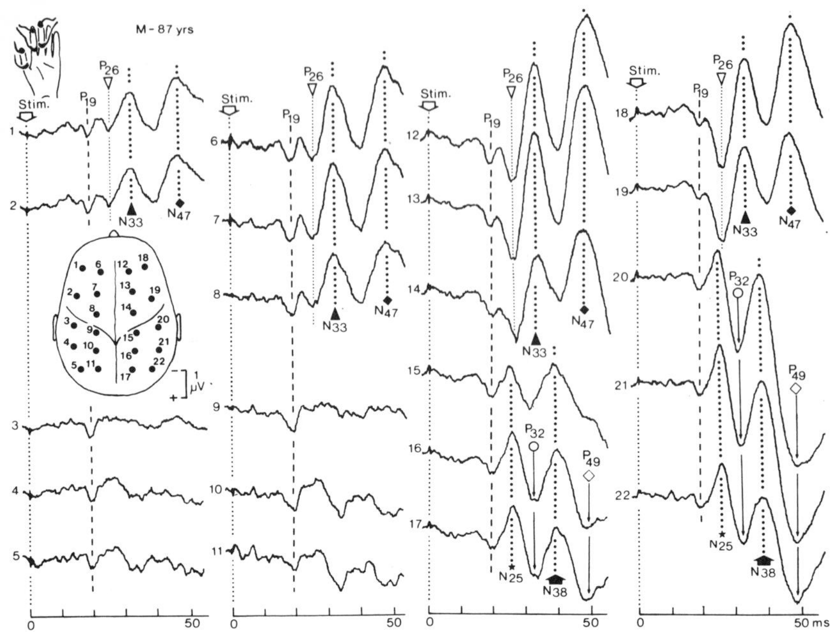

FIGURE 11. Scalp topography of SEP components elicited by finger stimulation in a healthy octogenerian of 87 years. Earlobe reference recording. The amplification is identical for all traces. The figurine indicates electrode positions. The components are labeled, not by the modal peak latency, but by the actual peak latencies in this experiment. Vertical lines with different design are drawn through the various components whereby the stable peak latencies of each component in its own scalp extent offer a sharp contrast with the clear latency transitions across the central fissure. Early components are lacking at the ipsilateral parietal scalp. The P_{14} (here P_{19}) far-field is present in all traces. (From Desmedt & Cheron.[21] By permission of *Electroencephalography and Clinical Neurophysiology*.)

are significantly larger in healthy octogenarians.[21] FIGURE 11 presents a topographical study of SEP components to finger stimulation (earlobe reference in this example) in a healthy male of 87 years. The labels of components here include, not the modal latency value, but the actual peak latencies of each recorded component.

With an earlobe reference, the scalp far-fields are virtually reduced to a P_{14} (here P_{19}) that appears in all 24 traces. The contralateral parietal region shows a large N_{20} (here N_{25}) followed by a characteristic "W" pattern of P_{32} and P_{49} at electrodes 15–17 and 20–22. There is no consistent early response at the ipsilateral parietal region: this lack of an ipsilateral cortical response for finger or distal upper limb stimulation was documented by Desmedt and Robertson (Reference 29, FIGURES 2, 4, 7) and it correlates with anatomical evidence of a lack of callosal connections for the primary somatosensory cortical areas representing the distal limb.[40] This has been a consistent finding in our series.[20,21,23] Electrodes in front of the central fissure record a positive P_{22} (here P_{26}) and two large peaks of negativity N_{33} and N_{47} (electrodes 12–14, 18, and 19). These frontal activities were also seen ipsilaterally with reduced amplitude (electrodes 1, 2, and 6–8).

SEP responses recorded at a frontal electrode were at one time interpreted by a "traveling wave" hypotheses.[10] We do not agree with this proposal because no gradual shift of components in time can be found along the antero-posterior scalp in the 24-electrode maps. The data of FIGURE 11 rather emphasize (together with extensive replication studies[9,23]) that each SEP component maintains its peak latency throughout its own scalp territory. The peak latencies of either the positive or the negative components are clearly different before or behind the central fissure, which appears to represent a sort of boundary between the SEP components' territories. We favor thus the hypothesis that prerolandic or postrolandic components are locally generated by distinct neural cortical systems and that these are activated sequentially by the incoming somatosensory excitation.[9,21,23]

The data are also incompatible with the hypothesis put forward by Broughton[7] who ascribes to a "deep dipole" in the central sulcus the parietal N_{20} and anterior P_{22} components. The onset latency of the prerolandic P_{22} is longer than the onset latency of the parietal N_{20}, and we found a mean delay of 0.62 msec for P_{22} onset in a series of 40 young adults (see the discussion in Reference 23). In fact, P_{22} is not a mirror image of N_{20}, as it should be if it was generated by the same deep dipole. Clear inversion of polarity across the central sulcus were not found in cortical recordings,[8,61] and a definite latency delay was documented very clearly by Papakostopoulos and Crow.[54] A decisive set of data has been provided by studies in patients with focal cortical lesions in whom complete dissociation and separate occurrence of either P_{22} or N_{20} were recorded.[49] Evaluation of these data in conjunction with studies in primates can be found elsewhere.[26]

Indeed the studies in human newborn and infants have shown that the parietal N_{20} phenomenon (which is not recorded in primates) is not an artifact from potential field distortions, but reflects a genuine human phenomenon that has to do with a presumably different organization of the parietal receiving areas in man.[27]

FIGURE 12 illustrates scalp maps of the four main early cortical components with their individual variations in extent. The extensive data collected emphasize the clear boundary in the SEP components' territories that corresponds to the Rolando fissure.[9,23]

CORTICAL GENERATORS OF N_{20} AND P_{22}

The SEP responses recorded in front or behind the central fissure are thus related to distinct cortical generators. Two possibilities for the activation of these generators can be considered:[21] (a) the afferent thalamocortical volley activates the primary parietal areas 3-1-2 and then area 5, and the precentral areas 4 and 6 are subsequently activated by cortico-cortical connections from areas 2 and 5; this would be in line with well-established anatomical pathways;[38–40] or (b) separate thalamocortical pathways with rapid CV convey the sensory inputs to the motor areas 4 and 6. This latter possibility receives the support of physiological data in primates,[6,45,58] but it still appears to raise some questions with respect to its actual anatomical basis.[38]

The rather brief difference in onset latencies for N_{20} and P_{22} that we found in man[23] and the persistence of the precentral responses after parietal lesions in patients[49] would seem to emphasize the second mechanism.

THE SPINO-CORTICAL TRANSIT TIME

Combined recordings of potentials along the peripheral nerve and at the scalp were initially used to estimate the central conduction time.[14] More direct evidence was then provided by the recording of neck potentials. However, these were generally recorded

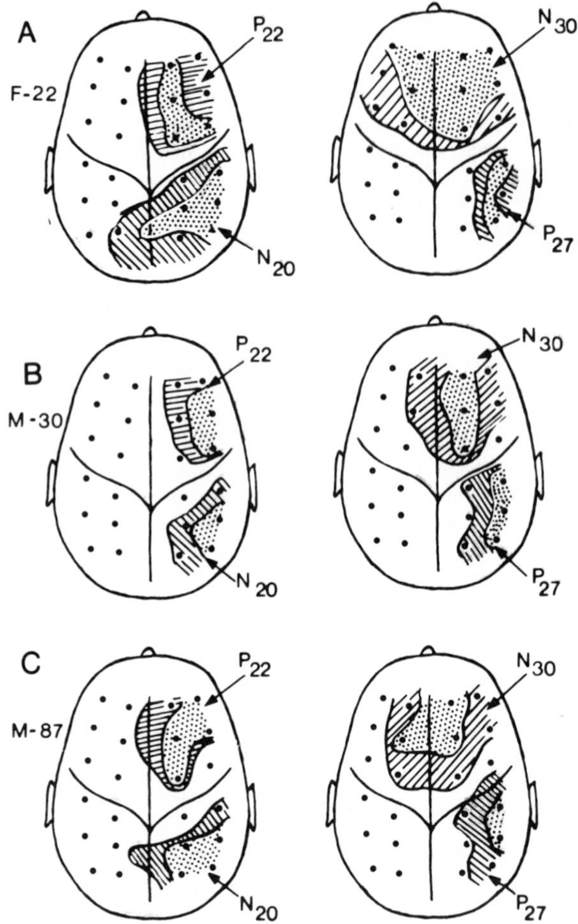

FIGURE 12. Maps of scalp distributions of the early SEP components recorded precentrally (P_{22} and N_{30}) and postcentrally (N_{20} and P_{27}) in 3 subjects: (A) female of 20 years; (B) male of 30 years; (C) male of 87 years, whose records are presented in FIGURE 11. The dotted areas correspond to the scalp focus where the component considered presents at least 75% of its maximum amplitude. The hatched area corresponds to the scalp extent over which the same component presents amplitudes between 50% and 75% of the maximum. (From Desmedt & Cheron.[21] By permission of *Electroencephalography and Clinical Neurophysiology*.)

with a frontal scalp electrode connected to grid 2 of the amplifier, and the peak of the waveform was taken as the spinal time reference.[1,31,32,37,42,59] Such a measure of the peak of this composite (see above) waveform would represent in fact the P_{14} peak contributed by the frontal reference (FIGURE 2 E–F) rather than the true spinal entry time. Moreover, the above studies considered the time interval to the peak of the N_{20} component. The actual time of arrival at the primary parietal cortex is indexed, we think, by the divergence of the N_{20} rising limb from the underlying widespread N_{18} (FIGURES 8 & 9) as can be measured by superimposition with the ipsilateral parietal trace or by considering the record with earlobe reference or by subtraction of the ipsilateral trace (FIGURE 9).

The problem with the method based on interpeak delay between "N_{14}" and N_{20} is that this by no means represents a spino-cortical time. We showed that the peak of "N_{14}" recorded between the posterior neck with a frontal scalp electrode reference connected to grid 2 of the amplifier is spuriously generated by the far-field P_{14} from the front, and does not represent any true spinal component (FIGURE 2 E–F).[20] Since the P_{14} far-field peak corresponds, we think, to the time of arrival of the afferent volley at the thalamus (TABLE 1), what is being measured in that method is in fact the time from arrival at the thalamus up to a full activation (not the onset of activation) of the N_{20} generator in parietal cortex. One should therefore not present the method as a spino-cortical interval, but as a rough estimate of a thalamo-cortical delay.

It is conceded that measures of peaks are easier to do and may be worthwhile under adverse conditions or for rapid testing as has to be done, for example, when monitoring patients during surgery or in intensive care units.[59] The actual significance of the measure should nevertheless be correctly appreciated along with the fact that the lower brainstem and spinal cord is not being tested by the measure. Such correct interpretation of the measure is essential to avoid errors, even for practical clinical purposes, when the data are to be compared with results of the brainstem auditory evoked potentials that is testing another (lower) part of the brain.

We emphasize therefore that the true spino-cortical transit must be estimated from the onset of the posterior neck N_{11}, recorded preferably with a noncephalic reference (FIGURE 1), to the onset of the parietal N_{20} (FIGURES 8 & 9) as was proposed by us[20,22] and further elaborated recently with the identification of the N_{18} phenomenon.[23] Under such conditions the spino-cortical transit time has a mean value of 6.4 msec (TABLE 1) which is longer than the values of 5.4 to 5.6 msec obtained for the interpeak "N_{14}" to N_{20}. The spino-cortical transit time corresponds to afferent conduction along the dorsal column-lemniscal pathway and is unaffected by lesion of the spinothalamic pathway.[36,51]

REFERENCES

1. ABBRUZZESE, M., E. FAVALE, M. LEANDRI & S. RATTO. 1979. Electrophysiological assessment of the central lemniscal pathway in man. Experientia **35:** 775–776.
2. ALBE-FESSARD, D., G. ARFEL, G. GUIOT, P. DEROME, J. HERRAN, H. KORN, E. HERTZOG, G. VOURCH & P. ALEONARD. 1963. Activités électriques caractéristiques de quelques structures cérébrales chez l'homme. Ann. Chir. **17:** 1185–1214.
3. ANZISKA, B. & R. Q. CRACCO. 1980. Short latency somatosensory evoked potentials: Studies in patients with focal neurological disease. Electroencephalogr. Clin. Neurophysiol. **49:** 227–239.
4. AREZZO, J., A. D. LEGATT & H. G. VAUGHAN. 1979. Topography and intracranial sources of somatosensory evoked potentials in the monkey. I. Early components. Electroencephalogr. Clin. Neurophysiol. **46:** 155–172.
5. AREZZO, J. C., H. G. VAUGHAN & A. D. LEGATT. 1981. Topography and intracranial sources of somatosensory evoked potentials in the monkey. II. Cortical components. Electroencephalogr. Clin. Neurophysiol. **51:** 1–18.
6. ASANUMA, H., K. D. LARSEN & H. YUMIYA. 1979. Direct sensory pathways to the motor cortex in the monkey: A basis of cortical reflexes. *In* Integration in the Nervous System, H. Asanuma & V. J. Wilson, Eds. pp. 223–238. Igaku-Shoin, Tokyo, Japan.
7. BROUGHTON, R. J. 1969. *In* Average Evoked Potentials. E. Donchin & D. B. Lindsley Eds. NASA SP-191: 79–84. U.S. Government Printing Office, Washington, D.C.
8. CELESIA, G. G. 1979. Somatosensory evoked potentials recorded directly from human thalamus and SM I cortical area. Arch. Neurol. (Chic.) **36:** 399–405.

9. Cheron, G. & J. E. Desmedt. 1982. Intersubject variability in the scalp topography of somatosensory evoked potential components recorded with a non-cephalic reference in man. Electroencephalogr. Clin. Neurophysiol.
10. Cracco, R. Q. 1980. Scalp-recorded potentials evoked by median nerve stimulation: subcortical potentials, traveling waves and somato-motor potentials. *In* Clinical Uses of Cerebral, Brainstem and Spinal Somatosensory Evoked Potentials. Prog. Clin. Neurophysiol. J. E. Desmedt, Ed. Vol. 7: 1–14. S. Karger, Basel, Switzerland.
11. Cracco, R. Q. & J. B. Cracco. 1976. Somatosensory evoked potentials in man: Far-field potentials. Electroencephalogr. Clin. Neurophysiol. **41:** 460–466.
12. Dawson, G. D. 1954. A summation technique for the detection of small evoked potentials. Electroencephalogr. Clin. Neurophysiol. **6:** 65–84.
13. Debecker, J. & J. E. Desmedt. 1964. Les potentiels évoqués cérébraux et les potentiels de nerf sensible chez l'homme. Acta Neurol. Belg. **64:** 1212–1248.
14. Desmedt, J. E. 1971. Somatosensory cerebral evoked potentials in man. *In:* Handbook of Electroencephalogr. Clin. Neurophysiol. A. Remond, Ed. Vol. 9: 55–82. Elsevier, Amsterdam.
15. Desmedt, J. E., Ed. 1977. Visual Evoked Potentials in Man: New Developments. Clarendon Press, Oxford, England.
16. Desmedt, J. E. 1977. Some observations on the methodology of cerebral evoked potentials in man. *In* Attention, Voluntary Contraction, and Event-Related Cerebral Potentials. Prog. Clin. Neurophysiol. J. E. Desmedt Ed. Vol. 1: 12–29. S. Karger, Basel, Switzerland.
17. Desmedt, J. E., Ed. 1980. Clinical uses of cerebral, brainstem and spinal somatosensory evoked potentials. Prog. Clin. Neurophysiol. Vo. 7. S. Karger, Basel, Switzerland.
18. Desmedt, J. E. 1981. Scalp-recorded cerebral event-related potentials in man as point of entry into the analysis of cognitive processing. *In* The Organization of the Cerebral Cortex. F. O. Schmitt, F. G. Worden, G. Adelman & S. D. Dennis, Eds. Chap. 19: 441–473. MIT Press, Cambridge, Mass.
19. Desmedt, J. E., E. Brunko, J. Debecker & J. Carmeliet. 1974. The system bandpass required to avoid distortion of early components when averaging somatosensory evoked potentials. Electroencephalogr. Clin. Neurophysiol. **37:** 407–410.
20. Desmedt, J. E. & G. Cheron. 1980. Central somatosensory conduction in man: neural generators and interpeak latencies of the far-field components recorded from neck and right or left scalp and earlobes. Electroencephalogr. Clin. Neurophysiol. **50:** 382–403.
21. Desmedt, J. E. & G. Cheron. 1980. Somatosensory evoked potentials to finger stimulation in healthy octogenarians and in young adults: Wave forms, scalp topography and transit times of parietal and frontal components. Electroencephalogr. Clin. Neurophysiol. **50:** 404–425.
22. Desmedt, J. E. & G. Cheron. 1981. Prevertebral (oesophageal) recording of subcortical somatosensory evoked potentials in man: the spinal P_{13} component and the dual nature of the spinal generators. Electroencephalogr. Clin. Neurophysiol. **52:** 257–275.
23. Desmedt, J. E. & G. Cheron. 1981. Non-cephalic reference recording of early somatosensory potentials to finger stimulation in adult or aging normal man: differentiation of widespread N_{18} and contralateral N_{20} from the prerolandic P_{22} and N_{30} components. Electroencephalogr. Clin. Neurophysiol. **52:** 553–570.
24. Desmedt, J. E., J. Debecker & D. Robertson. 1979. Serial perceptual processing and the neural basis of changes in event-related potentials components and slow potential shifts. *In* Cognitive Components in Cerebral Event-Related Potentials and Selective Attention. Prog. Clin. Neurophysiol. J. E. Desmedt, Ed. Vol. **6:** 53–79. S. Karger, Basel, Switzerland.
25. Desmedt, J. E., L. Franken, S. Borenstein, J. Debecker, C. Lambert & J. Manil. 1966. Le diagnostic des ralentissements de la conduction afférente dans les affections des nerfs périphériques: Intérêt de l'extraction du potentiel évoqué cérébral. Rev. Neurol. (Paris) **115:** 255–262.
26. Desmedt, J. E. & A. Galaburda. 1982. Somatosensory evoked potentials: Postrolandic early negativity is characteristic of man (but not primate) and unrelated to any unusual layout of cytoarchitectonic areas. Ann. Neurol.

27. DESMEDT, J. E. & J. MANIL. 1970. Somatosensory evoked potentials of the normal human neonate in REM sleep, in slow wave sleep and in waking. Electroencephalogr. Clin. Neurophysiol. **29:** 113–126.
28. DESMEDT, J. E., P. NOËL, J. DEBECKER & J. L. NAMÈCHE. 1973. Maturation of afferent conduction velocity as studied by sensory nerve potentials and by cerebral evoked potentials. *In* New Developments in Electromyography and Clinical Neurophysiology. J. E. Desmedt, Ed. Vol. 2: 52–63. S. Karger, Basel, Switzerland.
29. DESMEDT, J. E. & D. ROBERTSON. 1977. Differential enhancement of early and late components of the cerebral somatosensory evoked potentials during fast sequential cognitive task in man. J. Physiol. (London) **271:** 761–782.
30. DONCHIN, E., E. CALLAWAY, R. COOPER, J. E. DESMEDT, W. R. GOFF, S. A. HILLYARD & S. SUTTON. 1977. Publication criteria for studies of evoked potentials in man: report of a committee. *In* Attention, Voluntary Contraction and Event-Related Cerebral Potentials in Man. Prog. Clin. Neurophysiol. Vol. 1: 1–11. S. Karger, Basel, Switzerland.
31. EISEN, A. & K. ODUSOTE. 1980. Central and peripheral conduction times in multiple sclerosis. Electroencephalogr. Clin. Neurophysiol. **48:** 253–265.
32. GANES, T. 1980. Somatosensory conduction times and peripheral, cervical and cortical evoked potentials in patients with cervical spondylosis. J. Neurol. Neurosurg. Psychiatry **43:** 683–689.
33. GIBLIN, D. R. 1964. Somatosensory evoked potentials in healthy subjects and in patients with lesions of the nervous system. Ann. N.Y. Acad. Sci. **112:** 93–142.
34. GILLIATT, R. W. & T. A. SEARS. 1958. Sensory nerve action potentials in patients with peripheral nerve lesions. J. Neurol. Neurosurg. Psychiatry **21:** 109–118.
35. GRISOLIA, J. S. & W. C. WIEDERHOLT. 1980. Short latency somatosensory evoked potentials from radial, medial and ulnar nerve stimulation in man. Electroencephalogr. Clin. Neurophysiol. **50:** 375–381.
36. HALLIDAY, A. M. & G. S. WAKEFIELD. 1963. Cerebral evoked potentials in patients with dissociated sensory loss. J. Neurol. Neurosurg. Psychiatry **26:** 211–219.
37. HUME, A. M. & B. R. CANT. 1978. Conduction time in central somatosensory pathways in man. Electroencephalogr. Clin. Neurophysiol. **45:** 361–375.
38. JONES, E. G. 1982. The nature of the afferent pathways conveying short-latency inputs to primate motor cortex. *In* Brain and Spinal Mechanisms of Movement Control in Man. J. E. Desmedt, Ed. Raven Press. New York, N.Y.
39. JONES, E. G., J. D. COULTER & S. H. C. HENDRY. 1978. Intracortical connectivity of architectonic fields in the somatic sensory, motor and parietal cortex in monkeys. J. Compr. Neurol. **181:** 291–348.
40. JONES, E. G. & T. P. S. POWELL. 1969. Connexions of the somatic sensory cortex of the rhesus monkey. II. Contralateral cortical connexions. Brain **92:** 717–730.
41. KEEGAN, J. J. & F. D. GARRETT. 1948. The segmental distribution of the cutaneous nerves in the limbs of man. Anat. Rec. **102:** 409–437.
42. KIMURA, J., T. YAMADA & H. KAWAMURA. 1978. Central latencies of somatosensory cerebral evoked potentials. Arch. Neurol. (Chic.) **35:** 683–688.
43. KRITCHEVSKY, M. & W. C. WIEDERHOLT. 1978. Short latency somatosensory evoked responses in man. Arch. Neurol. (Chic.) **35:** 706–711.
44. KRNJEVIC, K. & M. E. MORRIS. 1976. Input-output relation of transmission through cuneate nucleus. J. Physiol. (London). **257:** 791–815.
45. LEMON, R. N. & J. VAN DER BURG. 1979. Short-latency peripheral inputs to thalamic neurones projecting to the motor cortex in the monkey. Exp. Brain Res. **36:** 445–462.
46. LORENTE DE NO, R. 1947. A study of nerve physiology. Studies from the Rockefeller Institute, New York **132:** Chap. 16.
47. MATTHEWS, W. B., M. BEAUCHAMP & D. G. SMALL. 1974. Cervical somatosensory evoked responses in man. Nature (London) **52:** 230–232.
48. MAUGUIÈRE, F. & J. COURJON. 1981. The origin of short-latency somatosensory evoked potentials in man. A clinical contribution. Ann. Neurol. **9:** 707–710.
49. MAUGUIÈRE, F., J. E. DESMEDT & J. COURJON. 1982. Dissociated loss or enhancement of the frontal or parietal components of somatosensory evoked potentials in patients with unilateral hemispheric lesion: Clinical and CT scan correlates. Ann. Neurol.

50. NAKANISHI, T., Y. SHIMADA, M. SAKUTA & Y. TOYOKURA. 1978. The initial positive component of the scalp-recorded somatosensory evoked potential in normal subjects and in patients with neurological disorders. Electroencephalogr. Clin. Neurophysiol. **45:** 26–34.
51. NOËL, P. & J. E. DESMEDT. 1975. Somatosensory cerebral evoked potentials after vascular lesions of the brainstem and diencephalon. Brain **98:** 113–128.
52. NOËL, P. & J. E. DESMEDT. 1980. Cerebral and far-field somatosensory evoked potentials in neurological disorders involving the cervical spinal cord, brainstem, thalamus and cortex. *In* Clinical Uses of Cerebral, Brainstem and Spinal Somatosensory Evoked Potentials. Prog. Clin. Neurophysiol. J. E. Desmedt, Ed. Vol. 7: 205–230. S. Karger, Basel, Switzerland.
53. OHYE, C., A. FUKAMACHI & H. NARABAYASHI. 1972. Spontaneous and evoked activity of sensory neurons and their organization in the human thalamus. Z. Neurol. **203:** 219–234.
54. PAPAKOSTOPOULOS, D. & H. J. CROW. 1980. Direct recording of the somatosensory evoked potentials from the cerebral cortex of man and the difference between precentral and postcentral potentials. *In* Clinical Uses of Cerebral, Brainstem and Spinal Somatosensory Evoked Potentials. Prog. Clin. Neurophysiol. J. E. Desmedt, Ed. Vol. 7: 15–26. S. Karger, Basel, Switzerland.
55. SMALL, D. G., M. BEAUCHAMP & W. B. MATTHEWS. 1980. Subcortical somatosensory evoked potentials in normal man and in patients with central nervous system lesions. *In* Clinical Uses of Cerebral, Brainstem and Spinal Somatosensory Evoked Potentials. Prog. Clin. Neurophysiol. Vol. 7: 190–204. S. Karger, Basel, Switzerland.
56. STARR, A. 1978. Sensory evoked potentials in clinical disorders of the nervous system. Ann. Rev. Neurosci. **1:** 103–127.
57. STOCKARD, J. J. & F. W. SHARBROUGH. 1980. Unique contributions of short-latency auditory and somatosensory evoked potentials to neurologic diagnosis. *In* Clinical Uses of Cerebral, Brainstem and Spinal Somatosensory evoked potentials. Prog. Clin. Neurophysiol. Vol. 7: 231–263. S. Karger, Basel, Switzerland.
58. STRICK, P. L. & J. B. PRESTON. 1978. Multiple representation in the primate motor cortex. Brain Res. **154:** 366–370.
59. SYMON, L., J. HARGADINE, M. ZAWIRSKI & N. BRANSTON. 1979. Central conduction time as an index of ischemia in subarachnoid haemorrhage. J. Neurol. Sci. **44:** 95–103.
60. WIEDERHOLT, W. C. 1980. Early components of the somatosensory evoked potential in man, cat and rat. *In* Clinical uses of cerebral, brainstem and spinal somatosensory evoked potentials. Prog. Clin. Neurophysiol. Vol. 7: 105–117. S. Karger, Basel, Switzerland.
61. WOOLSEY, C. N., T. C. ERICKSON & E. WARREN. 1979. Localization in somatic sensory and motor areas of human cerebral cortex as determined by direct recording of evoked potentials and electrical stimulation. J. Neurosurg. **51:** 476–506.
62. WOOLSEY, C. N., W. H. MARSHALL & P. BARD. 1942. Representation of cutaneous tactile sensibility in the cerebral cortex of the monkey as indicated by evoked potentials. Johns Hopkins Hosp. Bull. **70:** 399–441.

DISCUSSION OF THE PAPER

B. ANZISKA: What do you think the relationship is between the second positive or P_{11} potential recorded in scalp noncephalic recordings and N_{11} recorded in cervical spine scalp leads?

J. E. DESMEDT: I think they have the same generators. Do you agree with that?

ANZISKA: Yes.

T. ALLISON: A comment on the choice of the reference electrode. I agree that one can get spurious activity using an F_z reference when looking at the spinal components, but, on the other hand, if one wants to track the increased latency of N_{11} up the cord, you can do that quite nicely with an F_z reference. The increase in latency in the average subject is about 0.5 or 0.7 msec.

The great advantage of the F_z reference is that it is not nearly as sensitive to EKG or muscle artifact. For topographic studies, where one is interested in locating sources, a noncephalic reference is quite good. But for routine clinical purposes I think the F_z reference is preferable.

DESMEDT: It all depends on what you want to do. Whenever you mix up grid 1 and grid 2 contributions, I think you are making the problem difficult for yourself. However, in certain clinical situations it may be necessary in the presence of excess interference.

G. CELESIA: We did not find the polarity changes described by Broughton in any of our human cortical recordings. Similarly, Woolsey and Erikson did not find them. Furthermore, Goldring has shown that there are sensory units in the motor cortex. In man, it seems that the motor cortex and the sensory cortex function as a unit in which the sensory cortex is mostly sensory and the motor cortex is mostly motor. A simple dipole theory would appear to be an oversimplification.

DESMEDT: There is also evidence in the monkey showing direct cutaneous input to motor cortex. There is much to suggest this is more complex than once thought.

K. CHIAPPA: We, and Anziska and Cracco have recorded P_{13} and P_{14} in patients who meet all clinical criteria of brain death. We know from neuropathological experience that in these patients the pons and upper medula are necrotic. This suggests that these potentials are generated from below the level of the pons and the upper medulla.

DESMEDT: But could they be generated between the cunate nucleus and the upper medulla?

CHIAPPA: Yes, that is true, and we have to define then which parts of the medial lemniscus we are talking about.

SHORT-LATENCY SOMATOSENSORY EVOKED POTENTIALS TO MEDIAN AND PERONEAL NERVE STIMULATION: STUDIES IN NORMAL SUBJECTS AND PATIENTS WITH NEUROLOGIC DISEASE*

Roger Q. Cracco, Brian J. Anziska, Joan B. Cracco, George A. Vas, Paolo M. Rossini, and Paul J. Maccabee

Department of Neurology
State University of New York
Downstate Medical Center
Brooklyn, New York 11203

MEDIAN NERVE EVOKED POTENTIALS

Short-latency somatosensory evoked potentials (SSEPs) to median nerve stimulation which arise in subcortical and cortical structures have been recorded using scalp-noncephalic and ear reference leads and cervical spine–scalp leads.[1–10] Here we describe the results of investigations concerning these potentials carried out on normal subjects and patients with neurological disease. The purpose of these studies was to provide information concerning the nature of the generator sources of potentials recorded using the different methods and to determine the relative usefulness of these techniques in the evaluation of patients with different types of diseases affecting the nervous system.

Materials and Methods

Normal subjects and patients were sedated (if necessary) with oral diazepam (10 mg) or chloral hydrate (1.5 g). Stimulating electrodes consisting of tin discs (6 mm in diameter) were attached with collodion to the skin over the median nerve at the wrist. The cathode was placed 3 cm proximal to the anode. Stimulating pulses (duration: 0.2 msec) were delivered independently to the left and right median nerves, usually at rates of about 9/sec. The stimulus intensity was adjusted to produce a thumb twitch.

Recording electrodes identical to the stimulating electrodes were similarly attached to the scalp at C_3' and C_4' (2 cm directly caudal to the conventional International 10–20 placements), C_z and to both ears (A_1 and A_2). Recording electrodes were also attached to the skin over the seventh and second cervical spinous processes and over Erb's point bilaterally. Noncephalic reference (NC) electrodes were placed on the shoulders or hands. The impedance of all recording electrodes was maintained between 1000 and 3000 ohms.

Potentials recorded in scalp–ear (C_3'–A_1 with right median nerve stimulation and C_4'–A_2 with left median nerve stimulation), scalp (C_3', C_4', C_z)–NC (shoulder or hand contralateral to the side of stimulation) and Erb's point (ipsilateral to the side of stimulation)–NC leads were obtained from 43 normal subjects who ranged in age from 14 to 92 years. The scalp distribution of SSEPs was investigated in NC leads in 7

*This work was supported by Research Grant NS 12039 from The National Institutes of Health, U.S. Public Health Service.

0077–8923/82/0388–0412 $1.75/0

of these normal subjects by attaching electrodes widely over the scalp and comparing the amplitudes and peak latencies of each of the components at different scalp locations. These SSEPs were also obtained from 26 patients with definite multiple sclerosis,[18] 11 patients who fulfilled the clinical criteria for brain death, 25 patients with diffuse neurological disease of varying etiology and severity, and 31 patients with focal neurological disease. In patients with focal disease, the brachial plexus was involved in 2, the rostral cervical spinal cord in 2, the brainstem in 18, and the diencephalon in 5. Four patients had cerebral lesions involving somatosensory cortex. Lesions were localized on the basis of the clinical evaluation and confirmed in many of the patients by radiological procedures.

Cervical spine–scalp, cervical spine–NC and scalp–NC recordings were obtained and compared in 9 normal subjects and 20 patients to determine the relative contributions of the spine and scalp electrodes to potentials recorded in cervical spine scalp leads. Eight of these patients had intracranial lesions; 7, spinal cord lesions; 1, a brachial plexus root avulsion; and 4 were brain dead.

Input from the recording electrodes was led to differential amplifiers and the output was averaged by a computer and then recorded by an *xy* plotter. In most studies responses to 1000–4000 stimuli were averaged and the frequency response of the recording apparatus was 10–3000 Hz (−3dB). In 6 of the normal subjects C_3' or C_4'–F_z recordings were obtained using bandpasses of 5,150 and 300–3000 Hz. An analysis time of 20 or 40 msec was used often with a 5 msec delay between the shock and the sweep onset. Two or 3 averaged responses were superimposed to differentiate time locked from random components. Relative positivity at grid 1 resulted in a downward deflection in all recordings.

Results and Discussion

In scalp–NC leads 3 positive potentials were recorded with peak latencies of about 9 (P_9), 11 (P_{11}), and 13 (P_{13}) msec[3] (FIGURE 1). P_{11} was not consistently recorded in all normal subjects. P_{13} was bilobed in most normal subjects, the second component (P_{14}) peaking about 1 msec after the first. These were followed by a prominent negative potential peaking at about 20 msec (N_{20}) and a positive potential peaking at about 25 msec (P_{25}). One or two inflections were often observed on both the rising and falling phases of N_{20}. In C_3' or C_4'–F_z recordings performed using frequency bandpasses of 150 or 300–3000 Hz, Maccabee *et al.*[11] found that the N_{20}–P_{25} component was fractionated into up to 7 subcomponents suggesting that there are multiple generator sources for these potentials (FIGURE 2).

In topographic studies performed using NC leads,[3,6] P_9 and P_{11} were widely distributed over the scalp and were similar in amplitude and peak latency at all cephalic locations including the ears. They were poorly defined or absent in scalp–ear and scalp bipolar leads because they underwent cancellation (FIGURE 1). The P_{13}–P_{14} component was also widely distributed but was most prominent around the vertex. It was poorly defined in ear–NC leads and was recorded with reduced amplitude in scalp bipolar and ear reference recordings. P_{13} was better defined than P_{14} in scalp–NC leads, but P_{14} was often better defined than P_{13} in scalp-ear leads. This suggests a more rostral origin for P_{14} than for P_{13}.[12] The N_{20} component was bilaterally distributed but was most prominent at scalp locations overlying the specific somatosensory cortex contralateral to the side of stimulation (C_3' or C_4'). It was well defined in scalp bipolar and ear reference leads.

In scalp–ear and scalp–NC recordings, N_{20} and P_{25} often showed progressive increased in peak latency from anterior to posterior scalp locations in the sagittal plane and from contra- to ipsilateral to the side of stimulation in the coronal plane. In the sagittal plane these latency differences were greatest across the central sulcus[13] (FIGURE 3). These "traveling waves" may be the result of two or more generators which are activated closely but not simultaneously in time. The observation that these potentials were fractionated into multiple peaks in recordings peformed using narrow bandpass filters suggests such an interpretation.

In patients with focal and diffuse neurological disease, multiple sclerosis, and

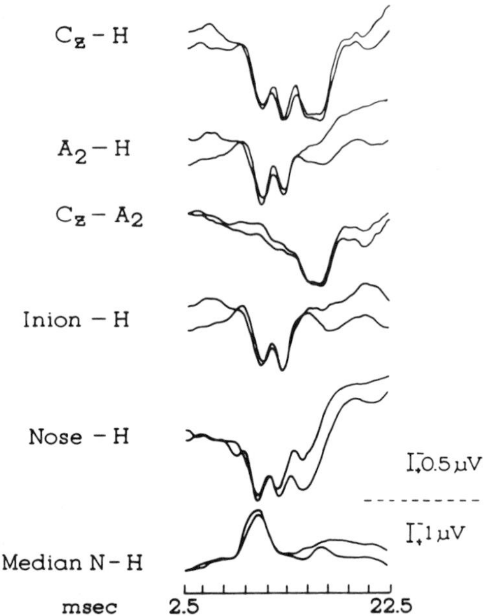

FIGURE 1. Evoked potentials to right median nerve stimulation recorded from the vertex (C_z), right ear (A_2), inion, nose, and over the median nerve just proximal to the axilla (median N) in reference recordings where reference electrode was placed on the dorsum of the left hand (H) or right ear (A_2). Two recordings are superimposed in each trace. In hand reference recordings, the response at the vertex consists of three positive potentials (P_9, P_{11}, and P_{13}–P_{14}). P_9 and P_{11} are similar in amplitude at the vertex, right ear, inion, and nose. They are not clearly defined in the vertex–ear lead. In hand reference recordings, P_{13}–P_{14} is greatest in amplitude at the vertex, smaller at the nose and not clearly evident at the ear or inion. It is well defined in the vertex–ear lead. The latency of onset of P_9 recorded from the scalp, nose, and ear is similar to that of the negative potential recorded over the stimulated median nerve 2 cm proximal to the axilla (bottom trace). (From Cracco & Cracco.[3] By permission of *Electroencephalography and Clinical Neurophysiology*.)

brain death, two criteria were used to define SSEP abnormality: absence of components consisently recorded in normal subjects (P_9, P_{13}–P_{14}, N_{20}) and interpeak latency differences (P_9–P_{11}, P_9–P_{13}, P_9–N_{20}, P_{11}–P_{13}, P_{11}–N_{20}, P_{13}–N_{20}) which were greater than 2.5 SD beyond the mean for normal controls. The P_{13}–P_{14} peak was treated as a single component since its bilobed configuration was not consistently seen in normal subjects. P_9, P_{11}, and P_{13}–P_{14} peak latencies were measured in C_3' and C_4'–NC leads. The N_{20} peak was measured in C_3' and C_4' ear recordings. When potential peaks could

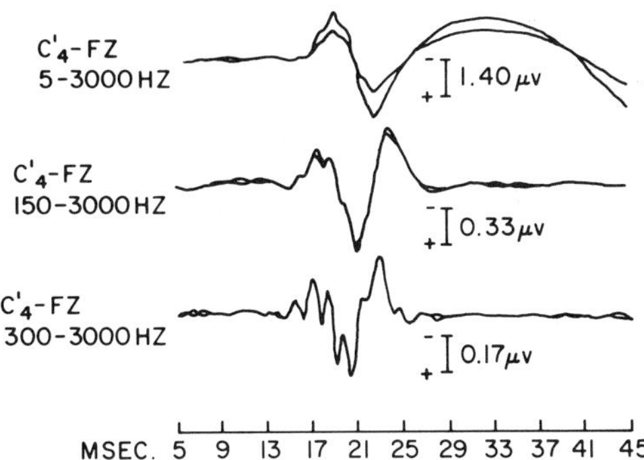

FIGURE 2. Effect of different frequency bandpass filters on SSEPs (C_4'–F_z recordings) to left median nerve stimulation. The analysis time is 40 msec and there is a delay of 5 msec between the shock and the sweep onset. Peak latencies of components are indicated. In the 5–3000 Hz bandpass recordings, inflections superimposed on the N_{20}–P_{25} component are apparent. These inflections are better defined in the 150–3000 Hz recordings. In the 300–3000 Hz recordings, 7 well-defined peaks are superimposed on the N_{20}–P_{25} potential.

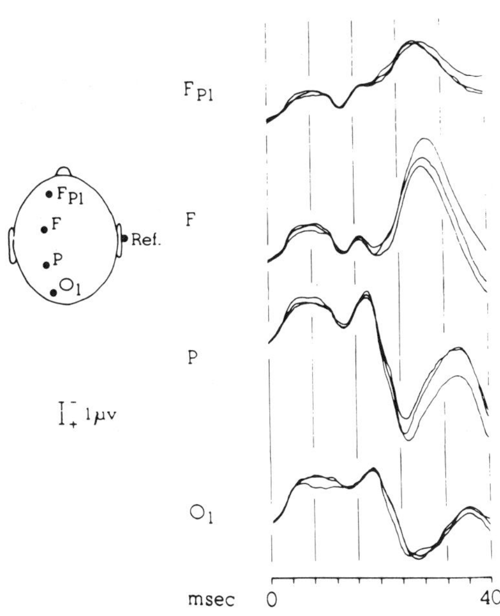

FIGURE 3. Electrodes F and P are placed equidistant between FP_1 and 0_1. Three recordings are superimposed in each trace. The initial positive potential peaks at 14 msec at all recording locations. The subsequent negative, positive and negative potentials (N_{20}, P_{25}, N_{35}) progressively increase in peak latency from front to back. These latency differences are greatest between F and P. (From Cracco.[13] By permission of *Electroencephalography and Clinical Neurophysiology*.)

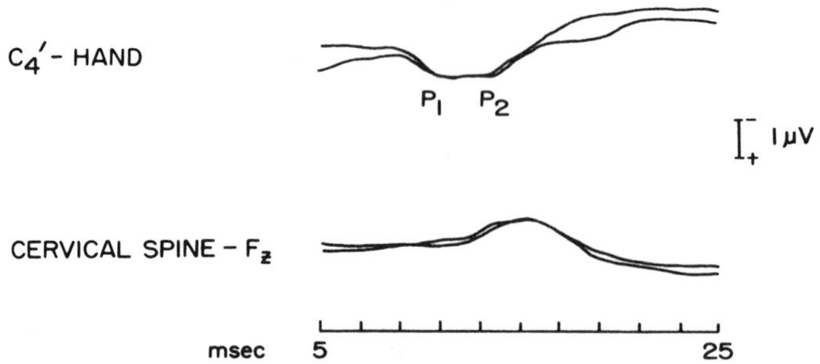

FIGURE 4. Evoked potentials in a 70-year-old male with a right medial medullary infarct. Left median nerve stimulation evoked only P_9 and P_{11} (labeled P_1 and P_2) in the C_4' hand lead. P_{13} is absent. The cervical spine-scalp lead yields a well-defined N_{13} potential. (From Anziska & Cracco.[15] By permission of *Electroencephalography and Clinical Neurophysiology*.)

not be precisely defined, then the peak latency was defined as the point of intersection of lines drawn over the ascending and descending slopes of the potential.[14]

In one patient with a severe chronic brachial plexus lesion situated distal to the dorsal root ganglion, no SSEPs were recorded. In another patient with root avulsion from C_5 through T_1, only P_9 was recorded which suggests this potential arises primarily in stimulated peripheral nerve fibers. The observation that the onset latencies of P_9 and the potential recorded over stimulated mean nerve fibers near the axilla were the same suggests that P_9 reflects the volley as it courses through the brachial plexus from axilla to spinal cord[3,12,15] (FIGURE 1).

Evidence based on a study[15] of 29 adult patients with focal lesions of the cervical spinal cord, brainstem, or cerebrum suggests the following origins for the subsequent potentials: P_{11} in the dorsal columns of the cervical cord, P_{13} primarily in brainstem lemniscal pathways (FIGURE 4) and P_{14} in brainstem or diencephalic lemniscal pathways. The volley arrives in cerebral cortical elements at/or before the peak of N_{20}.[15] The P_{11}, P_{13}, and P_{14} components most probably arise in fiber bundles and do not significantly reflect synaptic activity. This is because the anatomical orientation of cell bodies and dendrites in nuclei would be expected to generate close fields which might

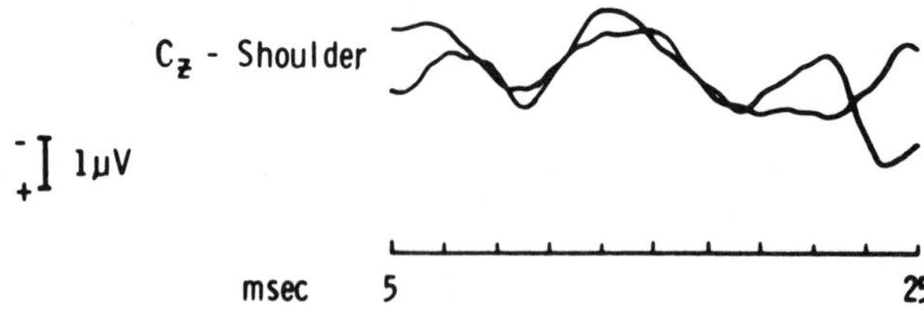

FIGURE 5. Patient with multiple sclerosis who had no abnormal sensory, brainstem or cerebellar findings on examination. Stimulation of the right median nerve elicits only a P_9 potential. (From Anziska *et al.*[6] By permission of *Electroencephalography and Clinical Neurophysiology*.)

not be recorded in the far field.[16] However, the synchronous discharge of axons which are oriented perpendicular to the scalp would be expected to generate an open field of electrical activity which would be recorded in the far field from the scalp as positive events. The short durations of these potentials also favor axonal rather than synaptic events. Arezzo *et al.*[17] have provided evidence in monkeys that suggests that these short-latency potentials arise primarily if not exclusively in fiber tracts.

SSEPs were abnormal in 24 of 26 patients with definite multiple sclerosis.[6,18] Absent components (P_{13}–P_{14} and/or N_{20}) were observed in 17 patients and prolonged peak latency differences in 13 patients (FIGURE 5). Sequential recordings obtained in 5 of 9 patients revealed loss of components or latency changes. Evoked potential abnormalities were often found in the absence of either brainstem signs or clinically evident disturbances of specific sensory systems (FIGURE 5). This probably reflects the multiplicity of lesions known to be present in patients with multiple sclerosis, many of which are not clinically evident.

In a study of 25 patients with diffuse neurological disease of varying etiology and severity,[19] abnormal SSEPs were recorded in 3 of 4 patients with dominant hereditary ataxia, 2 of 3 patients with Huntington's disease, 1 of 3 patients with amyotrophic lateral sclerosis, 1 patient with early Wilson's disease, 1 with CNS sarcoid, 2 with Jamaican neuropathy and 1 with subacute combined degeneration. SSEPs were normal in 2 patients with subacute sclerosing panencephalitis, 4 with anoxic encephalopathy and in the remaining 4 with miscellaneous disorders. Improvement in SSEP abnormality following treatment was observed in the patient with Wilson's disease. These observations demonstrate that SSEPs may be abnormal in patients with diffuse CNS disease as well as in patients with focal lesions and demyelinating disease. It seems that this method may prove useful in estimating prognosis and in evaluating the effect of therapy in some patients with diffuse CNS disease.

SSEPs were recorded in 11 patients who fulfilled the clinical criteria for brain death.[20] P_9 was recorded in all 11 patients, P_{13}–P_{14} was recorded in 5 patients, and N_{20} was recorded in 1 patient (FIGURE 6). SSEP interpeak latency differences were normal in most of these patients. Recordings performed using longer analysis times showed that all later potentials were absent. The observation that certain potentials which arise in central somatosensory pathways can be recorded in some of these patients suggests that these pathways are capable of conducting impulses in some patients who fulfill the clinical criteria for brain death.

In normal subjects, cervical spine–scalp leads yield potentials that are similar in peak latency but opposite in direction to the SSEPs recorded in scalp–NC leads. This would imply they are negative in polarity if one assumes the cervical electrode is more active than the scalp electrode for these components. Components recorded in cervical spine–scalp leads are referred to as N_9, N_{11}, N_{13}, and N_{14}. We compared cervical spine and scalp–NC recordings with cervical spine–scalp leads in normal subjects and patients with focal lesions to obtain information concerning the relative contributions of the cervical spine and scalp electrodes to potentials recorded in cervical spine–scalp leads.[12]

Cervical spine–NC leads yielded initially positive biphasic and triphasic potentials.[3,10,12] The initial positive component had a peak latency of 9 msec, which is essentially identical to P_9 (scalp–NC) and N_9 (cervical spine–scalp). The peak latency of the subsequent negative potential recorded in cervical spine–NC lead was about 13 msec, which was similar to P_{13} (scalp–NC) and N_{13} (cervical spine-scalp). Similar bi- or triphasic potentials have been recorded from the dura and surface over the cervical spinal cord of animals.[21] The initial positive potential reflects the approaching volley recorded in the far field. The first portion of the negative potential arises primarily in the dorsal columns and the subsequent portion reflects primarily spinal cord synaptic activity.

Based on their study of normal subjects and patients, Anziska and Cracco[12] concluded that N_9 (cervical spine–scalp) is actually a positive potential which is greater in amplitude over the scalp, where it is recorded as P_9, than it is over the caudal cervical spine, where it is recorded as the initial positive component of the bi- or triphasic potential. This potential arises chiefly in proximal segments of stimulated median nerve fibers. It is poorly defined in rostral cervical spine–scalp leads because the positive potentials are similar in amplitude at these two locations and undergo cancellation.[3,12] N_{11} is contributed to by both the initial portion of the negative potential recorded by the cervical electrode, which arises in the dorsal columns, and P_{11} recorded by the scalp electrode which is the far-field reflection of the dorsal

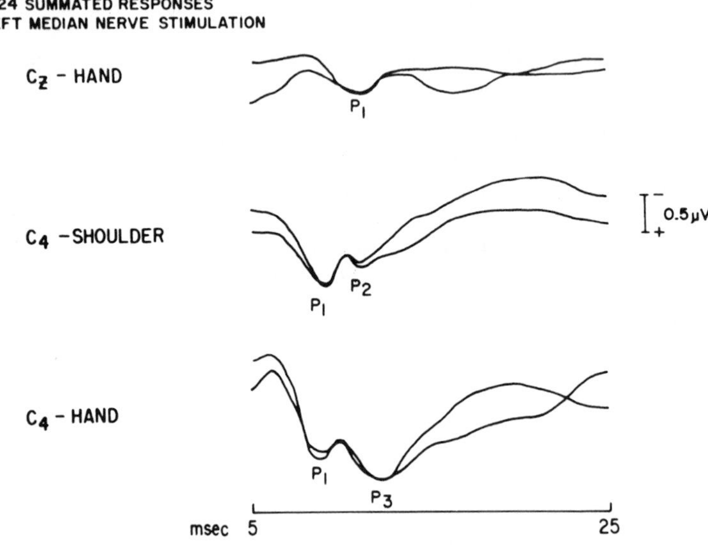

FIGURE 6. Recordings from three clinically brain dead patients. P_9, P_{11}, and P_{13}–P_{14} are labeled P_1, P_2, and P_3, respectively. Only P_9 is recorded in top trace. P_9 and P_{11} are recorded in middle trace, and P_9 and P_{13} in bottom trace. (From Anziska & Cracco.[20] By permission of *Archives of Neurology*. Copyright 1980, American Medical Association.)

column volley. N_{13} is also a composite potential that is contributed to by the major portion of the negative potential recorded by the cervical electrode, which reflects primarily spinal cord and perhaps caudal medullary synaptic activity, and P_{13} recorded by the scalp electrode, which reflects the medial lemniscal volley recorded in the far field. N_{13} (cervical spine–scalp) was recorded in 4 patients in whom P_{13} (scalp–NC) was not recorded which confirms the composite nature of N_{13} (FIGURE 4).[12,15] The "N_{14}" component was thought to be the P_{14} potential recorded by the scalp electrode since a component at this peak latency was never convincingly observed in cervical spine–NC recordings. P_{14} is thought to arise in brainstem or diencephalic lemniscal pathways. These conclusions are in general agreement with those of Desmedt and Cheron.[10]

Comparison of Recording Methods

Scalp-noncephalic recordings are sometimes noisier than cervical spine–scalp recordings. This results from the greater distance between the recording electrodes in scalp–NC leads and the greater likelihood for the NC electrode to record muscle or movement artifact. The N_{11}–N_{13} complex recorded in cervical spine–scalp leads is usually greater in amplitude than P_{11}–P_{13} recorded in scalp–NC leads. This results because of the summation of the negative spinal cord potential recorded by the neck electrode and the positive potentials recorded by the scalp electrode. The greater amplitude of this component in cervical spine–scalp leads is advantageous for clinical use but the multiple generator source of this composite potential may complicate the interpretation of brainstem lesions in the presence of an intact spinal cord (FIGURE 4).

In scalp–ear leads, the 9- and 11-msec potentials are not well defined because the scalp and ear electrodes are similarly active for these components and they undergo cancellation (FIGURE 1).

In the clinical evaluation of patients, a recording electrode is usually placed over Erb's point ipsilateral to the stimulated median nerve. In normal subjects, a well-defined predominantly negative potential is recorded. The Erb's point potential is important as a measure of input into the system. The latency of this potential may be subtracted from the latency of potentials recorded in other derivations (cervical spine–scalp, scalp–NC, scalp–ear). This provides information concerning transit times from Erb's point to the generator sources of potentials recorded in each of the other leads. This is preferable to using absolute latencies since variability due to differences in arm length and temperature are eliminated. Interpeak latency differences of potentials recorded in neck–scalp, scalp–NC, and scalp–ear derivations also provide information concerning transit times between the generator sources of the potentials, which are prolonged in some patients with neurological disease. Absence of potentials consistently recorded in normal subjects provides another indicator of abnormality.[6,9,12,15,19,20,22]

PERONEAL NERVE EVOKED POTENTIALS

Somatosensory evoked potentials (SEPs) to stimulation of nerves in the lower limbs have been recorded from the scalp of normal subjects.[23,24] They have also been recorded in patients with spinal cord dysfunction to provide information concerning the physiological integrity of spinal cord afferent pathways.[25,26] Additionally, these potentials have been utilized in patients undergoing surgery for scoliosis to monitor spinal cord function during this procedure.[27,28] However, the longer latency potentials that have been investigated in these patients are unstable and the definition of what constitutes a normal or an abnormal response has not been precisely defined. This has severely limited the clinical utility of this method.

We attempted to provide a clinically useful method for recording scalp SEPs to stimulation of nerves in the lower extremity.[29,30] Such a method would be expected to provide stable, well-defined potentials whose peak latencies can be reliably identified and measured. Since short-latency SEPs (SSEPs) to median nerve stimulation are more stable than the longer-latency potentials, SSEPs to peroneal nerve stimulation were investigated. The effect of different low-frequency filters on these potentials was determined and their scalp topography was investigated.[29,30] These studies provided information concerning the nature of their generator sources. Evoked potentials to peroneal nerve stimulation were also recorded from surface electrodes placed over the spine and scalp and the conduction characteristics of the response from cauda equina to cerebral cortex were investigated.[29]

Materials and Methods

Observations were made on 48 normal young adult subjects (28 females, 20 males; mean age: 25 years). Tin disc stimulating electrodes (6 mm in diameter) were attached to the skin over each peroneal nerve in the popliteal fossa with collodion and filled with conductive jelly. The cathode was placed 3 cm proximal to the anode. The intensity of stimulation (duration: 0.2 msec) was adjusted to produce dorsiflexion of the foot. Stimuli were delivered at rates of 9–10/sec.

Recording electrodes (6 mm diameter tin discs) were attached to the scalp and over the spine with collodion and filled with conductive jelly. Electrode impedance was maintained between 1 and 4 kΩ. Recordings were obtained with stimulation of one or both peroneal nerves. C_z–ear (A_1 or A_2) and C_z noncephalic shoulder reference (NC) recordings were obtained using frequency bandpasses of both 5–3000 Hz and 150–3000 Hz (−3dB). Peak latencies and interpeak latencies of short latency (15–40 msec) components were determined.

Scalp distribution studies were performed on 10 subjects. An array of 6–8 recording electrodes was placed in the mid-sagittal plane and in the central coronal plane. Close bipolar scalp recordings were obtained and compared with scalp–ear and scalp–FP_z recordings. These recordings were performed using bandpasses of both 5–3000 Hz (6 subjects) and 150–3000 Hz (10 subjects).

In 9 subjects, recording electrodes were attached to the skin over the L_3, T_{12}, T_6, and C_7 spinous processes. Recording electrodes were also placed 3–4 cm rostral to these locations and close bipolar recordings of the response to stimulation of either one or both peroneal nerves were obtained. The C_z–ear and C_z–FP_z scalp recordings were also performed. These recordings were done using bandpasses of both 5–3000 Hz and 150–3000 Hz. The onset latency of the first negative potential recorded at each spinal location and the peak latency of the first scalp recorded potential of presumed cerebral cortical origin were measured. Straight line distances were measured from the L_3, T_{12}, T_6, and C_7 spines to the scalp vertex and spine to scalp propagation velocities were determined.

In each recording the average response to 2000 stimuli was obtained by a computer and written out by an *xy* plotter. An analysis window of 5–45 msec after stimulus was used for scalp recordings and 5–25 msec or 5–45 msec for spinal recordings. The computer horizontal dwell time was 160 μsec. Relative positivity at grid 1 resulted in a downward deflection in all recordings. Two or three averaged responses were superimposed to differentiate time-locked from random activity.

Results and Discussion

Scalp SEPs

In 17 subjects C_z–ear and C_z–NC recordings were performed using a bandpass of 150–3000 Hz. Responses elicited by bilateral nerve stimulation were greater in amplitude but identical in peak latency to responses obtained with stimulation of one nerve. SSEPs (peak latency, 15–40 msec) to stimulation of one or both peroneal nerves consisted of up to 6 positive and 2 negative potentials (P_{1A}, P_{1B}, P_{2A}, P_{2B}, P_{3A}, P_{3B}, N_{1A}, N_{1B}) (FIGURE 7). The negative phases of the potentials preceding N_1 were sometimes prominent.

The P_{1A} and P_{1B} components were small potentials and were recorded in 15 of 17 subjects. P_{2A} was identified in 8 of these 17 subjects either as a well defined peak or inflection on the rising phase of P_{1B}. Components P_{2B} and P_{3A} were well defined in all 17 subjects and were usually greater in amplitude than the preceding potentials. P_{3B} appeared as a notch on the rising phase of N_1 in the 7 subjects in whom it was identified. N_1 was recorded in all 17 subjects and was usually the most prominent of these short-latency potentials. It was bilobed in 4 subjects (N_{1A}, N_{1B}). Except for variability in P_{3B} waveform and slight differences in N_1 peak latency, these potentials were stable in recordings performed on the same subject on different days. The observation that P_{2B}, P_{3A}, and N_1 were stable and well-defined potentials suggests they may be useful in the clinical evaluation of patients. Maccabee[31] has found that P_{3A} is relatively stable in recordings performed on patients undergoing corrective surgery for scoliosis.

In C_z–ear and C_z–NC recordings performed using a bandpass of 5–3000 Hz (13

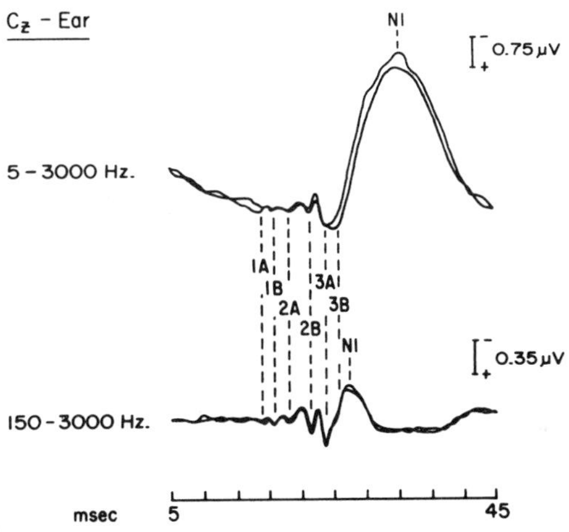

FIGURE 7. Effect of different bandwidth filters on SSEPs (C_z-ear recordings) to bilateral peroneal nerve stimulation. P_{1A} is not defined in the 5–3000 Hz recording. In the 150–3000 Hz recording, N_1 is reduced in amplitude and peak latency and P_{3B} is apparent as an inflection on the rising planes of N_1.

subjects) (Figure 7), all the aforementioned components could be identified in 3 subjects. In these subjects the peak latencies of all these short-latency components were similar using the two bandpasses except that the peak latency of N_1 occurred 4–10 msec earlier in recordings performed with the narrower bandpass. The amplitude and duration of N_1 were also considerably less in recordings performed using a 150–3000 Hz bandpass. In the remaining subjects P_{2A}, P_{2B}, and P_{3B} could clearly be identified in 5, 4, and 3 subjects, respectively; otherwise components were absent or fused to form broad potentials whose peak latencies did not coincide with those recorded with the narrower bandpass.

Topographic studies of SSEPs to stimulation of one or both peroneal nerves recorded in ear or NC reference leads using 150–3000 Hz filters revealed that P_{1A}, P_{1B}, and P_{2A} were widespread in their scalp distribution. Their short latency, positive polarity, and topography suggest these components are far-field events that arise in deep subcortical structures. They were absent or poorly defined in scalp bipolar recordings because they were similar in amplitude at various scalp locations and underwent cancellation. The onset latency of P_{1A} was similar to that of the evoked potential recorded over the caudal cervical spinal cord. This suggests that this component arises in rostral spinal cord afferent pathways. P_{1B} may arise in cervical spinal cord or caudal brainstem and P_{2A} in brainstem lemniscal pathways. P_{2B}, which was more restricted in its scalp distribution and maximal at frontal central scalp locations, may arise in brainstem lemniscal pathways or the thalamocortical radiations. P_{3A}, P_{3B}, and N_1 were most prominent at C_z–P_z recording locations. They had relatively restricted scalp distributions and steep potential gradients, which suggests that they arise in the cerebral cortex. The observation that these potentials often had an amplitude predominance at central parietal locations ipsilateral to the side of stimulation suggests the occurrence of cortical dipole generator sources in the parasagittal cortex contralateral to the side of stimulation which are oriented tangentially to the scalp recording electrodes.[24] The peak of P_{3A} may reflect the approximate time the volley arrives in cerebral cortical elements. The average interpeak latency difference of 8.5 msec between P_{1A} and P_{3A} is consistent with the expected transit time of a volley from rostral spinal cord to cerebral cortex.[10]

In recordings performed using low-frequency filters of 5 Hz, certain potentials were characterized by progressive increases in peak latency from anterior to posterior scalp locations in the sagittal plane and from contralateral to ipsilateral to the side of stimulation in the coronal plane. These "traveling waves" suggest the occurrence of multiple generators which are activated closely but not simultaneously in time.[13] These peak latency shifts were not apparent in recordings performed using low-frequency filters of 150 Hz. The progressive phase shifts of the first negative and the preceding major positive potential observed in recordings performed using a 5-Hz low-frequency filter resulted because subcomponents of these composite potentials, which were evident in 150-Hz low-frequency filter recordings, had different scalp distributions in both the sagittal and coronal planes. Shorter latency subcomponents of these composite potentials were relatively greater in amplitude at anterior scalp locations and at recording sites contralateral to the side of stimulation, while the longer latency subcomponents were more restricted to the vertex and central parietal regions ipsilateral to the side of stimulation.

Spine and Scalp SEPs

Responses recorded over the lumbar, thoracic, and cervical spines with a bandpass of 5–3000 Hz consisted of initially positive triphasic potentials sometimes followed by a negative potential. More complex potentials were sometimes recorded over the lower thoracic spine. In recordings performed using a bandpass of 150–3000 Hz these potentials were transformed into more complex wave forms and the peak latencies of some components were decreased. However, the onset latency of the initial negative potential obtained at all spinal recording locations was similar using both frequency filters. Cervical potentials were recorded with difficulty using either frequency filter because of their small signal size.

The peak latency of the scalp recorded P_{3A} was taken to reflect the approximate time the volley arrived in cerebral cortical elements. The propagation velocities and standard deviations of the responses from the L_3, T_{12}, T_6, and C_7 spinous processes to P_{3A} were 39.7 ± 1.4, 36.6 ± 1.8, 33.5 ± 1.3 and 22.9 ± 1.0 m/sec, respectively.

These studies show that, when both scalp and spinal evoked potentials are recorded in the same subject, information concerning the conduction characteristics of the response from cauda equina to cerebral cortex is provided. Even when responses over the rostral spinal cord cannot be recorded because of their small signal size, responses recorded over more caudal spinal regions may be used as latency indicators and the conduction characteristics from these locations to the stable scalp recorded SEPs may be determined. The generator sources of the evoked potentials recorded over the spine and the spinal afferent tracts which transmit the scalp recorded SEP may not be identical. Nevertheless, this method should provide useful information in the evaluation of patients with disorders affecting the cauda equina and spinal cord.

Summary

SSEPs to median nerve stimulation which arise in the brachial plexus, subcortical and cortical structures can be recorded from the scalp. Abnormalities of these potentials have been found in patients with demyelinating disease and focal or diffuse disease of the nervous system. SSEPs to peroneal nerve stimulation which arise in rostral spinal cord, brainstem, and cerebral structures have also been recorded from the scalp. These methods can be expected to provide useful information in patients with certain neurological disorders.

References

1. Cracco, R. Q. 1972. The initial positive potential of the human scalp recorded somatosensory evoked response. Electroencephalogr. Clin. Neurophysiol. **32:** 623–629.
2. Matthews, W. B., M. Beauchamp & D. G. Small. 1974. Cervical somatosensory evoked responses in man. Nature **252:** 230–232.
3. Cracco, R. Q. & J. B. Cracco. 1976. Somatosensory evoked potential in man: Far field potentials. Electroencephalogr. Clin. Neurophysiol. **41:** 460–466.
4. Jones, S. J. 1977. Short latency potentials recorded from the neck and scalp following median nerve stimulation in man. Electroencephalogr. Clin. Neurophysiol. **43:** 853–863.
5. Kritchevsky, M. & W. C. Wiederholt. 1978. Short latency somatosensory evoked potentials. Arch. Neurol. **35:** 706–711.
6. Anziska, B., R. Q. Cracco, A. W. Cook & E. W. Feld. 1978. Somatosensory far field potentials: Studies in normal subjects and patients with multiple sclerosis. Electroencephalogr. Clin. Neurophysiol. **45:** 602–610.
7. Nakanishi, T., Y. Shimada, M. Sakuta & Y. Toyokura. 1978. The initial positive component of the scalp recorded somatosensory evoked potentials in normal subjects and in patients with neurological disorders. Electroencephalogr. Clin. Neurophysiol. **45:** 26–34.
8. Yamada, T., J. Kimura & D. M. Nitz. 1980. Short latency somatosensory evoked potentials following median nerve stimulation in man. Electroencephalogr. Clin. Neurophysiol. **48:** 367–376.
9. Chiappa, K. H., S. K. Choi & B. R. Young. 1980. Short latency somatosensory evoked potentials following median nerve stimulation in patients with neurological lesions. *In* Progress in Clinical Neurophysiology. J. E. Desmedt, Ed. Vol. **7:** 264–281. S. Karger Publishing Co., Basel, Switzerland.
10. Desmedt, J. E. & G. Cheron. 1980. Central somatosensory conduction in man: Neural generators and interpeak latencies of the far field components recorded from neck and right or left scalp and earlobes. Electroencephalogr. Clin. Neurophysiol. **50:** 382–403.
11. Maccabee, P. J., E. I. Pinkhasov & P. Tsairis. 1981. Effect of different frequency filters on short latency somatosensory evoked potentials. (Abstract.) Electroencephalogr. Clin. Neurophysiol. **52:** 108 p.
12. Anziska, B. & R. Q. Cracco. 1981. Short latency SEPs to median nerve stimulation: Comparison of recording methods and origin of components. Electroencephalogr. Clin. Neurophysiol. **52:** 531–539.
13. Cracco, R. Q. 1972. Traveling waves of the human scalp recorded somatosensory evoked response. Electroencephalogr. Clin. Neurophysiol. **38:** 557–566.

14. STARR, A. & J. ACHOR. 1975. Auditory brain stem responses in neurological disease. Arch. Neurol. **32:** 761–768.
15. ANZISKA, B. & R. Q. CRACCO. 1981. Short latency somatosensory evoked potentials: Studies in patients with focal neurological disease. Electroencephalogr. Clin. Neurophysiol. **49:** 227–239.
16. LORENTE DE NO, R. 1947. A study of nerve physiology. Studies from the Rockefeller Institute. Vol. 132, Chap. 16. Rockefeller Institute, New York, N.Y.
17. AREZZO, J., A. D. LEGATT & H. G. VAUGHAN. 1979. Topography and intracranial sources of somatosensory evoked potentials in the monkey. Early components. Electroencephalogr. Clin. Neurophysiol. **46:** 155–172.
18. MCALPINE, D. 1972. Course and prognosis. *In* Multiple Sclerosis: A Reappraisal. 2nd ed it. D. McAlpine, C. E. Lumsden & E. D. Acheson, Eds. p. 202. Churchill-Livingston, Edinburgh, Scotland.
19. ANZISKA, B. & R. Q. CRACCO. 1981. Short latency somatosensory evoked potentials (SSEPs) in patients with diffuse neurological disease. (Abstract.) Electroencephalogr. Clin. Neurophysiol. (In press).
20. ANZISKA, B. & R. Q. CRACCO. 1980. Somatosensory evoked short latency potentials in brain dead patients. Arch. Neurol. **37:** 222–225.
21. ALLISON, T. & A. L. HUME. 1981. A comparative analysis of short-latency somatosensory evoked potentials in man, monkey, cat and rat. Exp. Neurol. (In press).
22. HUME, A. L., B. R. CANT & N. A. SHAW. 1979. Central somatosensory conduction time in comatose patients. Ann. Neurol. **5:** 379–384.
23. DAWSON, G. D. 1947. Cerebral responses to electrical stimulation of peripheral nerve in man. J. Neurol. Neurosurg. Psychiatry **10:** 134–140.
24. TSUMOTO, T., N. HIROSE & S. NONAKA. 1972. Analysis of somatosensory evoked potentials to lateral popliteal nerve stimulation in man. Electroencephalogr. Clin. Neurophysiol. **33:** 379–388.
25. HALLIDAY, A. M. & G. S. WAKEFIELD. 1963. Cerebral evoked potentials in patients with dissociated sensory loss. J. Neurol. Neurosurg. Psychiatry **26:** 211–219.
26. PEROT, P. L., JR. 1973. The clinical use of somatosensory evoked potentials in spinal cord injury. Clin. Neurosurg. **20:** 367–382.
27. NASH, C. L., R. A. LORIG, L. A. SCHATZINGER & R. H. BROWN. 1977. Spinal cord monitoring during operative treatment of the spine. Clin. Orthop. **126:** 100–105.
28. ENGLER, L. L., N. I. SPIELHOLZ, W. N. BERNHARD, F. DANZIGER, H. MERKIN & T. WOLFF. 1978. Somatosensory evoked potentials during Harrington Instrumentation for scolioses. J. Bone Joint Surg. **60:** 528–532.
29. ROSSINI, P. M., R. Q. CRACCO, J. B. CRACCO & W. J. HOUSE. 1981. Short latency somatosensory evoked potentials to peroneal nerve stimulation: Scalp topography and the effect of different frequency filters. Electroencephalogr. Clin. Neurophysiol. **52:** 540–552.
30. VAS, G. A., J. B. CRACCO & R. Q. CRACCO. 1981. Scalp recorded short latency cortical and subcortical somatosensory evoked potentials to peroneal nerve stimulation. Electroencephalogr. Clin. Neurophysiol. **52:** 1–8.
31. MACCABEE, P. J., E. I. PINKHASOV, P. TSAIRIS & D. B. LEVINE. 1981. Spinal and short latency scalp derived somatosensory evoked potentials during corrective spinal column surgery (Abstract). Electroencephalogr. Clin. Neurophysiol. (In press).

DISCUSSION OF THE PAPER

W. WIEDERHOLT: What makes you think that the voltage that is eliminated by raising the low end filter is any less important than the voltage that remains. Secondly, if it is not important and you can get rid of it, where do you think it is coming from?

R. Q. CRACCO: I do not know where the voltage that is eliminated is coming from, but I would think it is important. The generator sources of the eliminated voltage are less synchronous than those of the remaining potentials. The eliminated voltage might arise in relatively asynchronously firing fiber tracts. Synaptic activity might also contribute to it. The remaining potentials, however, are real because in occasional subjects all the potential peaks recorded with narrow bandpass filters are apparent in wide bandpass recordings.

J. AREZZO: I would like to comment on the multiple generators that Dr. Cracco has just described. We have just completed a study in the monkey on the initial cortical responses that arise in and around the pre- and post-central gyrus. Using both gross potential and multiple unit techniques, we found three generators that are active within 3 msec and are independent. The first component, peaking at 10 msec (P_{10}), is generated in area 3B and is oriented in the anterior posterior direction. The second component, peaking at 12 msec (P_{12}), is generated in areas 3, 1, and 2 on the crown. The third component, peaking at 13 msec (P_{13}), is generated in the pre-central gyrus, principally in the sulcus but also extending to the crown. The field distribution from these three generators is exceptionally complex and very consistent with some of the data you presented here.

BRAINSTEM AUDITORY EVOKED RESPONSES: INTRODUCTION

James Jerger

Baylor College of Medicine
and The Methodist Hospital
Houston, Texas 77030

The auditory brainstem reponse has been a viable clinical and research tool for only a decade. During that short period, however, it has led to an elucidation of possible generator sites. Great impetus to the clinical exploitation of the auditory brainstem response (ABR), especially in children, was provided by the important early paper of Hecox and Galambos in 1974 on the nature and maturation of the response in infants, and it is to Dr. Galambos that I think all of us owe a very important debt because of his continual reiteration of a fundamental principle that I think had a profound effect on my own view of these phenomena and hopefully will become a more pervasively accepted concept, namely, that anyone who is working with auditory brainstem response must be at the same time both an audiologist and a neurophysiologist. The ABR response cannot be interpreted without a thorough understanding of its nature from the standpoint of both disciplines.

During the past five years the clinical exploitation of ABR has been widespread in two major areas: first, in the neuro-audiologic evaluation of patients with suspected retrocochlear auditory disorders: and second, in the audiologic evaluation of infants and other difficult-to-test children.

In the first area, neuro-audiologic evaluation, the value of ABR in detecting, for example, acoustic tumors and early multiple sclerosis, has been dramatically documented. Used judiciously as part of a comprehensive test battery, ABR shows high sensitivity and quite amazing site specificity, but a high false-positive rate, especially in eighth nerve tumors, remains a persistent problem.

The successful application of ABR in pediatric populations has been particularly gratifying. It has truly provided a giant leap forward in this persistently difficult area. Present pediatric applications include newborn screening, evaluation of children in newborn intensive care units, the difficult-to-test child at any age, children with neurologic disorders with suspected nonorganic problems, and recent interest, of course, in the sudden infant death syndrome.

There are however, two lingering problems here: first, the continuing problem of frequency specificity; and second, the problem of evaluating the total auditory system to provide a true electrophysiologic analog to the behavioral response. Although much has been learned about the basic properties of the ABR, the problem of generator sites is still not entirely solved, and there is accumulating evidence that the simplistic view of a series of waves uniquely corresponding to specific neural structures must be abandoned in favor of a model incorporating complex and multiple generator sources.

In this section, these key issues—both basic and applied—will be addressed by a series of distinguished speakers.

DISTRIBUTION OF AUDITORY BRAINSTEM POTENTIALS OVER THE SCALP AND NASOPHARYNX IN HUMANS*

A. Starr and K. Squires

Department of Neurology
University of California, Irvine
Irvine, California 92717

Introduction

A dominant characteristic of scalp-recorded auditory brainstem potentials or responses (ABR) is their relative invariance over much of the surface of the human head, a fact accounted for by the "far-field" nature of the recordings.[1] A number of idiosyncracies in the scalp distributions of the various peaks have been noted, which may be relevant both for understanding the neural generators of these events and for defining recording arrays for clinical use.

Picton *et al.*[2] utilized a noncephalic reference and reported that wave I was largely restricted to the region of the mastoid and that waves I and III varied in polarity according to whether they were recorded from the mastoid ipsilateral or contralateral to a monaural stimulus. Waves I and III were negative in polarity for ipsilateral stimulation and positive in polarity for contralateral stimulation. Wave III, moreover, could be attributed to a theoretical source having both horizontal and vertical dipole components. Waves IV and V had positive polarities at all scalp locations and, while largest at the vertex, showed very little variation in amplitude over their electrode array.

Terkildsen *et al.*[3] also used a noncephalic reference and noted similar results. Wave I appeared as a negative deflection at the ipsilateral mastoid and as a positive deflection at the vertex. Waves II and III were a "single wave" at the ipsilateral mastoid and all subsequent peaks recorded from the vertex and mastoid were in phase.

Streletz *et al.*[4] reported that wave V recorded with a noncephalic reference was widely distributed over the scalp, but most prominent at the vertex. They also noted that, while wave I was recorded as a small positive potential at the vertex, it appeared as a larger negative potential at the mastoid ipsilateral to stimulation.

Stockard *et al.*[5] utilized an ankle reference and found that wave I was a negative deflection at the ipsilateral earlobe. Waves IV and V were positive at the earlobe. In a further comparison of waveforms obtained with referential recording between the vertex and ipsilateral and contralateral ears, both Stockard *et al.*[5] and Hixson and Mosko[6] confirmed earlier observations[7] that waves I and III are attenuated with a contralateral reference. Both studies defined that wave V was clearer using a contralateral rather than an ipsilateral earlobe as reference.

Van Olphen *et al.*,[8] however, reported no significant differences between the responses to ipsilateral and contralateral stimulation using the laryngeal prominence as a reference site, although no quantitative data were presented. In contrast to the previous studies, van Olphen *et al.*[8] also studied the anterior–posterior gradient of the ABR along the midline. While the maximum amplitudes were obtained at the vertex,

*This work was supported by Grant 11876 from the National Institutes of Health.

0077–8923/82/0388–0427 $1.75/0 © 1982, NYAS

sizable potentials were obtained as far anteriorly as the nasion, whereas at the inion all peaks were extremely small.

A map by Martin and Moore[9] of the ABR scalp distribution in humans is difficult to relate to the other studies because of the use of binaural stimuli. No asymmetries or polarity inversions were observed.

Thus, the various peaks of the ABR waveform differ in their relative amplitudes over the scalp, and some studies find that a few of the peaks have amplitude asymmetries in the lateral plane according to the ear of stimulation. In the present study we have made a systematic analysis of the amplitude distributions of all ABR peaks in both the lateral (coronal) and anterior–posterior (sagittal) scalp distributions using a noncephalic reference. A second purpose of the present experiments was to measure component latencies at several different electrode positions.

The definition of latency of the ABR components as a function of scalp derivation was not systematically investigated in these prior studies. Results from several animal species[11,12] have demonstrated that the latency of the ABR components do change across the scalp.

A third purpose was to study the ABR recorded from the human nasopharynx as reported by Martin and Coats.[10] They reasoned that nasopharyngeal electrodes might afford a unique view of the ABR since the electrode tip would be in close proximity to the presumed generators of the ABR and ideally positioned to distinguish lateralized differences according to the ear stimulated. They reported large differences in the amplitudes of two components, labeled NP3 (latency, about 4 msec) and NP6 (latency, about 7.5 msec), according to whether the stimulus was ipsilateral or contralateral to the nasopharyngeal electrode. They suggested that the NP3 potential may reflect near-field activity in the medial superior olive. In their study, however, the reference electrode was situated at the mid-forehead, which is now known to be an "active" site[8] and thus, the results do not allow for the definite separation of near-field potentials. In the study to be reported below, the nasopharyngeal electrode was referenced to a noncephalic site.

Finally, we compared the ABR derived from several recording arrays (C_Z referenced to a noncephalic site, C_Z referenced to either mastoid, and intermastoid recordings) to clarify the advantages and limitations of these various recording montages.

METHODS

Eleven subjects participated in the study. All were laboratory personnel between the ages of 21 and 35 years. Their audiograms were not defined and hearing was assumed to be normal based on their own evaluation and their normal thresholds to the click signals.

Measurement of the amplitude of the ABR was made on five of the subjects, all male. Silver cup electrodes were applied to scalp with collodion at nine locations. A mid-sagittal array was formed by electrodes at the nasion (N), vertex (C_Z), inion (I), a frontal location (F) midway between the nasion and vertex, and a parietal location (P) midway between the vertex and inion. A coronal array was formed by electrodes at each mastoid (M_1 and M_2), vertex (C_Z), and at points midway between the vertex and each mastoid (L_1 and L_2). A noncephalic reference electrode was fastened to the skin over the seventh cervical vertebra (C_{VII}) and a ground electrode was placed on the forearm.

The EEG was amplified 50,000 times with a bandpass of 0.35–10 kHz and recorded on FM tape (bandpass, 0 to 5 kHz) for later analysis. The final bandpass during analysis was 0.1–3 kHz.

During testing, the subject was supine in a sound-attenuating chamber. Monaural condensation clicks generated by activating TDH39 earphones with 0.1 msec pulses were presented at 10/sec rate and at an intensity of 65 dB above the sensory level (SL). The five electrodes making up the sagittal or coronal array were recorded simultaneous. The C_Z electrode was included in both arrays. Each ear was tested separately and replicate ABR waveforms averaged over 4096 trials were collected in a balanced order. The waveforms were displayed and amplitude measured from baseline (determined as the average value of the potential in a 3 msec period prior to stimulus presentation) to the peak of the wave. The latency at which each peak's amplitude was measured was derived from the ABR recorded at C_Z for each subject.

To facilitate presentation of the coronal-array data, the electrode positions are labeled according to their relationship to the stimulated ear. Thus, the ABRs recorded from M_1 and L_1 for left-ear stimulation and M_2 and L_2 for right-ear stimulation are labeled M_I and L_I (ipsilateral), whereas the recordings from M_1 and L_1 for right-ear stimulation and M_2 and L_2 for left-ear stimulation are labeled M_C and L_C (contralateral).

In a second study latency measures of the ABR were made on six different subjects (three men, three women). Electrodes were affixed at C_Z, M_I (mastoid ipsilateral to stimulus), M_C (mastoid contralateral to stimulus), and referenced to C_{VII}. A nasopharyngeal electrode was inserted in each subject and also referenced to C_{VII}. Monaural condensation clicks were presented at 11/sec and at 65 dBSL to each ear and two sets of 2000 trials were collected to obtain duplicate averages. Appropriate manipulation of this data in computer memory allowed the definition of potentials between C_Z–M_I, C_Z–M_C, and M_C–M_I. (For instance, to obtain the C_Z–M_I derivation, the potentials recorded between M_I–C_{VII} were subtracted from the potentials recorded between C_Z–C_{VII}.) Measures of polarity, latency, and amplitude of the various components were obtained from both the noncephalic referential recordings and the derived bipolar recordings.

RESULTS

The Amplitude of Auditory Brainstem Responses

The ABR waveforms over the scalp for one subject are shown in FIGURE 1. The grand-average ABR waveforms over two replications (8192 trials) for the right and left ears are superimposed. No significant amplitude differences were found as a function of whether the left or right ear was stimulated when the electrode sites were identified according to their relationship (ipsilateral or contralateral) to the ear stimulated. Six characteristic vertex-positive waves, labeled I–VI and five vertex-negative, waves labeled I_n to V_n, were identifiable at most electrode sites.

The mean amplitude of each peak are presented in FIGURE 2A. Wave I was positive in polarity at C_Z and over the scalp contralateral to the stimulated ear. At the ipsilateral mastoid, the polarity of wave I was negative (FIGURES 1 & 3). The difference between the amplitudes of wave I at the ipsilateral and contralateral mastoids was significant at the 0.05 level. Waves I_n, II, IV, and V were largest at C_Z, and were not significantly lateralized. Wave III was significantly larger at the contralateral mastoid and of opposite polarity than at the ipsilateral mastoid ($p < 0.01$). Waves II_n, III_n, and IV_n showed little amplitude variation across the array.

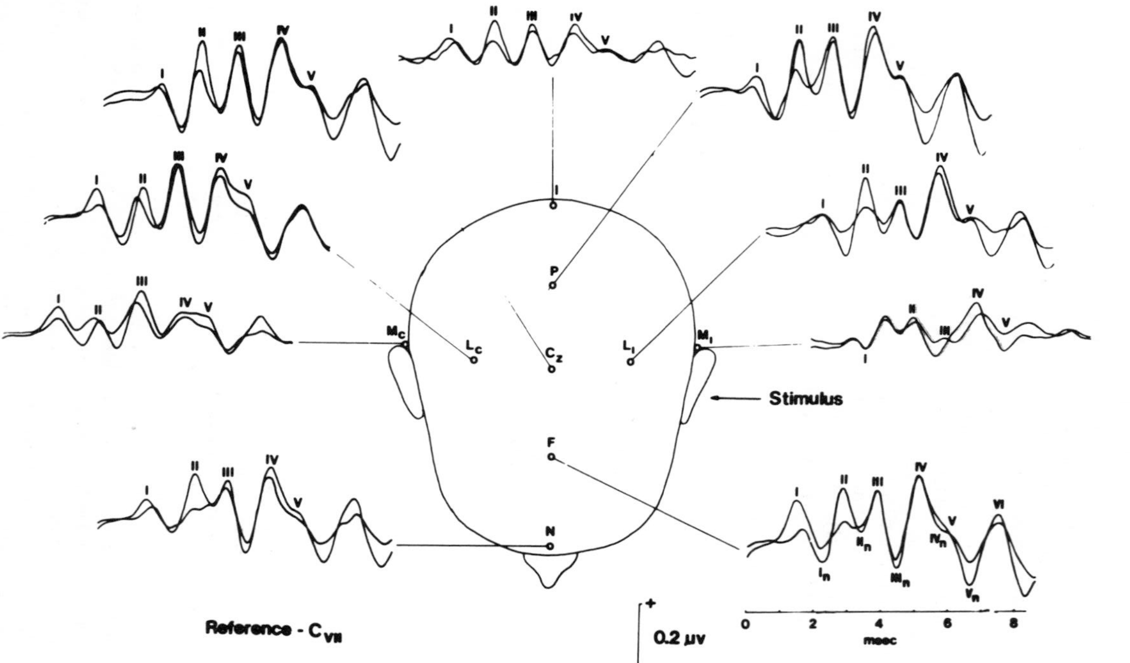

FIGURE 1. Auditory brainstem responses (ABRs) recorded from scalp electrodes in coronal and sagittal arrays referenced to a noncephalic site (seventh cervical vertebra, C_{VII}). The stimuli were monaural condensation clicks, 65 dBSL, presented at a rate of 10/sec. Duplicate averages of 4096 trials are presented. The designation of *I* and *C* in this and all subsequent figures refers to the electrode site as being ipsilateral (I) or contralateral (C) to the ear being stimulated (designated by arrow).

Wave VI was generally negative with respect to the prestimulus baseline and showed no significant amplitude variation. There was, however, an overall tendency to more negative voltages over the contralateral scalp for the latter portion (6 msec and later) of the ABR waveform. This can be seen as a trend in the graphs for waves V_n and VI, and as a significant difference between the voltages at the two mastoids for wave VI_n ($p < 0.05$) with the component being larger at the contra- versus ipsilateral mastoid.

The mean amplitudes for the sagittal array are shown in FIGURE 2B. The interpretation of amplitude of the ABR peaks in the sagittal distribution must be evaluated cautiously since the electrodes at the ends of the array vary in their

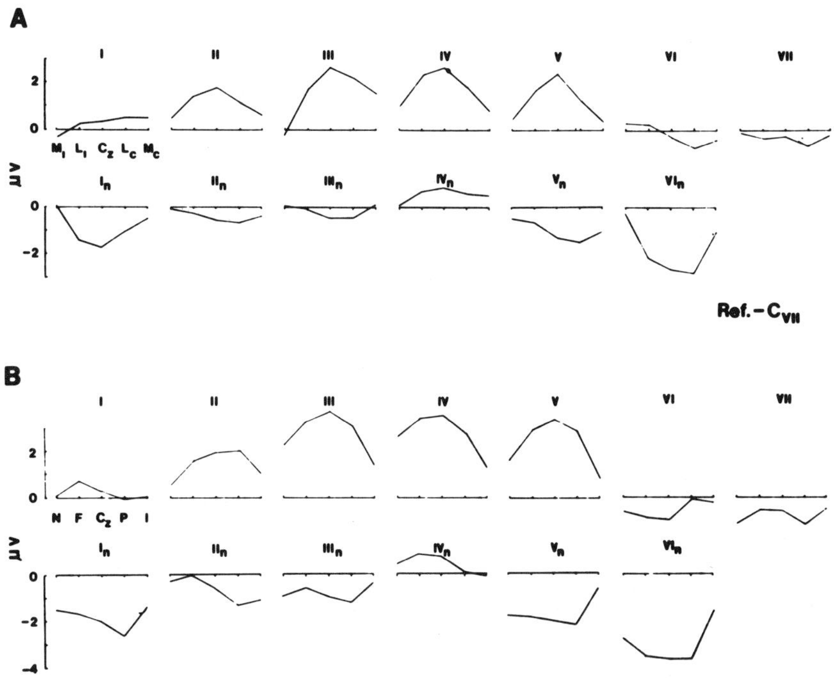

FIGURE 2. Amplitudes of auditory brainstem response components from different scalp locations. The data from the coronal array is in (A) and from the sagittal array is in (B). The amplitude scale is in tenths of microvolts.

longitudinal proximity to the reference electrode at the neck (C_{VII}). While the reference electrode (C_{VII}) was relatively inactive with respect to defining the ABR when recorded against an electrode even more remote from the scalp (first lumbar spinous process), it cannot be considered absolutely "referential" in recording the far-field reflections of the ABR. These considerations were not raised for the coronal array since the primary issue was the amplitude variation between electrodes equidistant from the "reference." Two general observations seem justified from the data of the sagittal array: first, wave I was largest over the fronto-central scalp, and decreased in amplitude posteriorly. For four of the subjects, the largest amplitude wave I occurred at the frontal electrode and for the fifth subject the largest wave I was recorded at C_z. Second, the amplitudes of the positive waves II through V and the negative waves III through VI were generally largest near the vertex.

The Latency of Auditory Brainstem Responses

The latencies of the ABR components change as a function of scalp location. TABLE 1A contains the mean values for three positions, (C_Z, M_C, M_I) referenced to a noncephalic site, C_{VII}. Components I_n, II_n, and III_n could not be identified consistantly at the ipsilateral mastoid (M_I), and IV_n could not be identified consistently at the contralateral mastoid (M_C). Percentage of subjects who demonstrated these components from these recording sites was 60% for I_n, 25% for II_n, 8.3% for III_n, and 33% for IV_n. All other components were identified in 90% or more of the trials. Examples of the waveforms from one subject and the grand average of all of the subjects are depicted in FIGURE 3. There are significant latency differences for wave II, III, III_n, IV, and V at these three sites (TABLE 2A). The magnitude of the mean latency differences at the three electrodes was quite small, ranging up to 0.24 msec. For each subject, wave IV at the ipsilateral mastoid occurred from 0.1 to 0.5 msec *before* wave IV recorded at the contralateral mastoid whereas wave V occurred from 0. to 0.6 msec *later* at the ipsilateral mastoid than at the contralateral mastoid. Viewed another way, the time separation between waves IV and V at the ipsilateral mastoid averaged 1.13 msec (range, 0.9–1.4 msec) while the separation of these waves at the contralateral mastoid averaged 0.5 msec (range, 0.3–0.7 msec). Such significant latency changes in the ABR components at the two mastoid recording sites can affect recordings using these sites as a "reference." For instance, in both Stockard *et al.*[5] and the present study, waves IV and V are more easily distinguished in recordings from C_Z–M_C than from C_Z–M_I (FIGURE 6). The enhancement of IV and V using the former derivation is

TABLE 1

MEAN LATENCY OF AUDITORY BRAINSTEM RESPONSE COMPONENTS (msec)*

	I	I_n	II	II_n	III	III_n	IV	IV_n	V	V_n	VI	VI_n
(A) Noncephalic Reference												
C_Z–C_{VII}	1.54	2.27	2.85	3.28	3.70	4.19	4.90	5.29	5.62	6.34	7.14	7.70
M_I–C_{VII}	1.53	†	2.81	†	3.80	†	4.74	5.32	5.77	6.39	7.16	7.72
M_C–C_{VII}	1.53	2.18	2.73	3.25	3.83	4.43	4.95	†	5.58	6.22	7.06	7.67
(B) Differential Recordings												
C_Z–M_I	1.52	2.28	2.73	3.18	3.68	4.23	4.98	5.27	5.53	6.24	7.13	7.78
C_Z–M_C	1.58	2.23	2.82	3.28	3.59	4.08	4.82	5.28	5.66	6.38	7.23	7.84
M_C–M_I	1.52	2.45	†	†	3.62	4.53	†	†	5.31	5.99	6.74	7.63
	$NP1_n$	NP1	$NP2_n$	NP2	$NP3_n$	NP3		$NP4_n$	NP4			
(C) Nasopharyngeal Recordings												
NP_I–C_{VII}	1.67	2.18	2.53	3.20	3.88	4.53		5.30	5.82			
NP_C–C_{VII}	†	†	†	†	3.65	4.40		5.20	†			

*Click intensity, 65 dBSL; 11/sec stimulus rate; condensation polarity; 6 subjects (3 men, 3 women); each ear stimulated monaurally. Electrode placements: C_Z = vertex; M_I = ipsilateral mastoid; M_C = contralateral mastoid; NP_I = ipsilateral nasopharynx; NP_C = contralateral nasopharynx; C_{VII} = seventh cervical spinous process.

†Components not evident in 10% or more of trials.

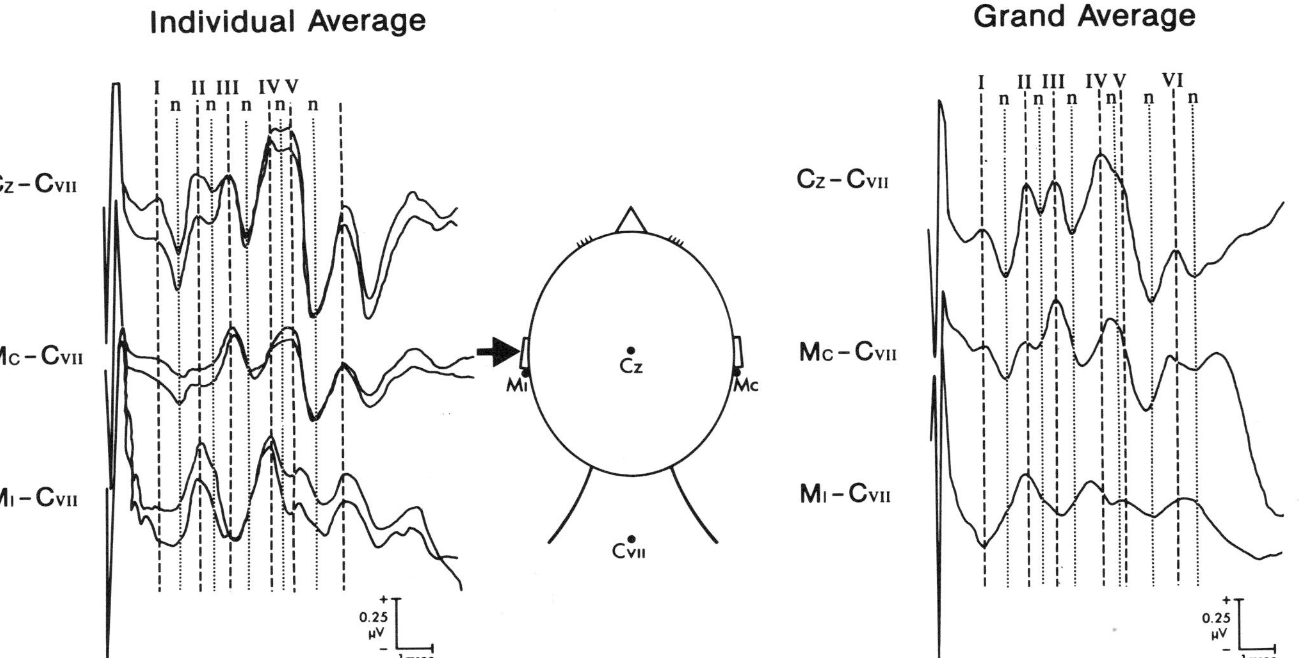

FIGURE 3. Auditory brainstem responses from both an individual subject and the grand average of six subjects recorded from C_Z, M_C, and M_I, all referenced to C_{VII}. In this and all subsequent figures, vertical lines descend through the components as defined at C_Z. Two separate averages of 2000 click trials are superimposed in the individual average while the grand average comprises 4000 click trials from each of the six subjects. Note the polarity and latency changes in these three derivations.

due to differences in the temporal occurrence of the components recorded at M_C and C_Z. Wave IV peaks at C_Z before it does at M_C whereas wave V peaks at C_Z later than it does at M_C. The effect of amplifying these differences in a C_Z–M_C recording is to enlarge the separation of waves IV and V. The difficulty of identifying *equivalent* peaks in differential recordings between two active sites is exemplified by the ABR derived from recording between the two mastoids, M_C referenced to M_I (FIGURE 6). During the time domain of the IV to V complex a component can be identified that falls intermediate between IV and V as identified at the vertex. There are no clear criteria to label such a peak at IV, IV_n, or V.

Nasopharyngeal Auditory Brainstem Responses

The brainstem potentials for the NP electrode referenced to the noncephalic electrode (C_{VII}) are in FIGURE 4 both for an individual and for the grand average of all subjects. The figure also contains the standard C_Z–C_{VII} recording. The positive peaks at the nasopharynx have been labeled NP1 through NP4 and negative subscripts have been added in the preceding trough to designate the corresponding negative peaks. The potentials to ipsilateral stimulation consist of a prominent positive–negative–positive sequence of waves (NP2, $NP3_n$, NP3) occurring in the time domain between components II_n and IV as recorded at the vertex.

TABLE 2

SIGNIFICANCE LEVELS OF LATENCY MEASURES OF AUDITORY BRAINSTEM RESPONSE COMPONENTS*

	I	I_n	II	II_n	III	III_n	IV	IV_n	V	V_n	VI	VI_n
(A) Noncephalic Reference												
C_Z–C_{VII} vs M_I–C_{VII}	ns	†	ns	†	0.04	†	0.006	ns	0.002	ns	ns	ns
C_Z–C_{VII} vs M_C–C_{VII}	ns	ns	0.05	ns	0.05	0.03	ns	†	ns	ns	ns	ns
M_I–C_{VII} vs M_C–C_{VII}	ns	†	ns	†	ns	†	0.006	†	0.03	ns	ns	ns
(B) Differential Recordings												
C_Z–M_I vs C_Z–M_C	ns	ns	ns	ns	ns	0.001	0.001	ns	0.06	0.03	ns	ns
	$NP1_n$	NP1	$NP2_n$	NP2	$NP3_n$	NP3		$NP4_n$	NP4			
(C) Nasopharyngeal Recordings												
NP_I–C_{VII} vs NP_C–C_{VII}	†	†	†	†	0.03	0.001		ns	†			

*See TABLE 1 for details. Values given are those of p (the probability of an incorrect hypothesis) as determined by the variance ratio, F. ns = not significant.

†F-ratios not determined because of insufficient data.

When the ear contralateral to the NP electrode was stimulated the potentials invert in polarity and shift to a slightly shorter latency (TABLE 1C). The NP brainstem potentials that occur before 3 msec and after 6 msec were greatly attentuated for contralateral stimulation so that consistent measures of peaks in those time domains were not possible. Subtracting the NP potentials to contralateral stimulation from those evoked by ipsilateral stimulation (bottom panel of FIGURE 4) resulted in a potential reflecting the time domain of differences in the two recordings.

In two subjects mapping of the nasopharynx was made to define the extent to which the latency and polarity of the NP potentials depend on electrode placement. Recordings were made when the electrode was inserted in the nasopharynx in the midline and when it was displaced laterally. This was done for each side of the

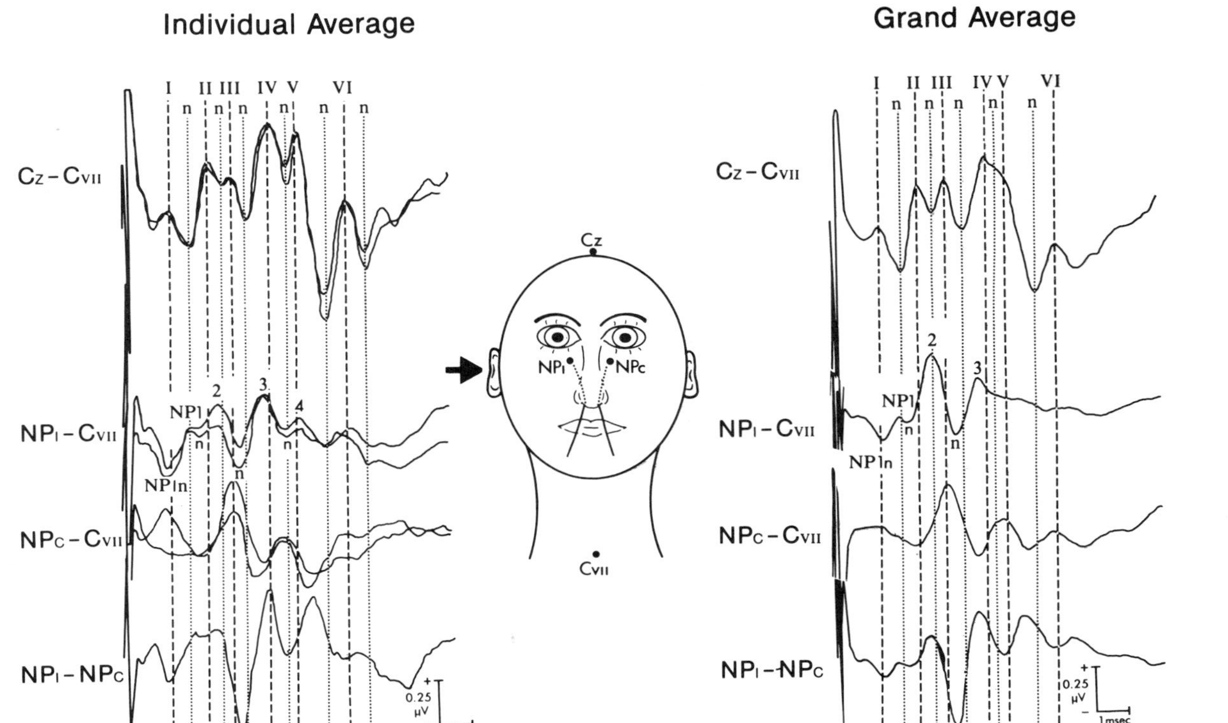

FIGURE 4. Auditory brainstem responses from both an individual subject and the grand average of six subjects recorded from C_Z, the ipsilateral and contralateral nasopharynx (NP_I and NP_C, respectively) all referenced to C_{VII}. The difference potential between the two nasopharyngeal recording sites is in the bottom trace (NP_I–NP_C). The vertical lines descend through the components defined at C_Z. The ipsilateral nasopharyngeal components have been labeled in sequence with arabic numerals and have the subscript *n* if the polarity is negative. Note the polarity and/or latency differences between the nasopharygeal recordings.

nasopharynx. We estimate the separation of the electrodes across the midline to be 5 mm and the separation between the lateral and midline nasopharyngeal placements to be 3–5 mm. The results from the two subjects corresponded, and recording from four nasopharyngeal sites from one of the subjects are shown in FIGURE 5. Note that the electrodes in the same half of the nasopharynx record quite similar events but that recordings from either side of the midline change as a function of the site of stimulation as described previously.

It was of considerable interest to determine whether the potentials recorded at the scalp and at the nasopharynx were manifestations of the same neural events. A comparison of the various waves from FIGURE 6 suggests a correspondence between several of the scalp and NP recorded waves.

In the early portion of the waveform the ipsilaterally evoked nasopharyngeal waves $NP1_n$, NP1, NP2, and $NP3_n$ had latencies approximating components I, I_n, II_n, and III recorded at the vertex, respectively. A major distinction between these two sites was that the components were of opposite polarity. In contrast, the polarity of the nasopharyngeal waves $NP1_n$ and $NP3_n$ to ipsilateral stimulation were the same as waves I and III recorded from the ipsilateral mastoid. Wave NP3 recorded from both the ipsilateral and contralateral nasopharynx corresponded most closely in latency to wave III_n recorded from the contralateral mastoid, and waves $NP4_n$ and NP4 were of similar latency to waves IV_n and V from the scalp. There were two events for which no correspondence could be defined: Wave $NP2_n$ from the ipsilateral nasopharynx occurred at a latency between waves I_n and II at the scalp and wave IV at the scalp had no corresponding component from the nasopharynx.

Auditory Brainstem Responses from Different Differential Scalp Montages

FIGURE 7 contains the ABR recorded in four arrays from one subject and the grand average of all subjects: C_Z–C_{VII} (vertical montage), C_Z–M_I and C_Z–M_C

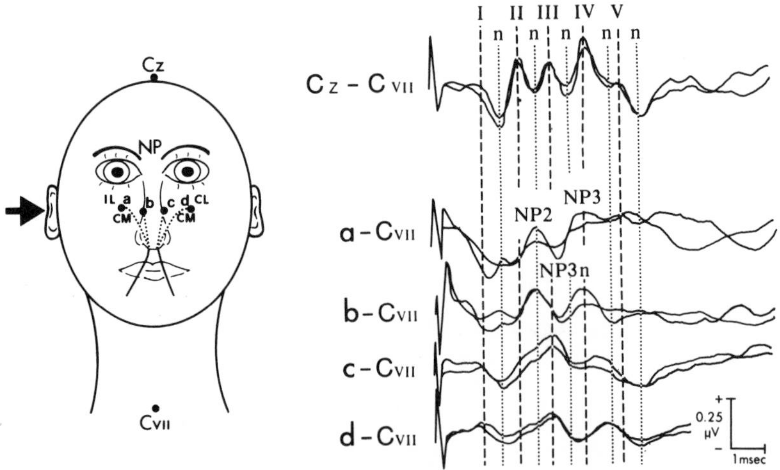

FIGURE 5. Auditory brainstem responses recorded from various positions within the nasopharynx referenced to C_{VII}. Note the similarity of the potentials recorded on each side of the nasopharynx but the change that occurs on crossing the midline.

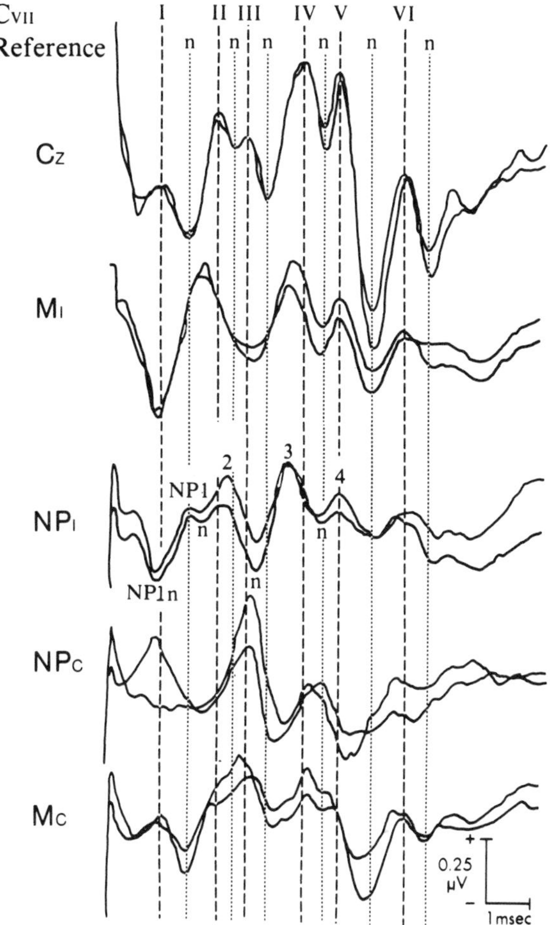

FIGURE 6. Auditory brainstem responses from an individual recorded from various scalp sites (C_Z, M_I, and M_C) and the nasopharynx (NP_I, NP_C) referenced to C_{VII}. The vertical lines descend through the components recorded at C_Z.

(diagonal montages), and M_C–M_I (horizontal montage). The mean latencies of the components are presented in TABLE 1B. Significance values for the latency differences between C_Z–M_C and C_Z–M_I are presented in TABLE 2B.

Wave I is easiest to define from the horizontal and one of the diagonal montages (C_Z–M_I). There may be no wave I when recording from the other diagonal montage (C_Z–M_C). Wave I_n is well seen in all recordings. Waves II and II_n are either indistinct or absent in the horizontal montage (M_C–M_I) and are best visualized in the vertical (C_Z–C_{VII}) and one of the diagonal (C_Z–M_C) derivations. Wave III is particularly small in one of the diagonal montages (C_Z–M_C) and is of high amplitude and broad dimension in the horizontal montage (M_C–M_I). This broadening of wave III reflects that the negative troughs surrounding III are both earlier and later in the horizontal derivation than in the other recording arrays. Wave III_n also occurs earlier in the

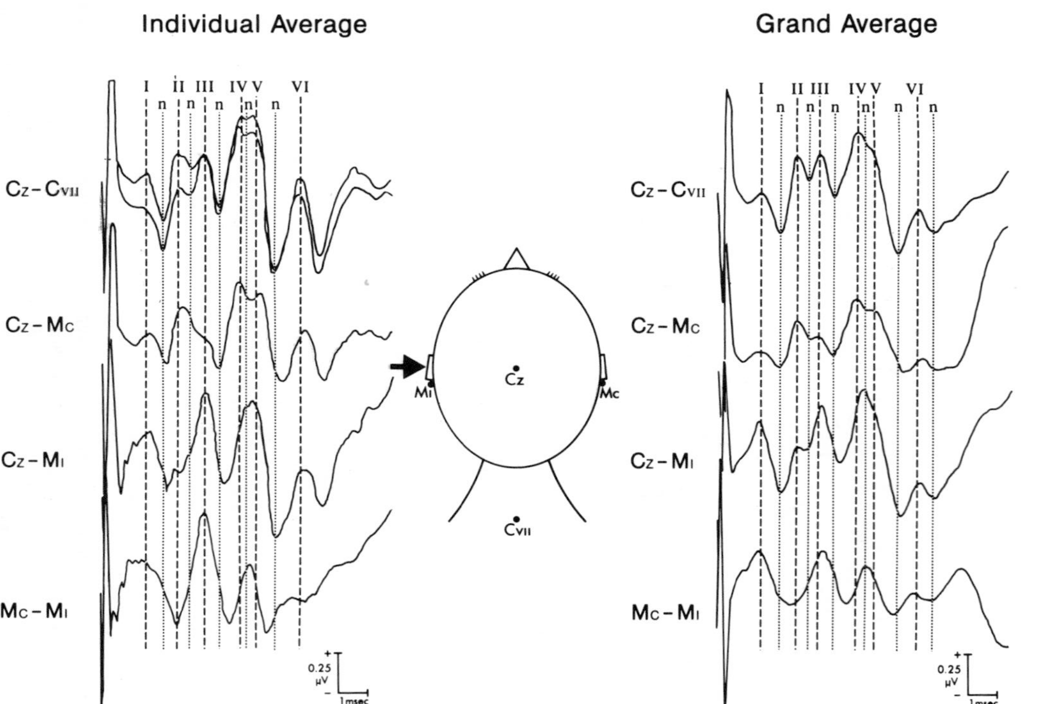

FIGURE 7. Auditory brainstem responses recorded from both an individual and the grand average of six subjects from several recording arrays; C_Z–C_{VII} (vertical), C_Z–M_C and C_Z–M_I (diagonal), and M_C–M_I (horizontal). The vertical lines descend through the components defined at C_Z–C_{VII}. Note that different components are recorded with different latencies and configurations in the various derivations.

C_Z–M_C montage than in the C_Z–M_I array. The IV/V complex is particularly segregated into separate waves in the diagonal array, C_Z–M_C, because wave IV occurs earlier and wave V occurs later in this array than in C_Z–M_I. In the horizontal montage (M_C–M_I), the IV/V complex appears as a low amplitude single component with a latency intermediate between the IV and V waves. V_n is well defined in the vertical and diagonal arrays and is of shortest latency from C_Z–M_I. Finally, wave VI is attenuated in the horizontal recording array (M_C–M_I).

Discussion

The aims of this study were: (1) to investigate further the distribution of amplitudes and latencies of the ABR over the scalp, particularly with regard to lateralization according to the stimulated ear; (2) to record ABRs referenced to a noncephalic site from the unique perspective provided by a nasopharyngeal electrode; (3) to define the ABR from several different montages that have clinical relevance. The findings with regard to these aims will be discussed for each of the ABR components.

Wave I. Recorded as a positive peak over most of the scalp, wave I has a negative polarity at the mastoid ipsilateral to stimulation. These results confirm those reported by several groups of investigators.[2–4] In addition to the lateral amplitude gradient, there is also an anterior–posterior gradient with the maximum mean wave I amplitude occurring frontally. At the nasion and inion the mean amplitude was essentially zero. The counterpart of wave I recorded from the ipsilateral nasopharynx ($NP1_n$) has a negative polarity. On the basis of these data, the wave I field can be idealized as vectors in both the coronal and sagittal plane, with an origin located near the ipsilateral mastoid and nasopharynx. The current clinical recording method of recording between C_Z, where wave I is positive, and the mastoid or earlobe ipsilateral to the ear stimulated where wave I is negative, clearly optimizes the detection of this component. In fact, wave I may be absent if recording vertex to the contralateral mastoid.

Wave I_n. Recorded with largest amplitudes at the midline, particularly over the parietal scalp, this component was also lateralized. It was defined as a negative peak at the contralateral mastoid whereas at the ipsilateral mastoid it was either not identified or appeared as a positive inflection between waves I and II. The temporal counterpart of wave I_n at the ipsilateral nasopharynx has a positive polarity (NP1). Thus, wave I_n recorded at the ipsilateral mastoid and nasopharynx appears as a small positive wave while it is recorded as a pronounced negative wave over most of the scalp. Its generators have vectors similar to that of wave I.

Wave II. A positive peak at all scalp locations, wave II is largest at the vertex and parietal electrode sites. Thus, there is a progression from wave I, which is negative at the ipsilateral mastoid and positive at the vertex, to wave II, which is positive at all sites, with a "transitional" wave I_n at the ipsilateral mastoid. This result has been noted by Terkildsen *et al.*[3]

The counterpart of wave II recorded from the ipsilateral nasopharnx ($NP2_n$) was negative in polarity. Wave $NP2_n$ was not reliably recorded for contralateral stimulation.

These results are compatible with a dipole orientation of the generator(s) of wave II in the sagittal plane, since the amplitude of wave II showed no lateral asymmetry and the largest amplitudes tended to be recorded over the posterior scalp. This suggestion is supported by the polarity reversal between the ipsilateral mastoid and the more anteriorly situated nasopharynx electrode.

Wave II_n. Wave II_n was negative at all electrode sites and largest over the posterior scalp, but could not be distinguished at the ipsilateral mastoid where it blended with wave III. The nasopharyngeal counterpart (NP2) was of opposite polarity. In many respects, wave II_n resembles wave II except in polarity and could be generated by similar mechanisms.

Waves III and $NP3_n$. The scalp-recorded wave III has its maximum amplitude at the vertex, however it is markedly lateralized to the contralateral scalp, in agreement with Picton *et al.*[2] and Stockard *et al.*[5] At the ipsilateral mastoid, the mean amplitude of wave III is actually negative. The nasopharyngeal counterpart of wave III in terms of latency is wave $NP3_n$. $NP3_n$ is also markedly lateralized and is negative for ipsilateral stimulation and positive for contralateral stimulation. The correspondencies between $NP3_n$ and III suggest the two events may be generated by similar processes. Wave III tended to be larger over the anterior than the posterior scalp. These results are consistent with the suggestion of Picton *et al.*[2] that wave III is the result of both vertically and horizontally oriented dipoles. The horizontal dipole reflected in wave III is situated medial to the ipsilateral nasopharyngeal electrode, and the vertically oriented dipole is directed slightly anteriorly. Moreover, the observation that the width of wave III is considerably broadened when recording from the horizontal plane (mastoid–mastoid) suggests a wide spatial extent of the horizontal dipole for the generation of this component.

Waves III_n and NP3. The scalp derived wave III_n was remarkable in its scalp distribution since it had no counterpart from the ipsilateral mastoid but did demonstrate a significant latency disparity between the vertex and contralateral mastoid. NP3 and $NP3_n$, were the most prominent peaks recorded from the nasopharynx. NP3 was markedly lateralized, being positive for ipsilateral stimulation. The significant latency disparity between NP3 to ipsilateral and contralateral stimulation is incompatible with a simple dipole source. NP3 to both ipsilateral and contralateral stimulation corresponds in latency to wave III_n at the contralateral mastoid. Wave NP3 is likely a reflection of near-field brainstem activity of generators located slightly medial to the nasopharyngeal electrode as suggested by Martin and Coats.[10] Wave III_n at the vertex may be a far-field reflection of this brainstem event.

Wave IV. Wave IV exhibited no significant amplitude lateralization, but tended to be largest over the anterior scalp. The latency of wave IV was earlier (mean: 0.24 msec; range: 0–0.5 msec) at the ipsilateral than contralateral recording sites. These results would be compatible with generators having vertical dipoles of short latency ipsilaterally and slightly longer latency contralaterally.

Wave V and V_n. Wave V was positive and V_n negative at all scalp locations and largest over the central and frontal sites. The latency of wave V was significantly delayed at the ipsilateral mastoid compared to the contralateral mastoid (mean difference: 0.27 msec; range: 0.1–0.6 msec). The values for wave V_n were similar (mean difference: 0.23 msec; range: 0–0.5 msec), but did not achieve statistical significance. The absence of wave V and V_n from the horizontal recording arrays (M_C–M_I) is consonant with vertically oriented dipoles for these components having significant latency disparity on both sides of the brainstem.

Wave VI and VI_n. There was a general trend toward a greater negativity over the contralateral scalp, which was a significant only for wave VI_n. This result is consistent with that reported by Martin and Coats.[10] The absence of wave VI in the horizontal array is compatible with vertically oriented generators.

The results of this scalp distribution study of the ABR in humans differ from similar mapping studies in monkey[11] and rat and cat.[12] In the animal studies there were significant changes in the number of ABR components as a function of scalp recording site. In the monkey, for instance, waves II and III at the vertex each segregate into two distinct components with different lateral scalp distributions. In the present study in humans, the components did not break into separate subcomponents, possibly because of the large volume of the human skull relative to the brainstem, making the definition of various subcomponents difficult. In agreement with the animal studies, there were significant latency shifts of some of the components over the scalp in the human subjects. The latency shifts were maximum for waves IV and V but could also be distinguished for short latency events (wave II). The presence of latency shifts suggest that the generator sources comprising some of these components may move within the brainstem. The nerve action volleys traveling along fiber pathways could be a source of such moving generators.

The definition of latency disparities over the scalp is emphasized by differential recording arrays.[5,6] However, the interpretation of the mechanism of such latency shifts with differential recordings is complex since both sites are "active." It may be that the generator sites for components identified in such differential arrays are relatively specific for each array.

Both depth recording and lesion studies in animals[13–15] suggest that there may be multiple generator sites within the brainstem for many of the components of the ABR. The results from both the present study and other studies in humans[2,3] suggest that the generator sources for the ABR components can have different orientations and dimensions. Wave III, for instance, has both vertical and horizontal dipoles. Moreover, the broadening of component III in the horizontal montage suggests that the horizontal dipole is spatially extensive. These data from scalp recordings can be interpreted as indicating that the generators for some of the ABR components in man are not discrete but, rather, are spatially distributed.

The clinical use of ABR techniques is still in its beginnings. There has been reasonable success utilizing a single diagonal recording array (vertex–ipsilateral earlobe), and it has been suggested that the definition of abnormalities may be increased by utilizing other recording arrays.[5] In the present study *different values* for latencies of ABR components were obtained from recordings over the scalp in a horizontal, vertical, and two diagonal planes as well as sampling close to the brainstem itself from the nasopharynx. It is obvious that other arrays could also be defined. Thus, criteria need to be developed to allow critical judgment for the selection of the appropriate recording array(s) for clinical applications. We suggest the use of a single array (vertex-ipsilateral mastoid) may be suitable for most applications. However, in those instances when components are difficult to recognize the use of additional recording arrays may resolve the ambiguities.

Since the preparation of the manuscript, several articles relevant to the issue of scalp distribution of auditory brain stem potentials and the vectors of their generation have appeared.[16–19] The data in these articles expand our knowledge of possible generators of auditory brain stem responses.

Summary

Auditory brainstem potentials were recorded from various scalp and nasopharyngeal sites referenced both to a noncephalic site and to certain scalp locations in normal humans. The distribution of amplitudes and latencies of the components were defined. There were significant amplitude, polarity, and latency asymmetries over the scalp in both referential and differential recordings. The data indicated that several of the ABR components have generator sources that are lateralized and move through the brainstem in particular orientations.

ACKNOWLEDGMENTS

Dr. Aaron Allen, Dr. John Polich, and Gary Geil provided valuable assistance in this study. We are indebted to Beverly Henderson for her care in collecting the recordings from the nasopharynx.

REFERENCES

1. JEWETT, D. L. & J. S. WILLISTON. 1971. Auditory evoked far-fields averaged from the scalp of humans. Brain **94:** 681.
2. PICTON, T. W., S. A. HILLYARD, H. I. KRAUSZ & R. GALAMBOS. 1974. Human auditory evoked potentials. I: Evaluation of components. Electroencephalogr. Clin. Neurophysiol. **36:** 179–190.
3. TERKILDSEN, K., P. OSTERHAMMEL & F. HUIS IN'T VELD. 1974. Far field electrocochleography, electrode positions. Scand. Audiol. **3:** 123–129.
4. STRELETZ, L. J., L. KATZ, M. HOHENBERGER & R. CRACCO. 1977. Scalp recorded auditory evoked potentials and sonomotor responses: An evaluation of components and recording techniques. Electroencephalogr. Clin. Neurophysiol. **43:** 192–206.
5. STOCKARD, J. J., J. E. STOCKARD & F. W. SHARBROUGH. 1978. Nonpathologic factors influencing brainstem auditory evoked potentials. Am. J. EEG Technol. **18:** 177.
6. HIXSON, W. C. & J. D. MOSKO. 1979. Normative Bilateral Brainstem Evoked Response Data for a Naval Aviation Student Population: Group Statistics. Publication No. 1262. Naval Aerospace Medical Research Laboratory, Pensacola, Fla.
7. STARR, A. & L. J. ACHOR. 1975. Auditory brainstem responses in neurological disease. Arch. Neruol. **32:** 761.
8. VAN OLPHEN, A. F., M. RODENBURG & C. VERWEY. 1978. Distribution of brain stem responses to acoustic stimuli over the human scalp. Audiology **17:** 511–518.
9. MARTIN, M. E. & E. J. MOORE. 1977. Scalp distribution of early (0 to 10 msec) auditory evoked responses. Arch. Otolaryngol. **103:** 626–628.
10. MARTIN, J. L. & A. C. COATS. 1973. Short-latency auditory evoked responses recorded from human nasopharynx. Brain Res. **60:** 496–502.
11. ALLEN, A. R. & A. STARR. 1978. Auditory brainstem potentials in monkey (*M. Mulatta*) and man. Electroencephalogr. Clin. Neurophysiol. **45:** 53–63.
12. PLANTZ, R. G., J. S. WILLISTON & D. L. JEWETT. 1974. Spatio-temporal distribution of auditory evoked far field potentials in rat and cat. Brain Res. **68:** 55–71.
13. ACHOR, L. J. & A. STARR. 1980. Auditory brainstem response in the cat. I. Intracranial versus extracranial recordings. Electroencephalogr. Clin. Neurophysiol. **48:** 154–173.
14. ACHOR, L. J. & A. STARR. 1980. Auditory brainstem responses in the cat. II. Effects of lesions. Electroencephalogr. Clin. Neurophysiol. **48:** 174–190.
15. BUCHWALD, J. S. & C. HUANG. 1975. Far-field acoustic response: Origins in the cat. Science **189:** 382–384.
16. BARRATT, H. 1980. Investigation of the mastoid electrode contribution to the brain stem auditory evoked response. Scand. Audiol. **9:** 203–211.
17. FITZGERALD, P. & T. PICTON. (in press). Temporal and Sequential Probability in Evoked Potentials Studies. Canad. J. Psychol.
18. INO, T. & K. MIZOI. 1980. Vector analysis of auditory brain stem responses in human beings. Arch. Oto-Rhino-Laryngol. **226:** 55–62.
19. TERKILDSEN, K. & P. OSTERHAMMEL. 1981. The influence of reference electrode position on recordings of the auditory brainstem responses. Ear and Hearing **2:** 9–14.

BRAINSTEM AUDITORY EVOKED POTENTIALS IN ADULT AND INFANT SLEEP APNEA SYNDROMES, INCLUDING SUDDEN INFANT DEATH SYNDROME AND NEAR-MISS FOR SUDDEN INFANT DEATH*

James J. Stockard

EEG and Evoked Potential Laboratory
University of California Medical Center
San Diego, California 92103

Department of Neurosciences†
University of California, San Diego
School of Medicine M-024
La Jolla, California 92093

INTRODUCTION

The proximity of the brainstem auditory pathway to pontomedullary respiratory centers suggested that brainstem auditory evoked potentials (BAEPs) might help to elucidate the pathogenesis of certain adult and infant sleep apnea (SA) syndromes. In preliminary investigations, we found BAEP abnormalities in a large proportion of adults with acute or progressive central SA (57%)[1] and in infants predisposed to recurrent apnea by metabolic disorders, status epilepticus, and congenital anomalies (32%).[2] In some of the patients with central BAEP abnormalities, those test abnormalities were the first objective evidence that brainstem lesions underlay the apnea syndromes. This high yield of BAEP abnormalities in patients who only subsequently, and often much later, proved to have structural brainstem pathology indicated that BAEPs might also shed light on the pathophysiology of the more common cryptogenic SA syndromes of adulthood such as chronic obstructive SA. It has been suggested that some adults with chronic obstructive SA syndromes may have carried with them a life-long proclivity to SA and survived a period in infancy during which they were at especially high risk for fatal SA, or sudden infant death syndrome (SIDS);[3] also, SA is considered the most common final pathway to SIDS.[4]

We hoped that BAEPs would provide insight into the pathophysiology of this related but poorly understood phenomenon of SIDS, which is the leading cause of mortality in the first year of life after the first week.[4] There is a growing consensus that SIDS represents failure of central respiratory control mechanisms during sleep,[5–8] and there are many reported cases of infants with abnormally prolonged apneic episodes during sleep subsequently dying of SIDS.[9–13] Episodes of prolonged sleep apnea, without obvious precipitants, which are associated with clinical evidence of hypoxemia, e.g., cyanosis and flaccidity, requiring resuscitation of the infant before spontaneous ventilation is resumed, are termed "near-miss-sudden-infant-death" (NMSID) episodes. There is considerable evidence that NMSID events may represent a *forme fruste* of SIDS and there have been numerous reports of survivors of

*This work was supported in part by University of California Research Committee Award R-F141-M through the Academic Senate of the San Diego campus.

†Address for reprint requests.

0077–8923/82/0388–0443 $1.75/0 © 1982, NYAS

NMSID episodes eventually succumbing to SIDS, further linking these phenomena.[10–15]

This link between NMSID and SIDS and the fact that many NMSID events and crib deaths appear to result from abnormally prolonged episodes of sleep apnea, along with our previous observations of abnormal BAEPs in adults with central SA syndromes, led us to investigate BAEPs in carefully defined cases of NMSID. By studying BAEPs in such cases and in apparently normal infants who subsequently died of SIDS, we hoped to be able to delineate some type of premonitory BAEP profile that would allow both identification of infants at high risk for SIDS, and preventive measures. We also hoped that application of this technique in such cases would help to elucidate the pathogenesis of SIDS and the postulated relationship between NMSID and adult SA syndromes.

Methods

Brainstem Auditory Evoked Potentials

Brainstem auditory evoked potentials were elicited from adults using 60-dBSL monaural clicks (60 dB sensation level or 60 dB above the perceptual threshold of the subject for the click) at a rate of 10/sec that were generated by passing a 100-μsec square wave through a shielded TDH-39 headphone which produced broadband clicks (2–4 kHz peak power) that were essentially monophasic with respect to their acoustic polarity (i.e., condensation clicks in our adult studies). When attainment of a perceived click intensity 60 dB above threshold required use of stimulus intensities greater than 70 dBHL (70 dB above the average normal hearing threshold in the recording environment), the nonstimulated ear was masked with white noise of 40 dBHL lower intensity than the clicks to prevent cross-hearing. BAEPs were recorded from tin–tin-chloride cup electrodes at the vertex (C_Z) and medial side of both the earlobe ipsilateral (A_i) and the (A_c) earlobe contralateral (A_c) to the click. A midline forehead electrode (F_{pz}) served as ground.

Two channels of bipolar EEG activity (C_Z–A_i and C_Z–A_c) were amplified 100,000 times, filtered (-6 dB points at 100 and 3000 Hz) and analyzed for 10.24–12.80 msec post-stimulus with a signal averager (Grass Model 10 or Nicolet CA-1000, 10 or 20 μsec sampling intervals). Each trial comprised an average of 2048 to 4096 individual responses and was replicated and plotted. Latencies of vertex-positive (ear-negative) peaks I through V and vertex-negative peaks I_N, III_N, and V_N[16] were measured to the nearest 10–20 μsec using a cursor with digital read-out. Peak-to-peak amplitude of the IV/V complex was measured from the most vertex-positive point of the complex, whether that was peak IV or V, to the most negative point in the subsequent trough of wave V_N. Peak-to-peak amplitude of wave I was similarly measured from the most vertex-positive/ear-negative point of the peak of wave I to the most vertex-negative/ear-positive subsequent point in the trough of wave I_N. These amplitudes were measured to the nearest 10–20 nV, and the ratio of the amplitudes of wave IV/V to that of wave I, the (IV/V):(I) amplitude ratio (AR), was calculated.

The two major BAEP measures of central auditory conduction were the wave I to wave V interpeak latency (I–V IPL) and the (IV/V):(I) AR defined above. The criteria for adequate reproducibility of these values in separate averages (trials) from the same ear were that the I–V IPL vary by less than 2% and the (IV/V):(I) AR by less than 5% between trials. That is, the maximum intertrial variability of the I–V IPL between two trials in the same recording session had to be 2% of the mean I–V IPL value obtained in that session, e.g., less than 80 μsec maximal intertrial variability in an adult subject or patient having a mean I–V IPL of 4.00 msec. Likewise, in an adult patient with a IV/V:I amplitude ratio of 2.00, the maximal allowable excursion in this dimensionless measure could be from 1.95 to 2.05 ($\overline{X} \pm 0.05\ \overline{X}$).

The I–V IPLs and IV/V:I amplitude ratios obtained from patients were compared individually and collectively (using *t*-tests) with normative data collected from a large control population of neurologically normal subjects without sleep apnea who were matched for age, gender, audiogram, and stimulation and recording technique. FIGURE 1 shows the distribution of I–V IPLs from the 100 neurologically and audiometrically normal adult subjects who served as controls in this study. The inter-ear asymmetry in the I–V IPLs ($|\text{I–V}|_{L\text{-}R}$) from these 100 normal subjects never exceeded 0.4 msec (range: 0.00–0.38), had a mean value ($\overline{X}$) of 0.191 msec with standard deviation (SD) of 0.218 msec and standard error of the mean (SE) of 0.022 msec. The mean IV/V:I amplitude ratio of the control group was 3.14; the SD, 1.05, and the SE, 0.011. The upper limit of normal for the I–V IPL was chosen to encompass both the entire range and mean + 3 SD for the appropriate quadrant of the adult population, for age and sex (FIGURE 2). For both the $|\text{I–V}|_{L\text{-}R}$ interaural latency difference and the IV/V:I amplitude ratio, the distributions were so non-Gaussian that the upper limit of normal (ULN) for the $|\text{I–V}|_{L\text{-}R}$ of 0.4 msec and lower limit of normal (LLN) for the (IV/V):(I) AR of 0.4 μV/μV were chosen because they encompassed the entire range of normal values. There were no age- or gender-related differences in these latter two BAEP measures which had to be accounted for in establishing limits of normality.

Similar methods were employed in the infant population with the exception that 57

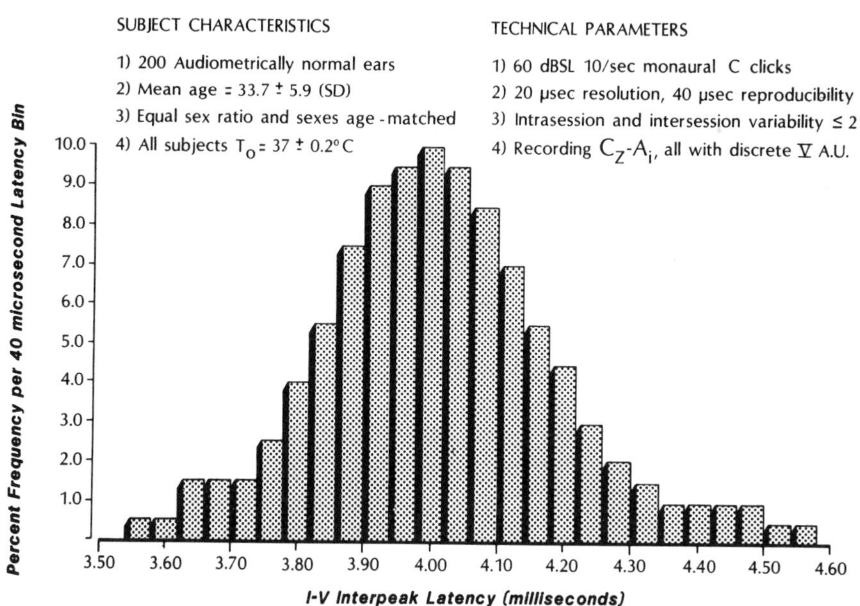

FIGURE 1. Gaussian distribution of I–V interpeak latencies (IPLs) from the 200 ears of the 100 control adult subjects who served as the normative data base for the central and mixed/obstructive sleep apnea (SA) patients who were adults. Pertinent subject and technical variables given above with abbreviations as defined in text and footnote for TABLE 1.

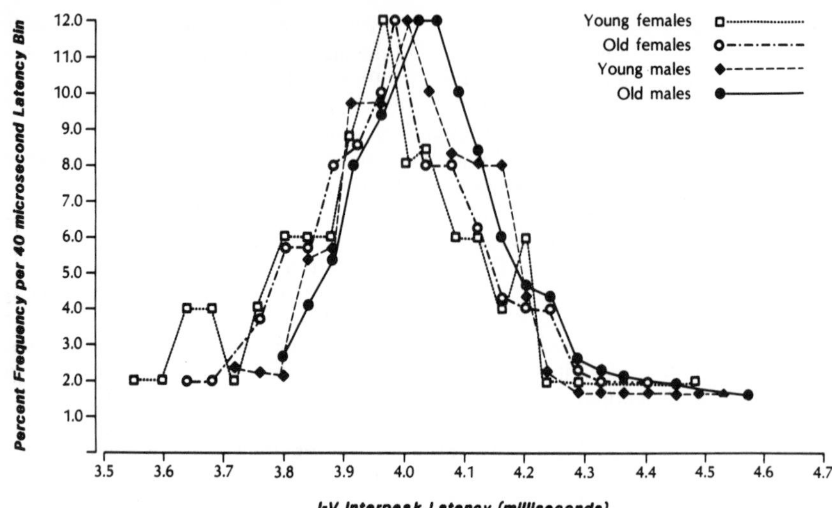

FIGURE 2. Gaussian distributions of the four subsets of the heterogeneous control population shown in FIGURE 1 which showed significant (ANOVA, $p < 0.05$) I–V IPL differences: "young" and "old" refer to below and above the age of 40 years, respectively; the above- and below-40 groups clustered into two discrete groups for both sexes, and there was no age-latency continuum within these two clusters.

dBHL‡ rarefaction clicks were employed rather than the 60 dBSL condensation clicks used in adults. The reason for the slight stimulus differences in the two age groups were: (1) condensation clicks allowed resolution of a discrete wave V (with a separate peak from wave IV) in all adult subjects in contrast to rarefaction clicks, which often fuse wave IV and V and preclude meaningful measurements of I–V interpeak latency;[17] (2) use of a fixed sensation level (SL) intensity in the adults compensated for variable reductions of effective stimulus intensity due to mild flat conductive hearing losses present in some of the adult controls and patients; (3) sensation level (SL) intensities obviously could not be employed in infants (as this involves subjective threshold determination); (4) rarefaction clicks were used in infants because of the greater response stability and better resolution of wave I they afford; because of greater separation between waves IV and V in infants compared to adults,[18] discrete waves V could always be resolved in the control and patient infants in this study despite routine use of rarefaction clicks.

The major BAEP measure of central auditory conduction that we analyzed was the I–V IPL, as we have found that inter-ear IPL differences are unreliable in this age group and amplitude ratios are too variable within and among normal infants to allow their meaningful diagnostic application in all but extreme cases.[19,20] Our normal I–V IPLs for the four different infant age groups studied in this investigation are presented in TABLE 2. All infant controls and patients were recorded when their rectal

‡The 70 dBHL dial setting on the stimulus generator resulted in an effective stimulus intensity of 57 dBnHL (57 dB above mean perceptual threshold for normal-hearing adults) in the infant recording environment. The most important measure of the click intensity employed was its physically constant peak-equivalent sound pressure level (peSPL) of 103 dB peSPL, with 0 dB peSPL = 2.04×10^{-4} dynes/cm^2.

temperatures were 36.5–37.5°C and virtually all infants were within a narrower range of 37.00 ± 0.2°C at the time of testing. BAEPs were obtained during wakefulness and quiet and active sleep and showed no significant differences in any amplitude or latency measures as a function of state. In contrast to the gender differences in BAEP measures of retrocochlear conduction which we originally described for adults,[16,18] no male–female BAEP differences existed in any of the infant populations in this study, in accord with previous investigations by us[17,19] and by others[21] in this age group (term to 4 months).

Classification of Sleep Apneas, SIDS, and NMSID

Central sleep apnea was defined as complete cessation of diaphragmatic and intercostal muscle movement associated with absence of nasal/oral airflow for greater than 10 seconds more than 30 times per hour of sleep (sleep–hour).

Mixed central and obstructive SA and purely *obstructive SA* were considered together in this study since all apneic episodes of the "mixed" type we recorded were predominantly obstructive in nature. *Mixed/obstructive SA* was said to occur when there was cessation of nasal/oral airflow for greater than 10 seconds more than 30 times per sleep-hour despite diaphragmatic contractions that either commenced with the onset of apnea (pure obstructive type of apnea) or developed later in the apneic episode and progressively strengthened until the episode ended. The methods by which obstructive versus central contributions to the SA could be reliably differentiated in all cases are outlined below.

Sudden infant death syndrome (SIDS) was defined as the unexpected, unexplained sudden death during sleep of a previously thriving infant between 3 weeks and 6 months of age which occurred in the absence of clear predisposing factors or discernible precipitants of cardiopulmonary arrest. *Near-miss-sudden-infant-death* (NMSID) events were defined as episodes of prolonged apnea during sleep in a previously thriving 3-week-to-6-month-old infant which result in generalized cyanosis and flaccidity and required resuscitation. Definition of NMSID episodes is necessarily largely subjective and the term "near-miss-SIDS" carries some inherent assumptions that can never be proven nor disproven; nonetheless, the NMSID group is the only available model for the antemortem study of SIDS and the term "near-miss" is almost uniformly used by workers in the field to describe episodes such as the one in the above definition.

Sleep apnea studies in adults were carried out using modified techniques of Guilleminault and co-workers.[22] Eight channels of EEG, two of EOG, a submental EMG (surface recording over genioglossus), and an ECG were continuously recorded prior to, during, and after the sleep sample, which was eight hours in duration in the majority of patients. During sleep, nasal/oral airflow (monitored with nasal thermistors and bias flow mask with pressure transducers), intraesophageal pressure (via balloon manometer), and chest wall and abdominal movement (with accelerometers) were recorded continuously. Blood oxygen saturation was also continuously measured with an ear oximeter, and PaO_2, $PaCO_2$, and arterial pressure were periodically determined through an indwelling radial artery catheter. Prolonged polygraphic monitoring similar to that used in the adult studies were carried out in two of the seven NMSID infants who had recurrent apneic episodes. Informed consent was obtained from all adult patients and the parents of all infants enrolled in the study, and the experimental protocol was approved by the Human Research Committees of both the Mayo Clinic and the University of California, San Diego, Medical Center.

RESULTS

Central SA (Adults)

Five males and three females ranging in age from 16 to 68 years presented with acute-onset or progressive central sleep apnea. In four of these cases, the BAEPs were well within normal limits as compared with age-matched sex-matched control data for the I–V interpeak latencies (FIGURE 2), the interaural I–V IPL symmetry, and (IV/V):(I) amplitude ratio norms given above. These patients all had mild central SA syndromes with only 31–45 apneic episodes per sleep-hour and only one had apneas greater than 20 seconds in duration. This latter patient, a 16-year-old male who had central SA or acute onset, had apneic episodes ranging 45–60 seconds in duration and an abnormally flat ventilatory response to inhaled CO_2. The patient had nystagmus, ataxia, and "brainstem atrophy" on computerized tomographic (CT) head scan. His sleep apnea did not progress and he was later diagnosed as having a Chiari malformation. The central SA also did not progress in the other three patients of this group (on four year follow-up) with normal BAEPs, and no cause for the central SA was found in any of them despite full work-ups for posterior fossa lesions.

TABLE 1
SUMMARY OF CENTRAL SLEEP APNEA (SA) CASES WITH ABNORMAL BRAINSTEM AUDITORY EVOKED POTENTIALS (BAEPS)*

Case	Sex/Age	Clinical Presentations	BAEP Abnormality	Ultimate Diagnosis
1	M 42	SA initial symptom; no definite stem signs at time of BAEP testing; SA noted by spouse.	Absent waves after III A.S.; ↓(IV/V):(I) A.D.	Biopsy-proven pontomedullary astrocytoma
2	M 66	SA then dysphagia first symptoms ⇢ spastic paraparesis, diplopia due to third nerve paresis over next year; SA and other symptoms resolved	Absent wave V A.U. ⇢ normalized 2 mos. later with ↓↓ symptoms after steroid therapy	Inflammatory and/or demyelinating brainstem process most severely involving midbrain
3	F 65	Ondine's curse (obligatory SA); nonprogressive, nonresolving medial medullary syndrome, L > R stem involvement	↓↓ amplitude of waves after II A.S., and after III A.D.	Pontomedullary infarct involving left pons and medulla more than right
4	M 68	Ondine's curse (obligatory SA); no other evidence of stem dysfunction when tested; later had fatal resp. arrest while asleep	↑ I–III A.S. with resultant ↑ I–V IPL; AD: normal BAEPs, A.D., ↑ III_N–V_N A.S. also (see text)	Left-sided and midline pontomedullary infarct, autopsy-proven

*A.S., in response from left ear; A.D., in response from right ear; A.U., in monaural responses from both ears; AR, amplitude ratio; IPL, interpeak latency; ↓, abnormally reduced, ↓↓, markedly reduced; ↑, prolonged.

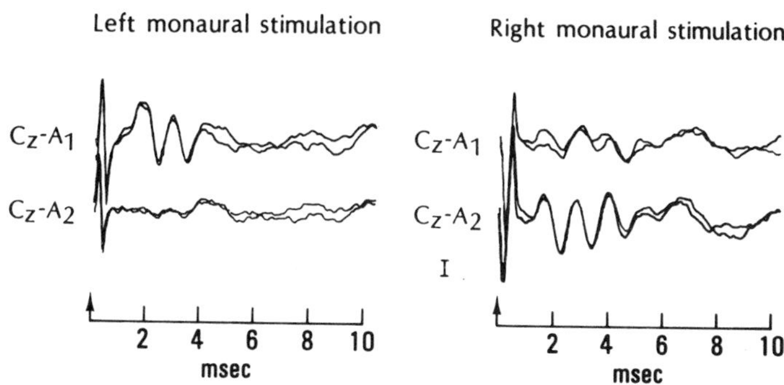

FIGURE 3. Brainstem auditory evoked potentials (BAEPs) in case 1 with central SA (compare TABLE 1 and text). The calibration unit is 0.1 μV. Simultaneous vertex (C_Z) to both left (A1) and right (A2) earlobe recordings shown. In this and subsequent figures, two separate averages of responses to 2048 clicks are superimposed (unless otherwise specified). Relative electropositivity at C_Z compared to the earlobe recording sites produces upward deflections in plots.

The remaining four patients in the central SA group had central BAEP abnormalities from one or both ears. Clinical information about these four cases is presented in TABLE 1 and the BAEPs from each case are shown in FIGURES 3–6. It is of interest that in two of these four cases (cases 1 and 2 in TABLE 1, FIGURES 3 & 4), SA was the first symptom of the brainstem lesion and was the only symptom of the brainstem lesion in a third (case 4 in TABLE 1, FIGURE 6). Clinical manifestations progressed in cases 1 and 4, resolved largely in case 2 (in association with normalization of the BAEPs, as seen in FIGURE 4), and stayed the same in case 3. The BAEP abnormalities in cases 1 and 4 (FIGURES 3 & 6) provided the first objective evidence that the SA syndromes reflected structural brainstem lesions; in case 2, other clinical signs had developed prior to BAEP testing, which indicated stem involvement and, in case 3, a medial medullary syndrome was present from the time of onset of the obligatory SA. Case 4 also had obligatory SA, or Ondine's curse,§ days before showing any other signs.

Mixed/Obstructive SA (Adults)

Twenty males aged 44–69 years (mean: 55.8 ± 1.63 years) presented with moderately severe, predominantly obstructive SA syndromes. In 14, the apneic episodes were purely obstructive at their outset and in the other six, the SA was of the mixed type but always predominantly obstructive. These episodes occurred 30 to 128 times per sleep-hour with a mean incidence in this group of 84.11 ± 7.12 per sleep-hour. BAEPs were normal from each ear of all 20 patients with I–V IPLs of 4.12 ± 0.037 msec, $|\text{I–V}|_{L\text{-}R}$ of 0.176 ± 0.039 msec, and (IV/V):(I) AR of 2.83 ± 0.12, all well within the normal limits for the corresponding means and dispersion of these values given in the Methods section. The similarity of the mixed/obstructive SA group with age-matched, sex-matched controls (the "older males" curve in FIGURE 2) can best be appreciated from the superimposition of the I–V IPL distributions of the

§After the sea nymph of German mythology whose curse of loss of automatic respiration resulted in the death of her unfaithful human lover when he fell asleep.

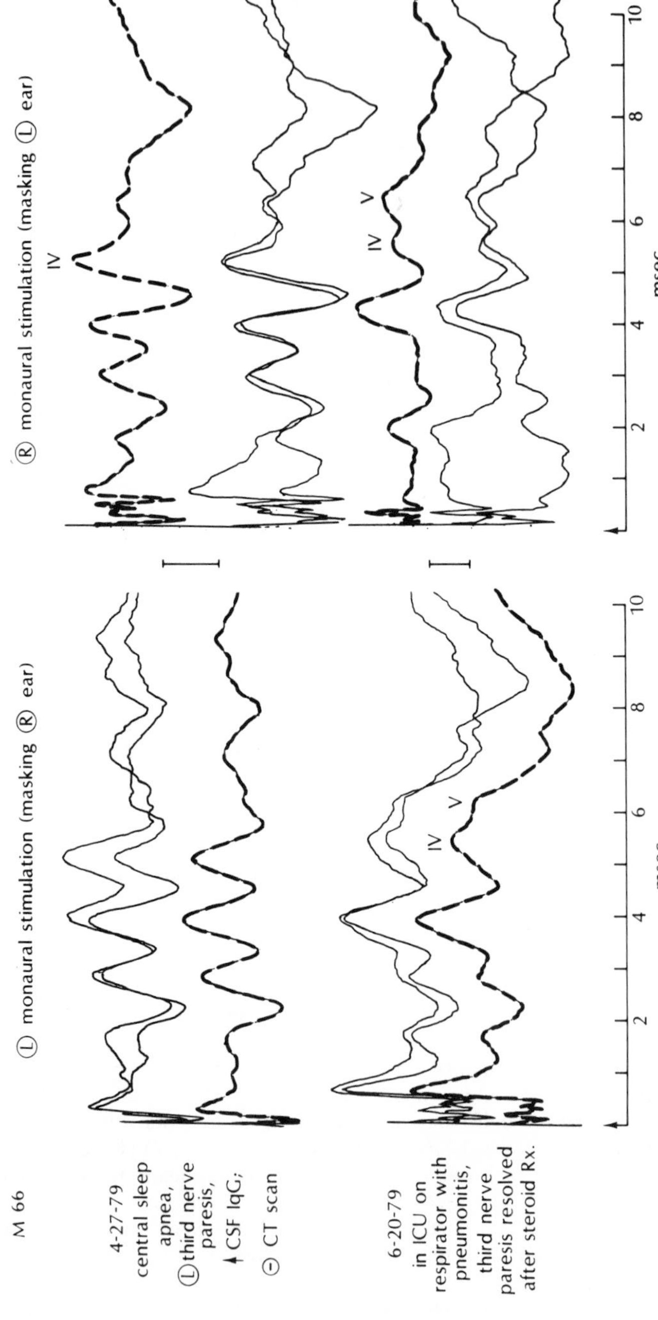

FIGURE 4. BAEPs in case 2 with central SA (compare TABLE 1 and text). The calibration unit is 0.2 μV. Dashed lines indicate additional averages of 4096 responses. Identification of BAEP components in this and all cases was accomplished by methods described elsewhere.[18]

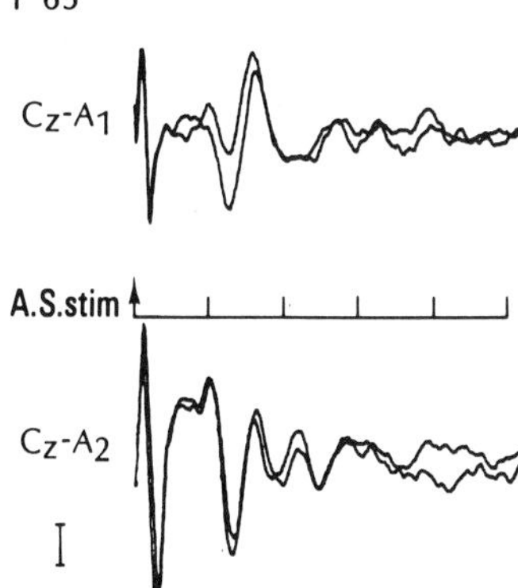

FIGURE 5. BAEPs in case 3 with central SA (compare TABLE 1 and text). The calibration unit is 0.1 μV. Recordings derivations (C_Z–A_i) and absence of cross-evoked BAEPs to contralateral monaural stimuli were verified in this and all cases.

two populations in FIGURE 7, which show frequency polygons normalized in percentages.

Sudden Infant Death Syndrome

We studied BAEPs in one case prior to the first and only NMSID episode (at 17 weeks of age) and a subsequent (at age 19 weeks) crib death, the only case of actual

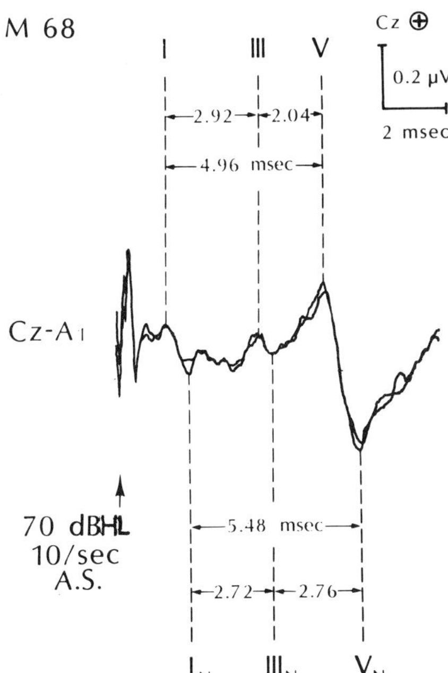

FIGURE 6. BAEPs in case 4 with central sleep apnea (Ondine's curse); compare TABLE 1 and text. Prolongation of the III_N–V_N IPL beyond the 2.62 msec upper limit of normal (ULN) for this value[18] indicates the intra-axial (brainstem) locus of the central auditory conduction disturbance suggested by conventional BAEP measures derived from the vertex-positive (ear-negative) response components. The I–III IPL prolongation (2.9 vs ULN of 2.6 msec in this age/sex group[16]) accounting for the I–V IPL prolongation in this case does not allow intra- vs extra-axial localization.

SIDS in which BAEPs have been recorded and, to our knowledge, the only case of NMSID in which BAEPs were performed prior to the first apneic episode. The infant was born at 28–29 weeks by dates and was of 29–30 weeks estimated conceptional age by Dubowitz scoring. He was a 1020-gram first-born twin and had APGAR scores of 8 and 9 at one and five minutes. The perinatal period was complicated by transient hyperbilirubinemia which peaked at 9.6 mg% and had fallen to within normal limits by age two weeks, and by respiratory distress syndrome requiring intubation and mechanical ventilation for the first two weeks. By the time of discharge from the nursery at age 37 weeks, his examination and clinical status were unremarkable and he was gaining weight well. BAEPs were first performed at this time for purposes of hearing screening and revealed normal-for-age I–V IPLs of 5.43 and 5.63 msec from the left and right ears, respectively, and prolonged-for-intensity wave I (FIGURE 8). The latter finding was consistent with mild conductive hearing loss as BAEP latency–intensity functions had normal slopes.[19] This is a common transient finding in normal prematures.[19,20] The infant did well until 17 weeks of age when, after developing an upper respiratory infection, he was found apneic and cyanotic in his crib and was resuscitated by his mother. He was hospitalized and monitored for apnea for the next ten days. Four days after discharge from the hospital, at the age of 19 weeks, he was found dead in his crib.

Near-Miss-Sudden-Infant-Death

In addition to the case described above, we studied BAEPs in six other NMSID cases between six and twelve days after the event. Data from these cases are

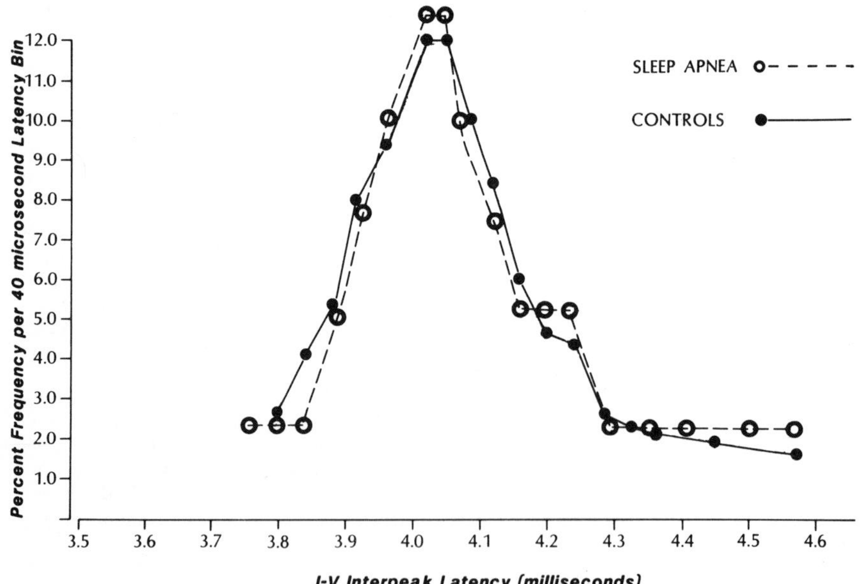

FIGURE 7. Near-perfect superimposition of the frequency polygons of the I–V IPL values falling in given 40 μsec bins for the 40 ears of the 20 patients with mixed/obstructive SA and the 50 ears of the 25 "old males" of the control population shown in FIGURE 2, the solid curve being the same as that in FIGURE 2.

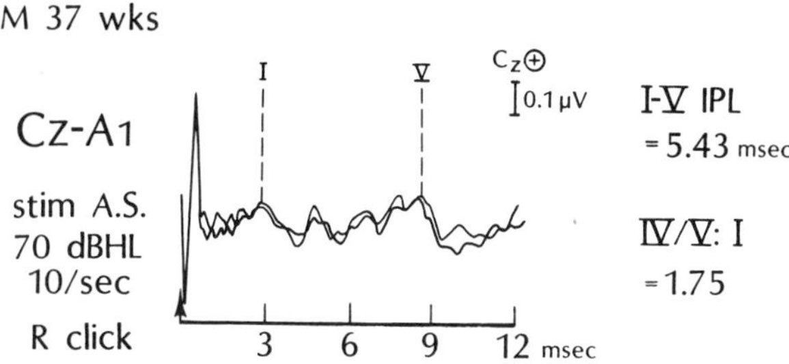

FIGURE 8. Normal-for-age I–V IPL from left ear in case 1 of TABLE 2 (see text) who survived a NMSID episode at age 17 weeks and died a crib death at age 19 weeks post-term. The above BAEPs were recorded 23 weeks earlier, at 37 weeks conceptional age (CA).

summarized along with the above case (case one) in TABLE 2. As can be seen in that Table, all NMSID survivors had normal BAEPs when they are considered as individuals compared to age-matched normal controls but all of them, except cases 1 (recorded prior to the first NMSID event) and 6 (recorded after only two episodes), had average I–V IPLs that fell 1.2 to 2.2 standard deviations to the right of the normal mean value for their age group. These high-normal IPLs were recorded from four infants (cases 2, 3, 5, 7) who each had multiple NMSID episodes requiring resuscitation prior to testing and a fifth (case 4) who had had the most severe single apneic episode of the group, one from which resuscitation was followed by several hours of unresponsiveness and mild acidosis. In case 2, who had the most NMSID episodes (more than 20) prior to testing, the high Z score in TABLE 2 partly reflects the relative youth of the patient with respect to the age range of the pooled normative data he is being compared with. However, at the age of 3.8 months, his I–V IPLs had not yet shown any of the expected normal shortening with maturation (FIGURE 9). Repeat BAEP testing at age one year after an eight-month period during which no apneic episodes occurred revealed normal-for-age I–V IPLs close to the mean normal value for that age, (mean I–V IPL for both ears was 4.18 msec vs 12-month norm of 4.23 ± 0.16 msec). Such divergence and convergence from normal age–IPL functions correlate well with apnea incidence in our experience with other high risk infants and appear to parallel, if not predict, the frequency and severity of subsequent apneas.[26]

Comparison of the averaged individual (mean of the two ears) I–V IPLs for the five infants in TABLE 2, who were between 2.5 and 3.5 months of age, gave a mean of 4.91 ± 0.136 msec versus 4.66 ± 0.067 msec for the 12 age-matched normal infants aged 3.0 ± 0.5 months. Comparison of these sample means in two-way analysis with 15 degrees of freedom gives $t_{15} = 6.94$, $p < 0.005$,¶ even though no individual value in the NMSID group falls outside of the normal limits obtained using the routine stimulation parameters (57-dBHL rarefaction clicks, 10/sec) and recording in the vertex to ipsilateral ear derivation. We attempted to increase the chances of recording BAEP abnormalities in the six infants studied after NMSID episodes by the following maneuvers, which may occasionally unmask BAEP abnormalities not evident when only the routine stimulation and recording parameters are employed: (1) by measuring stimulus-to-peak IPL changes as a function of decrements in stimulus intensity

¶Comparison of the two groups with a nonparametric test, the Mann-Whitney test, showed that the I–V IPLs from the five NMSID infants aged 3 ± 0.5 months ranked 1st, 2nd, 3rd, 4th, and 12th in length when pooled with the 12 age-matched control and patient values ($N = 17$). This gives $U_{(1), 5, 12} = 53$, $p < 0.01$, for significantly longer IPLs from NMSID infants versus age-matched normal infants.

TABLE 2

SUMMARY OF BAEP FINDINGS IN NEAR-MISS SUDDEN INFANT DEATH SYNDROME (NMSID) PATIENTS RELATIVE TO AGE-MATCHED NORMAL BAEP VALUES*

Case	Gender	Age at First NMSID Episodes (months)**	Age at Last NMSID Event Prior to Testing (months)**	Age at Testing (months)**	Left Ear I–V IPL	Right Ear I–V IPL	Mean IPL (Both Ears)	Z-Score re Normal Mean*	Follow-Up Information
1	M	4	—	−0.8 preterm (37 weeks C.A.)	5.43	5.60	5.52	+0.81†	Died of SIDS at 4.5 mos. (no autopsy)
2	M	1	2 3.5 3.5	2.5 3.8 12	5.16 5.21 4.14	5.20 5.19 4.22	5.18 5.20 4.18	+2.22‡ +2.31‡ −0.31¶	Recurrent NMSID with recorded obstructive SA for 1 mo. after first tested, during which BAEPs showed no maturation; no further NMSID, BAEP normal at 1 yr
3	F	2	2.5	3	5.18	4.84	5.01	+1.50‡	Multiple NMSID prior to testing, none in subsequent 5 mos.
4	F	3	3	3	5.00	5.00	5.00	+1.45‡	Single prolonged NMSID episode with unconsciousness, mild acidosis several hrs. following resuscitation

5	F	2	3	3.5	4.92	4.94	4.93	+1.15‡	Multiple NMSID events prior but none subsequently for 6 mos., then lost to follow-up
6	M	1	3.5	3.5	4.40	4.46	4.43	−0.98‡	Two NMSID events prior to testing, both mild; no subsequent NMSID
7	F	4	6	6.5	4.68	4.64	4.66	+1.18§	Sibling subsequently died of SIDS at age 1 mo.; patient had multiple NMSID episodes prior to BAEPs, none thereafter in 1-year follow-up

*The number of standard deviations from the age-matched normal mean value (mean + 2.33 SD = the one-tailed upper limit of normal, since I–V IPLs have a Gaussian distribution).[31] C.A.: conceptional age.

†For ages 36–37 weeks C.A., mean: 5.27 msec; SD: 0.31.[20]

‡For ages 2.5–3.5 months, mean: 4.66 msec; SD: 0.23 msec.

§For ages 5.5–6.5 months, mean: 4.46 msec; SD: 0.17 msec.

¶For ages 11–13 months (1 yr), mean: 4.23 msec; SD: 0.16 msec.

**Ages are given in months with respect to term. Only one infant (Case 1) was born prematurely and was also tested 3 weeks (−0.8 months) pre-term at a C.A. of 37 weeks.

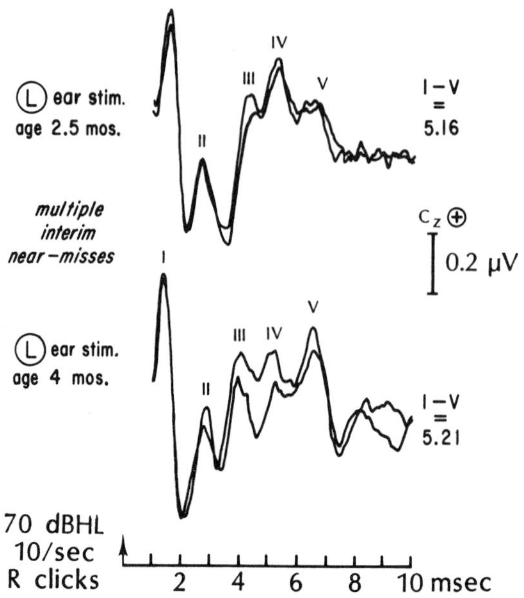

FIGURE 9. Lack of IPL shortening despite technically identical studies and equivalent peripheral auditory function in BAEP studies 5½ weeks apart in case 2 of TABLE 2. Infant had recurrent apneic episodes in the interim. This represents abnormal divergence from our control age-IPL functions even though IPLs never became abnormal when considered cross-sectionally at given ages. Longitudinal studies also showed complete convergence with normal IPL-age functions between ages 4 and 12 months in parallel with cessation of the NMSID and all apneic episodes.

from 70 dBHL to 30 dBHL (FIGURE 10); (2) by increasing rate of stimulus presentation from 10/sec to 80/sec to see if any rate-dependent IPL abnormalities could be elicited (FIGURE 11); (3) by recording simultaneously in C_z–A_i, C_z–A_c, and A_i–A_c (ear-to-ear) recording derivations (FIGURE 12) to see if derivation-specific abnormalities existed; (4) by changing the acoustic polarity of the clicks from initially rarefaction to initially condensation at high and low stimulus intensities (70 and 30 dBHL) and fast and slow (80 and 10/sec) rates to unmask possible polarity-dependent abnormalities or abnormal rate-intensity-polarity interactions. As was the case when using routine stimulation and recording parameters, there was no BAEP abnormality from either ear in any of the six infants when these maneuvers were employed.

The relationship between the number of prior episodes, the severity of the episodes, and the degree of deviation from the mean I–V IPL for normal age-matched controls was further assessed by comparing BAEPs of the five infants in cases 2–6 with those of eleven other age-matched infants (2.5–3.5 months) who had suffered recurrent or chronic hypoxemia/ischemia due to congenital heart disease (CHD), six of whom also had suffered chronic hypoxemia from respiratory distress syndrome (RDS) as premature infants. These infants had a mean I–V IPL of 4.82 ± 0.107, which does not differ significantly from the NMSID group but differs from that of age-matched normal controls at the same level of significance as the NMSID survivors did. Like the NMSID survivors, none of the infants with other causes of chronic hypoxemia/

ischemia met the criteria for BAEP abnormality when considered as an individual versus the age-matched normal subjects. Only when compared collectively against group normative data did the highly significant population differences emerge, as was the case for the NMSID infants. It was of interest that the infants with CHD or RDS all had I–V IPLs falling between 1 and 2 standard deviations above the mean, as did most NMSID surivivors.

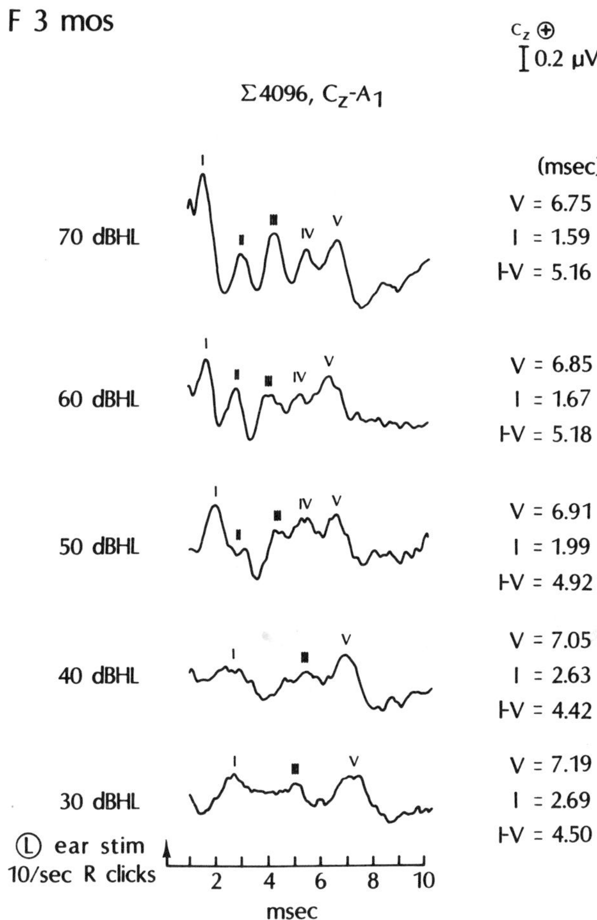

FIGURE 10. Normal latency-intensity functions for peaks I, III, and V and the I–V IPL[19,20,32] over a 40 dBHL range in 10 dB steps from a NMSID survivor. The stimulus-to-peak and interpeak latencies from the other ear at each intensity were virtually identical. Each average is the sum of two averages of 2048 responses that superimposed closely. This study was performed in a sound-attenuated room rather than the usual infant recording environment and the 103 dB peSPL click was actually 70 dBnHL as well as "70 dBHL" on the dial calibrated for use in soundproof rooms (see footnote‡). The 5.00 msec IPL in this case shown in TABLE 2, like the other cases and controls, is in response to a 103 dB peSPL click in an environment with a constant ambient noise level that raised normal adult thresholds by 13 dB.

Discussion

Elucidation of the potential role BAEPs might play in the understanding and/or prevention of SIDS is of especial importance now in the wake of highly publicized studies by Nodar, Orlowski, and Lonsdale stating that BAEPs are consistently abnormal in NMSID infants.[23,24] The conclusions of that study were the BAEPs should prove useful in identifying infants at risk for SIDS and that the abnormal BAEPs found in all NMSID infants implicated the brainstem (and brainstem respiratory center) as the area of dysfunction in SIDS.

The implications of these statements are so important that we initially tried to confirm them in a largely retrospective collaborative multicenter study of 28 NMSID infants.[2] Instead of the reported 100% prevalence of BAEP abnormalities, we found abnormalities in only 3 of 28 cases (11%) and, in each of those three cases, there was evidence that the abnormalities were related more likely to the result—than to the cause—of the severe apneic episodes.[2] Gupta, Guilleminault, and Dorfman[25] subsequently reported *normal* BAEPs in 100% (9/9) of NMSID cases and concluded, as we did, that single BAEP studies have no value in identifying infants at higher risk for SIDS. Furthermore, the Stanford group concluded that BAEPs have no value in

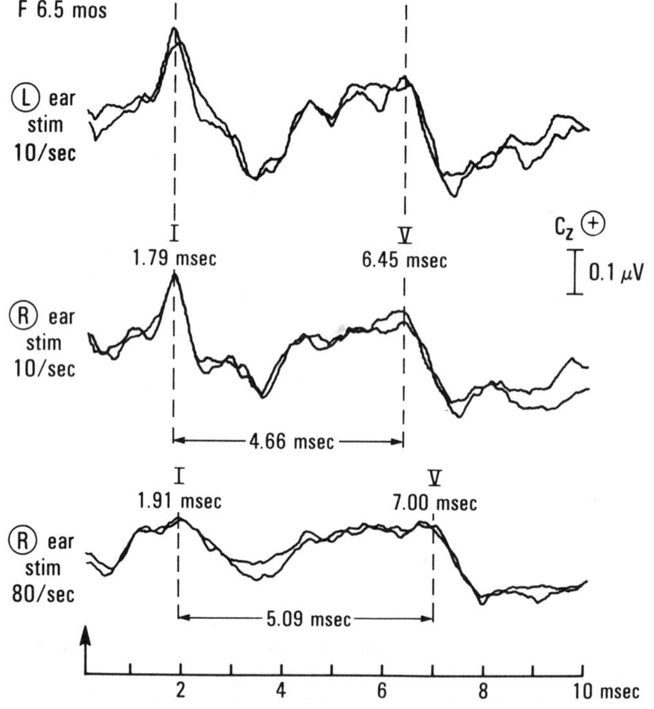

FIGURE 11. Normal I–V IPL at routine stimulus presentation rates and normal rate-dependent increase in I–V IPL with change in click rate from 10/sec to 80/sec (shown for right ear only). The mean increase in I–V IPL with change in rarefaction click rate from 10/sec to 80/sec in normal subjects in this age group is 0.89 msec[17] (SD = 0.10 msec), a value never reached by any of the NMSID survivors we tested. Rate-latency changes were also normal with condensation clicks, and were of lesser magnitude, for all subjects tested.

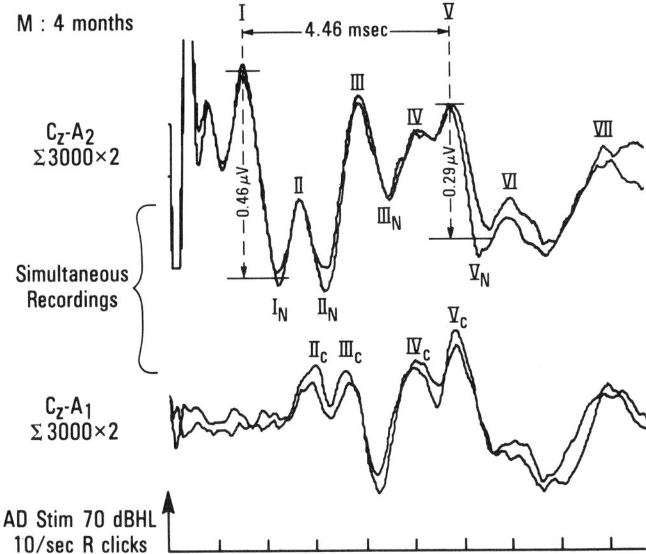

FIGURE 12. A normal I–V IPL and (IV/V):(I) amplitude ratio (AR) in an ipsilateral-ear (C_Z-A_i) recording derivation and normal amplitude and latency effects[16,18] on all BAEP components simultaneously recorded in C_Z-A_c from survivor of NMSID episode. Top pair of superimposed averages shows how the (IV/V):(I) AR is routinely measured in the C_Z-A_i (C_Z–A_2 in this case) derivation; (IV/V):(I) AR = 0.63 here (0.29 μV/0.46 μV).

studying the pathogenesis of SIDS either, although they performed no serial BAEP studies.[25]

The methods, findings, and conclusions of our investigation of BAEPs in NMSID differ strikingly from the findings of the Cleveland Clinic group[23,24] and are more in agreement with those of the Stanford study. The Standford group could not reconcile their diametrically opposed results *vis-à-vis* the Cleveland Clinic data of Orlowski, Nodar, and Lonsdale, except to note that the latter authors provided no quantitative data to support their claims about BAEP abnormality in NMSID cases. Equally noteworthy are the criteria for BAEP abnormality that Orlowski *et al.* did choose:[23] (1) stimulus-to-peak latency prolongation 0.2 msec longer than "normal" [norms not provided] for a given SL [sic] intensity in the infants tested; (2) intraear [sic] peak latency, [where the authors presumably mean interpeak latencies since peak-to-peak measurements between waves I, III, and V were cited as criteria for abnormality, but no limits of normality were specified]; (3) interear peak latency differences of wave V greater than 0.2 msec; (4) "unstable" responses, defined as poor intertrial reproducibility in successive averages; (5) amplitude reduction of a given peak by 50% when compared to the same peak obtained at the same SL [sic—assume the authors meant HL since SL is not ascertainable in infants, by definition]; (6) abnormal "wave shape," including "flattening" of the peaks, broad-based peaks, and additional peaks; (7) peak presence [not further elaborated on, but the implication being that all seven vertex-positive (ear-negative) BAEP components in the classical response must be present for the response to be judged normal].

BAEPs of all 15 infants with NMSID studied by Nodar *et al.*[24] were classified as abnormal because they met two or more of these seven response criteria. This is not surprising since, even when the totally unspecified criteria numbers 2 and 7, or the self-validating criteria 4 and 6 (the BAEPs are abnormal because of their abnormal wave shape, unstable because of their instability, etc.) are excluded, the remaining three arbitrary criteria are in no way *a priori* evidence of central auditory conduction disturbance. By far the most common cause of BAEPs meeting their criteria 1, 3, and 5 is peripheral auditory dysfunction, which is very common and often asymmetric as reflected in infant BAEPs.[20] Criterion number 4, if met, is almost certainly a reflection of inadequate technique rather than of any problem with the patient such as high-risk-for-SIDS, no more than it is evidence for brainstem demyelination in adults, as has been claimed by the same group.[26] Attribution of poor intertrial reproducibility to pathologic random "moment-to-moment differences" in BAEPs within trials[26] is a logical fallacy contradicted by some simple mathematical considerations.‖ Predictably, more than two-thirds (55/78 or 71%) of our normal infant population met two of the three criteria (numbers 1, 3, 5) of Orlowski *et al.*[23] and Nodar *et al.*[24] which could even serve as a basis for interlaboratory comparison. We are convinced by the absence of any NMSID episode or any other neurologic or otologic abnormality on 18 to 54 month clinical follow-up, that none of them was ever at higher risk for SIDS, or any otoneurologic or other problem.

Like Gupta *et al.*, [25] we chose to compare our NMSID infants to age-matched normal infants using rigorously defined quantitative BAEP measures that have actually been shown to reflect central auditory conduction and to specify limits of normality (and quantitative criteria for abnormality) of these measures. The methods of our study differed somewhat from that of the Stanford group, and although our conclusions were the same regarding the lack of predictive value of a single study regarding an individual infant being at higher risk for SIDS, our results and conclusions regarding BAEPs in the NMSID population differed slightly from those of the study Gupta *et al.*[25]

We do not yet rule out the potential value of BAEPs in this application, particularly if infant BAEPs are studied longitudinally—in which case our preliminary findings show at least a strong correlation between age-IPL *trends* in serial BAEPs and risk of apnea.[26] Also, the data presented here already indicate that BAEPs can be of value in characterizing certain aspects of the NMSID syndrome itself—even if they prove incapable of predicting risk for SIDS in individual cases. Gupta *et al.*[25] do not share even this limited optimism, based on their failure to find any *group* differences between NMSID survivors and normal "controls."

The nine NMSID survivors of that study ranged in age from 1 to 14 weeks when they had their first NMSID episode and had BAEPs recorded. The method of statistical analysis employed by these authors was a paired t-test, in which each NMSID survivor was compared with only a single age-matched control individual

‖Although the signal-to-noise ratio decreases with averaging in a roughly inverse proportion to the square root of the number of samples averaged, the ratio of any random variability of the signal to the variability of the noise always remains at least about 10^{-2} when averaging BAEPs and would usually be even lower.[28] Thus detection of random variability of the signal within—and thus between—single averages (trials) is unlikely with routine methods, and significant intertrial variability is best attributed to one of the remaining three possibilities which averaging under these circumstances can detect: (1) systematic variation in the signal, (2) systematic variation of the noise; (3) random variation in the noise. An example of the first would result if a patient became progressively hypothermic between successive trials or if repositioning of headphones between trials resulted in changes in effective stimulus intensity; the second if the patient had progressively more post-auricular EMG artifacts contaminating the average between trials; the third and most common situation results from changes in the amount of EMG contamination and other artifacts such as 60 Hz interference between trials.

selected from the community or hospital population. The control data were then pooled and the patient data pooled to form the basis for a group t-test comparison. Data from both ears of each subject were combined in both tests and treated statistically with the assumption that about twice the degrees of freedom (DF; $18 - 1 = 17$ DF vs $9 - 1 = 8$ DF) were created by the inclusion of both ears from each subject in the analysis (which is not the case).

We considered and then rejected these three statistical manipulations in analyzing our data; we considered them initially because the small numbers of NMSID survivors that one can study who are close to the same age makes it difficult to compare age-matched groups with the parametric** statistical analysis of choice, in this context, two-sample comparison of means by t-test.[29] However, use of the paired t-test between entirely different individuals is questionable here,[30] especially given the considerable interindividual variability of infant BAEP latencies as a function of determinants other than age.[31,32] Pooling data from a wide range of ages for group t-test comparison is also somewhat risky, as large differences between age-paired individuals could be averaged out by large differences in the opposite direction at other ages. Finally, inclusion of both ears of each individual in interindividual comparisons does not really double the DF or enhance the power of the t-test in proportion to the apparent increase in DF. Because of the tight linkage in latency values between the two ears of a given individual, e.g., as reflected by the fact that virtually all normal subjects have $|\text{I–V}|_{L\text{-}R} < 0.4$ msec,[32] inclusion of both subject ears as independent values injects the low intraindividual variability between ears into an analysis that should be testing only for interindividual variability. Spuriously low or high p values—either α/Type I or β/Type II errors—can result from such overestimation of the true degrees of freedom, which should obviously never exceed the number of individuals being compared in interindividual comparisons.

Thus, we averaged the two values from each ear of individual subjects in our patient and control groups and assigned only $N - 1$ degrees of freedom to each group rather than $2N - 1$, since there were only N individuals in each group. By then comparing two groups of individuals who were closely age-matched within groups and between groups (3.0 ± 0.5 months), we were able to find a significant†† population difference in I–V IPLs even though no individual patient's IPL fell outside of normal limits when considered as an individual *vis-à-vis* the control population. These differences in statistical approach may account for our different findings from Gupta *et al.* with respect to this one point—that as a group, BAEPs of NMSID survivors do differ from normal infants. This is not surprising given that these infants have had severe apneic episodes, most with associated bradycardia.

In the developing brain, it has been shown that the periventricular, brainstem, and especially the brainstem auditory structures are more vulnerable than other brain

**As noted previously, the nonparametric analysis of choice, the Wilcoxon-Mann-Whitney test yielded $U_{(1),5,12} = 53$, $p < 0.01$, for the significance of the longer I–V IPLs of the five NMSID survivors compared to the 12 normal age-matched infants. Other nonparametric tests gave even lower p values. No test employed yielded $p \geq 0.01$.

††$p < 0.01$ with the Mann-Whitney nonparametric test, and $p < 0.005$ with the two-tailed comparison of sample means by t-test. With regard to the latter, the small N of the two groups that were compared does not affect the validity of the comparison because of the Gaussian distribution of infant I–V IPLs[31] and the absence of any differences—apart from survival of NMSID episodes—of the two groups. The fewer DF are offset by the increased number of standard errors of the mean (SEM) required to achieve a given level of statistical significance for any difference between means; pooled variances including SEMs of both groups are employed. The variances of the NMSID subgroup when compared to the age-matched normals did not differ significantly, the variance ratio test giving $p\ (F_{11,4} = 1.69) > 0.050$ for the two groups.

regions to hypoxemic/ischemic insults.[33–37] This, of course, is in contrast to the pattern of selective regional vulnerability to anoxia in the adult brain in which the converse obtains, i.e., the most vulnerable areas being the cerebral (including hippocampi) and cerebellar cortices, with the adult brainstem being relatively resistant to anoxic insults.[37,38] We postulate that the absence of any population difference in BAEPs between our adult SA patients (FIGURE 7) and controls, despite the prominent oxygen desaturation associated with many of these episodes, reflects this lesser vulnerability of the adult brainstem to anoxia as compared to that of newborns and infants. Accordingly, the three adult patients with idiopathic mixed/obstructive SA syndromes on whom we have postmortem neuropathologic material have shown no abnormal brainstem histology, although Lehrman *et al.* did find midline brainstem gliosis in a patient with sleep apnea associated with Shy-Drager syndrome.[39]

The following observations in the present study also support the role of prior hypoxic/ischemic brainstem insults in causing these slight but highly significant BAEP differences seen in NMSID survivors: (1) the positive correlation within our NMSID group between the number and severity of apneic episodes and the degree of I–V IPL prolongation above the age-matched normal mean; (2) the progressively-long-for-age I–V IPLs (failure to mature between ages 2.5 and 3.8 months) in one infant while he continued to have severe apneic episodes; and then, conversely, by normalization (for age) of BAEPs over a subsequent eight month period in which no further apneic episodes occurred; (3) the very close similarity of BAEPs from NMSID survivors and those from survivors of recurrent hypoxic insults due to CHD and/or RDS—and the common BAEP differences of these groups versus age-matched controls; (4) that among seven patients having NMSID episodes, the two having I–V IPLs falling within one SD of the normal-for-age mean value were the only infants who had not had multiple or severe apneic episodes prior to BAEP testing.

Naeye *et al.*,[40] Guilleminault *et al.*,[10] and Takashima *et al.*[41] have all reported brainstem gliosis in SIDS victims and, in the latter study, the gliosis was found to be notably increased in the nucleus solitarius, dorsal motor nucleus of the vagus, the nuclei ambiguus and retroambigualis, and the reticular formation.[41] These areas are the primary sites for generation of respiratory rhythm,[42] and this finding raised the possibility that morphologic changes and secondary dysfunction of these brainstem respiratory centers were involved in the pathogenesis of SIDS. However, involvement of these particular brainstem regions also occurs in infants who have chronic or recurrent hypoxemia/ischemia due to congenital heart disease[41] and respiratory distress syndrome.[43] The question then becomes whether the brainstem gliosis in SIDS is not a nonspecific reaction to chronic hypoxemia associated with alveolar hypoventilation[7,9] and/or ischemia associated with bradycardia accompanying their frequent mixed and obstructive sleep apnea episodes.[9,10]

It seems plausible that the pathogenesis of SIDS involves a vicious circle in which hypoxemic/ischemic injury to brainstem respiratory nuclei from recurrent apnea results in more frequent episodes of apnea and bradycardia, which in turn further damage brainstem cardiorespiratory control centers. The BAEP differences we found between NMSID infants and controls may reflect such cumulative damage to adjacent brainstem auditory tracts, and this BAEP finding may be one more "marker" of chronic hypoxia such as those evidenced elsewhere in the brain[40,41,44–46] and throughout other organ systems.[47–54] As such, the BAEP differences we found in the NMSID population would best be considered nonspecific effects or epiphenomena of the syndrome rather than as directly related to the etiology of NMSID and SIDS.

We thus agree with Gupta *et al.*[25] that BAEP test results have no predictive value with respect to the risk of an individual infant being at risk for SIDS because, in our experience, they fall within normal limits in the vast majority of cases and were normal in one infant who subsequently died a crib death. We do not concur completely with Gupta *et al.* that BAEPs lack value in studying the pathophysiology of idiopathic infant apnea syndromes since we found collectively prolonged mean I–V IPLs in the NMSID population, which, for reasons outlined above, probably represent an electrophysiologic marker for the cumulative effects of chronic and/or recurrent hypoxic brainstem insult in this abortive form and possible precursor of SIDS. As pointed out by Gupta *et al.*, however, for the test to be clinically meaningful, a BAEP result which carries such grave prognostic implications should distinguish unequivocally between normal and at-risk infants—which BAEPs clearly have not done in their studies or ours, at least in single studies at one age purported to "screen" for risk of the syndrome.[23,24]

With such high stakes involved for the infant (life or death) and for the parents (false hope or inappropriate dread, the financial and psychologic burdens of home apnea monitoring, etc.), the consequences of either false-positive or false-negative test results could be disastrous. Other methods of determining risk-for-SIDS appear to hold more promise[7,9–11] than BAEPs and have greater therapeutic implications (in terms of prevention).[55] In summary, our findings suggest that BAEPs do have value in the study of the NMSID syndrome and that the technique may contribute to our understanding of SIDS itself. However, the test does not allow reliable discrimination of individual infants at higher risk for crib death from those who are not, as has been suggested.[23,24] Analysis of maturational trends in longitudinal BAEP studies may still prove useful in this regard, however, given the correlation between these trends and the occurrence/recurrence of apnea in high-risk infants.[26]

Acknowledgments

The author is grateful to Janet Stockard, Rochelle Hatleli, Marilyn Anderson, and Bruce Staats, M.D., for their assistance with data collection; to Dorie Kehew, for assistance in preparation of the manuscript; and to Frank Sharbrough, M.D., and the nursery staffs of St. Mary's Hospital, Rochester, Minnesota and University Hospital, San Diego, California for their support and cooperation in this study.

References

1. Stockard, J. J., F. W. Sharbrough, B. A. Staats & P. R. Westbrook. 1980. Brainstem auditory evoked potentials in sleep apnea. Electroencephalogr. Clin. Neurophysiol. **50:** 167P.
2. Stockard, J. J. & K. Hecox. 1981. Brainstem auditory evoked potentials in sudden infant death syndrome, "near-miss-for-SIDS," and infant apnea syndromes. Electroencephalogr. Clin. Neurophysiol. **51:** 43P.
3. Guilleminault, C., W. C. Dement & N. Monod. 1973. Syndrome "mort subite due nourrison": Apnée au cours du sommeil. La Nouvelle Presse Med. **2:** 1355–1358.
4. Naeye, R. L. 1970. The sudden infant death syndrome: A review of recent advances. Arch. Pathol. Lab. Med. **101:** 165–167.
5. Steinschneider, A. 1972. Prolonged apnea and the sudden infant death syndrome: Clinical and laboratory observations. Pediatrics **50:** 646–654.
6. Guilleminault, C., R. Peraita, M. Souquest & W. C. Dement. 1975. Apneas during sleep in infants: Possible relationship with sudden infant death syndrome. Science **190:** 677–679.
7. Shannon, D. C. & D. Kelly. 1977. Impaired regulation of alveolar ventilation and the sudden infant death syndrome. Science **197:** 367–368.

8. WEITZMAN, E. D. & L. GRAZIANI. 1974. Sleep and sudden infant death syndrome: A new hypothesis. Adv. Sleep Res. **1:** 327–344.
9. GUILLEMINAULT, C., R. ARIAGNO, M. SOUQUET & W. C. DEMENT. 1976. Abnormal polygraphic findings in near-miss sudden infant death. Lancet **1:** 1326–1327.
10. GUILLEMINAULT, C. & R. L. ARIAGNO. 1978. Why should we study the infant "near miss for sudden infant death"? Early Human Dev. **2/3:** 207–218.
11. SHANNON, D. C., D. H. KELLY & K. O'CONNELL. 1977. Abnormal regulation of ventilation in infants at risk for sudden-risk-death syndrome. New Engl. J. Med. **297:** 747–750.
12. GUILLEMINAULT, C., R. ARIAGNO, L. S. FORNO, L. NAGEL, R. BALDWIN & M. OWEN. 1979. Obstructive sleep apnea and near miss for SIDS: 1. Report of an infant with sudden death. Pediatrics **63:** 837–843.
13. BERGMAN, A. B. 1979. Sudden infant death syndrome in King County, Washington. *In* Sudden Infant Death Syndrome. A. B. Bergman, J. B. Beckwith & C. G. Ray, Eds. pp. 83–107. Univ. Washington Press, Seattle, Wash.
14. VALDES–DAPENA, M. A. 1980. Sudden infant death syndrome: A review of the medical literature 1974–1979. Pediatrics **66:** 597–614.
15. GUILLEMINAULT, C. & R. KOROBKIN. 1979. Sudden infant death: Near-miss events and sleep research. Some recommendations to improve comparability of results among investigators. Sleep **1:** 423–433.
16. STOCKARD, J. J., J. E. STOCKARD & F. W. SHARBROUGH. 1978. Non-pathologic factors influencing brainstem auditory evoked potentials. Am. J. EEG Technol. **18:** 177–193.
17. STOCKARD, J. E., J. J. STOCKARD, B. F. WESTMORELAND & J. F. CORFITS. 1979. Normal variation of brainstem auditory evoked potentials as a function of stimulus and subject characteristics. Arch. Neurol. **36:** 823–831.
18. STOCKARD, J. J., J. E. STOCKARD & F. W. SHARBROUGH. 1980. Brainstem auditory evoked potentials in neurology: Methodology, interpretation, clinical application. *In* Electrodiagnosis in Clinical Neurology. M. J. AMINOFF, Ed., pp. 370–413. Churchill-Livingstone, New York, N.Y.
19. STOCKARD, J. E. & J. J. STOCKARD. 1981. Brainstem auditory evoked potentials in normal and otoneurologically impaired newborns and infants. *In* Current Clinical Neurophysiology: Update on EEG and Evoked Potentials. C. E. HENRY, Ed. pp. 421–466. Symposia Specialists, Miami, Fla. and Elsevier/North Holland, Amsterdam.
20. STOCKARD, J. E. & B. F. WESTMORELAND. 1981. Technical considerations in the recording of the brainstem auditory evoked potential for neonatal neurologic diagnosis. Am. J. EEG Technol. **21:** 31–54.
21. COX, C., M. HACK & D. METZ. 1981. Brainstem-evoked response audiometry: Normative data from the preterm infant. Audiology **20:** 53–64.
22. GUILLEMINAULT, C. & M. SOUQUET. 1979. Sleep states and related pathology. *In* Advances in Perinatal Neurology. R. KOROBKIN & C. GIULLEMINAULT, Eds. pp. 225–247. Spectrum Publications, New York, N.Y.
23. ORLOWSKI, J., R. NODAR & D. LONSDALE. 1979. Abnormal brainstem auditory evoked potentials in infants with threatened sudden infant death syndrome. Cleveland Clin. Q. **46**(3): 77–81.
24. NODAR, R., D. LONSDALE & J. ORLOWSKI. 1980. Abnormal brainstem auditory evoked potentials in infants with threatened sudden infant death syndrome. Otolaryngol. Head Neck Surg. **88:** 619–621.
25. GUPTA, P. R., C. GUILLEMINAULT & L. J. DORFMAN. 1981. Brainstem auditory evoked potentials in near-miss sudden infant death syndrome. J. Pediatr. **98:**(5): 791–794.
26. STOCKARD, J. J., J. E. STOCKARD, R. COEN, A. MERRITT & D. A. TRAUNER. 1982. Correlation between evoked potential indices of brainstem maturation and clinical course of infantile apnea syndromes. Ann. Neurol. (Submitted).
27. NODAR, R. H. 1980. Brainstem auditory evoked potentials on 42 individuals with multiple sclerosis. *In* Evoked Potentials. C. Barber, Ed. pp. 581–586. University Park Press, Baltimore, Md.
28. JEWETT, D. L. & J. S. WILLISTON. 1971. Auditory evoked far-fields averaged from the scalp of humans. Brain **100:** 19–40.
29. O'BRIEN, P. C. & M. A. SHAMPO. 1981. Statistics for clinicians: Comparing two samples (the two-sample t-test). Mayo Clin. Proc. **56:** 393–394.

30. O'BRIEN, P. C. & M. A. SHAMPO. 1981. Statistics for clinicians: One sample of paired observations (paired t-test). Mayo Clin. Proc. **56:** 324–326.
31. STOCKARD, J. J. & J. E. STOCKARD. 1982. Statistical considerations in defining auditory brainstem response abnormalities in adults and infants. Proc. Int. ABR Standards Symp., Laguna Beach, Calif., Feb. 3–5, 1982, (in press).
32. STOCKARD, J. E., & J. J. STOCKARD. 1981. Technical and physiological variables affecting brainstem auditory-evoked responses in normal infants and adults. *In* Electrocochleography and Brainstem Electric Response Audiometry. E. Moore, Ed. Grune and Stratton, New York, N.Y.
33. GILLES, F. H. 1969. Hypotensive brain stem necrosis. Arch. Pathol. **88:** 32–41.
34. RANCK, J. B. & W. F. WINDLE. 1959. Brain damage in the monkey, *Macaca mulatta,* by asphyxia neonatorum. Exp. Neurol. **1:** 130–154.
35. SOKOLOFF, L. 1959. The action of drugs on the cerebral circulation. Pharmacol. Rev. **11:** 1–85.
36. HALL, J. 1964. On the neuropathological changes in the CNS following neonatal asphyxia. Acta Otolaryngol. Suppl. **188:** 331–339.
37. MYERS, R. E. 1973. Two classes of dysergia brain abnormality and their conditions of occurrence. Arch. Neurol. **29:** 394–399.
38. LAMPERT, P. W. 1961. The selective vulnerability of the brain to anoxia. Canad. Med. Assoc. J. **84:** 1172–1176.
39. LEHRMAN, K. L., L. GUILLEMINAULT, J. S. SCHROEDER, A. TIKLIAN & L. M. FORNO. 1978. Sleep apnea syndrome in a patient with Shy-Drager syndrome. Arch. Intern. Med. **138:** 206–209.
40. NAEYE, R. L. 1976. Brain-stem and adrenal abnormalities in the sudden-infant-death syndrome. Am. J. Clin. Pathol. **66:** 526-530.
41. TAKASHIMA, S., D. ARMSTRONG, L. BECKER & C. BRYAN. 1978. Cerebral hypoperfusion in the suddent infant death syndrome? Brainstem gliosis and vasculature. Ann. Neurol. **4:** 257–262.
42. MITCHELL, R. A. 1976. Control of respiration. *In* Pathophysiology. E. D. Frohlich, Ed. pp. 131–147. 2nd edit. Lippincott, Philadelphia.
43. BRAND, M. M. & A. BIGNAMI. 1969. The effects of chronic hypoxia on the neonatal and infantile brain: A neuropathological study of five premature infants with the respiratory distress syndrome treated by prolonged artificial ventilation. Brain **92:** 233–254.
44. TAKASHIMA, D. ARMSTRONG, L. E. BECKER & J. HUBER. 1978. Cerebral white matter lesions in sudden infant death syndrome. Pediatrics **62**(2): 155–159.
45. KOROBKIN, R. & C. GUILLEMINAULT. 1979. Neurologic abnormalities in near miss for suddent infant death syndrome infants. Pediatrics **64**(3): 369–374.
46. GADSDON, D. R. & J. L. EMERY. 1976. Fatty change in the brain in perinatal and unexpected death. Arch. Dis. Child. **51:** 42–48.
47. NAEYE, R. L. 1973. Pulmonary arterial abnormalities in the sudden-infant-death syndrome. N. Engl. J. Med. **289:** 1167–1170.
48. NAEYE, R. L. 1974. Hypoxia and the sudden infant death syndrome. Science **186:** 837–838.
49. NAEYE, R. L., R. FISHER, M. RYSER, *et al.* 1976. Carotid body in the sudden infant death syndrome. Science **191:** 567–569.
50. NAEYE, R. L., B. LADIS & J. S. DRAGE. 1976. Sudden infant death syndrome. A prospective study. Am. J. Dis. Child. **130:** 1207–1210.
51. NAEYE, R. L., P. WHALEN, M. RYSER, *et al.* 1976. Cardiac and other abnormalities in the sudden infant death syndrome. Am. J. Pathol. **82:** 1–8.
52. NAEYE, R. L. 1977. Placental abnormalities in victims of the sudden infant death syndrome. Biol. Neonate **32:** 189–192.
53. VALDEŚ-DAPENA, M. A., M. M. GILLANE, R. CATHERMAN, *et al.* 1976. Brown fat retention in sudden infant death syndrome. Arch. Pathol. Lab. Med. **100:** 547–549.
54. VALDEŚ-DAPENA, M. A., M. M. GILLANE, D. ROSS, *et al.* 1976. Extramedullary hematopoiesis in the liver in the sudden infant death syndrome. Arch. Pathol. Lab. Med. **103:** 513–515.
55. KELLY, D. H., D. C. SHANNON & K. O'CONNELL. 1978. Care of infants with near-miss suddent infant death syndrome. Pediatrics **61:** 511–514.

AUDITORY BRAINSTEM RESPONSE
AND
THE MASKING LEVEL DIFFERENCE

James Jerger, Maureen Hannley, and Victor Rivera

Department of Otorhinolaryngology and Communicative Sciences
Department of Neurology
Baylor College of Medicine
Houston, Texas 77030

INTRODUCTION

One of the most striking behavioral measures of binaural interaction in the auditory system is the masking level difference (MLD) effect.[1,2] When binaural noise masks a binaural tone, the degree of masking depends on the interaural phase relations of both the noise and the tone. If N_0 and N_π denote conditions in which binaural noise is, respectively, in phase and 180 degrees out of phase at the two ears, and if S_0 and S_π denote similar conditions in which a binaural tone is, respectively, in phase and 180 degrees out of phase at the two ears, then N_0 S_0 and N_π S_π are homophasic conditions, while N_0 S_π and N_π S_0 are antiphasic conditions.

Masking of the tone by the noise is greater under either homophasic condition than under either antiphasic condition. The difference between the masking effects under homophasic and antiphasic conditions is called the binaural masking level difference, usually abbreviated BMLD or, simply, MLD. It has been suggested that the MLD effect is a behavioral correlate of binaural interaction at the brainstem level,[3,4] but the specific structures mediating the effect have only been conjectured.[5]

A more direct electrophysiologic assessment of the brainstem structures thought to be crucial to binaural interaction is provided by the auditory brainstem response (ABR). In particular, Jewett's wave III has been linked to the superior olivary complex,[6–9] a brain stem site frequently associated with binaural interaction.[10–12]

Patients with multiple sclerosis provide an unique opportunity to study the effects of brainstem lesions on both ABR and MLD. Noffsinger *et al.*[13] studied changes in MLD in 61 patients. They found a variation from no observable change to virtual elimination of the MLD effect. Similarly, Robinson and Rudge[14] studied ABR in patients with multiple sclerosis. Results varied from no effect, through selective loss of late waves, to obliteration of all waves beyond wave I. Thus, multiple sclerosis has been shown to affect both MLD and ABR.

In this study, we compared ABR and MLD in 20 patients with multiple sclerosis in order to determine whether abnormality in ABR, especially delay or loss of wave III is related to reduction in the MLD effect. We reasoned that if the superior olivary complex is, indeed, a key structure in mediation of the MLD effect, then abnormality of wave III should be associated with concomitant reduction in MLD. In all cases, the disease was diagnosed neurologically on the basis of McAlpine's criteria.[15]

METHOD

The auditory brainstem response was recorded by conventional scalp electrode technique. Standard EEG cup electrodes (silver–silver-chloride) were affixed to vertex

0077–8923/82/0388–0466 $1.75/0 © 1982, NYAS

(active) and to each mastoid. The mastoid electrode on the side of sound stimulation was always the reference; the electrode on the opposite mastoid was always ground. The EEG signal was bandpass filtered from 300–3000 Hz, and preamplified by a factor of 200,000 (Grass preamplifier model P511). The auditory signal was the haversine transformation of a single cycle of a 5000-Hz sinusoid. The resultant transformed waveform is roughly equivalent to a half-cycle of a 2500-Hz sinusoid. Signals of alternating polarity were presented at a click hearing level (HL) of 70 dB at the rate of 20/sec. Successive responses to 2048 signals were signal-averaged (Nicolet, model 1010) and written out on an *X-Y* plotter for later analysis.

To measure the MLD effect the subject tracked threshold for a binaural 500 Hz sinusoid in the presence of binaural broadband thermal noise at an overall sound pressure level of 60 dB (re 2×10^{-5} Pa). The 500 Hz signal was periodically interrupted at the rate of 3 interruptions per second with a rise-decay time of 25 msec. The MLD effect was defined as the decibel difference between thresholds tracked under N_0 S_0 and N_0 S_π conditions. Under each condition, threshold was tracked by means of a Bekesy-type audiometry (Grason-Stadler, model E800) for a minimum of 24 pen reversals. Threshold was defined as the average of the 24 midpoints of the tracing.

The 20 subjects were divided into three groups on the basis of ABR responses. Latency of each component wave was measured from signal onset to the point of steep inflection at the termination of the positive peak of the wave. Latency was classified as abnormally delayed if it exceeded, by two standard deviations, normal values obtained with this specific instrumentation on a group of 10 young adults with normal hearing. Group A consisted of 5 subjects with normal ABRs from both ears. All component waves I through V were observed within their respective normal latency ranges. All absolute and all interwave latencies were within normal limits. Group B consisted of 7 subjects with delayed wave III from at least one ear, but with recordable wave III from both ears. Group C consisted of 8 subjects for whom wave III was absent from one or both ears.

RESULTS

FIGURE 1 shows the mean and range of MLDs for each group. In a group of 7 young adults with normal hearing, the average MLD obtained on this specific equipment and using the same test procedure was 9.7 dB, with a range from 8.7 to 10.8 dB. We note a systematic decline in the MLD effect as the ABR waveform becomes more abnormal. The mean MLD effect decreased from 8.7 dB in group A, to 5.7 dB in group B, and to only 1.9 dB in group C.

There was no overlap among the three groups. MLDs ranged from 0.3 to 3.6 dB in group C, from 4.3 to 7.7 in group B, and from 8.2 to 9.3 in group A. Furthermore, in group C, there was no significant difference in MLD between subjects with only one absent wave III ($N = 5$) and subjects with bilaterally absent wave III ($N = 3$). In the former group, MLD ranged from 0.3 to 2.6 dB. In the latter group, it ranged from 1.1 to 3.6.

In order to demonstrate that this systematic reduction in MLD is uniquely related to the ABR abnormality, it is necessary to rule out other factors that might be responsible for such changes. It is known, for example, that MLD is modified by peripheral hearing loss,[16,17] especially eighth nerve disorder. It is also possible that patient's age and/or duration of the disease might exert confounding influences on the MLD effect. TABLE 1 compares the three groups on each of these three potentially contaminating variables. There was little evidence of peripheral hearing loss in any

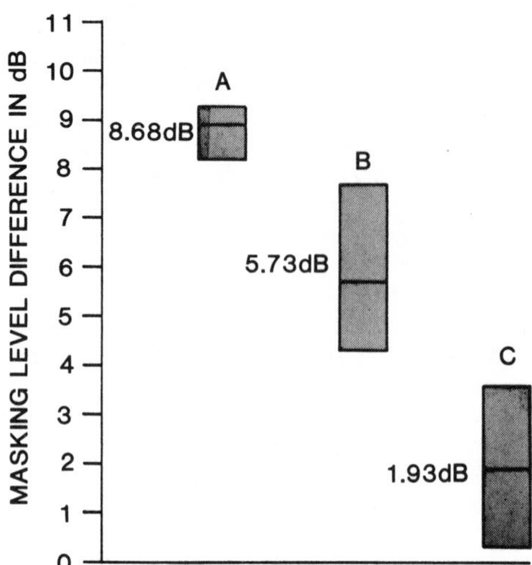

FIGURE 1. Mean and range of the masking level difference (MLD) effect in three groups of patients with multiple sclerosis. Group A: ABR normal in both ears. Group B: wave III delayed on one or both ears, but present bilaterally. Group C: wave III absent from one or both ears.

TABLE 1

DISTRIBUTION OF AGE, SEX, HEARING LEVEL, AND DURATION OF MULTIPLE SCLEROSIS (MS) IN THREE EXPERIMENTAL GROUPS

	Group*		
	A	B	C
Age (years)			
Mean	37.2	36.8	37.6
Range	26–48	25–48	24–54
Sex			
Male	0	2	4
Female	5	5	4
Hearing level (dB)†			
Right ear			
Mean	4.6	11.3	2.8
Range	0–10	0–20	– 5–10
Left ear			
Mean	5.6	9.0	4.2
Range	3–12	2–20	– 5–12
Duration of MS (years)			
Mean	8.1	7.9	8.0
Range	1–13	1–12	1–16

*Group A: ABR normal in both ears. Group B: wave III delayed in one or both ears, but present bilaterally. Group C: wave III absent from one or both ears.

†Average of hearing threshold levels at 500, 1000, and 2000 Hz measured by standard clinical audiometry.

group, and there were no substantial differences among the three groups on either age or duration of multiple sclerosis.

Finally, it is usually the case that abnormality of wave III is accompanied by concomitant abnormality of waves IV and V. Is it possible, then, that the reduction in MLD is related to the wave IV–V abnormality as well as the wave III abnormality? To answer this question, we searched our files for patients with normal wave III latency, but with delayed or absent wave IV–V. We found three such patients; two were cases of suspected multiple sclerosis, the third was a case of suspected vascular insult to the brain stem. In all three cases, wave III latency on each ear was within one standard deviation of the young adult norm for our instrumentation. On at least one ear of each patient, however, wave V was either significantly delayed (more than two standard deviations from the norm) or absent from the auditory brain stem response.

None of these three patients showed abnormal reduction in MLD. Actual MLD values ranged from 9.0 to 11.5 dB.

These findings strengthen our conclusion that the reduction in MLD observed in our groups B and C is related specifically to abnormality of wave III, rather than to more generalized abnormality of later waves, especially V.

Discussion

These data link abnormality of ABR and reduction of MLD. They suggest that when the latency and/or amplitude of ABR wave III is degraded, then MLD will also be affected. It does not necessarily follow, however, nor do we argue, that either measure is mediated solely at a single brainstem site. Cranford *et al.*[18] have shown, for example, that ablation of neocortex in cat has at least some effect on MLD, although Noffsinger *et al.*[13] found no effect on MLD in patients with temporal lobe disease.

The correspondence, however, between abnormality of wave III, and abnormality of MLD, supports the view that the two measures share common brainstem structures.

Acknowledgments

We acknowledge the contributions of Wini Breitbach and Connie Jordan to data collection.

References

1. Durlach, N. 1963. Equalization and cancellation theory of binaural masking-level differences. J. Acoust. Soc. Am., **35:** 1206–1218.
2. McFadden, D. 1975. Masking and the binaural system. *In* The Nervous System. D. B. Tower, Ed. Vol. **3:** 137–416. Raven Press, New York, N.Y.
3. Goldstein, D. P. & S. D. G. Stephens. 1975. Masking level difference: A measure of auditory processing capability. Audiology **14:** 354–367.
4. Quaranta, A. & G. Cerbellera. 1977. Masking level differences in central nervous system disease. Arch. Otolaryngol. **103:** 482–484.
5. Jeffress, L. 1972. Binaural signal detection: Vector theory. *In* Foundations of Modern Auditory Theory. J. Tobias, Ed. Vol. 11. Academic Press, New York, N.Y.
6. Jewett, D. L. 1970. Volume-conducted potentials in response to auditory stimuli as detected by averaging in the cat. Electroencephalogr. Clin. Neurophysiol. **28:** 609–618.
7. Lev, A. & H. Sohmer. 1972. Sources of averaged neural responses recorded in animal and

human subjects during cochlear audiometry (electrocochleogram). Arch. Klin. Exp. Ohr.-Nas. Kehlk, Heilk. **201:** 79–90.

8. Buchwald, J. S. & C. H. Huang. 1975. Far-field acoustic response: Origins in the cat. Science **189:** 382–384.
9. Koh, M. 1978. Changes of BSR by destruction of superior olivary nuclei in cats. Audiology (Japan) **21:** 645.
10. Moushegian, G., A. Rupert & M. A. Whitcomb. 1964. Medical-superior-olivary-unit response patterns to monaural and binaural clicks. J. Acoust. Soc. Am. **36:** 196.
11. Moushegian, G., A. Rupert & T. Langford. 1967. Stimulus coding by medical superior olivary neurons. J. Neurophysiol. **30:** 1239.
12. Moushegian, G., R. Stillman & A. Rupert. 1972. Characteristic delays in superior olive and inferior colliculus. *In* Physiology of the Auditory System. M. Sachs, Ed. pp. 245–254. National Ed. Consultants, Baltimore, Md.
13. Noffsinger, D., W. Olsen, R. Carhart, C. Hart & V. Sahgal. 1972. Auditory and vestibular aberrations in multiple sclerosis. Acta Otolaryngol. (Suppl.): 303.
14. Robinson, K. & P. Rudge. 1977. The early components of the auditory evoked potential in multiple sclerosis. *In* Auditory Evoked Potentials in Man. J. Desmedt, Ed. Karger, Basel, Switzerland.
15. McAlpine, D., C. Lumsden & E. Acheson. 1972. Multiple Sclerosis; A Reappraisal. Livingstone, Edinburgh, Scotland.
16. Olsen, W. & D. Noffsinger. 1976. Masking level differences for cochlear and brain stem lesions. Ann. Otol. **85:** 820–825.
17. Olsen, W., D. Noffsinger & R. Carhart. 1976. Masking level differences encountered in clinical populations. Audiology **15:** 287–301.
18. Cranford, J., J. Stramler & M. Igarashi. 1978. Role of neocortex in binaural hearing in the cat. III. Binaural masking-level differences. Brain Res. **151:** 381–385.

ANALYSIS OF CLICK-EVOKED BRAINSTEM AUDITORY ELECTRIC POTENTIALS USING HIGH-PASS NOISE MASKING AND ITS CLINICAL APPLICATION

J. J. Eggermont and M. Don

Department of Medical Physics and Biophysics
University of Nijmegen
Nijmegen
the Netherlands

House Ear Institute
Los Angeles, California

INTRODUCTION

Latencies form the essence of most brainstem auditory electric potential (BAEP) studies. Their use is quite diverse, as in studies on auditory development,[31] diagnosis of conductive versus sensorineural hearing loss,[20] detection of pontine angle tumors,[28] and differential diagnosis of brainstem abnormalities.[30] It is therefore worthwhile to investigate the factual meaning of latency, to determine which factors influence the latency of the various BAEP waves. As Stockard *et al.*[32] have investigated largely the influence of technical factors on latency, we will deal with the inherent audiological factors. Latency is best specified when a transient "click-like" stimulus is used. A click stimulus, however, is a broad-spectrum stimulus and thereby may stimulate the entire cochlear partition when presented at stimulus levels of 60–70 dB above the sensation level (SL) as commonly used.

As we know from the numerous studies of von Békésy[2] and from many others after him, a click stimulus impinging on the tympanic membrane sets up a traveling wave starting at a position between the cochlear windows and moving toward the apex of the cochlea. Recent measurements of traveling wave delay in animals[1] and humans[12,14] indicate that for the apical region of the cochlea this delay is quite appreciable. In the human ear, for instance, it takes about 3 msec for the traveling wave set up by the click to reach the 1 kHz region.[12] This traveling wave delay will enter in all latency measurements for the BAEP and its contribution may differ for clicks and tonebursts, may be intensity dependent, and may depend on the integrity of the cochlear receptor.

Thus, for wave I, representing the click-locked synchronous activity of fibers from the auditory nerve, latency depends to a large extend on which part of the cochlear partition is stimulated, the remainder being determined by synaptic delay and neural integration time. The later waves in the normal BAEP can be regarded in a simplified way as relayed and delayed replicas of wave I.[32] This means that the activity of the auditory nerve (representing a given part of the cochlear partition) that is represented in wave I also determines what is represented in waves III and V. A more realistic model assumes multiple origins of the various later waves. Another view could be that waves I, III, and V originate—directly or indirectly—from partly different regions on the cochlear partition.

It is our aim in the present review to examine two concepts:

First of all, what effects on the cochlear level might influence the latency behavior of the various waves? If the effect on the waves differs to some extent, does it have any

0077–8923/82/0388–0471 $1.75/0 © 1982, NYAS

impact on the wave I–wave V delay? Could this influence diagnostic procedures that are based on the I–V delay?[14]

Secondly, are all changes that one observes in the I–III or I–V delay caused by mechanisms central to the cochlea? Is it feasable or justifiable to speak about "central conduction time"[29] or "brainstem transmission time?"[19]

In order to separate peripheral and central influences on the BAEP latencies, a method is needed that specifies what parts of the cochlear partition contribute to the BAEP. A low-intensity toneburst can stimulate the cochlea in a restricted place,[4,11] but at these levels (<40 dB above the normal hearing level, HL) generally only wave V can be observed. We have adopted and substantially extended a method first introduced by Teas *et al.*[33] for the auditory nerve action potential in guinea pigs, and later on applied in human electrocochleography by Elberling[17] and Eggermont.[9,10]

This method in its present form uses high-pass noise masking to derive narrow-band contributions to the BAEP. Because the concepts of the high-pass noise masking technique have been questioned recently,[24] we will elaborate quite extensively on the validation of the derived narrow-band responses and recall results that have been obtained by electrocochleography. For the BAEP, these results apply to wave I without any restriction.

Realization

When a 60-dB SL click is presented monaurally to a human listener, one typically may record from surface electrodes on the vertex and the ipsilateral mastoid a series of up to seven waves (Figure 1A). Following the convention by Jewett *et al.*,[22] these waves are indicated by Roman numerals. In most recordings waves I, III, and V are generally very clear. The click, of which a smoothed acoustic waveform is shown, has a wide-band spectrum, i.e., contains energy up to about 10 kHz. Basically this click can stimulate the whole cochlear partition.

This 60-dB SL click is now masked using a wide-band noise masker, the level of which is adjusted so as to mask nearly completely the BAEP. The spectrum of the noise is largely similar to that of the click and both are in fact determined by the type of headphone used (Figure 1B).

Without any further changes in the spectral level, i.e., leaving the attenuator settings the same, the noise is passed through a high-pass filter or a system of cascaded high-pass filters. This filter system preferably has a very steep filtering slope, e.g., around 96 dB/octave. The spectrum of 6 kHz high-pass noise is shown in Figure 1C. Caution is needed in the way this has to be measured; one should either be able to use a measuring device such as a wave analyzer which is more selective than the filter used[8] or average spectra obtained by Fourier transformation of relatively long noise samples.[7]

Also shown in Figure 1C is the BAEP evoked by the click in the presence of this 6 kHz high-pass noise masker, i.e., generated in that part of the cochlea and connected higher centers that is not masked. The amplitude of this response is smaller and the latencies are longer than in the originally unmasked response.

The next situation (Figure 1D) shows a 3 kHz high-passed noise spectrum and the click response evoked in the presence of that noise. The part of the cochlea that contributes to the response is now smaller and located more apically than for the 6 kHz high-pass noise masking. Hence, its latencies will be longer by the increase in traveling wave delay.

By subtracting the spectra shown in Figure 1D and C one may obtain an impression of the difference in contribution from the cochlear receptor to the

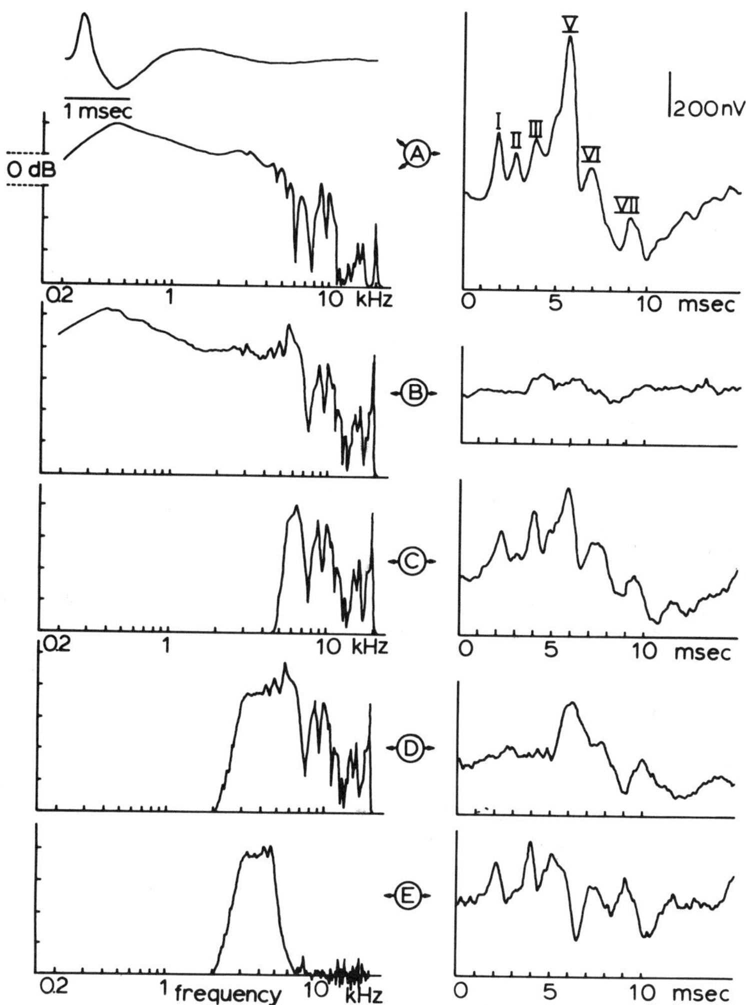

FIGURE 1. The technique of deriving narrow-band BAEP responses. In part (A) the click waveform, click spectrum, and 60 dB SL click-evoked BAEP are shown. In part (B) the spectrum of the wide-band masking noise is shown together with the nearly completely masked response. Part (C) shows the 6-kHz high-pass noise spectrum and the click-evoked BAEP in the presence of that noise. The same is illustrated in part (D) for 3 kHz high-pass noise. Part (E) finally shows an impression of the relative contributions of a narrow band, one-octave wide with a center frequency of 4 kHz, together with a narrow-band BAEP. This response is considered to originate from the area around 4 kHz. The slope of the filters used is 96 dB/octave. (From Don & Eggermont.[7] By permission of the *Journal of the Acoustical Society of America*.)

responses in (C) and (D). This response difference is shown in FIGURE 1E together with the difference in the spectra. The response shown is called a narrow-band response; the narrow band is characterized by the central frequency (CF), in this case around 4 kHz, and its width at −3 dB (about one octave). A narrow-band response represents synchronous activity from a limited region along the cochlear partition: a one-octave-wide narrow band corresponds to about 5 mm. It is possible to consider half-octave-wide narrow bands by corresponding adjustments of the cut-off frequencies of the filter.

Such narrow bands in a way result in a functional serial-sectioning of the cochlea, fibers within a narrow band have rather homogeneous properties and the narrow-band responses (especially for the auditory nerve) may be compared with single-nerve-fiber data.

A more detailed explanation of these methods has been published.[7,8,15,16]

Validation of the Narrow-Band Concept

Preservation of the Physical Noise Properties on the Cochlear Partition

From recent investigations on the vibration pattern of the basilar membrane by Mössbauer methods,[23,27] capacitive-probe techniques,[34] and laser illuminations,[25] it has become increasingly clear that the apical front of the traveling wave and of the excitation profile on the basilar membrane has a very large gradient. In frequency terms one may speak of around 100 dB/octave. These results have been obtained using rather high-frequency continuous tones as stimuli. Extrapolation to the low-frequency results of von Békésy[2] could indicate smaller slopes at more apical parts. On stimulation with, say, a 1-kHz tone, a traveling wave develops from the region of the cochlear windows and moves towards more apical regions. Thereby the vibrations of the basilar membrane increase in amplitude while the velocity of the traveling wave decreases; hence no energy is lost, until the resonance region is reached after which the energy contained in the wave is rapidly dissipated in a very small portion of the basilar membrane.[5]

When short tonebursts are used instead of continuous tones, a spectral broadening occurs which may affect the steepness of the slopes on the excitation profile. By applying the high-pass noise-masking technique to toneburst-evoked compound action potentials (AP) of the auditory nerve, one may for each tone-burst intensity determine the extension of the activation region. This is done by measuring the contributions to the subsequent half-octave narrow bands.[9,10] In this way, one obtains a so-called response area (FIGURE 2), in this case for 2 kHz toneburst stimulation. One observes for toneburst of 25 dB above threshold an activation region of around 1 octave. At 70 dB level, however, the toneburst activates the entire cochlear partition basally from around 700 Hz. What is important for us is the very minute spread of activity towards the region with low CFs, as revealed by the high-pass noise masking technique. The actual slope is very much near 96 dB/octave, i.e., near the value generated by the filter. This result is what one expects when the slope of the excitation profile toward the apex is *larger* than that of the filter. In that case the filter slope is the limiting factor.

From this result we expect that also the slope of the high-pass noise masker is preserved on the cochlear partition, at least for frequencies down to 2 kHz. Note that there is correspondence between results obtained in the guinea pig and those in the human ear. Elberling and Evans[18] have measured the effect of high-pass noise masking on single auditory nerve fibers and did not find an effect for CFs of 2 kHz of above. Below 1 kHz in the guinea pig there was some masking when this was not expected from the physical properties of the noise. This could point to an actual decrease in the slope of the apical part of the excitation profile, as suggested from an interpolation between von Békésy's results[2] and the results from the more recent experiments on basilar membrane movement. Whether this is also the case for the human ear has yet to be investigated.

From the shape of the response area (FIGURE 2), one observes that it is mandatory to use high-pass noise masking instead of low-pass noise masking or even bandstop masking *except* at relatively low stimulus levels (<50 dB).

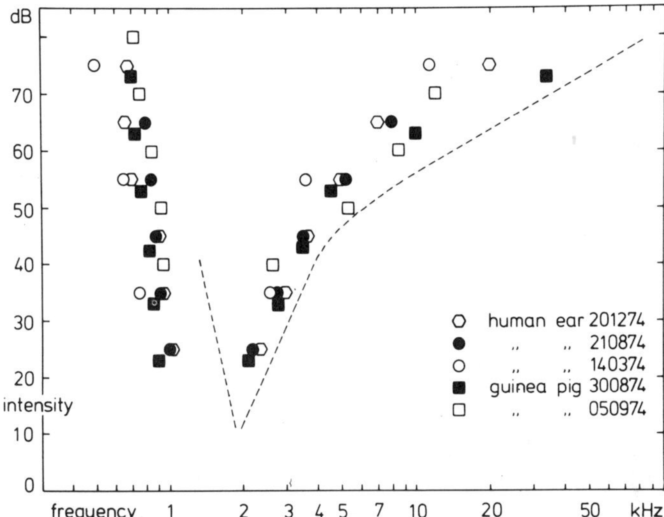

FIGURE 2. Response area for a short 2 kHz toneburst in the human and guinea pig ear. For a series of toneburst intensities the extend of the activation region in the cochlea is determined by high-pass noise masking. It appears that for higher stimulus levels there is considerable spread of activity towards the higher center frequencies (i.e., toward the base of the cochlea) but hardly any toward lower center frequencies. The slope of the low-frequency boundary in fact approaches 96 dB/octave, i.e., the actual filter slope. (From Eggermont.[9] By permission of the *Journal of the Acoustical Society of America.*)

Estimation of Traveling Wave Delay

Properties of the traveling wave propagation in the human cochlea can be estimated from the latency of the narrow-band action potentials evoked by click stimuli. These narrow-band responses show a gradually increasing latency for decreasing central frequencies (FIGURE 6). For a series of normal human ears, the onset latency of the narrow-band action potentials (NAP) is plotted in the upper part of FIGURE 3. This NAP onset latency consists of a central-frequency-independent part (synaptic delay and neural integration time) and a part depending on the central frequency (CF) of the narrow band. This frequency-dependent part is shown in the

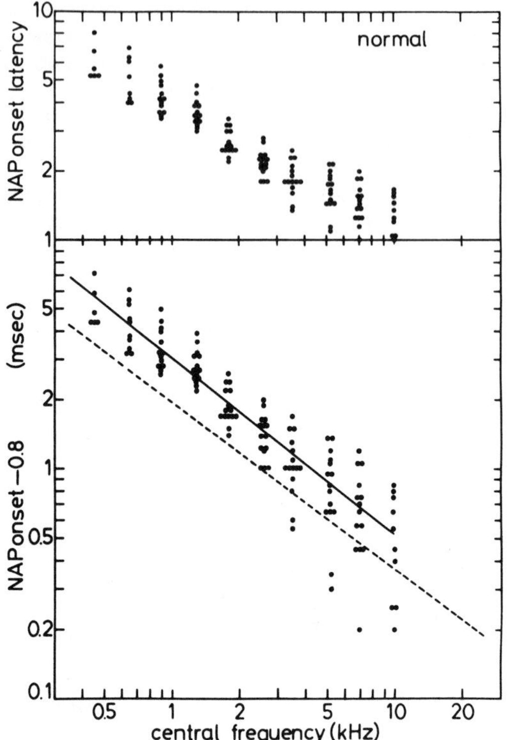

FIGURE 3. Narrow-band action potential (NAP) latencies for the normal human cochlea. In the upper part of the FIGURE the onset latencies are shown as a function of CF for 15 normal human ears. In the lower part of the FIGURE an asymptotic value of 0.8 msec is subtracted in order to determine the contribution of the traveling wave delay to the NAP latency. The dashed line represents a regression line determined from squirrel monkey single-unit data.[1]

lower part of the FIGURE and is obtained by subtraction of 0.8 msec. The drawn line is a power curve fit to the data points:

$$\tau_{CF} = 3.4 f_{CF}^{-0.77}$$

One could consider this as reflecting the group delay of the traveling wave produced by click stimulation.

Anderson *et al.*[1] calculated for single nerve fibers from the auditory nerve in the squirrel monkey a travel time from the slope of the phase-difference versus frequency curve. This is allowed in the case of a wave with a stimulus-frequency-independent velocity of propagation. In other words, Anderson *et al.*[1] obtained a phase delay. The dotted line represents the best fit to their data points:

$$\tau^{1}_{CF} = 1.95 f_{CF}^{-0.725}.$$

When we take into account the species difference and the fact that probably some dispersion occurs in the traveling wave propagation, the correspondence in slope for both situations is remarkable. We may conclude that the high-pass noise masking technique results in a reliable estimate of the traveling wave delay.

By calculating latency differences between adjacent narrow bands, one obtains the time it takes for the traveling wave to traverse that region. Using von Békésy's map[3] for the frequency–place relationship in the human cochlea, we have calculated the average velocity for each particular region. These results together with those of Elberling[17] and Zerlin[35] have been plotted in FIGURE 4. Zerlin's data were obtained by a psychophysical lateralization experiment using tones of slightly different frequency, the other data by electrocochleography. Subsequently, Parker and Thornton[26] have validated this for using narrow band BAEPs. The drawn line is calculated through the mean values supplied by the various investigators and follows:

$$v_t = 2.8 f_{CF},$$

i.e., the wavelength of the traveling wave is approximately 2.8 mm, being close to the width of a half-octave narrow band, and suggesting that if dispersion is present it is only very minor.

From the close correspondence of the masking data and the lateralization data, we draw additional support for the validity of the high-pass masking technique.

One of the consequences of a decreasing traveling wave velocity towards the apex of the cochlea is a gradual decrease in the amount of synchronization of the fiber activity that contributes to the NAPs for lower CFs. Since the click produces a traveling wave along the BM that slows down considerably towards the apex, and half-octave narrow bands still extend for the same number of mm along the basilar membrane, narrow bands with high CF will be traversed in less than 0.1 msec, while this may take up to 1 msec around 1 kHz. This will be reflected in the width of the

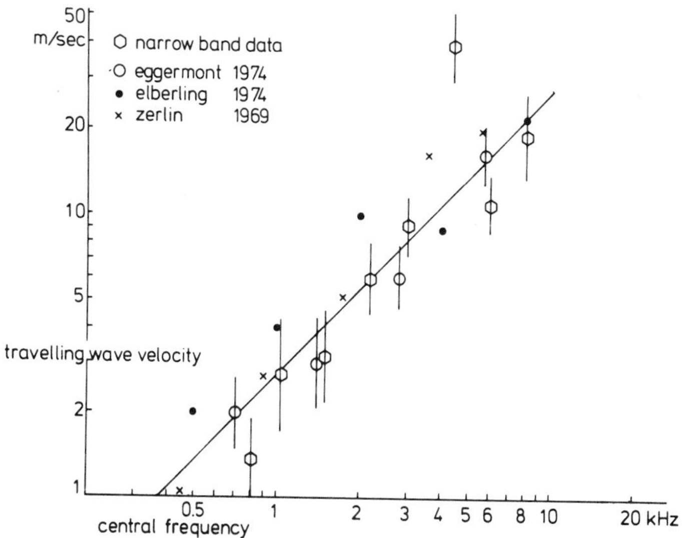

FIGURE 4. Estimation of traveling wave velocity in the normal human cochlea. Results from bandstop masking procedures (O) and narrow-band responses (●,O) are compared with results from a psychoacoustic lateralization study (×). The regression line is calculated using mean values (excluded is the one outlying value around 4 kHz). The vertical bars indicate one standard deviation. (From Eggermont.[9] By permission of the *Journal of the Acoustical Society of America.*)

NAPs as was determined[9] by calculating the regression for relative width (RW) with respect to a standard of 0.3 msec. It was found to be

$$\mathrm{RW} = 6.5\, V_I^{-0.5} = 3.9 f_{\mathrm{CF}}^{-0.5}.$$

This means that the relative width increases each 2 octaves by a factor of 2. When for each NAP about the same number of fibers contribute, assuming a constant innervation density and a sufficiently intense stimulus, one expects that increasing width parallels an amplitude decrease.

For a click level of 90 dB peak equivalent sound pressure level (p.e. SPL) (≃70 dB SL), the NAP amplitude has been measured for a group of normal ears and plotted in FIGURE 5. The average value for each CF is indicated by a hexagon, the dashed line drawn through the average values has a slope of −3 dB/octave. This means that the amplitude of the NAPs on average decreases by a factor of two each two octaves.

Combining this with the increased width of the NAPs, one concludes that the area under the NAP curve stays approximately constant for a very large part of the cochlear partition.

NARROW-BAND BRAINSTEM AUDITORY EVOKED POTENTIALS

A priori one expects that the latencies of waves III and V follow with a fixed delay the latency of wave I for the various narrow-band BAEPs. This is illustrated for a complete half-octave narrow-band analysis for a 60 dB SL click (FIGURE 6). On the left, the effect of high-pass noise masking with cut-off frequencies that are lowered in approximately half-octave steps is shown. On the right, the corresponding narrow band BAEPs are shown; one clearly observes the gradual increase in the wave V

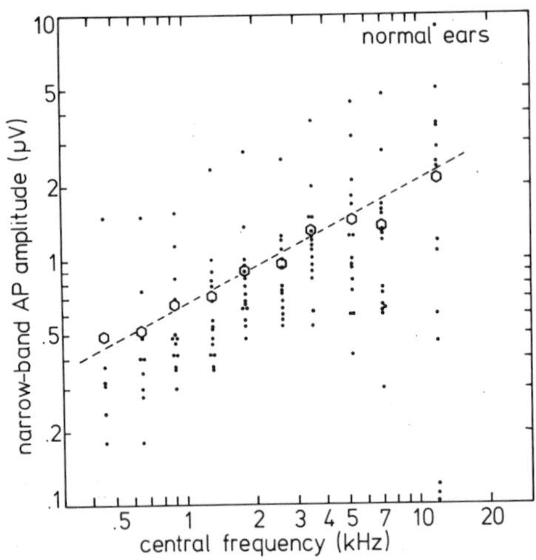

FIGURE 5. Narrow band action potential amplitudes for normal human ears. The mean values at each center frequency are indicated by hexagons, the dashed line having a slope of −3 dB/octave was visually fitted through the mean values. When median values are chosen, the trend remains the same. This indicates the dominance of contributions from the basal end of the cochlea to the compound AP.

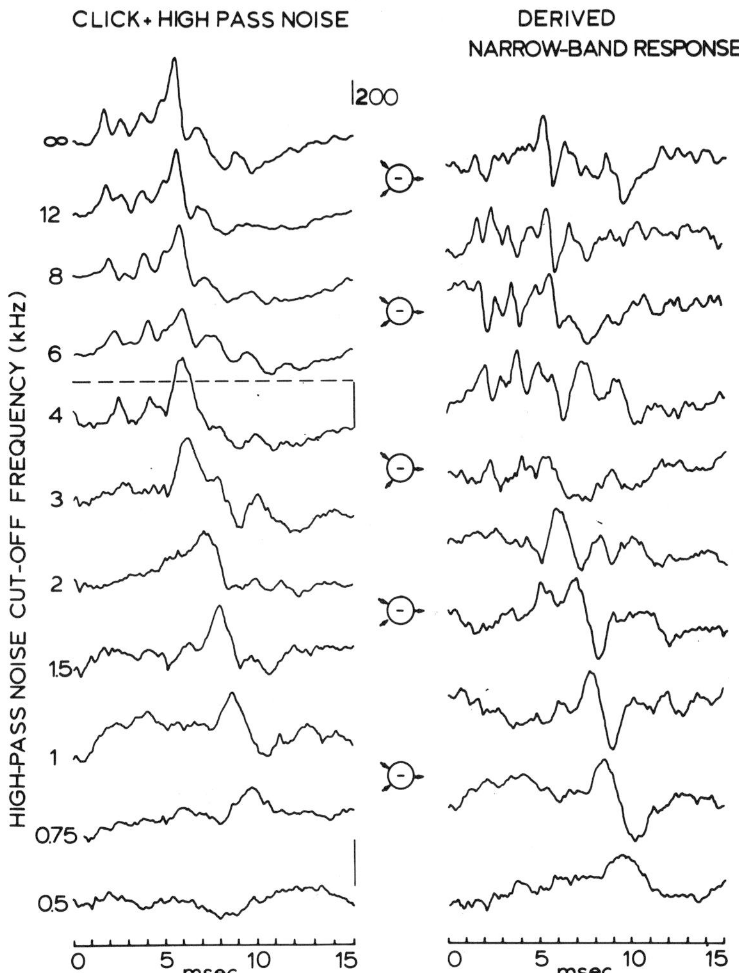

FIGURE 6. A complete half-octave narrow-band analysis for human BAEP. In the left part of the FIGURE the click-evoked response (60 dB SL) is shown for various high-pass noise masking conditions. By successive subtractions the narrow-band responses are derived (right-hand side). One observes the gradual latency increase as more and more high-frequency parts of the cochlea are masked. Surprisingly there is only a minor influence upon wave V amplitude (From Don & Eggermont.[7] By permission of the *Journal of the Acoustical Society of America.*)

latency from about 5 msec for a CF of 12 kHz (uppermost trace) to about 10 msec for a CF of about 425 Hz (lowermost trace).

For seven normal subjects, the latencies for waves I, III, and V are plotted for all CFs at which they could be discerned (FIGURE 7). It is clear that wave V can be detected more often at the lower CFs than wave III or wave I. It appears that the I–III and I–V delays are CF independent.[7]

The greater detectability for wave V, of course, will be reflected by its larger

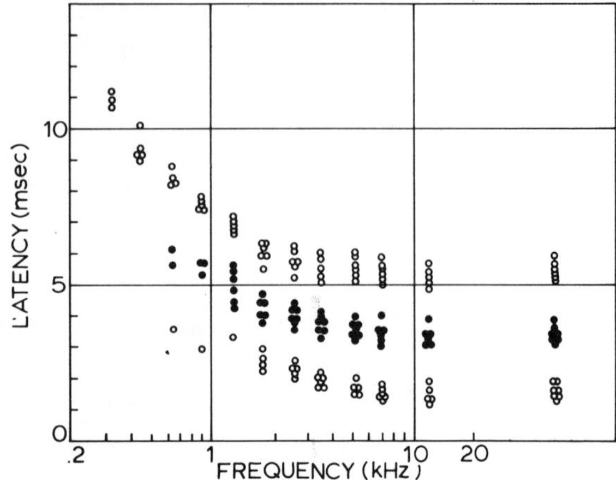

FIGURE 7. Pooled data for the latencies of waves I, III, and V in seven normal hearing subjects. The approximately parallel shift for the latencies is noted. The unmasked latencies are shown on the far right side. Click intensity was 60 dB SL. (From Don & Eggermont.[7] By permission of the *Journal of the Acoustical Society of America.*)

amplitude. The normalized (to the unmasked wave V amplitude) and averaged amplitudes for the three waves as a function of CF are shown in FIGURE 8. Waves I and III appear to have on average a lower amplitude by at least a factor 10 for CFs below 1 kHz. The wave V amplitude is nearly constant from say 400 Hz to 12 kHz.

This suggests that either the number of contributing neural units to wave V

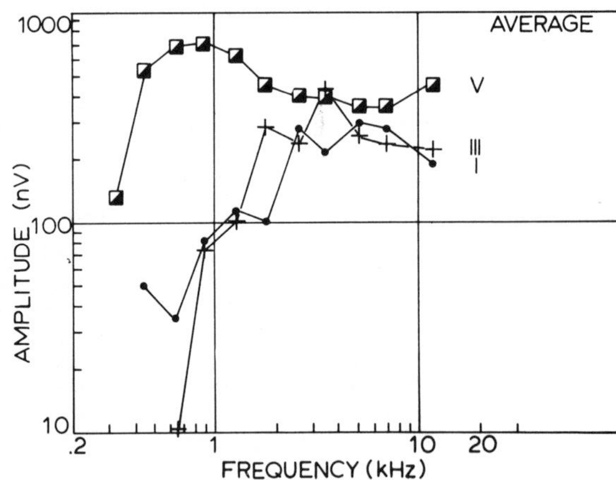

FIGURE 8. Mean normalized amplitude values for waves I, III, and V for 60 dB SL clicks for seven normal hearing subjects. The gradual decrease in amplitude for waves I and III especially below a central frequency of 1 kHz contrasts greatly with the gradual increase for the narrow-band wave V responses. (From Don & Eggermont.[7] By permission of the *Journal of the Acoustical Society of America.*)

increases for lower CF values (because wave V also broadens considerably towards lower CFs) or that wave V is inherently more resistant against desynchronization effects. This has been suggested also from results with variations in click rate[6] or from experiments with different rise-times for noise bursts.[21]

What is quite clear, however, from these results is that the BAEP contains contributions from nearly the whole cochlear partition and not only from the basal turn. The unmasked, 60-dB click BAEP represents only the activity from the basal two or three octaves as can be deduced by comparing the latencies of the unmasked BAEP waves with those of the narrow band responses.

To elaborate further on the amplitude differences as a function of CF, 13 normal hearing subjects were monaurally stimulated at click levels of 10–60 dB SL in 10 dB steps. All amplitude measures were again normalized to the unmasked wave V amplitude at 60 dB SL, and averaged. The resulting input–output curves obtained for

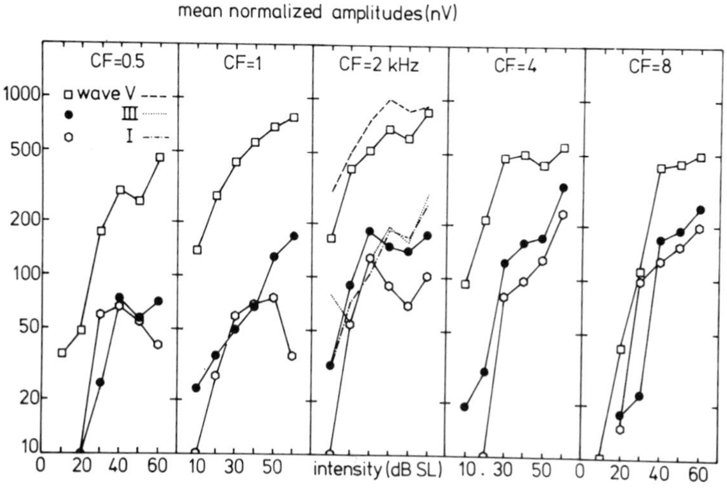

FIGURE 9. Mean normalized amplitudes for waves I, III, and V as a function of click intensity for five octave-wide narrow bands. The amplitude functions for the unmasked click response are shown by dashed, dotted, and dash-dot lines in the middle box. The data were obtained in 13 normal hearing subjects (From Eggermont & Don.[15] By permission of the *Journal of the Acoustical Society of America.*)

octave narrow bands are shown in FIGURE 9. We indicate a few differences: For CF = 0.5 kHz, the input–output curves for waves I and III clearly saturate above 40 dB SL, while wave V continues to increase in amplitude. For CF = 4 kHz and CF = 8 kHz, the wave V amplitude saturates above 40 dB, while the input–output curves for waves I and III show a monotonous increase. At intensity values of 10 or 20 dB, the largest contribution to the BAEP for *all waves* comes from the 2 kHz area, i.e., where the ear is most sensitive.

One of the crucial tests for the validity of the narrow band concept, in fact, originates from applying the method to ears having a hearing loss for a limited frequency range, e.g., a dip in the audiogram at 4 kHz. One such example is shown in FIGURE 10 where, at a click level of 50 dB HL (referring to the threshold in the normal population), the BAEP and its five octave narrow-band contributions are shown. The

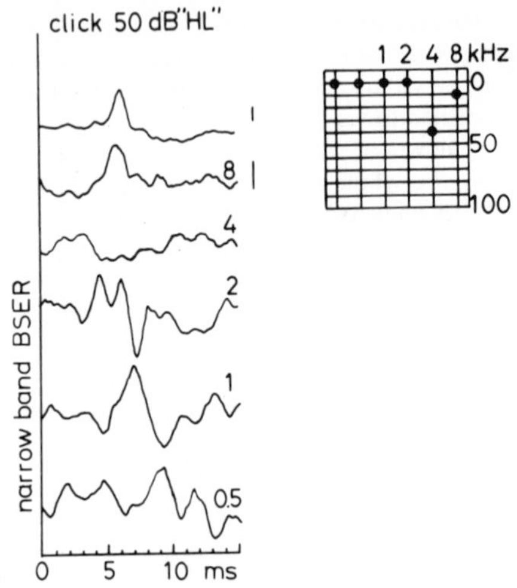

FIGURE 10. Narrow-band responses for a 50 dB HL click in a subject with a 4 kHz hearing loss (see audiogram). The upper trace represents the unmasked response. The lower traces show the octave-wide narrow band responses (different gain factor). One notes the absence of a contribution to the 4 kHz narrow band. By estimating in each narrow band the click level at which the response disappears one may obtain an objective audiogram.

audiogram shows a 40 dB hearing loss at 4 kHz and normal thresholds at the other audiometric frequencies. For the narrow band responses one observes the clear contributions to all narrow bands *except* 4 kHz. Incidentally, the unmasked BAEP threshold for this ear was 0 dB SL, not pointing to any abnormality. The audiogram as based upon thresholds for the various narrow bands agreed very well with the behavioral audiogram.[8]

From the application of BAEPs to the detection of acoustic neurinomas,[16,28] it has become clear that the latency of wave V at the tumor side generally is longer than for the contralateral ear; this latency originates from an increase of the I–V delay. It has been suggested that this increase in "central conduction time" is due to a decrease in the propagation of action potentials along the nerve tracts.

If this were the main reason, one should also expect that for narrow-band responses the difference between the two ears in either the wave V latencies or the I–V delays is the same as for the unmasked BAEP. Of course, cochlear factors can affect the observed latencies because of the hearing losses associated with tumors.

One case in which it is particularly evident that one cannot speak of central conduction time prolongation is shown in FIGURE 11. It shows the BAEP and its narrow band contributions in a subject with a tumor of the auditory nerve. In the upper trace the prolonged I–V delay is indicated; it amounts to 5.4 msec, i.e., about 1.4 msec longer than the average normal value. Considering the narrow band responses, one observes two facts: first of all, the I–V delays are shorter than in the unmasked BAEP and borderline normal. Secondly, wave I in the unmasked BAEP corresponds to wave I from the 8 kHz narrow band response, while wave V for the unmasked BAEP is based upon contributions from the 1 and 2 kHz regions.

From the first observation we conclude that the prolonged I–V delay is not due to an increased central conduction time. The second observation leads to the proposition that the tumor acts differently upon the high-frequency fibers in the auditory nerve and the low-frequency fibers, which are located more toward the center of the auditory nerve.

In cases of acoustic neurinomas, wave I is frequently absent (32%).[16] In these cases, a direct measure of central conduction time is not possible, but a comparison of wave V latencies for the various narrow band responses of the normal and the tumor ear may also provide evidence for the mechanism that causes a latency increase for the unmasked wave V. FIGURE 12 shows a case where the interaural wave V latency difference for the click BAEP is about 1.5 msec. Wave I is probably present in the tumor BAEP as well but is not discernable in the narrow band responses. The solid line curves represent the responses for the normal ear to 70 dB HL clicks and one observes that the latency of wave V for the unmasked response corresponds with those in the narrow band responses with a CF of 4 and 2 kHz. For the tumor ear (dashed line), however, the unmasked wave V latency corresponds with that of the 1 kHz narrow band response. Thus, clearly the wave V contributions for both ears originate from different regions on the cochlear partition. Furthermore, the wave V latencies for the 1 kHz narrow band responses in both ears are very nearly the same.

One is forced to conclude also in this case that the abnormally large interaural latency difference for wave V is due to different contributions from both cochleae to the BAEP. The difference in latency finds its origin on the cochlear partition. We think that a differential diagnosis may be based upon the amount of agreement between the unmasked BAEP and narrow band BAEP latencies as to the location and extension of the tumor.[36]

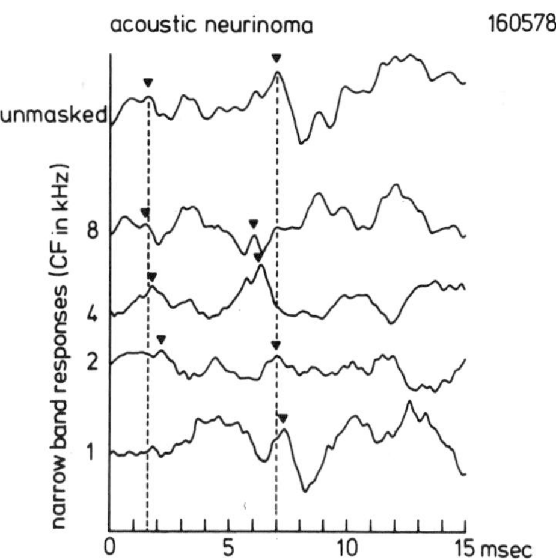

FIGURE 11. Narrow-band responses for an acoustic neurinoma patient. In the unmasked response, the waves I and V are indicated, suggesting an increased value: clearly a sign of abnormality. For various narrow-band responses it appears that the I–V delay is smaller than in the unmasked response. Note that the contributions to wave I in the unmasked response originate from the 8 and 4 kHz regions, while for wave V the contributions of the 2 and 1 kHz regions dominate. (From Eggermont *et al.*[16] By permission of the *Annals of Otology, Rhinology, & Laryngology.*)

Discussion

The usefulness of the narrow band concept is based on four assumptions: (1) The physical properties of the high-pass noise are preserved on the cochlear partition of the normal ear. (2) Remote masking from low-frequency components in the noise upon the more basal parts of the cochlear partition can be neglected in comparison with the masking effect of the high-frequency components. (3) There is no nonlinear interaction between the simultaneously present click and high-pass noise on the cochlear partition. (4) The previous three assumptions also hold for cases with a moderate cochlear hearing loss.

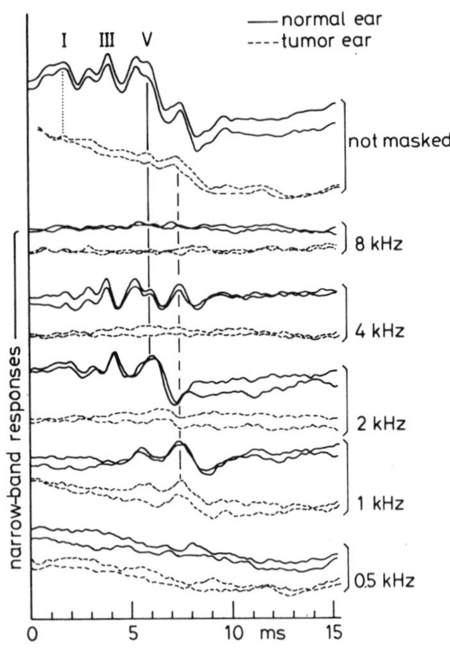

FIGURE 12. Narrow-band responses for the normal and the tumor ear in a patient. In the normal ear response the waves I, III, and V are indicated. The probable place of wave V in the tumor ear is indicated by the dashed line. The interaural latency difference between both ears is about 1.5 msec. When tracing back the narrow bands to where the contributions for the normal wave V originate, one finds the 4 and 2 kHz regions. In the tumor ear, however, the 1 kHz contribution is the most important. Note that the wave V latencies for the normal and the tumor ear in the 1 kHz narrow band are the same.

The first assumption has been amply documented in this paper and is generally justified: the steep filter slope appears to be preserved in the cochlear stimulation profile.

Remote masking effects generally are present especially at the higher masking levels. When one uses a low-pass filtered noise masker, sufficiently loud to mask a 70 dB SL click, a lowering of the cut-off frequency results in increasing amplitudes for the remaining response. However, hardly any changes are observed in the response latency (unpublished results). The gradual changing latencies found for high-pass noise masking (FIGURE 6) definitely cannot be explained by changes in remote masking. There might nevertheless remain a small effect upon amplitude.

The assumption about nonlinear interaction has been explored[13] and it appeared that the latency functions found for continuous masking and in forward masking were the same. The applicability to moderate hearing losses is justified by the accurate audiogram estimation based on the high-pass noise methods.[8] Still, some doubt exists in cases with cochlear hearing loss. It has been found[14] that for pathological cochleas the NAP latencies are shorter than for normal ears, in fact shorter than the calculated traveling wave delay along the cochlear partition. This could point to greatly changed basilar membrane mechanics or to a differing place-frequency relationship for pathological cochleas, i.e., a certain place is now tuned to a lower frequency. These effects will have no influence on audiogram estimation but may affect the I–V delay.

The separation of the "whole cochlea" BAEP into several narrow-band responses may offer a forceful method to distinguish central influences upon the BAEP latencies from those that in fact are due to changes in the contributions from various narrow bands to the compound response.

REFERENCES

1. ANDERSON, D. J., J. E. ROSE, J. E. HIND & J. F. BRUGGE. 1971. Temporal position of discharges in single auditory nerve fibers within the cycle of a sine-wave stimulus: Frequency and intensity effects. J. Acoust. Soc. Am. **49:** 1131–1139.
2. BÉKÉSY, G. VON. 1960. Experiments in Hearing. McGraw-Hill Book Company, New York, N.Y.
3. BÉKÉSY, G. VON. 1963. Hearing theories and complex sounds. J. Acoust. Soc. Am. **35:** 588–601.
4. DALLOS, P. & M. A. CHEATHAM. 1976. Compound action potential (AP) tuning curves. J. Acoust. Soc. Am. **59:** 591–597.
5. DE BOER, E. & R. MACKAY. Reflections on Reflections. J. Acoust. Soc. Am. **67:** 882–890.
6. DON, M., A. R. ALLEN & A. STARR. 1977. Effect of click rate on the latency of auditory brainstem responses in humans. Ann. Otol. Rhinol. Laryngol. **86:** 186–196.
7. DON, M. & J. J. EGGERMONT. 1978. Analysis of click-evoked brainstem potentials in man using high-pass noise masking. J. Acoust. Soc. Am. **63:** 1084–1092.
8. DON, M., J. J. EGGERMONT & D. E. BRACKMANN. 1979. Reconstruction of the audiogram using brainstem responses and high pass noise masking. Ann. Otol. Rhinol. Laryngol. **88** (Suppl. 57): 1–20.
9. EGGERMONT, J. J. 1976. Analysis of compound action potential responses to tonebursts in the human and guinea pig cochlea. J. Acoust. Soc. Am. **60:** 1132–1139.
10. EGGERMONT, J. J. 1976. Electrocochleography. *In* Handbook of Sensory Physiology. W. D. Keidel & W. D. Neff, Eds. Vol. 5(3): 625–705. Springer-Verlag, New York, N. Y.
11. EGGERMONT, J. J. 1977. Compound action potential tuning curves in normal and pathological human ears. J. Acoust. Soc. Am. **62:** 1247–1251.
12. EGGERMONT, J. J. 1979. Narrow-band AP latencies in normal and recruiting human ears. J. Acoust. Soc. Am. **65:** 463–470.
13. EGGERMONT, J. J. 1979. Compound action potentials: Tuning curves and delay times. *In* Models of the Auditory System and Related Signal Processing Techniques. M. Hoke & E. de Boer, Eds. [Scand. Audiol. (Suppl. 9): 129–139.]
14. EGGERMONT, J. J. 1980. Narrow-band AP studies in normal and recruiting human ears. *In* Psychophysical, Physiological and Behavioural Studies in Hearing. G. van den Brink & F. Bilsen, Eds. pp. 153–160. Delft University Press, Delft, the Netherlands.
15. EGGERMONT, J. J. & M. DON. 1980. Analysis of click-evoked brainstem potentials in humans using high-pass noise masking. II. Effect of click intensity. J. Acoust. Soc. Am. **68:** 1671–1675.

16. EGGERMONT, J. J., M. DON & D. E. BRACKMANN. 1980. Electrocochleography and auditory brainstem electric responses in patients with pontine angle tumors. Ann. Otol. Rhinol. Laryngol. **89**(Suppl. 75): 1–19.
17. ELBERLING, C. 1974. Action potentials along the cochlear partition recorded from the earcanal in man. Scand. Audiol. **3:** 13–19.
18. ELBERLING, C. & E. F. EVANS. 1979. Personal communication.
19. FABIANI, M., H. SOHMER, C. TAIT, M. GAFNI & R. KINARTI. 1979. A functional measure of brain activity: brainstem transmission time. EEG Clin. Neurophysiol. **47:** 483–491.
20. GALAMBOS, R. & K. HECOX. 1977. Clinical applications of the brainstem auditory evoked potentials. *In* Auditory Evoked Potentials in Man. [Prog. Clin. Neurophysiol.] J. E. Desmedt, Ed. Vol. **2:** 1–19. Karger, Basel, Switzerland.
21. HECOX, K., N. SQUIRES & R. GALAMBOS. 1976. Brainstem auditory evoked responses in man. I. Effect of stimulus rise-fall time and duration. J. Acoust. Soc. Am. **60:** 1187–1192.
22. JEWETT, D. L., N. M. ROMANAO & J. S. WILLISTON. 1970. Human auditory evoked potentials: Possible components detected on the scalp. Science **167:** 1517–1518.
23. JOHNSTONE, B. M. & A. J. F. BOYLE. 1967. Basilar membrane vibrations examined with the Mössbauer technique. Science **158:** 390–391.
24. KEVANISHIVILI, Z. S. 1980. Sources of the human brainstem auditory evoked potential. Scand. Audiol. **9:** 75–84.
25. KOHLLÖFFEL, L. U. E. 1972. A study of basilar membrane vibrations. III. The basilar membrane frequency response curve in the living guinea pig. Acustica **27:** 82–89.
26. PARKER, D. J. & A. R. D. THORNTON. 1978. Cochlear travelling wave velocities calculated from the derived components of the cochlear nerve and brainstem evoked responses of the human auditory system. Scand. Audiol. **7:** 67–70.
27. RHODE, W. S. 1971. Observations on the vibration of the basilar membrane in squirrel monkeys using the Mössbauer techniques. J. Acoust. Soc. Am. **49:** 1218–1231.
28. SELTERS, W. A. & D. E. BRACKMANN. 1977. Acoustic tumor detection with brain-stem electric response audiometry. Arch. Otolaryngol. **103:** 181–187.
29. STARR, A. 1978. Sensory evoked potentials in clinical disorders of the nervous system. Ann. Rev. Neurosci. **1:** 103–127.
30. STARR, A. & L. J. ACHOR. 1975. Auditory brainstem responses in neurological disease. Arch. Neurol. **32:** 761–768.
31. STARR, A., R. N. AMLIE, W. H. MARTIN & S. SANDERS. 1977. Development of auditory function in newborn infants revealed by auditory brainstem potentials. Pediatrics **60:** 831–839.
32. STOCKARD, J. J., J. E. STOCKARD & F. W. SHARBROUGH. 1978. Nonpathologic factors influencing brainstem auditory evoked potentials. Am. J. EEG Technol. **18:** 177–209.
33. TEAS, D. C., D. H. ELDREDGE & H. DAVIS. 1962. Cochlear responses to acoustic transients: An interpretation of whole nerve action potentials. J. Acoust. Soc. Am. **34:** 1438–1459.
34. WILSON, J. P. & J. R. JOHNSTONE. 1975. Basilar membrane and middle ear vibrations in the guinea pig measured by capacitive probe. J. Acoust. Soc. Am. **57:** 705–723.
35. ZERLIN, S. 1969. Traveling-wave velocity in the human cochlea. J. Acoust. Soc. Am. **46:** 1011–1015.
36. DON, M. *et al.* 1982. (to be published).

AUDITORY EVOKED BRAINSTEM RESPONSE ASSESSMENT IN OTOLARYNGOLOGY

Jun-Ichi Suzuki, Kazuoki Kodera, and Kimitaka Kaga

Department of Otolaryngology
Teikyo University School of Medicine
Tokyo, Japan

INTRODUCTION

Language and auditory skills are most effectively developed during the first few years of life. Early detection and diagnosis of hearing impairment is very important in formulating rehabilitation programs. In our ear, nose, and throat clinic, auditory brainstem response (ABR) audiometry is performed in children who are young, hyperactive, emotionally disturbed, or otherwise "untestable" by behavioral methods. The results of the ABR audiometry are utilized for fitting hearing aids.

In children, an analysis of the ABR can give indications of peripheral deafness, brainstem immaturity, brainstem lesions or normal brainstem conditions. Furthermore, the ABR provides useful information for making a diagnosis of brainstem lesions as well as hearing disorders. To illustrate this, four kinds of typical neurological diseases are studied, namely, perinatal asphyxia, cerebral palsy, brain tumor, and leucodystrophy. In this paper, the usefulness of ABR assessment in otolaryngology is demonstrated from the aspects of audiological and neurological diagnosis.

METHODS

Methods of ABR Audiometry for Audiological Application

Auditory stimuli consisted of tone pips with 5-msec rise–decay times (linear ramp up and down) without a plateau. The frequencies of the tone pips were 500, 1000, and 2000 Hz. They were generated and attenuated by an auditory stimulator (Dana Japan, DA 502-A) and delivered through a TDH 39 earphone at the rate of 26/sec. The tone pips were gated on and off at zero crossings. The polarity of the tone pips was alternately reversed. FIGURE 1 shows the frequency spectra of the tone pips.

The subjects were administered sedatives and were tested in a supine position during sleep in an electrically shielded and sound-attenuated room. Silver disc electrodes were attached with conductive paste to the center of the forehead near the hair line and to both mastoids. Responses were differentially amplified with the negative input (G1) connected to the test ear mastoid. The positive input (G2) was attached to the forehead and the ground to the non-test mastoid.

After amplification (Nihon-Kohden, RB-45030; a band-pass of 48–1000 Hz, 3 dB points of 6 dB/octave roll-off curves), 2000 responses were averaged by a computer (Sanei Sokki, 7T07). Averaging was initiated by the stimulus onset and continued for 30 msec.

Methods of ABR for Neurological Application

Auditory stimuli were clicks (one cycle of a 3-kHz sine wave) produced by a signal generator delivered through TDH-39 earphones. Clicks were given monaurally at

0077-8923/82/0388-0487 $1.75/0

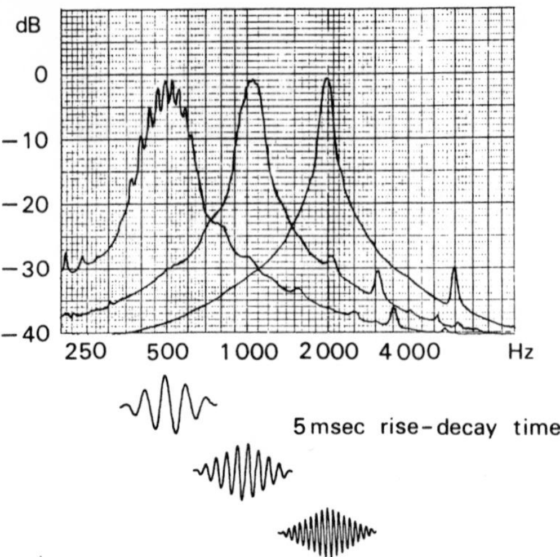

FIGURE 1. Frequency spectra of auditory stimuli. Auditory stimuli consisted of tone pips with 5-msec rise–decay times without a plateau. Frequencies of the tone pips were 500, 1000, and 2000 Hz.

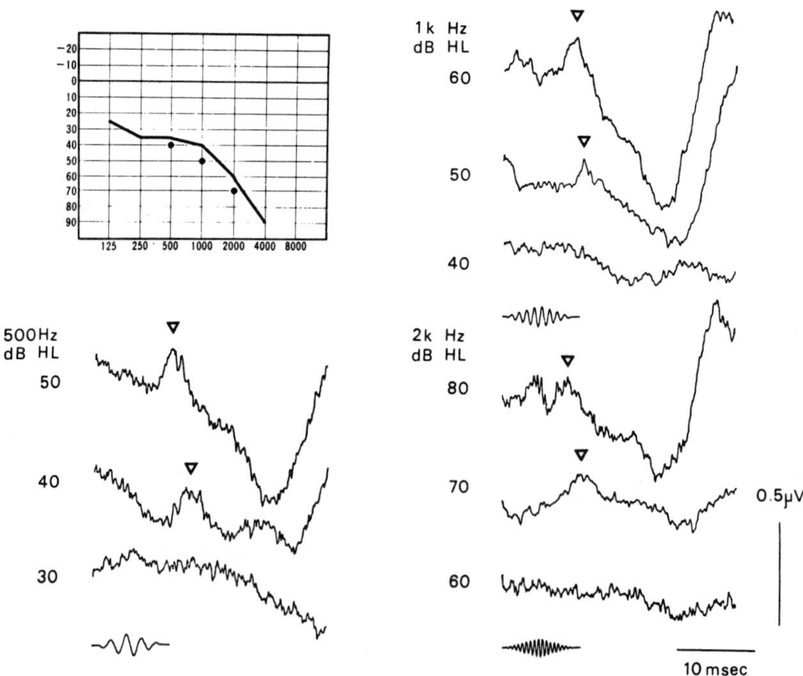

FIGURE 2. Auditory evoked brainstem response (ABR) recordings in a case with sensorineural hearing loss. An audiogram and ABR recordings evoked by tone pips at three speech frequencies are shown. ABR thresholds are indicated by filled circles in the audiogram.

intensities of up to 85 dB above the hearing level (HL) at the rate of 10/sec. The methods of the procedure and recording were the same as those for the audiological application above.

Results

Audiological Application

Correlations Between Thresholds by Pure-Tone Audiometry and ABR Thresholds

An audiogram and ABR recordings in a case with sensorineural hearing loss are shown in FIGURE 2. The hearing thresholds in pure-tone audiometry at 500, 1000, and 2000 Hz were 35, 40, and 60 dB HL, respectively. ABR thresholds at 500, 1000, and 2000 Hz were 40 dB, 50 dB and 70 dB HL, respectively.

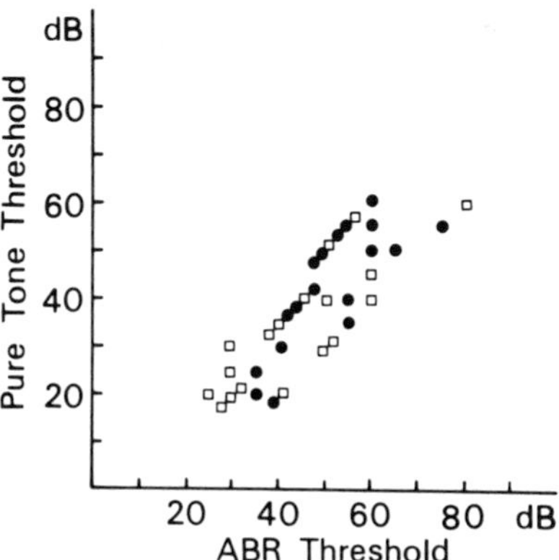

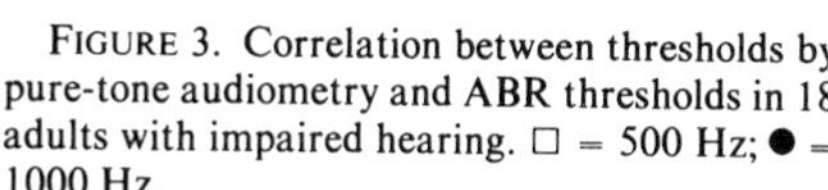

FIGURE 3. Correlation between thresholds by pure-tone audiometry and ABR thresholds in 18 adults with impaired hearing. □ = 500 Hz; ● = 1000 Hz.

In 18 adults with impaired hearing, pure-tone thresholds and ABR thresholds were compared. FIGURE 3 shows the correlation between the pure-tone thresholds and the ABR thresholds. The ABR thresholds were higher than the pure-tone thresholds by as much as 20 dB. The means of the differences at 500 and 1000 Hz were 10.9 and 7.3 dB, respectively. No clear difference related to frequency was observed.

Pure-tone thresholds and ABR thresholds were compared in 13 children with impaired hearing aged from 3 to 5 years. The pure-tone thresholds were obtained by repeated play audiometry. FIGURE 4 shows the correlation between the pure-tone thresholds and the ABR thresholds. The ABR thresholds were higher than the pure-tone thresholds by −15–10 dB.

The means of the differences at 500 and 1000 Hz were 1.7 dB and 0.5 dB, respectively.

Clinical Use of ABR Audiometry in Otolaryngology

Two illustrative cases with long-term follow-ups are presented in FIGURES 5 and 6. The first case visited the authors' ENT clinic at the age of 2 years 2 months with the complaint of speech retardation. Repeated COR audiometry and otological examina-

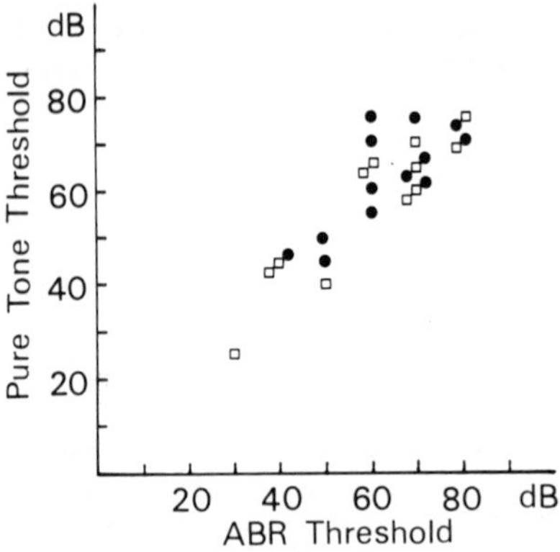

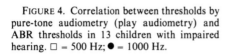

FIGURE 4. Correlation between thresholds by pure-tone audiometry (play audiometry) and ABR thresholds in 13 children with impaired hearing. □ = 500 Hz; ● = 1000 Hz.

tion indicated that the patient had sensorineural hearing loss. ABR audiometry was performed to evaluate the degree of hearing impairment correctly. According to the audiological tests, the patient's hearing aid was adjusted. Pure-tone audiograms taken when the patient was 7 years old demonstrated the reliability of ABR audiometry performed 5 years earlier.

The second case visited our ENT clinic at the age of 1 year and 8 months with the complaint of developmental retardation. On the basis of ABR audiometry at the age of 2 years 4 months, the case was diagnosed as suffering from moderate sensorineural hearing loss and auditory training was started using a hearing aid. Results of ABR

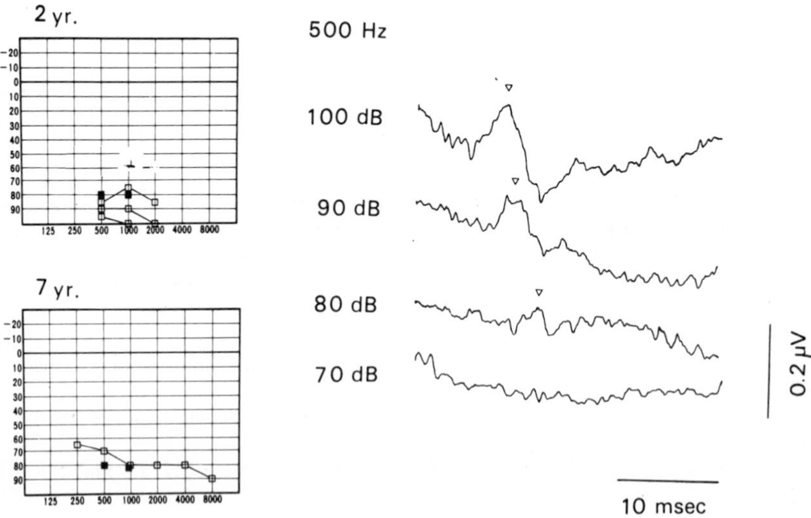

FIGURE 5. Audiograms and ABR recordings in a child with moderately severe sensorineural hearing loss. In the upper audiogram, results of repeated COR audiometry at the age of 2 years are shown by open squares. Filled squares indicate the results of ABR audiometry at 500 and 1000 Hz performed at the same age. In the lower audiogram, pure-tone thresholds confirmed by pure-tone audiometry at the age of 7 years are indicated by open squares. ABR thresholds obtained at the age of 2 are also by filled squares. ABR recordings at 500 Hz are shown on the right.

audiometry performed at 2 years and 4 months and 5 years and 8 months indicated the reliability of the test, as shown in FIGURE 6. As also shown in FIGURE 6, the results of COR audiometry as well demonstrate the reliability and usefulness of ABR audiometry in "hard-to-test" children with hearing impairment. At the age of 6, the patient was mute and his developmental quotient was 35.

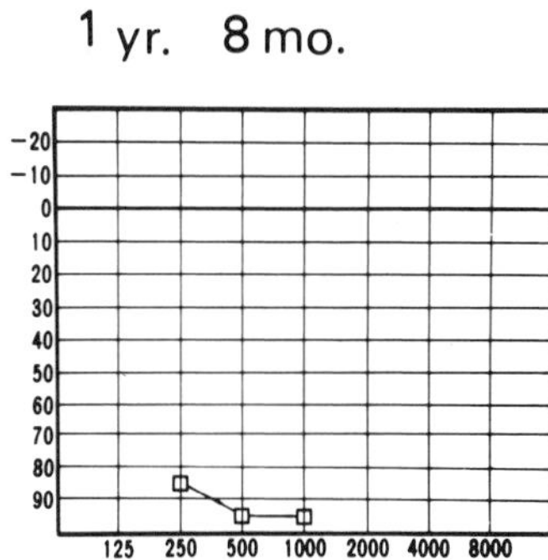

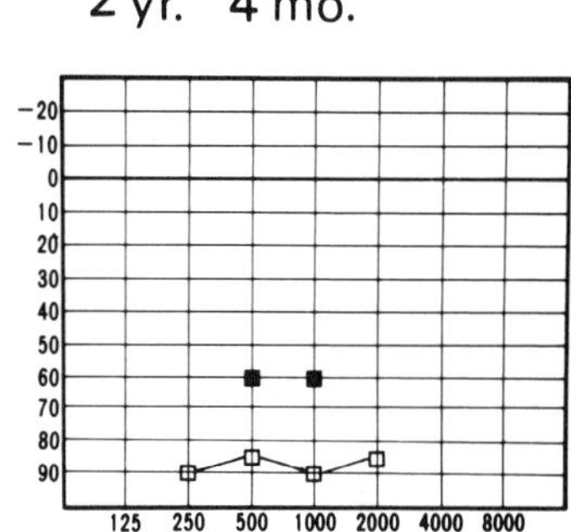

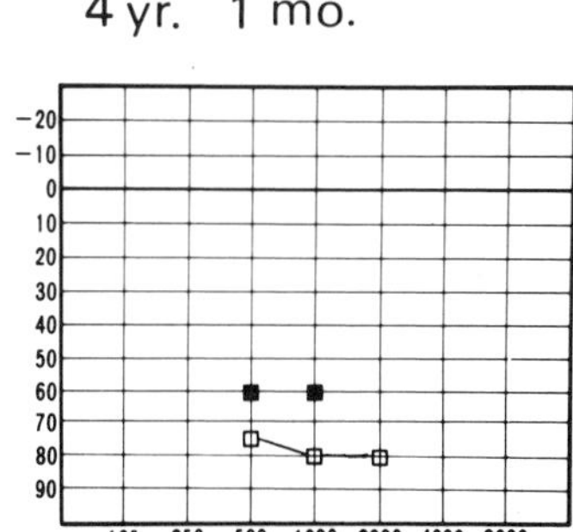

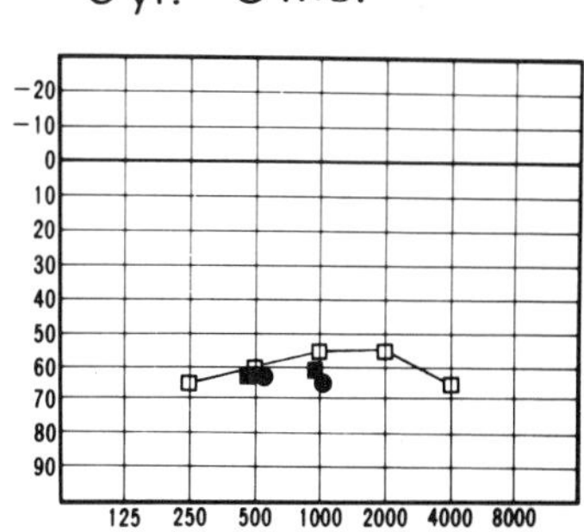

FIGURE 6. Audiograms in a child with moderate sensorineural hearing loss and severe mental retardation. Results of COR audiometry at each age are shown by open squares in four audiograms. ABR thresholds at the age of 2 years 4 months are indicated by filled squares. Filled circles indicate ABR thresholds at the age of 5 years 8 months.

Neurological Application

Normal Infants and Children

Developmental changes of ABR to clicks at 85 dB HL according to age are shown in FIGURE 7. Waveforms of the ABR are not well differentiated in neonates and young infants, but reach a mature pattern by 2 years of age. In FIGURE 8, changes in mean wave V thresholds of ABR with clicks and behavioral responses to 1- and 2-kHz pure-tone audiometry according to age are shown. Neonates had the highest wave V thresholds (28 dB), and adults had the lowest.

A similar pattern was observed for behavioral audiometric thresholds. Conventional audiometry was performed as follows: 1–3 months, behavioral observation audiometry; 4 months–2 years, conditioned orientation reflex audiometry; 3–4 years, play audiometry; and 5 years to adulthood, standard pure-tone audiometry.

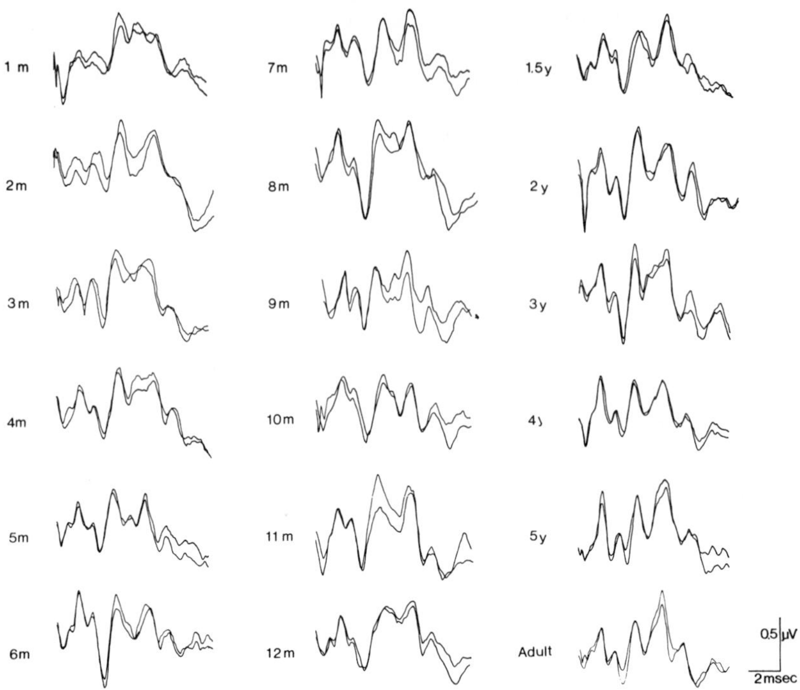

FIGURE 7. Configuration changes of auditory brainstem response at the 85 dB hearing level (HL). Each trace represents typical response for the age indicated.[5]

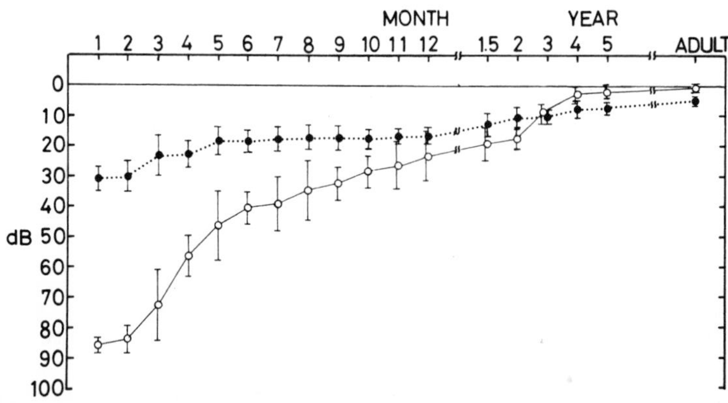

FIGURE 8. Changes in mean auditory brain stem response (ABR) thresholds for click and behavioral responses for 1- and 2-kHz pure tones with age.[5] The dotted line shows ABR thresholds and the solid line reflects behavioral sensitivity. ABR thresholds for the infants below the age of 4 month-old are now under further investigation and can be lower than those above.

The threshold functions determined by both methods converged with age. The functions crossed between the ages of 1½ and 2 years.

The latency age function to 85 dB stimuli is illustrated in FIGURE 9. There was a trend for all peak latencies to decrease with age. This effect was particularly pronounced for the later ABR components (III, IV, V, VI, and VII). Wave VII was recorded consistently only for ages of more than 3 years. Before that, wave VII had a low incidence of appearance.

Perinatal Asphyxia

ABRs in asphyxiated neonates are classified into (a) disappearance of early waves, (b) disappearance of later waves, (c) immature waves between waves II and V, (d) prolongation between wave V–I peak intervals, and (e) normal waves (FIGURE 10).[6] The cases who show disappearance of early waves frequently develop cerebral palsy of the athetoid type with deficient hearing at high frequencies. On the other hand, in cases manifesting disappearance of later waves, only minor motor disorder and clumsiness remain as sequelae. In spastic paresis, neonatal ABRs showing disappearance of early or later waves are rarely recorded. In general, immature waves and longer interwave intervals are the rule in asphyxiated groups, and such abnormal ABRs are sensitive to recovery trends from hypoxic encephalopathy.

Cerebral Palsy

Cerebral palsy is roughly classified into two types, spastic paretic and athetoid. In the spastic paretic type, ABRs are usually normal. However, in the athetoid type, ABRs frequently manifest threshold elevation, prolonged waves I and V and the absence of any responses. However, no brainstem lesion pattern is found. Kernicterus is a typical type of athetoid cerebral palsy. In FIGURE 11, the ABRs of 25 kernicterus patients at 85 dB HL, are shown. They are categorized into four groups: (A) normal threshold group (0–20 dB), (B) mild threshold elevation group (20–60 dB), (C) severe threshold elevation group (60–85 dB), and (D) very severe threshold elevation group (above 85 dB).[7]

Brain Tumor

Since the incidence of infratentorial tumors in children is three times that of supratentorial tumors and is higher than that in adults, ABRs can play an important role in diagnosing tumor sites. In FIGURE 12, three typical ABRs of patients with brain tumors that occurred at the lower, middle and upper brain stem, respectively, are illustrated. The abnormal ABRs showed partial disappearance of later waves.

Leucodystrophy

In FIGURE 13, ABR configurational changes with the progress of adrenoleucodystrophy are illustrated. This child was normal until the age of 5 years. His illness began with a gait disturbance, dysarthria and hearing difficulty. Later, spastic paralysis, serious deafness and blindness appeared. He died of respiratory failure 2 years after the onset.

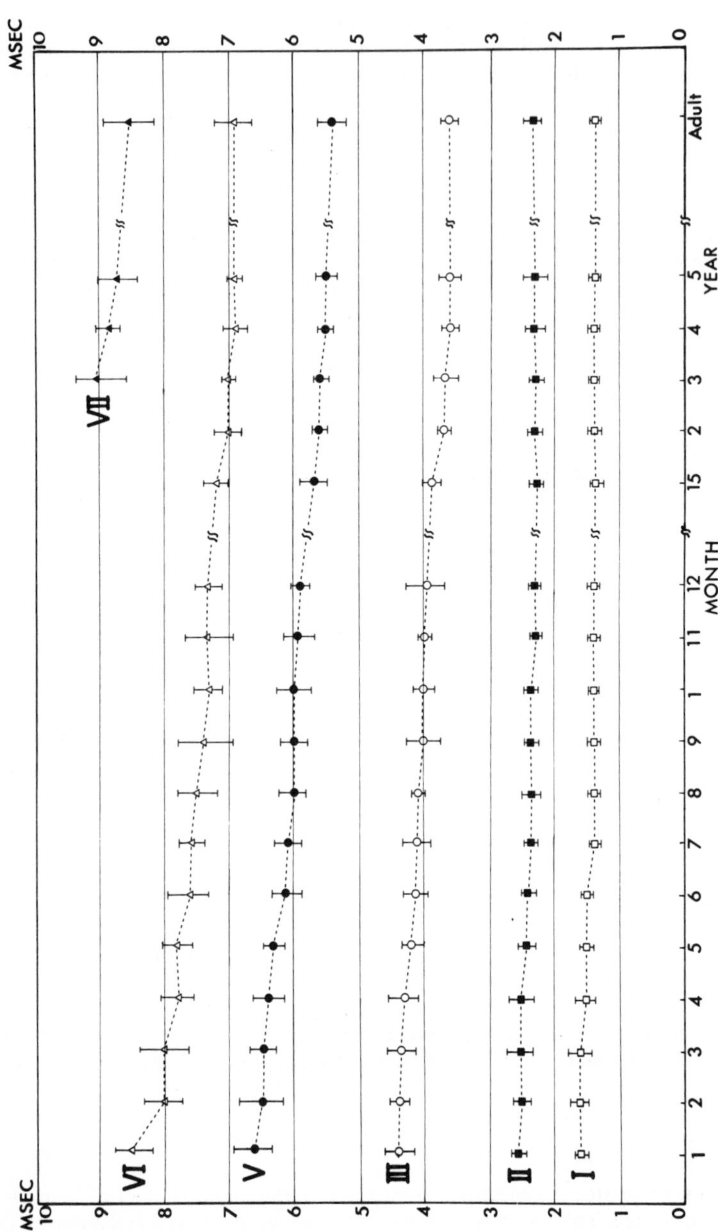

FIGURE 9. Peak latency changes of waves I, II, III, V, VI, and VII at 85 dB HL as a function of age.[5]

The ABR was normal at the onset, but later changed to an abnormal pattern.[8] Initially, there was lengthening of the wave V–I interpeak interval. As his general condition deteriorated, this was followed by the disappearance of the later components. At the terminal stage, only a prolonged wave I was recordable. The postmortem pathology revealed demyelination of the auditory nerves and remarkable neuronal loss in the auditory pathways of the brainstem; in addition, there was a variety of extensive degeneration throughout the cerebrum, in particular, complete degeneration of the white matter with secondarily occurring ganglionic cell changes. These data suggest that degeneration of the brainstem from the rostral to caudal levels occurred.

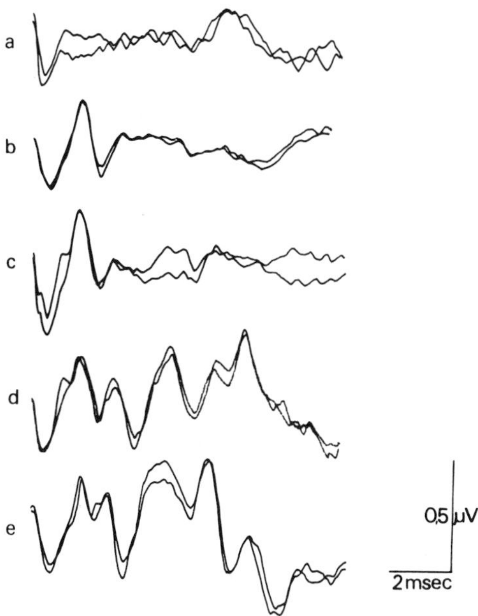

FIGURE 10. ABRs in asphyxiated neonates and infants.

DISCUSSION

Audiological Application

Objective audiometry by means of the auditory brain stem response (ABR) has been limited in its ability to estimate pure-tone thresholds, so long as brief clicks have been employed as stimuli. Although these acoustic stimuli with abrupt onsets do elicit well-defined and easily identified responses, they have such broad spectra, often dominated by the higher frequencies, that they cannot provide precise frequency-specific threshold estimates in the speech range.[1]

Tone pips—short sine-wave segments with rise and decay times of a few cycles—do appear to permit estimation of pure-tone thresholds in the speech frequencies with a degree of accuracy.[2] This is because, as shown in FIGURE 1, these stimuli have relatively narrow spectra and sufficiently rapid onsets to elicit a recognizable ABR. In

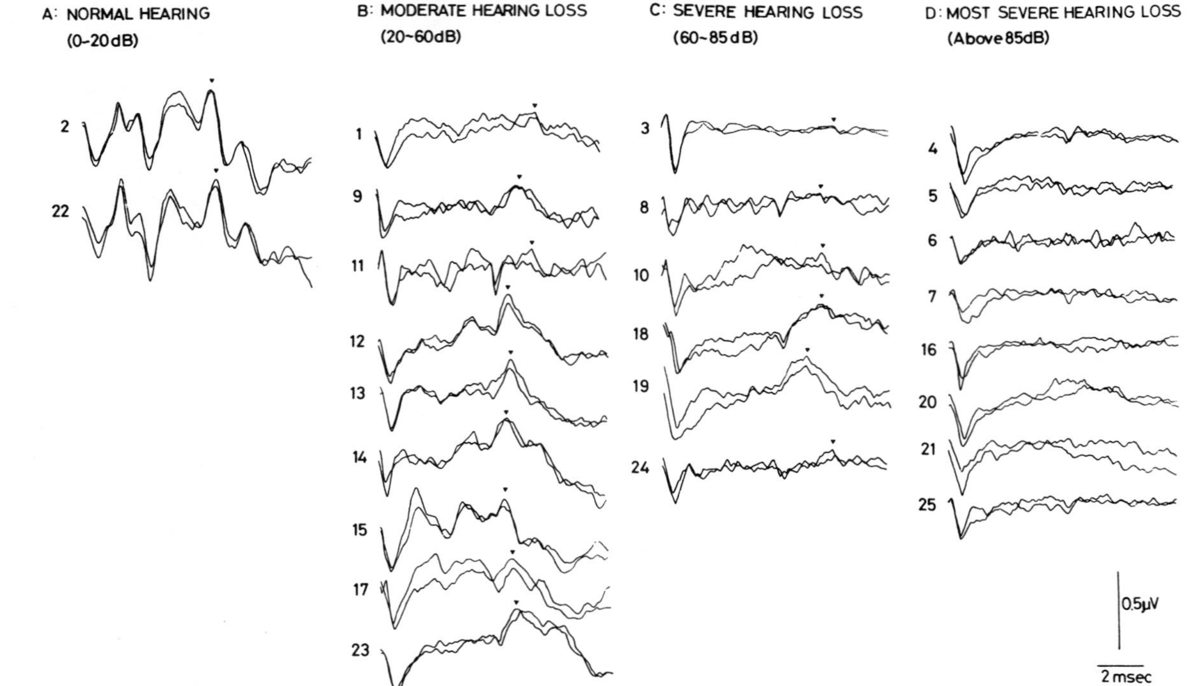

FIGURE 11. ABRs at 85 dB HL of 25 kernicterus cases, which are categorized into four groups: (A) normal threshold group (0–20 dB), (B) mild threshold elevation group (20–60 dB), (C) severe threshold elevation group (60–85 dB), and (D) very severe threshold elevation group (above 85 dB).[7]

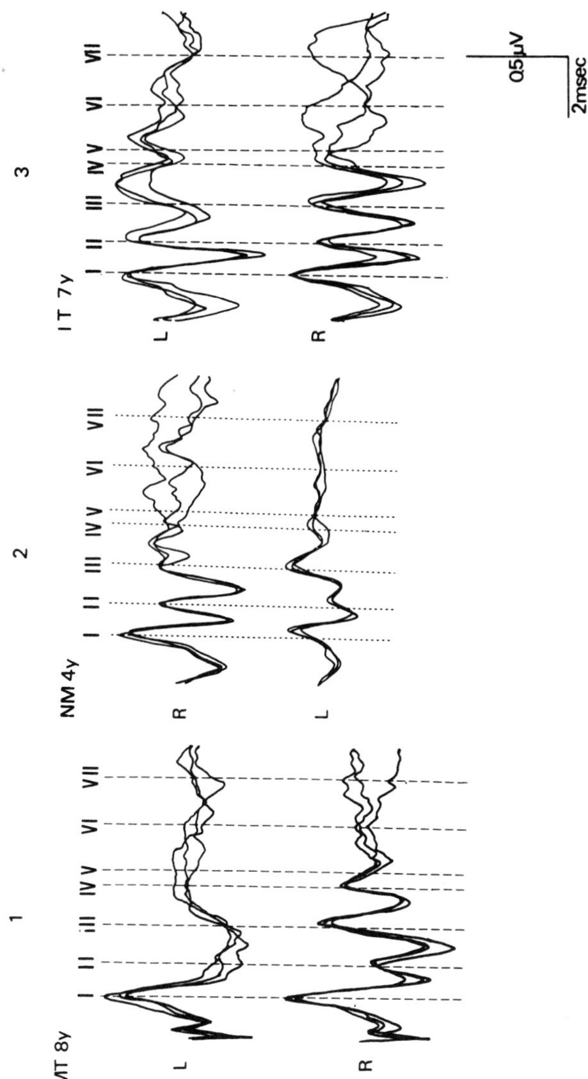

FIGURE 12. ABRs of brainstem tumor cases. Case 1: an 8-year-old-boy. Left pontine glioma. Case 2: a 4-year-old girl. Left pontine glioma. Case 3: a 7-year-old. Bilateral pontine glioma.

addition, the ABR thresholds are sufficiently close to the subjective thresholds of the tone pips.

In 18 adults with impaired hearing, as shown in FIGURE 3, the ABR thresholds were higher than the pure-tone thresholds by as much as 20 dB. If a pure-tone threshold is assumed to be 10 dB below a ABR threshold, the estimate should be correct within about 10 dB. This uncertainty is small enough to make ABR useful for objective audiometry.

As shown in FIGURE 4, in 13 children with impaired hearing, the mean difference

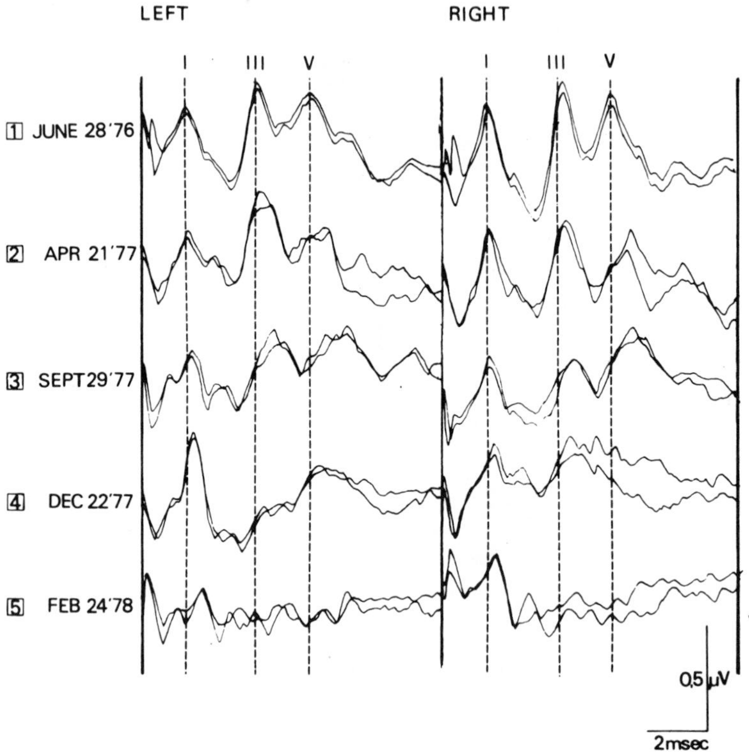

FIGURE 13. ABRs of an adrenoleucodystrophy case from the early (1) to the terminal stage (5).[8]

between ABR thresholds and pure-tone thresholds by play audiometry was less than 2 dB. If a pure-tone threshold is assumed to be 10 dB below a ABR threshold, the estimate might be higher than the correct value by 10 dB. However, thresholds by play audiometry are approximately 10 dB higher than "real" pure-tone thresholds. Therefore, results in children also indicate that ABR audiometry is reliable for objective audiometry in children.

Correct evaluation of hearing impairment in early life is very important and essential for the education of hearing handicapped children. ABR audiometry is a very useful and reliable method for objective audiometry as demonstrated by two cases.

Neurological Application

Our normative data on infants and children suggest that the ABR reflects brain stem maturation.[5] It is interesting to note the differences between the ABR threshold/age function and the behavioral threshold/age function. They perhaps reflect a difference in the rate of brain stem development and that of higher levels.

The reason for the decrease in the wave V threshold may be based on progressive myelination, which results in a greater degree of synchronization among impulses arriving at the wave V generators. This interpretation is supported by the result that peak latencies decreased with age. Such a finding would be predicted by an augmentation of nerve conduction velocity through myelination. Hecox has suggested that increases in dendritic arborization and fiber diameter growth might also underlie developmental changes of the ABR.[9] Both of these factors could enhance synchronization, as does myelination, by reducing the conduction time. But myelination does seem to be of prime importance since demyelinating diseases can reverse, to a large extent, the changes associated with development.

The direction of the developmental process seems to proceed rostrally. The fact that peak latencies of the earlier ABR components reached their adult values before the latter components is consistent with this view. This interpretation is also supported by the observation that behavioral response sensitivity, which depends on the maturation of cerebral function, was well above the ABR thresholds only for neonates and early infants. Since the ABR does seem to mature in the same direction as sensory information flows, it can potentially be used to monitor brain stem maturation.

The consequences of perinatal asphyxia range from death to various degrees of neurodevelopmental sensory or motor deficits. One of its well-known sequelae is sensorineural hearing impairment. Histopathological studies have provided evidence suggesting that the cochlea and brainstem of the human neonate is quite vulnerable to anoxia.[10] Our data on perinatal asphyxia suggest that ABR is useful as a screening test for cochlear and brainstem involvement in perinatal anoxia.

Although the association of athetosis and auditory dysfunction in kernicterus has been known for more than three decades, the identification of the lesion location has remained problematic. Currently, it is believed that either the cochlea, the brainstem, or both, are the lesion site. There are many reasons for the difficulty in establishing an auditory lesion location for kernicterus in infants and children. One is that hearing involvement itself seems to be a rather variable component of kernicterus. Another reason is that, until recently, there have been no reliable and objective means of testing hearing in infants—especially infants with psychomotor dysfunctions. Our data on kernicterus reveal that the deafness of kernicterus is mainly peripheral, and this fact gives rise to hope that its early detection and treatment can prevent later speech impairments.[7] In this respect, the ABR is well-suited as a means of early detection.

The definition of ABR abnormalities in our patients with midbrain and brainstem tumors raises the possibility that the measurement of ABRs may be of use in the early evaluation of patients with suspected tumors in this region.[11,12]

Our study of ABR in adrenoleucodystrophy also illustrated an important capability of ABR.[8] Not only can it help to identify the location of lesions, but it can also be used to evaluate the progress of degenerative diseases such as ALD.

Finally, it is emphasized that the above case studies illustrate the potential power of ABR as a diagnostic tool for hearing and brainstem disorders in pediatric audiology and neurology. Not only can ABR indicate the site of pathological processes, but it is capable of charting the progress of these processes as illustrated by premature infant, cerebral palsy, brainstem tumor, and adrenoleucodystrophy case studies.

REFERENCES

1. DAVIS, H. 1976. Principles of electric response audiometry. Ann. Otol. Rhinol. Laryngol. **85** (Suppl. 28).
2. DON, M., J. J. EGGERMONT & D. E. BRACKMAN. 1979. Reconstruction of the audiogram using brain stem responses and high-pass noise masking. Ann. Otol. Rhinol. Laryngol. (Suppl. 57): 1–20.
3. KODERA, K., H. YAMANE, O. YAMADA & J. SUZUKI. 1977. Brain stem response audiometry at speech frequencies. Audiology **16:** 469–479.
4. DAVIS, H. & S. K. HIRSH. 1979. A slow brain stem response for low frequency audiometry. Audiology **18:** 445–461.
5. KAGA, K. & Y. TANAKA. 1980. Auditory brain stem response and behavioral audiometry developmental correlates. Arch. Otolaryngol. **106:** 564–566.
6. KODAMA, K., E. KITAZUMI & K. KAGA. 1981. Auditory brain stem responses of asphyxiated infants in the perinatal period. Brain Devel. **13:** (in press).
7. KAGA, K., E. KITAZUMI & K. KODAMA. 1979. Auditory brain stem responses of kernicterus infants. Int. J. Ped. Otorhinolaryngol. **1:** 255–264.
8. KAGA, K., Y. TOKORO, Y. TANAKA, *et al.* 1980. The progress of adrenoleucodystrophy as revealed by auditory evoked responses and brain stem histology. Arch. Otorhinolaryngol. **228:** 17–27.
9. HECOX, K. & R. GALAMBOS. 1974. Brain stem auditory evoked responses in human infants. Arch. Otolaryngol. **99:** 30–33.
10. LEECH, R. W. & E. C. ALVORD. 1977. Anoxic-ischemic encephalopathy in the human neonatal period. The significance of brain stem involvement. Arch. Neurol. **34:** 109–113.
11. STARR, A. & A. E. HAMILTON. 1976. Correlation between confirmed sites of neurological lesions and abnormalities of far-field auditory brain stem responses. Electroencephalogr. Clin. Neurophysiol. **41:** 595–608.
12. STOCKARD, M. M. & V. S. ROSSITER. 1977. Clinical and pathologic correlates of brain stem auditory response abnormalities. Neurology **27:** 316–325.

GENERAL DISCUSSION

Moderators

James Stockard

University of California Medical Center
La Jolla, California 92103

James Jerger

Baylor College of Medicine
and The Methodist Hospital
Houston, Texas

K. CHIAPPA: Dr. Stockard, have you had a chance to evaluate the Cleveland Clinic data, and explain why they could get such a high abnormality rate? Are they using different measures of interpretation?

J. STOCKARD: They used seven criteria for abnormality, most of which were nonquantitative. Only three of the seven measures really allow even the possibility of interlaboratory comparison and they consist of a 0.2 msec prolongation of the peak latency of any of the seven vertex-positive BAEP waves; secondly, an asymmetry of these peak latencies of greater than 0.2 msec between the two ears; and thirdly, a 50% reduction of any wave amplitude compared to the corresponding wave elicited from the other ear. These criteria, as you know, are quite nonspecific with respect to central brainstem dysfunction, and, in fact, when we applied their criteria to 78 normal infants in our own series who were age-matched and sex-matched, 55 of 78, or 71%, of these normal infants met two out of three of the quantitative Cleveland Clinic criteria for brainstem abnormality as revealed by this test. I can assure you that none of those 55 infants were at high risk for anything. They were chosen because they were audiologically and neurologically normal and never had a near-miss episode, or anything else, for that matter. We followed them for up to 4 years now, so I think it was the nonquantitative nature of most of the criteria they used and the nonspecific nature of the quantitative criteria that they did use.

They did not have their own controls either, which does not help. There has been another study recently from the Stanford group which also found no abnormalities in 10 out of 10 near-miss for sudden infant death survivors. In 10 out of 10, they also found normal BAEP results. The Stanford group had slightly different conclusions, though, with respect to the near-miss-for-SIDS group as a population, and that was that they did not differ significantly with respect to their normal controls.

They used a paired *t*-test to match 1 of 10 individual survivors of near-miss episodes against only one control in each comparison. We took a different approach and used group *t*-tests for comparison of population means between our NMSID survivors and age-matched normal controls and the nonparametric Mann-Whitney U test and F-ratios to further compare the two groups, all tests utilizing information about every member of each group being compared. So I think there is a clear explanation for the subtle differences between our findings and those of the Stanford group. There is an even more obvious explanation, I think, for the differences between both of our groups' findings and those of the Cleveland Clinic group.

I. BODIS-WOLLNER: I would like to ask you if you think the anoxia or anoxemia might be the factor for the borderline BAEP anomaly that you can find in some subjects? Wouldn't you think that cortical evoked potentials, maybe in the somatosensory modality, would be a better test to use because, after all, the cortex is more sensitive to anoxia than the brainstem as far as I know of.

STOCKARD: That is true in adults but not in the perinatal period in which the converse obtains. In contrast to the adult hemispheral pattern of anoxic injury, neonates and infants show a different pattern of selective vulnerability to anoxia, which interestingly reveals that the brainstem auditory structures are among the most sensitive to both ischemia and hypoxemia in the entire brain, with the cochlear nucleus and the inferior colliculus being two of the most vulnerable.

I agree with you that somatosensory evoked potentials will probably be more useful than brainstem auditory evoked potential in assessment of these questions but I agree for a different reason. It is not because they are going to evaluate the cortex but because they are going to evaluate a more extensive length of subcortical afferent pathways in this age group.

Dr. Jerger, I just wanted to ask a question about your slide on 500 Hz stimuli compared to the clicks in terms of your estimates. Actually they were a little complex. Were you saying essentially that your 500 Hz thresholds were better predicted by your clicks than your 500 Hz pips?

J. JERGER: No, I am sorry. That slide is a bit confusing. We compared click predictions with thresholds at 2 kHz, and what we wanted to show was that that prediction was essentially unaffected by the audiometric contour, but that accuracy of prediction of the threshold at 500 Hz was accurate as long as the audiogram was rising or flat but that it became progressively too pessimistic as the audiogram began sloping downward from 500. The click data meant to show that the click prediction was not affected.

K. HECOX: Because we are in such desperate need of such comparisons between clicks and tone pips, do you know what the average discrepancies were for click behavioral thresholds versus 500 Hz and 2,000 Hz behavioral thresholds?

JERGER: You mean the distribution between prediction and the actuality? In these children, the prediction was quite good; the average error was perhaps on the order of no more than 4 to 5 db for the click and somewhat higher, perhaps 10 or 12 db, for the 500 Hz tone pip.

A. STARR: Dr. Jerger, did you mention whether the behavioral thresholds you did with the children were done with those very short duration pips, or with the usual standard long counts?

JERGER: No, the same stimuli. The behavioral threshold to the pip at a 20 solidus/sec rate.

CHIAPPA: Dr. Jerger, some of us might be a little bit happier if you could show us replications of tracings using the same stimulus parameters on the results. I noticed in your slides that the patients you showed all had abnormalities after wave III in addition; and, if that was the case all of the time, how do you know that its effect is at the level of III and not at the level of V? Occasionally you can find patients, MS patients, who have normal I–V separations and absent wave III. Did you have any patients that showed that effect?

JERGER: We found five in our series of MS patients who showed normal wave III's but absence or delay of IV-V, and in all of those cases MLD was quite normal so that we felt that it was not a question of generalized abnormality of all waves because they specifically had normal III's and a normal MLD but did not have a good IV-V.

M. HITERBOCKER *(Downstate Medical Center, Brooklyn, N.Y.):* Dr. Stockard, I was wondering whether you had an opportunity to consider the time of day of testing or relationship to the sleep or the state of sleep. Perhaps the diurnal rhythms or perhaps cortisol secretion or some other substance is impeding your almost normal results.

STOCKARD: Yes, we tested the infants during REM as well as non-REM sleep, as well as during wakefulness and there was no difference. We did not expect to find one since we had found no difference between those three states in our age-matched normal controls either. As far as time of day is concerned, it has been shown that there is a 0.2 msec per degree Celsius shift in wave V latency as a function of the circadian diurnal core temperature variation in humans. There is also a 0.1 msec per degree Celsius variation in the wave I latency. So you get a ±0.1 msec diurnal variation in I–V interpeak latency as a function of body temperature. We had our infants within 0.05°C at the time they were tested so the control for that variable was rigorous.

Of course, I should point out that the hypothesis, the null hypothesis, that there were no significant differences when individual infants were considered versus our normal age-matched control population, was quite adequately confirmed. It would be much more important to take those factors into account if one were saying that near-miss survivors did differ as individuals from a control population. And that, of course, was the contention of other groups.

R. GALAMBOS: We were shown some ABRs from people with multiple sclerosis that were rather badly smashed. For example, some of Jerger's recordings looked as if they had only wave I, and yet these people were responding to sounds really quite well; and of course Stockard has presented us in the past with a picture of an individual with only wave I who was audiologically normal. Dr. Starr, can you tell us how it can be that a person with virtually normal audiometric behavioral responses can be giving no brainstem response with the exception of wave I. This fact really sounds as if it is giving brainstem audiometry a pretty bad name. If someone with normal hearing can have no brainstem response, then what exactly is it that we are working with here?

STARR: Actually Dr. Galambos knows the answer to that. There are two ways to answer it. One is that all you need is perhaps one eighth-nerve fiber to work, or very few eighth-nerve fibers to be working and very few brainstem fibers to be working for hearing to be preserved, particularly if the changes occur slowly over time. For instance, you have studied cats where you sectioned 99.5% of visual optic nerve fibers, yet you can get very nice visual functions from them. That can be one interpretation.

The other is that the auditory brainstem responses measure a very limited portion of the auditory pathway and actually measure only those parts that depend upon synchrony of firing. To get the potentials into the averaging process, all of the nerve elements have to be going together, but, for hearing, you may be able to get along very well with asynchronous input because we have a time constant in the auditory system that you can take information over time—long tone pips—and integrate them and get a certain amount of hearing. However, those two possibilities are, in my mind, synchrony, and for that you do not need many fibers.

GALAMBOS: Dr. Starr, you left us with the impression, I think, that you believe that it is the synaptic events that are responsible for this brainstem response, and I had hoped that you would have brought out the other theory.

STARR: No, I actually think most of it probably comes from nerve fiber pathway activity because that could mostly account for the latency changes across the scalp. I do not want to leave out synaptic activity, though I think perhaps with some of the components there will be some evidence of synaptic activity, but I think most of the brainstem response is a series of traveling waves in nerve fibers giving us not only these amplitude differences, but also latency differences.

J.J. EGGERMONT: Yes, especially the last case I showed; it was a case of auditory nerve tumor, and on the tumor side there was only a wave V from the 1–2 kHz area, predominantly from the 1 kHz area actually. Although the thresholds for 8 and 4 kHz were not that much lower, there was no way for wave V to go through. That was the point, that even with a small wave I there were some problems. What we are looking at in the brainstem are synchronous responses, just the functional connections for probably short-latency fibers or fast forming fibers, which Dr. Galambos used to call time-keepers. That may be just a small part of the auditory system as well. And I think hearing and the brainstem are intimately related in subjects with normal brainstems and brains, and if you assess that the brain and the brainstem is normal you would only occasionally find no brainstem response and normal hearing. I would suggest that you check the patient again in that case. In normal hearing especially, or even in peripheral hearing loss, we use the high-pass masking technique, the subtraction technique, to derive audiograms, and they are very accurate. If you use tone pips and you have a properly shaped tone pip with good spectral content, then you have as Dr. Suzuki has shown here the capability of arriving at the correct audiogram.

Dr. Jerger had a little bit of a problem with his 500 Hz tone. Maybe it did not have good spectral properties, let's say only one or a little bit more than one period of 500-Hz signal. In that case, actually, a very broad spectrum exists. In fact, in steep audiograms the limiting factor is the slope of the spectrum of the tone pip, so that fits in all quite nicely.

The important point is what Dr. Stockard said, that the statement that the ABR is not necessarily related directly to hearing, is not completely correct. I think if you can rule out brainstem or brain abnormalities, it has a lot to do with hearing. At least it predicts hearing thresholds very accurately.

STOCKARD: I agree with you that the ABR is intimately correlated with hearing if you have a normal brainstem. Dr. Galambo's question which never was really addressed directly, was a case of a patient with multiple sclerosis with lesions intrinsic to the brainstem, and I do not think that the specific example that he gave has yet been addressed. It is intriguing and not explained by anything that anyone has said thus far really: a patient with multiple sclerosis who would have an eighth nerve action potential only—when we know that only oligodendriglia making central myelin are involved and not peripheral nerve or distal eighth nerve myelin. Thus why would the abnormality begin with wave II? Why would there be no central components, assuming that most of the portion of myelin covering the most proximal eighth nerve and all the distal nerve was not involved by MS?

I think the lesion to answer your question, in our three cases, was in the most proximal portion of the projections from the cochlear nuclei intrinsic to the brainstem, those projections which are involved in subserving this synchrony and phase comparison between the two ears that Dr. Starr alluded to. That would account for the presence of wave I only and no wave II, or of the subsequent components.

ALLEN LUGGET *(Einstein Medical College, Bronx, N.Y.):* I would like to echo Dr. Starr's comments about the ultimate nerve physiologic source of the brainstem response in studies in the monkey. While we can record peaks of many millisecond duration, I think potential within the structures of the brainstem auditory pathways, especially within those that have a degree of organization in laminar structures, namely, the inferior colliculus and the superior olivary complex and cochlear nucleus, we think specifically of the dorsal cochlear nucleus. These potentials are not recordable more than a few millimeters away from these structures, and it is the shorter duration potentials recorded within the fiber pathways that can be traced into the far field.

I would also like to carry the multiple generator identification back to the monkey equivalent of the human wave II, which in some cases can be distinguished as having two sources on surface recordings. One of these is the cochlear nucleus and the other one turns out to be the N2 component of the eighth nerve action potential, and one of the published cases—well most of the published cases of humans with acoustic neuronomas—only display a wave I. There is one case in the Starr-Hamilton paper, which displays both wave I and wave II, and several cases reported by Stockard *et al.* that show this. This may reflect the N1 and N2 components in that case, and this, in fact, was what Stockard *et al.* and, independently, Moller *et al.* postulated in recent publications.

STOCKARD: That is a good point. In addition to the Starr and Hamilton case, Drs. Chiappa and Goldie and we also have cases of complete brain death in which there is preservation of wave II, and we found that this wave II differs from the wave II seen in routine BAER testing and that it corresponds exactly with N2 of the compound auditory nerve action potential recorded simultaneously with electrocochleography. It is only seen at higher intensities and has a different field distribution from that of that wave II, obtained at lower intensities, which also has a slightly longer latency.

HECOX: One of my pet peeves is going on here, and I am astonished that Dr. Jerger is tolerating it, given his long-standing record of looking at the auditory system in manners other than those characterized as the audiogram. We are being told that it is astonishing that we are not predicting hearing by these tests—hearing, of course, being equivalent to an audiogram. I think that is a dangerous supposition that when we are trying to characterize complexities of hearing in the pathologic patient, particularly those with central auditory disorders, that we should have any hope or be at all surprised that there is a lack of congruity between audiometric thresholds and super-threshold behavioral measures, super-threshold BAER measures, and super-threshold any measures. That is a long-standing principle in animal work, and I think that is an error we need to stop making.

The other thing is that the brainstem response has a lot to do with hearing; but again, if you have evidence of brain disease, then all bets are off because many, many very important inner ear phenomena are very closely paralleled by BER activity traveling wave and frequency specific activity. Tuning curves have now been done, very nicely matched auditory nerve tuning curves, cochlear nucleus tuning curves which correlate with the BER, and it does have a lot to do with hearing. As always with clinical measures, one has to be a little bit cautious about what one says and what one tries to do with that particular window on the auditory system.

JERGER: Dr. Hecox makes a very good point, but while patients with multiple sclerosis typically have normal audiograms, they typically have far from normal auditory function, and I attempted to show one example of this in the masking level difference effect, which can be quite abnormal in the presence of normal audiogram.

EGGERMONT: Another example of a hearing threshold problem is that in a series of 43 tumors we found subjective hearing thresholds estimated with the tone pips and the electrocochleogram correlated very accurately with the hearing threshold. But if you looked at wave V, there was no correspondence at all. The absence of wave V, or the problems that you are having correlating wave V thresholds or the presence of wave V, is eliminated by looking at electrocochleographic thresholds.

If you looked at mere speech discrimination, and so on, and included these things, then you can be sure that if there are problems in finding or identifying wave V, you have a complex problem which is quite different. There are more problems with the higher functions, so I think threshold is mainly, at least in these cases, a peripheral phenomenon, and all the other things are probably at the brainstem level or maybe above that level dominantly. Then the brainstem response would not be a good estimate.

D. KURTZBERG: Dr. Jerger mentioned some lingering problems in his introductory remarks. If I am not wrong he mentioned false-positive results, do you care to elaborate about this?

JERGER: Yes, there is quite a variation in the reported prevalence of false-positive results in identifying acoustic tumors, for example, ranging from as low as 1% or 2% to as high as 30% in the work of Clemis and Magee. It is a muddy problem because the criteria for what constitute abnormality are not uniform across most of these studies, compounded by the fact that when the degree of peripheral sensitivity loss exceeds the limiting value, then the results can be noncontributory in the sense that absence of the response could be due to the severity of the peripheral hearing loss, and it is that constellation of findings that I refer to as the false-positive problem.

It was reamplified in a recent publication *Scandinavian Audiology,* which was a report of a symposium in Scandiavian countries summarizing their experience, and it was their uniform conclusion as well.

M. KLEIN: Dr. Eggermont pointed out that the brainstem response differed depending on what temporal frequency was the dominant stimulus. Could some of the other speakers comment on how the brainstem responses differ so according to the different temporal frequencies? I would be especially interested in Starr commenting on that. The importance of this fact is that it seems to wipe out the simplicity of the source story.

STARR: By temporal frequency I assume you mean different spectral components. The brainstem response in normals is dominated by the high frequency input so that the thing that Dr. Eggermont was referring to and Jerger was showing, those longer latency waves V from the apical parts of the cochlea, usually occur out of phase with each other and they cancel. The problem comes in though when you have significant cochlear damage or implied lesions of the eighth nerve; for instance, a tumor that selectively affects the high frequency portions of the cochlea. Then you are going to get problems in interpreting your evoked potentials, and that is one of the major issues really in the use of auditory brainstem potentials in neurological or central applications. The interpretation is confounded if you do not know very much about what is happening at the periphery.

WIEDERHOLT: Dr. Starr, you told us rather convincingly that in the identification of the generator sources that none of the experimental methods reveal very acceptable anatomical correlations. What would you suggest to do to precisely localize or identify these generators if you could set up the ideal experiment? Also, you said you felt that most of the activity recorded on the surface is probably generated in pathways. What is your evidence for that?

And, Dr. Stockard, in regard to the four or five adult patients you showed with the central nervous system apnea syndromes, from what I gather they were pretty sick and had rather serious brainstem pathology. What is the clinical value in those patients getting the BAERs?

STOCKARD: In two of the patients, there was no other evidence, not even corroborating clinical evidence, for brainstem pathology. The patients just suddenly presented with a central sleep apnea syndrome. The BAERs indicated that structural brainstem pathology was the basis for their syndromes and only up to a year later did they develop other signs of a brainstem lesion.

In contrast, our patients with central sleep apnea syndromes who have normal BAERs have up to five years of follow-up now and never have shown any evidence for neurologic lesions. It is a useful diagnostic screening test for patients whth central sleep apnea of later onset in life as it clearly and reliably differentiates those who have central sleep apnea on the basis of progressive, structural brainstem lesions from those who do not.

STARR: Each of the techniques we use has limitations. Initially when we used the technique, I was very enthusiastic about recording techniques in the deaf. I thought that would give the answer and there are limitations, of course, just like the new enthusiasm for the current density source that Dr. Vaughan talked about. I do not think that that is going to tell us what is really happening at the scalp, but it will tell what is happening in the layers beneath perhaps.

Each of the techniques has had limitations. If we put them all together, they would build up a set of arguments which all point to the fiber tracts. The ideal experiment would be to make a lesion that would have no remote effects and that would affect only the particular structure. Right now we are toying with demyelinating lesions in experimental animals in a way that we can control and which will affect the axons but not affect the cell bodies. But I think we are going to have problems with that too.

The evidence for fiber tracts is really a deductive one. I do not have any such evidence. If I did, I would have shown you the experiment. It is all deductive, and the evidence tilts toward the fiber tracts. But I am holding it back because the nasopharyngeal data showing that interesting polarity switch across the brainstem that cannot be accounted for by a fiber tract. It has to be a synaptic potential, but the issue is: Is that thing that I am recording in the nasopharynx being reflected up on the surface or not?

BODIS-WOLLNER: Returning to the paradox that Dr. Galambos so happily defined for this discussion, I have a feeling that we should take out data more seriously and trust them. And what I mean by that is that maybe we should ask questions as he did and go one step further and ask about the pathophysiology of the disease. My reasoning was summarized yesterday for visual evoked potentials where I tried to muster the arguments which point to the fact that in demyelination, a conduction velocity decrement in multiple sclerosis does not account for all the observed and well-documented phenomena in visual evoked potentials as a result of multiple sclerosis.

That does not mean that the explanations are all in, but I wanted to suggest first that we use our evoked potentials in a constructive way in terms of research and trust them. If we trust them without noise, we know that it is a fact. Then we should dare to ask the question: Is it really this pathophysiology which we thought was there? Therefore, I refer to the question from Dr. Stockard: What is the actual evidence for what you said about a single oligodendrocyte being involved in the MS patient who has this type of audiogram and auditory brainstem response?

STOCKARD: I did not say that. I indicated that by virtue of the known involvement of oligodendrocytes (versus Schwann cells) by the disease and therefore central instead of peripheral myelin by the disease, that a plaque of demyelination intrinsic to the brainstem would have to be held accountable for the loss of all waves after wave I in the three cases we reported with MS and with only wave I present. The other possibility is that the small amount of central myelin, supplied by oligodendrocytes, that covers the proximal part of the eighth nerve was involved. This cannot be ruled out in these cases. But the absence of *any* evidence of auditory nerve dysfunction on extensive audiologic testing makes this alternative explanation unlikely. That evidence pertains to what is already well known about the disease, not to a subset of MS patients who have BAER abnormalities.

BODIS-WOLLNER: There have been several other demonstrations, for instance, from Mary Bornstein, who is at Albert Einstein, showing changes at the dendrites. At that time the vogue was to pin the disease to autoimmune processes, but the demonstration of dendrite abnormalities is a fact.

JOSEPH DANTACITY: *(Medical College of New York, New York, N.Y.)*: Even our super-threshold tests are not always sensitive to eighth nerve or brainstem involvement. Even the tests that we do using normal speech discrimination testing are insensitive in many cases to eighth nerve lesions that are fairly large and do, in fact, affect brainstem evoked response measures. Dr. Starr, in regard to your mapping work, what effect would the plane of this potential have on the localization of those data rather than purely the specific proximity to the electrode with a plane of that potential?

STARR: By plane I assume you mean the vector, which is the most important thing. For instance, if you record between the ears in a horizontal plane, waves IV and V seem to disappear and are replaced by a component of lower amplitude that occurs intermediate between IV and V. Wave III becomes extremely broadened and we end up there with wave III of a dipole, and you think of it as a dipole. The dipole is in a horizontal and a vertical direction, and so the plane is the most important thing. This is the standard plane that we now use, the vertex to the ipsilateral mastoid.

If I had to do it all over again I would record vertex against the back of the neck, but I do not have to. I am going to stick with this technique. Clinically it works very nicely, but we are sampling only a limited portion of the vectors that way.

DANTACITY: The question I was specifically concerned with regards the mapping that you presented in trying to localize specific areas for the locus of potentials. Could the fact be that you generated a larger potential perhaps at the forehead even when you were using the back as a reference? Could that not be because you were checking electrodes for the neurons that were firing in that plane? And had you measured two potentials simultaneously you might be able to vector in on the source that way?

STARR: The sagittal array is very poor because the electrodes have a different relationship to the back of the neck, but on the coronal plane they are all equidistant. But I do not know the answer to your question. We use two electrodes, and I am sure we can do better.

CHIAPPA: Dr. Stockard mentioned one fact about central myelin going out onto the eighth nerve into the canal, and, in fact, Dr. Letterman of the Cleveland Clinic is studying a series of patients who have currently large multiple sclerosis plaques in the eighth nerve. I do not think we should be surprised at the difference between the brainstem auditory evoked abnormalities and conventional behavioral audiometry. As Dr. Jerger has shown in one technique, if you continue and test behavioral hearing in other ways—for example, Housler and Levine used interaural time discrimination—you can find behavioral abnormalities in all of the patients with multiple sclerosis who show central brainstem auditory evoked response abnormalities.

So I do not think that this divergence really exists when you use better tests. The false-positives in acoustic neuromas are largely a factor of not using interwave separation criteria. The studies that have used interwave separation criteria have had very low incidences of false-positives and very low incidences of false-negatives. There really is not a high incidence of false-positives with acoustic neuromas with this test.

Dr. Starr, I was very surprised to hear you say that midbrain lesions knock out waves IV and V, since one of the very few cases published in the literature in humans of midbrain lesions was published by you, and in fact the point you made in the case was that this midbrain lesion had not affected waves IV and V. Perhaps you have some new human, clinical pathological correlation data that you would like to share with us.

Also, I would disagree with your formulation of ipsilateral versus contralateral recordings showing different wave forms in a patient with a brainstem lesion and saying that this is evidence for different generators of the wave recorded in those two different derivations. Why could the explanation not be that the lesion has changed the generators in such a way that the potential field distribution is changed?

The question that Dr. Stockard and Dr. Desmedt raised of short-latency somatosensory evoked potentials in near-miss for SIDS infants has been investigated by us, and we have not found any abnormalities in the short-latency somatosensory evoked potentials in those infants.

Finally, I would like to take a difference of opinion with Dr. Jerger's initial comment with respect to the fact that you have to be an audiologist to interpret these things. As a neurologist I could say that you have to be a neurologist to interpret these things in diseases of the central nervous system. I do not think that either statement is necessarily true and both fields can interpret these brainstem auditory evoked responses.

STOCKARD: I think you have to be an otoneurologist actually to interpret them.

STARR: I think Dr. Chiappa is quite correct that another interpretation for that shift is that you have a shifting vector. When I was brought up, I thought of generators as a point source, but when I think of generators as a vector, then the different planes will record different vectors and in fact different generators. And you are quite correct that in the case that I reported of a midbrain lesion, if the midbrain lesion is restricted to the tectum, there will be no change in the IV-V complex; but if the midbrain lesion extends down into the tegmentum and gets into those fiber tracts, then you will get a change.

M. COHEN: *(City University of New York, New York, N.Y.)*: Dr. Stockard, do you have any information on the use of vitamin therapy in SIDS. I refer to a report by Orlowski.

STOCKARD: Lonsdale and Orlowski reported, in two cases, reversal of putative brainstem auditory evoked potential abnormalities in near-miss survivors by large doses of thiamine. Perusal of those two cases reveal that they were not abnormal centrally to begin with; that is, the BAEP findings did not reflect retrocochlear auditory dysfunction. The sorts of abnormalities that those patients had were much more likely to be attributable—just on the basis of the data presented itself—to technical and/or peripheral hearing problems. So I suspect that, given that was the most likely etiology for the so-called abnormalities to begin with, the correction of technical and/or peripheral factors was also the most likely etiology for their so-called reversal.

The two patients of Lonsdale who responded to thiamine (B_1) therapy had known disorders of thiamine metabolism–Leigh's disease in at least one case–and so even well-documented BAEP improvement would not have surprised me in these patients. Possible alterations in metabolism or deficiency of B_1 in other infantile apnea syndromes is also quite plausible, and needs to be systemically evaluated.

We have studied patients with Wernicke's encephalopathy both during the acute phase in which they were ophthalmoplegic and demented and ataxic, and after high-dose vitamin B_1 therapy and found in those cases who showed dramatic clinical improvement, absolutely no BAEP change in response to thiamine treatment.

COHEN: Dr. Eggermont, the brainstem potential has been described many times by several authors as a series of fat, short wavelets superimposed on the slow positive upswing of the baseline that peaks at wave V. There is some recent information as to a differential effect in MS patients on the slow-wave component as opposed to the earlier preceding pathways. Have you looked into this question with regard to deriving evoked potentials from narrow bands along the cochlea? And is it different from the slow-wave component as opposed to the short waves?

EGGERMONT: I did not look specifically for these effects. We use a filter setting of 100 Hz to 3 kHz, so I think most of the slow wave was missing. And I do not know if they behaved differently.

QUESTION: Dr. Suzuki, do you have any pathologic data to support the inference that your electrophysiologic changes are due to peripheral involvement of the auditory system in kernicterus or to hyperbilirubin encephalopathy?

Our studies in the newborn in the intensive care nursery population have, in a small group of patients, pointed towards a peripheral involvement with delays only being found in wave I, and then subsequent components. But I am wondering whether it is the cochlear or the auditory nerve that is being affected by hyperbilirubinemia?

I would like to ask Dr. Stockard if he has evaluated that particular risk factor.

SUZUKI: Well, we have no pathological data, but I think I made a conclusion that at least some of the cases had a peripheral lesion. We cannot exclude brainstem lesions, however: so this is a restriction on this test. But mostly, I think, people believe the hearing impairment in kernicterus must be from the brainstem lesions. The point of my report is that there are cases with lesions in the peripheral end organs.

STOCKARD: We have seen both peripheral-type and so-called central-type abnormalities separately in patients with kernicterus, a few of which had autopsies. In the cases in which the abnormalities were of the retrochochlear type they began with wave II, and the two patients had very heavy staining of the cochlear nuclei, which are thought to be selectively vulnerable to hyperbilirubinemia.

HECOX: This is a hard problem primarily because it occurs in the context, at least in the United States, of two entities, one of which is the asphyxiated acidotic infant, and we well know that the distribution of pathology in those cases is very much in the cochlear nucleus, inferior colliculus, and so on, and paralleling the metabolic rates in those areas. Thus it is very difficult to disentangle the effects of asphyxia from the effect of hyperbilirubinemia.

My own feeling about that is that it is probably only in those infants who have hemolytic syndromes who are term babies and have enormous elevations in their bilirubin, out of the context of any asphyxia, that we are going to answer the question.

I have seen a number of kernicteric babies, almost all of whom had some degree of peripheral loss with recruitment implying a cochlear disease not a nerve disease, the presence of recruitment there being helpful. Some of those babies had central auditory problems but almost all of them have had significant asphyxia at the same time. We have three children with hemolytic syndromes where the bilirubin level was greater than 30 and none of those have had any abnormalities centrally but have had marked abnormalities in the periphery, so our suspicion is that it is a cochlear impairment even though staining is surely demonstrable along the central auditory pathway.

S. JONES (*Medical Research Council, London*): I would like to bring up something that nobody else has mentioned very much so far, and that is a question of the polarity of the wide band click. We have not done very many brainstem responses but we were very impressed by the waveform difference betweeen a compression and a rarefaction click, and yet obviously, the click contains compression and rarefaction phases no matter which way it is, according to the manual.

But do you see waveform differences that are consistent in the individual latency shifts to the early components, although quite often wave V is of the same latency for the two polarities. These changes do not seem to be very consistent between individuals. I am wondering if any of the speakers has any information on that?

STOCKARD: We studied this extensively, so extensively in fact that I do not even know if I want to get into it. You may not have seen our article on this complex subject, which I shall send you in lieu of bogging down the discussion in this particular Pandora's box we opened several years ago.

CHIAPPA: That is true; you studied it before we did. We studied 600 neurologic 598patis of whom we found 20 patients with neurological diseases, mostly multiple sclerosis, who had the unusual finding of having completely normal brainstem evoked responses through wave V with one click polarity and having absolutely no wave V with the other click polarity. I am going to show some of these in the Roundtable Sessions later. We never saw this effect in 45 normals and I guess you have done more normals. The normals show subtle but statistically significant latency shifts and amplitude shifts but nothing as dramatic as this and, in fact, we also found in those patients that if you then reduced stimulus intensity, the previously absent wave V would suddenly appear. I am sure that Dr. Eggermont will have something to say about that. But it turned out that, of those 20 patients culled from a series of 600, 17 showed that effect such that the wave V was missing with rarefaction clicks but present and normal with condensation clicks, and the other 3, of course, were *vice versa*.

We have been trying to develop ways of determining whether this is a central or peripheral effect. Some of the patients who showed this effect had perfectly normal conventional audiometry, and beyond waiting for a patient who has a focal lesion, the only other way we can think of is perhaps using the binaural interaction waveform, which appears to be clearly a centrally produced effect only. If anybody has any other suggestions on this problem or we can bring it up again in the Roundtable Sessions.

STOCKARD: In those patients with MS or whatever in whom wave V disappeared as a function of acoustic phase, how can you be sure that wave V did not merely fuse with wave IV? This is a very characteristic change in waveform morphology as a function of click phase from condensation to rarefaction, which is the direction of change that resulted in disappearance of wave V in 17 of your 20 cases of phase-related "abnormality."

CHIAPPA: Because IV was present at both click polarities with absolutely the same latency and shape. When you see the figures, it is clear to see that what has happened is simply that wave V has reappeared. I do not think there is any question of that being the problem.

STOCKARD: What I am saying is that is what happens in normal subjects. In about 6% of normals you can get complete merger of wave V with wave IV, simply as a function of changing the click phase.

CHIAPPA: Yes, but what I am saying is that with one click polarity we have four peaks and the fourth peak has a latency of 5 msec; with the other click polarity we have five peaks, and the fourth one is still at 5 msec, and the fifth one is at 6 msec.

When I show you the figures and I think you will see how the problem is resolved.

M. ROCOL: *(Cleveland)*: Dr. Starr, in regards to your patient with a tectal lesion, does that mean that waves IV and V are generated entirely caudal to the colliculus? And would you care to comment on the generation of waves VI and VII as to where they might arise from?

STARR: I only have that one human case and I have not heard of any others. The only reason we knew the lesion was restricted to the tectum was because the surgeon went in and he actually took a biopsy of the inferior colliculus and there was an astrocytoma, and when the patient died it still had not spread down.

So I think that the wave IV/V comes from below the tectum. But I still think that they both come from the midbrain and probably from the lateral lemniscal fibers.

Waves VI and VII are very variable. That is the whole problem of trying to do the pathophysiology of it. I have some personal thoughts about VI and VII, but I think, regarding what Dr. Eggermont has said, they are another volley coming through the auditory system and not from an anatomically higher site.

JANET CAMP *(New York)*: Dr. Stockard, I am interested in your hypothesis that the slight prolongation of latency in the SIDS infants is a result of the hypoxic episode that they had experienced. I wondered if you had any chance to follow these infants up later to see whether this is a permanent or a transient effect and also whether you have had any opportunity to examine infants who have had an hypoxic birth, low Apgar scores, or whatever?

STOCKARD: That is a good question. We have followed the near-miss-for-SIDS survivors and infants who have recurrent apneic episodes. The latter will continue to diverge from a normal age latency function for I–V interpeak latency. Those NMSID survivors who do not have recurrent near-miss episodes will no longer diverge. We have not followed them long enough to see if they converge ultimately with the mean and reach normal values. On limited follow-up, they usually stay parallel with the normal latency–age function while 1–6 months of age and then begin to converge with normal age IPL functions; that convergence usually correlates with disappearance of apneic tendencies and abnormal periodic breathing during sleep.

As for your second question, there are people here who have studied that much more intensively than I. The question is, what are the effects of perinatal asphyxia in general? I think I would like to ask Dr. Hecox to comment on this. I could give you just one caveat. Unlike in the adult age group in which—in the clinical context of anoxic encephalopathy—these abnormalities have fairly straightforward and often unfavorable prognostic and diagnostic significance, all bets are really off in the perinatal age group. This is probably for a variety of reasons: differing ability to withstand the insult, greater plasticity, and larger neuronal reserve. Dr. Hecox, and I think Dr. Starr also, have seen cases such as we have seen in which patients have, as the result of acute hypoxemia, lost all brainstem auditory potentials transiently. Subsequently, they regained not only normal BAEPs but had normal neurologic outcome.

HECOX: I am going to discuss actually this afternoon some of the recovery function phenomena that you can encounter in that group so maybe we can defer that. Maybe Dr. Galambos has some other comments because he has as much experience as anybody in that area.

GALAMBOS: Yes, there is developing a very interesting and knowledgeable literature on the question you asked. There are several aspects to you question. In the study that we published a year and a half ago in *Pediatric Research* we had a hundred youngsters of whom 22 had really very low Apgar scores, and out of this 22 only 11, if you want to say only 11, had permanent damage, as measured by their brainstem response. They all had sensorineural hearing losses, and none of them had neurological disorder as evidenced by an increased I–V interval.

So here you could split two groups of asphyxic babies, approximately equal in external measures. What is the reason that half of them had abnormal ABRs and the other half did not? The answer, the best correlation we could get by going through their clinical histories, was that those babies with troubles were exactly the ones who had repeated bouts of acidosis during their postnatal history.

That is to say, they were roughly equally acidotic and asphyxic at birth, but then if they had other troubles like RDS syndrome, and so on, and were clinically very badly in trouble afterwards, then they were likely to be in our group of damaged people by ABR.

EGGERMONT: I would like to join the few comments about wave VI. There really is a wave VI. About 60% of wave VI, in a normal click response, comes from the basal portion of the cochlea and is actually wave V. If it shifted a little bit it might in fact reduce the whole thing because then you do not have a sharp peak anymore but just a broad thing filling up that negative area of the wave V.

I would ask everybody to be very careful about any statement about wave VI in just a normal click response because you do not know what you looking at, definitely not.

The next thing is about rarefaction and condensation phases in the narrow band responses. We have studied rarefaction and condensation phases and their effect. We should expect especially at the low central frequencies about a half period shift in latency for the rarefaction click and the condensation click because only one phase excites the nerve fibers. Now you do not find it, not in wave I, and not in wave V. The only thing that happens is that there are unexplained amplitude variations across the narrow bands. And these are more or less consistent in normal subjects. We do not find too much change that might be due to some unexplainable latency difference between I and V because I–V for rarefaction and I–V for condensation in the narrow bands are always the same. So it is definitely not. It must be some interacting effect due to different contributions from different parts of the cochlea. But to be conclusive about it, you would have to do about 50 normals to get out of this problem.

INFANT VISUAL DEVELOPMENT: EVOKED POTENTIAL ESTIMATES*

Samuel Sokol

Department of Ophthalmology
New England Medical Center
and Tufts University School of Medicine
Boston, Massachusetts 02111

INTRODUCTION

The visual evoked potential (VEP) can be elicited by either a flashing light or a pattern stimulus consisting of gratings or checks. A pattern stimulus is particularly useful in the study of the human visual system because it generates electrical signals that reflect, in part, the activity of the mechanisms that underlie the processing of pattern information from photoreceptors to visual cortex. For this reason the pattern VEP is useful as a noninvasive, objective measure of the maturing infant visual system. Two parameters of the pattern VEP can be analyzed: amplitude and peak latency (or phase). The purpose of this presentation is to discuss how each of these parameters reflects the maturation of visual acuity, contrast sensitivity, spatial-temporal interactions, and visual accommodation in human infants.

AMPLITUDE

Measurements of pattern VEP amplitude indicate that visual acuity develops rapidly during the first 6 months of life. Sokol and Dobson[1] used a checkerboard pattern stimulus that was phase reversed at 12 alternations per second (alt/sec) to record VEPs from infants. They found that by 6 months of age the peak of the amplitude–check size function occurred at a check size of 15 minutes of arc, which was in agreement with that obtained from adults (FIGURE 1). While these data suggest that the similar peaks in the infant and adult check size function reflects their similar acuity level, the peak of the check size function is not a direct estimate of infant (or adult) acuity. More directly, Sokol[2] used an extrapolation technique similar to the one described by Campbell and Maffei[3] and found that infant acuity reaches adult levels of 20/20 Snellen by 6 months. This technique consisted of extrapolating a straight line from the peak of the amplitude–check size function to 0 μV. FIGURE 2 shows how this technique is used to estimate the acuity of two infants between the ages of 3 and 6 months. Subject RD showed a shift in the peak of the amplitude–check size function from 24 minutes of arc at 3 months to 12 minutes at 5 months. Also, extrapolation of a straight line from each peak to 0 μV shows a shift in "threshold" from 5 min (20/100) to 1.5 min (20/30). Similar results occur with subject AC. FIGURE 3 shows that acuity measured by the extrapolation technique increases linearly between 2 and 6 months of age and reaches the adult Snellen equivalent of 20/20 between 6 and 7 months. Marg *et al.*[4] used the amplitude of the pattern onset–offset VEP to estimate infant acuity. They increased the spatial frequency of a square-wave grating until the signal did not

*This work was partially supported by the National Eye Institute (Research Grant EY-00926).

0077–8923/82/0388–0514 $1.75/0 © 1982, NYAS

differ from that obtained with a defocused pattern of 1-minute stripes. Marg and his colleagues found that acuity reached 20/20 Snellen equivalent by 5 months, which was a younger age than found by Sokol.[2] The fact that the checkerboard stimuli used by Sokol have different fundamental spatial frequency characteristics than the square-wave grating used by Marg may account for this difference.[5] The fundamental spatial frequency component of a checkerboard is oriented at 45° and 135° to the check pattern and is higher than the visual angle of the individual checks (in minutes of arc) by a factor of $\sqrt{2}$. Correction for this difference results in closer agreement

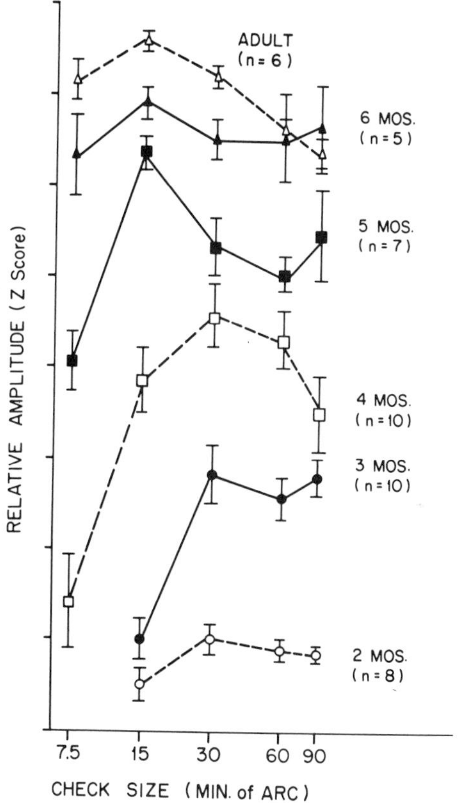

FIGURE 1. Relative amplitude of the pattern reversal VEP as a function of check size for infants and adults. Vertical lines indicate the standard error of the mean. Note the shift in the peak of the amplitude–checksize function between 2 and 6 months of age. (From Sokol.[2] By permission of *Vision Research*.)

between the two studies. Also, each study used different temporal rates of pattern presentation, which also may have influenced the results (see SPATIAL–TEMPORAL INTERACTION below). Harris *et al.*[6] recorded pattern VEPs from one 6-month-old infant and extrapolated VEP amplitude to the 100% contrast level. The cut-off occurred at 30 cycles/degree (c/deg), or a Snellen equivalent of 20/20. Finally,

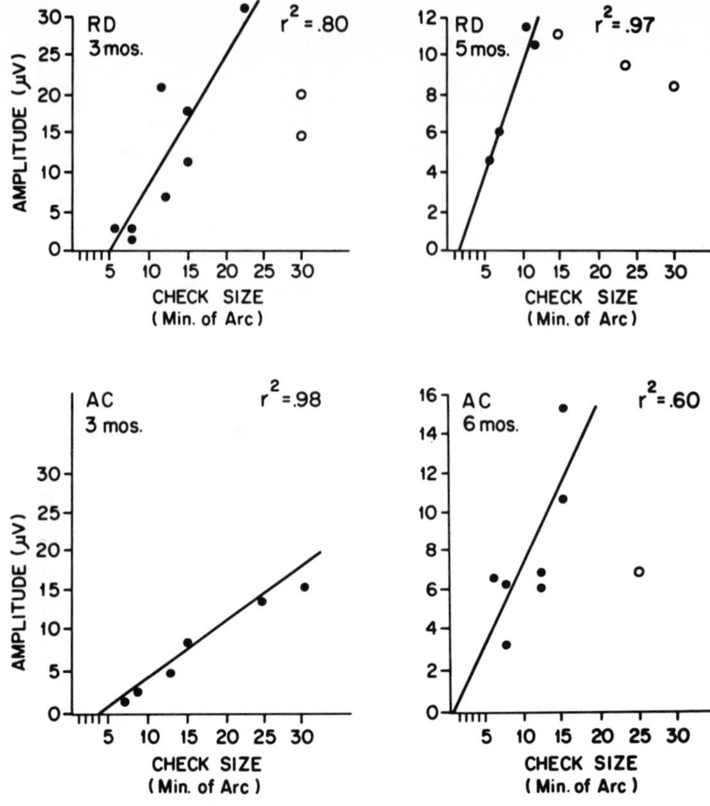

FIGURE 2. Amplitude of the checkerboard pattern VEP (microvolts) as a function of check size for 2 infants between 3 and 6 months of age. Using the method of least squares, regression lines were fit to the data points shown by the closed circles; data shown by the open circles were not included in the regression analysis. Note how the *x*-axis intercept shifts toward smaller checks as each infant gets older. (From Sokol.[2] By permission of *Vision Research*.)

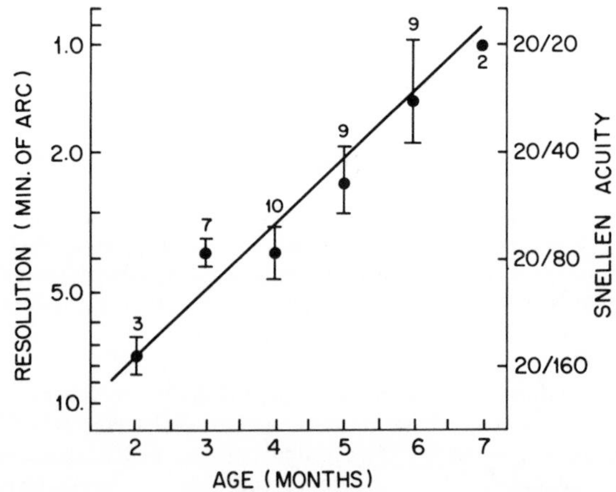

FIGURE 3. Acuity estimates of infants between 2 and 7 months of age derived from the extrapolation method shown in FIGURE 2. Vertical lines indicate ± 1 standard error of the mean; number of subjects is also shown. (From Sokol.[2] By permission of *Vision Research*.)

Pirchio *et al.*[7] also extrapolated the VEP amplitude to 100% contrast in infants between 2 and 10 months of age. They found that acuity improved linearly from approximately 20/200 at 2 months to 20/30 by 6 months and then remained stable between 6 and 10 months.

In addition to acuity estimates, contrast sensitivity functions for infants have also been derived from the amplitude of steady-state pattern VEPs.[6–8] The peak of the contrast sensitivity function shifts from around 1 c/deg at 2 months to adult values of 3–4 c/deg by 6 months. The maximum contrast sensitivity also increases progressively and reaches adult values by 12 months of age. Harris *et al.*[6] compared a VEP-derived contrast sensitivity function with a psychophysically determined curve from one 6-month-old infant. While they found that this particular infant was less sensitive than an adult at spatial frequencies higher than 2 c/deg, comparison of the VEP curve with the behaviorally determined contrast sensitivity curve from the same infant showed marked agreement between the two methods.

LATENCY

There are some advantages in using the latency of the pattern VEP rather than amplitude for both adults and infants. For example, the primary clinical value of the adult pattern VEP is that the variance of the absolute latency (in msec) is small, both within and, more importantly, between age-matched subjects. Therefore, fairly reliable decisions regarding latency normality can be made between age-matched normals and patients. The absolute amplitude (in microvolts), on the other hand, is of minimal value for comparisons between subjects, since one subject with entirely normal visual function may elicit a 15 μV VEP while another subject with equally normal visual function may elicit a signal as small as 5 μV (while both may have identical peak latencies). Amplitude comparisons between eyes of an individual can still be made since the absolute interocular difference in visually normal subjects is small; this is particularly useful in the detection of amblyopia.

In adults, the latency of the VEP elicited by checkerboard patterns and sinusoidal gratings varies as a function of check size and spatial frequency.[9–11] As check size decreases, or as spatial frequency increases, the peak latency of the VEP increases. FIGURE 4 shows that this effect also occurs with infants.[12] At 4–5 weeks of age, the latency of the first major positive component (P_1) for 30-minute checks is nearly 65 msec longer than the latency of P_1 obtained with 240-minute checks. As infants grow older there is a shift of the latency–check size function downward along the ordinate (shorter latency) and leftward along the absissa (smaller checks). For example, at 35–39 weeks, the P_1 latency difference between 30- and 240-minute checks is considerably smaller (12 msec) compared to that at 4–5 weeks (65 msec). However, at 35–39 weeks there is now a large difference (30 msec) between 7.5- and 60-minute checks.

FIGURE 5 presents the data of FIGURE 4 in a different format showing the rate of change in peak latency as a function of age for six different check sizes. By 40 weeks of age the mean P_1 latency for checks 30 minutes of arc and larger is equivalent to adult values. However, infant P_1 latency for small (7.5- and 15-minute) checks is still longer than adult latencies, showing a mean difference of nearly 40 msec for 15-minute checks. The peak latency of the pattern reversal VEP for checks 15 minutes and smaller continues to decrease up to 3 years of age. Spekreijse[13] has found that the latency of onset–offset VEPs obtained with 9-minute checks does not reach adult levels until 10 years of age.

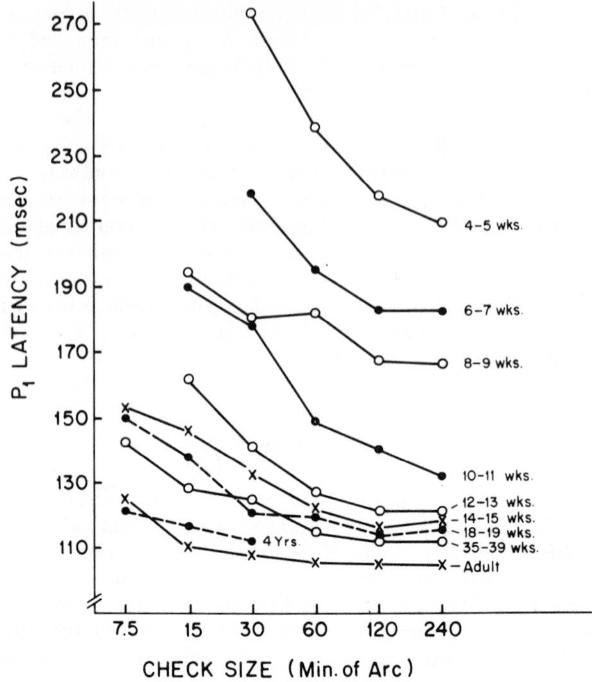

FIGURE 4. Mean P_1 latency (msec) as a function of check size for infants, 4-year-old children and adults. (From Sokol & Jones.[12] By permission of *Vision Research*.)

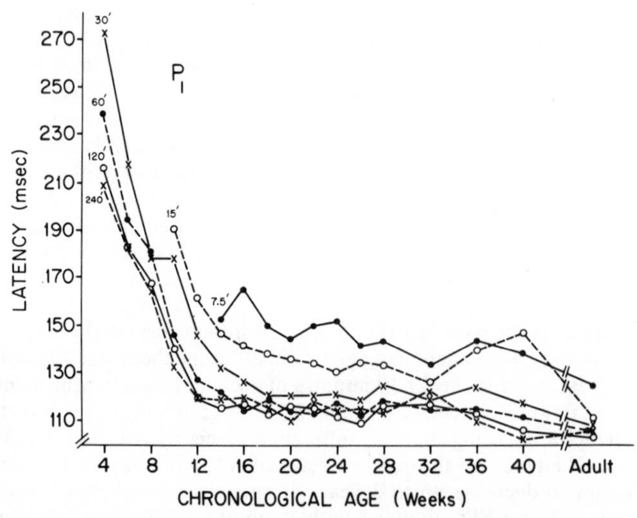

FIGURE 5. P_1 latency as a function of chronological age for check sizes ranging from 7.5 to 240 minutes of arc. (From Sokol & Jones.[12] By permission of *Vision Research*.)

Spatial-Temporal Interactions

Regan[14-16] has shown in adults that there is an interaction between check size and pattern reversal rate. For small checks (<15 minutes) the largest VEP amplitude is found at reversal rates of 5–7 alt/sec (2.5–3.5 Hz), while VEPs to large checks show a bimodal function, with one peak at 5–7 alt/sec and a second peak at 10 alt/sec. Regan has hypothesized that VEPs obtained with large checks reflect a mixture of contrast- and luminance-specific activity and that the second peak at 10 alt/sec is not unlike the large peak obtained at 10 flashes/sec when unpatterned flickering stimuli are used. This interaction of check size and alternation rate has important implications in the design and interpretation of pattern VEP experiments in infants since the choice of a particular alternation rate may lead to conclusions that are not necessarily valid for the visual system as a whole.

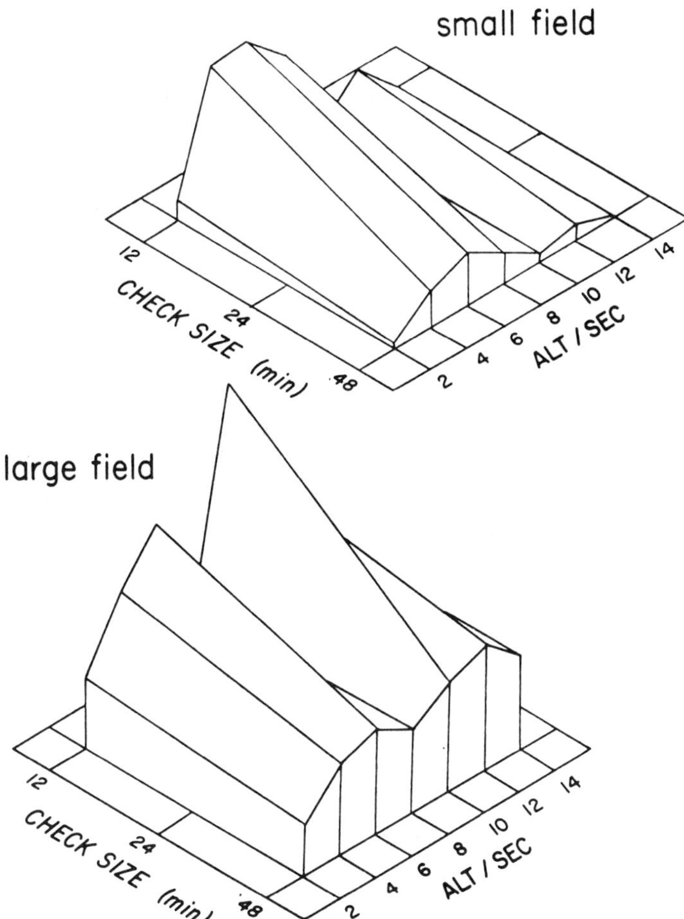

FIGURE 6. Three-dimensional surface representation of the effect of the interaction of alternation rate (x axis) and check size (y axis) on the amplitude (z axis) of the pattern reversal VEP in an adult. Amplitude is expressed in relative (z-score) units.[17]

In order to determine if infants show an interaction between check size and alternation rate, pattern VEPs were recorded from adults and from infants between 7 and 26 weeks for different combinations of check size and alternation rate.[17] FIGURE 6 shows a 3-dimensional surface representation of the effect of check size and alternation rate on VEP amplitude in an adult for two different field sizes. The large field subtended 17° by 22° and was the same field size used to test infants; the small field subtended 6° by 6°. These results confirm Regan's earlier findings. For the small field condition, 12-minute checks elicit the largest amplitude VEPs at 5–7 alt/sec. The results obtained with the large field showed a sharper resonance-like peak for 12 minute checks at 6 altsec and a large increase in the peaks at 10–12 alt/sec for both 12- and 48-minute checks.

FIGURE 7 shows data obtained from four infants in four age groups for the large field condition. At 7–10 weeks, VEPs were obtainable only for 48 minute checks and the function showed a peak at 4 alt/sec. At 11–14 weeks, records significantly different from noise were obtained for 24-minute as well as 48-minute checks; the temporal peak for both check sizes has now shifted toward faster alternation rates (8–10 sec), although there still is a "shoulder" at 4 alt/sec for 48-minute checks. At 15–18 weeks, VEPs were recorded for small (12 minutes) as well as large checks. The response surface at this age shows a further shift toward 10 alt/sec for large (48- and 24-minute) checks; the region encompassed by 12-minute checks is bimodal with peaks at 4 and 12 alt/sec. At 19–22 weeks, the region occupied by 12 minute checks is now distinctly bimodal with sharp peaks at 4 and 12 alt/sec, while the temporal characteristics of the surface representing 48-minute checks has only one peak at 10 alt/sec. The function obtained with 12 minutes after 15 weeks is most likely due to field size (FIGURE 6).

These results demonstrate that, as in adults, there is a spatial–temporal interaction of pattern reversal VEPs in infants. The interaction found in infants is not, however, identical to the adult interaction obtained under the same stimulus conditions. For example, if one compares the amplitude response surface obtained from 19–22-week-old infants (FIGURE 7) with the data from adults (FIGURE 6, large field), infants have a lower contrast-specific temporal peak (4 alt/sec) than adults (6 alt/sec). This suggests that pattern VEP studies of infants 12 weeks or older (which is when recordable signals are first obtained to checks smaller than 20 minutes) should be carried out at rates between 3 and 5 alt/sec. Therefore, we now routinely use 4 alt/sec in our evoked potential studies of infant vision.

LATENCY MEASUREMENTS OF INFANT ACCOMMODATION

We have recently used VEP latency to measure the amplitude of visual accommodation in infants. The mechanisms of visual accommodation can be activated in the human eye either by moving a target closer to the subject's eyes or by placing minus (concave) lenses of increasing power in front of the subject's eyes while keeping the target distance fixed.[18] When a minus lens is used, the subject overcomes the increase in minus power by increasing accommodation. This is accomplished by introducing an equivalent amount of plus power from the subject's own lens. When the total amplitude of the subject's accommodation has been reached and there is no longer sufficient plus power to neutralize the minus lens, the target will appear blurred. If, instead of a printed target, a reversing checkerboard pattern stimulus is presented and VEPs are recorded, the peak latency of the first major positive component (P_1) remains constant as long as the subject accommodates and then increases when the checks become blurred.[11] FIGURE 8 shows changes in P_1 latency obtained both when a

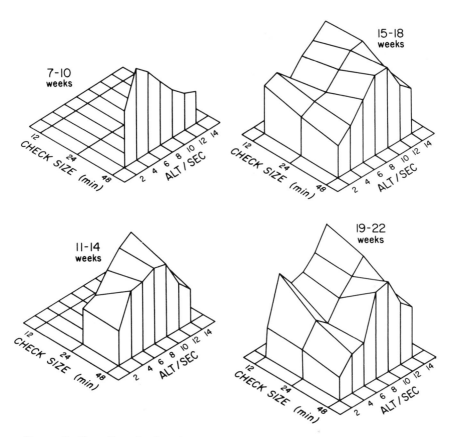

FIGURE 7. Three-dimensional surface representation of the effect of the interaction of alternation rate (x axis) and check size (y axis) on the amplitude (z axis) of the pattern reversal VEP in infants from 7 to 22 weeks of age. Amplitude is expressed in relative (z-score) units.[17]

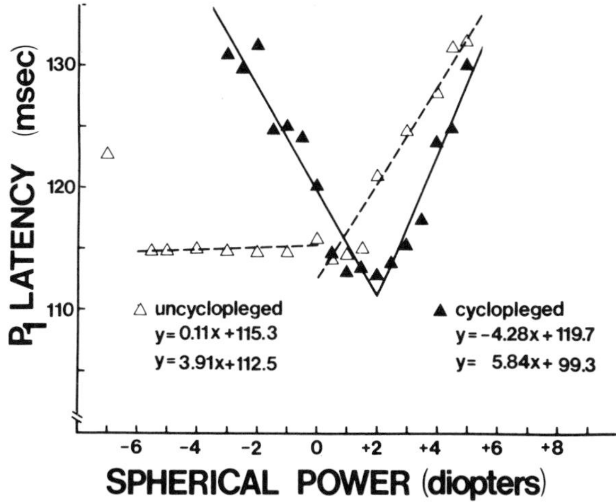

FIGURE 8. Increase in P_1 latency (msec) as a function of positive and negative spherical power in the freely accommodating eye (open triangles) and in the eye of the same subject with accommodation paralyzed (closed triangles). (From Sokol & Moskowitz.[11])

subject was able to accommodate and when accommodation was paralyzed. As long as the subject was capable of accommodating to increasing amounts of minus lens, there was no change in P_1 latency. However, when 8 diopters of minus sphere were used, the latency increased by nearly 10 msec. This increase correlated with the subject's clinically measured amplitude of accommodation. After cycloplegia, latency increased immediately and progressively as minus spheres were added. Positive lenses caused similar increases in latency under both cyclopleged and uncyclopleged conditions.

FIGURE 9 shows the correlation between a subjective report of blur (open circles) and a P_1 latency increase (closed circles) in two adults. Subject PH, a presbyope, had 1.75 diopters of accommodation and reported that the checks were blurred when minus 2 diopters (or more) of sphere were introduced. At the same time, P_1 latency increased beyond the 95% confidence limits (which were established by replicating 10 blocks of 64 sweeps of pattern VEPs with no lenses). Subject AM, who had 6.5 diopters of accommodation, reported total blur at 6 diopters; a significant increase in

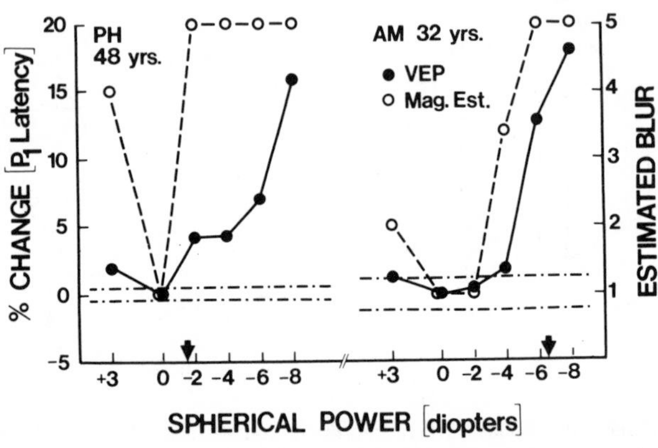

FIGURE 9. Percent increase in P_1 latency for 15-minute checks (closed circles) and subjective estimate of blur (open circles) as a function of spherical power. The percent increase is relative to the latency measured without a lens (O). The estimate of blur was based on a magnitude scale of 1 to 5: 1 = checks clear; 2 = checks clear, with initial transient blur; 3 = checks clear after some accommodative effort; 4 = checks clear with considerable accommodative effort; 5 = checks always blurred.

latency (beyond the 95% confidence limits) occurred at the same spherical power. On the basis of this correlation in adults, we have recently recorded VEPs from infants in order to determine their accuracy and total amplitude of accommodation.

FIGURE 10 shows VEP waveforms obtained with a series of minus lenses from 2 infants and one adult using 30-minute checks reversing at 3.75 alt/sec. Subject AP, an adult with 9 diopters of accommodation, showed no change in peak latency nor reported any subjective experience of blur when 8 diopters of minus sphere were used. The VEPs recorded from infant CC began to increase in latency at 4 diopters; at 8 diopters a recordable signal was still present but latency had increased even further. Infant NS showed a small latency increase at 4 and 6 diopters; at 8 diopters the VEP was not significantly different from a noise condition.

FIGURE 11 shows what proportion of the 11–13-week-old infants tested saw the checks clearly at varying amounts of blur. Infants whose peak latency for a given amount of blur fell within the 95% confidence limits for the no lens condition (which

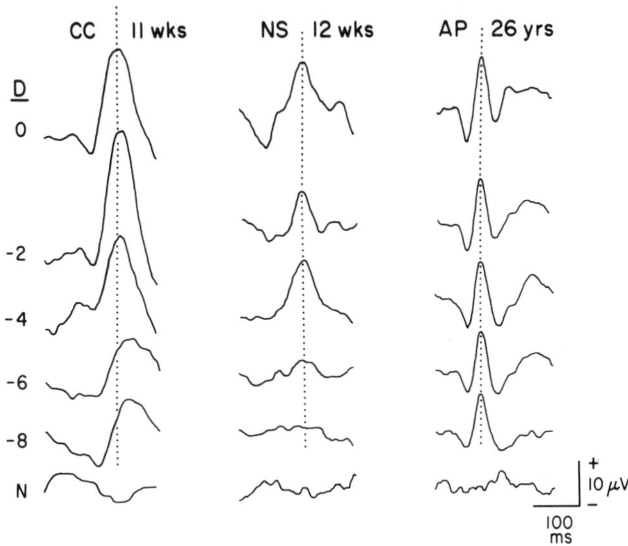

FIGURE 10. Pattern reversal binocular VEPs recorded from 2 infants and 1 adult using lenses of increasing amounts of minus sphere. 30-minute checks alternating at 3.75 reversal/sec were used to evoke cortical potentials.

was based on 10 replications of the no lens condition for 3 infants) were categorized as seeing the checks clearly. All of the infants categorized as "blurred" had P_1 peak latencies that were beyond the 95% confidence limits for the no lens condition. As shown in FIGURE 9, the upper level of the 95% confidence limits coincided with the point at which adults first reported blur. The bar graph shows that 75% of the infants tested were able to meet an accommodative demand of −3.33 D [a −2.00 lens at a distance of 0.75 m (−1.33 D)]. When −5.33 and −7.33 diopters of accommodation were required, the proportion of infants who were able to maintain "clear" vision dropped to 50%. Less than 20% were able to accommodate to 9.33 diopters (a value

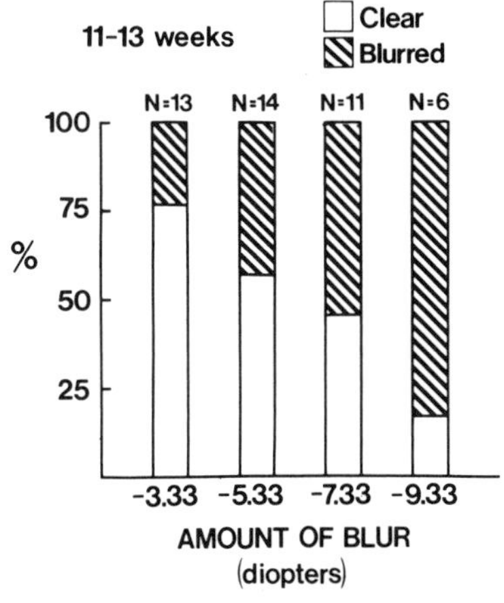

FIGURE 11. Bar graph indicating the percentage of infants who were able to accommodate to blur produced by minus lenses.

not difficult to maintain for a young emmetropic adult). These data suggest that infants either cannot, or will not attempt to, induce significant amounts of accommodation. This is probably of no consequence to an emmetropic infant. However, infants with a significant amount of hyperopia may have difficulty in accommodating since they have to overcome their hyperopia in addition to accommodating for a particular distance.

CONCLUSION

The development of the human visual system as measured by the pattern VEP progresses rapidly during the first year of life and then slowly thereafter. Measurements of the amplitude of both the transient and steady-state pattern VEP indicate that visual acuity and contrast sensitivity reach the "adult" equivalent between 5 and 12 months of age[1,2,4,6,7] On the other hand, measurement of the peak latency of transient pattern evoked cortical potentials and the spatial–temporal interactions of the pattern VEP show changes in the visual system up to at least 2–3 years of age[12,17] and in some cases as long as 10 years.[13] These gradual changes in the temporal nature of the VEP could relate directly to specific aspects of the developing visual system or they may be related to a more general, nonsensory maturation of the central nervous system. There is anatomical evidence that suggests that the slow latency change is modality specific. For example, Hickey[19] found a differential rate of growth in the cells of the human lateral geniculate nucleus, with cells in layers 3–6 (parvocellular) developing faster than cells in layers 1 and 2 (magnocellular). It is not until at least two years of age that all cells have reached adult levels. Further, Magoon and Robb[20] have found that myelin density along the optic nerve and optic tract continues to increase dramatically during the first two postnatal years, which contradicts many of the earlier studies which claimed that myelination was complete by 6 months of age. Finally, behavioral studies of children with a history of strabismus have shown that the critical period for the development of binocularity continues until 2–3 years of age.[21,22] Thus, the electrophysiological, anatomical, and psychophysical data from infants and young children indicate that the critical period for maturation of the visual system continues until at least 3 years of age.

ACKNOWLEDGMENT

I thank Dr. V. Leo Towle for his helpful comments and suggestions, particularly with regard to FIGURES 6 and 7.

REFERENCES

1. SOKOL, S. & V. DOBSON. 1976. Pattern reversal visually evoked potentials in infants. Invest. Ophthalmol. **15:** 58–62.
2. SOKOL, S. 1978. Measurement of infant visual acuity from pattern reversal evoked potentials. Vision Res. **18:** 33–39.
3. CAMPBELL, R. W. & L. MAFFEI. 1970. Electrophysiological evidence for the existence of orientation and size detectors in the human visual system. J. Physiol. **207:** 635–652.
4. MARG, E., D. N. FREEMAN, P. PELTZMAN & P. J. GOLDSTEIN. 1976. Visual acuity development in human infants: evoked potential measurements. Invest. Ophthalmol. **15:** 150–153.

5. KELLY, D. H. 1976. Pattern detection and the two-dimensional Fourier transformation: Flickering checkerboards and chromatic mechanisms. Vision Res. **16:** 277–287.
6. HARRIS, L., J. ATKINSON & O. BRADDICK. 1976. Visual contrast sensitivity of a 6-month-old infant measured by the evoked potential. Nature (Lond.) **264:** 570–571.
7. PIRCHIO, M., D. SPINELLI, A. FIORENTINI & L. MAFFEI. 1978. Infant contrast sensitivity evaluated by evoked potentials. Brain Res. **141:** 179–184.
8. ATKINSON, J., O. BRADDICK & J. FRENCH. 1979. Contrast sensitivity of the human neonate measured by the visual evoked potential. Invest. Ophthalmol. Visual Sci. **18:** 210–213.
9. PARKER, D. M. & E. A. SALZEN. 1977. Latency changes in the human visual evoked response to sinusoidal gratings. Vision Res. **17:** 1201–1204.
10. JONES, R. & M. J. KECK. 1978. Visual evoked response as a function of grating spatial frequency. Invest. Ophthalmol. **17:** 652–659.
11. SOKOL, S. & A. MOSKOWITZ. (In press.) Effect of retinal blur on the peak latency of the pattern evoked potential.
12. SOKOL, S. & K. JONES. 1979. Implicit time of pattern evoked potentials in infants: an index of maturation of spatial vision. Vision Res. **19:** 747–755.
13. SPEKREIJSE, H. 1978. Maturation of contrast EPs and development of visual resolution. Arch. Ital. Biol. **116:** 358–369.
14. REGAN, D. 1973. Evoked potentials specific to spatial patterns of luminance and color. Vision Res. **13:** 2381–2402.
15. REGAN, D. & W. RICHARDS. 1973. Brightness contrast and evoked potentials. J. Opt. Soc. Am. **63:** 606–611.
16. REGAN, D. 1978. Assessment of visual acuity by evoked potential recording: Ambiguity caused by temporal dependence of spatial frequency selectivity. Vision Res. **18:** 439–443.
17. MOSKOWITZ, A. & S. SOKOL. 1980. Spatial and temporal interaction of pattern-evoked cortical potentials in human infants. Vision Res. **20:** 699–707.
18. SLOANE, A. E. 1970. Manual of Refraction. 2nd edit. Little, Brown and Company, Boston, Mass.
19. HICKEY, T. L. 1977. Postnatal development of the human lateral geniculate nucleus: Relationship to a critical period for the visual system. Science **198:** 836–838.
20. MAGOON, E. H. & R. M. ROBB. 1981. Development of myelin in human optic nerve and tract: A light and electron microscopic study. Arch. Ophthalmol. **99:** 655–659.
21. HOHMANN, H. & O. D. CREUTZFELD. 1975. Squint and the development of binocularity in humans. Nature **254:** 613–614.
22. BANKS, M. S., R. N. ASLIN & R. D. LETSON. 1975. Sensitive period for the development of human binocular vision. Science **190:** 675–677.

SPINAL SOMATOSENSORY EVOKED POTENTIALS: MATURATIONAL AND CLINICAL STUDIES*

Joan B. Cracco and Roger Q. Cracco

Department of Neurology
State University of New York
Downstate Medical Center
Brooklyn, New York 11203

Somatosensory evoked potentials which arise in the cauda equina and in spinal cord afferent pathways have been recorded in man from surface electrodes attached to the skin over the spine.[1,2] They have also been recorded from spinal epidural and intrathecal electrodes.[3–7] The advantage of epidural and intrathecal recordings is that the responses are considerably larger and either no (intrathecal) or few (epidural) responses need to be averaged. The obvious disadvantage is the invasiveness of these procedures. Because the technique is noninvasive, skin or surface recordings would be preferred. Unfortunately, potentials obtained by this method may be small and difficult to record, particularly over rostral cord segments in adults. Nevertheless, surface recordings of these potentials have provided a noninvasive method for studying human spinal cord maturation.[8,9] They have also provided a method for evaluating spinal cord function in patients with localized spinal cord lesions and in patients with diffuse diseases involving the nervous system.[10–12] Recently spinal potentials obtained from electrodes placed in spinous processes or interspinous ligaments have been combined with scalp recorded somatosensory evoked potentials in monitoring spinal cord function during surgical procedures on the spine.[13,14]

The following discussion deals primarily with evoked potentials to lower limb stimulation recorded from surface electrodes placed over the spine of normal adults, infants, and children and children with diseases affecting the spinal cord and cauda equina.

METHODS

Recording disc electrodes were attached to the skin over the spine. Because of the small signal size low and stable electrode impedances were maintained during recording sessions. Bipolar or reference recordings were obtained from various spinal locations. Reference recordings often yielded larger responses; however, they were often considerably noisier than bipolar recordings.[8,15] Since recordings were often obscured by myogenic activity, this was minimized by recording when the subject was drowsy or sleeping. The frequency responses of the recording apparatus was about 10–3000 cps. Analysis times were 20–40 msec, and 1000–4000 responses were averaged. If potentials were being recorded only over the cauda equina or caudal spinal cord, fewer responses had to be averaged. These potentials are larger than those recorded over rostral cord segments and technically satisfactory recordings have been obtained by averaging fewer than 100 responses.[16]

Either the peroneal nerve in the popliteal space or the tibial nerve at the ankle was

*This work was supported by a research grant from the National Foundation and by Research Grant NS12039 from the National Institutes of Health, U.S. Public Health Service.

0077–8923/82/0388–0526 $1.75/0 © 1982, NYAS

stimulated at an intensity that produced a visible muscle twitch. Stimulation of purely sensory nerves, such as the saphenous or the sural nerves was not utilized since this yields only small potentials over the cauda equina and caudal spinal cord.[17,18] The stimulus may be triggered by the EKG to reduce EKG artifact. However, faster rates of stimulation (7–10/sec) were used in these studies with little interference from EKG artifact.

Simultaneous stimulation of multiple nerves, such as both peroneal or both tibial nerves, increased the signal size[10,19] (FIGURE 1). This is important since the major problem with this method is the small size of these potentials, particularly over the rostral spinal cord in adults. Raising stimulus intensity also increases the amplitude of the response, but this may not be feasible because of discomfort to the subject.[17,18]

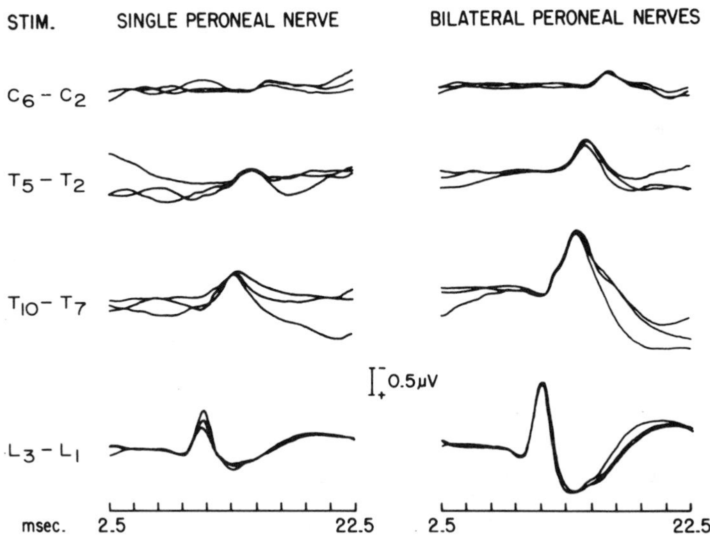

FIGURE 1. Comparison of spinal potentials evoked by stimulation of single and bilateral peroneal nerves. In this and all subsequent figures, designated electrode placement refers to spinous process level rather than spinal cord segment. Two or three recordings are superimposed in each trace. Relative negativity in grid I (caudal electrode) results in an upward deflection. With bilateral stimulation potentials are greater in amplitude and better defined at all recording locations than they are with single nerve stimulation.

DISCUSSION

These potentials progressively increased in latency from lumbar to cervical recording locations (FIGURE 1). Similar potentials have been recorded from epidural and intrathecal electrodes in man.[3–7] In bipolar surface leads over the lumbar spine, the response consisted of initially positive triphasic potentials. This is seen when recording an impulse traversing a nerve trunk in volume and would be consistent with potentials arising in the roots of the cauda equina. In reference recording over the cauda equina, this initial triphasic potential was often followed by a second negative potential or inflection (FIGURE 2). This second potential is thought to arise in post-synaptic ventral root fibers or to reflect a volume conducted response arising from the caudal spinal cord.[20–22]

In bipolar leads over the lower thoracic spine (overlying caudal spinal cord segments) the response was greater in amplitude and duration and usually more complex in configuration than in more rostral or caudal leads. In infants and young children a large positive negative diphasic potential followed by a broad negative and then, at times, a positive potential was recorded over this area (FIGURE 3).[8,9] Investigations of similar potentials recorded over caudal cord segments in animals suggest that the initial diphasic potential arises in the intramedullary continuations of dorsal root fibers and the subsequent potentials reflect synaptic and postsynaptic activity, which is associated with local reflex mechanisms rather than with propagation of the response to rostral cord levels.[20,23] In older subjects less complex potentials were recorded over this area (FIGURE 3). The configuration of the response over the cauda equina and caudal cord segments was different in bipolar and reference recordings (FIGURE 2).

Over the rostral cord the response in both children and adults, recorded in both

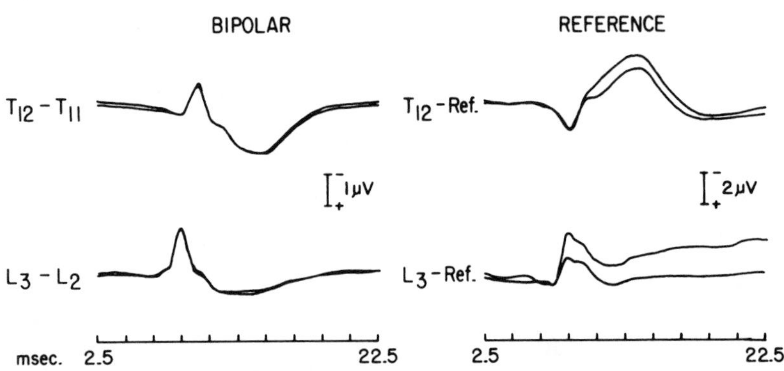

FIGURE 2. Comparison of close bipolar with iliac crest reference recordings of the response obtained over the cauda equina and caudal spinal cord in a normal child. Although the onset latency of potentials (at L_3 and T_{12}) are similar, the response configuration is considerably different in bipolar as compared with reference recordings. The second negative component recorded over L_3 in the reference recording is not apparent in the L_3–L_2 bipolar recording. The response recorded over T_{12} in reference recordings is considerably broader than that recorded in the T_{12}–T_{11} bipolar recording.

bipolar and reference leads, consisted of small initially positive triphasic potentials with poorly defined positive phases, which progressively decreased in amplitude rostrally (FIGURE 3).[2,8,9] This amplitude decrement probably reflects temporal dispersion of impulses and the greater distance between the skin recording electrode and the spinal cord at rostral recording sites. Studies in animals suggest that these potentials arise in multiple rapidly conducting afferent pathways, including the dorsolateral columns, which lie primarily ipsilateral to the stimulated nerve.[17,20,23] These animal studies also suggest that the peripheral nerve fibers that mediate the potentials recorded at all spinal levels are primarily muscle nerve rather than cutaneous nerve afferent fibers.[17]

These responses were generally larger and more complex in infants than in older subjects.[8] In both animals and man, the amplitude, duration, and waveform complexity of the spinal response increased from surface (skin) to depth (dura) (FIGURE 4).[10,17] Therefore, the greater amplitude and complexity of the response in the young child

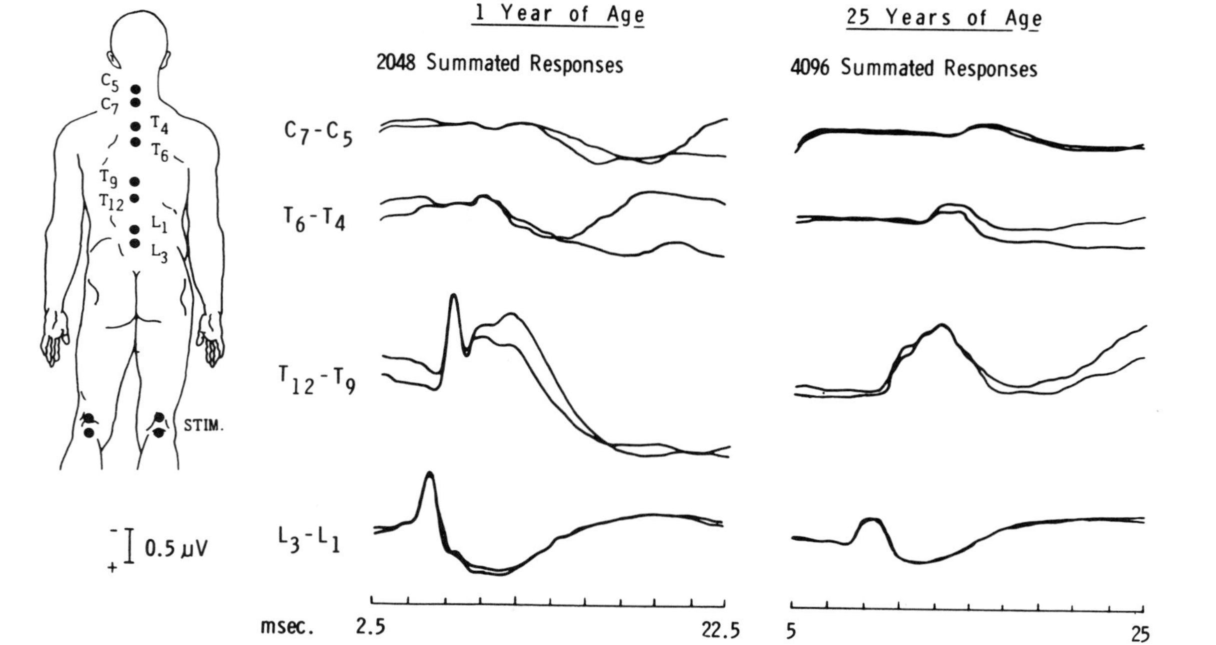

FIGURE 3. Comparison of bipolar recordings of the spinal response to peroneal nerve stimulation in a 1-year-old infant and an adult. Over the cauda equina (L_3 spine) the response in both the infant and the adult consists of triphasic potentials with poorly defined initial positive phases. In the infant, the response over caudal spinal cord (T_{12} spine) consists of a positive negative diphasic potential followed by a broad negative positive diphasic potential. In the adult it consists of a broad negative potential with 2 or 3 inflections. The response over rostral spinal cord in both the infant and adult consists of small initially positive triphasic potentials with poorly defined positive phases. (From Cracco *et al.*[9] By permission of *Electroencephalography and Clinical Neurophysiology*.)

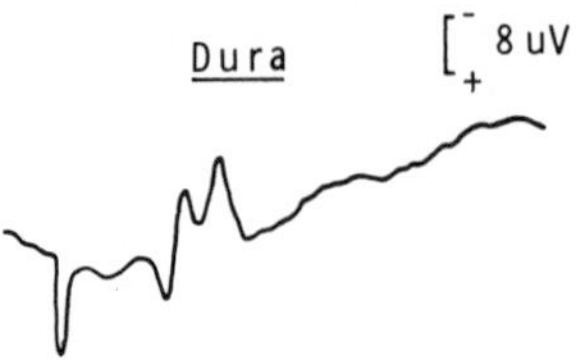

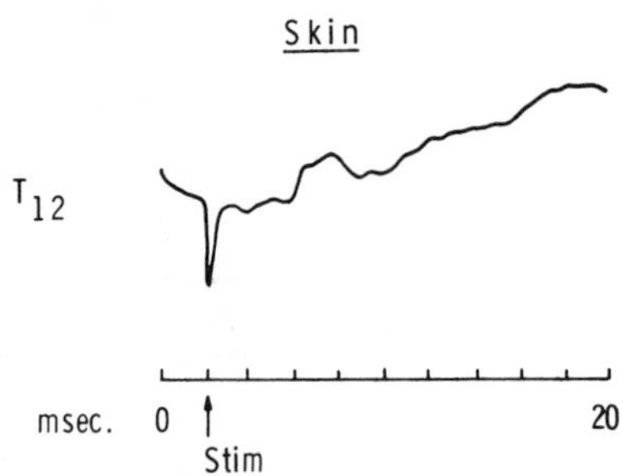

FIGURE 4. Comparison of the spinal response to sciatic nerve stimulation in a cat taken from skin and dura over the lower thoracic region. The response increases in amplitude and waveform complexity from surface to depth.

may reflect, in part, the shorter distance between the spinal cord and skin recording electrodes.

The onset of the negative potential at each recording site was used to determine conduction velocity of the fastest fibers contributing to the response. However, it should be noted that small nonpropagated potentials that precede the propagated spinal potential have been observed in animals over the thoracic cord (FIGURE 5).[20,23] Therefore, in the calculation of conduction velocities it is important to differentiate propagated from nonpropagated events.

The conduction velocity up the spine was nonlinear; it was slower over caudal spinal cord segments than over the cauda equina or rostral spinal cord.[2,9] This decrease in the speed of conduction over caudal cord segments probably reflects branching of dorsal root fibers and synaptic activity since this is the region where these fibers undergo synaptic contact in Clark's column and other nuclei.

In adults the mean overall conduction velocity of the responses from mid-lumbar to lower cervical recording locations (L_3 to C_7 spines) was about 70 m/sec. Segmental conduction velocities were about 65 m/sec over peripheral nerve and cauda equina (point of peripheral nerve stimulation to L_3 spine), 50 m/sec over caudal spinal cord (T_{12} to T_6 spines), and 85 m/sec over rostral spinal cord (T_6–T_7 spines).[9] Jones and Small found similar overall conduction velocities of surface recorded spinal potentials of about 65 m/sec in the adult.[15] Intrathecal recordings of spinal potentials have disclosed considerably slower conduction velocities (about 37 m/sec) between the lumbar and cervical spinal enlargements.[6] The reason for these differences in conduction velocities is uncertain. It may be related to the fact that these different techniques record potentials that arise in different spinal afferent tracts. In animals it has been noted that onset latency of the response recorded from surface leads over rostral thoracic segments is usually several tenths of a millisecond less than it is in dural recordings.[17,23,24] However, this alone would account for only slight differences in conduction velocity.

Both overall and segmental conduction velocities increased with maturation. In the newborn infant they were about half of the adult values. Peripheral conduction velocities were within the adult range by 3 years, whereas velocities over the spinal cord did not reach adult values until the fifth year (FIGURE 6).[9] This suggests that maturation of rapidly conducting spinal afferent pathways proceeds at a slower rate than maturation of rapidly conducting peripheral sensory fibers. Similar maturational findings were obtained by Desmedt[25] who used the scalp-recorded evoked response to median nerve stimulation. Conduction velocities in median nerve reached adult values between 12 and 18 months of age, whereas conduction velocity within central lemniscal pathways did not reach adult values until 5–7 years of age. This increase in the speed of conduction that accompanies maturation in peripheral nerve and spinal cord is probably related to the increasing fiber diameter and progressive myelination that accompanies maturation. However, explanations for the differential rate in maturation of peripheral and central afferent pathways can only remain speculative at this time. Precisely how such factors as increasing fiber diameter, synaptogenesis, and myelinogenesis contribute to these maturational changes is uncertain.

In a study of a group of infants and children with myelodysplasia on whom spinal and cerebral evoked potentials were recorded, there was often good correlation between the evoked potential abnormalities and the clinical status of the patients.[26,27] In a few patients with either myelomeningocele or occult dysraphism it was possible to diagnose caudal displacement of the spinal cord; in these patients the large complex spinal potential which is normally recorded over the lower thoracic spines was recorded over lumbar spinous process (FIGURE 7). In some children with myelomeningocele a positive potential was recorded in leads immediately rostral to the lesion. This potential progressively decreased in amplitude but did not change in latency rostrally (FIGURE 8). This is a nonpropagated volume-conducted potential that is consistent

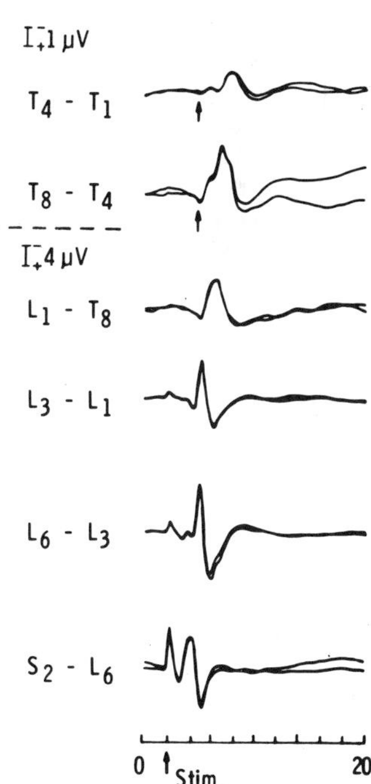

FIGURE 5. Bipolar surface recordings of sciatic nerve evoked potentials in a monkey. Nonpropagated positive–negative potentials that decrease in amplitude but do not change in latency rostrally are recorded from thoracic leads (arrows). (From Feldman *et al.*[23] By permission of the *Annals of Neurology*.)

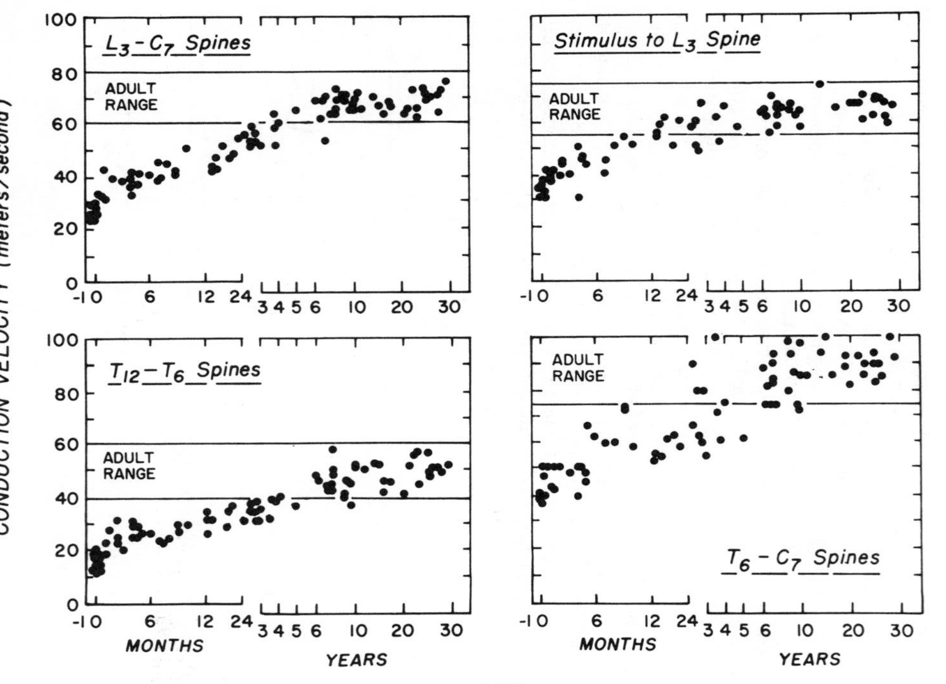

FIGURE 6. Relationship between age and overall conduction velocity (L_3–C_7 spines) and segmental conduction velocities over peroneal nerve and cauda equina (stimulus to L_3 spine), caudal spinal cord (T_{12}–T_6 spines), and rostral spinal cord (T_6–C_7 spines). Note the change in age scale (from months to years) on the abscissa. Infants from − 1 to 0 months were premature. All conduction velocities progressively increase with age. The overall velocities are in the adult range by 5 years of age. Velocities over peroneal nerve and cauda equina increase rapidly during the first year of life and most values are in the adult range by 3 years. However, velocities over both caudal and rostral spinal cord are not in the adult range until 5–6 years of age (From Cracco *et al.*[9] By permission of *Electroencephalography and Clinical Neurophysiology*.)

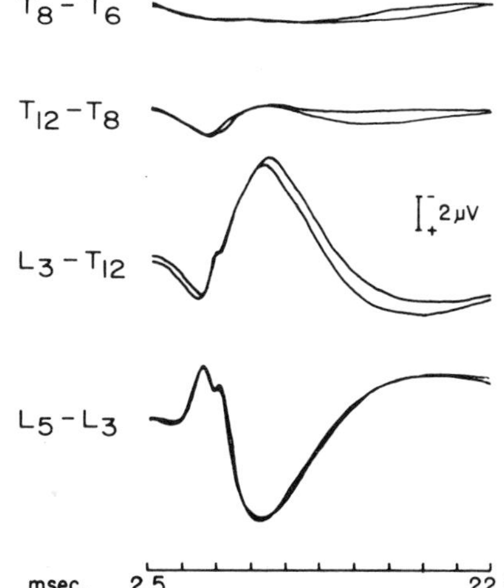

FIGURE 7. Spinal responses in a 3-year-old child with thoracolumbar myelomeningecele. The large complex response that is recorded over T_{12} to T_9 spines in normal children is present over L_3 spine in this child, suggesting caudal displacement of the spinal cord.

with physiological transection of neural pathways. Similar positive potentials have been recorded rostral to spinal cord transections in animals (the killed end effect).[17,20,23,28]

Spinal and scalp-recorded SEPs were investigated in a group of children with degenerative diseases of the central nervous system.[11] The conduction velocity over peroneal nerve was normal. Responses were recorded over the spinal cord in most

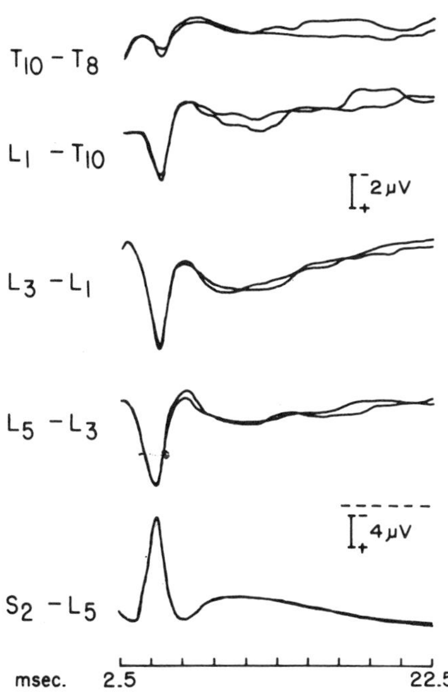

FIGURE 8. Spinal responses in a 6-year-old child with a lumbosacral myelomeningocele. A prominent predominantly negative triphasic potential is recorded over S_2. However, rostral leads yield only positive potentials that do not change in latency but decrease in amplitude rostrally.

patients but these potentials were considerably greater in duration in many of these patients than they were in age-matched controls. This suggests the occurrence of increased temporal dispersion. Conduction velocities over the cord were slowed in most of these patients, but this slowing was greater over rostral spinal cord segments (T_6–C_7 spines) than over the caudal cord (T_{12}–T_6 spines) (FIGURE 9). This selectively greater involvement of rostral cord segments may be due to a central "dying-back process." Although this term is generally used to describe peripheral nerve degeneration in which the earliest changes occur in the distal axons, this process has also been found in the central extentions of axons within the spinal cord.[29] The scalp-recorded short latency evoked potentials to median nerve stimulation that arise within and rostral to the brainstem were absent in most of these patients as were the longer-latency potentials of cerebral cortical origin. One patient with myoclonus showed enhancement of cerebral potentials. In this patient the peroneal nerve evoked negative potential recorded over the caudal spinal cord was followed by a positive potential that was much greater in amplitude and duration than that seen in normals. (FIGURE 10). A similar enhancement of this positive potential has been observed following strychnine administration in animals.[20] This strychnine-enhanced potential is thought to reflect activation of polysynaptic spinal cord segmental reflexes.[30] Therefore, in some patients with myoclonus there may be a pathological enhancement of spinal cord polysynaptic systems.

Spinal and peripheral nerve conduction velocities were also found to be slowed in some clinically asymptomatic juvenile diabetics.[12] Peripheral nerve or spinal conduction velocity alone was slowed in some patients, while in others both peripheral and spinal conduction velocities were slowed. As in the children with CNS degenerative disease, spinal conduction velocity over rostral spinal cord segments was chiefly affected.

Recording of both scalp and spinal evoked potentials can also provide information concerning the conduction characteristics of the response from cauda equina to

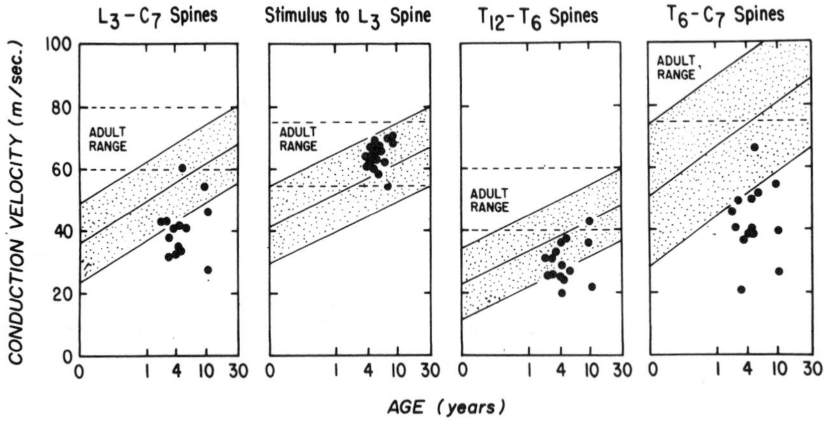

FIGURE 9. Distribution of the overall spinal conduction velocities (L_3–C_7 spines) and the segmental conduction velocities over peroneal nerve and cauda equina (stimulus to L_3 spine), caudal spinal cord (T_{12}–T_6 spines) and rostral spinal cord (T_6–C_7 spines) in the patients with degenerative disease. The shaded area represents the regression line with 95% confidence limits for these conduction velocities in 95 normal subjects. The range of conduction velocities in normal adults is indicated by the dashed lines (From Cracco *et al.*[11] By permission of *Electroencephalography and Clinical Neurophysiology.*)

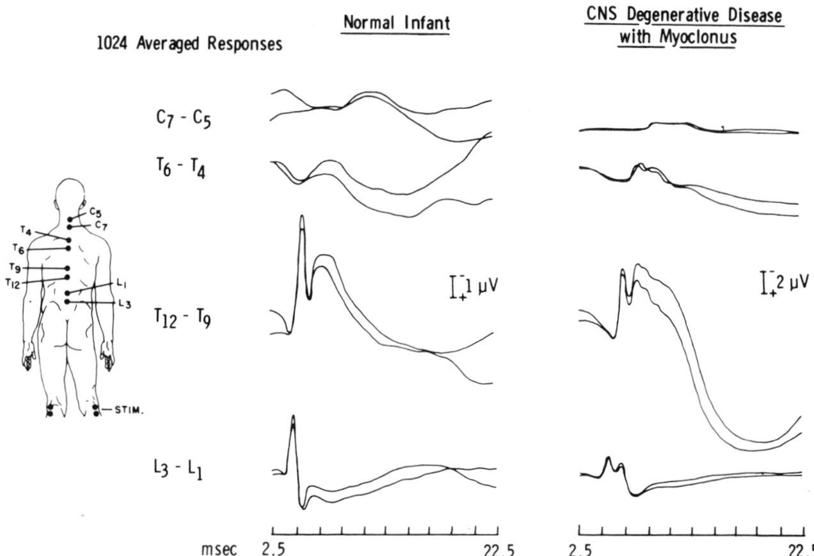

FIGURE 10. Comparison of spinal evoked potentials in a normal infant and a child with degenerative disease and myoclonus who had enhancement of longer-latency scalp-recorded SEPs. The positive potential that follows the large complex negative potential recorded over the caudal spinal cord (T_{12}–T_9 spines) is much larger in the patient. (Note the calibration difference.) (From Cracco *et al.*[11] By permission of *Electroencephalography and Clinical Neurophysiology.*)

cerebral cortex.[31] When responses over the rostral spinal cord cannot be recorded because of their small signal size, potentials recorded over more caudal spinal regions may be used as latency indicators, and the conduction characteristics from these spinal locations to the stable scalp recorded SEPs can be determined. It should be noted, however, that the generator sources of the spinal potentials may not be the same as the spinal afferent tracts that transmit the scalp-recorded SEP. Nevertheless, this method should still provide useful information in the evaluation of patients with cauda equina or spinal cord pathology.

Dorfman has devised a method by which conduction velocities within spinal cord afferent pathways may be indirectly estimated.[32] He used the scalp-recorded SEPs to stimulation of a nerve in the lower extremity and the upper extremity as latency indicators and subtracted out the estimated peripheral conduction time by determining the F-wave latencies for the stimulated peripheral nerves. Overall spinal conduction velocity (55 m/sec) as determined by this method was somewhat slower than the velocity calculated from surface recordings. He found slowing in the speed of conduction within spinal cord afferent pathways in normal elderly subjects, in some patients with multiple sclerosis, and also in some diabetics.[33–35]

SUMMARY

Somatosensory evoked potentials recorded from surface electrodes over the spine provides a noninvasive method for study of maturation of the human peripheral and central nervous system. This technique can also provide important pathophysiological information in some patients with localized spinal cord lesions or diffuse diseases of the nervous system.

ACKNOWLEDGMENT

The authors are indebted to Ms. Nancy Gillum for typing the manuscript.

REFERENCES

1. LIBERSON, W. I., M. GRATZUR, A. ZALES & B. GRABINSKI. 1966. Comparison of conduction velocity of motor and sensory fibers determined by different methods. Arch. Phys. Med. **47:** 17–23.
2. CRACCO, R. Q. 1973. Spinal evoked response: Peripheral nerve stimulation in man. Electroencephalogr. Clin. Neurophysiol. **35:** 379–386.
3. MAGLADERY, J. W., W. E. PORTER, A. M. PARK, & R. D. TEASDALL. 1951. Electrophysiological studies of nerve and reflex activity in normal man. IV. The two neuron reflex and identification of certain action potentials from spinal roots and cord. Bull. Johns Hopkins Hosp. **88:** 499–519.
4. CACCIA, M. R., E. UBCALI & L. ANDREUSSI. 1976. Spinal evoked responses recorded from the epidural space in normal and diseased humans. J. Neurol. Neurosurg. Psychiatry **39:** 962–972.
5. ERTEKIN, C. 1976. Studies in the human evoked electrospinogram. I. The origin of the segmental evoked potentials. Acta Neurol. Scand. **53:** 2–30.
6. ERTEKIN, C. 1976. Studies in the human evoked electrospinogram II. The conduction velocity along the dorsal funiculus. Acta Neurol. Scand. **5:** 21–38.
7. SHIMOJI, K., M. MATSUKI & H. SHIMIZU. 1977. Wave form characteristics and spatial distribution of evoked spinal electrogram in man. J. Neurosurg **46:** 304–310.
8. CRACCO, J. B., R. Q. CRACCO & L. J. GRAZIANI. 1975. The spinal evoked response in infants and children. Neurology **25:** 31–36.
9. CRACCO, J. B., R. Q. CRACCO & R. STOLOVE. 1979. Spinal evoked potential in man: A maturational study. Electroencephalogr. Clin. Neurophysiol. **46:** 58–64.
10. CRACCO, R. Q., J. B. CRACCO, R. SARNOWSKI & A. B. VOGEL. 1980. Spinal evoked potentials. *In* Progress in Clinical Neurophysiology, J. E. Desmedt, Ed. Vol. 7: 87–104. Karger, Basel, Switzerland.
11. CRACCO, J. B., V. V. BOSCH, & R. Q. CRACCO. 1980. Cerebral and spinal somatosensory evoked potentials in children with CNS degenerative disease. Electroencephalogr. Clin. Neurophysiol. **49:** 437–445.
12. CRACCO, J. B., S. CASTELLS & E. MARK. 1980. Conduction velocity in peripheral nerve and spinal afferent pathways in juvenile diabetics (Abstr.). Neurology **30:** 370–371.
13. LUEDERS, H., A. GURD, J. HAHN, J. ANDRISH, G. WEIKER & G. KLEM. 1981. A new technique for intraoperative monitoring of spinal cord function. Spine (in press).
14. MACCABEE, P. J., E. I. PINKHASOV, P. TSAIRIS & D. B. LEVINE. 1981. Spinal and short latency scalp derived somatosensory evoked potentials during corrective spinal column surgery (Abstr.). Electroencephalogr. Clin. Neurophysiol. (in press).
15. JONES, S. J. & D. G. SMALL. 1978. Spinal and sub-cortical evoked potentials following stimulation of the posterior tibial nerve in man. Electroencephalogr. Clin. Neurophysiol. **44:** 299–306.
16. DIMITRIJEVIC, M. R., L. E. LARSSON, D. LEHMKUHL & A. M. SHERWOOD. Evoked spinal cord and nerve root potentials in humans using a noninvasive recording technique. Electroencephalogr. Clin. Neurophysiol. **45:** 331–340.
17. SARNOWSKI, R. J., R. Q. CRACCO, H. B. VOGEL, & F. MOUNT. 1975. Spinal evoked response in the cat. J. Neurosurg. **43:** 329–336.
18. MCKAY, W. B. & G. L. GALLOWAY. 1979. Technological aspects of recording evoked

potentials from the cauda equina and lumbosacral spinal cord in man. Am. J. EEG Technol. **19:** 83–96.

19. CRACCO, R. Q., J. B. CRACCO, & B. J. ANZISKA. 1979. Somatosensory evoked potentials in man: Cerebral, subcortical, spinal and peripheral nerve potentials. Am. J. EEG. Technol. **19:** 59–81.
20. CRACCO, R. Q. & B. EVANS. 1978. Spinal evoked potential in the cat. Effects of asphyxia, strychnine, cord section and compression. Electroencephalogr. Clin. Neurophysiol. **44:** 187–201.
21. DELBEKE, J., A. J. MCCOMAS & S. J. KOPEC. 1978. Analysis of evoked lumbosacral potentials in man. J. Neurol. Neurosurg. Psychiatry **41:** 293–302.
22. PHILLIPS, L. H., II, & J. R. DAUBE. 1980. Lumbosacral spinal evoked potentials in humans. Neurology **30:** 1175–1183.
23. FELDMAN, M. H., R. Q. CRACCO, P. FARMER & F. MOUNT. 1980. Spinal evoked potential in the monkey. Ann. Neurol. **7:** 238–244.
24. HAPPEL, L. T., H. J. LEBLANC & D. G. KLINE. 1975. Spinal cord potentials evoked by peripheral nerve stimulation. Electroencephalogr. Clin. Neurophysiol. **38:** 349–354.
25. DESMEDT, J. E., P. NOEL, J. DEBECKER & J. NAMECHE. Maturation of afferent conduction velocity as studied by sensory nerve potentials and cerebral evoked potentials. *In:* New Developments in Electromyography and Clinical Neurophysiology. J. E. Desmedt, Ed. Vol. 2: 52–63. Karger, Basel, Switzerland.
26. CRACCO, J. B., R. Q. CRACCO & L. J. GRAZIANI. 1974. The spinal evoked response in infants with myelodysplasia. (Abstract). Neurology **24:** 359–360.
27. CRACCO, J. B. & R. Q. CRACCO. 1979. Somatosensory spinal and cerebral evoked potentials in children with occult spinal dysraphism (Abstr.). Neurology **29:** 543.
28. ROSSINI, P. M., F. GRECO, L. DEPALMA & L. PISANO. 1980. Electrospinogram of the rabbit. Eur. Neurol. **19:** 409–413.
29. SPENCER, P. S. & H. H. SCHAUMBERG. 1977. Ultrastructural studies of the dying back process. IV. Differential vulnerability of PNS and CNS fibers in experimental central-peripheral distal axonopathies. J. Neuropath. Exp. Neurol. **36:** 300–320.
30. BERNHARD, C. G. & W. KOLL. 1953. On the effect of strychnine asphyxia and dial on the spinal cord potentials. Acta Physiol. Scand. **29** (Suppl. 106): 30–41.
31. ROSSINI, P. M., R. Q. CRACCO, J. B. CRACCO & W. J. HOUSE. 1981. Short latency SEPs to peroneal nerve stimulation: Scalp topography and the effect of different frequency filters. Electroencephalogr. Clin. Neurophysiol. **52:** 540–552.
32. DORFMAN, L. J. 1977. Indirect estimation of spinal cord conduction velocity in man. Electroencephalogr. Clin. Neurophsiol. **42:** 26–34.
33. DORFMAN, L. J., T. M. BOSLEY & K. L. CUMMINS. 1978. Electrophysiological localization of central somatosensory lesions in patients with multiple sclerosis. Electroencephalogr. Clin. Neurophysiol. **44:** 742–753.
34. DORFMAN, L. J. & T. M. BOSLEY. 1979. Age related changes in peripheral and central nerve conduction in man. Neurology **29:** 38–44.
35. GUPTA, P. R. & L. J. DORFMAN. 1981. Spinal somatosensory conduction in diabetes. Neurology **31:** 841–845.

DEVELOPMENTAL DEPENDENCIES OF THE HUMAN BRAINSTEM AUDITORY EVOKED RESPONSE

Kurt Hecox*

Department of Neurology and Waisman Center
University of Wisconsin
Madison, Wisconsin 53706

Robert Burkard

Department of Communicative Disorders and Waisman Center
University of Wisconsin
Madison, Wisconsin, 53706

INTRODUCTION

The technical and methodological challenges to the study of sensory development are multiple. Measurements must often be performed on a small scale, the phenomena under study are in constant evolution, there is increased behavioral variability, and the relative contributions of genetic versus experiential factors must be defined. For these reasons, the introduction of a rapid technique able to circumvent the vagaries of behavioral instability without the sacrifice of quantification is usually enthusiastically received. It is not surprising, therefore, that the introduction of the human brainstem auditory evoked response (BAER), as a quantitative measure of auditory maturation, is rapidly gaining in popularity. This chapter reviews the BAER as applied to the study of auditory development. There are three main divisions in this article. The first is a description of the age dependence of response parameters, the second examines the effect of stimulus variables on response parameters, and the third summarizes selected issues surrounding the application of the brainstem auditory evoked response to the study of human auditory development.

RESPONSE VARIABLES

Of the many possible response measures, only four have been widely used: amplitudes, amplitude ratios, latencies, and interpeak latencies (IPL), as shown in FIGURE 1. Scalp topography, another response variable, is gainingin popularity. While there is considerable topographic data available on adults, only limmited information has been collected from infants. Each of these response properties is discussed in the following paragraphs.

Amplitudes

There are few systematic studies of the amplitude of each of the components of the brainstem auditory evoked potential as a function of age. The earliest studies were by Lieberman and Sohmer,[1] who suggested that infant and adult responses were submicrovolt in amplitude, the largest responses occurring in infants, the smallest in

*Address for correspondence: Neurology Department, Waisman Center, 1500 Highland Avenue, University of Wisconsin, Madison, Wisc. 53706.

0077-8923/82/0388-0538 $1.75/0 © 1982, NYAS

neonates, with adults assuming an intermediate value. Standard deviations in this study were large and no statistical evaluation was presented. Unfortunately the Lieberman study provided little detail on the age distribution of the infants, so that developmental trends are difficult to discern. A subsequent article by Salamy, Fenn, and Bronshvag[2] describes the age dependency of the amplitude of these responses in a more consistent fashion. In their study, eighth nerve amplitudes (wave I) increased until nearly 3 months, thereafter decreasing through adulthood. Wave III showed a

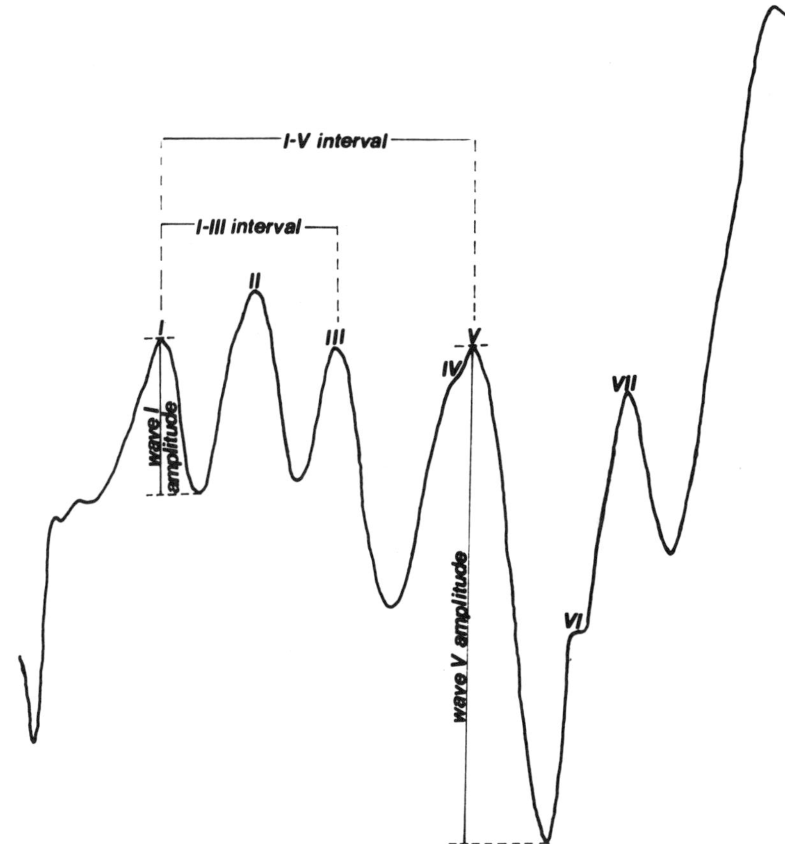

FIGURE 1. A schematized BAER depicting the response parameters discussed in this presentation is shown above. The various peaks are labeled with roman numerals I to VII by the convention of Jewett.

similar trend, peaking at about 3 months, while waves II, V, and VI peaked at 5 years, only to decrease again in adults. Our own normative values suggest that waves I, III, and V peak at about 1 year, again decreasing slightly in adults. Salamy *et al.*[2] also presented data on the frequency with which components could be identified as a function of age, showing that wave II has the lowest identification rate and that adult identification rates are achieved by 12 months for all waves.

The number of variables contributing to changes in response amplitude is large. These include such technical factors as electrode impedance, the orientation of the referential recording system with respect to the orientation of the dipole, and the amount of contamination by muscular activity. Biologic variables such as the scalp potential, the thickness of the calvarium, and the impedance properties of the intervening brain tissue also affect response amplitude. At present there seems little hope of separately assessing the contribution of each possible variable. Additional interpretative difficulties may arise with respect to amplitudes as a function of age since it must be assumed that the stimulus is constant across ages. The meaningful investigation of central changes necessitates holding constant the peripheral input. A simple conpensation for differences in peripheral sensitivity may not be adequate, as will become apparent in subsequent sections of this chapter. There have been no serious attempts to use "equivalent" stimuli in investigating the age dependency of response amplitudes. For these and many other reasons, Starr and Achor[3] suggested in 1975 that amplitude ratios be used as a dependent variable, in lieu of absolute amplitudes.

Amplitude Ratios

There are many possible amplitude ratios in a response complex containing seven components. However, only the ratio of wave V to wave I has been widely used. The reasons for focusing on the ratio of these two components are both practical and theoretical. The practical reason is that wave V is the most stable and generally the largest component. Thus, wave V is the primary dependent variable in a variety of analyses for the diagnoses of peripheral and central auditory disorders. The separation of peripheral from central auditory disorders is facilitated by the additional analysis of the peripheral nerve response. Disorders of the middle or inner ear should be reflected in abnormalities of the eighth nerve action potential (wave I) while disorders of central auditory structures should result in abnormalities of those components that follow wave I. Therefore, waves I and V are the most critical for clinical and research applications.

Numerous investigators have described the age dependency of the V/I amplitude ratio and all agree that the amplitude ratio is smaller in newborns than adults.[4–6] Typically the V/I amplitude ratio is about 3 in the adult while it is approximately 1 in the neonate [FIGURE 2]. Adult values are attained sometime following 12 months of age with little consensus as to the precise age of maturation. As with absolute amplitude, there are several factors that might contribute to developmental changes in amplitude ratios. Electrode impedance, cephalic skin potentials, calvarial thickness, and so on, as described above, should equivalently affect both waves I and V. It appears that the number of technical nonneuronal factors that may influence the amplitude ratio is smaller than that for absolute amplitudes. Nevertheless, the problem of stimulus equivalence as a function of age is important for the interpretation of changes in ratios since the peripheral generators for waves I and V may not be the same. Both Klein and Teas,[7] and Don and Eggermont[8] question the assumption of identical peripheral generators for waves I and V, on the basis of masking studies. Thus, age-dependent differences in the effective stimulus could interact to elicit different percentages of the total population of potential generators for each of the waves.

Since the percentage change in amplitude of wave I is less than that for wave V, the decreased amplitude ratio seen in the neonate is due primarily to the diminished absolute amplitude of wave V. The mechanisms for such changes may be complex.

Central or peripheral changes in synchrony could account for variations in response amplitude. Changes in either the dipole orientation of each of the waves or the number of units with sufficient maturity to respond could account for amplitude ratio differences. Undoubtedly, multiple mechanisms are involved, and the relative contribution of each to the observed effects awaits more sophisticated methodology.

Several other technical factors also affect the interpretation of pathologic or developmental changes. Many stimulus variables influence amplitude ratios, including intensity, rate, and stimulus frequency.[9,10] In addition, "norms" for amplitude ratios

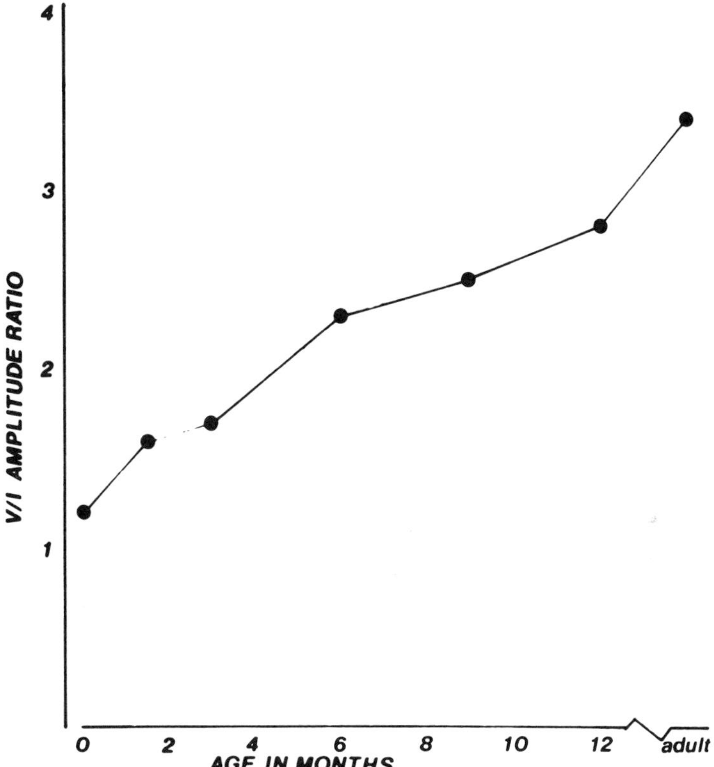

FIGURE 2. The relationship between the amplitude of wave V and wave I is shown as a function of age. Note the monotonically increasing function that peaks sometime beyond the fifth year of life.

must take into account the rather marked kurtosis of the underlying distributions. There is considerable asymmetry in the distributions, with little spread of values below the mean, especially for infants. The sensitivity of amplitude ratios to changes in stimulus conditions and the high false-positive rate of pathologic diagnosis at high signal intensities has led to the abandonment of amplitude ratio as a clinical measure in many centers. This is unfortuante since, when care is taken in the selection of appropriate stimulus conditions, amplitude ratios provide considerable insight into pathophysiologic mechanisms.

Scalp Topography

Preliminary data obtained in our laboratory suggests that there is a significant interaction of electrode montage, response components, and age. Data obtained from adults have demonstrated that wave I is maximally recorded ipsilateral to the ear of stimulation, either at the mastoid or earlobe. Wave V is best seen centrally, with maximal amplitude from C_Z to F_PZ.[11–13] While wave I is consistently larger measured from mastoid to mastoid compared to vertex to mastoid (the standard montage), wave V is virtually never larger in the mastoid-to-mastoid montage, compared to the standard recording configuration in adults. In contrast, mastoid-to-mastoid recordings of wave V are larger than vertex to mastoid responses in nearly 40% of newborns and infants less than 8 months of age [FIGURE 3]. There is very little difference in the relative amplitude of wave I as a function recording configuration for the same pediatric population. Explanations for these effects are several, including the possibility that the generators of each response component may not be the same across ages. Alternatively, the dipoles responsible for the generation of waves V and I may have a constant anatomic substrate, but as maturation advances their geometric orientation shifts, producing a change in their surface-recorded properties. In all cases, such major shifts in scalp topography make the interpretation of age-dependent changes in amplitude ratios more complex and the investigation of the physiological substrate of age-dependent amplitude changes more difficult.

Latencies

As shown in FIGURE 1 the brainstem auditory evoked potential consists of a series of components designated by Roman numerals I–VII. All of these components show an age dependency in their latency. The rate of maturation and the age at which each peak attains adult values varies in a consistent fashion. The more rostral (longer latency) the response, the longer is its time course of development. Wave I approximates adult latencies by 3 months, while the wave V response attains adult values between 1 and 2 years of age. Waves II–IV attain maturity at intermediate ages, although waves III and IV nearly parallel the development of V, and wave II is essentially identical to wave I in its developmental timetable. Waves VI and VII, in our experience, have such variable morphology and are so unpredictably elicited, even in adults, that a discussion of their maturation will be omitted.

The anatomic origins and physiological mechanisms responsible for the decreasing latency of all wave components with increasing age is unknown. A variety of factors have been postulated, including the resolution of middle ear transduction abnormalities, the maturation of the most basal portions of cochlea, increased synaptic efficiency of transmission, improved myelination, and superior synchrony of responses with increasing age. The relative contribution of each of these factors to the observed changes is unknown and will require careful studies in animal models.

Interpeak Latencies (IPL)

Because of differences in the rate of maturation of the waves, interpeak latencies must also depend on age. Thus, the mean interpeak interval between waves I and V is 5.1 msec in the neonate and only 4.0 msec in the adult.[14–17] Interestingly, the time course of maturation of the I–III interval is longer than that for the III–V interval. This observation is contrary to the general rule that the more rostral the generator or interwave interval, the more immature its response properties. The interpretation of maturational changes in interpeak latencies is that central conduction times (velocities) are being directly measured, and reflect improved myelination. This assumes serial generators, an assumption that is not entirely borne out by masking studies, nor by animal lesioning studies of the underlying generators. Fortunately, the "serial" approximation is useful for clinical studies. Thus, clinical abnormalities of brainstem white matter seem to produce abnormalities in conduction time as would be predicted

AMPLITUDE HORIZ. > VERT

	I	III	V	VI	V/I
>8 mos	81%	70	0	60	15
<8 mos	70%	90	40	22	15

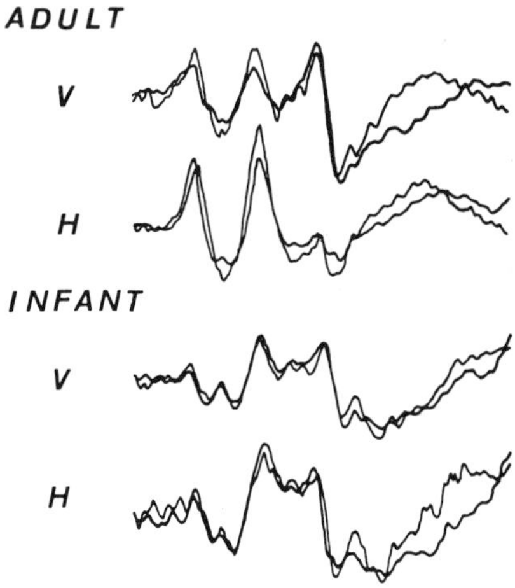

ADULT

V

H

INFANT

V

H

FIGURE 3. The relative amplitude of each wave component for horizontal (mastoid to mastoid) versus vertical (vertex to mastoid) electrode montages. The values in the boxes are the percentages of responses that are larger for that electrode configuration. Note that wave V is virtually never larger in the horizontal records for adults, although it is often greater in the horizontal recordings from the neonates.

secondary to impaired myelination. Empirical evidence suggests that interpeak latencies and their abnormalities are useful and reliable indices of pathology impinging upon the brainstem auditory pathways.

In summary, each of the above described response properties—amplitudes, amplitude ratios, latencies, and interwave intervals—show an age dependency. The rate of maturation depends on the component of interest. The mechanisms responsible for normal maturation have received relatively little attention, and most data has been collected at a descriptive level. The implications of shifts in scalp topography are multiple and should affect the interpretation of both amplitude and amplitude ratio changes. In addition, surprisingly little information is available on other aspects of the age dependency of response properties, including their spectral content, the slopes of the waves, and the area under each of the waveforms. The limitation of response variables to the set described to date will undoubtedly prove too restrictive in the future.

Stimulus Dependencies

Intensity

Response latency decreases as signal intensity increases for both infants and adults. The slope of the latency–intensity function shows a small age dependency such that the rate of change in the neonate is approximately 7μsec per dB slower than that seen in the adult.[18,19] In addition, the distribution of latency–intensity functions is skewed to lower values in neonates than in adults.

Few studies have described the absolute threshold for eliciting the brainstem evoked potential across age although there is general agreement that the discrepancy between studies and neonates is no greater than 20 dB.[20,21] The only available study comparing electrophysiologic and behavioral thresholds noted that by 5 months of age essentially all subjects had thresholds of 20 dB or less, but it was not until approximately 3–4 years of age that subjects demonstrated the 10 dB thresholds seen in the adult.[22] This same study examined behavioral estimates of threshold simultaneous with the evoked potential measures. Not until nearly one year of age did behavioral results approximate levels achieved by electrophysiologic techniques. Initially (neonates) behavioral estimates of sensitivity were 50 to 60 dB greater than the BAER thresholds. The amount by which neonatal BAER threshold exceeds the adult BAER threshold is 15–20 dB in several studies.[20–22] The threshold value of intensity depends on many variables, including frequency, rise time, rate of repetition, and signal duration. In addition, response identification is dependent upon recording parameters, such as the number of sums, and subjective variables, such as background activity, factors that both contribute to signal-to-noise ratio changes. Insufficient effort has been made to optimize signal-to-noise ratios for these low level responses and there have been few investigators who have collected more than the routine 2000 to 4000 sums. Most attempts at threshold estimation have been performed for click stimuli, although several studies of adults using tone-burst brainstem evoked potentials have been published.[23,24] Little developmental information is available on the interaction of stimulus frequency and intensity.

Spectrum

The spectrum of a stimulus is sensitive to changes in duration, rise–fall time, and component frequency. Decreasing rise–fall times, decreasing durations, and decreasing frequency interact to produce increasingly dispersed acoustic energy. Perhaps a more fundamental notion is that stimulus purity (narrowness of the acoustic spectra) is not equivalent to frequency specificity. The stimulus spectrum is related to the acoustic properties of the stimulus while frequency specificity is related to physiological responses. While related, they may not necessarily by equivalent. The multiple determinants of frequency specificity were first demonstrated by Teas, Eldredge, and Davis[25] for the eighth nerve response and subsequently reconfirmed for the BAER by Eggermont and Don.[8] For example, increasing stimulus intensity results in a basalward shift of the cochlear region from which a response is derived, a result not predicted by changes in signal spectrum. Near threshold there is a systematic relationship between signal frequency and place along the basilar membrane. As stimulus intensity is raised the region of basilar membrane motion spreads, particularly in the basal direction. One of the consequences of this phenomena is that stimulation with low frequency signals produce responses from a large portion of the basilar membrane, introducing uncertainty into the frequency specificity of the response.

Compared to click-evoked responses, tone-burst-evoked potentials have higher thresholds and longer latencies,[23,24] although these effects have not been clearly separated from the influence of differences in rise–fall time. There have been no systematic studies of tone pips (filter transients) as a function of age, or of bandpassed noise bursts. The paucity of data in this area hardly permits speculation of underlying mechanisms. There are at least two unpublished theses[26,27] in the area of tone-burst-evoked potentials in neonates, each demonstrating that it is possible to elicit well-defined evoked potentials at moderate intensities for a 500 Hz signal and at low intensities for 1600 and 4000 Hz signals. A number of laboratories are actively investigating this problem, and more data will surely be available within the next few years.

Rise–Fall Time

In adults, increases in rise time result in increased latencies with minimal effects on response amplitude.[28] These effects are found for both noise and tone burst stimuli.[29-31] FIGURE 4 illustrates the increase in latency with increasing rise time for neonates and adults. Interestingly, the smaller values of rise time produce a smaller effect for neonates than adults. This result is somewhat surprising in that less rapidly rising stimuli are usually more poorly processed the more immature the organism. There are a number of possible explanations for this phenomenon, the most likely of which is that the neonate has a relative immaturity of the basal region of the cochlea that prevents the generation of short latency responses. Preliminary evidence suggests that decreases in response latency with decreasing rise time imply a basalward shift in the region of response generation (electrophysiologically), and the inability to shorten latency as rise time decreases would reflect impairment of the most basal portions of the inner ear. Other alternatives are that impulsive stimuli with rapid rise times cannot be faithfully transduced in the neonate because of the increased compliance of the middle ear system, or that there are impedance differences in the basilar membrane. At present, there is little way to distinguish amongst these possiblities or other alternatives. Preliminary data obtained in our laboratory suggests that the frequency impairment hypothesis is the most tenable explanation.

Duration

Increasing signal duration results in increasing loudness and decreasing thresholds. These well-known psychophysical effects are not, however, paralleled in the

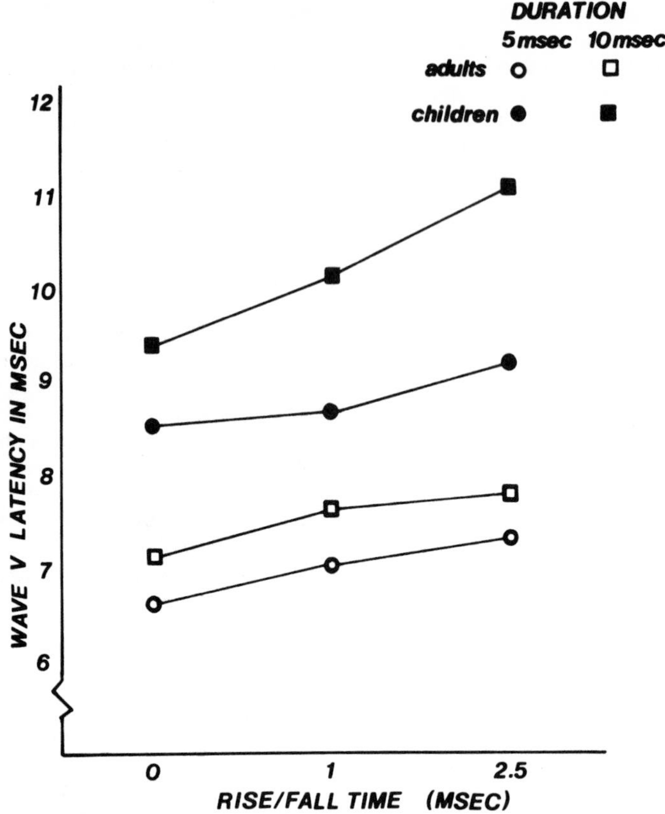

FIGURE 4. Wave V latency as a function of rise/fall time is shown for noise-burst signals. Note that latency increases with increasing rise/fall times and that the magnitude of the effect interacts with age.

amplitude or latency of the BAER.[28,31] The relative insensitivity of the BAER to duration is a reflection of its fundamental character: It is an onset response. When care is taken to control for possible effects of forward masking, by separately manipulating signal *on* and *off* time, *on* time has minimal effects on response amplitude and only slightly prolongs response latency (and then only when a critical *off* time has been reached).[28] The age dependency of this phenomenon is poorly described, although preliminary data from our laboratory shows differences between adults and neonates. There appears to be minimal effect of signal *on* time for the neonate until a critical value of *off* time is reached, after which the neonates' responses essentially disappear. Whether this reflects increased sensitivity to forward masking or some other post-stimulatory process is not known. Although commonly seen in neonates, this phenomenon is never seen in adults. No reports of neonatal responses to tone burst stimuli of varying duration could be located.

Masking

To improve the frequency specificity of the BAER, several masking paradigms have evolved including band-passed masking, tonal masking, and notched-noise masking. With the exception of the high-pass masking paradigm, no information is available on the age dependency of these methods. The preliminary information available on high-pass masking suggests that neonates are less sensitive to high-pass maskers than adults.[18] Many fundamental questions remain unanswered in the area of masking, including the effect of subtotal masking on response latencies and amplitudes. The physiological consequences of masking and its rate of spread are also undefined in both adult and pediatric populations. Since the primary motivation for masking paradigms is improved frequency resolution in the hearing-impaired, it is critical to demonstrate that such signal paradigms do not interact with pathologic diagnosis. The theoretical advantage of masking paradigms for frequency specificity has not yet been translated to an empirical demonstration of their superiority in the hearing impaired. The normative data available are insufficient to permit the use of these methods in clinical or developmental assessment.

Binaural Interaction

Controversy exists concerning the reality and significance of binaural interaction responses in the brainstem auditory evoked potential.[32,33] These responses are defined by comparing summed monaural responses to responses obtained under binaural conditions—any resulting difference being an interaction response. While "crossover" has been seen in these conditions and remains an important effect to be controlled, binaural interaction can be demonstrated at intensity levels below those of transcranial conduction.[32] Binaural interaction can also be seen in the infant (FIGURE 5) and

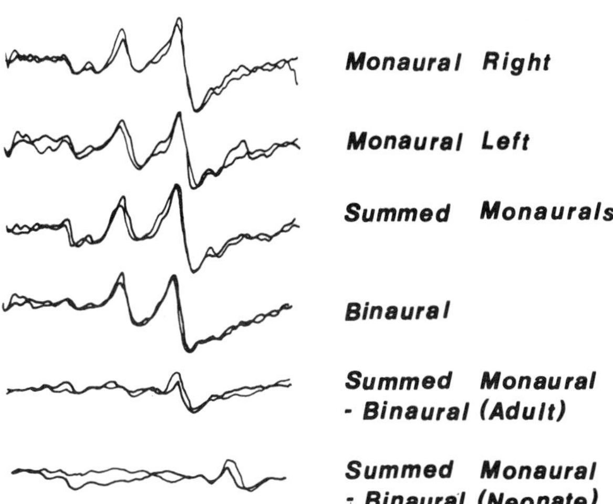

FIGURE 5. This figure illustrates the method by which binaural interaction is defined. The interaction response is the difference between the summed monaural responses and the binaural response. The latency of this "difference" wave depends on age, as shown above.

its latency is prolonged by an amount comparable to the latency prolongations seen in monaural stimulation. Aside from the documentation of its occurrence, there is little information available on the parametric dependency of binaural interaction, or on the earliest age at which such interactions appear. This is indeed unfortunate since this is a promising area for psychophysical–electrophysiological correlations and the investigation of their age dependencies. No doubt there will be increased interest and activity in this area in the near future, particularly as it relates to critical periods for the development of binaural interaction in the unilaterally hearing-impaired.

Rate

Rate of stimulation has been investigated for both adults and infants with consistent results.[35–39] The average change in latency with changing rate of stimulation can be predicted by multiplying the change in rate of stimulation (rate of test minus the base rate) by 0.006 msec for adults. The correction factor for neonates is 0.14 and at 6 months of age it is 0.009. In an investigation of the age dependency of this phenomenon, no second-order or third-order effects were found, suggesting that rate interacts linearly with intensity.[39] Latency values can therefore be converted from one rate to any other in a linear fashion. The mechanisms responsible for such changes are unknown although fatigue seems unlikely given the lack of interaction with signal intensity. Increased effects of adaptation, differences in locus of generation of the evoked potential along the basilar membrane, and differences in recovery processes at the level of the synapse are all possible mechanisms that have been invoked in other contexts. It is noteworthy that the latency shifts seen in wave V as a function of rate cannot be accounted for by latency changes in wave I in adults or neonates.[39] Both rate changes and their age dependency are the result of peripheral and central contributions. Several investigators have suggested that increased rates of stimulation provide a means of clinically "stressing" the auditory system to increase the yield of patholgoic findings.[41,42] Our experience has been that approximately 4% of patients with central auditory pathology will exhibit these rate abnormalities alone. While multiple sclerosis and closed head injury have been the most common disorders exhibiting rate dependency to the pathology, we have seen it in several clinical contexts.

Adaptation

In this paradigm a train of signals is presented followed by a pause of varying duration and the sequence of signals is repeated. The responses are then summed according to their position in the train of stimuli. The dependent variable in this paradigm is a decrease in response amplitude or increase in response latency as a function of position in the stimulus train. While this paradigm is frequently used in sensory physiology, there is only one published report of its application to the BAER.[43] In 1976, Don, Allen, and Starr[43] reported that the adaptation exhibited by the BAER was virtually complete by the fourth stimulus in a signal train. In that study it was not possible to directly investigate the relative contribution of peripheral versus central factors. Based upon preliminary data, adaptation in the infant is larger than that in the adult although, like the adult, it is essentially complete by the fourth trial. There is, therefore, an age dependency to adaptation although insufficient numbers of infants have been collected to establish the precise age at which adult responses are obtained. This particular paradigm may be of interest in clinical applications since the diagnosis of retrocochlear disease based on behavioral measurements is facilitated by the demonstration of abnormally rapid adaptation. Thus, adaptation is a reliable electrical phenomenon seen in all adults to date and exhibits age dependency, thus holding promise as a scientific and clinical tool.

Click Phase

Click signals may be condensation or rarefaction in phase. These phases are reported to produce different response morphologies and latencies, at least for waves II, IV, and VI of the BAER. The only report of the developmental dependency of click phase is in Stockard *et al.*[44] The effect of click phase was said to be enhanced in newborns as compared to adults, although mean latency differences did not exceed 0.13 msec. Rarefaction clicks appeared to produce longer interpeak intervals and shorter wave I's for both adults and newborns. It appears that click phase has a small but significant effect on response morphology and latency and the magnitude of this reponse may be age dependent.

Special Problems

Following the introduction of any new response, there is a period of intense investigation of its fundamental properties. Only after these properties are sufficiently well defined is the response used as a dependent variable in pursuit of scientific and clinical answers. The BAER is now entering the latter phase of application. Among these applications several deserve special mention, either because of their importance in clnical medicine or in basic science. The selection of these "special problems" reflects the biases of the authors and is neither exhaustive nor objective.

Recovery of Function

One of the most fascinating intersections between basic science and clinical medicine is recovery from central nervous sytem damage. The definition of conditions that may facilitate or accelerate recovery from acute insults to the brain has been a matter of great interest for decades. One of the greatest obstacles in the study of recovery of function is the poor correlation between animal models and human diseases. Additionally, quantification has been elusive in models of human disorders. In this regard, the BAER may play a unique role since it can provide quantitative information on the status of peripheral transduction and the integrity of central auditory pathways. The earliest examples of "recovery" were reported in patients with multiple sclerosis[41] and central pontine myelinolysis[45] in whom behavioral and electrophysiologic indices paralleled one another. Documentation of recovery of neurophysiologic function in pediatric patients has been primarily in hypoxic–ischemic encephalopathy. Following hypoxia, it is not rare (although it is uncommon) to see improvement of IPLs, the re-emergence of responses, and normalization of rate dependencies. Interestingly, in the experience of the major author, at least, marked abnormalities of amplitude ratios have rarely normalized least follwoing episodes of hypoxia. Improvements in amplitude ratios can, however, be seen following removal of posterior fossa tumors or following the relief of raised intracranial pressure. The reversilitiy of peripheral auditory pathology has not been as well documented. This is because improvements in response could be secondary to the reversal of middle ear disease. The solution to this problem is to use bone-conducted BAERs once adequate infant normative data has been collected. Another form of reversible central auditory pathology is in inborn errors of metabolism. We have seen two instances, each of maple syrup urine disease and hyperphenylalinemia, in which normalization of serum metabolic parameters was associated with normalization of the BAER. There is little doubt that responses may change significantly with respect to degree of pathology in longitudinal studies of patients with CNS disease. Whether these changes are predictive of ultimate outcome or may provide insight into pathophysiological mechanisms is an exciting but unanswered question.

Critical Periods

It seems inevitable that the BAER will be used in the investigation of critical periods in auditory development. Prior to the advent of BAER testing, it was not possible to define the degree or locus of auditory pathology noninvasively in the very young patient. Therefore, the impact of amplification devices and age at which the infant experiences a "normal" sound enviroment has not been studied. As a result, there is little information on the critical age for the placement of hearing aids and there is even less data on the critical period for the development of binaural hearing. The ease with which the BAER is administered and its high level of repeatability make it an appropriate measure in this area. Animal work on critical periods has recently begun to increase also. Thus, the investigation of critical periods not only promises to yield information of interest to basic scientists, but also carries with it important implications for the management and care of the hearing-impaired infant.

Premature Infants

The investigation of maturation in the premature population remains a thorny problem. The earlier the gestational age, the more prolonged the response latencies, and the smaller the response amplitudes.[16, 46–49] These findings have been verified by numerous investigators, and the slope of the age–latency function is approximately 0.2 of a msec per week of gestational age for wave V. The earliest age at which responses can be elicited is between 28 and 30 weeks. There has been no attempt to evaluate the interaction between stimulus variables and prematurity, with the exception of rate, which shows rather striking dependency on gestational age.[50] The premature infant would seem to present an exceptional opportunity for the study of the early maturation of the auditory system. It is not at all uncommon in a 26–28 week premature infant to be unable to elicit a wave V complex. Within one to two weeks the complex will appear as a poorly defined waveform, which becomes better synchronized, larger in amplitude, and shorter in latency as the infant approaches term. The clinical period for the age at which binaural interaction first occurs could certainly be investigated in this population.

One of the major problems in the establishment of norms for this age group relates to the definition of gestational age. Despite years of investigation and the use of multiple methods, there is still no generally agreed upon method of determining gestational age with less than a two week error. As shown by Galambos and Despland,[50] the BAER can be demonstrated to change from week to week for individual patients. It would appear therefore that validation of the BAER would depend on criteria that are less sensitive than the BAER itself. Until satisfactory external criteria are defined for determining gestational age, it seems unlikely that age-specific norms for this population will be more than approximate. This does not preclude their clinical utility but only results in wider standard deviations. A related but distinct issue is the question of whether there is a *normal* premature infant. Even when the premature infant is born without respiratory distress and the clinical course is without significant medical complications, the premature infant enters the world with immaturity of many organ systems, including liver, kidney, brain, and endocrine. The potential effects of the immaturity of these organ system on CNS function is undefined. Furthermore, virtually all such infants are exposed to ototoxic agents and are placed in high noise environments. Again, it is difficult to believe that such infants should be considered normal by any criteria. Despite these conceptual pitfalls the investigation of responses in the premature population will surely continue. The very high rate of peripheral auditory dysfunction (about 10%)[19,51] defined for this population, and the possibility of more precisely determining neurologic prognosis are sufficiently important and alternative measures sufficiently inadequate that the BAER will undoubtedly constitute a major clinical and investigational tool in the premature population.

Frequency Selectivity in Hearing Loss

One of the most common applications of the BAER is to the diagnosis of hearing impairment in infants. While other auditory evoked potentials have been used in the past for this same purpose, none have achieved the reliability and validity of the BAER. The use of the BAER as an index of auditory pathology in the premature and infant populations (especially in intensive care units) has resulted in rather striking changes in the statistics of auditory impairment. In contrast to the approximately one per thousand incidence of significant auditory damage found in "normal" newborn screening, there is an incidence of nearly 10% for significant bilateral auditory impairment in graduates of neonatal intensive care units.[19,51] The limited follow-up information available suggests that approximately half of these infants will have persistent auditory impairment sufficient to require amplification. This makes hearing impairment one of the most common chronic disabilities in neonatal intensive care unit graduates. The reason for discrepancies between the statistics derived from BAER measures and other tests of auditory function is not clear although the ease with which the neurologically impaired patient can be tested by BAER is surely an important factor. In our experience the incidence of hearing loss is directly proportional to the incidence of significant neurological impairment. It is in this neurologically impaired population that behavioral and even impedance measures are the most difficult to obtain and interpret, often requiring months to years of observation. Also, many of the youngsters who might have been identified as hearing-impaired following discharge from the neonatal intensive care unit have been placed in institutional facilities where testing of sensory function is not routinely performed. The scarcity of resources available for the care of these children will undoubtedly raise important ethical issues since it would be expected that many of these severely neurologically impaired patients would gain little functional benefit from amplification although the cost and time for the fitting of aids may be great.

The diagnosis of hearing impairment in cooperative adults is not an area in which the BAER has had significant impact. While helpful in the diagnosis of eighth nerve tumors, the presence of central auditory pathology on the BAER cannot distinguish between vascular disease, tumors, or demyelinating disorders. The BAER is primarily helpful in that a normal BAER (by central nervous system criteria) makes exceedingly unlikely that any impairment is secondary to an eighth nerve tumor.

One of the major areas of discussion among practitioners of BAER audiometry is the choice of an optimal stimulus. Because the click contains many frequencies, attempts have been made to choose a signal that more selectively stimulates restricted regions of the basilar membrane. There are several ways in which this can be done but there have been few comparative studies of the diagnostic utility of clicks versus tone bursts versus clicks in high-pass noise versus tone bursts in high-pass noise versus tone bursts in notched-noise, and so on. Until such direct comparisons are available we feel that it is premature to base testing decisions on the theoretical possiblity of improved results using tone-burst stimuli or complex masking paradigms. While tone-burst signals should surely improve the accuracy of evoked potential testing, the gain in accuracy versus the potential pitfalls inherent in the use of such signal paradigms is unknown. Those conditions in which the BAER is poorest in predicting audiometric patterns are predominantly low frequency impairment and 4 kHz notches (as commonly seen in noise-induced hearing loss). In the former case the use of impedance audiometry is critical since the vast majority of such losses are secondary to middle ear disease and are reflected in abnormalities of impedance measures; in the latter case, we have not seen a single infant proven to have a 4 kHz notch over the past six years. Thus the groups in which click-evoked responses most poorly predict audiometric abnormalities in adults are not a large problem in infants and young children. It is for these reasons that we caution the practitioner from immediately accepting the more time-consuming and costly use of tone-burst or noise-masking audiometry. Finally, each of the more sophisticated stimulus conditions has as its goal the definition of auditory thresholds. In our experience one of the most useful aspects of the BAER is the ability to define abnormally steep latency–intensity functions (electrophysiologic recruitment) since that information can be critical to the selection of an appropriate amplification device. Information on electrophysiologic recruitment cannot be derived from a "threshold" measure.

Neurologic Applications

The localization of neurologic disease is often difficult in patients unable to cooperate with the traditional neurologic exam. Thus, adjunctive measures of localization are most critical for the pediatric population. The clinical literature on brainstem syndromes has a heavy emphasis on vascular disease as seen in older adults. Yet, a number of pediatric patients evaluated for "unknown neurologic syndromes" have in our experience demonstrated clear brainstem pathology not apparent on their neurologic exam. A subgroup of infants with universally poor prognosis with hypoxic–ischemic encephalopathy have also been identified using the BAER.[52] These post-hypoxia patients are essentially indistinguishable from other patients on the basis of their clinical, historic, or neurologic examination. The list of disorders with unsuspected brainstem pathology includes a large proportion of those patients with chromosomal disorders (excluding Downs syndrome), those with malformation syndromes of nonchromosomal nature, and in progressive neurologic disorders including "peripheral neuropathies."

Correlations between abnormalities of the BAER and autism[53,54] and other forms of psychopathology have been made. Results of such measurements have been variable and in our opinion clarification of those studies awaits a more rigid and objective criterion for the diagnosis of the particular form of psychopathology. Our experience has not been very positive with regard to the diagnosis of unsuspected brainstem disorders in patients with "learning disabilities." Whether the prospective analysis of patients demonstrated to have early signs of brainstem dysfunction may show long-term abnormalities of learning despite normalization of their BAER is an interesting but untested hypothesis.

One of the major challenges in the near future will be to refine BAER testing to perform differential diagnosis of neurological disease. Thus, attempts must be made to define recording and stimulating conditions that selectively elicit abnormalities secondary to tumors, vascular disease, or myelin disorders.

Relation to Other Measures of Auditory Responsivity

The brainstem auditory evoked response is just one subportion of the auditory evoked potential (AEP). The middle, long latency components, P300 response, the CNV, and even longer latency potentials are just a few of the electrical phenomena that have been studied in different auditory information processing tasks.[55,56] There is little data available on the application of each of these epochs and paradigms to the diagnosis of neurological disease or to the description of normal human development. Some information is available on the infant middle-latency responses[57,58] as well as the longer-latency N_1–P_2 response, as a function of age.[59,60] Over the past year, we have collected responses from as many of these response categories as time permitted on pediatric and adult patients with known or suspected neurologic disorders. The spectrum of results is striking and it is quite clear that significant neurologic and audiologic information can be obtained by supplementing BAER testing with other responses. For example, we have seen patients with no brainstem auditory evoked responses but virtually normal long-latency responses to the same stimuli. The explanation for this phenomenon is unclear although possiblities include differences in the frequency specificity of each time epoch, differences in neuronal subpopulations along the classical auditory pathway, or different but parallel pathways for different categories of surface recorded potentials. This is just one example of the many interesting combinations of evoked potential results that can be defined by more broadly sampling from the entirety of the auditory evoked potential. Thus, both in clinical applications and in the study of normal developmental processes the collection of a BAER must not be viewed as a comprehensive approach to the issues and substrates of human auditory development. It is unlikely that any single measure will prove comprehensive in scope. Instead, multiple measures of auditory function will be needed to adequately describe the subtleties of human auditory development.

Acknowledgments

Of the many direct and indirect contributors to the studies cited herein, Barbara Cone, Terry Hieber, Lynn Weatherby, Don Deegan, Mary Malischke, and Janice Cunniff deserve special thanks.

References

1. Leiberman, A., H. Sohmer & G. Szabo. 1973. Cochlear audiometry (electrocochleography) during the neonatal period. Dev. Med. Child Neurol. **15:** 8–13.
2. Salamy, A., E. Fenn & M. Bronshvag. 1979. Ontogenesis of human brainstem evoked potential amplitude. Dev. Psych. **12**(5): 519–526.
3. Starr, A. & L. J. Archor. 1975. Auditory brainstem responses in neurological disease. Arch. Neurol. **32:** 761–768.

4. SALAMY, A., C. M. MCKEAN, G. PETTETT & T. MENDELSON. 1978. Auditory brainstem recovery processes from birth to adulthood. Psychophysiology **15**(3): 214–220.
5. HECOX, K. & B. CONE. Applications of brainstem auditory evoked responses to pediatric neurological diseases. Neurology (in press).
6. CONE, B., K. HECOX & T. FINITZO-HIEBER. 1977. The brainstem auditory evoked response in neonates: A narrative study. Trans. Am. Acad. Ophthalmol. Otolaryngol. **84:** 188.
7. KLEIN, A. J. & D. C. TEAS. 1978. Acoustically dependent latency shifts of BSER (Wave V) in man. J. Acoust. Soc. Am. **63**(6): 1887–1895.
8. EGGERMONT, J. J. & M. DON. 1980. Analysis of the click-evoked brainstem potentials in humans using high-pass noise masking. II. Effect of click intensity. J. Acoust. Soc. Am. **68**(6): 1671–1675.
9. STOCKARD, J. J., J. E. STOCKARD & F. W. SHARBROUGH. 1978. Non-pathologic factors influencing brainstem auditory evoked potentials. **18:** 177–209.
10. PRATT, H. & H. SOHMER. 1976. Intensity and rate functions of cochlear and brainstem evoked responses to click stimuli in man. Arch. Otorhinlaryngol. **212**(2): 85–92.
11. PICTON, T. W., S. A. HILLYARD, H. I. KRAUSZ & R. GALAMBOS. 1974. Human auditory evoked potentials. I: Evaluation of components. Electroenceph. alor. Clin. Neurophysiol. **36:** 180–190.
12. GOFF, W. R., T. ALLISON & W. LYONS, *et al.* 1977. Origins of short latency auditory evoked potentials in man. *In* Auditory Evoked Potentials in Man. Psychopharmacology Correlates of Evoked Potentials. Progress in Clinical Neurophysiology. J. E. Desmedt, Ed. Vol. 2: 30–43. Basel, Karger, Switzerland.
13. MARTIN, M. E. & E. J. MOORE. 1977. Scalp distribution of early (0 to 10 msec) auditory evoked responses. Arch. Otolaryngol. **103**(6): 326–328.
14. SALAMY, A. & C. M. MCKEAN. 1976. Postnatal development of human brainstem potentials during the first year of life. Electroencephalogr. Clin. Neurosphysiol. **40**(4): 418–426.
15. GOLDSTEIN, P. J., A. KRUMHOLZ, K. FELIX, D. SHANNON & R. F. CARR. 1979. Brainstem evoked response in neonates. Am. J. Obstet. Gynecol. **135:** 622–628.
16. STARR, A., R. N. AMLIE, W. H. MARTIN & S. SANDERS. 1977. Development of auditory function in newborn infants revealed by auditory brainstem potentials. Pediatrics **60**(6): 831–839.
17. SALAMY, A., C. M. MCKEAN & F. B. BUDA. 1975. Maturational changes in auditory transmission as reflected in human brainstem potentials. Brain Res. **96**(2): 361–366.
18. HECOX, K. 1975. Electrophysiological correlates of human auditory development. *In* Infant Perception. L. Cohen & P. Salapateck, Eds. Vol. 2: 151–191. Academic Press, New York, N.Y.
19. DESPLAND, P. A. & R. GALAMBOS. 1980. Use of auditory brainstem responses by premature and new born infants. Neuropaediatrie **11**(2): 99–107.
20. SCHULMAN-GALAMBOS, C. & R. GALAMBOS. 1979. Brainstem evoked response audiometry in newborn hearing screening. Arch. Otolaryngol. **105:** 86–90.
21. MOKOTOFF, B., C. SCHULMAN-GALAMBOS & R. GALAMBOS. 1977. Brainstem auditory evoked responses in children. Arch. Otolaryngol. **103:** 38–43.
22. KAGA, K. & Y. TANAKA. 1980. Auditory brainstem response and behavioral audiometry: Developmental correlates. Arch. Otolaryngol. **106**(9): 564–566.
23. SUZUKI, T., Y. HIRAI & K. HORIUCHI. 1977. Auditory brainstem responses to pure tone stimuli. Scand. Audiol. **6**(1): 51–56.
24. PICTON, T. W., J. OUELLETTE, G. HAMEL & A. D. SMITH. 1979. Brainstem evoked potentials to tone pips in notched noise. J. Otolaryngol. **8:** 289–314.
25. TEAS, D., D. ELDREDGE & H. DAVIS. 1962. Cochlear responses to acoustic transients: And interpretation of whole nerve action potentials. J. Acoust. Soc. Am. **34:** 1438–1459.
26. CONE, B. 1978. The human auditory evoked response. Unpublished thesis. Univ. of Texas, Dallas, Tex.
27. FRYE-OSIER, H. 1981. Tone burst early and middle latency evoked potentials. Unpublished thesis. Univ. of Wisconsin, Madison, Wisc.
28. HECOX, K., N. K. SQUIRES & R. GALAMBOS. Brainstem auditory evoked responses in man. I. Effect of stimulus rise fall time and duration. J. Accoust. Soc. Am. **60**(5): 1187–1192.
29. SUZUKI, T. & K. HORIUCHI. 1981. Rise time of pure-tone stimuli in brainstem response audiometry. Audiology **20:** 101–112.

30. Kodera, K. H., Yamane & O. Yamoda. 1977. Brainstem responses evoked by tone bursts in humans: Effects of rise time and frequency. Audiology (Japan) **20:** 151–157.
31. Brinkmann, R. D. & M. Scherg. 1979. Human auditory on- and off-potentials of the brainstem. Scand. Audiol. **8:** 27–32.
32. Dobie, R. A. & C. I. Berlin. 1979. Binaural interaction in brainstem evoked responses. Arch. Otolaryngol. **105:** 391–398.
33. Van Olphen, A. F., M. Rodenburg & C. Verwey. 1978. Distribution of brainstem responses to acoustic stimuli over the human scalp. Audiology **17:** 511–518.
34. Ainslie, P. J. & J. R. Boston. 1980. Comparison of brainstem auditory evoked potentials for monaural and binaural stimuli. Electroencephalogr. Clin. Neurophysiol. **49:**(3–4): 291–302.
35. Fujikawa, S. M. & B. A. Weber. 1977. Effects of increased stimulus rate on brainstem electric response (BER) audiometry as a function of age. J. Am. Audiol. Soc. **3**(3): 147–150.
36. Pratt, H. & H. Sohmer. 1977. Correlations between psychophysical magnitude estimates and simultaneously obtained auditory nerve, brainstem and cortical responses to click stimuli in man. Electroenephalogr. Clin. Neurophysiol. **43**(6): 802–812.
37. Kodera, K., O. Yamada, H. Hamane & J. I. Suzuki. 1978. Effects of number and interstimulus interval of tone pips on fast responses. Audiology **17**(6): 500–510.
38. Harkins, S. W., T. M. McEvoy & M. L. Scott. 1979. Effects of interstimulus interval on latency of the brainstem auditory evoked potential. Int. J. Neurosci. **10**(1): 7–14.
39. Hecox, K. E., A. C. B. Lastimosa, B. Mokotoff & R. Sandlin. Human brainstem auditory evoked potential: Rate effects. (Submitted for publication).
40. Stockard, J. J. & V. S. Rossiter. 1977. Clinical and pathologic correlates of brainstem auditory response abnormalities. Neurology **27**(4): 316–325.
41. Hecox, K., B. Cone & P. Cooper. 1977. Optimal stimulus selection in evoked response studies. Proc. San Diego Biomedical Symp. **16:** 67–69.
42. Yagi, T. & K. Kaga. 1979. The effect of the click repetition rate on the latency of the auditory evoked brainstem response and its clinical use for neurological diagnosis. Arch. Otorhinolaryngol. **222:** 91–97.
43. Don, M., A. R. Allen & A. Starr. 1977. Effect of click rate on the latency of auditory brainstem responses in humans. Ann. Otol. Rhinol. Laryngol. **86**(2): 186–195.
44. Stockard, J. E., J. J. Stockard, B. R. Westmoreland & J. L. Corfits. 1979. Brainstem auditory–evoked responses: Normal variation as a function of stimulus and subject characteristics. Arch. Neurol. **36:** 823–831.
45. Stockard, J. J., V. S. Rossiter & W. C. Weiderholt. 1976. Brainstem auditory evoked responses in suspected central pontine myelinolysis. Arch. Neurol. **33:** 726–728.
46. Schulman-Galambos, C. & R. Galambos. 1975. Brainstem auditory-evoked responses in premature infants. J. Speech Hear. Res. **18:** 456–465.
47. Salamy, A., T. Mendelson, W. H. Tooley & E. R. Chaplin. 1980. Differential development of brainstem potentials in healthy and high-risk infants. Science **210**(4469): 553–555.
48. Morgan, A. & B. Salle. 1980. A study of brainstem evoked responses in prematures. Acta. Otolaryngol. **89**(3–4): 370–375.
49. Galambox, R. & P. Despland. 1980. The auditory brainstem response (ABR) evaluates risk factors for hearing loss in the newborn. Pediatr. Res. **14:** 159–163.
50. Despland, P. A. & R. Galambos. 1980. The auditory brainstem response (ABR) as a useful diagnostic tool in the intensive care nursery. Pediatr. Res. **14**(2): 154–158.
51. Marshall, R. E., T. J. Reichert, S. M. Kerley & H. Davis. 1980. Auditory function in newborn intensive care unit patients revealed by auditory brainstem potentials. J. Pediatr. **96**(4): 731–735.
52. Hecox, K. & B. Cone. Prognostic importance of brainstem auditory evoked responses after asphyxia. Neurology (in press).
53. Sohmer, H. & M. Student. 1978. Auditory nerve and brainstem evoked responses in normal, autistic, minimal brain dysfunction and psychomotor retarded children. Electroencephalogr. Clin. Neurophysiol. **44:** 380–388.
54. Student, M. & H. Sohmer. 1978. Evidence from auditory nerve and brainstem evoked responses for an organic brain lesion in children with autistic traits. J. Austism Child. Schizophrenia **8**(1): 13–20.

55. Picton, T. W., S. A. Hillyard, H. I. Krausz & R. Galambos. 1974. Human auditory evoked potentials. I: Evaluation of components. Electroencephalogr. Clin. Neurophysiol. **36:** 179–190.
56. Picton, T. W., D. L. Woods & G. B. Proulx. 1978. Human auditory sustained potentials. II. Stimulus relationships. Electroencephalogr. Clin. Neurophysiol.
57. Mendel, M. I., C. D. Adkinson & L. A. Harker. 1977. Middle components of the auditory evoked potentials in infants. Ann. Otol. Rhinol. Laryngol. **86**(3): 293–299.
58. Wolf, K. E. & R. Goldstein. 1980. Middle component AERs from neonates to low-level tonal stimuli. J. Speech Hear. Res. **23**(1): 185–201.
59. Ohlrich, E. S. & A. B. Barnet. 1972. Auditory evoked responses during the first year of life. Electroencephalogr. Clin. Neurophysiol. **32:** 161–169.
60. Ohlrich, E. S., A. B. Barnet, I. P. Weiss & B. L. Shanks. 1978. Auditory evoked potential development in early childhood: A longitudinal study. Electroencephalogr. Clin. Neurophysiol. **44:** 411–423.

EVENT-RELATED POTENTIALS IN THE EVALUATION OF HIGH-RISK INFANTS*

Diane Kurtzberg

Departments of Neuroscience and Neurology
Rose F. Kennedy Center for Research
in Mental Retardation and Human Development
Albert Einstein College of Medicine
Bronx, New York 10461

Premature and low-brith-weight (LBW) infants represent the principal source of neonatal mortality and of later neurobehavioral disorders. Very-low-birth-weight (VLBW) infants, weighing 1500 grams or less at birth, have the poorest prognosis for mortality and morbidity due to the high incidence of severe physiological disturbances associated with the immaturity of vital body systems. With the introduction of neonatal intensive care, mortality of VLBW infants has been reduced substantially, even for infants less than 1000 grams.[1,2] The impact of neonatal intensive care upon neurobehavioral outcome remains unclear, however, due to the complex and variable pathophysiology of the VLBW newborn and the equally complex effects of the various modalities of therapeutic intervention.

The incidence of specific visual and auditory deficits among low-birth-weight infants is significantly greater than the normal full-term population. In addition to the well-documented visual impairment due to retrolental fibroplasia among those infants receiving ventilatory assistance with increased oxygen, follow-up studies reveal that from 6%–39% of LBW infants have significant refractive errors in the first few years of life compared with estimates of less than 2% in the normal preschool population.[3–5] Strabismus is seen in 6%–20% of LBW infants compared with 0.1%–3% in the normal population.[3,4,6] Hearing disorders occur in 2%–12.4% of LBW infants compared with an incidence of 0.26% among children born at term.[3,5]

The implementation of reliable clinical methods for evaluating the neurobehavioral and sensory development of high-risk infants is of clear importance not only for early detection of deviant infants to permit appropriate therapeutic interventation, but also for identifying the significant etiological factors that lead to later neurobehavioral impairment.

Electrophysiological evaluation of the high-risk neonate provides a direct source of information on the integrity of sensory pathways and brain structure during the course of development. Recordings of event-related cortical potentials also offer a unique insight into the development of brain function during infancy.

Evaluation at 40 Weeks Post-Conceptional Age

In the high-risk newborn population it is most useful to screen for visual, auditory, and more general cerebral dysfunction at 40 weeks post-conceptional age (PCA), which corresponds to term gestation. This is the earliest time at which a comparison can be made between the high-risk infant and the normal full-term population.

*This work was supported by Grants HD 11562, HD 01799, and MH 06723 from the U.S. Public Health Service and by a grant from the United Cerebral Palsy Foundation.

0077-8923/82/0388-0557 $1.75/0 © 1982, NYAS

Low-birth-weight infants are likely to be still in the hospital at 40 weeks PCA where it is feasible to carry out electrophysiologic testing. Unless a thorough neonatal neurobehavioral evaluation is carried out, infants with subtle sensory deficits or brain damage may be overlooked, only to return for evaluation when obvious developmental delays are noticed later by the parents, pediatrician, or the schools. Thus, the earliest possible detection of cerebral dysfunction is of great importance.

There are some difficulties inherent in neonatal evaluation. The neural activity recorded in these young infants is rather labile, so fluctuations in event-related potential (ERP) morphology and amplitude across runs are often seen. The mechanisms that underlie this variability of response are not well understood. Moreover, the brain responses of normal full-term infants appear to differ in maturational status at birth, so one must define very carefully the normal range for each ERP measure used to evaluate infants at risk.

Event-Related Potential Methods Available for Neonatal Assessment

Several techniques are available to evaluate neonatal visual and auditory function. The visual ERP assessment methods to be discussed include the visual evoked potential (VEP) to flash stimulation and pattern reversal and the cortical potentials associated with visual scanning. The auditory ERP include the brainstem auditory evoked potential (BAEP) and cortical AEP to tones and speech sounds.

VEP—Flash

The most practical technique for assessment of visual function in the neonatal period is the VEP to light flash stimulation. Although the flash VEP has been considered too variable for use in the evaluation of neonatal visual status,[7] measures can be taken to optimize its recording. Although these points may appear obvious they merit emphasis. Strict control of the technical aspects of stimulation (e.g., consistent head position with respect to light source) lessens this variability. Proper electrode placement is a particularly important consideration. Most studies have used a single or at most a few electrodes. However, electrode placements based upon estimation of intracranial anatomy from external landmarks may not be optimal for picking up responses from the cortical area of interest. This problem is especially important in young infants, whose brains are small and whose ERP components are often sharply circumscribed in scalp localization. This locus varies somewhat across babies due to individual differences in the position of cortical areas with respect to bony landmarks. In the adult, the relatively broad scalp distribution of most ERP components, due to thicker and less conductive tissues between the generator and scalp electrodes, renders precise positioning of electrodes less critical. By contrast, in infants, major differences in ERP waveshape may be observed over a 2 cm distance so that a component can be completely overlooked when a single electrode is utilized. Thus, even when the ERP of interest is known to be generated within a localized brain region, a small cluster of electrodes is desirable in order to optimize the recording. This is especially important in studies of abnormal subjects in whom waveform deviance is suspected.

Repeated runs are mandatory to evaluate the response variability and to enhance the likelihood of observing a normal response. The individual averages may vary in morphology, but as is the case with behavioral evaluations of the neonate, it is valid to judge normality based on an optimal response and to identify only consistently abnormal responses as deviant.

Since the flash VEP is not greatly affected by sleep state[8] and can be recorded with the eyes closed, it is especially useful for screening newborn infants who are difficult to maintain in an alert state. The validity of the VEP at term to light flash stimulation as a measure of visual and cerebral function in the neonatal period will be discussed later.

VEP—Pattern Reversal

VEP to pattern-reversal stimulation has certain advantages over the VEP to diffuse light flash in the assessment of visual function in infancy. Since the total amount of luminance is held constant, the VEP reflects the activity of mechanisms underlying pattern vision, not merely light sensitivity. In addition, objective measures of visual acuity can be obtained by systematically varying the element size within the pattern.[9] One serious disadvantage of this method when used with very young infants is the necessity for fixation on the visual display, which requires an alert and attentive baby. Although maintaining this optimal state during a testing session is possible in many infants, it is especially difficult in those high-risk infants in whom an assessment of visual function is most important. There are also substantial individual differences in maturational status of the visual system at term, so it is not possible to say whether a poorly defined resonse is deviant because of visual system dysfunction or immaturity. Thus, the clinical usefulness of pattern-reversal VEP in the newborn is questionable. In our experience, however, by 2 or 3 months of age most infants show clearly defined pattern VEP with similar morphology. Therefore, follow-up recording after 2 months with the more specific pattern-reversal VEP in infants who had displayed deviant flash VEP at term is recommended.

Cortical Potentials Associated with Visual Scanning

The third ERP technique that is useful for assessing infant visual function is the recording of brain potentials associated with active visual scanning. Normal visual behavior comprises a series of fixational pauses interrupted by saccadic eye movements, which repeatedly shift the direction of gaze to salient features of the visual scene. A direct index of the neural mechanisms underlying active vision is obtained by averaging the brain activity in association with the saccadic eye movements.

In the infant at term, the eye movement potentials recorded over the occipital region are relatively simple, consisting of a negative-positive complex (FIGURE 1). This is the classic "lambda" complex, which reflects both the change in retinal stimulation caused by movement of the eyes across a pattern field and visual input from each new fixation.[10] As with the pattern VEP, there is substantial intersubject variability in morphology and peak latency of the eye movement potentials recorded at term. Since the recording of these potentials requires that the infants be alert and actively scanning the target, these individual variations most likely reflect differences in maturational status of the visual system rather than effects of attention or arousal. By contrast, the eye movement potentials recorded at 2 months of age are quite similar in all infants. The components are more sharply defined and peak at shorter latencies. As with the pattern VEP, the clinical usefulness of recording potentials associated with visual scanning is doubtful before 2 or 3 months of age. In addition to the occipital lambda complex, responses arising from temporal and parietal regions are present in some normal infants as early as 40 weeks PCA. The significance of these components

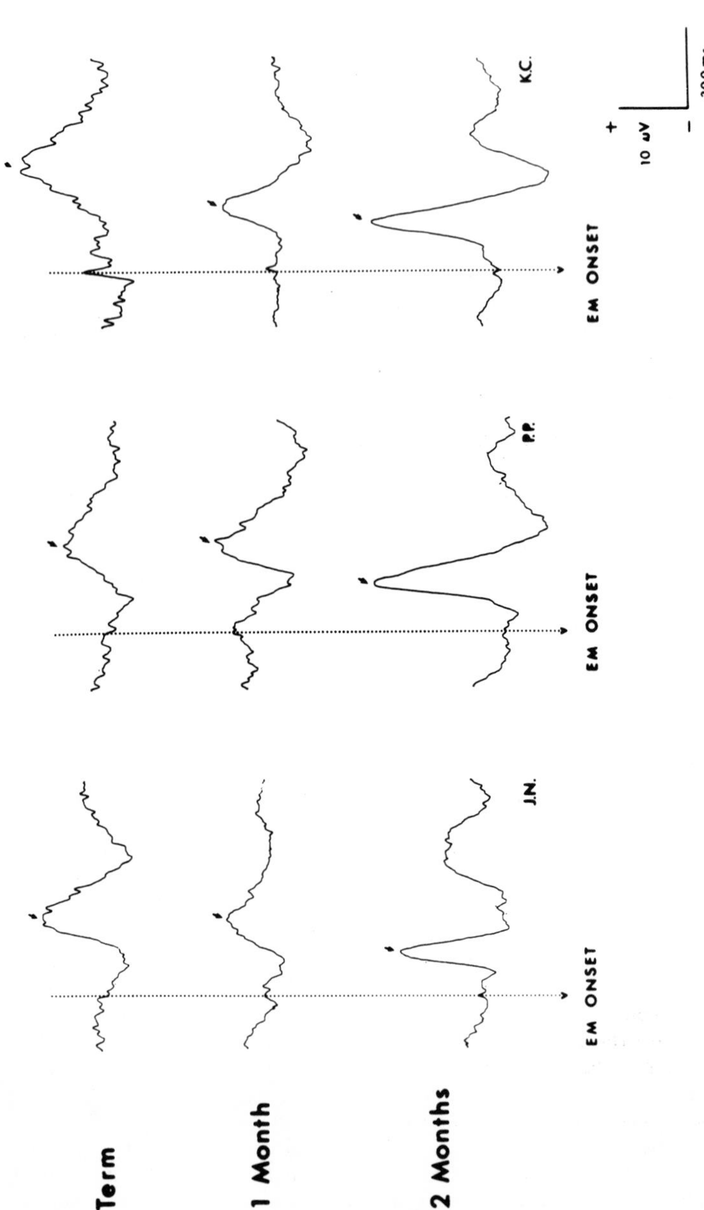

FIGURE 1. Occipital lambda response recorded from three infants at term, 1 month, and 2 months of age. Note the greater intersubject variability at term, which diminishes by 2 months.

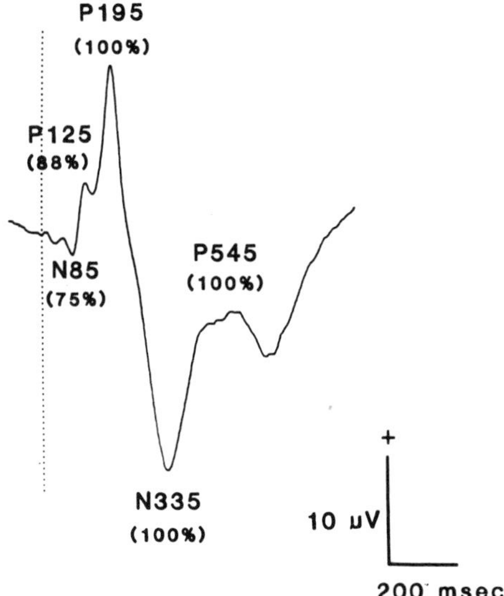

FIGURE 2. Normal VEP to flash recorded at term. Five major peaks are identified and the percentage of occurrence in the normal full-term population of each component is noted. In this and subsequent figures stimulus onset is indicated by a dotted line.

in the infant is yet to be determined, but they have been implicated as indices of higher spatial and verbal processing in older children and adults.[11]

The Clinical Usefulness of the Flash VEP at Term

The flash VEP recorded in the infant at term is complex, and all components may not be present in normal full-term infants' responses. Therefore, only the latency and amplitude characteristics of those components that are invariably present can provide normative standards for evaluating the responses of infants at risk. The occipital VEP comprises five major components of which three are present in all normal full-term responses (FIGURE 2). The first major negativity in our sample has a mean latency of 85 (± 8.2) msec and is present in 75% of the children. The first positivity peaks at 125 (± 26.4) msec and appears in 88% of the sample. The second major positivity, peaking at 195 (± 21.3) msec is present in all infants. Ellingson[7] also found this component to be the least variable in his studies of the newborn VEP. The negativity and positivity that follow peak at 335 (± 49.5) msec and 545 (± 60.9] msec and are also present in all normal infants.

For a response to be considered normal, the three invariably present components of the VEP must be observed over both right and left occipital areas at latencies within 2 SD of the normal control mean. Since absolute amplitude varies a great deal among normal full-term infants, we use amplitude only for the definition of deviant hemispheral asymmetries. Deviant responses are classified into three categories:

(1) Atypical waveshape (FIGURE 3), in which the major components of the VEP are absent or do not conform in configuration and latency characteristics to the normal response, bilaterally.

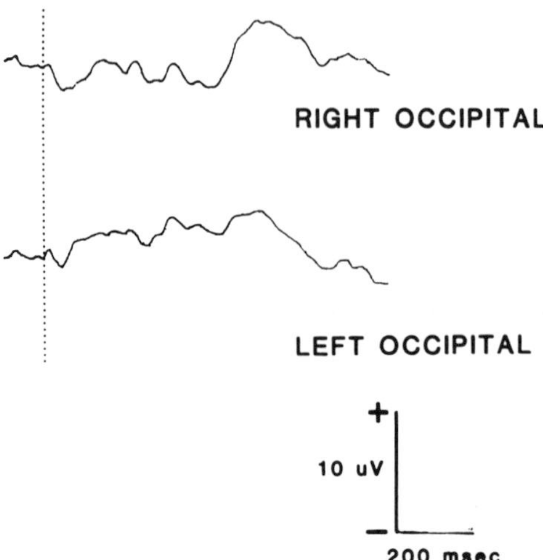

FIGURE 3. Deviant VEP recorded from a very-low-birth-weight (VLBW) infant at 40 weeks post-conceptional age (PCA) depicting bilaterally atypical waveshape. Note the absence of the three major components appearing in normal full-term infants.

(2) Significant hemispheral asymmetry in either morphology or amplitude (FIGURE 4).

(3) An immature response (FIGURE 5) which is inappropriate in morphology and latency for the infant's age, but consistent with responses that are normal during the preterm period.

In a group of 79 VLBW infants under study in a comprehensive follow-up program, 40 babies (51%) demonstrated normal flash VEP at 40 weeks PCA. Fourteen (18%) displayed hemispheric asymmetries of occipital responses. These included asymmetrical waveshape in 8 and significant amplitude asymmetries in 6. Nineteen babies (24%) showed immature responses, and 6 (7%) demonstrated atypical responses.

Of interest is the concordance between the flash VEP and structural characteristics of the brain as assessed by CT scan, which each of the infants in the study received at 40 weeks PCA. CT scans were classified into four groups: (1) normal; (2) diffuse periventricular low density (D-PVLD) extending from zones contiguous with the frontal and occipital horns around and above the bodies of the lateral ventricles; (3) D-PVLD with an associated abnormality such as intraventricular hemorrhage, enlarged ventricles or porencephalic cyst; and (4) scans with lesions only, without areas of low density.

Although the incidence of normal CT scans in this group of VLBW infants is only 24%, 88% of the normal CT group had normal VEP at 40 weeks. The incidence of D-PVLD either with or without associated lesions was 92% in infants with asymmetric occipital VEP, 100% in babies with immature responses and 80% in babies with atypical responses. That the VEP at 40 weeks may be sensitive to D-PVLD is not altogether surprising since the geniculo-calcarine tract passes through the periventricular white matter. It appears that the VEP may differentiate cases with a greater involvement of the subcortical visual pathways, a finding of potential prognostic significance.

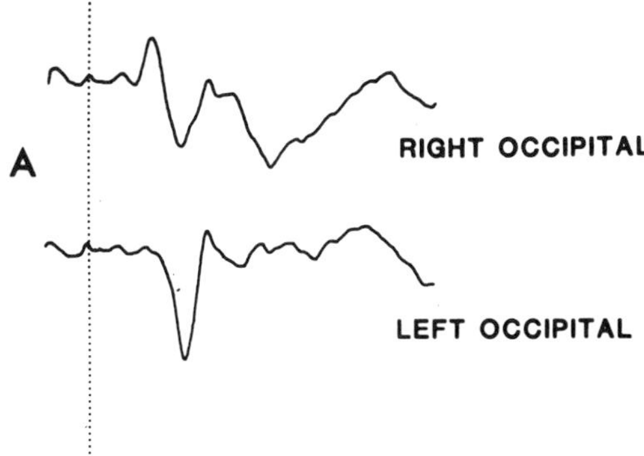

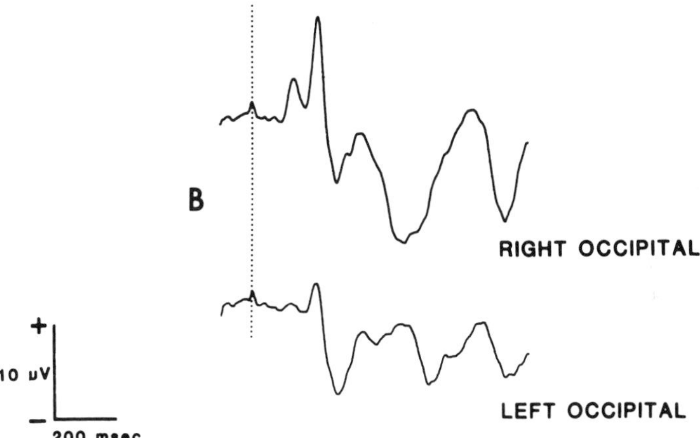

FIGURE 4. Deviant VEP to flash recorded from two VLBW infants at 40 weeks PCA illustrating morphology (A) and amplitude (B) asymmetries.

The validity of the term VEP to flash in the evaluation of visual function was assessed by follow-up electrophysiological studies of VEP to flash and pattern reversal and cortical potentials associated with visual scanning, at monthly intervals through 6 months, and bimonthly to 12 months of age (corrected for term gestation).

An infant was considered normal if no persistent deviancies during each 6-month period for the three visual ERP measures were seen. An infant showing persistent abnormalities in two of the three ERP measures was considered deviant. FIGURE 6 illustrates for each term VEP group the incidence of deviant visual ERP during the first and second 6 months of life.

Of the babies with normal flash VEP at 40 weeks PCA, 92% had normal visual

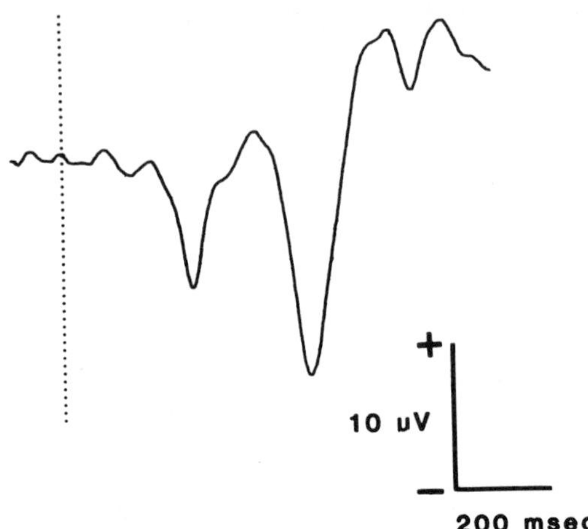

FIGURE 5. Deviant immature VEP recorded from a VLBW infant at 40 weeks PCA. Note the absence of large positivity at 200 msec. This reponse is typical of normal preterm infants at 35 weeks PCA.

ERP during the first 6 months of life and 97% were normal at one year. Abnormalities persisted during the first year of life in more than 50% of the asymmetric and immature groups and in more than 75% of the atypical group. These findings support the value of the VEP at 40 weeks PCA for screening of visual function, since an infant who shows a normal flash VEP will most likely remain normal, even when tested with more specific electrophysiologic measures. Conversely, babies with deviancies at term can be retested during the first few months of life to assess the persistence and nature of the apparent dysfunction.

In addition to the electrophysiologic evaluations, each infant was given standard neurological examinations during the first year of life, which were categorized as normal, suspect, or abnormal. The examination results at 7 months and 1 year for normal and deviant VEP groups are illustrated in FIGURE 7.

Although the incidence of deviant neurological findings is high among these VLBW babies, the probability of having a normal neurological examiniation at 7 and 12 months of age is greatest for those infants with normal term VEP. It is noteworthy that the incidence of abnormal neurological examinations decreases from 7 to 12 months of age, especially among the infants in the deviant 40 week VEP group. Yet the incidence of normal 12 month neurological exams is 54% in the normal VEP group contrasted with only 17% in the deviant group.

The results of developmental follow-up examinations using the Bayley Scales at 7 and 12 months are illustrated in FIGURE 8. A Mental Development Index (MDI) score of less than 85 is considered deviant. Those infants with normal VEP at 40 weeks have the highest probability of scoring within the normal range at both 7 and 12 months. As with the neurological evaluation, the number of infants with abnormal MDI scores decreased from 7 to 12 months.

Thus, the three follow-up measures (electrophysiologic, neurological, and developmental evaluations) concur to varying degrees in that the probability of having normal follow-up at one year is highest for those babies with normal 40 week PCA VEP. It appears that the flash VEP recorded at term provides a valid measure of both visual and more general cerebral status. The ease of recording the flash VEP taken together

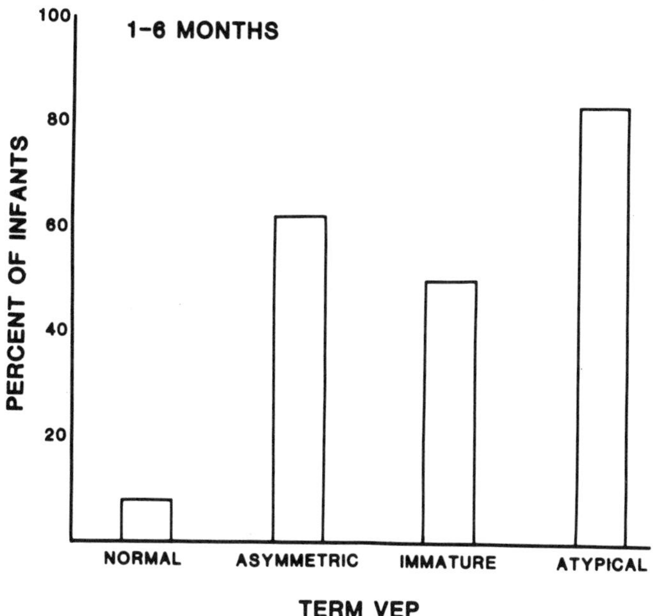

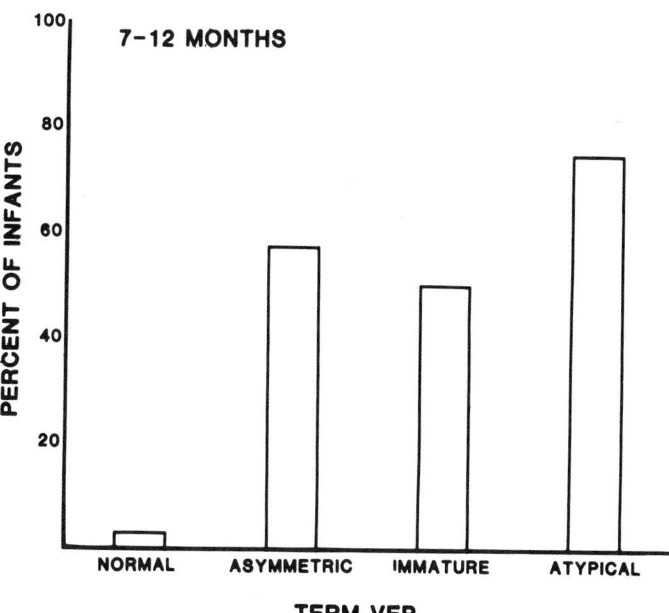

FIGURE 6. Incidence of deviant visual ERP (VEP to flash and pattern reversal, cortical potentials associated with visual scanning) during the first and second 6 months of life for each VEP group at term.

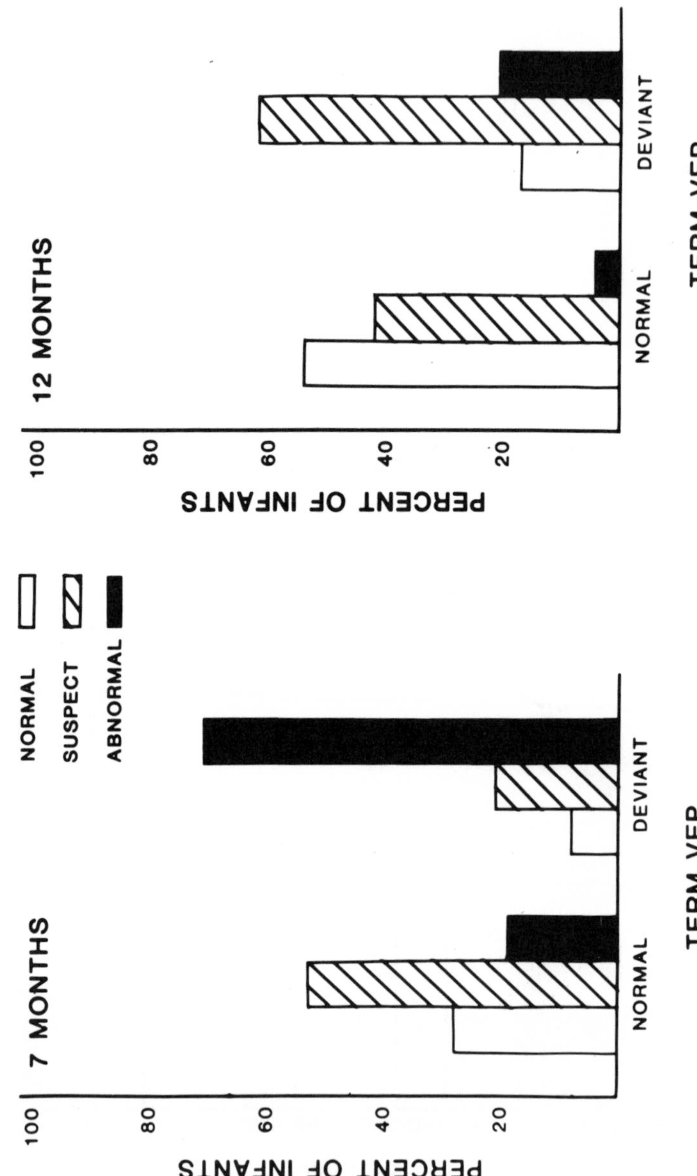

FIGURE 7. Neurological examination results at 7 months and 12 months of age for normal and deviant VEP groups.

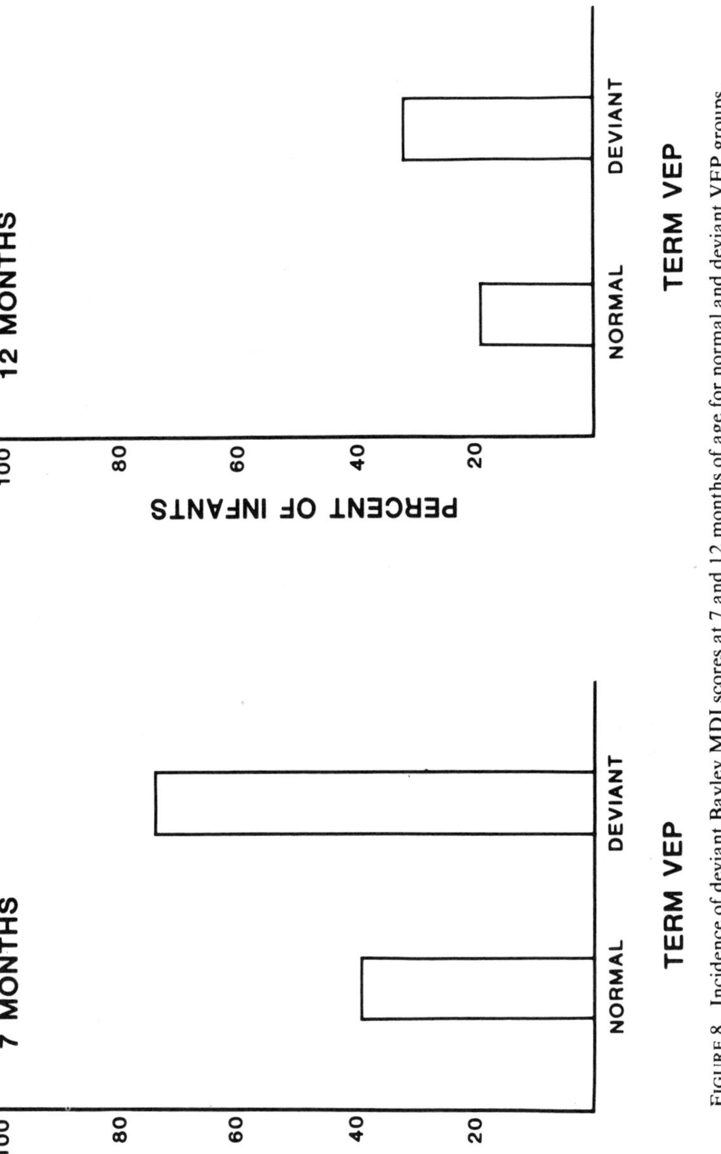

FIGURE 8. Incidence of deviant Bayley MDI scores at 7 and 12 months of age for normal and deviant VEP groups.

with its sensitivity in detecting potential deviant visual function during the first year of life appear to make it an extremely useful device for screening purposes in the newborn period.

I now briefly consider auditory system evaluation. The availability of specific electrophysiologic techniques to assess the functional integrity of both subcortical and cortical auditory structures provides an opportunity to identify the infant with potential deficits in peripheral or central auditory processing at a very early age.

Brainstem Auditory Evoked Potentials (BAEP)

The clinical usefulness of BAEP recording in the evaluation of both auditory and brainstem function in adults is well established.[12,13] The application of this technique to screen high-risk newborn infants for auditory dysfunction has disclosed an incidence of hearing deficits ranging from 2%–16% among infants in neonatal intensive care units.[14–18] Our experience parallels these findings. In 54 VLBW infants tested at 40 weeks PCA, 76% showed normal BAEP bilaterally. Among the 13 infants with abnormal 40 week responses, 7 had normal BAEP on repetition one month later. Thus, the incidence of persistent deviant BAEP in this group of VLBW infants was 11%.

Cortical Auditory Evoked Potentials

That an infant shows a normal BAEP at 40 weeks provides some evidence of the integrity of subcortical auditory pathways. The BAEP, however, does not assess the functional status of the mechanisms associated with more complex aspects of auditory processing at the cortical level. Because of the technical simplicity and reliability of BAEP recording, the use of cortical AEP for assessment purposes has been relatively neglected. Cortical responses, however, permit the assessment of higher order auditory processing, especially those aspects that are necessary for language acquisition.

To this end, we have recorded AEP to tones and speech sounds in our cohort of VLBW infants. AEP to consonant-vowel syllables differing in voice onset time (/DA/, /TA/) and 800 Hz tones are being recorded at 40 weeks PCA and at follow-up.

The AEP at term recorded from an electrode placed at midline halfway between F_Z and C_Z consists of four major waves, with latency characteristics that vary according to the stimulus (i.e., voice onset time) (FIGURE 9). A positive-negative-positive waveform followed by a slow negativity comprises the normal full-term response. In some VLBW infants at 40 weeks PCA, immature responses, which consist of a surface negative response without the prominent positive wave, are seen. This immature response resembles the AEP recorded in healthy premature infants at 35 weeks PCA. Of the VLBW infants in our sample, 20% had immature AEP at term.

The topography of the AEP to speech sounds and tones is complex (FIGURE 10). There is a differential maturation of responses recorded from the midline in contrast to those seen in more lateral placements. At term, the lateral response is most often characterized by a large negativity that becomes positive by two months of age. This presumably reflects differential rates of maturation in two distinct generators, one most probably located in the superior temporal plane that projects its field upward toward the vertex,[19,20] and another, later maturing generator located along the lateral surface of the temporal lobe. The appropriate development of mature responses from this secondary auditory area, which includes the classic Wernicke's area in the adult,

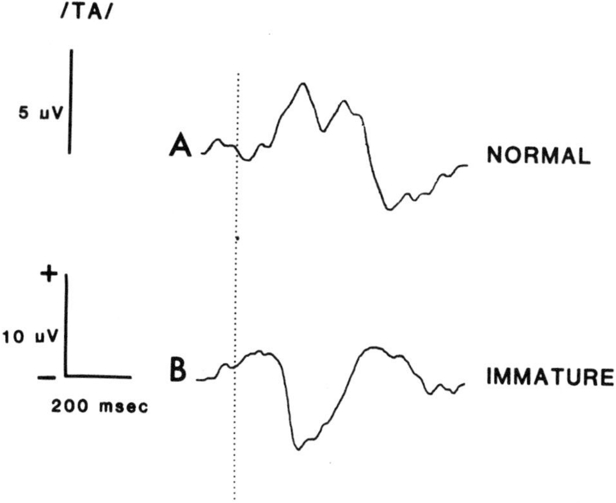

FIGURE 9. Cortical auditory evoked potentials to the speech sound /TA/ recorded from a normal full-term infant (A) and a VLBW infant at 40 weeks PCA (B). The normal response is characterized by a surface positive complex in contrast to immature response, which is surface negative and typical of that recorded in the preterm period. Electrode placement is halfway between F_Z and C_Z referred to O_Z.

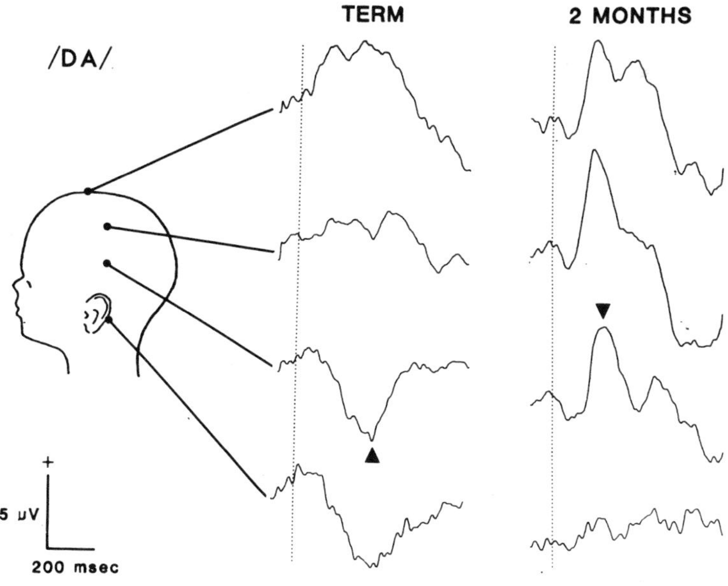

FIGURE 10. Coronal distribution of AEP to the speech sound /DA/ recorded at term and at 2 months. Note the surface negative response recorded over lateral temporal cortex, which becomes positive by 2 months of age.

may be crucial for higher level auditory processing and language acquisition. After 2 months of age, the responses change very little in morphology, but the components peak at progressively shorter latencies.

The clinical usefulness of the cortical AEP to tones and speech sounds has yet to be demonstrated. However, the maturational status of the two distinct auditory cortical areas can be evaluated, at term for the midline response, and by 2 months of age for the more lateral potentials. Judging by the neurobehavioral outcome of the infants who show an immature VEP at term, we would expect that babies displaying similar age-inappropriate responses to auditory stimulation would be at higher risk for subsequent neurobehavioral dysfunction.

Conclusions

The usefulness of 40 week post-conceptional age electrophysiologic testing of visual and auditory function and evaluation of cerebral status rests with the possibility for identifying individual infants with actual or potential neurological damage.

The ERP methods that are presently most useful for a term evaluation are the VEP to flash and the BAEP. Further study is needed, but it is likely that the cortical AEP to tones and speech sounds will yield information on the integrity of cortical auditory processing centers. Subsequent electrophysiological re-evaluation during the first year of life in infants with deviant term visual and auditory ERP permits the identification of those children with persistent abnormalities for whom appropriate therapeutic measures are required.

Acknowledgment

The author wishes to acknowledge the collaboration of Herbert G. Vaughan, Jr., Judith Kreuzer, Pamela Hilpert, Cecelia McCarton-Daum, Holly Ruff, Alice Edelman, and Allan Danziger in the Low Birth Weight Infant Follow-Up Program, and to thank Linda Murphy for preparation of the manuscript.

References

1. Stewart, A. L. 1977. Prognosis for infants weighing 1000 grams or less at birth. Arch. Dis. Child. **52:** 97–104.
2. Yu, V. Y. H. & E. Hollingsworth. 1979. Improving prognosis for infants weighing 1000 grams or less at birth. Arch. Dis. Child. **55:** 422–426.
3. Kitchen, W. H., M. M. Ryan, A. Rickards, A. B. McDougall, F. A. Billson, E. H. Keir & F. D. Naylor. 1980. A longitudinal study of very low-birthweight infants. IV: An overview of performance at eight years of age. Develop. Med. child Neurol. **22:** 172–188.
4. Kumar, S. P., E. K. Anday, L. M. Sacks, R. Y. Ting & M. Delivoria-Papadopoulos. 1980. Follow-up studies of very low birth weight infants (1,250 gram or less) born and treated within a perinatal center. Pediatrics **66:** 438–444.
5. Bjerre, I. & E. Hansen 1976. Psychomotor development and school-adjustment of 7-year-old children with low birthweight. Acta Paediatr. Scand. **65:** 88–96.
6. Schulte, F. J. & E. Sterrert. 1978. Hearing defects in preterm infants. Arch. Dis. Child. **53:** 269–270.
7. Ellingson, R. J., G. H. Lathrop, T. Danahy & B. Nelson. 1973. Variability of visual evoked potentials in human infants and adults. Electroencephalogr. Clin. Neurophysiol. **34:** 113–124.

8. BARNET, A. B., S. L. FRIEDMAN, I. P. WEISS, E. S. OHLRICH, B. SHANKS & A. LODGE. 1980. VEP development in infancy and early childhood. A longitudinal study. Electroencephalogr. Clin. Neurophysiol. **49:** 476–489.
9. SOKOL, S. 1978. Measurement of infant visual acuity from pattern reversal evoked potentials. Vision Res. **18:** 33–39.
10. KURTZBERG, D. & H. G. VAUGHAN, JR. 1977. Electrophysiological observations on the visuomotor system and visual neurosensorium. *In* Visual Evoked Potentials in Man: New Developments. J. E. Desmedt, Ed. pp. 314–331. Clarendon Press, Oxford.
11. KURTZBERG, D. & H. G. VAUGHAN, JR. 1979. Maturation and task specificity of cortical potentials associated with visual scanning. *In* Human Evoked Potentials—Applications and Problems. D. Lehmann and E. Callaway, Eds. pp. 185–199. Plenum Press, New York, N.Y.
12. STOCKARD, J. J. & V. S. ROSSITER. 1977. Clinical and pathologic correlates of brain stem auditory response abnormalities. Neurology **27:** 316–325.
13. CHIAPPA, K. H., J. L. HARRISON, E. B. BROOKS & R. R. YOUNG. 1979. Brainstem auditory evoked responses in 200 patients with multiple sclerosis. Ann. Neurol. **7:** 135–143.
14. DESPLAND, P. A. & R. GALAMBOS. 1980a. Use of the auditory brainstem responses by prematures and newborns infants. Neuropädiatrie **11:** 99–108.
15. DESPLAND, P. A. & R. GALAMBOS. 1980b. The auditory brainstem response (ABR) is a useful diagnostic tool in the intensive care nursery. Pediatr. Res. **14:** 154–158.
16. COX, C., M. HACK & D. METZ. 1981. The brainstem-evoked response audiometry: Normative data from the preterm infant. Audiology. **20:** 53–64.
17. GALAMBOS, R. & P. A. DESPLAND. 1980. The auditory brainstem response (ABR) evaluates risk factors for hearing loss in the newborn. Pediatr. Res. **14:** 159–163.
18. SALAMY, A., T. MENDELSON, W. H. TOOLEY & E. R. CHAPLIN. 1980. Differential development of brainstem potentials in healthy and high-risk infants. Science **210:** 553–555.
19. VAUGHAN, H. G., JR. & W. RITTER. 1970. The sources of auditory evoked responses recorded from the human scalp. Electroencephalogr. Clin. Neurophysiol. **28:** 360–367.
20. VAUGHAN, H. G., JR., W. RITTER & R. SIMSON. 1980. Topographic analysis of auditory event related potentials. *In* Motivation, Motor and Sensory Processes of the Brain: Electrical Potentials, Behaviour and Clinical Use. Prog. Brain. Res. **54:** 279–285. Elsevier, Amsterdam.

SIMULTANEOUS ELECTRORETINOGRAMS AND EVOKED POTENTIALS

John C. Armington

Department of Psychology
Northeastern University
Boston, Massachusetts 02115

Simultaneous recordings of the human electroretinogram (ERG) and visual evoked cortical potential (VECP) are frequently made in an effort to monitor the processing of the visual information as it travels from the receptors to the cortical level. This goal is not always achieved with high success, however, because the recordings performed at the two sites may give information of an entirely different character. For example, unless proper precautions are taken, the evoked potential will reflect photopic activity initiated by stimulation of the central part of the retina while the electroretinogram will be scotopic and arise from stimulation of the peripheral retina. The complexity of the evoked potential is well recognized, but the electroretinogram is not as simple a phenomenon as it seems either. Thus, the complex functioning of each of these two potentials must be considered in any study comparing the actions of different levels of the visual system.

It is clear that some of the confusing results that exist in parallel studies of the evoked potential and the electroretinogram can be brought into a better relation with one another when attention is given to the various stimulus conditions under which they were recorded. The present discussion will review some of stimulus conditions that influence recordings, and several examples will be given of their important bearing on the type of result produced. Specific mention will be made of problems arising from *stray light* within the eye, the *temporal* rate of stimulation, the *level of the eye's adaptation* and *stimulus pattern.* Some attention will also be given to the problem of *response measurement.*

Stray light. Stray light is a particular concern in the case of the electroretinogram.[1] When it is elicited by stimuli that flash regardless of whether they are patterned or not, the electroretinogram is generated by virtually all of the retina acting in concert. This happens because a significant proportion of the light entering the eye that is intended for the image area is actually distributed to much broader regions. Some is scattered before it reaches the retina, and some is reflected without being absorbed within the image area. When the stray light flashes it will participate in initiating the electroretinogram, and hence, it will be difficult to follow the action of a limited sector of the visual field. When stimuli are directed to the center of the visual field, the evoked potential will be less influenced by stray light than the electroretinogram.[2] The two responses are thus apt to provide information regarding separate parts of the visual field rather than the transfer of information from one level of the visual system to another. The problems of stray light may be reduced by delivering carefully tailored background light to all regions outside of the image area,[3] but a much simpler solution is to use patterned stimuli that are presented in temporal phase alternation.[4] When using this latter approach, however, attention must also be given to the effects of different temporal rates of stimulation on the responses.

Temporal Rate. When presenting stimuli in phase alternation, it is important to remember that the stimulus actually is flickering within small local patches of the total stimulus area. Because of this, one must not attribute effects to pattern when

0077–8923/82/0388–0572 $1.75/0 © 1982, NYAS

they can be explained in terms of flicker alone. Rather complicated properties of the electroretinogram may be revealed with flicker. The amplitude of the electroretinogram does not simply decrease as the temporal rate is increased; there is a range where it actually grows larger.[5] Best has shown that certain flickering stimuli produce subharmonics;[6] in this case every other flash in a flickering train produces a large B wave. The flashes in between produce small ones. Flicker lets components of the electroretinogram, which would be masked by large scotopic B waves if single flashes were used, stand out more clearly. Thus, although changes can be seen in the ordinary electroretinogram when the optic nerve is severed,[7] they are much more evident with stimulus alternation.[8] Whether the advantages of stimulus alternation in this case are due to temporal factors or to pattern, *per se*, is not yet known. Flicker also exerts effects on the visually evoked potential, but these have been less well worked out than for the electroretinogram.

Adaptation. Adaptational variables interacting with those of stray light and of temporal rate further complicate the interpretation of simultaneous recordings. When flashing lights of relatively high intensity are used, and when they are imaged on a central part of the retina, the evoked potential will be photopic, and its spectral sensitivity will approximate the photopic CIE curve. The electroretinogram, however, will be scotopic because it will be produced chiefly by the peripheral retina through the action of stray light.[9] This problem, as already mentioned, may be corrected by using alternating stimulus patterns.[2] FIGURE 1 shows spectral curves for the VECP and the ERG at a high and a low level of light adaptation. The alternating stimulus was presented in the near periphery of the visual field. The data, connected with straight lines, are compared with smooth photopic and scotopic psychophysical functions. A Purkinje shift is evident with the shift from rod to cone vision. Yet, it must still be recognized that although the spectral curves for the evoked potential and the electroretinogram are similar, they are not identical. The evoked potential exhibits more sensitivity to the long wavelengths than does the electroretinogram. Even under carefully selected conditions, a perfect match is not obtained.

Pattern. The data relating the evoked potential and the electroretinogram to pattern vision may be the least simple of all. The striped or checkered patterns that are used as stimuli in vision research are described in terms of their spatial frequency. The results of experiments with pattern are frequently presented as plots of amplitude against spatial frequency. FIGURE 2 summarizes typical results.[10] The electroretinogram is usually described as responding best to low spatial frequencies and as dropping off in amplitude as the pattern becomes finer. The evoked potential, on the other hand, responds best to a two-dimensional pattern of intermediate spatial frequency and decreases in amplitude at both low and high frequencies. Actually, there is some evidence that the electroretinogram may also drop off at very low frequencies under particular testing conditions.[11] The effect may depend on the rate at which the stimulus pattern is alternated. Korth has recently shown that the oscillatory component of the electroretinogram possesses a degree of spatial tuning.[12] When oscillations are superimposed on the B wave, they too could help produce a tuning effect. The important point for the present discussion, however, is that any potential peak in the response versus spatial frequency relation appears at a much lower spatial frequency for the electroretinogram than for the visually evoked cortical potential as may be seen in FIGURE 2.

Mitchell Brigell and the present author have recently obtained spatial frequency data at a high (or photopic) and a low (or scotopic) level of light adaptation.[13] Specific positions on the retina were tested extending from a centrally fixated region out to a region approximately 10 degrees in the visual periphery. The stimuli appeared to the subjects as a series of rings of approximately the same thickness but of differing radius

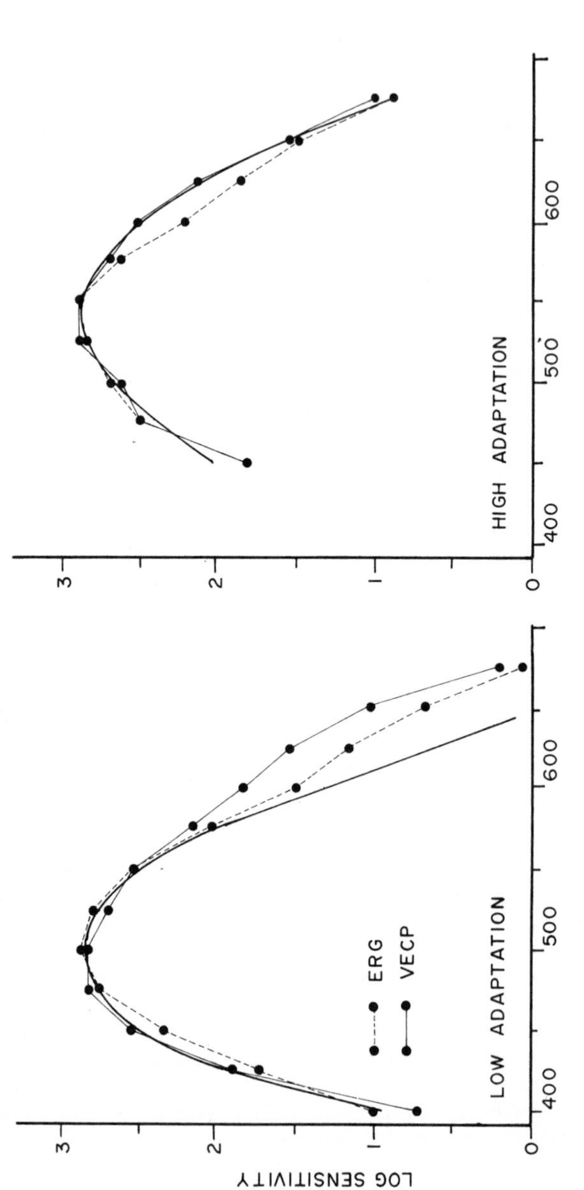

FIGURE 1. Spectral sensitivity of the electroretinogram and the evoked potential at a photopic and a scotopic level of adaptation. (After Korth & Armington.[2])

(FIGURE 3). The rings were composed of checks whose spatial frequency could be set by the experimenter. Thus, it was possible to compare the action of the near periphery with that of the central visual field. As is true of all stimuli presented in temporal phase alternation, the luminance of the checks determined both the level of stimulation and the level of the eye's adaptation.

Examples of recordings are shown in panel A at the left of FIGURE 4. At the high level of stimulation the electroretinogram was characterized by a small A wave, a sharp photopic B wave, and a rounded late negative potential. At the low level the

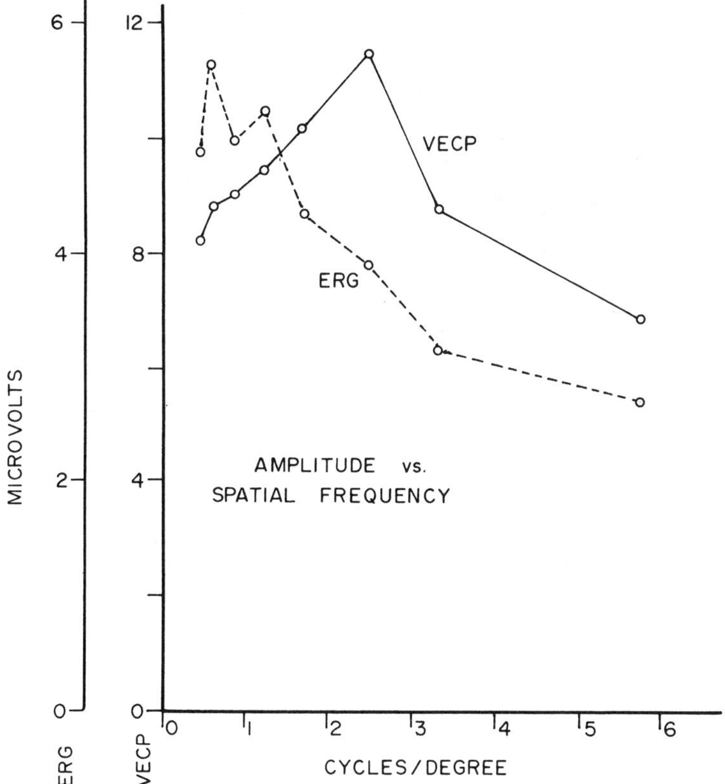

FIGURE 2. Amplitude of the electroretinogram and the evoked potential as a function of the spatial frequency of the test pattern. A large test area was used. (After Armington.[10])

electroretinogram was too small for accurate measurement. In fact, at the low level, the electroretinogram is so insensitive that it can be elicited only with stimuli subtending relatively large areas.[14] At the high level of stimulation, the evoked potential showed a clear, upward positive potential (labeled P) followed by a negative downward wave (N). These same waves appeared at the low level, but with increased latency. The double arrows indicate how peak to peak amplitude measures were made of the responses.

FIGURE 5 plots the amplitude of the evoked potential and of the electroretinogram

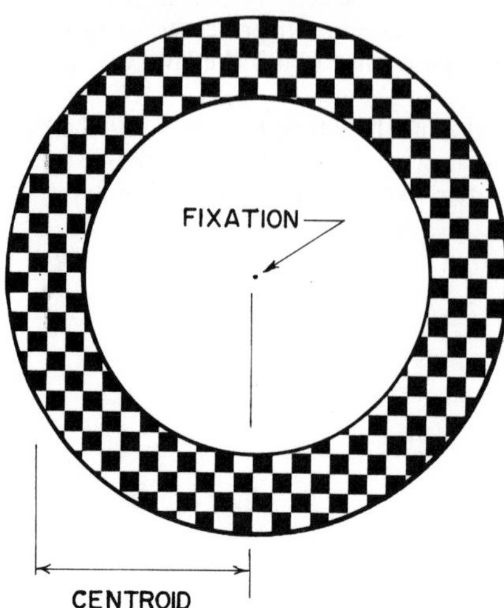

FIGURE 3. Test pattern for investigating pattern response. The centroid defined the position of the stimulus as measured from the fovea in units of visual angle.

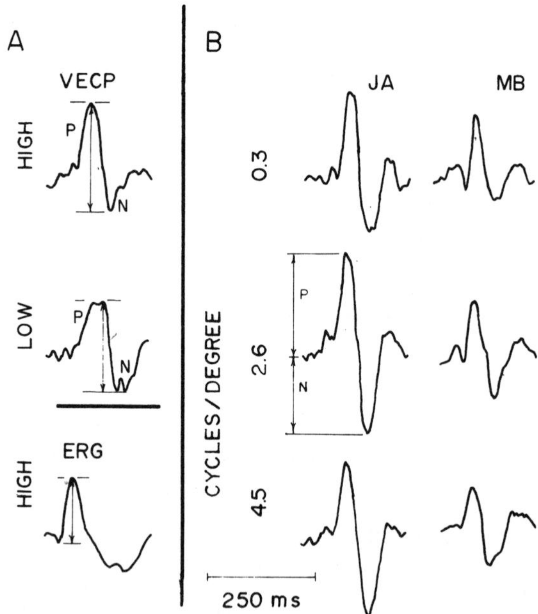

FIGURE 4. Typical responses to checkered stimuli. Panel (A) shows evoked potentials from subject JA for a high (photopic) and a low (scotopic) level of adaptation. A high level ERG is also shown. Panel (B) compares evoked potentials for two subjects at the high level of adaptation and for three spatial frequencies.

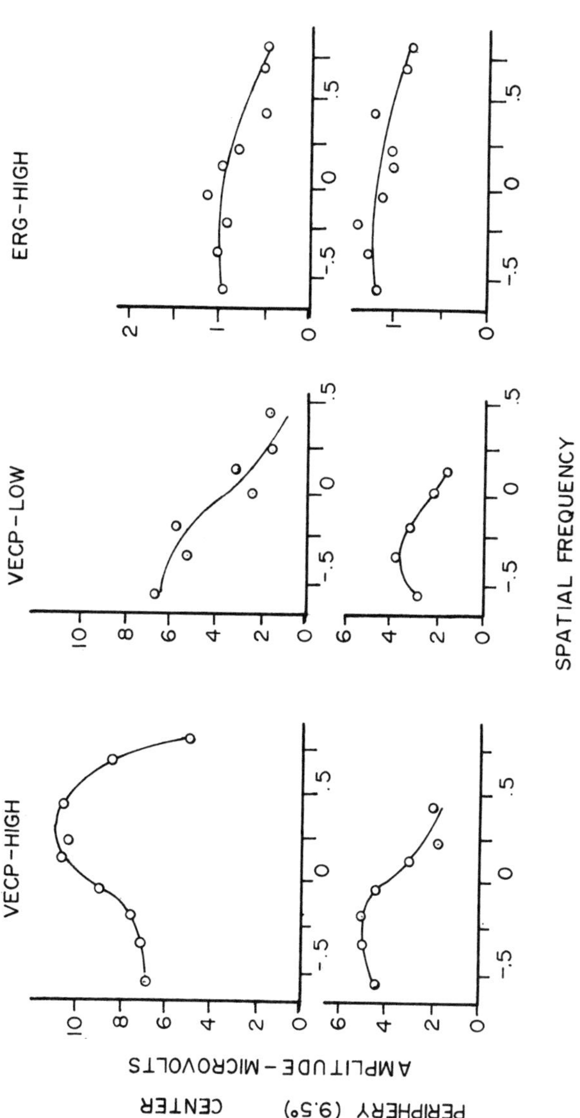

FIGURE 5. Amplitude versus spatial frequency plots for a central and a peripheral stimulus position. (After Armington & Brigell.[13])

as a function of the spatial frequency of the stimulus for a central and for a peripheral position. Data are shown for the evoked potential for both levels of stimulation and for the electroretinogram at the high level only. The spatial frequency that produces the largest evoked potential varies with retinal location and the level adaptation. When the stimulus was directed to the central retina and the eye was adapted to the high level of stimulation, relatively fine patterns produced the largest response. Coarse patterns were needed at the low level. Stimuli of intermediate texture were optimum in the peripheral position. These changes may be accounted for in terms of scotopic and photopic action and in the changing sizes of receptive fields.

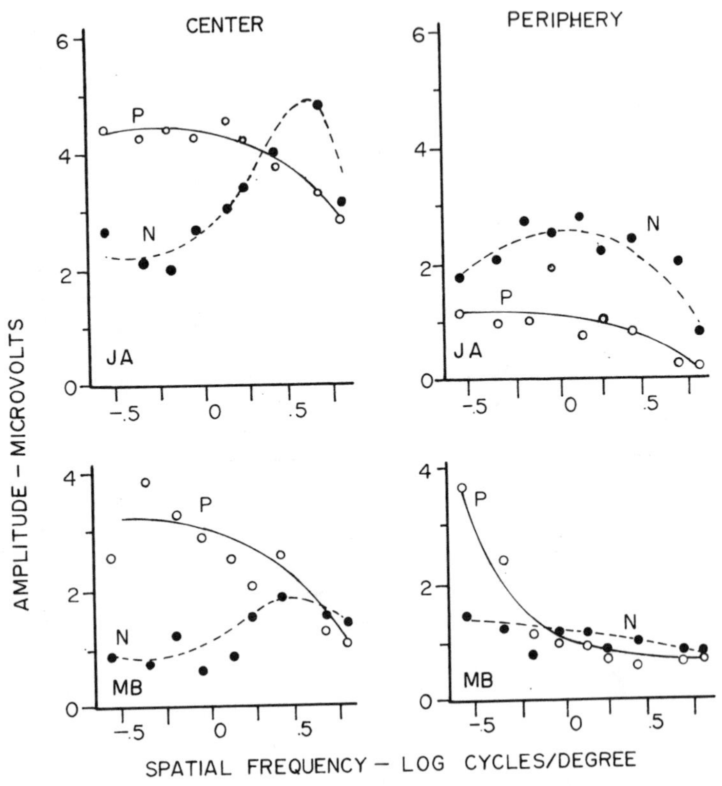

FIGURE 6. Changes in the amplitude of the positive and negative peaks of the VECP in two subjects and at two retinal locations. High level of adaptation.

The electroretinogram showed essentially the same characteristics for all conditions over which it could be measured. It was always large with the coarse frequencies and dropped off as the stimulus became very fine.

Response Measurement. It must be noted that the measured relative sensitivity of the VECP to patterns of various texture may be contingent on other variables in addition to the level of adaptation and the retinal location of the stimulus. Thus, the temporal rate of stimulation is likely important. The manner in which the response is measured must be considered also. The figures just examined were based on peak-to-peak measures. In a related series of experiments, the components, marked P and N, were measured from the baseline as indicated in panel B of FIGURE 4. The records for two subjects are shown. Plots of amplitude versus spatial frequency derived from these measures are shown in FIGURE 6. Subject JA's potentials are larger overall than those of MB, but more important is the difference in waveform. Wave P is relatively large in subject MB while wave N is large in JA.

The plots (FIGURE 6) present a picture of considerable complexity. Certain trends can be seen, however—particularly for the central stimulus position. P is favored by low spatial frequencies; N is favored by middle and high frequencies. Relatively simple plots such as seen in FIGURE 4 are possibly a composite of several spectral processes. In any event, it is clear that responses may be measured so as to reveal either diverse or similar properties both with respect to pattern sensitivity and individual functioning.

These data and many others purporting to trace visual activity from stimulus to the retina and from thence to the cortex illustrate the problems that can be encountered. Elaborate transformations take place at all functional levels. One or another visual subsystem may be reflected by any given measure. Progress is to be made only through a careful consideration of stimulus, recording, and measuring procedures. When this is done, many of the apparent conflicts in experimental results and their interpretation will be resolved.

REFERENCES

1. ARMINGTON, J. C. 1974. The Electroretinogram. Academic Press, New York, N. Y.
2. KORTH, M. & J. C. ARMINGTON. 1976. Stimulus alternation and the Purkinje shift. Vision Res. **16:** 703–711.
3. AIBA, T. S., M. ALPERN & F. MAASEIDVAAG. 1967. The electroretinogram evoked by the excitation of human foveal cones. J. Physiol. **189:** 43–62.
4. RIGGS, L. A., E. P. JOHNSON & A. M. L. SCHICK. 1964. Electrical responses of the human eye to moving stimulus patterns. Science **144:** 567.
5. VAN DER TWEEL, L. H. 1961. Some problems in vision regarded with respect to linearity and frequency response. Ann. N.Y. Acad. Sci. **89:** 829–856.
6. BEST, W. & K. BOHNEN. 1957. Untersuchungen über das Elektroretinogramm des Menschen bei verwendung farbiger lichtreize. Bibl. Ophthalmol.: 77–86.
7. FEINSOD, M., H. ROWE & E. AUERBACH. 1971. Changes in the electroretinograms in patients with optic nerve lesions. Doc. Ophthal. Mol. **29:** 169–200.
8. MAFFEI, L. & A. FIORENTINI. 1981. Electroretinographic responses to alternating gratings before and after section of the optic nerve. Science **211:** 953–955.
9. ARMINGTON, J. C. 1966. Spectral sensitivity of simultaneous electroretinograms and visual responses. *In* Clinical Electroretinography. H. M. Burian & J. H. Jacobson, Eds. pp. 225–233. Pergamon, Oxford, England.
10. ARMINGTON, J. C. 1978. Potentials accompany eye movement. *In* Frontiers in Visual Science. S. J. Cool & E. L. Smith III, Eds. pp. 460–470. Springer, New York, N.Y.
11. SOKOL, S. & K. JONES. 1977. The human electroretinogram and visually evoked potential as a function of changes in contrast, spatial frequency, and rate of alternation. Invest. Ophthalmol. Vis. Sci. (Suppl.) **16:** 158.
12. KORTH, M. 1981. Human fast retinal potentials and the spatial properties of a visual stimulus. Vision Res. **21:** 627–630.
13. ARMINGTON, J. C. & M. BRIGELL. 1981. Effects of stimulus location and pattern upon the visually evoked cortical potential and the electroretinogram. Int. J. Neuroscience. **13:** 169–178.
14. ARMINGTON, J. C. 1976. Spectral sensitivity of low level electroretinograms. Vision Res. **16:** 31–35.

CLINICAL AND EXPERIMENTAL EVIDENCE THAT THE PATTERN ELECTRORETINOGRAM (PERG) IS GENERATED IN MORE PROXIMAL RETINAL LAYERS THAN THE FOCAL ELECTRORETINOGRAM (FERG)

G. B. Arden, Vaegan, and C. R. Hogg

Electrodiagnostic Clinic
Moorfields Eye Hospital
London, EC1V 2PD
United Kingdom

It is widely recognized that the evoked response of the visual cortex consists of various components and that the resulting waveforms depend critically on the parameters of the stimulus used. Thus, the visually evoked potential (VEP) produced by contrast reversal of a periodic pattern without net luminance change, is considerably more sensitive to changes in visual capacity than is the potential evoked by a flash increase of luminance in a featureless field. We wish to report that the same is true for the electroretinogram (ERG).

In ophthalmic practice, the ERG is usually recorded with ganzfeld stimuli, in eyes well adapted to the dark. The response is large (approx 400 μV) and late because it is dominated by the effects of light of lower intensity scattered widely within the eye.[1] Hence it mainly reflects the activity of the peripheral, scotopic system. Gross abnormalities of dark adaptation and photoreceptor function can be readily determined. However, of the main components, the leading edge of the *a* wave represents receptor activity and the *b* wave is known to be largely produced in the distal retinal layers by retinal glia (the Müller cells),[2] in response to changes in extracellular potassium.[3,4] Although it remains a valuable clinical tool, such recordings have revealed little about the details of retinal physiology. It is commonly found that optic atrophies,[5] and amblyopia[5,6] produce no change in the ERG. Furthermore, it is often very difficult to detect any abnormality in the presence of gross macular degeneration, for, although the retina may be locally nonfunctional, the fovea comprises a very small area, in relation to the retina as a whole.

Attempts have been made to record from focal retinal areas by keeping the stimulus luminance at levels such that the reflected stray light will not exceed the threshold for the unstimulated parts of the retina.[7] The responses seen are small and the total luminous flux must of necessity be kept so low that the extremely small signals that are obtained can only be recorded in a physiological preparation. Photopic focal ERGs can be recorded in man if the flickering stimulus is presented within a large, bright surround.[8] The responses obtained are much smaller and faster than the usual ERG. When patients have localized retinal lesions, focal ERGs are particularly relevant, but if the lesion is macular, the necessary fixation cannot be maintained. Jacobson, Sandberg, Effron, and Berson[5] improved the technique by stimulating with a hand-held ophthalmoscope.

If any pattern with a 0.5 mark/space ratio is reversed in contrast there is no associated, nonfocal, stray light modulation because the mean luminance remains constant. Several groups have recorded potentials to such stimuli, hoping to isolate specifically photopic ERGs from defined retinal areas. (See References 9 for a historical and 10 for a more recent review). In such stimuli, at each change, the

0077–8923/82/0388–0580 $1.75/0 © 1982, NYAS

illumination of many small areas of retina is increasing and, simultaneously, that of an equal number of others is decreasing. All previous workers have presumed that the sum, on alternate triggers, of the focal ERG responses to *on* and *off* (hereafter referred to as the FERG) is identical to the response to pattern contrast reversal (the PERG).[11–16] In one study the responses to the two patterns were directly compared,[14] but the FERGs and PERGs were indistinguishable because, in the absence of extensive computer averaging with artifact rejection, the records were noisy. Even in our own records, before we merge over many replications, the two responses are quite similar in shape and timing. Linear spatial summation and macular overrepresentation are characteristics of both responses.[9,10]

We wish to report here that, when we carefully compared the FERG and PERG in man, we found significant timing differences, similar to those first observed in the pigeon retina.[17] Furthermore, we can record a response to the raster of a blank TV screen, with properties similar to a high frequency FERG, which is much easier to record. We have already reported that the PERG is frequently unilaterally reduced in untreatable human amblyopes.[10,18,19] We can now report that both the FERG and raster-evoked ERG need not be affected in amblyopic eyes and other eyes with reduced PERGs. This clinical separation of the responses occurs in conditions where it is normally accepted that only the inner retinal layers and the optic nerve can be damaged, such as toxic, traumatic, congenital, and degenerative optic atrophy and optic neuritis. It is best seen in unilateral diseases. The PERG asymmetry can not be detected for several weeks after an acute optic nerve injury, as if time for retrograde degeneration was involved. In cats, experimental optic nerve section has a similar, slowly developing effect on the PERG.[20] These results, which are the first to clinically distinguish FERGs and PERGs, cannot be conclusive without histological evidence. It now seems very likely, however, that a much greater proportion of PERG than FERG generation is proximal in the retina. This result further strengthens the argument that there could be a retinal basis for human amblyopia.[21,22] If, as seems likely, the PERG mainly reflects the activity of the ganglion cell layer, it will prove to be an important additional tool for the electrophysiological disection of the visual pathway.

METHODS

We have previously described our methods and the advantages they offer.[10,18,19] ERGs were recorded, usually bilaterally, with gold-on-mylar electrodes[23] hooked between the lid and sclera and referenced to the eyebrow with a central forehead ground, common to both eyes. In the course of the investigation we stopped the routine use of Benoxinate drops since these tended to produce more irritation than the electrodes. Responses were amplified with a standard EEG machine (0.3–50 Hz bandpass), whose output was fed to a computer (PDP 11/10). Programs for data collection and analysis included on-line artifact rejection and facilities for merging replications within sessions and across subjects. No traces illustrated have been digitally filtered after averaging.

We used our TV pattern generator system.[24,25] The standard clinical PERG test used checkerboard stimuli, which keep mean luminance almost constant. Contrast [defined as $(L_{max} - L_{min})/(L_{max} + L_{min})$] was 86%, the maximum possible. Squares (checks) with sides subtending 1 deg were alternated at 2 Hz (i.e., 4 reversals sec^{-1}) for 8 replications of 250–500 sweep averages. There were 4 replications with checks of 0.5 and 0.25 deg per side at 4 Hz. The screen subtended 22 × 16 deg at 114 cm. Mean luminance was 50 cd m^{-2}. The FERG stimulus was generated by increasing the check size until one check covered up to three quarters of the screen. If the "spatial frequency" is further reduced, most monitors cannot maintain constant contrast and mean luminance. We covered the screen with a large, opaque white card with a 16-cm-diameter circular aperture in it, which, at 57 cm, the closer of the two testing distances we employed in these circumstances, still subtended 16 deg while only exposing the uniform area of the screen. The surround was illuminated by two 100–watt "anglepoise" lights to each side of the subject plus normal room lighting and had a reflectance that was at least equal to the screen and was on occasions considerably higher. We repeated previous work[8] and confirmed that the various levels of surround illumination isolated the focal ERG. Where the PERG and FERG were directly compared, the same surround was used with both stimuli. The PERG is unaffected by surround luminance. Responses were also recorded to the TV raster, triggered from the frame signal, with various stimuli, as described in more detail below.

We used two types of TV monitor. One could be driven at 50 or 59.6 Hz by linking two pins on the synchronizing pulse generator in the pattern generator (i.e., we can run our TV at UK or USA frame rates). The other was a special purpose monitor, designed to run at 100 Hz, which could similarly be driven at 100 or 119 Hz. Our standard pattern generator drives the faster monitor when fitted with an oscillator crystal which runs at twice the normal frequency. The fast display had no discernable flicker. The higher frequency on either monitor was useful because we could arrange, by external associated circuitry, to "frame lock" the reversal time, and also to have brief appearance or disappearance events, at a fundamental frequency which was not a subharmonic of the power line frequency. This circuitry included an associated DC external video signal to the pattern generator which kept the mean screen luminance constant in blank and high-contrast pattern phases, where it would otherwise have been outside the linear range.

The PERG in particular has some discernable structure, with shoulders on the rising and falling phases, but we have not analyzed these in more detail. It can be seen clearly when the pattern transition time is 0.6 msec[15,16] rather than the 10 or 20 msec of the fast and slow TVs, respectively. We normally only measure the gross features of each trace; the amplitude from the *a*-wave-like dip to the *b*-wave-like peak and the time to peak (sometimes called latency). These features can be clearly seen in FIGURE 1. In clinic studies on uniocular disease we can eliminate random variability between trials and interindividual variability by taking the amplitude ratio of the affected eye to the fellow eye for each trial and testing for each individual whether the mean ratio differs significantly from 1.0. All normal mean ratios lie between 0.9 and 1.0, even when the worst eye of normals is taken to be the eye with the smallest mean PERG. We can thus be sure, for each individual, if the response from one eye is reduced.[10,18,19] In experiments we normally treat each averaged PERG record as a separate item for statistical analysis and then merge the traces to see the statistically established pattern directly (FIGURE 1).

RESULTS IN NORMAL SUBJECTS

Well-defined PERG and FERG traces, merged over many replications on both eyes of three subjects are shown in FIGURE 1. With high-contrast stimuli, the FERG is, in man, on average about 4 msec earlier than the PERG. This difference persists at all temporal frequencies of stimulation although the amplitude/stimulus-frequency characteristics of the two responses are otherwise identical.[10,17] Changing contrast distinguishes between the responses (FIGURE 1). The PERG peak time stays fixed while the FERG moves earlier as contrast is reduced, thus increasing the time difference to 10 msec. If the spatial frequency used to evoke the PERG were to be higher, the timing difference would be more pronounced. When luminance is reduced in one eye by filters, over a 2.2. log unit range, the times to peak of the PERG and FERG both become progressively longer.

The transition from the FERG to the PERG can be appreciated by changing the pattern size. FIGURE 2 shows that when 1 check fills the aperture in the 16-deg-diameter bright surround, a large early response occurs. If the aperture is filled

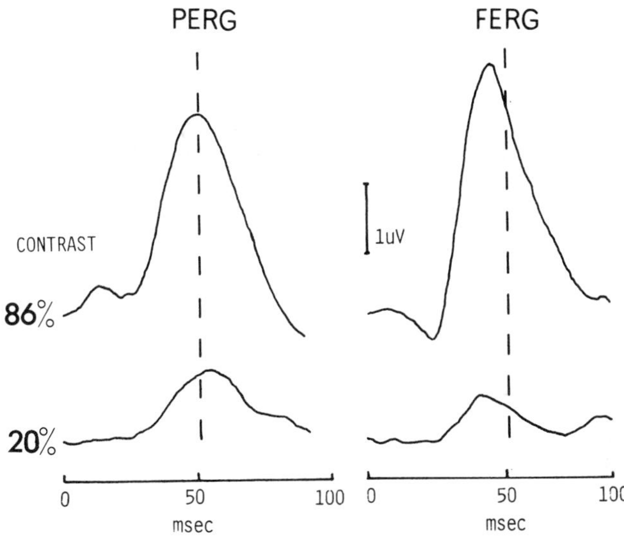

FIGURE 1. Direct demonstration of the difference between the focal ERG (FERG) and the pattern ERG (PERG) at high and low contrasts. Traces from a merge of the results from three normal observers (750 sweeps at 86% contrast and 2000 at 20% contrast for each). There is a difference between FERG and PERG time to peak which increases at lower contrasts because the FERG gets earlier while the PERG is unchanged. The shape difference, seen clearly only when many replications at high contrast are merged, is real, but has not been quantified.

with very large checks, positioned so that light and dark areas are equal, the response decreases in amplitude but remains early. At a check size of 2 deg per side a cusp occurs. The amplitude rises and the peak time increases by about 5 msec. In our normal working range (between 1 to 0.25 deg side^{-1}) the time to peak remains constant and amplitude again declines. When the checks are very small (0.165 deg side^{-1} or at even higher spatial frequencies, not illustrated here) the increase in time to peak of the human PERG is very pronounced. With 20-cycles deg^{-1} bars in a 5-deg-diameter field, we have recorded PERGs with a time to peak of 110 msec.[10]

Similar results, over a smaller range, have been obtained by Korth[15,16] and Armington (this conference). The very fine gratings that can be used and the characteristically delayed times to peak of the responses elicited, argues strongly that the PERG is not due merely to local luminance change.

The PERG response amplitude depends on the contrast, the size of the pattern elements, and the retinal area stimulated.[10] Although with a 16 × 22 deg field the largest response is obtained with checks of side subtense between 1 deg and 30 min or bars of about 2 cycles deg^{-1} (FIGURES 2 & 3), the voltage produced is only approx 0.007 $\mu V\ deg^{-2}$. The contribution of the 1.5-deg-radius paramacular area is 10 times as large.[10] As the spatial frequency of the pattern is increased, the proportion of the response generated by the paramacular and macular areas increases.[10] As spatial frequency increases the macular response declines very little in amplitude.[10] Furthermore, the PERG response to the entire field is, within the limits of measurement accuracy, equal to the sum of the responses to the center and surround. All three of these effects can be seen in FIGURE 3.*

RASTER RESPONSES

Ordinary TV pictures flicker slightly because the frame rate is slightly below critical flicker fusion frequency. We looked for ERGs and cortical responses evoked by this stimulus. FIGURE 4 shows that ERGs can be recorded when the ordinary monitor is run at 59.6 Hz, asynchronous to the power line frequency. The pattern was the bright phase of our normal FERG stimulus (approx 96 cd m^{-2}). A fast transient artifact from the TV vertical deflection coils could not be entirely removed by shielding in the clinic. We therefore broke each recording of 1000–2000 sweeps into blocks of 250. On alternate blocks the amplifier polarity was reversed and the screen covered by opaque white card. The averaging epoch was 50 msec and the trigger pulses came so quickly that the recordings were made rapidly. The smallest responses in FIGURE 4 are about 1 μV. They are approximately sinusoidal but contain additional high-frequency components.

This observation raises an interesting problem. The fly back time of the TV is short and the screen is dark for about 7% of the duty cycle. Each point on the screen is only bright for a very brief instant. Expressed as phase angle, the time between the trigger and the illumination of a particular region of the screen can take any value between 0 and somewhat less than 360 deg. Any line in the top of the screen would be 180 deg out of phase with that produced by another line in the bottom half, excluding a small central strip. If the latent period of the ERG to the raster is independent of retinal position, the responses from all parts of the screen, except that central strip, will cancel each other.

This is so, as can be seen in FIGURE 4. The raster response therefore seems to be a

*Linear spatial summation and macular overrepresentation are also characteristics of the FERG, as Brindley and Westheimer have shown,[8] and thus these features do not distinguish it from the PERG. It is interesting, in view of the results above, that when they used a stimulus of 2 deg diameter (the smallest from which they could record a response), the focal ERG had two peaks. The first occurred at 38 msec and was the same as found for other stimuli of larger subtense. The second peaked at approx 51 msec. This suggests that when the spot is so small that it had a pattern structure with a fundamental component of sufficiently high spatial frequency, a PERG (in our terminology) was also evoked. The result is similar to that obtained in visually evoked potential (VEP) recording. When the pattern size is 1 deg or more, the response is dominated by the change in luminance, rather than contrast.

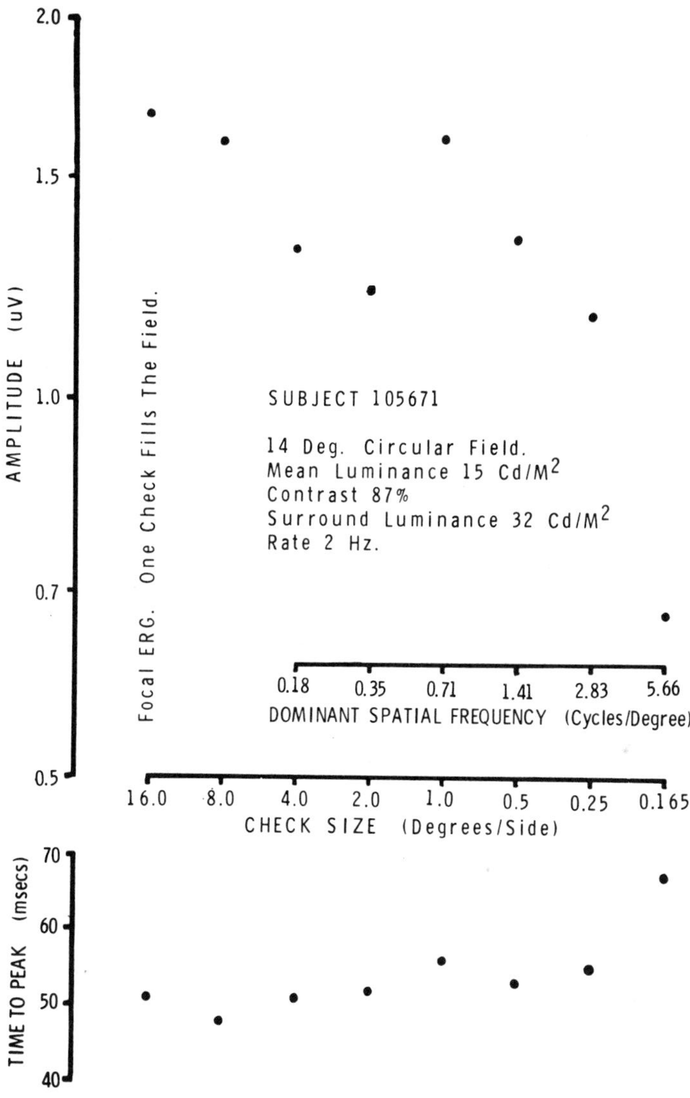

FIGURE 2. Change in amplitude and peak time of the response recorded when the stimulus, seen through an aperture in a large bright surround, varies from a blank screen to progressively smaller checks. The coarse stimuli were carefully centered to avoid any associated luminance change. All responses are the average of several replications. Note the double peak in the amplitude function and the later times to peak with smaller checks. As discussed in the text, these changes support the argument that the spatial summation area for the FERG is 2 deg or more, at which stage it adds to the PERG. With checks of smaller subtense there will be no net luminance change within the FERG summation area and so the PERG appears as a distinct response.

fast focal response from a small area. It is relatively easy to gaze at a blank TV screen and not make eye movements. Our subjects find it much more difficult to maintain fixation when the screen is flashing. We have therefore used raster responses clinically to look at focal ERGs in patients with asymmetrical PERGs (see below).

Raster-evoked ERGs are much larger (approx 10 μV) if recorded in the dark or with a +40 diopter lens before the eye, which further supports the view that, as recorded in FIGURE 4, the responses are focally generated. We attempted to record

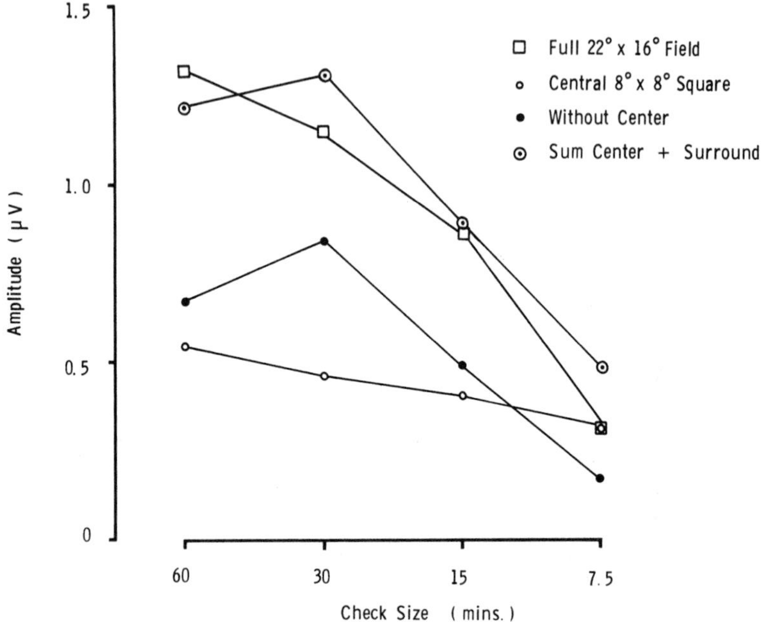

FIGURE 3. PERG responses generated by separate retinal areas using checkerboards with squares (checks) of various sizes. The responses to the full 16 × 22 deg field (□), the central 8 × 8 deg square (○), and the peripheral area remaining when the central 8 × 8 deg square is occluded (●) are shown. The algebraic sum of the responses from the two part fields is also indicated (⊙). Note that at all check sizes linear spatial summation holds to within the limits of experimental error, and that there is little attenuation of the paramacular response as check size decreases. The peripheral response is markedly attenuated over the same range of stimulus sizes and the maximum response is with 0.5-deg side^{-1} checks.

raster responses using the fast TV run at 119 Hz. No response could be recorded to the raster, but alternating screen brightness on every frame (i.e., 59 Hz stimulation) did produce a response. The raster ERG thus disappears at about the frequency where flicker can no longer be seen. The raster ERG to a standing checkerboard is about half the amplitude of that to a blank screen, as expected from the decrease in luminance. If the screen is then phase-reversed at about 4 Hz, the raster ERG and PERG can be recorded simultaneously if asynchronous two channel averaging is available, although

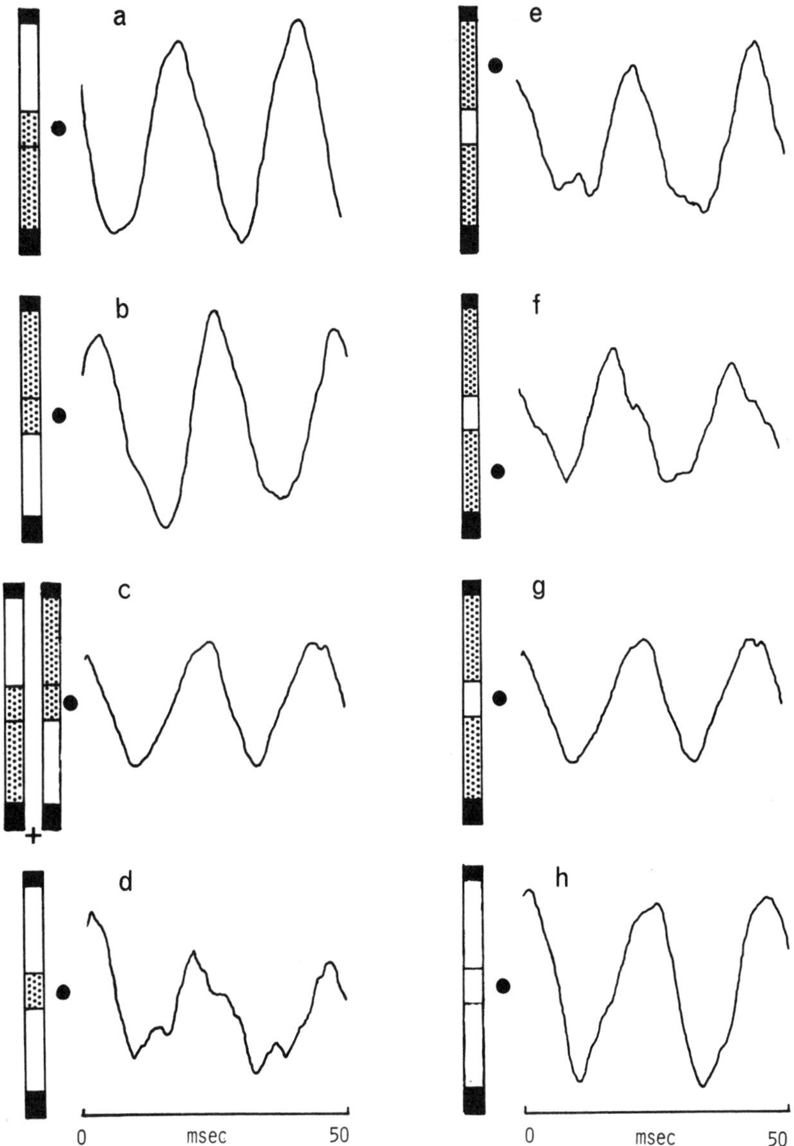

FIGURE 4. Demonstration that the response elicited by the raster is evoked by the luminance change from a narrow central strip of the screen. The patterned stripe beside each trace represents the TV frame beginning with the trigger at the top. Dark areas represent fly-back when the screen is blank. Dotted areas are the horizontal strips of screen occluded by white card. White areas are the part of the screen visible. The large black dots are at the level of the fixation point. (A) and (B) show that phase is caused by delay after the trigger and depends on the part of the screen exposed. These two responses practically cancel each other when algebraically summed (C) or simultaneously accumulated (D). This residue of incomplete cancellation (C or D) has timing similar to those from the center of the screen (E, F, or G). The response depends on screen position and not fixation (E, F, or G) and the full screen response (H) equals the sum of the center strip (E, F, or G) and the residue after cancellation (C or D). For further discussion, see the text.

this is inconvenient because of the controls needed to remove the electromagnetic artifact.†

PERG Reduction in Optic Nerve Disease

We frequently find the PERG is reduced in cases of multiple sclerosis, optic neuritis, glaucoma, and congenital, toxic, or traumatic optic atrophy where the disease process is initially limited to the optic nerve. In some cases the damage may be progressive and bilateral. However, in 12 patients it appears that there was, at a well-documented moment, a single, acute, monocular episode, as established by subjective report and clinical evidence including visual acuity and VEP changes. The ratio of the PERG amplitudes of the affected eye to that of the unaffected eye is plotted against log time since onset in Figure 5. It can be seen that, except for one case, there is no reduction during the acute stage even though form vision may be entirely absent in the affected eye. As vision returns, 2–7 weeks after the attack, the VEP reappears with a significant delay. At the same time, in 4 out of 5 cases, the PERG in the affected eye is larger. At about 10 weeks and later, the PERG is always reduced in the affected eye. These results are for the largest checks, but very similar ratios are obtained with all three check sizes. Since all parts of the retina are likely to be equally affected by retrobulbar neuritis, this result is quite understandable.

All the significant features of the above graph are clearly exemplified in Figure 6. These are the merged traces for one patient, a 37-year-old woman, who, when first seen, complained of unilateral loss of vision of 3 weeks duration. There was pain on moving the eye and no history of a previous attack. Her vision had at this point recovered from a period of no light perception (OS) to 6/4, 6/60. The pattern VEP was normal in the right eye and small and very late in the left. The ERGs, as tested in our standard clinical routine, were equal in both eyes. The PERG in the affected eye is significantly larger, at all square sizes, than in the normal eye. When the patient returned some 7 weeks later, vision had improved to 6/5, 6/9 (part), and the VEP was still classically small and delayed. The PERG was now smaller in the affected eye.

It is interesting that very recent unilateral insults to the optic nerve do not produce asymmetries. Our results might suggest a hypernormal phase before the decline in the PERG, perhaps because neurones on the point of degenerating become hyperexcitable.

The rate at which the PERG is reduced in cats after optic nerve section is similar.[20] We cannot comment on the observation that responses to the highest discernable spatial frequencies disappear more slowly[20] becasue our stimuli (1.0–0.25 deg side^{-1} checks) are all relatively coarse and were equally affected. Our clinical data and physiological experiments[20] both seem consistent with the PERG being affected by retrograde degeneration of the optic nerve. The change appears to be rather too rapid

†We have not been able to record consistently recognizable cortical VEPs to the raster in our subjects, although a "visual hum" can be detected in some subjects.[26] No such 50 Hz waveform can be detected with monocular stimulation and electrodes placed at C_1 and C_3 so that central and retinal raster responses are not generated in the same place.

The PERG recorded to the 100 Hz set differs only in that it is slightly larger and over 5 msec faster, as expected from the decreased composition time and brighter screen. Frame locking, asynchronous from the power line frequency, is not critical. With it we can control the luminance changes which otherwise accompany high-contrast pattern appearance and disappearance. The recorded responses resemble those to pattern reversal but are of about half the amplitude, as is expected from the reduced contrast change.

for any transsynaptic retrograde degeneration, but firm conclusions will require more adequate histology.

Glaucoma is a common and severe condition where it is believed the optic nerve is primarily affected. However, it is rarely unilateral and gross asymmetries are uncommon. It is much less frequent in younger patients. We have examined 9 patients in their 40s and 50s with obvious asymmetries on routine clinical examination. In none was the field loss extensive or visual acuity severely depressed. We took care that they

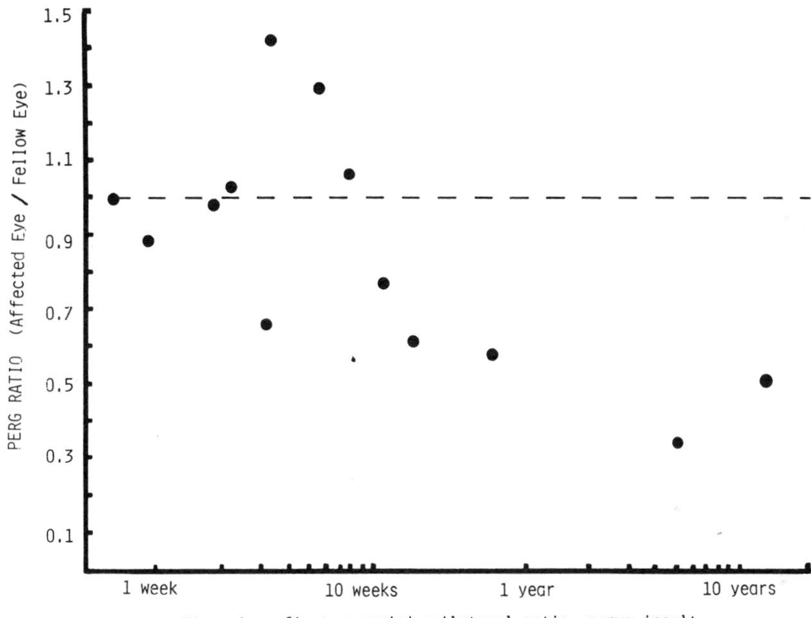

FIGURE 5. Ratio of PERG amplitudes of the affected eye to those of the unaffected eye in 12 cases of unilateral optic nerve insult as a function of log time since the first recorded attack. The longest suffering case has congenital optic nerve hypoplasia. The others have optic neuritis. One case seen twice (FIGURE 6). The curiously large responses in the affected eye in the 2–7 week period may represent hyperactivity following damage. More convincing is the rapid decrement by 10 weeks at latest . It would seem that this time is sufficient for retrograde degeneration, but too rapid for transynaptic degeneration, although the normal mamalian physiology is unknown. The results with 0.5- and 0.25-deg side^{-1} checks were very similar and no systematic effect of spatial frequency could be detected. The sizes we used are too coarse to demonstrate the slower high-frequency loss found in cats.[20]

had not used mitotic medication for over 24 hours so that their pupils were of normal size and that both were equal. In 7 out of 8, the response is reduced in the eye with the worst visual acuity (mean amplitudes, 1.4 and 1.7 μV) but, more strikingly the PERG amplitudes in both eyes are highly correlated ($r = 0.9$, $p < 0.005$) and for 6 out of 9 the responses in both eyes were under 2 μV (mean over all eyes, 1.6 μV). This series will be described in more detail elsewhere. The PERG is also substantially reduced bilaterally in cases of known multiple sclerosis with bilateral involvement.

PERG Reduction in Amblyopia

Amblyopia is a usually monocular condition of reduced visual acuity associated with squint, hypermetropia, or anisometropia. It only develops in childhood and, if not promptly treated, becomes irreversible.[22,27] Hitherto, it has not been possible to demonstrate a retinal defect in human amblyopes, though this has been suspected. Retinal changes are produced in an animal model.[21] Ikeda suggested that reduced activity was sufficient to arrest the development of retinal ganglion cells. However, five studies have now found that stimulus deprivation has no retinal effect (Reference 28; for a review see Reference 22). Nevertheless, independent studies have confirmed her report that blur[29] and large angle, nonalternating strabismus[30,31,45] do reduce the spatial resolution of central retinal ganglion cells. Murray Sherman[46] has suggested that deprivation affects Y-cells more than X-cells. We have argued that the retinal change might only occur in eyes where the image is always blurred (i.e., high spatial frequencies are selectively removed from the visual array). Y-cells will still respond to a blurred image, but the X-cells would be deprived. Possibly Y-cells then have a competitive advantage for synapses in the retina, just as the input from a nondeprived eye does in the cortex.[22] Neurological losses in human amblyopes could then be retinal, cortical, or both, depending on the inducing conditions.[22]

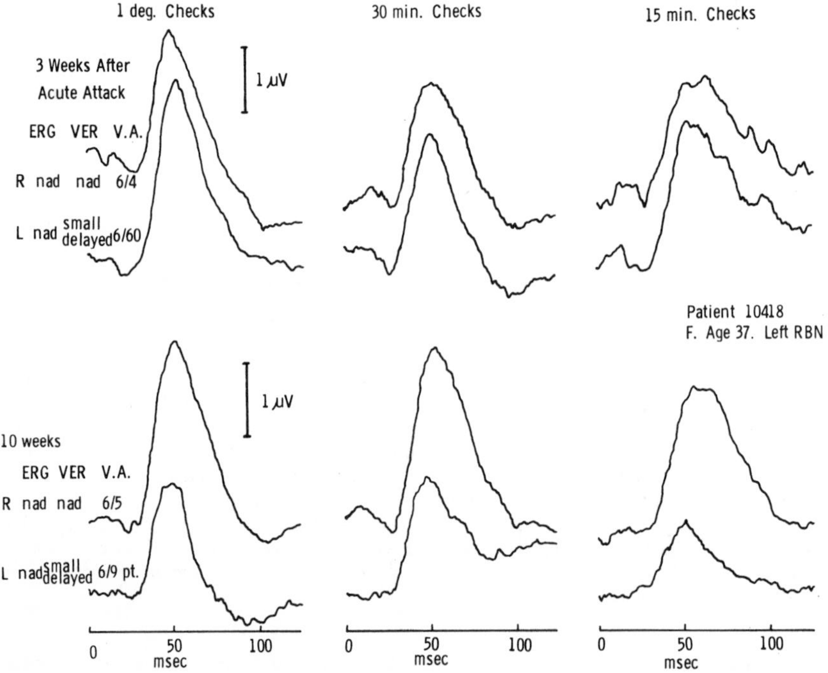

FIGURE 6. Ratio of PERG amplitudes, (affected eye)/(unaffected eye), in a case of optic neuritis first seen 3 weeks after the first attack. At this stage, although the visual acuity and VEPs from the affected left eye were very poor, the PERG was slightly larger at all check sizes. When the visual acuity had recovered, 7 weeks later, the PERG was substantially reduced. Traces are the merged averages of 4–8 replications at each check size.

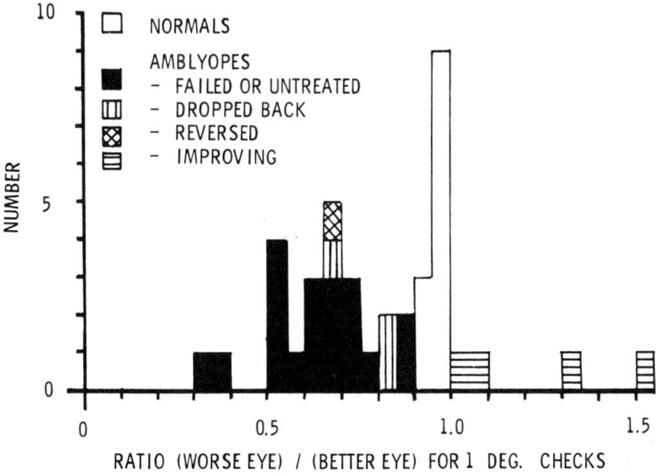

FIGURE 7. Histogram of the frequency of occurrence of various ratios of PERG amplitudes, (worse eye)/(better eye), for 27 amblyopes (shaded areas) and 12 normals (white areas) tested with 1-deg side^{-1} subtense checks. The worse eye for the normals was taken to be that with the smallest mean PERG amplitude so as to bias against a positive result. Nevertheless the amblyopes as a group have an average ratio less than the normals. Since the test was replicated many times on most patients it was possible to test whether the ratio, for each individual, differed from unity. Thus it is apparent that the amblyopes fall into two groups: 23 for whom the PERG was reduced in the amblyopic eye, and 4 for whom it was not. The amblyopes were of most common clinical types and the only clinical feature we could discover which separated the two groups was that those with ratios of 1.0 or greater (horizontal stripes) were responding to treatment. The others had failed to improve (black) or had dropped back (vertical stripes). One case (hatched), described in the text, had the conventional pattern although the now strabismic and amblyopic eye had been the better one prior to occlusion and strabismus surgery at age 4. This case of reversed retinal defect must have developed slowly because conventional occlusion did not reduce the response amplitudes in the better eyes of the other children.

It is well established that the FERG is not affected in human amblyopes.[5,6] Lawwill[32,33] found that, for 4 patients, the PERGs of normal and amblyopic eyes were equal. Sokol and Nadler[9] did find an asymmetry in 3 cases. We independently tested 14 amblyopes and found that the PERG was reduced in 12.[18,19] Results on a larger sample, and a number of experiments to control for potential artifacts have recently been reported elsewhere[10] and will be only briefly reviewed here.

Our basic result is illustrated in FIGURE 7. This is a histogram of the frequencies with which various ratios of PERG amplitudes (affected eye/unaffected eye) are encountered in amblyopes and normals. The amblyopes which were referred to us are by no means a representative sample. Typical case details, which have been already published[18] show clearly that the sample included most types of amblyopia, but was biased towards older, more cooperative children, and a large proportion were sent to us after it had been found that orthoptic treatment had failed to produce a sustained increase in visual acuity. All normal ratios fall in the range 0.9 to 1.0. All amblyopes in our biased sample fall outside that range. The amblyopic eye's response is larger for 4 patients who have maintained a positive response to treatment but not for 3 children who, despite an earlier substantial improvement, had regressed at the time of testing.

In all other cases[20] the PERG from the worse eye is reduced. For this group the relative amplitude in the amblyopic eye gets smaller with check size, as it should be if the PERG was sensitive to an amblyope's spatial resolution deficit (FIGURE 8). In one amblyopic eye the inactive area corresponded to the monocular scotoma.[18]

Arden and Barnard[35] have already reported that moderate occlusion therapy temporarily delays and reduces the VEP in the occluded eye. After prolonged occlusion the change can be very long lasting. Almost all our children have been

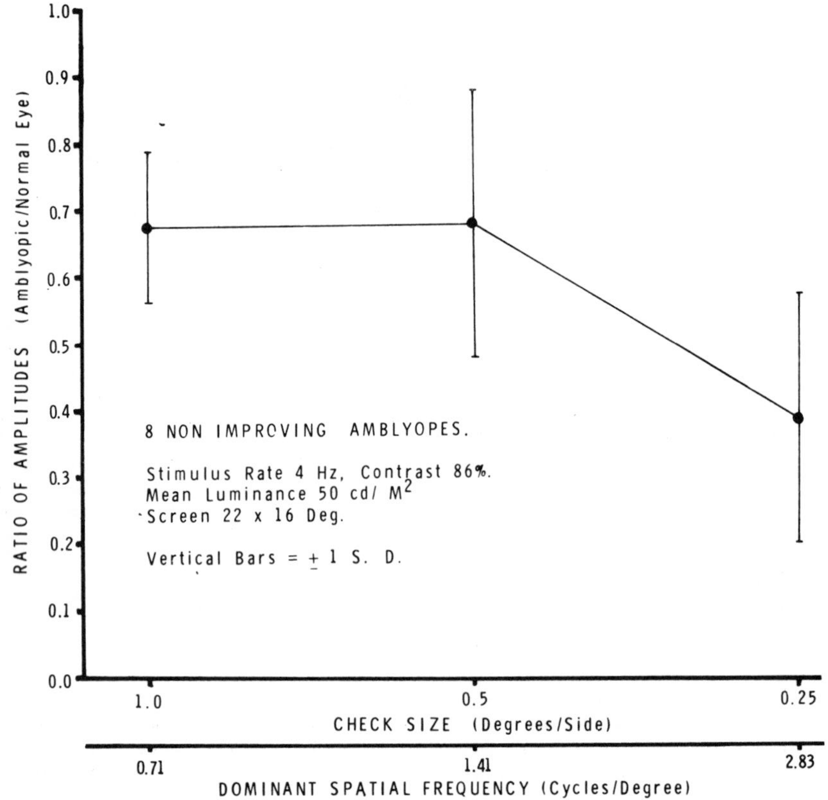

FIGURE 8. Change in the ratio of PERG amplitudes, (amblyopic eye)/(fellow eye), as check size decreases. The patients were 8 nonimproving amblyopes, each of whom show a significant reduction with the largest checks. The decreased ratio with small checks is what would be expected if the central retina, subserving high spatial resolution, was selectively depressed in human amblyopia.

extensively occluded. We therefore examined the PERGs from the better, treated eyes. They are of normal amplitude and time to peak.[10] The amblyopic eye's response in the four cases shown as responding to treatment is slightly small but in the normal range. There is one interesting exception, shown as reversed amblyopia in FIGURE 7:

> This boy had a 40 diopter left convergent strabismus with visual acuity 6/18, 1/60 when first seen at 3.5 years of age. By occlusion he was induced to alternate and the visual acuity

improved to 6/12, 6/12 at age 4 when he underwent corrective surgery. As sometimes happens in such cases, the strabismus reversed, he became a right eccentric fixator and the vision in the previously better eye deteriorated. At age 5.5 years, after left occlusion, the VEPs from both eyes were delayed. When we saw him again at 8.5 years of age, after further surgery, his visual acuity was 6/24, 6/4, the VEPs were still bilaterally delayed, and the PERG from the now squinting and amblyopic right eye was reduced.

This case clearly demonstrates that the retinal change in human amblyopia can involve deterioration, rather than simple arrest of development and that the retina has a critical period which extends throughout infancy. During this time very prolonged strabismic deprivation causes a deterioration of retinal visual acuity while adequate stimulation can restore normal growth and is therapeutic.

If this pattern continues, the PERG ratio would be important for monitoring amblyopia treatment and could also have prognostic significance for recovery. An extensive clinical program is underway but we have so far seen few recovering cases and followed none through treatment.

PERG Loss in Deprivation Amblyopia

As discussed above, total deprivation in animals only affects the cortex. It is therefore of some interest to examine the PERGs in cases of uncomplicated, unilateral deprivation amblyopia with well-documented histories and adequate optical correction[36] to see if any retinal change may be detected.

As can be seen in the top two traces of Figure 9, the PERG in children with early-onset, long-lasting deprivation is more profoundly disturbed than it is in the other types of amblyopia that we have seen. This finding is consistent with a simultaneous report by Levi (this conference). The traces from the deprived eyes show no evidence of any main positive peak at 50 msec, and all that remains resembles the slow base-line shifts. We do not know the origin of this slow component. It may be a residual luminance response or an eye movement artifact. The later the beginning of the stimulus deprivation and the shorter its duration, the more normal the PERG. Thus, in the third case in Figure 9, a boy deprived for 10 months at age 6.5, the PERGs are of normal shape and timing but of an amplitude less than a third of those in the fellow eye. The fourth example shows PERGs present and bilaterally equal after a relatively early onset (age 13 months) bilateral cataract which was kept as brief as possible (7 months) and very carefully managed. The lower traces for this child are obtained with ordinary spectacle correction worn in a trial frame to show that these losses cannot be due to the contact lenses. The fifth case is an ordinary nonamblyopic high myope in whom again the contact lens has no effect. The same results were obtained with 2 D over or under correction added in a trial frame so it is unlikely that the residual refractive errors which occur in these aphakic corrections is critical.

Clinical Separation of the FERG and PERG

In 5 out of 7 of the cases where both the FERG and PERG have been recorded and there is no evidence of receptor dysfunction, we have found that the PERG alone is asymmetrical. Three cases are illustrated in Figure 10. These results are unlikely to occur if the unilateral PERG reduction is an eye movement artifact. They support the idea that the PERG is generated more proximally in the retina than the FERG.

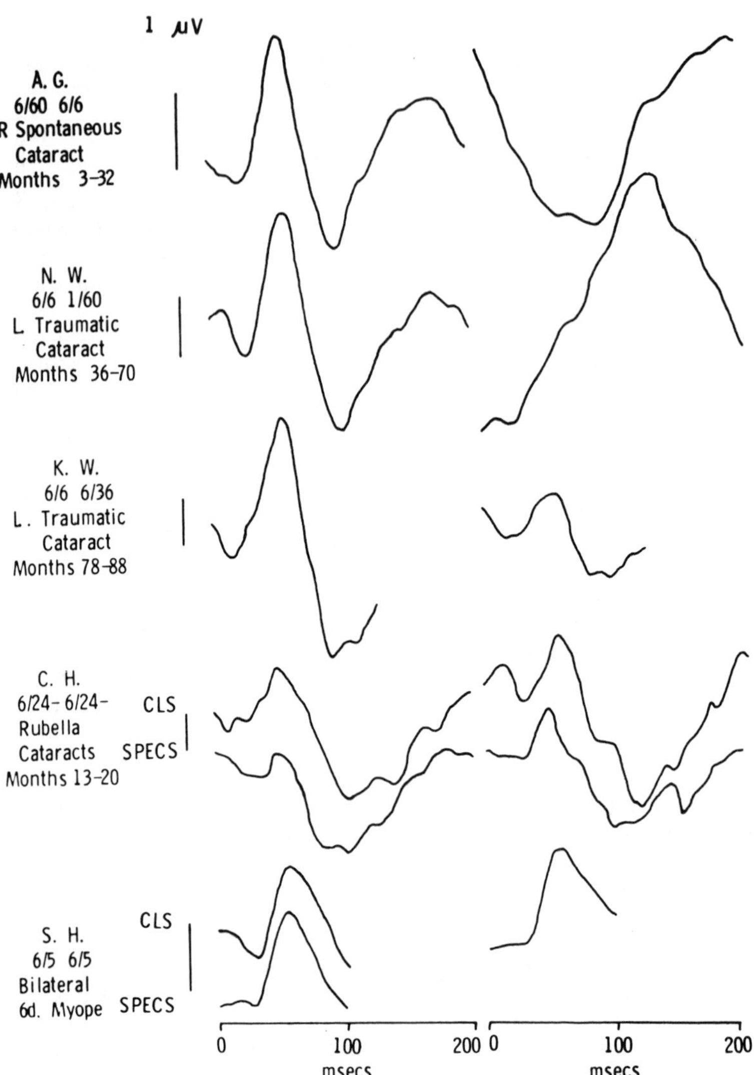

FIGURE 9. PERG reduction in deprivation amblyopia. In these top 3 unilateral cases the better eye is shown on the left. These cases are arranged in order of age of onset. Clearly the PERG is extremely abnormal in early onset, long deprivation cases (see also Levi in this conference). Even 10 months deprivation at age 6.5 reduces the PERG to less than 1/3 that in the fellow eye, though peak time and shape are then normal. These results were the same with 2 diopters over or under correction supplementing the usual soft contact lens these children wore to overcome their aphakia. Their residual refractive error, determined by streak retinoscopy and subjective refraction where possible, was less than 2 diopters, so blur cannot explain the response abnormalities. The fourth patient is a bilateral aphake, who has retained useful vision equally in both eyes, and clearly also has small but normal PERGs, because of relatively rapid treatment despite cataract onset at 13 months. The last case is a nonamblyopic bilateral high myope. Both of the last two patients establish that the reduced PERGs are not an artifact of contact lens correction. Their responses are unchanged with full spectacle correction.

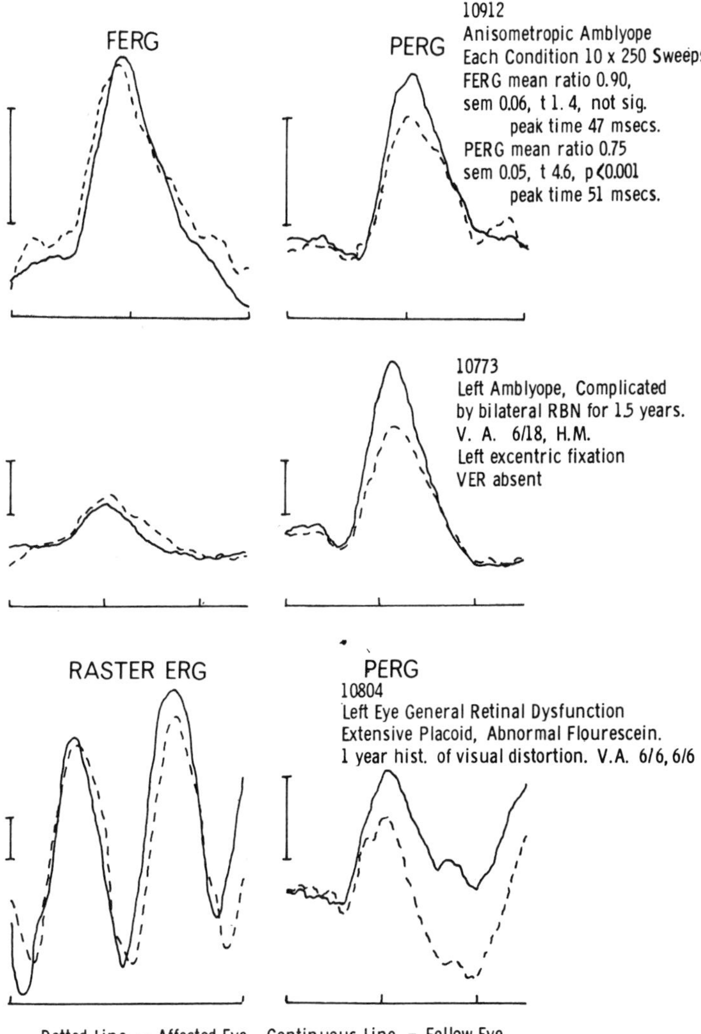

FIGURE 10. Normal symmetrical FERGs associated with asymmetrical PERGs in two patients. This is the most common result in those cases where both responses have been tested and receptors are not abnormal. The first case was a volunteer. Many replications of both conditions were made and statistically analysed. Only the PERG is reduced. As well as providing a clinical separation to support the experimental distinction between the two responses described above, the difference between the two is also strong evidence that the PERG reduction is not an eye movement artifact. In a third case the response to the raster has been recorded instead of the FERG because it is easier and appears to be a fast focal ERG. Again only the PERG is asymmetrical.

Controls for Artifacts

All our amblyopes, most of whom were children, were carefully screened by the Orthoptic Department at Moorfields Eye Hospital to exclude any superficially similar disease with a known organic basis. Where squint was present, the fixation target on the stimulating television screen was moved by half the angle of squint, so that the stimulus fell on comparable areas of each retina. Only patients with strabismus angles below 20 deg were tested. Precautions were also taken to prevent any other known source of error, such as extremely asymmetrical electrode placement relative to the entrance pupil, excessive tear secretion, large movements of the recording electrode produced by spasm of the orbicularis, or spikes associated with excessive neck muscle tonus.

We have mainly investigated uniocular disease where the fellow eye acts as a control, especially when the ratio between the eyes is the main variable. We are confident that refractive error could not cause the sort of asymmetries we have seen in our patients. They were carefully refracted and wore their full correction in properly fitted frames, not the lesser added lens as is sometimes prescribed for the amblyopic eye. FIGURE 11 shows that 1.5 D is the minimum degree of blur necessary to cause a

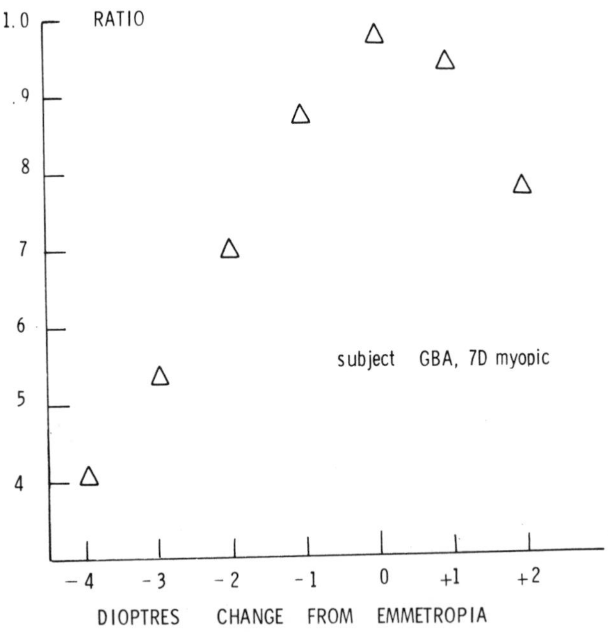

FIGURE 11. Ratio between the amplitudes of the PERGs in the two eyes as a function of the degree of refractive error caused by spectacle lenses which deviate from full correction, when the accommodation of one eye is paralyzed by Cyclopentolate 1%. Although there is a clear effect of blur, a clinically significant reduction (over 10%) requires substantially more blur than ever occurs in any of our subjects, who were optically corrected to within 0.5 diopter. Moreover, GBA is a high unilateral myope who habitually suppresses at this testing distance. With optimal correction, the myopic eye gives a response of normal amplitude, although it is not used to fixate.

significant effect on its own when the accommodation in one eye is paralyzed and the refraction artificially varied with lenses. This degree of blur is unlikely in our patients.

It has been suggested that abnormal eye movements may occur and reduce the responses in the amblyopic eyes. However, during binocular fixation, amblyopic eyes are as steady as the fixing eye and it appears that the wandering fixation movements of amblyopic eyes are a consequence of the poor vision during monocular fixation.[37] We have rigorously tested our eye movement artifact rejection system. The PERGs are completely unchanged if the subject makes eye movements during recording that mimic, in rate and amplitude, the largest we might encounter (FIGURE 12). The software rejects all channels when an artifact appears on any one. Moreover, the

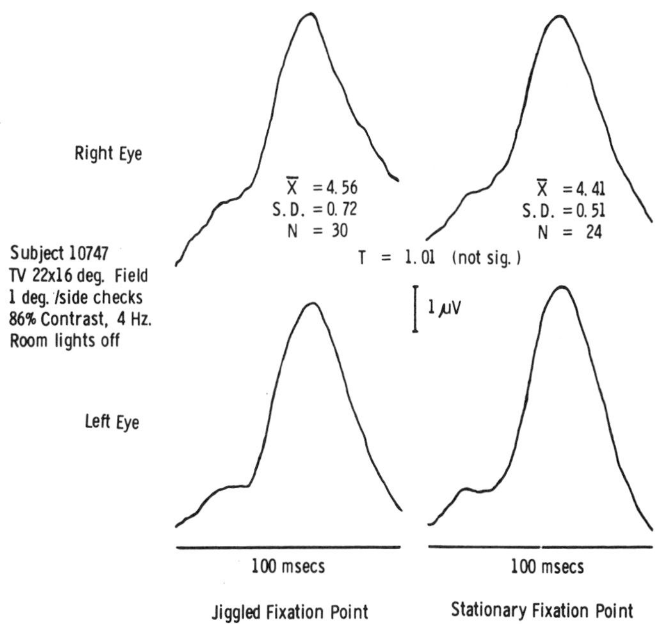

FIGURE 12. Test of the artifact rejection system. Trials with a steady central fixation point were alternated with those in which the experimenter jiggled the fixation point so as to mimic, in rate and extent, the most abnormal fixation movements likely to be encountered. The gain was set to reject 20–250 sweeps in the fixation state. Apart from causing the rejection of 4000–8000 sweeps and hence slowing averaging, eye movements do not affect PERG recording.

pattern of PERG loss seen in optic neuritis does not occur when the visual acuity is reduced during the acute phase; the loss occurs only after visual acuity has recovered. The subject for the experiment on refractive error also clearly has an optimally refracted ratio of 1.0, despite being a unilateral high myope who habitually suppressed the treated eye. Again there is no PERG reduction merely because the eye is not used for fixation. Finally, as reported above, we have recorded both FERGs and PERGs in several subjects. The FERG does not drop with the PERG, as it should if eye movements or malfixation caused the asymmetry (FIGURE 10).

Do PERGs Parallel Ganglion Cell Activity?

The PERG can be generated with isoluminant chromatic contrast.[11] There are no reports of color-coded amacrine cells. If this were universally true, PERG generators would be very likely to occur parallel to the amacrines, at the bipolar layer, or later.

It is interesting that changes in spatial frequency of the stimulating pattern are very effective in altering the time to peak of the PERG in man (Figure 2).[10,15,16,38] No other stimulus parameter, except mean luminance of the display, has a similar effect. Nor does the time to peak change in the diseases we have studied. It is evident that the slower responses to higher spatial frequencies must be generated by increasingly more central retinal areas (Figure 3). The timing of the positive wave of the normal PERG is very similar to the period when cat retinal gangion cells, stimulated at similar mean luminances, are discharging[39–41] and the actual waveform of the PERG to low and high spatial frequencies is similar to the envelope of spike discharges of brisk-transient (Y) and brisk-sustained (X) type cat retinal ganglion cells, respectively.[10,40] While one cannot take too seriously this correspondence in waveform between cat ganglion cell discharge and human gross potentials, both the PERG and the FERG are sufficiently rapid to be synchronous with neuronal activity.

In man, the PERG, but not the FERG, has a time to peak which does not change with contrast (Figure 1). The amplitude increases almost linearly with contrast in the range 5%–86%. The function, on linear coordinates appears to pass through zero. In this range the responses are so small that contrast thresholds could not be established. We have checked this relationship very thoroughly. Seven people were tested with 1-deg side^{-1} checks at 4 Hz and at 13 contrast levels in 0.1 log unit steps. There were at least 3 records of 500 sweeps at every stimulus level on each testing day. One subject was tested 6 times. No shift in peak time could be established either in the merged traces, over all eyes and sessions, or in the statistical analysis of the raw data. Holden and Vaegan[34] have found that, in the pigeon, both ganglion cells and the PERG also behave in this way. For stimuli where each check extends over one third of the diameter of the receptive field or less, the findings in the pigeon are consistent with findings in the cat.[39] In such circumstances, the total luminous flux on the field will be constant. In some circumstances,[42] cat retinal ganglion cells may behave more like our FERGs.

Discussion

All previous workers have accepted that the site and mode of generation of the FERG, PERG, and ERG were all identical.[11–16,40] The PERG has accordingly been under-utilized clinically and experimentally. Recently a few studies have suggested[18–20] or implied[43] that the PERG reflected proximal retinal activity but have not tested the alternative hypothesis. Thus, at present, theories exist which place the PERG generators at both sides of the retina. This is the first study to find clinical and experimental differences between the PERG and FERG in man and to describe a large, homogeneous clincial series with unilateral and bilateral optic nerve insult where the PERG is clearly reduced. The conclusion that the PERG is certainly generated in the very proximal retinal layers compliments similar findings from physiological studies in the pigeon and is consistent with personal communications we have had from a number of other workers [Maffei, Sherman, Levi (this conference), A. Fiorentini, and T. H. Kirkhamin].

The properties described above suggest that the PERG is a direct index of the functioning of the inner retinal layers. It could well represent the activity of any spatially organized retinal neurones such as amacrine or ganglion cells or the synaptic potentials feeding onto such cells. The results on patients with optic nerve disease strongly support the view that abnormalities in third-order neurones readily affect the PERG. If this is so, the results on amblyopes, in particular, take on added importance because we now have direct, objective, physiological evidence that there can be, in man, a retinal, neural abnormality, due to inadequate stimulation during development, which can cause a visual acuity loss. This abnormality must lie at or before the ganglion cell layer and can develop slowly at ages as late as 4 years.

Sherman's data (this conference) suggests that the PERG is never totally lost when the optic nerve is entirely destroyed but it is certainly reduced. Since he records monocularly, looks for clear records, and does not take the ratios of simultaneous traces, his data sampling procedure is different from ours. Hence his results are consistent with our FIGURE 5 but not with the cat experiments.[20] Only future research can decide between these alternatives. We feel that both the FERG and PERG that we record contain "luminance" and "pattern" components in varying proportions depending on spatial frequency. No stimulus that has been used clinically is likely to elicit a response that is unique to a single cellular mechanism.

Whatever the conclusion, it is clear that the PERG is an index of proximal retinal neural activity at a very early stage in the visual pathway. The cortical VEP has to be supplemented by recordings that may represent lateral geniculate nucleus activity.[44] The PERG appears to complete the chain. Perhaps we will now be able to record evoked potentials noninvasively from each level of the visual system.

SUMMARY

A TV monitor was used to evoke either a pattern ERG to a contrast-reversing checkerboard (PERG), or a focal ERG to alternate increases and decreases of luminance of the blank screen within a bright surround (FERG). Both responses are small (approx 2 μV) and fast (approx 50 msec to peak) and are similar in several other properties. However, they differ in timing and respond differently to changes in contrast. Each frame of a TV picture evokes a "raster ERG," even though the screen is blank. The response is focal and specific to a small central strip of the screen. It is simpler to record than the FERG, where the whole screen is flashing. Because the FERG summation area is about 4 deg, small squares (checks) reversing in contrast produce little luminance response. In 5 of 7 cases where the PERG is unilaterally reduced, the FERGs or raster responses were not affected. Thus clinical evidence also suggests that the PERG may be a separate phenomenon to the FERG and produced at a different site. Toxic, traumatic, congenital, and degenerative diseases of the optic nerve reduce the PERG. The comparison is most easily made in unilateral disease. Ten weeks after an optic nerve insult, the PERG becomes reduced in the affected eye as if retrograde degeneration was occurring. In 27 amblyopes of various types, the PERG was reduced in 23 where orthoptic treatment had failed. In 4 patients responding to treatment, PERGs of the amblyopic eye were as large as, or larger than, those of the fellow eye. The loss is greater with smaller checks. Retinal changes do occur after age 4 but so slowly that responses in heavily occluded eyes are not reduced. An additional level in the visual pathway is thus accessible to evoked potential investigation.

ACKNOWLEDGMENTS

The authors wish to thank their hospital colleagues who served as experimental subjects and as patients, Mr. P. Fells, who screened and referred the younger patients, Ms. Elizabeth Waddell, who summarized their histories, and Prof. I. McDonald, who screened and referred most of the optic nerve cases. Jerry Nelson referred us to some relevant neurophysiological papers. Equipment has been provided by The Medical Research Council and by Moorfields Eye Hospital Local Research Grants. Vaegan is supported by the Golden Nugget Fellowship.

REFERENCES

1. FRY, G. A. & S. H. BARTLEY. 1935. The relation of stray light in the eye to the retinal action potential. Am. J. Physiol. **111:** 335–340.
2. MILLER, R. F. & J. E. DOWLING. 1970. Intracellular responses of the Müller (glial) cells of mudpuppy retina; Their relation to the *b*-wave of the electroretinogram. J. Neurophysiol. **33:** 323–341.
3. KAROWSKI, C. J. & L. H. PROENZA. 1977. Relationship between Müller cell responses, a local transretinal potential, and potassium flux. J. Neurophysiol. **40:** 244–259.
4. DICK, E., & R. F. MILLER. 1978. Light-evoked potassium activity in the mudpuppy retina: Its relationship to the *b*-wave of the electroretinogram. Brain Res. **154:** 388–394.
5. JACOBSON, S. J., M. A. SANDBERG, M. H. EFFRON & E. L. BERSON. 1979. Foveal cone electroretinograms in strabismic amblyopia. Comparison with juvenile macular degeneration, macular scars and optic atrophy. Trans. Ophthal. Soc. U.K. **99:** 353–356.
6. BURIAN, H. M. & T. LAWWILL. 1966. Electroretinographic studies in strabismic amblyopia. Am. J. Ophthalmol. **61:** 422–430.
7. NELSON, R., E. ZRENNER & P. GOURAS. 1979. Pattern stimuli reveal spatial organization in the electroretinogram. *In* Proc. 16th Symp. of the Int. Soc. for Clinical Electrophysiology of Vision, Marioka, 1978. Y. Tazawa, Ed. pp. 161–169. [Japan. J. Ophthalmol.]
8. BRINDLEY, G. S. & G. WESTHEIMER. 1965. Spatial properties of the human electroretinogram. J. Physiol. **179:** 518–537.
9. SOKOL, S. & D. NADLER. 1979. Simultaneous electroretinograms and visually evoked potentials from adult amblyopes in response to a patterned stimulus. Invest. Ophthalmol. Vis. Sci. **18:** 848–855.
10. VAEGAN, G. B. ARDEN & C. R. HOGG. In press. Properties of normal electroretinograms evoked by patterned stimuli in man. Abnormalities in optic nerve disease and amblyopia. Doc. Ophthalmol. Proc. Series (19th ISCEV Conf., Zurich, 1981).
11. RIGGS, L. A., E. P. JOHNSON & A. M. L. SCHICK. 1964. Electrical responses of the human eye to moving stimulus patterns. Science **144:** 567.
12. JOHNSON, E. P., L. A. RIGGS & A. M. L. SCHICK. 1966. Photopic retinal potentials evoked by phase alternation of a bar pattern. *In* Clinical Electroencephalography. H. M. Burian & J. H. Jacobson, Eds. [Vis. Res. (Suppl. 1): 75–91.]
13. MILLODOT, M. & L. A. RIGGS. 1970. Refraction determined electrophysiologically: Responses to alternation of visual contours. Arch. Ophthalmol. **84:** 272–278.
14. SPEKREIJSE, H., O. ESTÉVEZ & L. H. VAN DER TWEEL. 1973. Luminance responses to pattern reversal. Doc. Ophthalmol. Proc. Series (10th ISCERG Symp.) **2:** 205–211.
15. KORTH, M. 1980 Luminosity functions of human electroretinogram wavelets evoked with pattern reversal stimuli. Invest. Ophthalmol. Vis. Sci. **19:** 810–816.
16. KORTH, M. 1981. Human fast retinal potentials and the spatial properties of the visual stimulus. Vis. Res. **21:** 627–630.
17. VAEGAN. 1981. Pigeon pattern electroretinograms differ from focal electroretinograms. J. Physiol. **319:** 75P.
18. ARDEN, G. B., VAEGAN, C. R. HOGG, D. J. POWELL & R. M. CARTER. 1980. Pattern ERGs are abnormal in most amblyopes. Trans. Ophthal. Soc. U.K. **100:** 453–460.
19. ARDEN, G. B., R. M. CARTER, C. R. HOGG, D. J. POWELL & VAEGAN. 1980. Reduced

pattern electroretinograms suggest a preganglionic basis for non-treatable human amblyopia. J. Physiol. **308:** 82P–83P.
20. MAFFEI, L. & A. FIORENTINI. 1981. Electroretinographic responses to alternating gratings before and after section of the optic nerve. Science **211:** 953–955.
21. IKEDA, H. & M. J. WRIGHT. 1975. Amblyopic cells in the lateral geniculate nucleus in kittens raised with surgically produced squint. J. Physiol. **256:** 41P–42P.
22. VAEGAN, G. B. ARDEN & P. FELLS. 1982. Amblyopia: Some possible relations between experimental models and clinical experience. *In* Paediatric Ophthalmology. D. Taylor & K. Wybar, Eds. Marcel Dekker, New York, N.Y.
23. ARDEN, G. B., R. M. CARTER, C. R. HOGG, I. M. SIEGEL & S. MARGOLIS. 1979. A gold foil electrode: Extending the horizons for clinical electroretinography. Invest. Ophthalmol. Vis. Sci. **18:** 421–426,
24. ARDEN, G. B., D. FAULKNER & C. MAIR. 1977. A versatile television pattern generator for visual evoked potentials. *In* Visual Evoked Potentials In Man: New Developments. J. E. Desmedt, Ed. Chap 3: 90–109. Clarendon Press, Oxford, England.
25. FAULKNER, D. 1978. A new television stimulator for contrast sensitivity and evoked response testing. J. Physiol. **275:** 7P–8P.
26. VAN LITH, G. H. M., G. W. VAN MARLE & S. VIJFVINKEL-BRUINENGA. 1979. Interference of 50 Hz electrocortical potentials evoked by TV systems. Br. J. Ophthalmol. **63:** 779–781.
27. SCHAPERO, M. 1971. Amblyopia. Clifton Book Co., Philadelphia, Pa.
28. BLAKEMORE C. & F. VITAL-DURAND. 1979. Development of the neural basis of visual acuity in monkeys. Trans. Ophthal. Soc. U.K. **99:** 363–368.
29. DUFFY, F. H., G. D. MOWER & J. L. BURCHFIEL. 1981. The role of a defocused image in strabismic amblyopia. Invest. Ophthalmol. Vis. Sci. (ARVO Abstr. Suppl.) **20:** 187.
30. CHINO, Y. M., M. S. SHANSKY & D. HAMASAKI. 1980. Development of receptive field properties of retinal ganglion cells in kittens raised with convergent squint. Exp. Brain Res. **39:** 313–320.
31. SINGER, W., M. W. VON GRAUNAU & J. RAUSCHECKER. 1980. Functional amblyopia in kittens with unilateral exotropia: 1. Electrophysiological assesment. Exp. Brain Res. **40:** 294–304.
32. LAWWILL, T. 1974. The bar pattern electroretinogram for clincial evaluation of the central retina. Am. J. Ophthalmol. **78:** 121–126.
33. LAWWILL, T. 1974. Pattern stimuli for clinical ERG. *In* Proc. 11th ISCERG Symp. E. Dolt & J. T. Pearlman, Eds. pp. 353–362. Dr. W. Junk, The Hague.
34. HOLDEN, A. L. & VAEGAN. Personal communication.
35. ARDEN, G. B. & W. M. BARNARD. 1979. Effect of occlusion on the visually evoked response in amblyopia. Trans. Opthal. Soc. U.K. **99:** 419–426.
36. VAEGAN & D. TAYLOR. 1979. The critical period for deprivation amblyopia in children. Trans. Ophthal. Soc. U.K. **99:** 432–439.
37. CIUFFREDA, K. J., R. V. KEYNON & L. STARK. 1980. Increased drift in amblyopic eyes. Br. J. Ophthalmol. **64:** 7–14.
38. ARMINGTON, J. C., T. R. CORWIN & R. MARSETTA. 1971. Simultaneously recorded retinal and cortical responses to patterned stimuli. J. Opt. Soc. Am. **61:** 1514–1521.
39. CLELAND, B. & C. ENROTH-CUGELL. 1970. Quantitative aspects of gain and latency in the cat retina. J. Physiol. **206:** 73–91.
40. CLELAND, B. G., W. R. LEVICK & K. J. SANDERSON. 1975. Properties of sustained and transient ganglion cells in the cat retina. J. Physiol. **228:** 649–680.
41. LEVICK, W. R. 1973. Variation in the response latency of cat retinal ganglion cells. Vis. Res. **13:** 837–853.
42. SHAPLEY, R. M. & J. P. VICTOR. 1978. The effect of contrast on the transfer properties of cat retinal ganglion cells. J. Physiol. **285:** 275–298.
43. MAFFEI. L. & F. W. CAMPBELL. 1970. Neurophysiological localisation of the vertical and horizontal vertical coordinates in man. Science **167:** 386–387.
44. SIEGFRIED, T. In press. Early wavelets in the VECP. Doc. Ophthalmol. Proc. Series (19th ISCEV Symp., Zurich, 1981).
45. MITCHELL, D. Personal communication.
46. SHERMAN, M. Personal communication.

PATTERN ELECTRORETINOGRAMS: GENERAL DISCUSSION

Ivan Bodis-Wollner, *Moderator*

Mount Sinai School of Medicine
City University of New York
New York, New York 10029

VAEGAN: The question to which I would specifically like to address myself is: Can we resolve more explicitly where the site of origin of this response is?

We have heard Dr. Maffei suggest possibly a ganglionic cell basis for this and Dr. Sherman suggested some of his clinical data were incompatible with the ganglion cell origin, suggesting it was in the more proximal retina, but not the ganglion cells themselves.

Dr. Armington and most other workers who have worked with the pattern electroretinogram have had a quite different assumption, which has not always been specifically tested, that is, that the pattern electroretinogram was merely a way of getting the photopic components of the ordinary electroretinogram. This is a response that is generated in the distal retina, and the idea is that if you have an alternating pattern then what you have are little local areas all going bright and dark simultaneously. But you are still recording the same thing that you would be if you summed the *on* and *off* responses of the ERG, and the ERG is generally believed to be generated by the distal retina.

Dr. Armington has already pointed out to you that you can certainly get rid of stray light effects with alternating patterns, and that is a very good reason for using them. But there is another way of getting focal responses.

Brindley and Westheimer showed that if you surround your stimulating area by a bright surround you can cut out the very late waves of the ERG and you will finish up with a very small early component, which is much tinier and much earlier than the late wave is usually considered to be. The main B wave, as Dr. Armington pointed out, is a late wave which reflects a scotopic function mainly due to stray light.

Brindley and Westheimer usually use about 10% of their stimulus. There you have the late B wave and here is a very early B wave which Brindley and Westheimer and we shall call the focal ERG.

All the ERGs have been recorded with bright surrounds at about the same level of luminance, and all our responses will be ones that are the sums of alternate *on*'s and *off*'s, and we have been able to separate this sort of focal ERG from the pattern ERG in two ways:

The first is experimental. The most remarkable thing about the pattern electroretinogram is it has an amazingly fixed peak time. We have been able to manipulate the timed peak of this response in only two ways. One is a relatively trivial way, that is, with luminance, and the other is, I will show you later, with spatial frequency, in humans only. They have a very well organized macula with different cell populations.

The pattern electroretinogram has a very peaked time as we measure it, and just to give you a difference from Dr. Armington we only measure the conventional A wave to B wave peak, although we do not know if these are actually A and B waves.

You can see that statistically there is no difference at high and low contrasts with the pattern electroretinogram. We have done this very carefully. The focal electroretinogram is earlier than the peak time of the pattern electroretinogram. We get the same sort of difference with the pigeon except, in the pigeon, the focal electroretinogram moves back by 10 msec. Otherwise the experiment is comparable.

So, we have here experimental separation of the two responses. Now, we have some clinical work which seems to suggest, as Dr. Maffei and others have already told you, that there is a ganglionic basis for this response.

We have recorded different reductions in the pattern electroretinogram in a large number of diseases and these include alcohol amblyopia, degenerative optic nerve disease, congenital optic nerve disease, and traumatic disruption of the optic nerve. In this last case, we saw the response was entirely absent, but I will not make much of that data because the subject had very poor fixation in both eyes. In early glaucoma there is a high correlation between the degree of glaucoma and the degree of loss. The response was just below 2 μV, where our average is 2½ in six of nine. In seven of the eight where the glaucoma is asymmetrical, the smallest response was in the eye with the worst field and there appears to be a 0.9 correlation between both eyes, which is highly significant.

These responses are for one degree per side checks, but, we get exactly the same results down to a quarter degree per side. This is not a high enough spatial frequency to see the high spatial frequency Professor Maffei showed. Dr. Sherman's case would come in about here, but that is because he recorded both eyes separately and looked for nice, clear responses. If he had recorded both eyes simultaneously and taken the average, this case would probably drop down a bit.

So, this data shows that the pattern and focal responses, even if there is very long optic nerve damage, are never totally separate, but, there is a component that drops out with degeneration.

I would like to stress the origin of the pattern electroretinogram in more detail, but, I think time will prevent me. However, one thing I would like to point out is that the pattern electroretinogram, as I have said, is extremely fixed in its time to peak. This is, according to Dr. Shapley, unlike the behavior of ganglion cells, which are supposed to change the time of peak for up to 20 msec at low contrast. We are not confident this happens with checkerboard pattern reversal.

Using microelectrode recordings in pigeon retinas, we have found that, with various contrasts, the time of peak stays absolutely constant with checkerboard phase reversal. So, I am not sure from this whether the experimental evidence is that the pattern responses are ganglionic. The literature will wait for that. It is certain that there may be a bit of luminous component in the pattern and a bit of pattern component in the luminous. Nevertheless, there is something in both that is different and we can get it in both experimental and clinical separations.

L. Maffei: I am glad to see that most of you in the crowd are in agreement with my physiological data and I would like to point out in a general remark that it is important to distinguish between the physiological clinical use of a given electrical wave and what one can get out in terms of information in the field of physiology. I think that the origin of the ERG of the different waves are of concern. It is very rare in my opinion to reach to the animal experiment where the condition can be more clearly established and repeatable, and in this context I would like to ask Dr. Veagan if he has tried in the pigeon to cut the optic nerve, which is the obvious experiment to do, and see whether the ERG is changing or not, or if he has done other experiments in which it would be very useful to hear the physiological properties of the ERG of the pigeon?

Vaegan: That is a rather opening question. I shall not talk too much about pigeon research because for the very most part there is hardly any difference between the pigeon results and the human. They are extremely close and very similar and this is most unlike most of the features of the pigeon retina and the human retina.

We can, with the pigeon, where the recording techniques are much more stable, record a spetial frequency rate per degree which is higher than the behavior we measured in the resolution of the pigeon.

We have done depth profiles through the retina in the pigeon retina with microelectrodes and we find exactly the same differences at the retinal surface. We find that the focal response stays positive and rolls over fairly quickly. But, at a different rate, the pattern response rolls over and changes at a different rate and a different manner than the focal response so that at different depths in the retina the two responses will be even more different. But, when you get very deep in the retina to the inner nuclear layer, the two responses have become identical and that is found at all positionings of the stimulus.

We have also tried pattern manipulation in the pigeon to try to work out whether the response is dominated by spatial frequency components or by edges or whether a stimulating receptive field centers it. It is rather tricky. Basically we tried to match the fundamental frequency components of the checks, bars and sine waves. We find that if the retina was responding to the fundamental spatial frequency components, then after we matched the fundamentals, all these patterns should give the same response. In actual fact, the bars and the sine waves give exactly the same response, which shows that edges are not important. But, nevertheless, the checkerboard response shows a reasonance, and that is at about a half a degree per side checks. After that there is a consistent difference between the checks, the bars, and sine waves. This is consistent with the predictions you would make if checkerboards are more optimally stimulating the center surround organization than are the sine waves and bars.

Other than that, we have not been able to do simple manipulations such as cut the optic nerve because the pigeon head is just too fragile. There would be too much damage.

J. SHERMAN: I would like to ask Dr. Maffei if it is possible that the flash ERG and the flicker ERG (which were recorded without a bright surround) might be coming from the periphery or the area outside of the area centralis of the cat whereas the pattern-reversal ERG is coming from the area centralis. Hence, one was present and one was not present.

MAFFEI: The experiment was done like that. The flash was tested before and after in the way that I described. So, the flash is a usual flash and the pattern ERG was a total view, but, we did not do any localized stimulus where the luminance is concerned.

You implied that only the central part of the retina could be involved, at least, in the early part. I think that is probably true.

I have an additional comment: The pattern ERG that we have in the cat and in the human must be definitive. In the cat any recording is all right. In the human, before recording the ERG we were obliged to put the reference electrode in the other eye because if we did not do that and put the reference electrode on the forehead or any other place, in many instances we picked up the visual evoked potential. So, I feel that this is a very important point because for years we tried to record the ERG and always we had interferences with the visual evoked potential.

J. C. ARMINGTON: I think the position of reference electrodes is always a problem. If you have a subject, such as myself, with alpha waves that are over a hundred microvolts, if you put the electrodes in the back of your head, and if you put an electrode on each eye and record, it looks just like anyone else's EEG. Out comes the alpha and all those other things.

SHERMAN: Some people might want to look at the pattern ERG as being a far-field potential from the cortex. Now, that is certainly possible. I would like to point out, however, that in the seven cases of optic nerve disease that I presented yesterday the pattern ERG was present, but, the pattern VEP was extinguished. In these instances, there is no way one can argue that the pattern ERG that was present is a reflection of the pattern-reversal VEP, which was extinguished.

M. TRICK: Why I have done is attempt to look at the effect of spatial frequency on the human pattern-reversal ERG and at the same time to look at the effect of temporal frequency. Essentially what I have done is replicate Dr. Armington's study for various temporal frequencies using various spatial frequencies.

In doing that I find, first of all, that, as long as we look at the human at spatial frequencies higher than about 0.25 cycles per degree, we can get, using those points, a very good fit to visual accuity. I used 14 emmetropic subjects, and, depending upon temporal frequencies, the range of visual accuities I predicted with a linear regression to noise level (rather than to zero) ranged from 20/60 to 20/25 over these normals with the best prediction being obtained with 1.8 or 3.75 Hz temporal frequency.

Now, I would also like to comment relative to what has been reported as the fall-off or the temporal spatial tuning question relative to the pattern-reversal ERG.

In doing my 14 normals, I noted that in some cases I did get a fall-off in amplitude for large checks and in other cases I did not get that fall-off for large checks. It disturbed me immensely that I was getting a lot of differences.

I went and repeated these experiments using just one temporal frequency on a number of subjects, and in this case I used a number of different blurs and I blurred the image to the eyes by up to 12 diopters. I found in those cases when there was blur, that as long as I was using checks of higher spatial frequency than 0.25 cycles per degree I got a rapid fall-off in the amplitude of the response as a function of blur, but, for large checks such as 4-degree checks I got no significant difference in the amplitude of the response when I blurred the subject.

So, I think this is an interesting point that we need to look at a little more extensively and it may be relevant to one of Dr. Spekreijse's comments earlier this week about the local luminance changes. Conceivably, in that case we may be picking up some luminance changes.

J. NELSON: I would like to address this to Dr. Vaegan. The results that you reported in amblyopia are in many respects similar to a number of recent reports in experimental amblyopia in cats in that ganglion cells are reported to have abnormal spatial resolutions in kittens reared with experimental amblyopia. I wonder if, maybe, you would like to expand on where you think the anomaly in amblyopia lies via the ERG anomalies that you have reported.

VAEGAN: The reason that we looked at the deprivation amblyopes in particular, as some of you might be aware, is because at present in the literature there is some degree of controversy about whether retinal effects occur in animal models of amblyopia, and up until very recently Dr. Ikeda was the only one to promote the idea that there might be a retinal change in amblyopia.

However, deprivation amblyopia does not seem to work in animals. We certainly find deprivation amblyopia does not produce retinal change in animals and there are at least five studies where deprivations failed to have an effect.

We, certainly, find that the majority of untreatable, amblyopes, or amblyopes failing to respond to treatment, have a retinal change, and it now looks a lot more like these cats and a lot more as if blur might be the important factor.

Now, as far as deprivation amblyopes, I believe that they are to a very large degree deprived rather than blurred because they were very rapidly corrected, i.e., in most cases very, very soon after surgery and contact lenses were put in. The optics were always good apart from the time they had cataracts. So, they seem to be totally different than the cats.

Most of the cat work to date has been a very short deprivation, and I think that if one starts to look carefully at long deprivation one will find retinal changes even in cats. My feeling is that a lot of the amblyopia that we are seeing may well have a retinal basis. There may be some cortical things as well and I am not prepared to say that you cannot have both.

S. FRICKER: We have approached this from rather a crude clinical viewpoint, but we have been doing ERGs on a somewhat different basis using the cross-correlation method. This allows you to consider the frequency of stimulation in a little different manner. For instance, if you want to get photopic ERGs you stimulate the system rather rapidly, and by doing the cross-correlation process you get a waveform that looks rather like a photopic ERG, but, you are putting in fairly high frequencies. We usually do it with flashes with a maximum of 50, an average of 25, and so on. So, in the course of a minute's test we get on the order of 1600 flashes in. We get a lot of signals and noise improvement, and we get nice clean signals. This is just what we want clinically because we want clean signals with minimum testing time and the ability to work through a lot of noise.

This is fine for flash VERs too. When we come to alternating checkerboards, we have to slow down a little bit because we are constrained, as most of you are I guess, by the TV screen. So, we have adopted a rather sloppy method in which our maximum rate of alternation is 25; average, 12½; and it can be lower. It can be anywhere basically from nothing up to 25. This is further compounded by the fact that you do not have the basic timing for the cross-correlation method linked to the 60 Hz or the TV set or you get into trouble.

The TV change is liable to have a bit of uncertainty in it—an average ±9 msec compared with the way you are doing the timing for your correlation.

In addition, when the checks change, of course, we have different periods of time in which something is black and, then, later on when it is white. We have the subject sitting there looking at this and the subject is probably, as most of our subjects are, moving around. He is not absolutely fixed, some of them less fixed than others.

The net result of all of this is that the stimulus on the retina is a very variable sort of thing. It is variable in time. It is not *off* equally as *on* for any given element. The elements themselves may change, and we have this added factor of frequency put into it.

When we first started doing the pattern VERs we looked at the pattern ERGs, and, I must say, on listening to some of the papers I would like to believe that there is a strong ganglion cell component. I am rather more inclined to agree with Dr. Spekreijse that there is a luminous component.

We have obtained quite variable results. We get small noisy responses which do not have fixed latencies, but, which seem to vary quite a bit. In spite of our noise suppression with the system, which is really rather effective, the responses tend to be variable and noisy. Sometimes we get them and at other times we do not. We wrote them off, perhaps rather too quickly, as being luminance variations. Possibly, that is a mistake and we should look at them some more. But the implications from our results with a fairly large number of clinical patients is that these are slippery things to get one's hands on. But, if you want to put in the added factor of timing of frequency, or temporal variations, consider the cross-correlation method because this factor is built in. You will have low and high frequencies, and you can change it and you can also get a lot of noise reaction too.

NELSON: In regard to where these new B waves are coming from, I think those of you who have not seen or read the Nelson, Zrenner, and Gouras paper, which was in the Morioka 16th ISCEV Symposium, should do so. That paper shows that if you take a profused cat eye cup and sprinkle it with little dots completely covering it and do an ERG, you will get what looks like a classical clinical ERG. Then, you take your dots and you make them larger and make the spaces between the dots less, but, keeping the luminance completely constant, the ERG B wave slowly diminishes. When you reach a critical dot size, the ERG B wave disappears completely. If you keep the process up, you reach a point finally where a small much faster little B wave appears. Now, this is certainly in line with some of the results that we have been hearing today, and this paper is about two years old.

BODIS-WOLLNER: What did you imply for those of us who might have read the paper, but, did not quite get the point?

NELSON: Gouras's notion is that there is spatial interaction at the B wave level, but, in a completely different part of the retina than we have recognized, that is, at the ganglion cell level.

BODIS-WOLLNER: In other words, those interactions would cancel out the B waves?

NELSON: Yes. He has a model for receptive field antagonism.

BODIS-WOLLNER: Are you suggesting, therefore, that there are certain spatial patterns which if they produce a pattern ERG would not produce a B wave and, therefore, the origin of the pattern ERG could be separated by this method?

NELSON: That is right. It is certainly something to think about in terms of the different B waves we all seem to be dealing with now.

SHERMAN: The first time I started to record these simultaneous pattern-reversal ERGs and VEPs in a small series of patients with bilateral optic nerve disease with acuities of 20/400 and worse, I could not get the ERGs.

You have to remember that this response is typically 2 or 3 μV. These patients with poor vision have poor fixation. The poor fixation itself is going to wipe out the 2 or 3 μV response. You really have to be very, very cautious to say that the pattern-reversal ERG is not there just because you do not get it the first time you try.

Now, in the seven patients that we reported yesterday we were very careful to make sure the patients were fixating carefully. Certainly, if you have a patient with monocular optic nerve disease it is much easier. The patient can fixate with the good eye through a mirror at a point out in space and you can stabilize fixation in the poor eye with the stimulus falling on the central retina. In that last case I presented yesterday, the blind eye was stimulated by the checks, while the good eye fixated. The normal or near normal pattern ERG was recordable from the blind eye using this simple trick.

VAEGAN: If I could make a point about technique too for those of you who are interested in doing this type of thing. The practice in our lab is to use the Arden gold foil electrode, obviously because it is Dr. Arden's lab. This is a very comfortable electrode to use, produces no rotations to the subject, and is very sensitive.

The other important aspect of our data collection is that we always record from both eyes simultaneously and take the ratio between the eyes as the critical variable.

Finally, our averaging techniques depend heavily on very good artifact rejection programs, and even those artifact rejection programs do not work unless we turn the gain of our system up to a point where we are rejecting somewhere between 20% to 50% of our traces.

You have to make sure that your artifact rejection system is really working well and is being pushed to get rid of all of the noise.

ROUNDTABLE SESSION: THE TRUE BINOCULAR VISUAL EVOKED POTENTIAL: INTRODUCTION

Ivan Bodis-Wollner

Departments of Neurology and Ophthalmology
Mount Sinai School of Medicine
New York, New York 10029

One of the primary aims of this conference has been to emphasize the two-way advantage of applying physiological concepts to the clinical usage of EP measurements. One of the areas where it may be expected that clinical data will add to our understanding of basic physiology is the study of binocular vision. How do the two eyes work together? As far as we know from morphological and single cell studies in the primate, the first stage in the visual system where information from the two eyes converges is in the cortex. Thus, truly binocular visual evoked potential abnormalities may be directly relatable to cortical pathology. The words *truly binocular* require clarification. By this, I mean that a monocular impairment alone cannot explain the presumed binocular dysfunction, but rather that one must postulate an abnormality at or beyond the site of binocular interaction. Therefore, to define a true binocular VEP it is necessary to clarify the rules regulating the cortical interaction of the two monocular inputs.

The situation is not as simple as it would first appear. It has been traditionally assumed that lesions behind the chiasma will affect both eyes, or, conversely, that the two pathways from both eyes converge behind the chiasma. Anatomically, they are in close proximity. Functionally, however, the two monocular pathways do not converge into a single structure until the first and probably only the second cortical relay neuron. Single cell recordings of the last 15 years in the visual cortex of cats and monkeys have amply demonstrated this fact. Thus, a truly binocular abnormality definitely localizes a lesion to the cortex, while a monocular abnormality may or may not be caused by a retrochiasmal lesion. If the lesion is large, such as a tumor, the classical concept stating that a monocular abnormality represents prechiasmal pathology still holds, but this is precisely the condition which hardly requires EP technology for diagnosis. Rather, EP technology would be most important for identifying small-sized lesions or selective involvement of long fiber tracts. The anatomy is further complicated because, as far as we know, both from anatomical and functional studies, parallel pathways representing fast and slow conducting fibers of the optic nerve (probably related to a subdivision of X and Y ganglion cells) remain independent up to the cortex. A division of these pathways must also consider their predominant linear versus nonlinear properties. Thus, even the simplest physiologically relevant assumptions concerning the binocular neurons in the cortex must take into account two separate pathways of two separate eyes, which raises the number of permutations. Further complications arise from the fact that two homologous areas of the two retinae in perfect unison superimposed on each other would not enable a cortical neuron to detect depth. Neurons that respond to retinal disparity, so-called "disparity-tuned neurons," are needed to detect stereoscopic depth,[1,2] and these cannot receive their information from the first binocular correlated neurons, as was first pointed out by Hubel and Wiesel.[3] Binocular correlators (zero disparity) and disparity-tuned neurons responding to stereoscopic depth are anatomically segregated. While it is controversial,[4] there is a possibility that, as first suggested, the latter predominate in area 18.[5] These anatomical facts are relevant to topographical studies of EPs.

Thus it is not trivial to discuss binocular interaction. It has been assumed that, for instance, binocular summation is evidence of binocular interaction. However, it is clearly possible that two monocular signals summate not in a single neuron, but rather at a recording electrode as a perfect algebraic device. In order to be sure that the two signals interact, not in the recording apparatus, but in the brain, one needs to show clear evidence that the presumably binocular signal could not have been produced in a passive electrical device having two inputs. Several methods have been proposed,[6,7] and many of the participants in this roundtable have found their own solution to the problem. Questions were posed to several members on the panel concerning the explicit or implicit linear versus nonlinear assumptions underlying binocular VEP studies. The questions that were asked each of them were geared to their own work, and we hoped that they would point to strengths and/or weaknesses in their own arguments. Consequently, in this roundtable discussion the applications of true binocular VEP recordings in amblyopic children to neurological patients with cortical pathology, and the question of linear versus nonlinear signal analysis of binocular responses will be discussed.

References

1. Barlow, H. B., C. Blakemore & J. D. Pettigrew. 1967. The neural mechanism of binocular depth discrimination. J. Physiol. **193:** 327–342.
2. Nikara, T., P. O. Bishop & J. D. Pettigrew. 1968. Analysis of retinal correspondence by studying receptive fields of binocular single units in cat striate cortex. Exp. Brain Res. **6:** 353–372.
3. Hubel, D. H. & T. N. Wiesel. 1970. Stereoscopic vision in macaque monkey. Nature **225:** 41–42.
4. Poggio, G. F. & B. Fischer. 1977. Binocular interaction and depth sensitivity in striate and prestriate cortex of behaving rhesus monkey. J. Neurophysiol. **40:** 1392–1405.
5. Ferster, D. 1981. A comparison of binocular depth mechanisms in areas 17 and 18 of the cat visual cortex. J. Physiol. **311:** 623–655.
6. Julesz, B., W. Kropfl & B. Petrig. 1980. Large evoked potentials to dynamic random-dot correlograms and stereograms permit quick determination of stereopsis. Proc. Natl. Acad. Sci. USA **77:** 2348–2351.
7. Bodis-Wollner, I., M. C. Barris, L. H. Mylin, B. Julesz & W. Kropfl. 1981. Binocular stimulation reveals cortical components of the human visual evoked potential. Electroencephalogr. Clin. Neurophysiol. **52:** 298–305.

SPATIAL FREQUENCY LIMITATIONS IN BINOCULAR NEURONS: VISUAL EVOKED POTENTIAL EVIDENCE*

Ken Nakayama, Patricia Apkarian, and Christopher W. Tyler

Smith Kettlewell Institute of Visual Sciences
San Francisco, California 94115

What can the visual evoked potential (VEP) tell us about the tuning characteristics of binocular versus monocular visual mechanisms? In particular, are binocular mechanisms tuned to the same range of spatial frequencies as monocular mechanisms?

To start, let us review some relevant information from other sources. First there are data from animals. In the monkey, which is the most appropriate animal to compare with man, the geniculate afferents terminate mostly in layer IV. Most cells in the visual cortex are binocularly driven, but layer IV cells are monocular, and in addition they have the smallest receptive fields of all cortical cells.[1] These findings might suggest that monocular neurons are tuned to higher spatial frequencies than binocular neurons. Whether the receptive field sizes translate into a difference in spatial frequency tuning functions is not clearly established, however. Binocular cells with large receptive fields (complex cells, for example) may be driven by subunits which themselves have small receptive fields.[2] As a consequence, the spatial frequency tuning function may not differ between cells having different receptive field sizes. Thus, the results of single unit research are as yet inconclusive.

In terms of human psychophysics, it is possible to compare the contrast sensitivity functions for monocular and binocular viewing. They both have the same shape, with the sensitivity of the monocular curve being about 1.4 times less than that for the binocular curve.[3] This might suggest that the spatial frequency tuning for binocular and monocular mechanisms was identical, but this conclusion would be unwarranted. If the binocular system had higher thresholds, for example, both sets of thresholds may be determined by monocular mechanisms alone, revealing nothing about binocular mechanisms. In fact, the ratio of binocular to monocular sensitivity is close to $\sqrt{2}$, a ratio that would be predicted if the threshold detection in the binocular situation were due to probability summation of independent monocular mechanisms. Thus, these threshold results do not argue for or against a different spatial frequency tuning for each type of mechanism. Moreover, they suggest that the use of contrast threshold psychophysics may be inappropriate to examine the difference between monocular and binocular mechanisms. More apt would be to examine the question with stimuli that were suprathreshold and would be sure to stimulate binocular mechanisms. This can be done either using different types of psychophysical technique or using visual evoked potential techniques.

Psychophysical experiments using such supratheshold stimuli favor the view that binocular mechanisms are tuned to lower spatial frequencies. First, there is an increase in perceived spatial frequency of random noise on closing one eye. For example, random noise viewed monocularly appears to be composed of smaller texture "grains" than the same noise viewed binocularly.[4,5]

Secondly, stereoscopic tilt from a spatial frequency between the two eyes[6] does not

*This work was supported in part by National Institutes of Health Grants R01 EY01582, R01 EY02124, GRS 5S01, and RR05566 and by the Smith Kettlewell Eye Research Foundation.

occur for sinusoidal gratings above about 15 cycles per degree, suggesting a fairly low upper spatial frequency cut-off, much lower than the contrast sensitivity cut off for the detection of gratings (~60 cycles/degree).

To examine whether a similar conclusion might be drawn from evoked potentials, we have measured the VEP amplitude as a function of several variables, in monocular and binocular viewing. For example, we have made many measurements of VEP amplitude versus spatial frequency. The evoked potential results are very different from the threshold psychophysics described above. Rather than showing an across-the-board increase in VEP response when comparing binocular versus monocular viewing, the differences are highly dependent on spatial frequency.

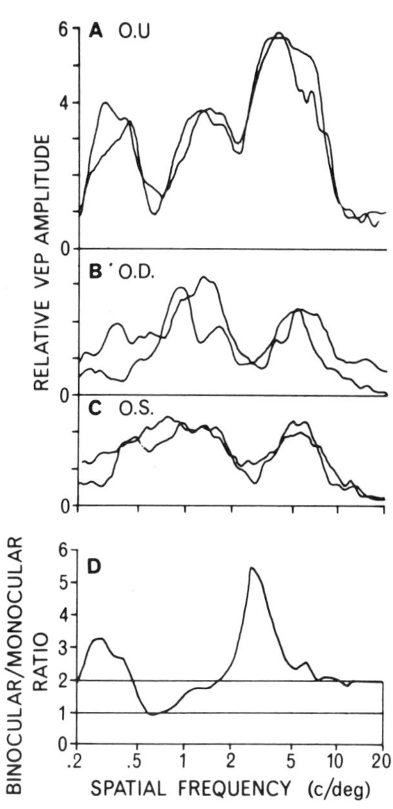

FIGURE 1. Comparison of binocular and monocular responses obtained with the steady-state VEP using the electronic sweep method.[7] This enables a complete VEP spatial frequency tuning function to be obtained in 20 seconds or less. Stimulus field was a 20° by 15° rectangle. Stimulus was a counterphase-modulated vertical sine-wave grating with a contrast of 80%. Reversal and recording frequency was 28 Hz. Two separate spatial frequency functions were obtained for each condition and are superimposed. The binocular response alone is seen in (A). The right and left eye response is seen in (B) and (C), respectively. The ratio of binocular to mean monocular response (averaged across the two eyes) is plotted as (D). Binocular neural facilitation is indicated by portions of the function above a binocular/monocular ratio of 2 (upper horizontal line in D).

As an example, consider the spatial frequency tuning for binocular and monocular viewing as determined by our technique of sweeping spatial frequency.[7] FIGURE 1 shows the binocular response (A), the right and left eye response alone (B and C), and the ratio of the binocular response to mean monocular responses (D). If the binocular response were the linear summation of monocular responses (assuming the same phase for each eye, which is indeed the case), the binocular-to-monocular ratio should remain at 2.0 (skirting the upper solid line in FIGURE 1D). This is not the case, however. There are spatial frequency regions where the actual binocular response is far greater than that predicted by linear summation alone (those regions in FIGURE 1D that are above the upper horizontal line). Such areas represent clear evidence of neural facilitation, a finding that has been commonly seen in single unit work[8,9] but rarely found with evoked potentials. What is critical in this regard is that binocular facilitation occurs at only some spatial and temporal frequencies and one must vary both over a wide range to see it reliably.[10] Because these regions of binocular facilitation are independent of the occurrence of monocular peaks, we presume that they represent the activity of binocular neurons.

Here we are interested in the range of spatial frequencies over which the binocular facilitation occurs. If it occurs systematically at a lower spatial frequency than the sharply tuned monocular peaks, it would support the view that binocular mechanisms are indeed tuned to lower spatial frequencies. In the case presented in FIGURE 1, the highest spatial frequency peak of binocular facilitation (2.5 c/deg) is still clearly below the point where the highest monocular spatial frequency peaks (5.0 c/deg).

FIGURE 2 shows a summary histogram of six such experiments, showing the highest spatial frequency peak of the monocular response in comparison to the highest frequency of the binocular facilitation peaks. Because binocular facilitation is not evident under all circumstances,[10] we have selected for comparison those spatial

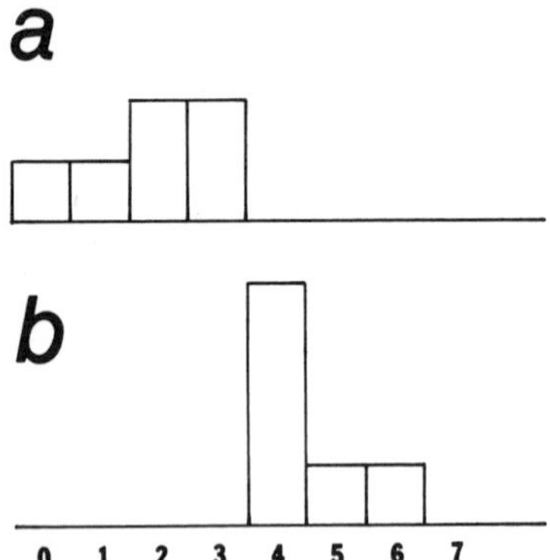

FIGURE 2. Histogram showing the center spatial frequency of the upper spatial frequency peak for binocular facilitation (a) in comparison to the upper frequency amplitude peak obtained with monocular stimulation (b). Data taken from four observers, using temporal frequencies where binocular facilitation is evident.

frequency curves (choosing the right temporal frequencies) where binocular facilitation is clearly evident. Note that the peak frequency of the facilitation effect is shifted down, such that the mean monocular frequency is about twice the mean frequency for binocular facilitation.

In conclusion, the evoked potential results are consistent with the view that binocular mechanisms are tuned to lower spatial frequencies than monocular mechanisms. They suggest that at some neurophysiological level, the neural mechanisms that are selectively activated by binocular stimuli are more apt to be driven by visual neurons that perform a coarser analysis of the visual image. Conversely, it suggests that the higher spatial frequency mechanisms responsible for the appreciation of fine spatial detail might not be involved in some stereoscopic mechanisms.

REFERENCES

1. HUBEL, D. H. & T. N. WIESEL. 1968. Receptive fields and functional architecture of monkey striate cortex. J. Physiol. **195:** 215–243.

2. MOVSHON, J. A., I. D. THOMPSON & D. J. TOLHURST. 1978. Receptive field organization of complex cells in cat's striate cortex. J. Physiol. **283:** 79–99.
3. CAMPBELL, F. W. & J. C. ROBSON. 1968. Application of Fourier analysis to the visibility of gratings. J. Physiol., **197:** 551–566.
4. MACKAY, D. M. 1961. Visual effects of non-redundant stimulation. Nature **180:** 739–740.
5. TYLER, C. W. 1975. Observations on binocular spatial frequency reduction in random noise. Perception **4:** 305–309.
6. BLAKEMORE, C. 1970. A new kind of stereoscopic vision. Vision Res. **10:** 1181–1200.
7. TYLER, C. W., P. APKARIAN, D. LEVI & K. NAKAYAMA. 1976. Rapid assessment of visual function: An electronic sweep technique for the pattern VEP. Invest. Ophthalmol. Vis. Sci. **18:** 703–713.
8. PETTIGREW, J. D., T. NIKARA & P. BISHOP. 1980. Binocular interactions on single units in cat striate cortex. Exp. Brain Res. **6:** 391–410.
9. BARLOW, H. B., C., BLAKEMORE & J. D. PETTIGREW. The neural mechanism of binocular depth discrimination. J Physiol **193:** 327–342.
10. APKARIAN, P., K., NAKAYAMA, & C. W. TYLER. 1981. Binocularity in the human visual evoked potential: Facilitation, summation, and suppression. EEG Clin. Neurophysiol. **51:** 32–48.
11. TYLER, C. W., P. APKARIAN & K. NAKAYAMA. 1978. Multiple spatial frequency tuning of electrical responses from the human visual cortex. Exp. Brain Res. **33:** 1978.

DISCUSSION OF THE PAPER

I. BODIS-WOLLNER: Dr. Nakayama, one of the questions, of course, is *how much* facilitation versus no facilitation is evidence of binocular interaction, whether or not an algebraic interaction can be used as evidence of binocular interaction.

K. NAKAYAMA: Well, we demonstrated a nonalgebraic interaction. In other words, if you put one plus one together and it equals four or five, this cannot be explained by the addition of electricity. I see that as evidence of binocular neuronal interaction.

I am not saying that that is the only type of binocular interaction. It turns out that many types of binocular interaction in the literature are inhibitory and this technique can miss that.

I. BODIS-WOLLNER: Couldn't one not say precisely that where the two eyes' input produces a lesser response than either eye alone there is binocular interaction?

NAKAYAMA: I would definitely say that is true as well.

COMMENT: I would like to make a comment. Even if you ascribe to the view that you should be able to predict things about the evoked potential from single-cell work, which I find a difficult jump to make conceptually, why couldn't it be, for example, that disparity detectors which are facilitated at the single-cell level have source-sink patterns that are oriented differently than the monocular cell source-sink patterns, and that, in fact, excitation of a disparity detector could lead to what appears to be a decrease, not a facilitation, in the VEP? In other words, why should one equate facilitation with the existence, with the activation, of disparity detectors?

NAKAYAMA: At this moment I did not equate it with the activation of disparity detectors. I just said that it is a very narrowly defined concept of binocular interaction.

In other words, any time you see a nonlinearity it seems to me that it involves some type of neuronal interaction, not independence. It means that two signals are combining in the brain in some form. That is all I really meant to say, and this is one way of picking out a very specific type of nonlinearity and Dr. Bodis-Wollner pointed out another one. As I said, it is casting out a net that is fairly restrictive. It does not include all types of binocular interaction and, also, it does not presume to say exactly what the underlying mechanism is, but, I do not think one can argue with the view that this indicates a neural convergence.

DO VISUAL EVOKED POTENTIALS STUDIES REVEAL AMBLYOPIC ABNORMALITIES NOT READILY APPARENT IN PSYCHOPHYSICAL TESTS?

Dennis M. Levi

University of Houston
College of Optometry
Houston, Texas 77004

The visual evoked potential (VEP) has been widely studied in humans with naturally occurring amblyopia. There are various motivations for such studies. Among these, perhaps the most frequent factors have been the following:

(1) To determine the utility of the VEP as a diagnostic test for amblyopia and concomitant anomalies of binocular vision.
(2) To assess prognosis and monitor treatment.
(3) To attempt to localize normal and abnormal function.
(4) To study mechanisms of amblyopia.

The recent use of patterned stimuli for eliciting VEPs of amblyopes has proven quite useful in the diagnosis of amblyopia, and in monitoring treatment in infants and young children.[1,2] More recently, the application of rapid techniques, e.g., the spatial frequency "sweep" technique, in conjunction with "0 voltage extrapolation" have allowed a more quantitative approach to the measurement of acuity, and the assessment of treatment in amblyopes.[3–6]

Appropriate manipulations of various stimulus parameters (e.g., spatial frequency, temporal frequency, and contrast) may also be of value in identifying normal and abnormal binocular vision.[7–11]

The VEP has also provided useful information regarding the mechanisms of amblyopia. VEP studies are consistent with the notion originally suggested by Wald and Burian, that amblyopia represents primarily an abnormality of the "form sense," which is present at or prior to primary visual cortex. In this regard, it is consistent with a large body of psychophysical data.[12–16]

The purpose of the present paper is to identify an amblyopic abnormality which is present in the VEP, and which is not readily apparent in psychophysical tests. A universal finding among VEP studies of amblyopia is that with small (high spatial frequency) patterned stimuli, amblyopes show a reduction in the amplitude of the VEP of the amblyopic eye compared to that of the nonamblyopic eye.[17–19] FIGURE 1A shows the VEP spatial frequency tuning curves of the amblyopic (squares) and nonamblyopic (circles) eyes of M.M. These results were obtained in response to sinusoidal gratings with a contrast well above threshold (44%). Below (FIGURE 1B) are the psychophysical contrast sensitivity functions (1/threshold contrast) for each eye of the same observer. While the exact shapes of the electrophysiological and psychophysical functions differ somewhat, it is clear that the amblyopic process is similarly reflected in the two functions.

FIGURE 2 shows the results of a quite different kind of psychophysical experiment, i.e., interocular successive contrast matching.[20] The abcissa shows the contrast of a 10

*This work was supported by a grant (RO1 EY01728) from the National Eye Institute.

0077–8923/82/0388–0615 $1.75/0 © 1982, NYAS

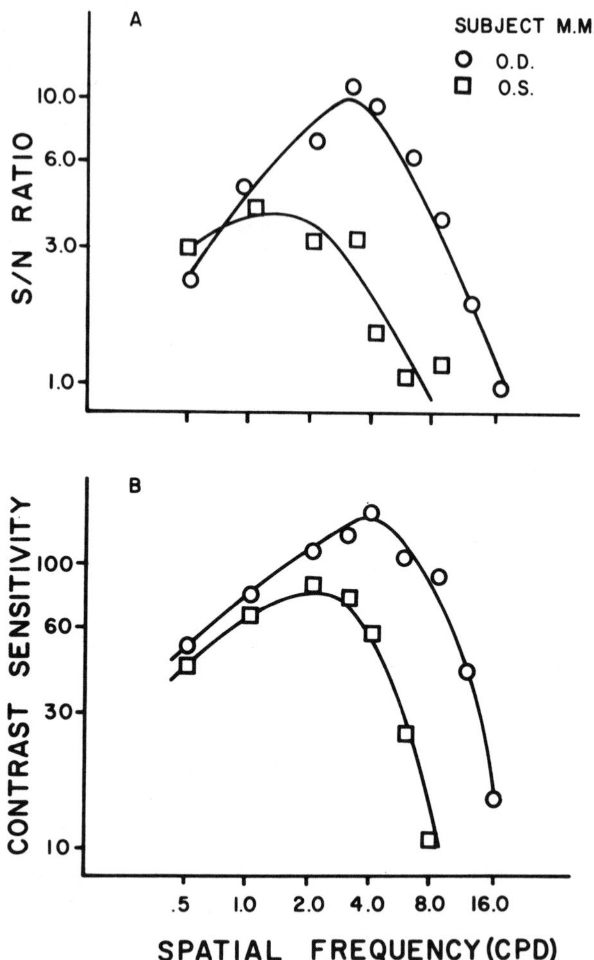

FIGURE 1. (A) VEP spatial frequency tuning functions for the nonamblyopic (circles) and amblyopic (squares) eyes of patient M.M.; stimulus contrast was 0.44. (B) Psychophysical threshold contrast sensitivity functions for the nonamblyopic (circles) and amblyopic (squares) eyes of subject M.M. (From Levi & Harwerth.[18] By permission of *Investigative Ophthalmology and Visual Science.*)

cycle/degree standard grating, presented to the fovea of the nonamblyopic eye. The ordinate shows the contrast of a variable test grating (of the same spatial frequency) presented to the fovea of the amblyopic eye, which has the same apparent contrast as the standard. The threshold for each eye is represented by the half-filled symbols. The diagonal line represents points of equal physical contrast. Data are shown for two amblyopic observers, an anisometropic amblyope (D.S., circles) and a strabismic/anisometropic amblyope (J.V., squares). For both observers, the threshold data show an elevation of threshold (reduced sensitivity) of the amblyopic eye. If this threshold abnormality were present over the entire range of suprathreshold contrast tested, than all of the suprathreshold data would fall on a line with a slope of 1, passing through the threshold symbol. However, it is clear that the threshold abnormality was not maintained in the suprathreshold contrast matching results. For observer J.V., when the grating is just visible to the amblyopic eye, it has normal apparent contrast. For anisometropic amblyope D.S., the threshold abnormality is evident for suprathreshold stimuli within about a log unit of threshold; however, at the higher contrast levels (>30%) the data fall close to the diagonal line, suggesting that there is little if any

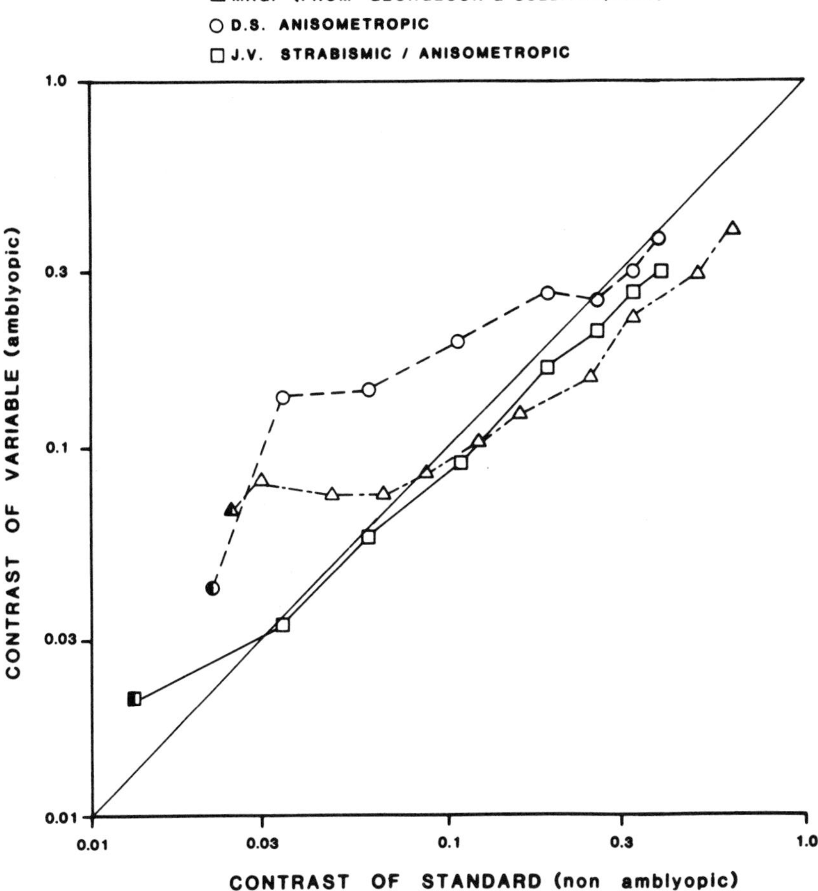

FIGURE 2. Apparent contrast of a 10 cycle/degree grating for two amblyopic observers, D.S. (circles) and J.V. (squares). Contrast of a sinusoidal grating (the variable) seen by the amblyopic eye which matched a grating of the same frequency seen by the nonamblyopic eye. Each point is the mean of 6 reversals of a staircase, and the error bars are equal to or smaller than the symbols. Thresholds for the variable and standard stimuli are indicated in two dimensions by the half-filled symbols. The diagonal line represents physical equality of contrast. The triangles are data of MAG (from Georgeson and Sullivan)[21] showing apparent contrast in peripheral vision, i.e., the contrast of a 10 c/deg grating seen in peripheral vision which matched a grating of the same frequency seen foveally.

abnormality in apparent contrast in the high contrast range. Thus it appears that there is substantial suprathreshold compensation for the reduced contrast sensitivity evident at threshold. Herein lies the paradox: the VEP to suprathreshold stimuli shows abnormalities in amblyopia that are not evident in suprathreshold psychophysical experiments (although they do appear to parallel the losses evident at threshold).

Georgeson and Sullivan[21] have suggested the normal visual nervous system corrects for contrast losses caused by blurring of the image early in the visual process. They showed that compensation for the attenuation of threshold contrast sensitivity occurs under a variety of conditions, including high spatial frequency stimuli, peripheral viewing and scotopic conditions. The triangles in FIGURE 2 are data of MAG (from Georgeson & Sullivan,[21] text figure 6). The results show the contrast of a grating (the variable) seen in peripheral vision which matched a grating of the same frequency seen foveally. It is interesting to note that contrast compensation for the peripheral retina is quite similar to that seen in the amblyopic fovea. Georgeson and Sullivan[21] suggested that cortical spatial frequency selective channels can adjust their gain to compensate for earlier optical and neural attenuation, thus achieving a deblurring of the image. Thus, the spatial frequency channels receiving their input

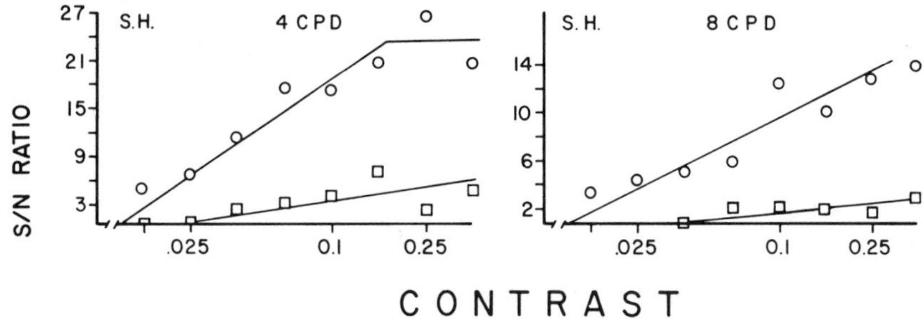

FIGURE 3. VEP signal/noise ratio vs stimulus contrast for gratings of 4.0 (Left) and 8.0 (Right) cycles per degree for the nonamblyopic (circles) and amblyopic (squares) eyes of S.H.

from an amblyopic eye would have a higher gain than those receiving input from a nonamblyopic eye. While this is precisely what is observed in psychophysical studies based upon contrast matching[20] or magnitude estimation,[22] VEP studies show just the opposite, i.e., the gain of the VEP versus contrast function of the amblyopic eye is lower than that of the nonamblyopic eye (FIGURE 3).[18]

How can the abnormalities evident in the VEP of amblyopes to suprathreshold stimuli be explained in the light of their apparently normal suprathreshold contrast perception?

One possible explanation is that the suprathreshold compensation for the amblyopic anomaly occurs at a higher level in the visual system, i.e., after the site of generation of the VEP, and is therefore not evident in the VEP results. While this possibility cannot be ruled out, Georgeson and Sullivan[21] have argued convincingly that the mechanism of contrast constancy exists within the system of spatial frequency and orientation selective channels in the visual cortex.

The VEP represents the pooled activity of many thousands of neurons. In animals deprived of normal visual experience by lid suture, experimental unilateral strabismus or anisometropia, cortical neurons that are normally responsive to visual stimulation of either eye lose their functional input from the deprived eye.[23–26] Recent evidence suggests that the pattern VEP of monocularly deprived cats is extremely sensitive to shifts of ocular dominance, even when the monocular deprivation periods were restricted.[27] Thus, it is possible that the abnormalities evident in the VEP to suprathreshold stimuli simply represent a reduction in the number of responsive neurons rather than changes in the sensitivity of the neurons themselves.[28]

While there are probably other plausible accounts for the discrepancy between the psychophysical and electrophysiological responses of amblyopes, the data nevertheless point out the need to exercise caution in extrapolating from VEP data to suprathreshold psychophysics (or vice versa). On the other hand, the results strongly suggest that the VEP to suprathreshold spatial stimuli may provide a very useful diagnostic tool for humans with naturally occurring amblyopia.

References and Notes

1. Arden, G. B., W. M. Barnard & A. S. Mushin. 1974. Visually evoked responses in amblyopia Br. J. Ophthalmol. **58:** 183–192.
2. Sokol, S. 1977. Visual evoked potentials to checkerboard pattern stimuli in strabismic amblyopia. *In* Visual Evoked Potentials in Man: New Developments. J. E. Desmedt, Ed. pp. 410–417. Clarendon Press, Oxford, England.
3. Regan, D. 1977. Rapid methods for refracting the eye and for assessing visual acuity in amblyopia, using steady-state visual evoked potentials. *In* Visual Evoked Potentials in Man: New Developments. J. E. Desmedt, Ed. pp. 418–426. Clarendon Press, Oxford, England.
4. Tyler, C. W., P. Apkarian, D. M. Levi & K. Nakayama. 1979. Rapid assessments of visual function: An electronic sweep technique for the pattern visual evoked potential. Invest. Ophthalmol. Vis. Sci. **18**(7): 703–713.
5. Sokol, S. 1978. Measurement of infant visual acuity from pattern reversal evoked potentials. Vis. Res. **18**(1): 33–40.
6. Note that the use of grating patterns will typically result in an overestimate of Snellen visual acuity in amblyopes, both psychophysically and in VEP studies. Levi, D. M. & R. S. Harwerth. 1978. A sensory mechanism for amblyopia: Electrophysiological studies. Am. J. Optom. Physiol. Optics **55**(3): 163–171.
7. Srebro, R. 1978. The visually evoked response: Binocular facilitation and failure when binocular is disturbed. Arch. Ophthalmol. **96:** 839–844.
8. Amigo, G., A. Fiorentini, M. Pirchio & D. Spinelli. 1978. Binocular vision tested with visual evoked potentials in children and infants. Invest. Ophthalmol. Vis. Sci. **17:** 910–915.
9. Lennerstrand, G. 1978. Binocular interaction studied with visual evoked responses (VER) in humans with normal or impaired binocular vision. Acta Ophthalmol. **56:** 628–637.
10. Apkarian, P. A., C. W. Tyler & K. Nakayama. 1981. Binocularity in the human visual evoked potential: Facilitation, summation and suppression. Electroencephalogr. Clin. Neurophysiol. **51:** 32–48.
11. Under certain conditions amblyopes show binocular interactions in the VEP which are in many respects similar to those of normal observers. Thus caution should be exercised in the use of the VEP to screen for anomalies of binocular vision. Apkarian, P., D. M. Levi C. W. Tyler. 1981. Binocular facilitation in the visual evoked potential of strabismic amblyopes. Am. J. Optom. Physiol. Optics **58:** 820–830.
12. Lombroso, C. T., F. H. Duffy & R. M. Robb. 1969. Selective suppression of cerebral evoked potentials to patterned light in amblyopia exanopsia. Electroencephalogr. Clin. Neurophysiol. **27:** 238–247.
13. Spekreijse, H., L. H. Khoe & L. H. van der Tweel. 1972. A case of amblyopia: Electrophysiology and psychophysics or luminance and contrast. *In* The Visual System. G. B. Arden, Ed. pp. 141–156. Plenum Press New York, N.Y.
14. Sokol, S. & B. Bloom. 1973. Visually evoked cortical responses of amblyopes to a spatially alternating stimulus. Invest. Ophthalmol. **12:** 935–939.
15. Levi, D. M. & R. E. Manny. 1981. The pathophysiology of amblyopia: Electrophysiological studies. Ann. N. Y. Acad. Sci. This volume.

16. For a recent review, see VON NOORDEN, G. K. 1980. Burian and von Noorden's Binocular Vision and Ocular Motility. 2nd edit. pp. 237–249.
17. LEVI, D. M. 1975. Patterned and unpatterned visual evoked responses in strabismic and anisometropic amblyopia. Am. J. Optom. and Physiol. Optics **52**(7): 455–465.
18. LEVI, D. M. & R. S. HARWERTH. 1978. Contrast evoked potentials in strabismic and anisometropic amblyopia. Invest. Ophthalmol Vis. Sci. **17:** 571–575.
19. LEVI, D. M. & J. W. WALTERS. 1977. Visual evoked responses in strabismic and anisometropic amblyopia: Effects of check size and retinal locus. Am. J. Optom. Physiol. Optics **54**(10): 691–698.
20. The paradigm was essentially the same as that described by Hess and Bradley (HESS, R. F. & A. BRADLEY. 1980. Contrast perception above threshold is only minimally impaired in human amblyopia. Nature **287:** 463). The paradigm was modified so that the observers task was to judge whether the apparent contrast of the variable stimulus was higher or lower than that of the standard. Depending upon the observers response the contrast or the variable stimulus was varied according to a staircase until the apparent contrast of the variable and the standard could not be distinguished. The paradigm will be described in greater detail elsewhere [(LEVI, D. M. & D. S. LOSHIN. (in preparation)].
21. GEORGESON, M. A. & E. D. SULLIVAN. 1975. Contrast constancy: Deblurring in human vision by spatial frequency channels. J. Physiol. **252:** 627–656.
22. MANNY, R. E. & D. M. LEVI. 1981. The temporal modulation sensitivity function in amblyopia: Spatio-temporal interactions. Invest. Ophthalmol. Vis. Sci (in press).
23. WIESEL, T. N. & D. H. HUBEL. 1963. Single cell responses in striate cortex of kittens deprived of vision in one eye. J. Neurophysiol. **26:** 1003–1014.
24. BLAKEMORE, C., L. J. GAREY & F. VITAL-DURAND, 1978. The physiological effects of monocular deprivation and their reversal in the monkey's visual cortex. J. Physiol. **283:** 223–282.
25. BAKER, F. H., P. GRIGG & G. K. VON NOORDEN. 1976. Effects of visual deprivation and strabismus on the response of neurons in the visual cortex of the monkey, including studies on the striate and prestriate cortex in the normal animal. Brain Res. **66:** 185–208.
26. BLAKEMORE, C. & H. EGGERS. 1978. Animal models for human visual development. *In* Frontiers in Visual Science. S. J. Cool & E. L. Smith, Eds. Springer-Verlag, Berlin.
27. SNYDER, A. & R. SHAPLEY. 1979. Deficits in the visual evoked potentials of cats as a result of visual deprivation. Exp. Brain Res. **37:** 73–86.
28. MITZDORF, U. & W. SINGER. 1978. Basic patterns of synaptic activity in visual cortex of normal and monocularly deprived cats: A current source density analysis of electrically evoked potentials. J. Physiol. **284:** 120p.

DISCUSSION OF THE PAPER

S. FRICKER: I think it is wonderful that we can obtain these graphs and functions from 18-month-old noncooperative squalling children in the space of about ten seconds. We cannot do that. You have a problem, really, in dealing with this when the subject of amblyopia comes up because, almost invariably, when we get results presented, they are not from the typical clinical patient who we find very difficult to test clinically sometimes; and these lab-type tests which are fine for your lab partners and some of the lab technicians rarely seem to be quite as applicable in the actual clinical situation.

I would suggest that the VEP is a very blunt tool indeed. It is blunt for a cooperative 25 or 30 year old amblyope, and it is even blunter for an 18 month old nontrained amblyope.

O. BRADDICK: Well, I concur entirely with Dr. Fricker's opinion of the 18 month old, but, I am rather startled by his implicit suggestion that the typically amblyopic patient is an 18 month old, squalling or otherwise, in that amblyopia is defined as a loss of visual accuity and so on, and very, very few 18 month olds or younger babies outside the Infant Visual Lab have had their accuities tested.

I would have thought the typically amblyopic patient was more likely to be a six year old who is a rather different kettle of fish. I have no doubt that these data are not from six year olds either, but I think one would have more prospects of gaining this sort of data from a child in that age range if it was thought to be useful to do so.

COMMENT: I think Dr. Fricker is quite right in pointing out the difficulties in gaining a lot of data in clinical patients, but I think one of the points I would like to make is that it is a pretty universal finding that if you can measure a pattern VEP in an amblyopic eye with a relatively small pattern in an amblyopic patient, then, it is almost always reduced in amplitude by comparison to the other eye. And in that sense there is some value, provided you can rule out optical factors and other disease processes.

COMMENT: Well, I think that is true. We do have problems in being certain that youngsters are looking at a pattern. A flash is one thing, but, a pattern is something else. You just do not know what they are doing.

The other thing I would like to suggest is I do not quite agree with you that the six year old is the typical one. Clinically, you can get responses out of youngsters from age two plus or minus a little bit with pictures. In clinical practice, you can use the preferential looking technique, which I expect many of you are familiar with. It is being used on a somewhat wider scale now, at least in an experimental way, and this is applicable to youngsters up to the age of one year. It is being extended to work with somewhat older children.

Again, when you listen to people talking about this, you get the feeling that you can just do the test in three minutes and that is it. In fact, it usually takes longer, but this is a nonelectrical, more psychological type of test; however, it actually seems to be working out moderately well in some cases.

So, in many clinical situations, if you have the choice of having to fund a VEP Lab or Preferential Looking Lab to do clinical work, you might be inclined to go towards the Preferential Looking Lab rather than the VEP unfortunately.

I. BODIS-WOLLNER: Dr. Levi, after all you somehow did not answer the question, which was simply: Do you think that if you had tested binocular VEP rather than monocular you would have gotten a closer match to your psychophysical studies?

D. LEVI: No.

BODIS-WOLLNER: Do you think that the binocular VEP, a truly binocular one as you tried to define it, would tell you something different than testing each eye separately?

LEVI: Very much so, and, in fact, I think the kind of stimuli we have heard about from Dr. Julesz and Dr. Braddick in particular may have great value. That is, stimuli which test directly binocular correlation and directly binocular disparity processing may be of great value. It may tell us different things about patients with amblyopia.

BRADDICK: I think, particularly with relation to the animal model, one thing we should keep very clear in all this discussion of amblyopia is the clinical definition. The aspect that is approached by Dr. Levi's work, is the issue of the acuity and, maybe also, the relative contrast sensitivity between the two eyes.

Most, not all, but most of the animal work related to animal models as proposed to be connected to amblyopia have been concered with issues of binocularity and the ocular dominance histogram. So, one has two defects which are, obviously, not independent: a lot of binocularity and a unilateral loss of acuity.

IS LINEAR ANALYSIS SUFFICIENT TO REVEAL ABNORMALITIES OF THE VISUAL SYSTEM UNDER DICHOPTIC STIMULATION CONDITIONS?*

Stephen J. Fricker and Mark C. Kuperwaser

Howe Laboratory of Ophthalmology
Massachusetts Eye and Ear Infirmary
Boston, Massachusetts 02114

Measurement of evoked potentials with low signal-to-noise ratios usually is carried out with conventional signal averaging. For visual evoked potentials repetitive stimuli are used, with timing intervals of the order of one-half to one second. Thus the first few hundred milliseconds of the averaged response can be viewed as the visual system's impulse response to a single stimulus. Because of the anatomic intermixing of the nerve transmission pathways from the two eyes, this type of measurement has to be carried out testing one eye at a time (monocular test condition). Some use has been made of steady-state VER testing, using fixed rate repetitive stimuli (e.g., flash stimuli at 30 Hz). In such cases it is possible to stimulate each eye independently by using two sets of stimuli at slightly different frequencies.[1] Thus the system is being tested under steady-state conditions with two fixed but different frequencies. While this does allow some degree of resolution of the steady-state responses from the two eyes, a more general approach to obtaining dichoptic VER measurements would be to stimulate each eye separately using independent stimuli covering a range of frequencies. Provided that the timing of the two sets of stimuli is randomized, then cross-correlation analysis allows the recovery of VER responses representing the impulse-response of the visual system to each of the two sets of noiselike stimuli. It is of interest to ask whether linear cross-correlation analysis of such dichoptically induced VERs demonstrates the same type of behavior of the visual system as observed under conditions of monocular stimulation, and whether the dichoptic test results suggest the existence of a significant degree of cross-coupling between the two visual pathways.

For some years now we have been applying cross-correlation techniques to the measurement of VERs (and ERGs) in human subjects.[2,3] For VER measurements either flash stimuli or video presentations of alternating checkerboards have been used. Regardless of the type of stimulus used the timing of the stimulus occurrence—flash or pattern-alternation—is controlled by a pseudorandom binary sequence, so that the intervals between stimuli vary in a statistically definable manner. The use of two independent pseudorandom binary sequences to control two separate stimulus presentations permits "true" dichoptic stimulation of the two eyes.[4] With this test arrangement the subject views two similar but independently driven displays—one for each eye. The amplified occipital signal, measured from a single pair of midline electrodes, then is cross-correlated with each of two independent reference waveforms, corresponding to the two independent pseudorandom binary sequences used to control the stimuli. The two cross-correlation VERs obtained in this manner from the occipital signal represent the first-order responses of the visual system to the stimuli presented to each eye.

Using a "true" dichoptic stimulation system, we can investigate differences in the

*This work was supported in part by the Massachusetts Lions Eye Research Fund, Inc.

0077–8923/82/0388–0622 $1.75/0 © 1982, NYAS

VER responses when both eyes are tested concurrently, as opposed to testing each eye individually. If any presumed cross-coupling (linear or nonlinear) between the visual pathways from the two eyes is present, then one might expect to obtain significantly different results when the two eyes are stimulated dichoptically, compared with the results obtained when each eye is tested alone.

The following two examples are presented to demonstrate the results obtained from linear cross-correlation processing under dichoptic stimulation conditions. In each case the subjects gave different VERs from the two eyes; in one case this was the result of using different colored stimuli, and in the other case the different VERs obtained were the result of unilateral optic neuritis.

Case 1

In this example different dichoptic VERs were obtained from the two eyes of a normal individual by placing a colored filter in front of one of the two independent alternating checkerboard stimuli. Figure 1 shows the right and left eye VERs obtained with 40-minute check size when the right eye was always viewing a plain black-and-white display, while the left eye's display could have various colored filters placed in front of it. The top two traces, Figure 1 (a) and (b), show the results when the left eye and the right eye each viewed a black-and-white display. The three other pairs of results were obtained when the display for the left eye had placed in front of it a blue filter (c,d), a green filter (e,f), and a red filter (g,h). Different delays are induced in the left eye's VER response by the filters, most pronounced for the red filter. Similar relative latencies were observed when the left eye viewed the black-and-white display, while the right eye viewed its display through the various colored filters.

Case 2

This young woman had a history of optic neuritis affecting the left eye, with visual acuity dropping to 20/80. At the time of the test, approximately 4 weeks later, she had regained 20/20 vision and there were no clinical indications of any residual problem. Both monocular and dichoptic VERs were recorded using alternating checkerboard stimuli subtending 40 minutes of arc; the results are shown in Figure 2 (a–g). The first two traces (a,b) show the VERs from the right and left eyes tested in the usual monocular fashion. There is an obvious timing difference, with the VER from the left eye delayed with respect to that from the right eye. The algebraic sum of the two VER responses is shown in (c). Traces (d) and (e) show the VERs for the right and left eyes under dichoptic test conditions, that is, with each eye viewing its own independently controlled alternating checkerboard display. The signal amplitudes are slightly lower than those obtained when each eye was tested alone, but the general waveshapes and relative timing differences are quite similar to the monocular test results. The algebraic sum of the dichoptic VERs from the two eyes is shown in (f). The last tracing (g) was obtained by cross-correlating the occipital response with a reference waveform obtained by summing the two independent reference waveforms corresponding to the timing of the two sets of stimuli. It is evident that the VER obtained in this fashion (g) is very similar to the algebraic sum of the two dichoptic VERs (f). This raises obvious questions about the amount of nonlinearity in the system, and the degree of possible interaction between the two visual pathways.

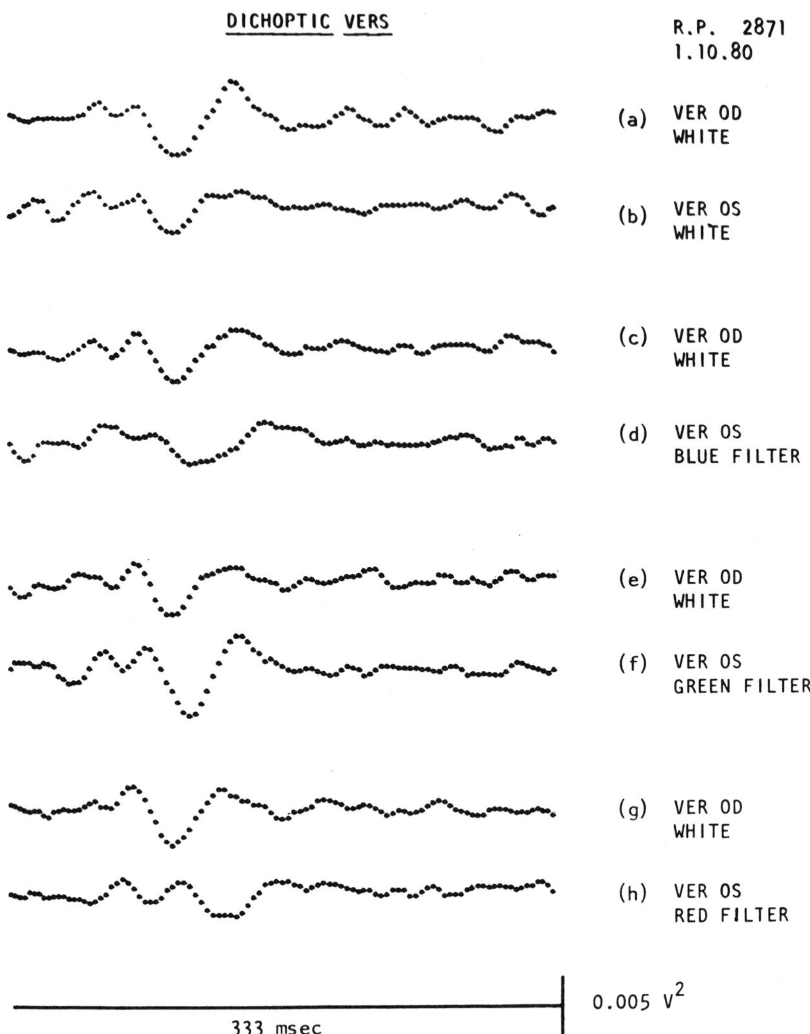

FIGURE 1. Case 1: Dichoptically induced visual evoked responses (VERs) using various colored stimuli.

DISCUSSION

The use of colored filters with pattern-shift stimuli for VERs provides a simple means for inducing latency variations in the responses. This is of value in investigating interactions of the visual pathways under dichoptic stimulation conditions. Subjective impressions of viewing dichoptic displays with different color stimuli for each eye are varied. Often one sees one color or the other exclusively for varying periods of time, or a variable mosaic of patches of color. Intuitively it might be expected that some significant amount of interaction would be revealed by the visual processing of signals from the two eyes under such dichoptic testing conditions. However, the linear, first-order, cross-correlation VERs shown in FIGURE 1 indicate that changing the latencies of a VER response from one eye by varying the color of the stimuli does not apparently affect the response from the other eye. Evaluation of second-order kernels might reveal some degree of nonlinear behavior or interaction,[5] but the inherently low

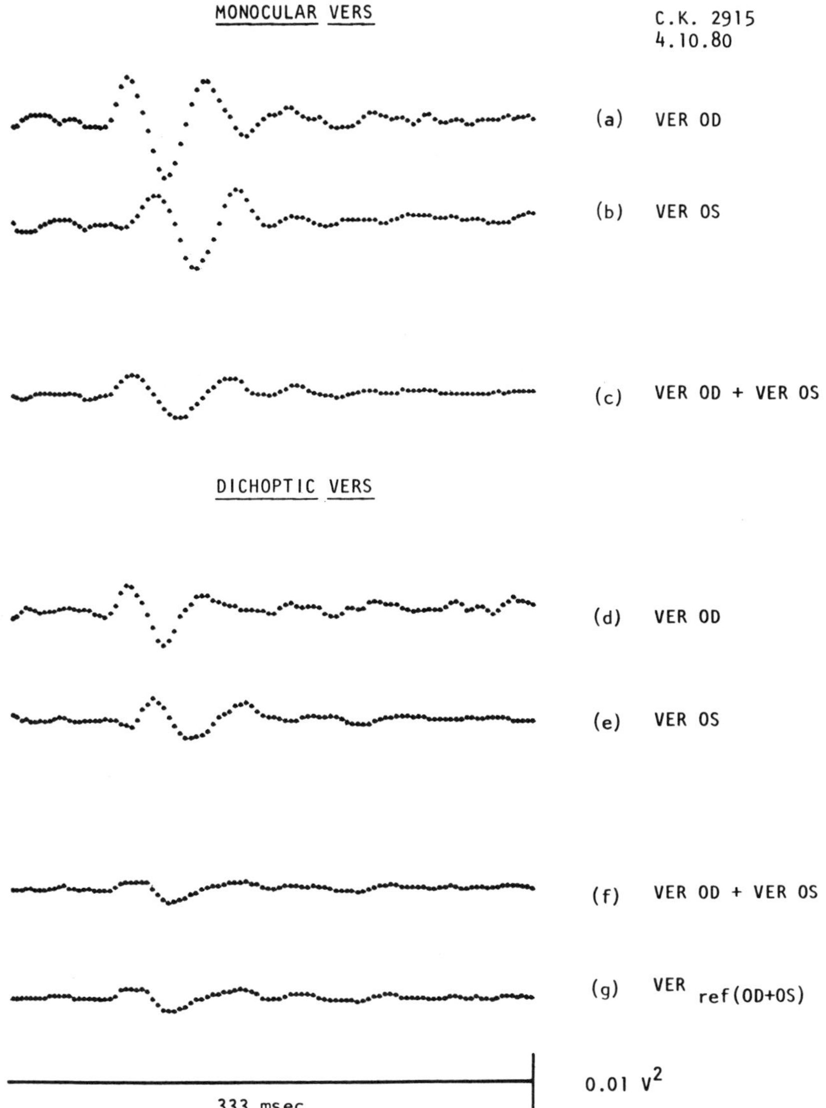

FIGURE 2. Case 2: Monocular and dichoptic VERs in a case of recovered unilateral optic neuritis.

signal-to-noise ratio of the occipital signal might make the results of such processing difficult to interpret.

The second example, a case of clinically recovered optic neuritis, showed that linear processing of the dichoptically induced VERs gave results similar to those obtained with monocular testing of one eye at a time. The fact that the algebraic sum of the two dichoptically induced VERs (FIGURE 2f) appears similar to the VER obtained using the summed reference waveforms for the two independent sets of stimuli (FIGURE 2g) implies that the linear portion of the visual system response, as obtained by first-order cross-correlation, is not affected to any major degree by dichoptic stimulation. Here again, second-order kernels might reveal some degree of nonlinear behavior or interaction, but the fact that the first-order response is essentially unchanged under conditions of dichoptic stimulation implies that the visual system is capable of processing two sets of stimuli with relatively little overall interaction between the visual pathways from the two eyes.

REFERENCES

1. REGAN, D. 1976. Latencies of evoked potentials to flicker and to pattern speedily estimated by simultaneous stimulation method. Electroencephalogr. Clin. Neurophysiol. **40:** 654.
2. FRICKER, S. J. & J. J. SANDERS. 1974. Clinical studies of the evoked response to rapid random flash. Electroencephalogr. Clin. Neurophysiol. **36:** 525–532.
3. FRICKER, S. J. & J. J. SANDERS. 1975. A new method of cone electroretinography: the rapid random flash response. Invest. Ophthalmol. **14:** 131–137.
4. FRICKER, S. J. & M. C. KUPERWASER. 1979. Use of pseudorandomly timed stimuli and crosscorrelation techniques for visual evoked response measurements. Computers in Ophthalmology, IEEE Catalog No. 79 CH1517-2C:242–249.
5. MCCANN, G. D. & P. Z. MARMARELIS. Eds. 1975. Proc. 1st Symp. on Testing and Identification of Nonlinear Systems. California Institute of Technology, Pasadena, Calif.

DISCUSSION OF THE PAPER

H. SPEKREIJSE: I think that your data very nicely demonstrate what Maffei and I were hinting at, that in such a situation you are studying monocular VEPs and not binocular VEPs at all. What is happening is that when you have two different stimuli in the two eyes either by spatial frequency or by time course difference, 50% of the time you have a response from the left eye and 50% of the time you have a response from the right eye.

R. SREBRO: I would like to take issue with what Dr. Spekreijse just said because I think we have been demonstrating for the last couple of years that you can use a paradigm where you are stimulating the same spatial and temporal frequencies except you create a large interocular luminance difference between the two eyes by simply putting a neutral density filter in front of one eye and varying the density of the filter, creating the delay that you are talking about between the two eyes while still keeping the same spatial and temporal characteristics. When you use a large enough neutral density filter, the amplitude of the response under the dichoptic situation is significantly less than the algebraic sum of the two monocular responses. This implies to me that it is not a simple linear interaction, or it cannot be accounted for simply in terms of a waveform interaction, but that is is it is some type of a binocular interaction.

J. NELSON: Throwing a colored filter in front of one eye and leaving white light for the other eye is not a very satisfactory manipulation. We know from neurophysiology, notably, from Zeki's work in monkeys, and some other labs as well, that color-selective cells in monkey cortex tend to be monocular. So, it would be then more interesting to put colored filters in front of both eyes and, indeed, the proper finesse here is to adjust not for red and black or green and black stimulation, but for chromatic contours.

In this case, one might then be selectively stimulating a population of cortical neurons that is monocular, and the prior observation of a reduction in amplitude might disappear.

I think the conclusion of my remark, then, is that there is a problem in stimulus control in clinical applications; and in looking at solutions for this problem one can take cues from the organization of the brain as revealed by neurophysiology in trying to plan the organization of one's experimental program.

WOULD AN EVALUATION OF BINOCULARITY USING ALGEBRAIC INTERACTION BETWEEN THE TWO EYES BE MODIFIED USING THE PSEUDORANDOM BINARY SEQUENCE METHOD?

Richard Srebro

Department of Ophthalmology
Southwestern Medical School
Dallas, Texas 75235

The phrase "algebraic interaction" refers to experiments in which a stimulus, S1, is presented to eye 1 (with eye 2 covered), a stimulus, S2, is presented to eye 2 (with eye 1 covered), and stimuli S1 and S2 are simultaneously presented to eyes 1 and 2, respectively. Visually evoked potentials (VEPs) are recorded for all three conditions and compared. Their amplitudes, or the amplitudes of derived components, are used to calculate an index of summation, for example, as the ratio of amplitude for S1 and S2 simultaneously presented to the sum of amplitudes for S1 and S2 individually presented. An index value different from 1 is taken as evidence for binocular interaction, or, more rigorously, binocular interaction is demonstrated when the VEP to simultaneous presentation of S1 and S2 is *not* equal to the sum of the VEPs to individual presentation of S1 and S2. There are two cases of the experimental paradigm. In case one, S1 and S2 are identical both in the temporal and spatial domains. In case two, S1 and S2 differ either in the temporal domain or in the spatial domain or in both domains.

For subjects with normal binocular function, the case one paradigm yields diverse results that depend both on the experimental conditions and on the method of data analysis. Summation indices ranging from 0.7 to 1.3 are frequently reported.[1–4] But one group of investigators[5] report index values as high as 2.5. Defocusing the image in one eye[2,3] or placing a prism base up in front of one eye[4] causes a reduction in the value of the index of summation. Subjects with abnormal binocular function have lower values of the index of summation than do normal subjects under the same test conditions.[4,6]

Experiments using the case two paradigm have generally been concerned with spatially dissimilar stimuli, and in subjects with normal binocular function the results suggest that binocular interaction occurs.[3,7–9] Under these conditions, simultaneous presentation of S1 and S2 may result in a temporally unstable perception called "retinal rivalry." The term "suppression" is used to label the subject's report that only one of the stimuli is seen even though both are presented. VEPs collected during epochs of suppression result in a summation index near 0.5. In one experiment,[7] S1 and S2 were not only spatially dissimilar, but also 90° out of phase temporally, and the phase of the VEP appeared to shift commensurately with the report of which stimulus was suppressed. All these experiments have been taken as providing evidence that suppression has a VEP counterpart, i.e., inhibition of the VEP of the suppressed eye by an unknown mechanism. However, at least one study[10] failed to confirm this result and a critical examination of the VEPs published by Cobb *et al.*[7] suggests phenomenon not compatible with the hypothesis that total inhibition of the VEP from the suppressed eye occurs.

All of the above studies including both case one and case two paradigms, have several limitations:

(1) The stimuli S1 and S2 are temporally correlated. Thus the contributions from the two eyes to the VEP during simultaneous presentation of S1 and S2 are confounded. Using stimuli that are out of temporal phase but at the same frequency for S1 and S2 does not suffice to sort the contributions to the VEP from the two eyes because it cannot be assumed that binocular interaction does not itself cause phase shifts. The inability to sort the contributions from the two eyes leads to ambiguities in interpreting results. For example, a summation index value near 0.5, as reported in the experiments concerning suppression, could occur if the VEP to simultaneous presentation of S1 and S2 contains equal contributions from each eye and each contribution is about half as large as the VEPs to individual presentations of S1 and S2. Thus the conclusion that the VEP due to the suppressed eye is zero, i.e., the conclusion that is usually drawn, is not the only one that can be drawn.

(2) The time course of S1 and S2 is usually a simple periodic pulse train or a sinusoid. Since the VEP is a nonlinear function of the evoking stimulus, such simple stimuli may not provide a rich enough ensemble of stimulus conditions to reveal the system complexity. Latency differences between the VEPs of the two eyes, as is sometimes seen in multiple sclerosis, appears to have an effect on algebraic summation.[11] It is plausible that this finding may generalize to normal subjects so that an adequate description of algebraic summation may require the examination of a range of timing differences between S1 and S2.

(3) In part, the use of spatially dissimilar stimuli is a matter of expediency in that the subject's perceptual report is used as a marker for "eye." However, suppression may also occur for spatially similar stimuli. In fact, subjects with abnormal binocular function suppress when presented with stimuli which in normals would ordinarily produce fusion and stereopsis rather than retinal rivalry.[12] Thus conclusions based on experiments on suppression in retinal rivalry in normals may not extrapolate to suppression in abnormals.

These limitations can, to a considerable extent, be circumvented by using relatively wide band temporally uncorrelated stimuli for S1 and S2. One way to construct such stimuli is to use a pseudorandom sequence to construct S1 and another different pseudorandom sequence to construct S2. If the periods of these two sequences are coprime, i.e., different prime numbers, they will be uncorrelated. Probably the simplest way to sort the contributions of the two eyes to the VEP is by means of a frequency domain analysis because the frequency components as well as the sum and difference frequencies of S1 will all be different from those of S2. Judicious selection of the sequence type and periods would ease problems of spectral resolution and permit a practical procedure. The wide band nature of these stimuli and the fact that they are uncorrelated would provide a rich ensemble of stimulus conditions, including timing differences between the two eyes, and thus the substrate for the expression of binocular interaction.

References

1. Campbell, F. W. & L. Maffei. 1970. Electrophysiological evidence for the existence of orientation and size detectors in the human visual system. J. Physiol.**207:** 635–652.
2. White, C. T. & L. Bonelli. 1970. Binocular summation in the evoked potential as a function of image quality. Am. J. Opt. Arch. Am. Acad. Opt. **47:** 304–309.
3. Fiorentini, A., L. Maffei, M. Pirchio & D. Spinelli. 1978. An electrophysiological correlate of perceptual suppression in anisometropia. Vis. Res. **18:** 1617–1621.
4. Srebro, R. 1978. The visually evoked response: Binocular facilitation and failure when binocular vision is disturbed. Arch. Ophthalmol. **96:** 839–844.

5. Apkarian, P. A., K. Nakayama & C. W. Tyler. 1981. Binocularity in the human visual evoked potential: Facilitation, summation, and suppression. Electroencephalogr. Clin. Neurophysiol. **51:** 32–48.
6. Amigo, G., A. Fiorentini, M. Pirchio & D. Spinelli. 1978. Binocular vision tested with visual evoked potentials in children and infants. Invest. Ophthalmol. Vis. Sci. **17:** 910–915.
7. Cobb, W. A., H. B. Morton & G. Ettlinger. 1967. Cerebral potentials evoked by pattern reversal and their suppression in visual rivalry. Nature **216:** 1123–1125.
8. Ciganek, L. 1971. Binocular addition of the visual response evoked by dichoptic patterned stimuli. Vis. Res. **11:** 1289–1297.
9. Harter, M. R., W. H. Seiple & M. Musso. 1974. Binocular summation and suppression: visually evoked cortical responses to dichoptically presented patterns of different spatial frequencies. Vis. Res. **14:** 1169–1180.
10. Martin, J. I. 1970. Effects of binocular fusion and binocular rivalry on cortical evoked potentials. Electroencephalogr. Clin. Neurophysiol. **28:** 190–201.
11. Hoeppner, T. J. 1980. Binocular interaction in the visual evoked response: temporal factors. J. Neurol. Sci. **47:** 49–58.
12. Schor, C. M. 1977. Visual stimuli for strabismic suppression. Perception **6:** 583–593.

SPATIAL FREQUENCY EFFECTS ON BINOCULAR INTERACTION ASSESSED WITH VISUAL EVOKED POTENTIALS*

Gunnar Lennerstrand and Peter Jakobsson

Department of Ophthalmology
University Hospital
S-581 85 Linköping
Sweden

We are conducting studies on binocular interaction to grating stimulation in the human visual evoked potentials (VEP) and this preliminary report describes the effects of changes in spatial frequency of the stimulus. Neurophysiological studies in animals have shown that ganglion cells in the retina,[1,2] and neurons in the lateral geniculate body[3] and in the striate cortex[4,5] exhibit selective sensitivity to spatial frequency. In animals with normal binocular functions, signals from the two eyes converge and activate the same cells in the striate cortex, provided the spatial characteristics of the stimuli to the individual eyes are similar.[6] It is reasonable to assume that these conditions apply also to the human visual system. This would explain results of psychophysical experiments in humans, showing that the interaction between the responses from the two eyes is largest when they are stimulated with identical grating patterns, and that the interaction decreases with increasing difference in spatial frequency content of the two stimuli.[7] These studies have indicated a selectivity of human binocular interaction of approximately two octaves around the conditioning spatial frequency.

Electrophysiological evidence of these mechanisms in humans might be obtained in recordings of visual evoked potentials to grating stimulation. It has previously been demonstrated that binocular interaction might be evaluated in VEP to checkerboard stimulation and that subjects with poor binocular functions can be identified with this method.[8] It was also shown that binocular VEP was dependent on the spatial content of the stimulus to the two eyes, much in the same way as indicated by the psychophysical experiments.[9] However, VEP binocular interaction to the grating stimulation, used in the psychophysical experiments, has not been previously studied.

The experiments were performed on two subjects with normal monocular and binocular vision. Vertical grating patterns with sinusoidal luminance profile and 50% contrast were generated on the screens of two identical oscilloscopes (Tektronix model 5440 with P31 phosphor). The mean luminance was 25 cd/m^2. The spatial frequencies of the patterns on the two screens could be varied independently. With the aid of a prism in front of one eye, the images of the two screens could be fused, although the patterns stimulated the individual eyes separately. The two patterns reversed at slightly different temporal frequencies (1.9 and 2.1 reversals per sec), allowing transient VEP for each eye to be recorded separately from the other, even during binocular viewing.[8] The signals were amplified and 128 sweeps averaged in the conventional manner.

In both subjects the VEP to sinusoidal grating stimulation consisted of a

*This work was supported by grants from the Swedish Medical Research Council (4751), the Research Committee of Östergötlands läns landsting and the Carmen and Bertil Regnér Foundation.

monophasic, positive deflection. The latency of the response varied slightly with the spatial frequency of the stimulus and was shortest at frequencies around 4 cycles/degree (c/d). Measurable monocular responses were obtained over a frequency range between 1 and 8 c/d.

Binocular VEP interaction was calculated as the ratio in percent between response amplitudes for each eye during binocular and monocular stimulation. A binocular/monocular ratio of 100% indicated no binocular interaction; a ratio of 50% or less was seen at full binocular interaction.[8-10]

FIGURE 1A shows the results in subject BL for grating stimuli of constant spatial frequency of 4 c/d to one eye and a varying frequency to the other. Maximal binocular interaction was obtained with identical grating stimuli to both eyes. Binocular interaction had vanished when one eye viewed a 4 c/d pattern and the other a 1 c/d or

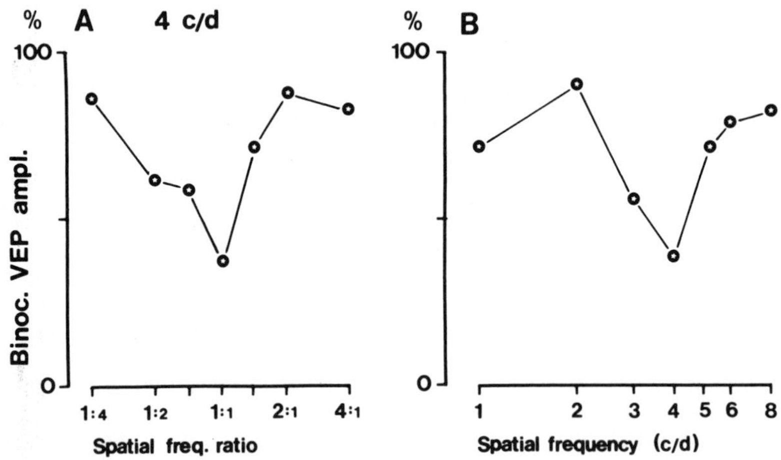

FIGURE 1. Binocular interaction in subject BL: (A) The effect of varying the spatial frequency of the grating stimulus to one eye while keeping the other constant at 4 c/d. On the abscissa the relation between the frequencies of the two stimuli are given. On the ordinate the binocular/monocular ratio in percent has been plotted. Each point represents the mean value for the two eyes. (B) Interaction for grating stimuli of identical spatial frequency to both eyes. Mean values for the two eyes are plotted.

an 8 c/d pattern. Reduced binocular interaction was observed at spatial frequency combinations between 1 and 4 c/d, and 4 and 8 c/d.

FIGURE 1B shows that in subject BL the binocular interaction for identical patterns to the two eyes varied with the spatial frequency of the stimulus and that the interaction was largest for spatial frequencies around 3–4 c/d.

These studies indicate that binocular interaction in the VEP can be used as a means of determining the spatial frequency specificity of human visual mechanisms in an objective manner. The specificity found with VEP was about the same as that obtained in psychophysical pattern adaptation and contrast sensitivity experiments. However, in contradistinction to the psychophysical results, binocular VEP interaction was seen only over a limited range of spatial frequencies. This might be due to experimental conditions imposed by the VEP technique, e.g., the temporal factors in the stimulation. We will explore such aspects further.

References

1. Enroth-Cugell, C. & J. G. Robson. 1966. The contrast sensitivity of retinal ganglion cells of the cat. J. Physiol. (Lond.) **187:** 517–522.
2. Ikeda, H. & M. J. Wright. 1972. Receptive field organization of 'sustained' and 'transient' retinal ganglion cells which subserve different functional roles. J. Physiol. (Lond.) **227:** 769–800.
3. Lehmkuhle, S., K. E. Kratz, S. C. Mangel & S. M. Sherman. 1980. Spatial and temporal sensitivity of X- and Y-cells in dorsal lateral geniculate nucleus of the cat. J. Neurophysiol. **43:** 520–541.
4. Campbell, F. W., G. F. Cooper & C. Enroth-Cugell. 1969. The spatial selectivity of the visual cells of the cat. J. Physiol. (Lond.) **203:** 223–235.
5. Movshon, J. A., I. D. Thompson & D. J. Tholhurst. 1978. Spatial summation in the receptive fields of simple cells in the cat's striate cortex. J. Physiol. (Lond.) **283:** 53–77.
6. Bishop, P. O. 1973. Neurophysiology of binocular single vision and stereopsis. *In* Handbook of Sensory Physiology. R. Jung, Ed. Vol 7(3)A: 255–301. Springer Verlag, Berlin.
7. Blake, R. & E. Levinson. 1977. Spatial properties of binocular neurones in the human visual system. Exp. Brain Res. **27:** 221–232.
8. Lennerstrand, G. 1978. Binocular interaction studied with visual evoked responses (VER) in humans with normal or impaired binocular vision. Acta Ophthalmol. (Kbh.) **56:** 628–637.
9. Lennerstrand, G. 1978. Some observations on visual evoked responses (VER) to dichoptic stimulation. Acta Ophthalmol. (Kbh.) **56:** 638–647.
10. Jakobsson, P. & G. Lennerstrand. 1981. A comparison of different VEP methods for the assessment of binocular vision. Doc. Ophthalmol. Proc. Series **27:** 337–344.

Discussion of the Paper

I. Bodis-Wollner: It is interesting to consider your data for some of us who are interested in nonlinear interactions in the binocular VEP.

You are essentially separating the two eyes by using one temporal frequency here and one temporal frequency there. In Dr. Braddick's and Dr. Julesz's techniques the separation for the two eyes uses color tags, red and green, and it just shows us that we have different methods and different problems. Some people may have color defects that have nothing to do with binocular vision. So, it is pretty difficult sometimes to make sure that there is adequate separation.

O. Braddick: Dr. Bodis-Wollner, you provoked me there. The issue of separating the two eyes is quite different in these two cases. In Dr. Lennerstrand's technique, he is trying to separate the responses to the two eyes by labeling the frequency. In our technique, all we are doing is separating the two eyes in the sense of putting different stimuli to them. I think I should emphasize that the use of the color has nothing at all to do with the color vision of the patient. It is simply an optical way of determining by those pairs of filters what stimuli actually get to the two eyes. The patient could be completely color blind and it would not make any difference to the degree of separation.

Bodis-Wollner: In the color blind patient the monocular contrasts would be different in the two eyes.

H. Spekreijse: Dr. Lennerstrand, on the basis of two different temporal frequencies, then, you have a constant consideration of rivalry. So, there is no monocular separation at all.

BODIS-WOLLNER: In your opinion, rivalry is not a binocular mechanism, Dr. Spekreijse?

SPEKREIJSE: I think the data you were showing are not contradictory to such a conclusion. So, the reduction will be the strongest when you are using the optimal spatial frequency.

BODIS-WOLLNER: In other words, the way you want to look at binocular interactions is to avoid any question of dominance. Is dominance a problem in amblyopia, Dr. Levi?

D. LEVI: It is a potential problem. We have made studies psychophysically in which we have looked at masking, that is, the effect of a grating in one eye on the contrast in the other eye. In fact, we found masking functions that looked quite similar to Dr. Lennerstrand's in both normal and amblyopic observers, provided the masking grating had sufficient contrast above threshold so that it is expressed in those terms.

Our interpretation from the masking data is that, perhaps, that represents some kind of inhibitory interaction between the two eyes which seems to be present in at least some strabismic and amblyopic patients whereas there may be an absence of other kinds of binocular interaction such as stereosis and binocular summation.

SPEKREIJSE: I agree completely that the amblyopic eye may be the dominant eye if you give it enough contrast and if you give the normal eye low contrast.

L. MAFFEI: I think in testing binocular vision if we wish to relate to physiology what we want to have is two frequencies which are actually equal or very, very similar because otherwise we are going to study some particular interaction, probably, inhibitory, which probably has nothing to do with what we want to study. Otherwise, as Dr. Spekreijse was pointing out, we are in a completely nonphysiological situation.

So, my question is, what is the logic of the experiment because I do not see any logic for any application in the study of binocular vision.

BODIS-WOLLNER: If I may reemphasize I think the logic in Dr. Lennerstrand's experiments is to try to achieve a condition where, while both eyes are open, you stimulate each eye separately.

Now, there are different techniques as in the random correlogram where one achieves a separation with colors, and I think the logic, here, was to separate it tagging it with frequency.

COMMENT: The method is not important. The only thing of importance is whether there is stabile monocular perception or not. And in the condition when there is not, then, I think we have to be fairly cautious because you are probably studying, well, complicated interaction problems, not particularly monocular interaction.

SREBRO: I am not sure I agree with that. I think Dr. Lennerstrand's technique is very interesting because I think he has one control. Well, let me just step back. I think his technique is very similar to the one that I presented. It is bringing out a nonlinearity to the system. You assume independence of the two eyes and show that when you combine the stimulation you get nonindependence. There is non algebraic summation.

Now, if I understand Dr. Spekreijse correctly, it is that you have some kind of rivalrous situation which is very abnormal, and the point I would like to raise is that the experiment in which Dr. Lennerstrand did the spatial frequency tuning, I think, is a situation where one would expect to see a considerable amount of rivalry when you have spatial frequencies. And this is a situation where he had not a particularly large effect.

BODIS-WOLLNER: The point is that linear versus nonlinear properties of binocular neurons have not been very well explored in neurophysiological experiments and I am sure that this is an explorable—in fact, very important—area for future studies.

CHRONOTOPOGRAPHICAL STUDY OF THE PATTERN-EVOKED RESPONSE AND BINOCULAR SUMMATION*

Nicole Lesevre

Laboratoire d'Electrophysiologie et Neurophysiologie Appliquee (L.E.N.A.)
Hôpital de la Salpêtrière
75651 Paris Cedex 13
France

I have been asked to comment on the following question: Can topographical analysis of VEPs allow us to evaluate the participation of monocular versus binocular neurons of the visual cortex? In fact, most topographical analyses of the VEP that have been realized in our laboratory by means of spatiotemporal maps, in particular, those for visual field studies, have only concerned binocular stimulation responses. However, with C. Bourdy—a psychophysicist researcher working at the Laboratoire de Physique Appliquée du Muséum in Paris—we have recently undertaken a study regarding binocular interactions by both psychophysical and electrophysiological approaches. The main purpose of this study was to evaluate in normal humans the influence of various parameters such as contrast level, luminance level, and ocular dominance, on binocular summation, in order to determine the best experimental conditions under which maximal summation can be obtained. In regard to the electrophysiological aspect of this study, which is the only one this paper will deal with, it must be noted that most human VEP studies on binocular interactions have chiefly taken into account amplitude values obtained from only but a very limited number of recording sites on the scalp, in spite of several results obtained long ago with flash stimuli showing that different electrode placements over the visual cortex gave rise to differing degrees of binocular summation effects in the VEP.[1] Therefore, although this was not the principal aim of this study, we undertook a topographical analysis of the binocular versus monocular pattern-reversal evoked response, assuming that changes in scalp distribution of each VEP component in relation to the degree of summation or interocular transfer should help us to understand the underlying neurophysiological mechanisms involved in human binocular summation, and, in particular, to evaluate the participation of monocular versus binocular neurones depending on the various visual areas activated.

The summation performances were measured through a brightness matched experiment[2] for the psychophysical study; in the electrophysiological study they were evaluated by comparing amplitude, latency, and scalp distribution characteristics of the various components of the pattern-reversal response[3] obtained from monocular stimulations and from fused binocular viewing with dioptic stimulations. The same experimental conditions in regard to contrast and luminance levels were used for both psychophysical and electrophysiological studies. The visual stimuli were 5° black and white checkerboards electronically generated on two oscilloscopes, made up of 12-minute checks and seen by an haploscopic device enabling the two eyes to be stimulated independently. Five subjects with normal binocular vision[4] participated in

*This work was achieved with the help of the Institut National de la Santé et de la Recherche Médicale (INSERM) Grant 78.2002.6.CA 413.

0077–8923/82/0388–0635 $1.75/0 © 1982, NYAS

the experiment. Each subject participated at least to five different sessions. During each session the mean luminance level remained constant and the contrast level was varied from 6% to 70% (8 different contrast levels were tested during each session). From one session to the other the luminance level was varied by means of filters placed in front of each stimulus (pattern) in order to have either the same or different luminance levels for each eye.[5]

The EEG was recorded from 9 active electrodes referred to linked ears, 4 cm apart,

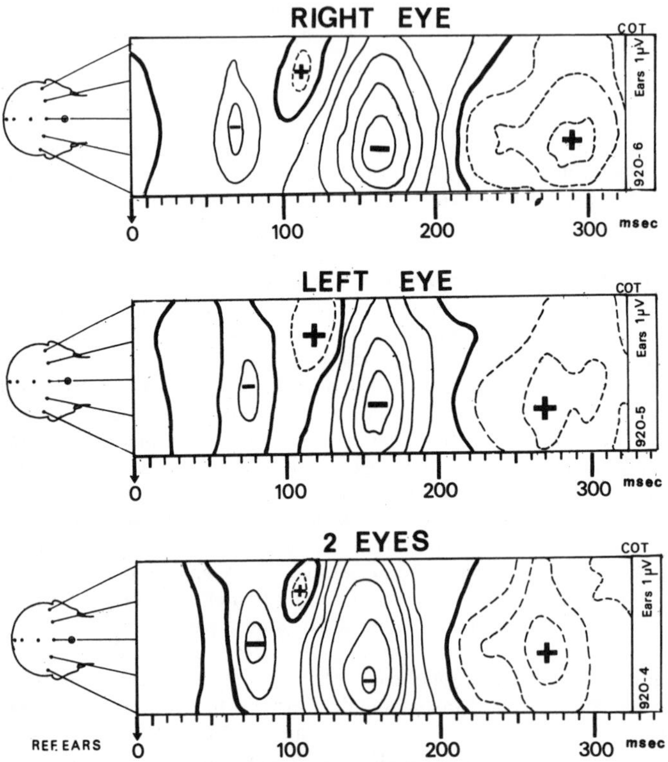

FIGURE 1. Averaged pattern-reversal response obtained on the transversal part of the montage from one same subject, under the same luminance and contrast level conditions, for right and left monocular stimulations and for binocular stimulation. Linked earlobes reference. Amplitude: 1 μV between two successive isopotential curves of positive (broken lines) or negative (plain thin lines) polarity. The sign + or − corresponds to the location of the peak along the transverse axis.

forming a "cross" montage; the most posterior electrode of the longitudinal axis was placed 1 cm lower than the inion; the transversal axis crossed the midline 3 cm above the inion and extended 8 cm away from the midline. The horizontal and vertical eye movements were simultaneously recorded in order to control fixation.

Whatever the experimental conditions concerning luminance and contrast levels, the responses either to binocular or to monocular stimulation were made up, for the subjects, of 3 main components: P100, N140, and P200 (labeled according to their polarity and mean latency across conditions and subjects); one subject out of 5 had in addition a relatively important N60 (which in fact peaked between 70 and 85 msec according to contrast and/or luminance levels) (FIGURE 1). The *topographical organization* of this response, i.e., the location of the peak of each component on both axes of the "cross" montage, was slightly different from one subject to an other, but for the same subject it was remarkebly stable accross the various experimental conditions tested in this study (except on the longitudinal axis, for different luminous levels). In particular the location of the peaks did not vary in relation to binocular versus monocular stimulation, nor according to the stimulated eye on both axes of the "cross" montage utilized in this study. For binocular and for both monocular conditions, on the transverse maps P100, N140, and P200 peaked on the midline for 3 subjects out of 5, whereas the 2 others showed an asymmetrical organization of their response. This asymmetry was not related to the stimulated eye nor to binocular versus monocular stimulation: for one subject, both N140 and P200 were always shifted to the left hemisphere no matter the stimulated eye whereas P100 was inconstantly shifted to the right (FIGURE 1 and upper maps of FIGURE 2). For the other subject, P200 and, to a lesser extent, N140 were shifted on the contrary to the right hemisphere (FIGURE 2, lower maps). Such individual interhemispheric distributions were statistically highly significant since they were observed for each subject under at least 25 out of 30 different experimental conditions concerning luminance and contrast. No relation between these asymmetrical responses and eye dominance could be found in the case of these two individuals since both had a right eye dominance (FIGURE 2). On the longitudinal axis, the location of N140 and P200 varied slightly but significantly in relation to luminance level (peaking more anteriorly for the lowest luminance level) but not in relation to binocular versus monocular stimulation. Therefore, and this is the first answer to the question I was asked, the analysis of the peaks locations of the pattern-reversal response, as observed on the maps, did not help to differentiate monocular versus binocular responses, at least within the range of topographical variation investigated in this study, i.e., with a limited number of electrodes, relatively large interelectrode distances and a montage made up of only two (vertical and horizontal) axes.

But a spatiotemporal analysis is not limited to the delimitation of peaks locations; it permits, in particular, a better differentiation of the various components of the response which might overlap in space and time. This study demonstrated that component N140 was—at least under our experimental conditions—the only component of this pattern-reversal response that evidenced binocular summation: indeed, N140 was the only component for which not only the increase in amplitude, but also, above all, the decrease in latency observed for binocular responses compared to monocular responses were statistically highly significant. This N140 decrease in latency was observed under all luminance and contrast conditions tested and had, across conditions, an average value of 10 msec. However, in order to investigate whether this summation effect varied or not in relation to contrast and/or luminance level, a summation index K, taking into account the peak latencies of N140, was calculated in such a way that K would increase with the amount of summation.[6] This index was plotted against the various contrast levels tested in each luminance condition. The results of this electrophysiological analysis (which were confirmed by the psychophysical measures) showed that the highest latency summation index was found for the medium contrast levels (i.e., 30% to 50% contrast, depending on subjects), whatever was the luminance (FIGURE 3): indeed, a statistically significant tendency[7] for this latency summation index to decrease for the highest and for the lowest contrast values tested was observed for the 5 subjects (FIGURE 3). However, no

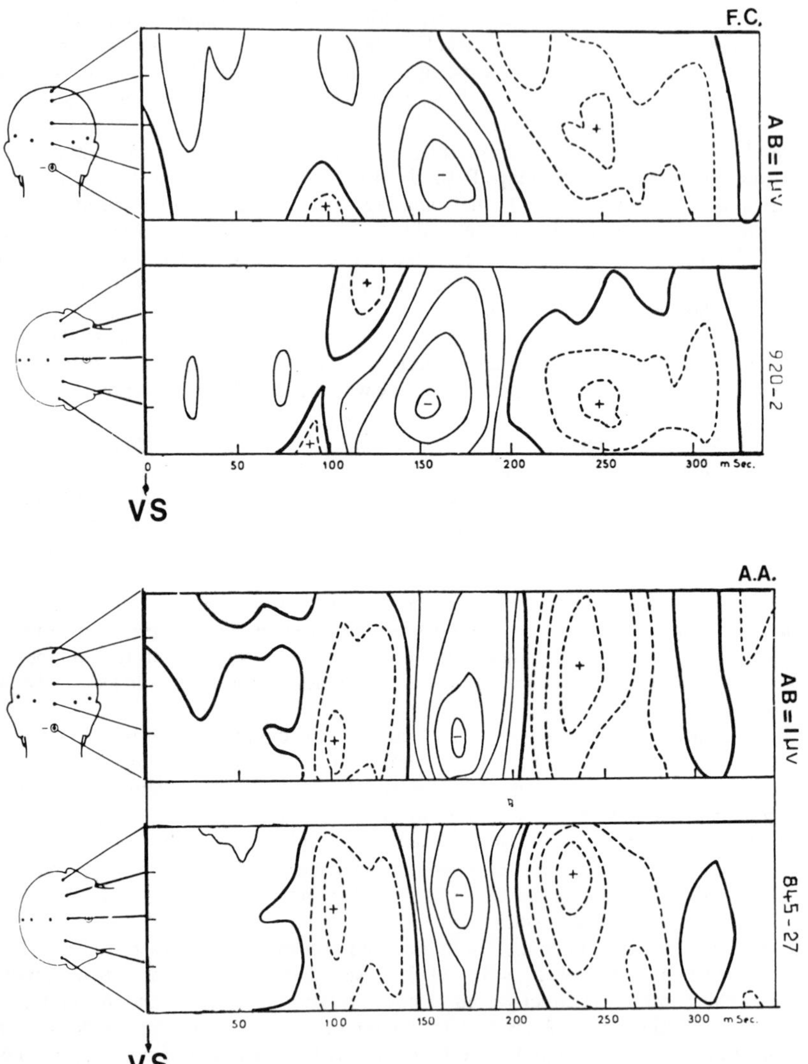

FIGURE 2. Averaged pattern-reversal responses obtained on both parts of the cross montage by stimulating the left eye, for two different subjects, both with right eye dominance (upper maps: same subject as that of FIGURE 1 seen under other luminance and contrast conditions). Linked earlobes reference, 1 μV between successive isopotential lines.

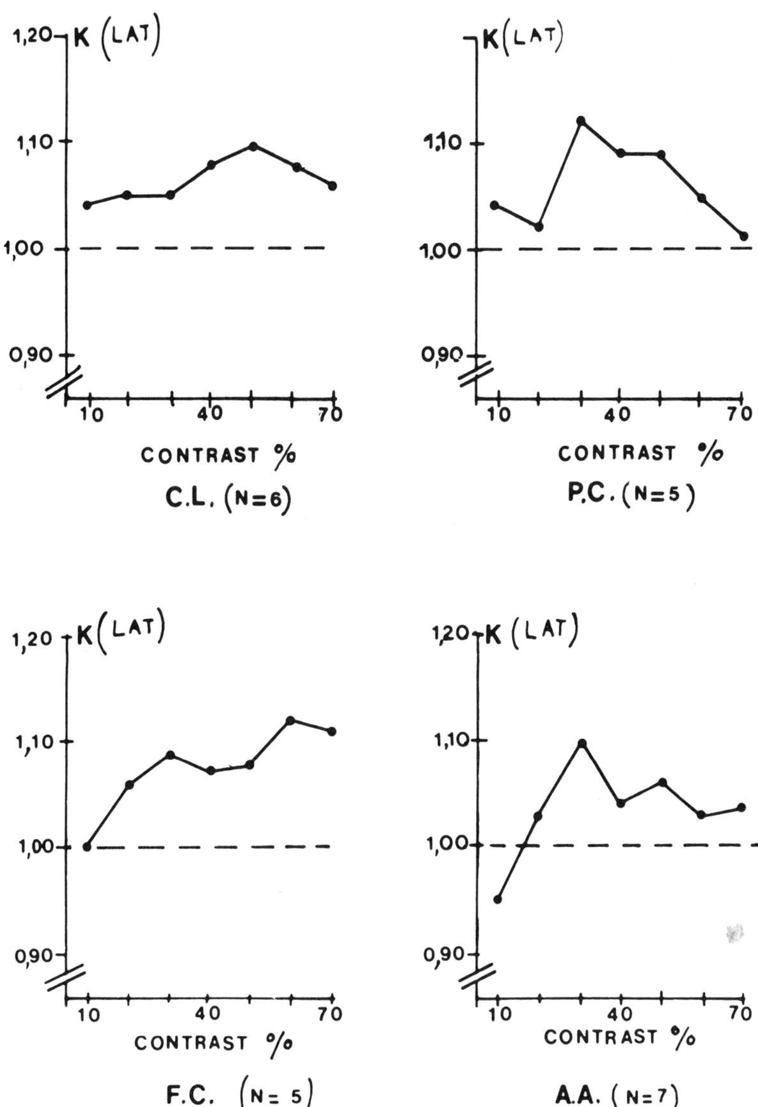

FIGURE 3. Summation index K calculated from the peak latencies of component N140,[6] plotted against contrast levels (abscissa). Four different subjects. For each contrast level the summation index K was averaged across different luminance conditions (5 to 7 according to subjects). There is a summation effect when $K > 1$. For statistical signification.[7]

significant relations were seen when a similar summation index taking into account the amplitudes of N140 was plotted against contrast values. Another result that must be noted is that the N140 summation index, be it of latency or of amplitude, was not significantly higher when both eyes were stimulated by targets of equal luminance than when targets of different luminance levels were presented to each eye. In the latter case, some subjects (3 out of 5) had a higher latency summation index when the higher luminance target was presented to their dominant eye. But this last finding would have to be confirmed with more subjects and by testing other aspects of eye dominance before drawing conclusions about this eye dominance effect on binocular summation.

In summary, the chief result of this pattern-reversal response study concerning binocular interactions was that the significant effect of contrast level on binocular summation was mostly reflected by component N140's latency characteristics, whereas it was only slightly reflected by its amplitude changes, and not at all, in spite of what we had expected, by its scalp distribution, at least under our spatial experimental conditions.

We should like to end by the two following comments:

(1) Usually the degree of binocular summation is inferred only by comparing differences in amplitude between monocular and binocular VEPs. Our spatiotemporal analysis has shown that significant differences in latency were also encountered and that these latency differences reflected the incidence of contrast level on binocular summation much better than did amplitude differences. This finding concerning the importance of latency characteristics is in agreement with data obtained long ago in cats[8] showing that the latency of unit responses to flashes in visual cortex were shorter for binocular than for monocular stimulation; however, no such differences were obtained at that time for the surface evoked potentials.

(2) We have shown that, at least under our experimental conditions, binocular summation was chiefly reflected by component N140. This component has been described by Jeffreys,[9] under the label CII (however, it must be noted that the CII of Jeffreys has a much shorter latency, around 100 msec), as related to contour-specific processes whereas P100 (Jeffreys' CI, which peaks around 75 msec) would be chiefly related to contrast specific processes. According to this author,[10] the former component shows more interocular transfer of the VEP attenuation caused by prior exposure to a similar pattern than the latter and therefore would more likely originate in extrastriate cortex where, at least in monkeys, a greater amount of binocularly driven cells have been found than in the striate cortex. Our findings concerning binocular summation as being better reflected by N140 than P100 are thus quite in agreement with Jeffreys' results. However, chronotopographical results from our EP visual field studies[11] favor an extrastriate origin for both components. Besides, we have shown that N140, which is thus probably more related to binocularly driven cells than P100, is not only related to contour processes but also to contrast level changes, at least as far as binocular interactions are concerned.

References and Notes

1. Perry, N. W., D. G. Childers & E. McCay. 1968. Vis. Res. **8:** 567–573.
2. This brightness matching experiment was inspired by that utilized by Fry and Bartley in order to show that border contrast was the main element producing Fechner's paradox (Am. J. Ophthalmol. 1933. **6:** 687–693). Part of the psychophysical results obtained with this method utilizing the Fechner's paradox has been presented by Bourdy and Lesevre at

the European Conference on Visual Perception (October 1979, Noordwijkerhout, Holland).

3. Each pattern reversal occured every 620 msec; each averaged evoked potential was obtained from 150 successive reversals. The averaging process was computed off line after elimination of responses occurring during an eye movement.
4. Binocular vision was studied by means of various tests in order to evaluate heterophoria, aniseiconia, and stereoscopic vision. Eye dominance was appreciated by a "sighting" test: 3 subjects showed a right eye dominance, the two others a left eye dominance.
5. The highest luminance level tested was 4 cd/m2. From one session to the other, series of neutral filters of density .1, 0.2, 0.5, and 0.9 were utilized. Before each session the subject was dark adapted for 20 minutes.
6. This N140 latency summation index was obtained as follows: K (lat) = [lat (RE) + lat (LE)]/2 lat (B), where lat (RE) and lat (LE) are the peak latencies of the right eye and of the left eye stimulation responses and lat (B) is the latency obtained to the fused binocular viewing stimulation.
7. The decrease of the summation index for the lower contrast levels is statistically significant ($P < 0.05$: Wilcoxon matched pairs signed-rank test) for the 5 subjects; the decrease for the higher levels is significant for 3 subjects out of 5.

8. Burns, B. D., W. Heron & B. Grafstein. 1960. Am. J. Physiol. **198:** 200–204.
9. Jeffreys, D. 1977. The physiological significance of pattern visual evoked potentials. In Visual Evoked Potentials in Man: New Developments. J.E. Desmedt, Ed. pp. 134–183. Clarendon Press, Oxford.
10. Smith, A.T. & D. A. Jeffreys. 1979. Evoked potential evidence for differences in binocularity between striate and prestriate regions of human visual cortex. Exp. Brain. Res. **36:** 375–380.
11. Lesevre, N. & J. P. Joseph. 1979. Electroenceph. Clin. Neurophysiol. **47:** 183–203. *See also* N. Lesevre. Chronotopographical analysis of the human evoked potential in relation of the visual field. This volume.

STIMULUS CONTROL IN VISUAL EVOKED POTENTIALS AND BEHAVIORAL ASSESSMENT OF INFANT VISION

Oliver Braddick and Janette Atkinson

Department of Experimental Psychology
University of Cambridge
Cambridge, England

Visual function in infancy (e.g., acuity, contrast sensitivity, binocular function) may be assessed either by the ability of stimuli to elicit a visual evoked potential (VEP) from the infant, or from behavioral manifestations of discrimination between stimuli, most commonly "preferential looking" (PL).[1] PL requires the infant to have reasonably good head and oculomotor control, and to sustain attention to the stimuli long enough for a psychophysical procedure to be completed. This limits the age range over which PL measures are feasible and reliable to about 1–9 months. In principle, VEPs can be recorded at any age. However, a VEP that is a meaningful guide to visual function depends on the infant maintaining fixation on the stimulus. The practicality of VEP assessment is therefore limited by the same problems of fixation and attention which limit the PL method. The critical issue in stimulus control in this field is not control of the external stimulus, but control of the stimulus that is maintained on the infant's retina.

We have found a number of techniques to be of value in such control. The first is to use an essentially homogeneous display which subtends as large a visual angle as possible, so that to some degree the stimulation is independent of the direction of the infant's fixation. The large-screen projection video system we have used to stimulate "cyclopean" binocular VEPs is an example.[2] Short viewing distances (e.g., 40 cm) aid the attempt to obtain large visual angles and also serve to maximise infants' attention, but create technical problems in achieving spatial frequencies that test the acuity of any but the youngest infants. A second, critical, aspect of technique is to monitor infants' fixation continuously, and to interrupt recording when the infant's fixation or attention is in doubt. A desirable refinement for this purpose would be to be able to discard the VEP signal on the sweep where the shift of fixation occurred; most available signal averagers simply allow an operator to prevent triggering on the *next* sweep. A third technique is to superimpose on the test stimulus some visual pattern which holds the infant's attention more reliably than the test stimulus itself. Provided the attention-holding stimulus bears no temporal relation to the test stimulus it should not contribute any spurious signal to the averaged VEP. One example we have used is to optically superimpose on the test stimulus the face of an adult observer, by means of a half-reflecting mirror.[3] This observer attempts to actively attract and hold the infant's attention and is also ideally placed to monitor fixation.

These techniques in combination make it possible to record VEPs that can give useful measures of visual function on neonates,[4] and on infants through the age range where PL data is also available. Rather little VEP data is available from infants of 9 months and over: at this age attention is extremely labile and it is also difficult to obtain any usable amount of recording time that is free of artifacts due to gross bodily movement.

If a VEP is obtained from an infant with central fixation, say as a measure of acuity, does this yield the same information as a PL acuity determination? PL depends on the infant making a shift of fixation towards the stimulus, which implies that it is initially in extrafoveal vision. Published PL acuity data have used stimulus positions separated by up to 45°, which might be taken to imply that the acuity being measured is that up to 22° in the infant's peripheral vision. This is an oversimplification, since the infant does not remain fixating on the midline, and part of the behavior used in the assessment occurs when he is fixating the stimulus. Indeed the figure of 45° separation comes from Teller's laboratory,[1] where the infant is actually directed towards each stimulus position in turn. However, there is no doubt that a major part of the discrimination is based on first fixations,[9] and thus must depend on extrafoveal vision.

We have recently explored the effect of varying the eccentricity of the stimulus (from 3° to 10°) on PL-assessed accuity in 1-, 2-, and 3-month infants, and do not find any consistent advantage when testing with less eccentric stimuli, at any age. It seems likely that two factors are trading off: the more separated stimuli produce more unambiguously detectable shifts of fixation, but the infant's attention is less readily attracted by them. (This latter is not necessarily a reflection of retinal variations in acuity; it is found also with stimuli that are not demanding for acuity).[10] It is possible that with truly foveal stimuli, a higher acuity would be found. (The highest reported values for 1-month acuity, though not for older infants, are from Banks and Salapatek[5] who are the only workers to have used stimuli that actually extend to the midline).

It would be risky to conclude that VEP measures reflected foveal acuity, while PL measures reflected parafoveal. Dobson and Teller's review[6] concluded that, age for age, VEP acuities were somewhat higher than PL; however, (a) there are criterion issues associated with each kind of measurement; (b) infants' acuity measures at any one age show striking individual variation, and the very few individuals we tested by both techniques showed closely comparable results;[3,4] and (c) the comparison has not generally been made under comparable conditions of pattern, luminance, screen size, and temporal frequencies. Even if VEP acuities do turn out to be higher than PL, eccentricity may not be the critical factor. We do not know how infant acuity varies with eccentricity. If, as has been reported, the fovea is structurally immature in the first months,[7,8] it may not be the area of maximum acuity. This may, however, be one problem where the VEP is the preferred method to yield the information. The fixation shifts intrinsic to PL make stimulus control, in the sense of achieving specific eccentricities, very difficult with that technique. It may prove possible to test the infant's pattern VEP with an annular stimulus, determine whether it is as foveally dominated as the VEP of the adult, and perhaps even to measure the acuity of extrafoveal regions.

References

1. Teller, D. Y. 1979. The forced-choice preferential looking procedure: A psychophysical technique for use with human infants. Infant Behav. Devel. **2:** 135–153.
2. Braddick, O., J. Atkinson, B. Julesz, W. Kropfl, I. Bodis-Wollner & E. Raab. 1980. Cortical binocularity in infants. Nature **288:** 363–365.
3. Harris, L., J. Atkinson & O. Braddick. 1976. Visual contrast sensitivity of a 6-month-old infant measured by the evoked potential. Nature **264:** 570–571.
4. Atkinson, J., O. Braddick & J. French. 1979. Contrast sensitivity of the human neonate measured by the visual evoked potential. Invest. Ophthalmol. Vis. Sci. **18:** 210–213.
5. Banks, M. S. & P. Salapatek. 1978. Acuity and contrast sensitivity in 1-, 2-, and 3-month old human infants. Invest. Ophthalmol. Vis. Sci. **17:** 361–365.
6. Dobson, V. & D. Y. Teller. 1978. Visual acuity in human infants: A review and comparison of behavioral and electrophysiological studies. Vis. Res. **18:** 1469–1483.

7. MANN, I. C. 1964. The Development of the Human Eye. British Medical Association, London.
8. ABRAMOV, I., J. GORDON, A. HENDRICKSON, L. HAINLINE, V. DOBSON & E. LABOSSIERE. 1981. Postnatal development of the infant retina. (abstract). Invest. Ophthalmol. Vis. Sci. **20** (3, Suppl.): 46.
9. ATKINSON, J., O. BRADDICK & K. MOAR. 1977. Development of contrast sensitivity over the first 3 months of life in the human infant. Vis. Res. **17:** 1037–1044.
10. HARRIS, P. & A. MACFARLANE. 1974. The growth of the effective visual field from birth to seven weeks. J. Exp. Child Psychol. **18:** 340–348.

STIMULUS PARAMETERS AND VISUAL EVOKED POTENTIAL DIAGNOSIS

John Camisa

New York Association for the Blind
New York, New York 10021

Mount Sinai School of Medicine
City University of New York
New York, New York 10029

Ivan Bodis-Wollner

Mount Sinai School of Medicine
City University of New York
New York, New York 10029

In the last few decades, studies in physiology, electrophysiology, and psychophysics have made it clear that the mammalian visual system is a composite of subsystems. Subsystems of the direct retino-cortical pathways that are distinct, both functionally and anatomically, are to a large degree, stimulus specific.

It is generally understood by vision scientists that the visual evoked potential reflects the contribution of stimulus-specific pathways. However, in the clinical application of visual evoked potentials, it is often tacitly assumed that there is a single normal latency which can be a measure of conduction velocity in the visual pathways. Yet, as we know, certain features of the stimulus are detected by distinct subgroups of neurons, which transmit the sensory information via pathways arriving at different cortical projection sites and structures. Therefore, any one visual evoked potential recording may reflect only those cortical structures sensitive to the particular configuration of features that comprise the VEP stimulus. In fact, the latency of the major positive peak varies depending on stimulus conditions. This is of clinical importance.

We know, for instance, that the VEP latency in normals is dependent on such stimulus factors as check size,[1] spatial frequency,[2] mean luminance,[3] and other less well explored parameters. Presumably, this change in the "normal" VEP latency with changes in these stimulus dimensions is accounted for by the differential stimulation of separate channels by the varying combinations of these stimulus factors. This raises an important question in the clinical application of the VEP: Can we assume that a patient's VEP classified as normal under one set of stimulus conditions would be found normal under a different set of conditions?

The question is hardly trivial when one considers that even a cursory glance at the clinical VEP literature for the last five years reveals as many combinations of stimulus parameters (check size range, 10′–60′; luminance range, 25–1025 cd/m^2; field size range, 4°–30°) as there are reported studies. If a patient's diagnosis was the same regardless of the stimulus conditions, then one could conclude that the given pathology affects all visual pathways indiscriminantly. If, on the other hand, a patient's diagnosis depended on the set of stimulus conditions used, then one could infer that the given pathology affects only a specific pathway. This *would* be of physiologic and clinical interest. In fact, the answer to the question "is one set of stimulus parameters sufficient for classifying a patient's VEP as normal?" is no.

We have found that, while the specific orientation of the grating has no appreciable affect on the *normal* VEP latency, it can have a determining affect on the classification of *patient* latencies.[4] In our laboratory for clinical diagnostic purposes we use both a 4° circular display of vertical sinusoidal gratings with a spatial frequency of 2.3 c/deg at a mean luminance of 0.7 log footlamberts (ft-L) and a 4° circular display of checks (14′ sq) at a mean luminance of 0.7 ft-L. Classification of patient latencies as normal or abnormal is accomplished by comparing individual patient latencies with those of the appropriate normal control group. Patient latencies falling outside the 95% probability elipse generated from the control group right and left eye latencies are classified as abnormal. We decided to test a number of patients with both vertical and horizontal gratings. Of the 22 patients we tested, 8 were classified as definite MS, 7 were probable MS and 7 were classified as possible MS. We found that the VEP latencies of 13 of the 22 patients were classified as abnormal when the vertical gratings were used. Of the 9 patients classified as normal with vertical gratings, 6 were classified as abnormal when horizontal gratings were used.

Over half of the patients we tested had an orientation—dependent VEP abnormality. Using gratings in two orientations rather than just one increased our diagnostic yield an average of 25% across the three diagnostic categories of MS. With finding orientation dependent losses in MS it seems reasonable, therefore, to explain the lower diagnostic yield obtained with checkerboard patterns in terms of stimulus complexity. The checkerboard is a very complex stimulus, and, therefore, it probably optimizes the response of a larger population of neurons than would any single orientation of sinusoids.

In addition to orientation, we have found that the luminance of the VEP stimulus also has a significant affect on the VEP diagnostic yield in MS. We recorded VEPs in a group of 18 MS patients with the identical checkerboard pattern set at two different luminance levels (1.7 and 0.7 log ft-L). We found that eight of the patients (44%) classified as normal with high luminance checks were classified as abnormal when the lower luminance checks were used. We repeated the measurements using both the low luminance check pattern and a grating pattern of the same mean luminance and identical spatial frequency as Fourier fundamental of the checkerboard. Both the check and the gratings stimuli diagnosed 28 of the 39 patients (72%) as abnormal. However, five of the patients categorized as normal with the grating stimulus were found abnormal when low luminance checks were used. Conversely, one patient, whose low luminance check VEP was normal, was found to be abnormal when the grating was used. With the joint information from the low luminance check and grating stimuli, 33 of the 39 (85%) MS patients were classified as abnormal.

As we mentioned earlier, there are as many combinations of these stimulus conditions as there are reports on VEP diagnosis in the literature, even though a comparison of diagnostic yield does not vary greatly. Given these differences in methodology, one wonders if the same patient would be given the same VEP diagnosis if tested in each of these laboratories. In our own lab, then, we have shown the importance of at least two stimulus parameters. It is possible that other parameters such as field size, spatial frequency or contrast may also have an important role in the effectiveness of VEP diagnosis in MS. If one takes into account what we know about stimulus specificity of the visual pathways, we could, in addition to the increase in the diagnostic yield, gain some insight into the physiology of disease.

References

1. Sokol, S. & A. Moskowitz. 1981. Effect of retinal blur on the peak latency of the pattern evoked potential. Vis. Res. (in press).
2. Parker, D. M. & E. A. Salzen. 1977. Latency changes in the human visual evoked response to sinusoidal gratings. Vis. Res. **18:** 1201–1204.

3. Cant, B. R., A. L. Hume & N. A. Shaw. 1978. Effects of luminance on the pattern visual evoked potential in multiple sclerosis. Electroencephalog. Clin. Neurophysiol. **45:** 496–504.
4. Camisa, J., L. H. Mylin & I. Bodis-Wollner. 1981. The effect of stimulus orientation on the visual evoked potential in multiple sclerosis. Ann. Neurol. **10:** 532–539.
5. Asselman, P., D. W. Chadwick & C. D. Marsden. 1975. Visual evoked responses in the diagnosis and management of patients suspected of multiple sclerosis. Brain **98:** 261–282.
6. Collins, D. W. K., J. L. Black & F. L. Mastaglia. 1978. Pattern reversal visual evoked potential. J. Neurol. Sci. **36:** 83–95.
7. Halliday, A. M. 1976. Visually evoked responses in optic nerve disease. Trans. Ophthal. Soc. U.K. **96:** 372–376.
8. Lowitzsch, K., U. Kuhnt, C. Sakman, K. Maurer, H. C. Hopf, D. Schott & K. Thater. 1976. Visual pattern evoked responses and blink reflexes in assessment of MS diagnosis. J. Neurol. **213:** 17–32.
9. Matthews, W. B., D. G. Small & E. Pountney. 1977. Pattern reversal evoked potential in the diagnosis of multiple sclerosis. J. Neurol. Neurosurg. Psychiatry **40:** 1009–1014.
10. Regan, D., B. A. Milner & J. R. Heron. 1976. Delayed visual reception and delayed visual evoked potentials in the spinal form of multiple sclerosis and in retrobulbar neuritis. Brain **99:** 43–66.

VISUAL EVOKED RESPONSES AND RETINAL ECCENTRICITY

Gastone G. Celesia and J. Todd Meredith

Department of Neurology
William S. Middleton Memorial Veterans Hospital
University of Wisconsin
Madison, Wisconsin 53705

The effect of stimulation of discrete areas of the retina on pattern-reversal visual evoked potentials (VEP) was studied in 16 normal volunteers. The stimulus consisted of a constant luminance 2°18′ square field containing 16 checks of 34′30″ reversing at a frequency of 500 msec. The luminance of the field was 20 footlamberts. As shown in FIGURE 1, stimulation at the fixation point evoked a reproducible potential characterized by two major positive–negative deflections. Similar evoked potentials were elicited when the stimulus was located within the 2° isopter.

The amplitude of the response declined rapidly as we moved from the fixation point along the horizontal, vertical, and oblique meridians, and usually no response was detected outside the 4° isopter (FIGURE 2).

In four subjects, the smallest size of field needed to evoke a reproducible response at fixation was determined. The size of the field varied slightly in each individual, ranging from 6′54″ to 20′42″ containing four individual checks of 3′47″ and 10′21″, respectively. The responses to these small fields consisted of a small, broadened, positive potential occasionally followed by a prominent negative wave. Although no responses with stimuli of 2°18′ were obtained outside the central 4° isopter, evoked responses could be elicited outside the foveal region by the use of larger stimuli. The smallest field capable of evoking a potential along the horizontal meridian at 8° and 14° isopters were calculated. Increasingly larger fields and larger checks were needed to produce an evoked potential as we moved from the fixation point to the 14° eccentricity (TABLE 1). The amount of striate complex activated by these pattern-reversal stimuli at the three retinal eccentricities was calculated based on the magnification factor of Cowey and Rolls[1] and that of Virsu and Rovamo,[2,3] respectively. The cortical magnification factor M is the amount of striate cortex in square millimeters corresponding to 1° of arc in visual space. The values shown in TABLE 1 were obtained by multiplying the field size in degrees of visual angle per M factor at the specified eccentricity. At the fixation point the visual stimulus will activate both striate cortices, therefore the value at this point was obtained by multiplying the size of the field per 2 M.

The data in TABLE 1 indicates that the visual stimuli at the three eccentricities studied activated approximately 7 square millimeters of striate cortex. This suggests that visual stimuli are equally effective in producing an evoked potential if they are M-scaled. The only value not agreeing with this hypothesis is the value at the fixation point based on the M-factor of Rovamo and Virsu.[2] The M-scaling of these authors is much smaller than the M-scaling of Cowey and Rolls.[1] The latter authors calculated the M factor from phosphenes produced by electrical stimulation of human cortex, while Rovamo and Virsu extrapolated their values from psychophysical methods.

In summary, the amplitude distribution of evoked responses to a 2°18′ stimulus in relation to retinal eccentricity correlates well with: (1) the decline in cone density in relation to retinal eccentricity;[4] (2) density distribution of human ganglion cell

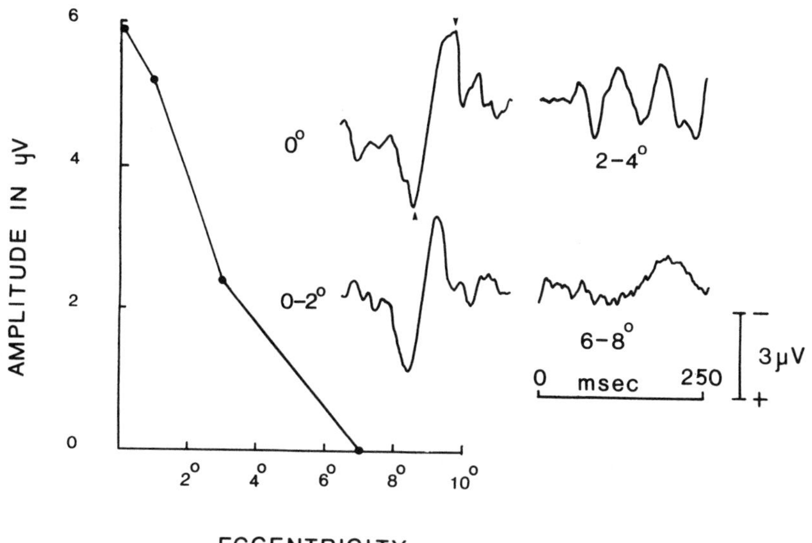

FIGURE 1. Relationship between amplitude of the major deflection P_1–N_1 (indicated by two arrows) and retinal eccentricity along the horizontal meridian.

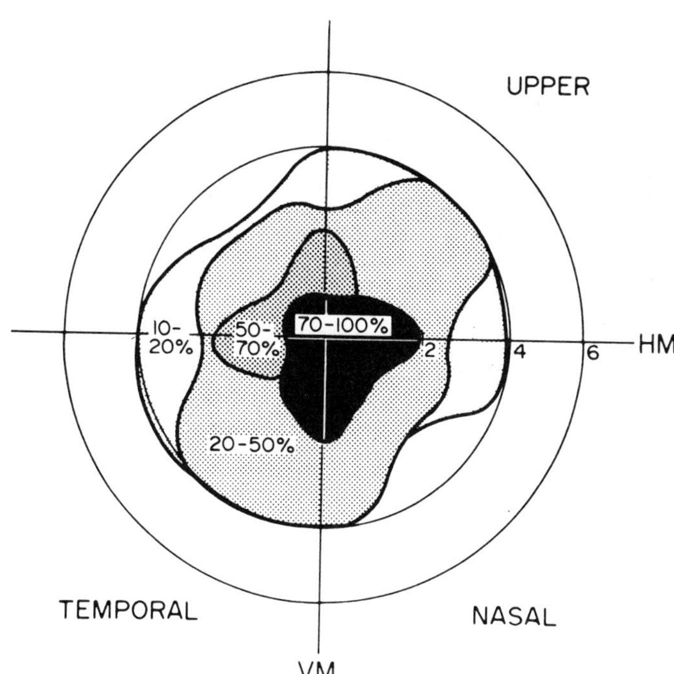

FIGURE 2. Relationship between normalized amplitude of evoked responses to a 2°18′ stimulus and various retinal eccentricities. Note the rapid amplitude decline outside the 2° meridian.

TABLE 1

SIZE OF STMULI REQUIRED TO PRODUCE AN EVOKED RESPONSE AND ESTIMATED SIZE OF STRIATE CORTEX ACTIVATED

	Fixation Mean ± SD	8° Nasal Mean ± SD	14° Nasal Mean ± SD
Field size	13′17″ ± 5′4″	3°18′ ± 34′	5°45″ ± 56′
Check size	6′35″ ± 2′5″	44″39″ ± 10′	1°43′ ± 14′
Cortex activated (mm^2)			
Using *M* factor of Rovamo & Virsu	3.42 ± 1.46	7.62 ± 1.54	7.13 ± 1.16
Using *M* factor of Cowey & Rolls	6.68 ± 2.84	6.93 ± 1.86	6.67 ± 1.09

population along the horizontal axis;[5] and (3) the decline of visual acuity in relation to eccentricity.[6]

Outside the fovea, the visual stimulus, to be effective, needs to reach the visual discrimination threshold and needs to stimulate sufficient numbers of receptors and ganglion cells. Our data are further supportive of the invariance principle of Rovamo *et al.*[7] stating that photopic stimuli presented anywhere in the visual field are equally effective if the stimuli are equivalent in terms of calculated cortical projection images and number of cortical neurons activated.

REFERENCES

1. COWEY, A. & E. T. ROLLS. 1974. Human cortical magnification factor and its relation to visual acuity. Brain Res. **21:** 447–454.
2. VIRSU, V. & J. ROVAMO. 1979. Visual resolution, contrast sensitivity and the cortical magnification factor. Exp. Brain Res. **37:** 475–494.
3. ROVAMO, J. & V. VIRSU. 1979. An estimation and application of the human cortical magnification factor. Exp. Brain Res. **37:** 495–510.
4. ØSTERBERG, G. 1935. Topography of the layer of rods and cones in the human retina. Acta Ophthalmol. (Suppl.) **61:** 1–102.
5. VAN BUREN, J. M. 1963. The Retinal Ganglion Cell Layer. Charles C. Thomas, Springfield, Ill.
6. WEYMOUTH, F. W., D. C. HINES, L. M. ACRES, J. E. RAAF & M. C. WHEELER. 1928. Visual acuity within the area centralis and its relation to eye movements and fixation. Am. J. Ophthalmol. **11:** 947–960.
7. ROVAMO, J., V. VIRSU & R. NÄSÄNEN. 1978. Cortical magnification factor predicts the photopic contrast sensitivity of peripheral vision. Nature (London) 271:54–56.

RADIAL SPATIAL PATTERNS AND MULTIFREQUENCY TEMPORAL PATTERNS: POSSIBLE CLINICAL APPLICATIONS*

Floyd Ratliff

Rockefeller University
New York, New York 10021

INTRODUCTION

There is no "best method" of stimulus control or of data analysis in the study of visual evoked potentials (VEPs). The problems are so varied, both in basic research and in clinical research, that no one method can possibly be best suited to all applications. For research purposes, the ideal solution is to have at hand extremely versatile methods of stimulus control—methods that permit adjustments to be made along several dimensions so as to maximize the effectiveness of the stimulus and to permit wide exploration of stimulus variables. It is also essential to have powerful and sophisticated methods of data analysis so as to maximize the amount of information that can be extracted from the limited data available. These are the directions we have chosen to go in the development of a Visual Stimulator/Data Analyzer for both basic and clinical research. Ultimately, we hope to develop much simpler instruments and techniques for specific clinical applications.

Two major problems in the study of VEPs are the low signal-to-noise ratio and the high variability of responses in time. These problems are especially critical in clinical situations where recording conditions are much less favorable than in the laboratory and repeated measurements, under stable conditions and over long periods of time, cannot be made. Therefore, it is very important that there be a good "impedance match" between the method of stimulus control, the physiological process under investigation, and the method of data analysis to be used. This discussion focuses on two of our attempts to improve this match. One concerns the use of radial spatial patterns, the other concerns the use of multifrequency temporal patterns.

RADIAL SPATIAL PATTERNS

Some of the most widely used patterned stimuli are homogeneous periodic patterns such as sinusoidal or bar gratings and checkerboards. One disadvantage they have, which we are attempting to surmount,[1] is that the spacing of their elements cannot be adjusted to stimulate maximally many different parts of the retina at the same time. If they are fine enough to match the resolving power of the fovea, they are too fine to match the resolving power of more peripheral regions. If they are coarse enough to match the resolving power of the more peripheral regions, then they are too coarse to match the resolving power of the fovea.

More effective stimuli can be provided by inhomogeneous radial patterns whose dimensions increase with distance from the fixation point and thus more closely match

*This work was supported by the National Institutes of Health and by the Esther A. and Joseph Klingenstein Fund.

0077-8923/82/0388-0651 $1.75/0 © 1982, NYAS

the functional characteristics of the visual system at different distances from the fovea. Two phases of a radial stimulus are shown at the top of FIGURE 1, a "windmill" and a "dartboard." Such stimuli can be contrast reversed in the same way as the usual parallel gratings or checkerboards. (Some earlier versions of radial stimuli used pattern onset only.[2,3]) In the first experiments to be described, the maximum contrast of all segments was 45%. The segments in the central disc and in the second annulus were sinusoidally contrast reversed at a frequency of 4.19 Hz. The segments in the first annulus and in the third annulus were static, and were fixed at a contrast of 45%. Consequently, the overall stimulus pattern alternated between the windmill and dartboard forms. At the bottom of FIGURE 1, the contrast of the static segments was set at 0%, equal to the mean luminance of the entire pattern.

Dimensions of these particular examples were chosen arbitrarily. However, there are rational neurophysiological and psychophysical bases for making a particular

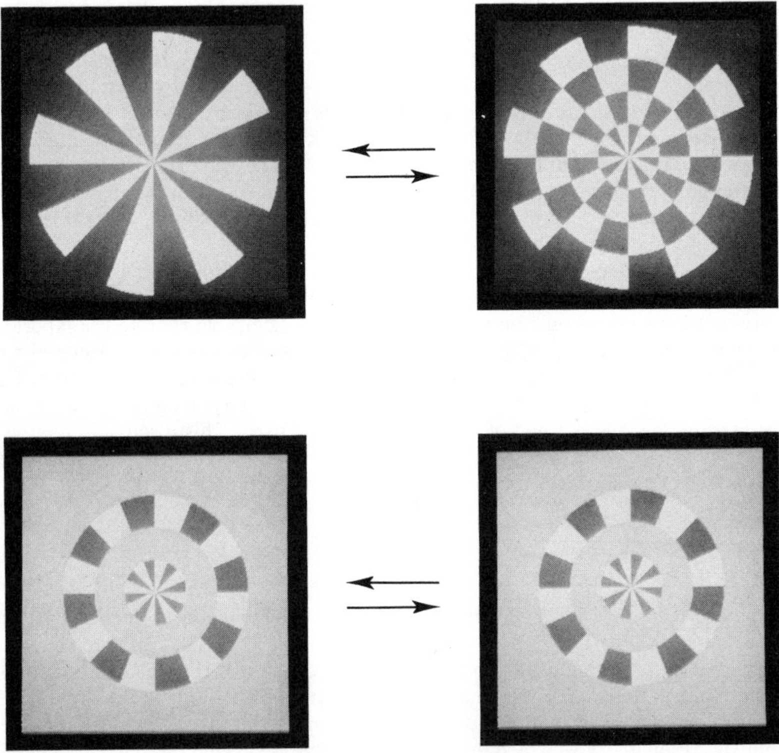

FIGURE 1. (Top) The windmill phase (left) and the dartboard phase (right) of a radial pattern. The segments in the first and third annuli were static. The segments in the central disc and second annulus were contrast reversed with a sinusoidal wave at 4.19 Hz. Thus the pattern alternated between the windmill and dartboard phases. Maximum contrast was 45%. (Bottom) The static segments in the first and third annuli were set at 0% contrast (equal to the mean luminance of the remainder of the pattern). With contrast reversal of the segments in the central disc and second annulus, the pattern alternated between the two phases shown. The diameter of the pattern subtended a visual angle of 4°, viewed binocularly at a distance of one meter. Mean luminance, 30 ft-L.

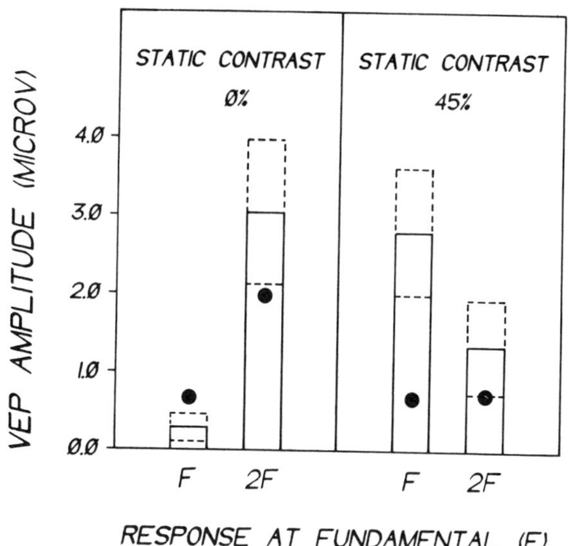

FIGURE 2. Amplitudes of average fundamental and second harmonic components of VEPs from eight normal subjects (bars) and one epileptic patient (filled circles). Dotted bars indicate one standard deviation. See text and FIGURE 1 for details of stimuli used.

choice: e.g., sizes of receptive fields[4] and sizes of so-called perceptive fields[5] at different distances from the fovea. (See also the reports by C. Tyler and by G. Celesia in this Roundtable Session.)

Some preliminary results obtained from normal subjects and from an epileptic patient, using our windmill-dartboard stimulus, are shown in FIGURE 2. (The patient was referred to us by Dr. Gale Solomon, Neurology Department, New York Hospital-Cornell Medical School.) When the static segments were "off" (set at 0% contrast) as shown at the bottom of FIGURE 1, the normal subjects (bar graphs) showed a large second harmonic and little or no fundamental response, as expected. When the static segments were "on" (contrast set at 45%, equal to the dynamic pattern) the normal subjects (bar graphs) showed a large fundamental response and a strong suppression of the second harmonic which we attribute to lateral interactions.[6-8] The epileptic patient (filled circles) showed a near-normal response when the static contrast was "off" (set at 0%). The response was clearly abnormal, however, when the static contrast was "on" (set at 45%). Although there was the expected second harmonic suppression, there was little or no response at the fundamental frequency. This is indicative, we believe, of some interference with lateral interactions in the visual pathways. It would be foolhardy, however, to draw any firm conclusions on the basis of this one experiment. Nonetheless, the results are very provocative.

A major advantage of this dartboard stimulus is that the amplitude of the response is much larger than with conventional homogeneous stimulus patterns such as sine-wave gratings and rectangular checkerboards. A disadvantage is that good fixation is required to keep the pattern centered (often a difficult problem with patients). The advantages and disadvantages must be weighed with respect to particular applications; there is no general rule that can be followed in every instance.

MULTIFREQUENCY TEMPORAL PATTERNS

It is important to record the VEP over as short a time span as possible in both clinical and basic research. Many patients, for one reason or another, cannot attend to the stimulus for a long period of time. And even if the patient or normal subject can attend to the stimulus, there is usually considerable variability in the response over long periods.

Several techniques have been developed to shorten the recording time. The pros and cons of some of them have been reviewed by Regan.[9] All of them save time, but at some cost. For example, "sweeping" through a range of frequencies may obscure phase information, and the inevitable "compression" of data by a sweep technique may obscure important maxima and minima in the response. We are using a method which presents several (usually eight) frequencies of contrast reversal simultaneously—a "sum-of-sinusoids" rather than a single sinusoid. A disadvantage of this

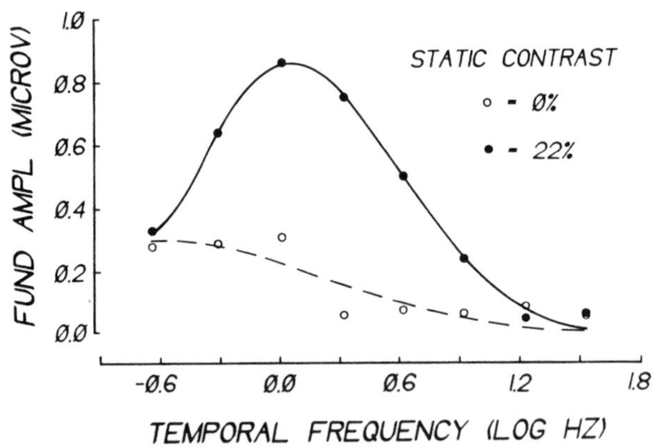

FIGURE 3. Amplitudes of VEPs at eight frequencies of contrast reversal, presented simultaneously as a sum-of-sinusoids. Same stimuli as in FIGURE 1. (Open circles) Contrast of static segments, 0%; maximum contrast of each of the eight sinusoids in the sum-of-sinusoids driving the dynamic segments was 11%. (Filled circles) Contrast of the static segments 22%; maximum contrast of each of the eight sinusoids in the sum-of-sinusoids 11%.

technique is that the total contrast in the stimulus must be divided among the several sinusoids; therefore the response to each of the several low-contrast sinusoids will necessarily be much smaller than the response to a single high-contrast sinusoid alone. To compensate for this requires several repetitions. In the worst case, the total recording time may be as long as with single sinusoids. But even if no time is saved (which is usually not the case) the method has the advantage that responses to *all* temporal frequencies are obtained under exactly the same conditions in each measurement. Variations from one measurement to the next are therefore less likely to affect responses at different frequencies selectively.

The method also has the advantage that proper choice of the frequencies of these eight temporal sinusoids yields combination frequencies at sums and differences that are distinct from the eight fundamental frequencies.[10] This provides a better signal-to-noise ratio than conventional methods of nonlinear analysis and permits easy separation of the first-order (linear) and second-order (non-linear) components of the

VEPs.[11-13] The amplitudes of the responses at the eight fundamental frequencies gives the first-order linear component. The joint frequency responses to pairs of the fundamental frequencies yield the second-order nonlinear component. This second-order frequency "kernel" is closely related to the Fourier transform of the second-order Wiener kernel.[14,15]

A sum-of-sinusoids contrast reversal of the same stimulus patterns (FIGURE 1) used in the previous experiment shows how the response is measured simultaneously across the whole spectrum of relevant frequencies. As before, the segments of the central disc and the second annulus were contrast reversed, but this time with the sum-of-sinusoids (contrast of each of the eight sinusoids at 11%). The segments of the first and third annuli were held constant (at a contrast of either 0% or 22%). When the static annuli were "off" (0% contrast) there was an insignificant response at the fundamental input frequencies (FIGURE 3, open circles), but there were strong second-order responses (not shown here). When the static annuli were "on" (contrast 22%) there was a substantial response at the fundamental input frequencies (filled circles) and the second-order responses were much attenuated.

Although the responses at each frequency in the sum-of-sinusoids technique are necessarily small (with a contrast of only 11% per sinusoid), the difference between the two conditions illustrated here is measurable and obviously significant.

SUMMARY AND CONCLUSIONS

Signal-to-noise ratios can be improved significantly by using radial stimuli with spatial characteristics that are closely matched to the functional characteristics of the visual system. The gain in the strength of the signal facilitates the recording of responses to several temporal frequencies simultaneously. Such a combination of an inhomogeneous radial spatial pattern (which is closely matched to the spatial inhomogeneity of the visual system) and a multifrequency temporal signal (which can elicit a wide range of first-order linear and second-order nonlinear responses) might provide a useful "screening" stimulus for clinical research. Once any abnormality was detected, the range of spatial and temporal variations in the stimulus could then be narrowed down so as to focus on that one particular problem.

ACKNOWLEDGMENTS

The exploratory experiments reported here were carried out in collaboration with Vance Zemon. I thank Norman Milkman and Gary Schick for technical assistance.

REFERENCES

1. MILKMAN, N., G. SCHICK, M. ROSSETTO, F. RATLIFF, R. SHAPLEY & J. VICTOR. 1980. A two-dimensional computer-controlled visual stimulator. Behav. Res. Meth. Instr. **12:** 283–292.
2. OCHS, A. L. & M. J. AMINOFF. 1980. Visual evoked potential elicited by circular gratings. Arch. Neurol. **37:** 308–309.
3. BARBER, C. & N. R. GALLOWAY. 1976. A pattern stimulus for optimal response from the retina. Doc. Ophthalmol. Proc. **10:** 77–86.
4. CLELAND, B. G., T. H. HARDING & U. TULUNAY-KEESEY. 1979. Visual resolution and receptive field size: Examination of two kinds of cat retinal ganglion cell. Science **205:** 1015–1017.

5. Ransom-Hogg, A. & L. Spillman. 1980. Perceptive field size in fovea and periphery of the light- and dark-adapted retina. Vis. Res. **20:** 221–228.
6. Ratliff, F., V. Zemon & J. D. Victor. 1980. Visual evoked potentials: Spatial interaction of dynamic and static sinusoidal gratings. J. Opt. Soc. Am. **70:** 1598.
7. Ratliff, F. & V. Zemon. 1982. Some new methods for the analysis of lateral interactions that influence the visual evoked potential. Ann. N.Y. Acad. Sci. This volume.
8. Zemon, V. & F. Ratliff. 1982. Visual evoked potentials: Evidence for lateral interactions. In preparation.
9. Regan, D. 1980. Speedy evoked potential methods for assessing vision in normal and amblyopic eyes: Pros and cons. Vis. Res. **20:** 265–269.
10. Victor, J. D., R. M. Shapley & B. W. Knight. 1977. Nonlinear analysis of cat retinal ganglion cells in the frequency domain. Proc. Natl. Acad. Sci. USA **74:** 3068–3072.
11. Shapley, R., Y. T. So & J. Victor. 1978. Nonlinear systems analysis of retinal ganglion cells and visual evoked potentials in the cat. J. Opt. Soc. Am. **68:** 1427.
12. Ratliff, F., J. D. Victor & R. M. Shapley. 1978. Nonlinear analysis of visual evoked potentials in the human. J. Opt. Soc. Am. **68:** 1427.
13. Ratliff, F. 1980. Form and function: Linear and nonlinear analyses of neural networks in the visual system. *In* Neural Mechanisms in Behavior: A Texas Symposium. D. MacFadden, Ed. pp. 72–142. Springer-Verlag, New York, N.Y.
14. Victor, J. D. 1979. Nonlinear systems analysis: Comparison of white noise and sum of sinusoids in a biological system. Proc. Natl. Acad. Sci. USA **76:** 996–998.
15. Victor, J. D. & B. W. Knight. 1979. Nonlinear analysis with an arbitrary stimulus ensemble. Q. Appl. Math. **37:** 113–136.

PROBLEMS OF STIMULUS CONTROL IN THE MEASUREMENT OF PEAK LATENCY OF THE PATTERN VISUAL EVOKED POTENTIAL*

Samuel Sokol

Department of Ophthalmology
New England Medical Center
and Tufts University School of Medicine
Boston, Massachusetts 02111

With the widespread clinical use of the peak latency of the first major positive component of the pattern-reversal VEP for the evaluation of optic nerve function,[1–4] it is increasingly important that strict control over both the pattern stimulus presentation and the state of the patient be maintained. There are a number of nonpathologic factors, albeit simple ones, that must be monitored in order to be certain that "abnormal" prolongation of latency is a true indication of pathology of the visual pathways rather than an artifact of poor control. This becomes particularly important when other clinical findings are equivocal. Examples of some of these factors are: check size, pupil size, uncorrected refractive error, and age.

FIGURE 1 shows that with the exception of N_2, the peak latency of the major components of the transient pattern-reversal VEP decrease as check size increases and finally asymptotes for checks larger than 20–30 minutes of arc.[5] Therefore, if checks smaller than 20–30 minutes of arc are used adult patients should be seated at a table with a chin rest in order to maintain a constant viewing distance from the pattern stimulus since any variation in distance will in turn effect check size. While it is difficult to have children use a chin rest, their distance from the pattern stimulus should nevertheless be closely monitored.

Another factor that affects latency is pupil size; small diameter pupils reduce retinal illuminance. For example, as pupil diameter is decreased from 9 mm to 1 mm there is a peak latency increase of nearly 20 msec. This effect is independent of check size.[6] A practical implication of a variation in pupil size on VEP latency is that results obtained from patients with miotic pupils, e.g., those with glaucoma, cannot be compared to normals. This problem can be avoided by acquiring normative data with an artificial pupil that is equivalent to the smallest pupil size one would expect to encounter in a clinical situation.

A third factor that can lead to a deceptive increase in latency is retinal blur caused by an uncorrected refractive error.[7] Figure 2 shows the effect of minus and plus spherical lenses on the P_1 latency of the monocularly recorded VEP from two cyclopleged ametropic subjects who viewed a checkerboard pattern stimulus comprised of 12 minute checks at a distance of one meter. Subject JM, a 4.75 diopter hyperope, had the shortest P_1 latency (and the sharpest subjective vision) at $+4.75$ diopters. Subject RP, a -4.00 diopter myope produced the shortest P_1 latency at -3.00 diopters, which is in agreement with the predicted value at a test distance of one meter.

FIGURE 3 demonstrates how in a patient recovering from bilateral optic neuritis, an abnormally prolonged latency was found to be the result of an uncorrected refractive

*Parts of this work were supported by a research grant (EY-00926) from the National Eye Institute.

0077–8923/82/0388–0657 $1.75/0 © 1982, NYAS

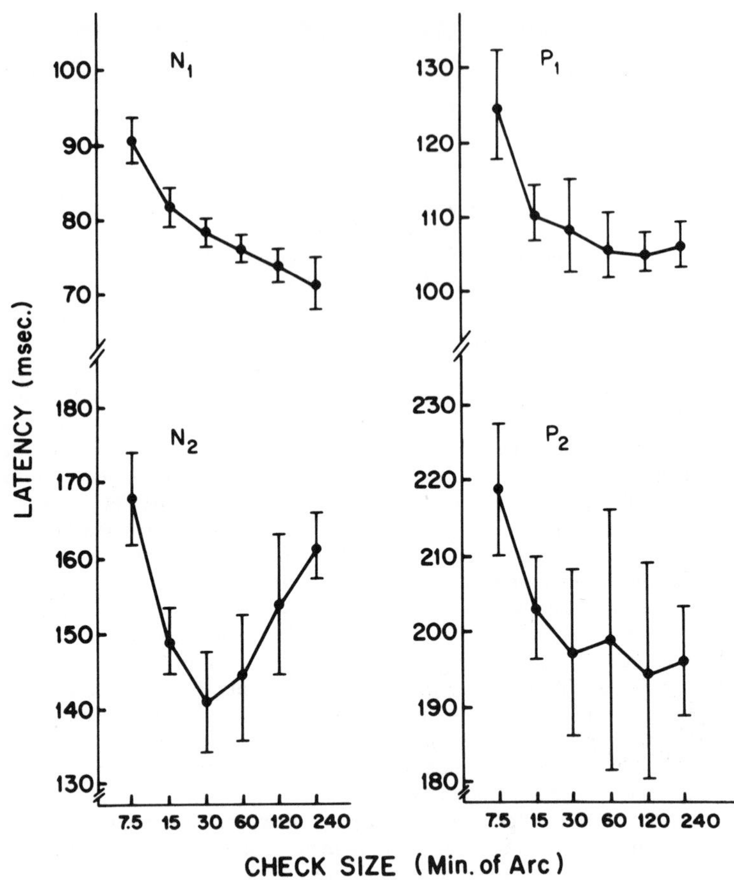

FIGURE 1. Peak latency of the four major components of the pattern-reversal VEP as a function of check size. The vertical lines indicate 1 standard error of the mean. N_1, P_1, and P_2 decrease as check size increases. N_2 shows an inverted U-shaped function, with shortest latency at 30 minutes and increases for smaller and larger checks.

error. The patient was a 13-year-old male who presented with complaints of pain on lateral and upward gaze for two weeks and blurred vision for one week in the right eye. Acuity was 20/50 in the right eye and 20/25 in the left eye. Visual fields showed a centrocecal defect in the right eye and an enlarged blind spot in the left eye. A bilateral red–green color vision defect was demonstrated and a relative afferent pupillary defect was present in the right eye. There was bilateral disc swelling. The vision in the right eye soon improved to 20/25 without correction. Shortly thereafter vision fell to counting fingers at 3 feet in the left eye. At this point serial VEPs were recorded from the patient over the next 85 weeks. Initially, VEPs were nonrecordable in the left eye (FIGURE 3, open circles) and abnormally prolonged in the right eye (closed circles). Within 6 weeks the peak latency for the right eye returned to normal and the latency for the left eye, now recordable, was abnormal. At 11 weeks the latency for the right and left eye were within normal limits but the difference between eyes was abnormal. At 18 and 25 weeks the latency and visual acuity for the left eye were abnormal. However, on examination the patient was found to have a hyperopic refractive error in the left eye of +3.00 diopters. In addition, the patient had asymmetric accommodation. Uncyclopleged, accommodation was 13 diopters in the right eye and only 10 diopters in the left eye. When retested at 32 weeks with correction, the VEP latencies obtained from the left eye were normal. Thus, a latent hyperopia brought out by an accommodative paresis possibly related to the patient's optic neuritis, caused a reduction in acuity and an abnormal prolongation of latency. With correction, both the visual acuity and latency returned to normal. While a prolonged latency in the presence of normal visual acuity is not the result of a refractive error the combination of reduced acuity and prolonged latency might therefore be due to a refractive error and not demyelinization. Moreover, this artifact is more likely to occur with small checks. Therefore, this problem can be minimized by using large (e.g., 50 minute) checks.

Another factor which affects the latency of the pattern reversal VEP is age. In general, latency increases with age, more rapidly for small checks than large checks.[8-9] For example, when 48 minute checks are used, P_1 latency increases from a mean of 104 msec at 20 years of age to 111 msec at 70 years of age. For 12 minute checks the rate of latency increase is more rapid; rising from a mean of 112 msec at 20 years to 124 msec at 70 years of age. Subjects over 50 years of age with normal visual acuity

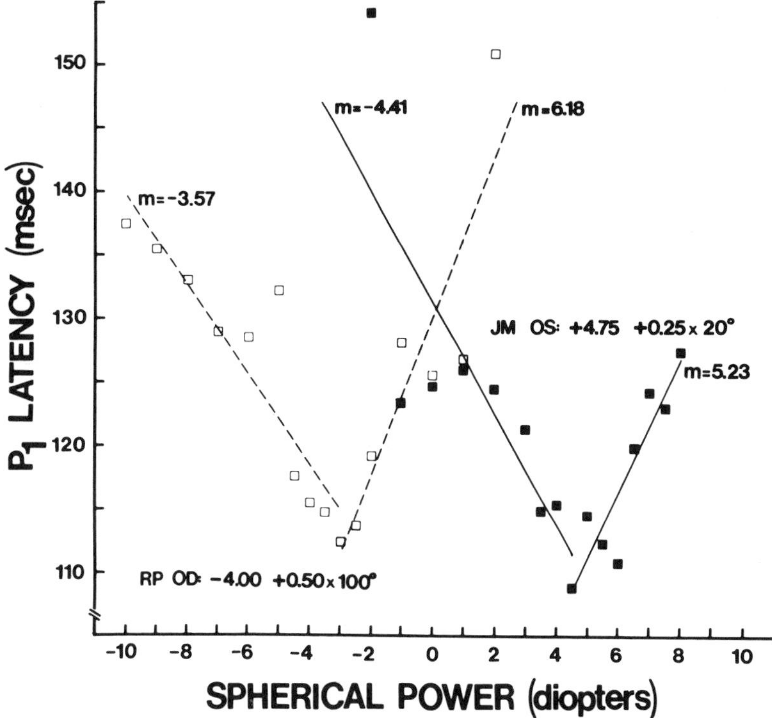

FIGURE 2. P_1 latency as a function of spherical power (diopters) for 12 minute checks for a myope (subject RP) and a hyperope (subject JM). Both subjects were cyclopledged. (From Sokol & Moskowitz.[7] By permission of *Vision Research*.)

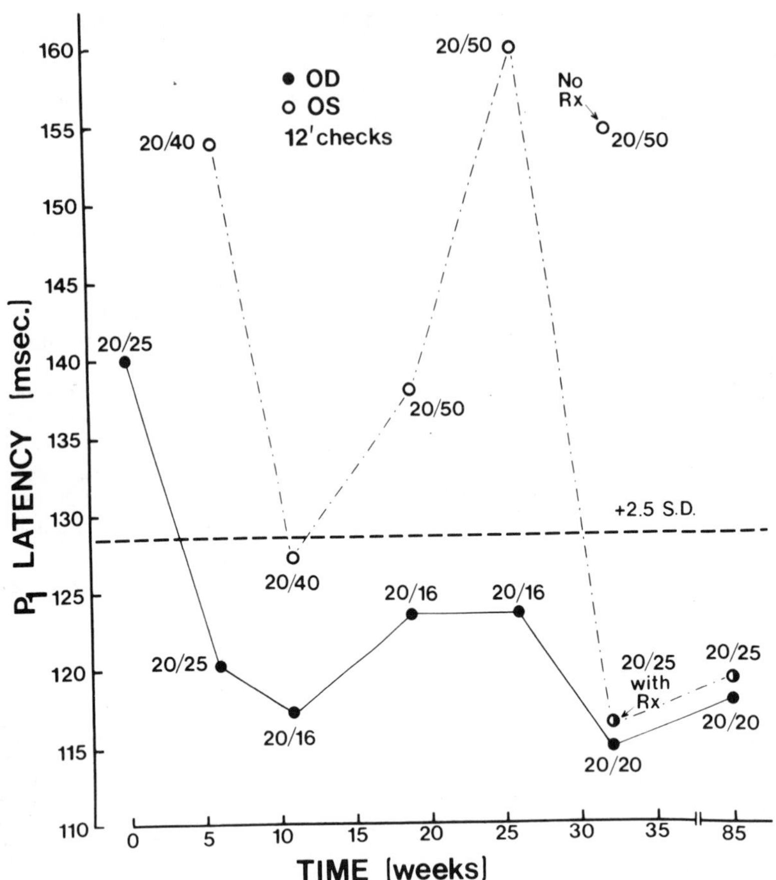

FIGURE 3. P_1 latency recorded over a period of 85 weeks from a patient recovering from bilateral optic neuritis. Note the decrease in latency that occurred when the patient was corrected (Rx) for a latent hyperopia of +3.00 diopters.

and no ocular pathology can elicit VEPs with latencies as long as 130 msec. This differential increase in the latency for large and small checks probably reflects a differential change in the capacity of the human visual system to process spatial frequency information.

In summary, check size, pupil size, refractive error and age will effect the latency of the pattern-reversal VEP and contributions from these factors could conceviably cause artifactual results. However, with the exception of pupil size, the effect is minimal with checks larger than 30 minutes of arc. Thus, these problems can be avoided by only using large checks. However, in my judgement this is not the best solution since information that arises from foveal receptors will be sacrificed if small checks are not used. An alternative is to use both small and large checks and maintain rigid adherence to control of both stimulus and patient parameters.

ACKNOWLEDGMENT

I thank J. W. Gittinger, Jr., M.D. who examined and referred the patient for VEP testing.

REFERENCES

1. HALLIDAY, A. M., W. I. MCDONALD & J. MUSHIN. 1972. Delayed visual evoked response in optic neuritis. Lancet (**1**): 982–985.
2. HALLIDAY, A. M., W. I. MCDONALD & J. MUSHIN. 1973. Visual evoked responses in diagnosis of multiple sclerosis. Br. Med. J. (**4**): 661–664.
3. HALLIDAY, A. M., E. HALLIDAY, A. KRISS, W. I. MCDONALD & J. MUSHIN. 1976. The pattern-evoked potential in compression of the anterior visual pathways. Brain **99:** 357–374.
4. CAPPIN, J. M. & S. NISSIM. 1975. Visual evoked responses in the assessment of field defects in glaucoma. Arch. Ophthalmol. **93:** 9–18.
5. PORCIATTI, V. & G. P. VON BERGER. 1981. Visual potentials evoked by pattern stimulation with different spatial frequencies in retrobulbar neuritis. *In* Visual Pathways: Electrophysiology and Pathology. Proc. 18th Int. Soc. Clin. Electrophysiol. Vis. (I.S.C.E.V.) Symp. H. Spekreijse & P. A. Apkarian, Eds. pp. 67–76. Dr. W. Junk, The Hague.
6. SOKOL, S., A. DOMAR, A. MOSKOWITZ & B. SCHWARTZ. 1981. Pattern evoked potential latency and contrast sensitivity in glaucoma and ocular hypertension. *In* Visual Pathways: Electrophysiology and Pathology. Proc. 18th I.S.C.E.V. Symp. H. Spekreijse & P. A. Apkarian, Eds. pp. 79–86. Dr. W. Junk, The Hague.
7. SOKOL, S. & A. MOSKOWITZ. 1981. Effects of retinal blur on the peak latency of the pattern evoked potential. Vis. Res. (in press).
8. CELESIA, G. G. & R. F. DALY. 1977. Effects of aging on visual evoked responses. Arch. Neurol. **34:** 403–407.
9. SOKOL, S., A. MOSKOWITZ & V. L. TOWLE. 1981. Age-related changes in the latency of the visual evoked potential: Influence of check size. Electroencephalogr. Clin. Neurophysiol. **51:** 559–562.

PROPERTIES OF LOCALIZED PATTERN EVOKED POTENTIALS*

Christopher W. Tyler and Patricia A. Apkarian

Smith-Kettlewell Institute of Visual Sciences
San Francisco, California 94115

The Netherlands Ophthalmic Research Institute
Amsterdam, the Netherlands

INTRODUCTION

The synchronous pattern evoked potential has been shown to have tremendous specificity to a number of stimulus parameters.[1,2] For example, changes in spatial or temporal frequency by as little as 10% can sometimes affect response amplitude by a factor of two or more, even when no corresponding change occurs in the psychophysical threshold function. A similar specificity may be used as a tool to study the properties of localized responses from small, homogeneous regions of cortex.

The question of localization has usually been approached by attempting to compute the position in the brain corresponding to the source of the scalp potential by means of large numbers of electrodes on the scalp. The most extensive effort in this direction has been made by D. H. Fender and his group at California Institute of Technology, but they have not so far used highly localized stimulation conditions.[3]

RETINAL LOCALIZATION AND THE VISUAL METRIC

We have taken a different approach toward the localization question. We have attempted to optimize the stimulus and recording conditions so as to record only from a small region of cortex lying immediately beneath the recording electrode. We have then studied the properties of this local population of neurons with other stimulus parameters. Thus, we are concerned not so much with spatial localization as with function isolation of small groups of neurons by retinal and scalp localization. The retinal localization is achieved simply by reducing the stimulating area to a small patch in the peripheral field.

In order to know what area of cortex is stimulated by a given retinal area, one has to take into account the retina-to-cortex mapping. The best data available from studies of this mapping in monkey cortex suggest that the mapping corresponds approximately to a radial-to-linear transform.[4,5] This transform is supported by human visual acuity data.[6-8] A schematic version of the transform is shown in FIGURE 1, where each major division on the retina corresponds approximately to one square centimeter of primary visual cortex. One of these areas has been divided into 100 subareas, each of which correspond to the estimated size of an elementary hypercolumn unit analyzing all orientations and each eye's input. This map represents schematically the best current estimate of human striate cortex organization.[9]

Most previous evoked potential work has not taken the retina-to-cortex mapping into account (review by Riggs and Wooten).[10] A typical procedure has been simply to

*This work was supported by the Smith-Kettlewell Eye Research Foundation and National Institutes of Health Grants EY 3622 and 3884, GRS 5S01, and RR 5566.

0077-8923/82/0388-0662 $1.75/0 © 1982, NYAS

move a stimulating patch to different positions in the retina without changing its size. It has therefore been common to state that most of the pattern evoked potential is generated by the central retina. This result can be better conceptualized in terms of the radial mapping function (FIGURE 1). The dashed-line circles indicate a stimulus of 8° diameter in central vision, a second one shifted by one diameter, and a third shifted by eight diameters in eccentricity. If we assume that hypercolumns in the human striate cortex are about 1 mm^2 in area,[9] the number of hypercolumns stimulated at each eccentricity can be estimated. As a result of the radial mapping, the central 8° area will stimulate about half the cortex or 3200 hypercolumns. The one diameter

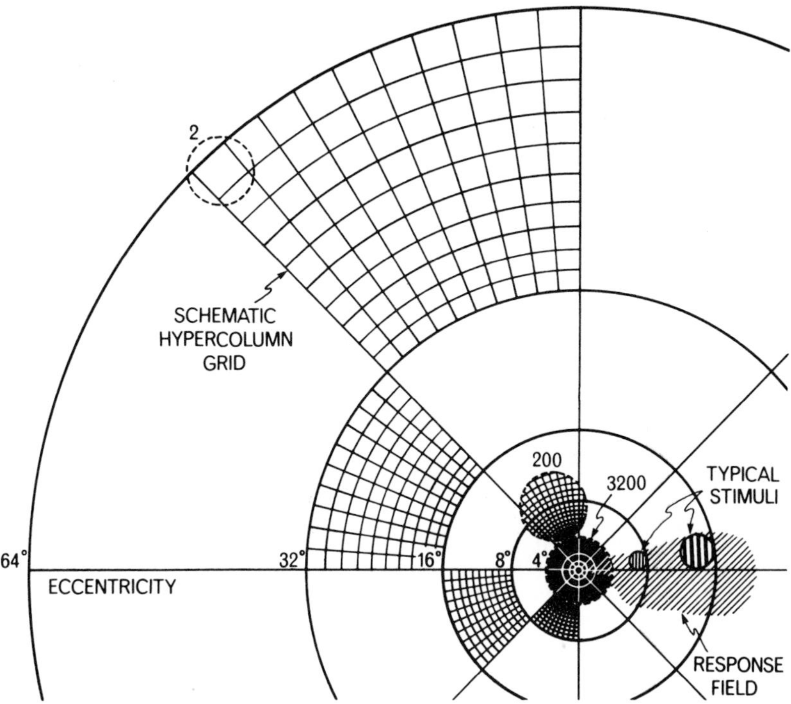

FIGURE 1. Schematic metric of human striate cortex. Radial grid indicates retinal format for projection to linear regions in striate cortex. Each small "square" projects to the estimated area of one hypercolumn structure in the cortex. For other details see text.

shift will reduce the cortical area stimulated to about 2 cm^2 or 200 hypercolumns (a reduction of 16:1) and the eight diameter shift will reduce the area to about 1.5 × 1.5 mm or 2 hypercolumns. This procedure indicates that there need be no specialization for the central retina for pattern evoked potentials beyond the radial transform. Even if each area of cortex has similar responses to its retinal area, the radial mapping will produce a massive preferential response from the central retina in investigations with a fixed field size.

We therefore conducted a study of the effect of retinal eccentricity taking the radial mapping into account. The field sizes for each eccentricity are shown in FIGURE

2 (left panel) and are quadrant annuli with the outer diameter twice the inner diameter. Thus, each field should stimulate about 2 cm^2 of primary visual cortex. It will also stimulate unknown amounts of other cortical areas, but presumably the total amount of cortex stimulated would be about the same for each field. Thus we expected that the amplitude of response for each area should be comparable, although some differences might be expected due to the variable arrangement of the cortical convolutions with respect to the electrodes. The stimuli were a uniform field and a 1 cycle per degree (c/deg) counterphase grating with a high temporal frequency of 80 reversals per second (rps). The pattern evoked potential (PEP) amplitudes obtained are shown in FIGURE 2 (right panel). As can be seen, they are all of comparable size, with no preference for the foveal or macular regions. Thus by scaling the stimulating area according to the radial mapping from retina to cortex, we find that roughly

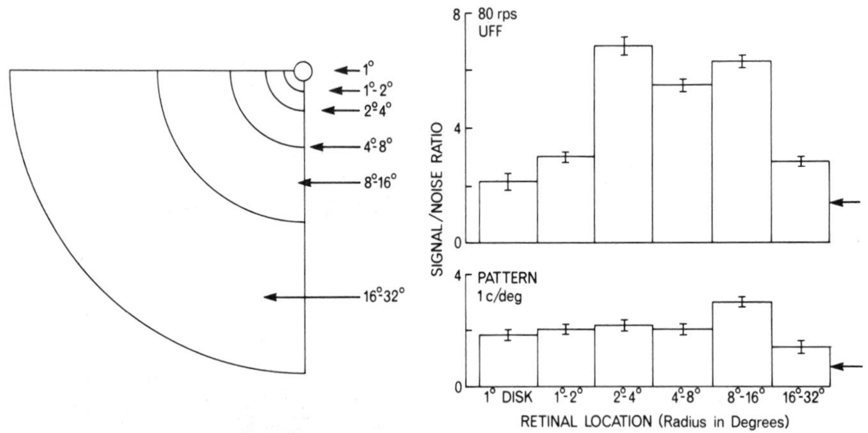

FIGURE 2. (Left) Retinal regions of stimulation for equivalent cortical response, from the fovea to 32° peripheral. (Right) Synchronous evoked responses at 80 reversals per second (rps) (40 Hz stimulation rate), for uniform field flicker (upper histogram) and 1 c/deg pattern stimulation (lower histogram). Noise level indicated by arrow. Note that responses are all of similar amplitude, in support of the radial transform from retina to cortex.

equivalent responses are obtained as far peripherally as 32°. This suggests that equivalent evoked potentials can be recorded from regions of cortex subserving any part of the retina, rather than being specialized for macular regions.

SCALP LOCALIZATION

In a second phase of the localization study, we wished to record the characteristics of the most localized regions of cortex available to our technique. In addition to retinal localization, we used novel types of electrode configuration and recording technique. Although all electrical recording is actually performed between two leads, when using a configuration referred to an electrode on a presumably inactive part of the head, such as the earlobe, it is conventional to call it a monopolar configuration. When both electrodes are on active parts of the scalp, the configuration is designated bipolar. Our localization configuration is based on three electrodes, and hence, we call it a tripolar configuration. The three electrodes are in a line with the active electrode at the center, and the two outer electrodes connected in parallel to the reference channel. In the present work, the three electrodes were closely spaced only 1.5 cm apart.

The advantage of this configuration is that it gives maximum response when the peak of the potential distribution on the scalp lies directly under the central electrode and a strong attenuation for other positions of the distribution. Given the type of potential distribution likely to be generated by a localized dendritic field,[11] the response from the tripolar configuration falls close to zero for potential distributions peaking further than 1 cm away from the central electrode. Broad potential fields derived from more distant sources or a spread of neural activity are also strongly attenuated. Thus, the tripolar electrode configuration provides a high degree of localization in isolating the response from only those local cortical regions giving rise to a potential field peaking within about a 2 cm range on the scalp.

Properties of Local Cortical Responses

Our localization study consisted of measurement of the response to stimulation of a series of small retinal areas located in different regions of the visual field. The retinal areas were circular with a diameter of 1° (in regions between 2° and 4° eccentricity), 2° (from 4° to 8° eccentricity), 4° (from 8° to 16° eccentricity), and 8° (from 16° to 32° eccentricity). Thus the retinal areas stimulated very small cortical regions estimated to vary from a diameter of 4 to 6 mm on the primary visual cortex (or between 12 and 30 hypercolumns), and probably equivalent amounts in the other cortical regions. According to the schematic representation, this corresponds to about 2 cm^2 of primary cortex (together with some presumably equivalent regions of other cortical representations). Under optimal conditions we have obtained reliable responses from an area estimated to stimulate only a 2.5 mm diameter region of striate cortex (5 hypercolumns). We found that with the electrode configuration shown in the inset all the recordable responses occurred within an octant sector from 1°–20° eccentricity (Figure 1). The limited range of response (as described below) is consistent with the notion that the tripolar electrode configuration records only from cortical regions lying immediately beneath the electrodes, although we have no direct means of verifying this supposition at present.

The technique used to obtain information rapidly and to avoid slow state changes in the brain was an electronic spatial frequency sweep method. The brain response to counterphase grating alternation was recorded with a phase-insensitive synchronous filter at a relatively high temporal frequency (28 Hz), while the spatial frequency of the bars in the grating was swept from low to high in a 20 second sweep. This technique is based on an optomechanical device by Regan[12] for which we substituted a more flexible electronic version.[13] The system provides a direct read-out of the entire spatial frequency tuning of the PEP in 20 secs, and thus, even with replications, allows rapid estimates of the spatial frequency tuning to be obtained for a large number of positions of the stimulus in the field.

Almost all the localized fields gave tunings that consisted of a single spatial frequency peak or two separate spatial frequency peaks. Some typical examples are shown in Figure 3 (left, single peaks; and right, double peaks). Thus it seems that local regions of the field give rise to PEP responses limited to a narrow range of bar sizes in the stimulus. This is consistent with recent neurophysiological findings that receptive field properties in a local cortical area are similarly limited to a narrow range of bar sizes.[14,15]

Restricted retinal stimulation also gives rise to a narrow region of response to temporal frequencies of stimulation. FIGURE 4 shows a contour map of response amplitude over spatial frequency and temporal frequency for a single retinal location of 2° diameter at 5° horizontally to the left of the fixation point. (Note that this is a contour map of the stimulus/response characteristics and not of the response profile across the head). The response is largely concentrated at 28 Hz and 5 c/deg, with very little activity for other stimulus ranges.

These responses from local fields suggest that each region of retina has a simple

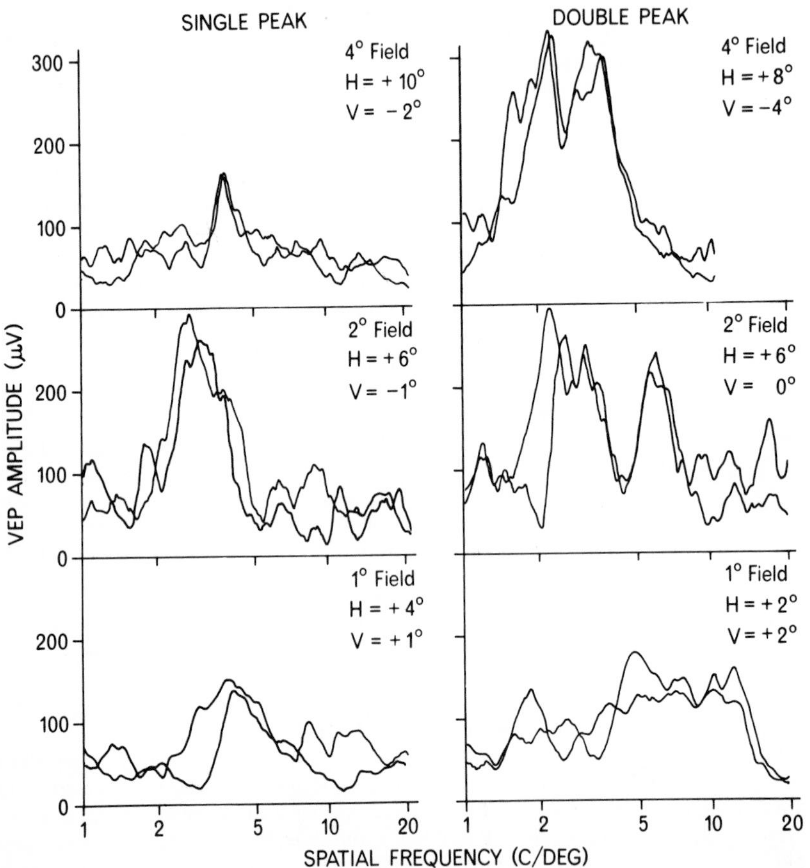

FIGURE 3. Examples of synchronous evoked potentials with localized stimulus swept in spatial frequency from 1 to 20 cycle/deg. Field size and retinal location are indicated for each condition. Two sweeps are shown for each condition, to indicate replicability. Note that the response is limited to a narrow band of spatial frequencies, with noise fluctuations outside these bands. Left column shows responses which exhibited a single peak, while right column shows examples that had a double-peaked form.

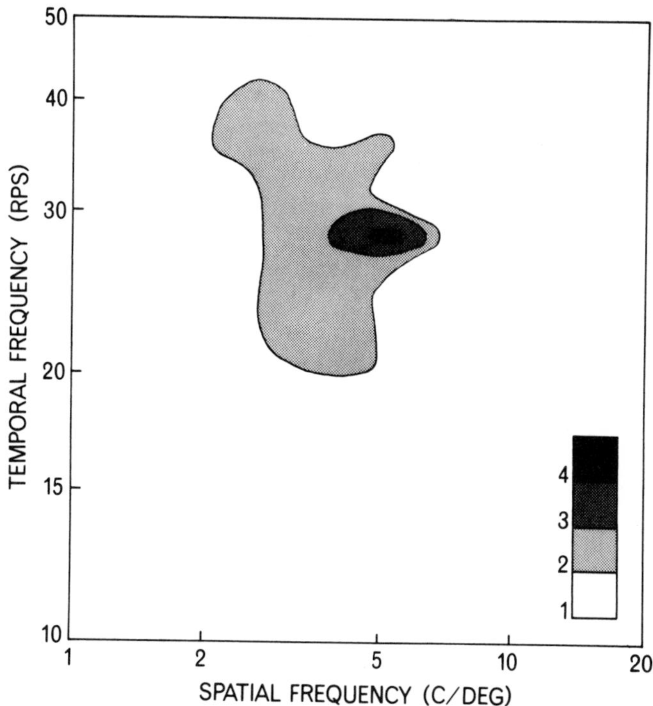

FIGURE 4. Spatiotemporal frequency contour map of the synchronous pattern evoked potential response characteristics from a restricted region of stimulation. Temporal frequency axis is magnified by a factor of 4 relative to spatial frequency axis (double log coordinates). Bar inset indicates signal-to-noise ratios at each grey level. Note that with stimulation localized to 2 cm^2 of striate cortex the response is concentrated in a narrowly tuned region around 28 Hz and 5 c/deg.

resonant-like response in the cortex, and that by selective recording techniques responses can be obtained from one or two local populations of neurons. It is therefore of interest to study the change in response characteristics across retinal regions. To do this, we measured the spatial frequency at which the one or two peak responses from each retinal region occurred. These peak spatial frequencies are plotted in FIGURE 5 in terms of the spatial period (bar width) on the retina as a function of the eccentricity of the retinal region stimulated.

DISCUSSION

The main feature of the data is a strong association between eccentricity and peak spatial period. As the stimulating patch was moved farther from the fovea, the peak spatial period of the response progressively increased. If these results were entirely attributable to the radial transform from retina to cortex, the data would fall on the solid line in FIGURE 5, and the data conform well to this function. (The actual regression line intercepts the x axis at $-2.6°$, compared with an average of 3.0° for data in the literature.

A secondary feature of the data is that the spatial frequency peaks appear to fall into two groups above and below the solid line. This tendency is clarified in the histogram at inset, which depicts the deviation of the peaks from the theoretical function, plotted in terms of the spatial frequency per millimeter of cortex after taking the radial transform into account. The histogram clearly shows a bivariate distribution, with the two peaks about a factor of two apart. Thus, although these two peaks are not found in the response from every retinal location, it appears that there are two distinct populations of neurons giving rise to the responses, tuned to different bar sizes separated by a factor of two.

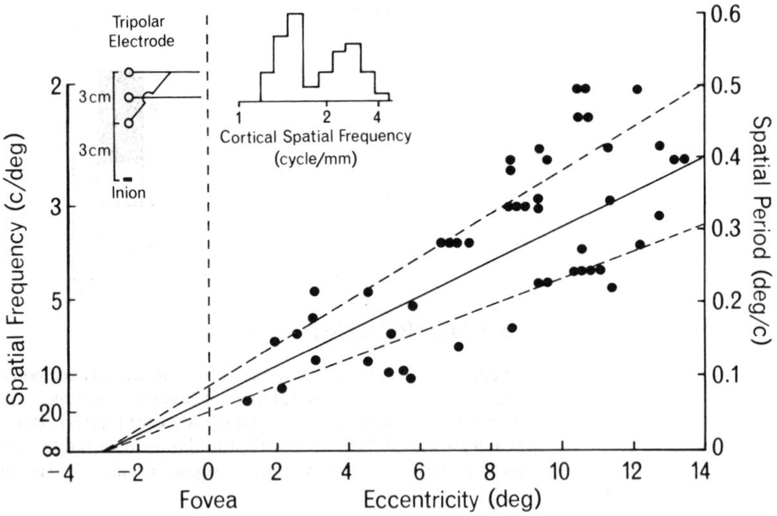

FIGURE 5. Peak retinal spatial frequency of localized responses (filled circles) as a function of retinal eccentricity, obtained from a large number of spatial frequency sweeps from stimuli within the hatched area of FIGURE 1. Left inset indicates tripolar electrode location on scalp. Full line shows constant cortical spatial frequency based on retina/cortex mapping function. Histogram of peaks in terms of cortical spatial frequency (right inset) shows bimodal distribution. Mean of each peak of the cortical spatial frequency distribution is shown as dashed lines in main figure. Note that distribution of response peaks is well described by the theoretical functions.

The fact that many retinal locations give rise to only one spatial frequency peak means that it is very unlikely that the two populations of neurons are co-adjacent in the same cortical regions. The functional isolation implies a spatial separation of the response fields in the cortex. Such a spatial separation could occur if either the "dipoles" for the two spatial frequency peaks were at different orientations, or the two populations are in separate cortical representations. Of these alternatives, the second appears much more likely. The highly convoluted structure of the cortex under the skull presumably ensures that only a patchy distribution of each cortical representation gives rise to recordable responses at a given localized recording site. The separation of the distributions of the two sets of responses thus probably implies that they arise in different cortical regions, for which the best candidates are the first two visual areas, V1 and V2 (corresponding to Brodman's areas 17 and 18).

This speculation is supported by the study in cat visual cortex by Movshon *et al.*[15] who measured the spatial frequency tunings of a large number of neurons of both simple and complex type in V1 and V2. While the simple and complex types had almost identical distributions of peak spatial frequency, neurons from V1 and V2 showed largely nonoverlapping distributions, with peak spatial frequencies separated by about 1.5 octaves (a factor of about 3) at each retinal eccentricity. Thus in cat, the spatial frequency separation occurs between retinal areas rather than between simple and complex cell types. Similar data are not yet available for monkey, but the present results suggest that we are now in a position to study separately the characteristics of responses arising from different visual areas in human, by first selecting the spatial frequency at which the recording is obtained.

Previous work in identifying components of the pattern evoked potential has been based on the scalp distribution of identifiable response peaks. Unfortunately, the two groups who have attempted this analysis[16–18] have come to opposite conclusions on this score. Furthermore, the different scalp distributions of their responses can also be explained on the basis of differing response properties (e.g., rectifying versus nonrectifying) of neurons within the same cortical areas. It therefore appears that this synchronous recording technique is the first to provide a functional separation of the responses from the different visual areas. This opens up an exciting range of possibilities for the localization of functional organization within the human visual system.

References

1. Tyler, C. W., P. Apkarian & K. Nakayama. 1978. Multiple spatial frequency tuning of electrical responses from the human visual cortex. Exp. Brain Res. **33:** 535–550.
2. Apkarian, P. A., K. Nakayama & C. W. Tyler. 1981. Binocularity in the human visual evoked potential: Facilitation, summation and suppression. Electroencephalogr. Clin. Neurophysiol. **51:** 32–48.
3. Fender, D. & T. P. Santoro 1977. Spatiotemporal mapping of scalp potentials. J. Opt. Soc. Am. **67**(11): 1489–1494.
4. Whitteridge, D. & P. M. Daniel. 1961. The representation of the visual field on the calcarine cortex. In The Visual System: Neurophysiology and Psychophysics. R. Jung & H. Kornhuber, Eds. pp. 222–228. Springer-Verlag, Berlin.
5. Hubel, D. H. & T. N. Wiesel. 1974. Uniformity of monkey striate cortex: A parallel relationship between field size scatter and magnification factor. J. Comp. Neurol. **158:** 295–306.
6. Weymouth, F. W. 1958. Visual sensory units and the minimal angle of resolution. Am. J. Ophthalmol., **46:** 102–113.
7. Rovamo, J. & V. Virsu. 1979. An estimation and application of the human cortical magnification factor. Exp. Brain Res. **37:** 495–510.
8. Schwartz, E. L. 1980. Computational anatomy and functional architecture of striate cortex: a spatial mapping approach to perceptual coding. Vis. Res. **20:** 645–669.
9. Hitchcock, P. F. & T. L. Hickey. 1979. Banding pattern in human striate cortex as demonstrated by reduced silver stain. Invest. Ophthalmol. Vis. Sci. (Suppl.) **17:** 157–8.
10. Riggs, L. A. & B. R. Wooten. 1972. Electrical measures and psychophysical data on human vision. *In* Handbook of Sensory Physiology. Vol. 7(4). Human Psychophysics D. Jameson and L. M. Hurvich, Eds. pp. 690–731. Springer-Verlag, Berlin.
11. Jeffreys, D. A. 1971. Cortical source locations of pattern-related visual evoked potentials recorded from the human scalp. Nature **229:** 502–504.
12. Regan, D. 1975. Recent advances in electrical recording from the brain. Nature **253:** 401–407.

13. TYLER, C. W., P. A. APKARIAN, K. NAKAYAMA & D. M. LEVI. 1979. Rapid assessment of visual function: An electronic sweep technique for the pattern VEP. Invest. Ophthalmol. Vis. Sci. **18:** 703–713.
14. MAFFEI, L., & A. FIORENTINI. 1973. The visual cortex as a spatial frequency analyzer. Vis. Res. **13:** 1255–1267.
15. MOVSHON, J. A., I. D. THOMPSON & D. J. TOLHURST. 1978. Receptive field organization of complex cells in cat's striate cortex. J. Physiol. **283:** 79–99.
16. JEFFREYS, D. A., & V. G. AXFORD. 1972. Source locations of pattern-specific components of human visual evoked potentials. I. Component of striate cortical origin. Exp. Brain Res. **16:** 1–21.
17. JEFFREYS, D. A. & J. G. AXFORD. 1972. Source locations of pattern-specific components of human visual evoked potentials. II. Component of extrastriate origin. Exp. Brain Res. **16:** 22–40.
18. HALLIDAY A. M. & W. F. MICHAEL. 1970. Changes in pattern-evoked responses in man associated with the vertical and horizontal meridians of the visual field. J. Physiol. **208:** 499–513.

SCALP AND CORTICAL RECORDINGS OF INITIAL SOMATOSENSORY CORTEX ACTIVITY TO MEDIAN NERVE STIMULATION IN MAN*

Truett Allison

Neuropsychology Laboratory
Veterans Administration Medical Center
West Haven, Connecticut 06516

Department of Neurology
Yale University School of Medicine
New Haven, Connecticut 06510

Progress is being made in clarifying the neural origins of potentials generated in the human somatosensory system by stimulation of peripheral nerves. However, there is disagreement concerning the identification and origins of the initial activity generated in the somatosensory cortex following median nerve stimulation. Here I review some current work relevant to this issue.

In both scalp and cortical surface recordings, the predominant potentials in the 20–30 msec latency range are contralateral and consist of an N20–P30 sequence recorded in the parietal area and a P20–N30 sequence recorded in the frontal area. We have concluded, as did Broughton,[1] that these potentials are generated by a single contiguous dipolar source layer located in the posterior bank of the central sulcus in area 3b (FIGURE 1A). Various lines of evidence favoring this conclusion have been discussed elsewhere.[2,3] One important type of evidence comes from cortical surface recordings like those shown in FIGURE 1B. The N20–P30 sequence was recorded posterior to the central sulcus, and anterior to it potentials of similar latency and waveform but of opposite polarity (P20–N30) were recorded. These potentials are large (50–100 μV) and have steep spatial gradients confined to the region of the central sulcus, strongly suggesting that this activity is generated either by one bank of the central sulcus or by oppositely activated generators in the pre- and post-central gyri.

It is also relevant to consider scalp recordings of these potentials (FIGURE 2). FIGURE 2A is a scalp topography at the peak of P30–N30. Note the general similarity of the potential field to that obtained from the cortical surface, but note also that the amplitude peaks are near P_4 and F_4, well posterior and anterior to sensorimotor cortex. FIGURE 2B shows a similar result for N20 and P20. A third example is shown in FIGURE 2C; again note that the N20–P30 and P20–N30 potentials (here labeled according to the actual latencies in each subject) are largest in parietal and frontal regions, respectively. In other words, there is a large discrepancy between scalp and cortical recordings in the location of maximum amplitude of these potentials. In cortical recordings they are near the central sulcus, whereas in scalp recordings they are well anterior and posterior to it. Any hypothesis regarding the generators of these potentials needs to account for this disparity.

FIGURE 3 summarizes two models currently proposed to account for these potentials. FIGURE 3, lower left, illustrates the model of FIGURE 1A and for

*This work was supported by the Veterans Administration and by the National Institute of Mental Health (Grant 05286).

0077-8923/82/0388-0671 $1.75/0 © 1982, NYAS

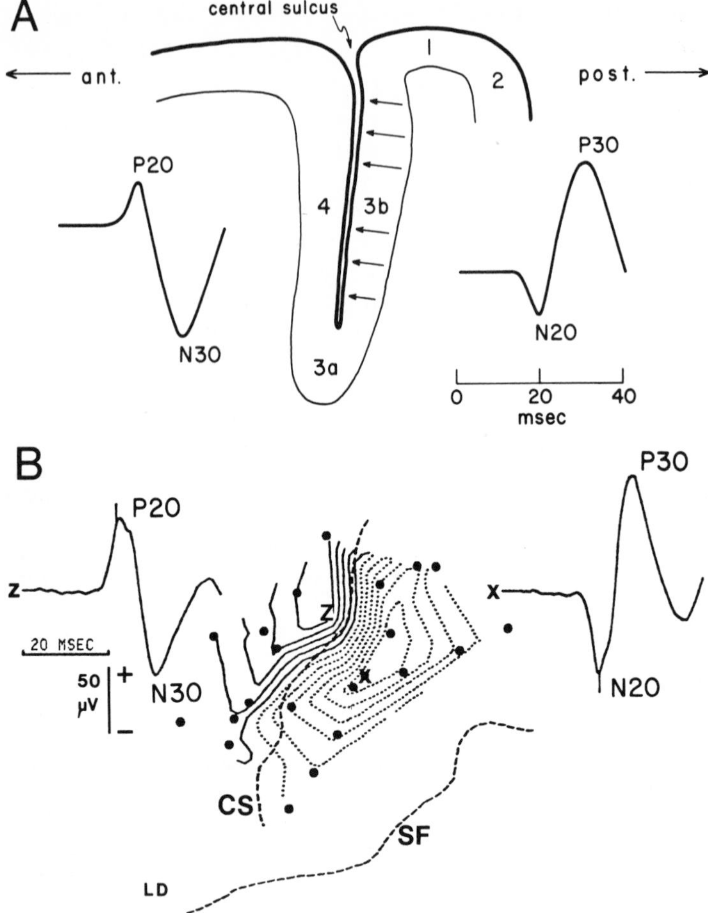

FIGURE 1. (A) Schematic short-latency potentials recorded at locations anterior and posterior to the central sulcus, and their hypothesized source layer in the posterior bank of the central suclus in area 3b. (B) Cortical surface recordings from electrode locations in relation to central sulcus (CS) and Sylvian fissure (SF) determined from photographs made during surgery in an awake but sedated patient undergoing removal of a left frontal lobe tumor; right median nerve stimulation. A computer-generated isopotential map (solid lines denote positive potential; dashed lines, negative) shows the distribution of activity at the latency (20 msec) indicated by the cursors.

convenience may be called a single-source model. Others concluded that these potentials are generated separately in post- and pre-central cortex;[4–7] this may be called a dual-source model (FIGURE 3, lower right). FIGURE 3, top row, shows normalized potential fields calculated in the saggital plane at the cortical surface and at the scalp. Both models have been adjusted to produce the observed highly focal cortical surface potentials (FIGURE 1B) whose positive and negative peaks are located approximately 5° (about 8 mm) anterior and posterior to the central sulcus. Note that

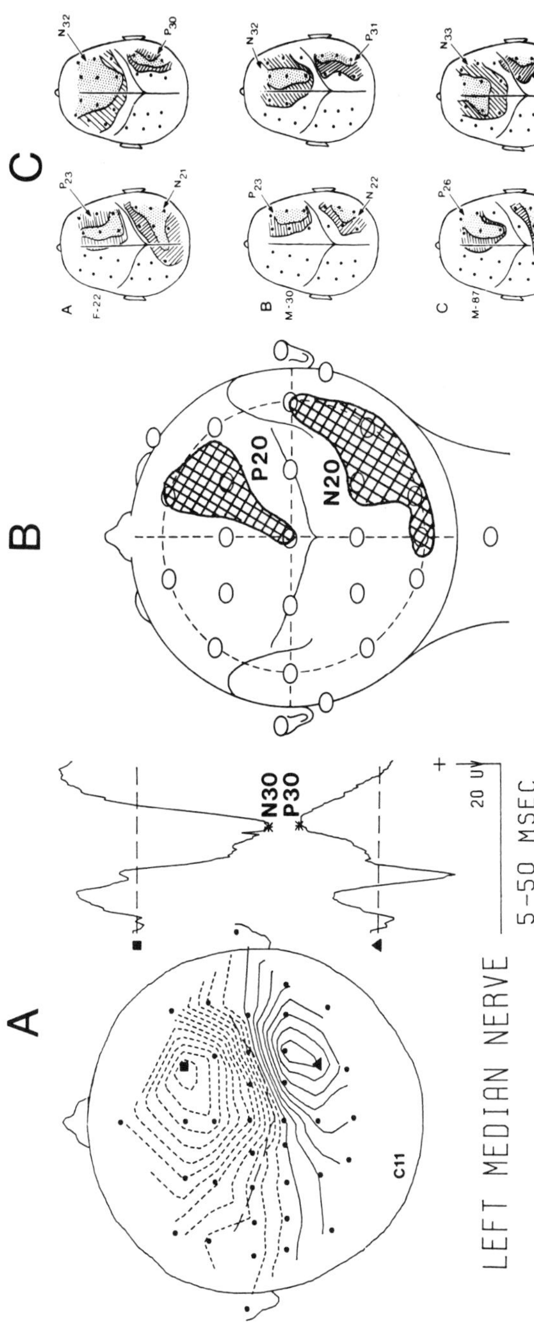

FIGURE 2. Scalp topography of potentials evoked by left median nerve stimulation. (A) Isopotential plot at peak (asterisk) of P30–N30 potentials; traces are from F_4 (square) and P_4 (triangle). (B) Topography of N20 and P20 potentials. Hatched area is enclosed by 50% of maximal amplitude isopotential line; group data from 12 subjects. (Adapted from Goff *et al.*[2]) (C) Topographies of N20–P20 and P30–N30 potentials in three subjects. (From Desmedt & Cheron.[6] By permission of *Electroencephalography and Clinical Neurophysiology*.)

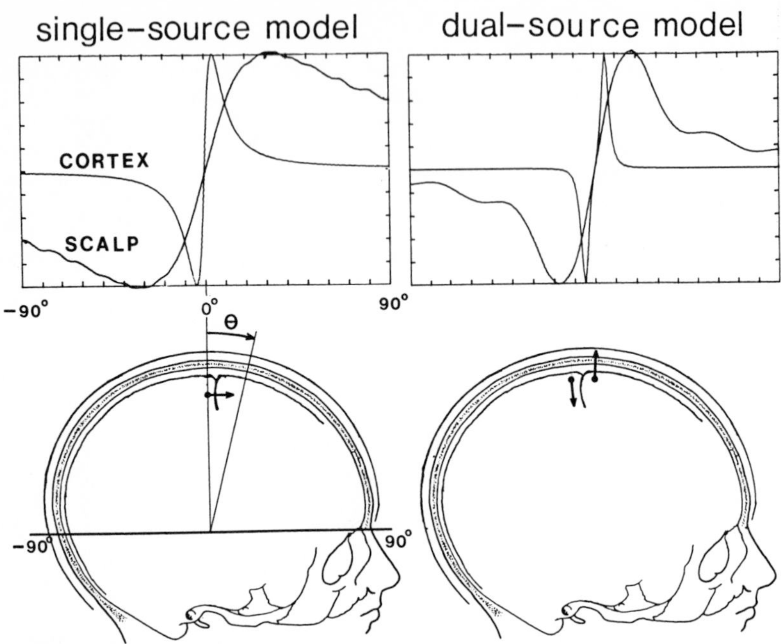

FIGURE 3. Computer-predicted cortical surface and scalp field potentials produced by two hypothetical sources: a single-source model assuming a tangential dipolar source located in area 3b, and a dual-source model assuming pre- and post-central radial dipolar sources. Cortical potentials are based on a homogeneous single-sphere head model (simulating craniotomy conditions), while scalp potentials are based on a 3-sphere head model (simulating intact skull and scalp conditions); see Darcey[18] for details. Simulation is for the N20–P20 case; for the P30–N30 components the calculated potential fields would have the same shape but opposite polarity.

the single-source model predicts scalp amplitude peaks that are displaced posteriorly and anteriorly by about 30° compared with the cortical recordings; these locations correspond roughly to P_4 and F_4 and agree well with the scalp topographies of FIGURE 2. In contrast, the dual-source model predicts surface and scalp foci which are displaced by only about 12°. To account for the observed scalp topographies it is necessary to move the locations of the two generators to cortex underneath P_4 and F_4, in which case they will no longer produce an appropriate cortical topography. This is also highly unlikely since cortical surface recordings, both in man and monkey,[1,3,4,8–12] show that large potentials in this latency range are recorded only from somatomotor cortex. We therefore conclude that the single-source model is better able to account for both the observed cortical surface and scalp topographies.

The most common objection to the single-source model is that there are often differences in latency of 1–2 msec between the peaks of the precentral and postcentral waveforms (e.g., FIGURE 2C). Hence, it is argued, the potentials cannot be generated by a single dipolar source buried in the central fissure. How damaging is this objection? The recording of FIGURE 4A is from area 1 of the postcentral gyrus and shows that the surface positivity (P12) is about 2 msec later than the peak of the negativity recorded in white matter. In the recording of FIGURE 4B the surface positivity is about 2 msec earlier than its negative counterpart in white matter. These laminar recordings are representative of those reported by others. The observed latency differences indicate either that concurrent activity generated outside area 3b (for example, the human P25 and related activity generated in areas 1 and 4[3,9]) makes some contribution to the area 3b potential fields, or that some feature of the spatiotemporal activation of area 3b neurons itself is responsible for the imperfect polarity inversion from surface to white matter. In either case, similar differences of 1–2 msec in peak latency between the anterior ("surface") and posterior ("white matter") waveforms in human recordings are not sufficient grounds to reject the single-source model as the primary generator of the N20–P20 and P30–N30 potentials. A crucial test of the contribution of area 3b to these potentials would be to record transcortically from area 3b in man. This cannot be done, but in monkeys such

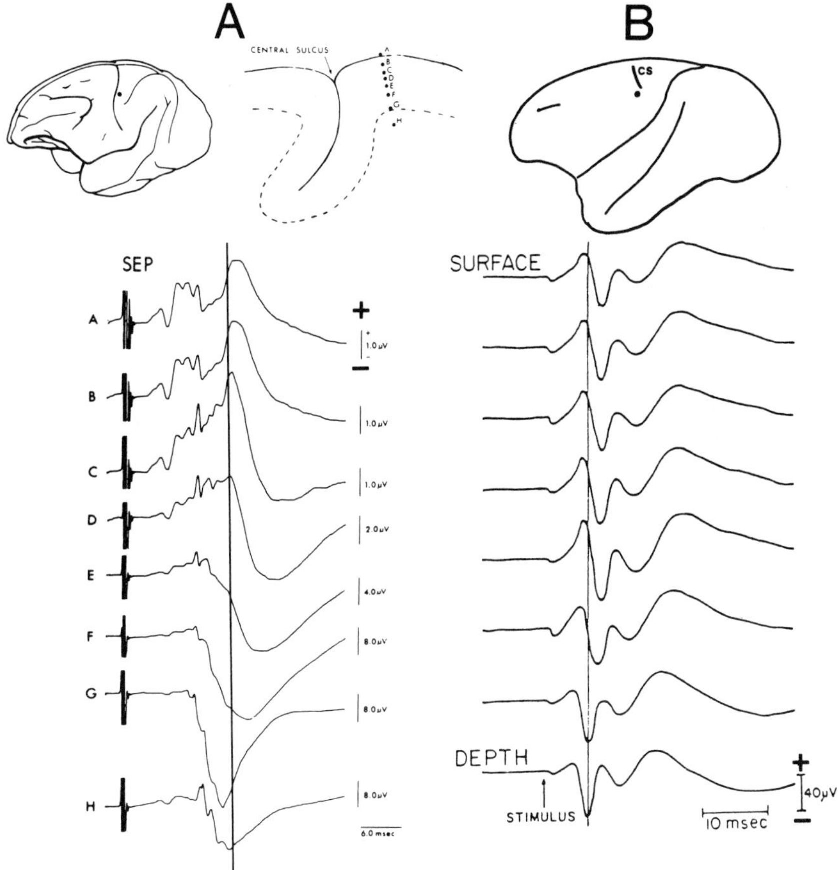

FIGURE 4. Laminar recordings from the hand area of the somatosensory cortex in monkeys. (A) Recordings from area 1 of rhesus monkey; isolatency line at 16 msec. (After Arezzo *et al.*[11]) (B) Recordings from area 3b of squirrel monkey (in this species area 3b extends anterior and lateral to the shallow central sulcus, CS). Isolatency line at 7 msec. (After Wood & Allison.[17])

recordings show polarity inversion of frontal P10–N20 ("surface") and parietal N10–P20 ("white matter") potentials which likely correspond, respectively, to the human P20–N30 and N20–P30 sequences.[11,12]

The dual-source model must postulate neurophysiological processes that simultaneously produce frontal and parietal surface potentials of opposite polarity. Various patterns of neuronal excitation and inhibition could be assumed which would produce such potential fields. However, the single-source model need only assume a polarity inverting positive-negative sequence of the type reported in animal studies (e.g., FIGURE 4) recorded from the surface of area 3b (and hence anterior to the central sulcus) and its polarity-inverted counterpart in white matter (and hence posterior to the sulcus).

Both the single-source and dual-source models assume that N20 is generated in somatosensory cortex. However, some groups have suggested that N20 is of subcorti-

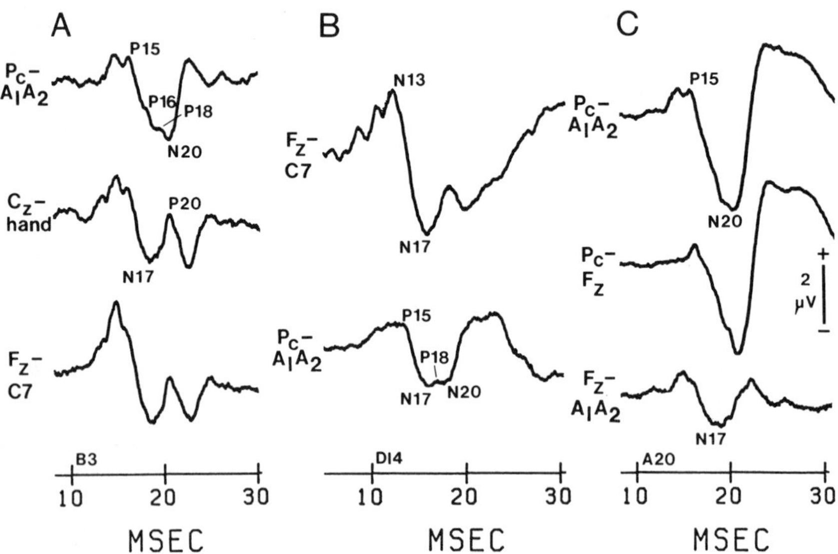

FIGURE 5. Scalp recordings in three subjects to illustrate differentiation of short-latency parietal scalp negativity into N17 and N20 fractions.

cal origin.[4,13–15] FIGURE 5A, top trace, shows a typical recording from a parietal scalp-linked ear derivation. P15 is followed by a scalp negativity which terminates in N20 and on which is superimposed two deflections labeled P16 and P18. In a C_Z–hand derivation, a scalp-negative potential is seen, which in most subjects has a peak latency between P16 and P18 and is thus labeled N17 (FIGURE 5A, middle trace). In this derivation N17 is followed by a small P20, since C_Z is anterior to the central sulcus. It is difficult to obtain good scalp topographies of N17 using a noncephalic reference because of muscle and stimulus artifacts, but it appears to have a broad scalp distribution as suggested by its presence at F_Z (FIGURE 5A, bottom row). FIGURE 5B, lower trace, is an example of a subject in whom N17 is unusually large (in this case even larger than N20) in the P_C–A_1A_2 derivation; here the scalp negativity clearly appears to be a mixture of the N17 and N20 potentials. In the subject of FIGURE 5C,

N17 is not apparent in a P_C–A_1A_2 derivation, but its presence can be inferred as follows: If it is the case that N17 is essentially equipotential at frontal and parietal scalp, a trans-Rolandic derivation (FIGURE 5C, middle trace) will eliminate N17 but sum the N20–P20 potentials; subtracting the middle from the upper trace yields an equivalent F_Z–A_1A_2 recording that reveals N17 (FIGURE 5C, lower trace). Thus the parietal scalp negativity can be decomposed into N17 and N20 fractions. Kimura at this conference reached the same conclusion.[16] Because of its broad scalp distribution and shorter latency, N17 is likely of subcortical origin, whereas N20 is of cortical origin as discussed above. N17 may contribute to the N2 potential described as having a broad scalp distribution and surviving lesions of somatosensory cortex.[14] Some authors label as N20 the potential here labeled N17. It is likely that differing opinions concerning the topography, origin, and pathological alteration of "N20" are in part due to the failure of workers, ourselves included, to distinguish what appear to be two separable parietal scalp-negative potentials in the 15–25 msec latency range.

In summary: (a) In cortical surface recordings, which are dominated by locally generated potentials, a dipolar source in area 3b provides a parsimonious explanation of the N20–P20 and P30–N30 potentials. (b) In scalp recordings, the initial potentials generated by the area 3b source (N20–P20) are superimposed upon an earlier, broadly distributed negativity that is probably of subcortical origin.

I thank T. Darcey, W. R. Goff, G. McCarthy, P. D. Williamson, and C. C. Wood for their collaboration in much of the work described here.

REFERENCES

1. BROUGHTON, R. J. 1969. *In* Averaged Evoked Potentials. E. Donchin & D. B. Lindsley, Eds. NASA SP-191:79–84. U.S. Government Printing Office, Washington, D.C.
2. GOFF, G. D., Y. MATSUMIYA, T. ALLISON & W. R. GOFF. 1977. Electroencephalogr. Clin. Neurophysiol. **42:** 57–76.
3. ALLISON, T., W. R. GOFF, P. D. WILLIAMSON & J. C. VANGILDER. 1980. *In* Progress in Clinical Neurophysiology. J. Desmedt, Ed. Vol. **7:** 51–68. Karger, Basel, Switzerland.
4. CELESIA, G. G. 1979. Arch. Neurol. (Chic.) **36:** 399–405.
5. CRACCO, R. Q. 1980. *In* Progress in Clinical Neurophysiology. J. Desmedt, Ed. Vol. **7:** 1–14. Karger, Basel, Switzerland.
6. DESMEDT, J. E. & G. CHERON. 1980. Electroencephalogr. Clin. Neurophysiol. **50:** 404–425.
7. PAPAKOSTOPOULOS, D. & H. J. CROW. 1980. *In* Progress in Clinical Neurophysiology. J. Desmedt, Ed. Vol. **7:** 15–26. Karger, Basel, Switzerland.
8. JASPER, H., R. LENDE & T. RASMUSSEN. 1960. J. Nerv. Mental Dis. **130:** 526–537.
9. STOHR, P. E. & S. GOLDRING. 1969. J. Neurosurg. **31:** 117–127.
10. WOOLSEY, C. N., T. C. ERICKSON & W. E. GILSON. 1979. J. Neurosurg. **51:** 476–506.
11. AREZZO, J., A. D. LEGATT & H. G. VAUGHAN, JR. 1979. Electroencephalogr. Clin. Neurophysiol. **46:** 155–172.
12. AREZZO, J., H. G. VAUGHAN, JR. & A. D. LEGATT. 1981. Electroencephalogr. Clin. Neurophysiol. **51:** 1–18.
13. FUKUSHIMA, T., Y. MAYANAGI & G. BOUCHARD. 1976. Electroencephalogr. Clin. Neurophysiol. **40:** 481–490.
14. GOLDIE, W. D., K. H. CHIAPPA, R. R. YOUNG & E. B. BROOKS. 1981. Neurology **31:** 248–256.
15. KRITCHEVSKY, M. & W. C. WIEDERHOLT. 1978. Arch. Neurol. **35:** 706–711.
16. KIMURA, J. 1982. Ann. N.Y. Acad. Sci. This volume.
17. WOOD, C. C. & T. ALLISON. 1981. Canad. J. Psychol./Rev. Canad. Psychol. **35:** 113–135.
18. DARCEY, T. 1979. Ph.D. Dissertation, California Institute of Technology. University Microfilm Order No. 7920351.

Discussion of the Paper

R. Cracco: Many of the potentials you show are characterized by progressive changes in peak latency over the scalp which suggests, at least to me, that these potentials arise from multiple generators which are activated closely but not simultaneously in time.

T. Allison: Most of us look at peak latencies. We tend to do it easily and naturally by visual inspection. However, whenever the instantaneous voltage is contributed to by multiple sources as often occurs with both the cortical and subcortical SEPs, the significance of peak latency must be interpreted with caution.

THE CONTRIBUTION OF AFFERENT FIBER TRACTS TO THE SOMATOSENSORY EVOKED POTENTIAL*

Joseph C. Arezzo and Herbert G. Vaughan, Jr.

Departments of Neuroscience and Neurology
Rose F. Kennedy Center for Research
in Mental Retardation and Human Development
Albert Einstein College of Medicine
Bronx, New York 10461

The short-latency components of the somatosensory evoked potential (SLSEP) provide an index of the integrity of the neuraxis from peripheral nerve to cortex. The full clinical utility of this technique is dependent on the degree to which each component can be ascribed to an anatomical structure and related to a portion of the ongoing neuroelectric activity in this region. Identification of the intracranial and intraspinal sources of the surface SLSEP components is a formidable task, which requires a three-dimensional analysis of the field potential distributions and the simultaneously recorded multiple unit activity. For a description of the techniques of source identification see the review by Vaughan in this volume. The results of our mapping studies in the monkey, following both median and peroneal nerve stimulation have been previously reported[1–4] and the following is a brief summary and discussion of these results.

Using a noncephalic reference, a series of five small positive components, which precede the initial cortical activity, were identified overlying the contralateral sensorimotor cortex. The latencies of the components are dependent on the length of the conduction pathways and varied with the stimulated nerve, point of stimulation and size of the animal; however, the response waveform was similar following either hindlimb or forelimb stimulation. On the basis of timing and topography the source of SLSEP1 was localized to the proximal portion of the peripheral nerve, including the dorsal root fibers. The dorsal roots are angled relative to the ascending spinal tracts and SLSEP1 is therefore seen as a "near-field" component overlying the spinal zone of root entry and as a "far-field" potential overlying rostral spinal and brainstem sites. At the vertex this potential is small and detected only as an inflection on the rising edge of SLSEP2. SLSEP2 is recorded as a traveling wave overlying the spinal cord rostral to the segments of root entry and as a fixed latency component above the level of the dorsal column nuclei. This component reflects propagated activity of the most rapidly conducting axons within the ascending spinal tracts, principally the dorsal columns. The administration of the neurotoxin acrylamide, which initially damages the preterminal axons in the gracile nucleus, distorts SLSEP2 following peroneal nerve stimulation, but leaves the response to median nerve stimulation unaffected. SLSEP3 and SLSEP4 are recorded as large potentials in the ventral thalamus, but each of these components can be traced with decreasing latency to caudal sites in the brainstem. The topography of the components suggest that they represent sequential activity in the mesencephalic and diencephalic portions of the medial lemniscus. SLSEP5 can be traced at increasing latencies from the thalamus to the sensorimotor

*This work was supported by Grants HD 01799, MH 06723, and ES 02168 from the U.S. Public Health Service.

0077–8923/82/0388–0679 $1.75/0 © 1982, NYAS

cortex and thus represents the activity within the thalamocortical radiations. The peak latency at the surface is the same as that of the activity in the proximal portion of the pathway as it enters the cortex. SLSEP5 is recorded only over the contralateral sensorimotor region and is seen as a positive potential at both pre- and post-central recording sites. The initial cortical activity is generated in area 3b within the posterior bank of the central sulcus. Because of the orientation of this region, its field is volume conducted in an antero-posterior direction. Additional cortical activity is initiated within milliseconds in Brodmann's areas 1, 2 and 4.

Our results indicate that the principal and probably the sole sources of the phasic surface-recorded SLSEP are synchronized volleys of action potentials within spinal, brainstem, and cerebral afferent fiber tracts. The common orientation of axons within fiber bundles, and the relative synchrony of activity within the largest diameter fibers of these pathways results in directionally coherent sources yielding volume-conducted potentials that are maximum in amplitude along vectors defined by the orientation of the fiber tracts. An advancing wavefront of axonal depolarization is seen as a monophasic positivity in recordings distal to the tract.[5] These brief positive components can be traced with decreasing amplitude to the surface of the brain. In contrast, graded post-synaptic potentials of dendritic and somatic origin are confined to the immediate vicinity of the relay nuclei. These post-synaptic potentials can be large in the near field, but, because of the lack of a common geometry, current flows diminish rapidly with distance from their generators[6] and they are therefore, not a significant source of surface-recorded SLSEP. The major contribution of afferent fiber tracts to the short-latency components of subcortical origin has been confirmed in the auditory system.[7]

The generation of surface-recorded short-latency EP components within fiber tracts has several important implications for their analysis and interpretation in clinical applications:

(1) Afferent fiber tracts exhibit varying geometry with regard to a fixed surface recording site. As the orientation of the fiber tract changes along its course, the contribution of specific portions of the pathway to the surface SLSEP is enhanced or diminished. Thus, for example, the decussation of the medial lemniscus in the medulla, having an oblique orientation, makes little or no contribution to the vertex potential, whereas activity in the rostral portion of the pathway ascending towards the thalamus is volume conducted to the dorsolateral surface of the brain. It is possible, due to changes in direction of a single fiber tract for its activity to be manifest at the surface as "components" with disparate timing and distribution.

(2) Potentials recorded within or adjacent to fiber tracts have latencies that vary with the position of the recording electrode along the pathway. The surface-recorded short-latency EP components represent summation of a traveling wave weighted by the proximity of the generating segments to the recording site. The peak latency of a surface component corresponds to the depolarization at the proximal end of an ascending tract, whereas the component onset reflects volume-conducted activity from the caudal segments of the pathway. Thus, the timing of activity seen in the near field is not identical to the potentials recorded distal to the termination of the fiber tract. These time relationships can only be observed by the systematic tracing of components along the course of the tracts that generate them to the surface.

(3) The segregation of the compound action potentials due to differential conduction velocity of axons within each tract results in temporal overlap of potentials from sequential segments of the afferent pathways which summate in far-field recordings. A cardinal feature of the somatosensory system is the great length of the afferent pathways. As a result the compound activity differentiates into a number of subcom-

ponents. In the rostral segments of the brainstem pathways, the activity can continue for up to 10 msec following the initial discharge and consists of between 10 and 12 positive subpeaks. Because of the presence of this sequence of subcomponents within different segments of the afferent pathways, each of the surface SLSEP components is at once coincident with potentials generated within spinal, brainstem, and diencephalic sources. For this reason, temporal coincidence between surface and intracranial potentials is an insufficient criterion for source identification. Since intracranial recordings in man are necessarily limited to a relatively small number of recording sites, these studies must be interpreted with extreme caution.

(4) The longer latency components of the caudal generators are volume conducted with approximately the same amplitude gradient and vectors as the initial discharge at each source. Most of these components are, however, masked by the temporally overlapping activity at the surface. Damage to rostral structures can eliminate or alter the contribution of the pathways central to the lesion, while unmasking concurrent activity generated within more caudal structures.

(5) With added distance of the active source from the recording site the amplitude gradients of the far-field components are increasingly shallow and decreasingly affected by differences in tissue impedance. Thus, the amplitude of the initial components generated within the cervical cord shows only slight changes in intracranial recordings from pons to cortical surface and only a 10% reduction from cortex to scalp. By contrast, the potentials generated within thalamocortical radiations undergo rapid decrement to the cortical surface and are diminished by over half their amplitude from cortical surface to scalp. This effect tends to reduce the contribution of the thalamocortical radiations to scalp-recorded SLSEP relative to the more caudal far-field potentials.

(6) Accurate representation of the SLSEP waveform requires the use of a noncephalic reference. Since far-field components are attenuated very slowly with increasing distance from their source, SLSEP1 and SLSEP2 are approximately isopotential over the entire dorsal surface of the head. Therefore, the use of a neck, ear, or frontal reference precludes the identification of SLSEP1 and diminishes or distorts the remaining components as well.

Although a cephalic reference site has been employed in some studies of the human SLSEP, this practice cannot be recommended in view of the difficulty of interpreting the waveform derived from differential recordings from two active sites. The wrist contralateral to the stimulated nerve is a more appropriate reference since this location is distant from the active tissue and is perpendicular to the axis of maximal field potentials volume conducted within the neuraxis.

The foregoing considerations have important implications for the clinical application and interpretation of short-latency evoked potential components in both the somatosensory and auditory modalities. It is evident from our detailed intracranial studies in the monkey that simple interpretations of short latency components in terms of single intracranial generators are unwarranted. Further studies of the impact of localized lesions on the short-latency EP components are required to adequately define the effects of specific intracranial pathology on EP morphology.

Acknowledgment

The authors wish to acknowledge the valuable assistance of James Barna, Chester Freeman, and Linda Murphy.

References

1. Arezzo, J. C., A. D. Legatt & H. G. Vaughan, Jr. 1979. Topography and intracranial sources of somatosensory evoked potentials in the monkey. I. Early components. Electroencephalogr. Clin. Neurophysiol. **46:** 155–172.
2. Arezzo, J. C., H. H. Schaumburg, P. S. Spencer & H. G. Vaughan, Jr. 1980. A novel approach for the detection of toxic distal axonopathy: The brainstem somatosensory evoked potential. Ann. Neurol. **8:** 96.
3. Arezzo, J. C., H. G. Vaughan, Jr. & A. D. Legatt. 1981. Topography of intracranial sources of somatosensory evoked potentials in the monkey. II. Cortical components. Electroencephalogr. Clin. Neurophysiol. **51:** 1–18.
4. Arezzo, J. C., H. H. Schaumburg, H. G. Vaughan, Jr., P. S. Spencer & J. Barna. 1981. Hindlimb somatosensory evoked potentials in the monkey: The effects of distal axonopathy. Ann. Neurol. (in press).
5. Woodbury, J. W. 1965. Potentials in a volume conductor. *In* Neurophysiology. (edit.) T. C. Ruch, H. D. Patton, J. W. Woodbury & A. L. Towe, Eds. pp. 85–91. Saunders, Philadelphia, Pa.
6. Lorente dé Nó, R. 1947. A Study of Nerve Physiology. Studies from the Rockefeller Institute, Vol. 132, Chap. 16. New York, N.Y.
7. Legatt, A. D. 1981. Short-latency auditory evoked potentials in the monkey. Ph.D. dissertation submitted to the Albert Einstein College of Medicine, Yeshiva University.

Discussion of the Paper

C. Wood: You ascribe these short-latency potentials to fiber tract activity and not to post-synaptic potentials. I find it very hard to believe that the entire waveform of P5.3, 6.2, and 7.2 including the envelope, that is to say the entire difference between baseline and peak, is not contributed to by post-synaptic activity.

J. C. Arezzo: The results of our intracranial studies indicate that the contribution of post-synaptic potentials to the short latency waveform, recorded in the far field, is negligible. For the most part, post-synaptic potentials are generated in cellular aggregates which possess a closed field configuration. There is little volume conduction of the activity beyond the boundaries of active tissue. What you refer to as the *envelope* is indeed the summation of components from disparate sources, but the activity is always traceable to propagated action potentials in fiber tracts. The longer latency volleys from caudal pathways contribute to the total deviation from baseline underlying the initial components from the rostral generators.

THE SOMATOSENSORY EVOKED POTENTIAL IN PATIENTS WITH SEVERE HEAD INJURY: OUTCOME PREDICTION AND MONITORING OF BRAIN FUNCTION*

Richard P. Greenberg,† Pauline G. Newlon, and Donald P. Becker

Division of Neurological Surgery
Medical College of Virginia
Richmond, Virginia 23298

Brain dysfunction and/or damage following severe human head trauma is caused both by primary mechanical tissue injury that occurs at the moment of impact and by secondary brain insults that may occur at different time periods following the head injury. Secondary insults that may affect the brain can be either intracranial, such as mass lesions, elevated intracranial pressure (ICP), brain ischemia, and meningitis, or systemic, such as septicemia, pneumonia, hyponatremia, and so on. It is often difficult in head injury patients to clinically evaluate brain dysfunction, predict outcome, or assess the efficacy of various treatments (barbiturate coma, hypothermia, etc.) because of the disseminated areas of CNS compromise, the various mechanisms of injury, and the temporal dispersion of secondary insults. The clinician's inability to communicate with comatose head trauma patients adds to the problem of evaluating brain function and limits the informative value of the neurological examination.

We have previously reported our results using multimodality (visual, auditory, and somatosensory) evoked potentials (MEP) recorded early after head injury and then serially as a clinical adjunct for prediction of final patient outcome and to monitor brain function.[1-3] In this communication, we evaluate the use of somatosensory evoked potential (SEP) data alone to forecast final outcome and evaluate the effects of secondary clinical insults on the brain function of 109 comatose, severe head injury patients. We also compare the prognostic value of SEP with that of the clinical examination, intracranial pressure (ICP) measurements, and computerized tomographic (CT) scan data in these patients.

Clinical Materials and Methods

Patient Population and Management

A total of 109 comatose, severely head injured patients, 24 female and 85 males, ranging in age from 5–70 (mean age: 30) who were admitted to our intensive care unit between 1977 and 1980, comprised the study group. On admission, all patients were comatose. None made any verbal response or motor movement to verbal command, but all exhibited evidence of neurological function, i.e., spontaneous respirations or better. Forty-two patients had space-occupying lesions, 40 of which required surgery. As previously described, patients were prospectively entered into a head injury center clinical management protocol and received standardized treatment.[4]

*This work was supported in part by Grant NS-12587 from the National Institute of Neurological and Communicative Diseases and Stroke (NINCDS).

†Recipient of NINCDS Teacher Investigator Award K07-NS346.

0077–8923/82/0388–0683 $1.75/0 © 1982, NYAS

Neurological Examination, ICP, CT Scan

The emergency room neurological examinations were performed by a neurosurgeon, while follow-up neurological examinations were performed by neurologists. The clinical features employed in data analysis were age, Glasgow coma score (GCS), pupillary response to light, eye movements, motor posturing, flaccidity, and surgical mass lesions.[5] Patient outcomes were classified at six months or more in survivors according to the following categories: good recovery, moderately disabled, severely disabled, vegetative, and dead.[6]

All patients had an emergency CT scan as well as protocol serial CT scans on the same post-injury days as their neurological examinations. The two CT scan categories employed in data analysis were derived as follows: patients with intracranial high-density lesions either extracerebral or intracerebral comprised one group while all others, i.e., patients with low density abnormalities or normal CT scans, formed the other group.[5]

The intracranial pressure was continuously monitored in all patients for 3–7 days post-injury. Patients were divided into two ICP groups: those with normal ICP measurements and those with ICP elevations that required treatment.[5]

Somatosensory Evoked Potentials

SEP and MEP were recorded from 109 patients after appropriate medical and surgical emergency care was completed.

Specific stimulating and recording techniques utilized to obtain SEP in comatose head injury patients have been described in detail elsewhere.[1,2] Briefly, cortical SEP were recorded with a filter bandpass of 1–1,000 Hz from left and right parietal scalp electrodes (P_3 and P_4) referenced to linked ears in response to median nerve stimulation produced by a 0.5 msec constant-current depolarization at 1 pulse per second. The left and right median nerves were stimulated separately, and the pulse intensity was clinically determined by observing the onset of thumb twitch and raising the stimulus intensity 2 mA above threshold. An average of 256 or 512 stimulations were needed to achieve adequate signal-to-noise ratios in the intensive care unit. The first 250 msec period of brain electrical activity post-stimulus was analyzed.

Short-latency SEPs were recorded from 69 patients with a filter bandpass of 30–3000 Hz from an active electrode positioned on the posterior cervical region at C_7 referenced to the vertex (C_z), while an Erbs point potential was obtained from an active electrode positioned at Erbs point and referenced to C_7. The initial 20 msec period was examined and 2,048 trials averaged for analysis.

Somatosensory evoked potentials obtained in each study period for all patients were analyzed off-line by an investigator who had no knowledge of the patient's condition or final outcome. Each response was evaluated for the presence or absence of abnormalities and assigned a grade ranked I to IV, based on a clinical head injury functional grading system developed earlier and described in detail elsewhere.[1–3] Briefly, an SEP was assigned a grade of I, II, III, or IV, if it satisfied the criterion for normal, mildly abnormal, severely abnormal or absent electrical activity, respectively. These grades of electrical abnormality have been shown to correlate with increasing neuronal dysfunction.[1,3]

The grades assigned to SEP show a rough correspondence with certain neuronanatomical subdivisions. For example, an SEP from a comatose head injury patient is considered grade I if the electrical activity generated by structures in the brainstem primary and secondary cortex is consistent with that from normal volunteers. A grade II response consists of brainstem and primary cortical wave components with the absence of secondary cortical potentials beyond 50 msec. This type response suggests that information is reaching primary cortical areas but may not be further processed by the brain. Shifts in latency and decrements in amplitude of wave peaks occurring between 15 and 50 msec are often observed in grade II SEP when compared to the grade I category. Grade II SEP abnormalities are considered mild and forecast recovery of the comatose patient. Grade III SEP are characterized by normal brainstem potentials and by the SEP primary cortical component—the so-called N_{20} wave. No later electrical activity can be found. We interpret this as a severely abnormal response, indicative of failure of cortical function. The most severely abnormal SEP, grade IV, that we obtain in comatose head injury patients consists of either no brainstem and cortically generated electrical activity or only the so-called P_{15} wave. We believe that the P_{15} wave is generated by structures in the brainstem medulla.[2] The grade IV SEP is characteristic of significant diencephalic and/or brainstem dysfunction.

Results

Prognostic Value of SEP

A significant relationship exists between the graded SEP data obtained within 48 hours of head injury in comatose head injury patients and the outcome of these patients at one year ($p < 0.001$, chi-square test); that is, as the severity of the SEP abnormality increases, so does mortality (Table 1A). Of the comatose patients, 81% (62/76) who had, at worst, mildly abnormal MEPs, i.e., grade I and II, realized a return to normal life or only moderate disability. Likewise, the outcome of 76% (23/30) of the patients who had a severely abnormal SEP, i.e., grade III or IV, was poor. Differences in outcome can also be discerned between the 4 SEP grades. For example, the percentage of patients attaining good/moderate recovery was 88%, 72%, 35%, and 0% for SEP grades I, II, III, and IV, respectively.

A comparison of the prognostic value of the somatosensory evoked potential and multimodality evoked potential (visual, auditory, and somatosensory) data is presented in Table 1. There is no significant difference between the predictive accuracy of SEP data alone and that of multimodality evoked potential data.

When only the short-latency components (<15 msec) of the SEP were utilized to forecast outcome, the accuracy of prediction dropped significantly from that achieved when longer latency SEP components (to 250 msec) were also analyzed. With short-latency SEP components alone, a good/moderate recovery was forecast correctly in only 58% of cases compared to 88% for SEP components with longer latency.

Effect of Secondary Insults

The data indicate that death and poor outcome occurs in patients with any SEP grade—even those with minimal SEP neuroelectric abnormalities (grade I included 3 deaths, 2 severe/vegetative; grade II, 6 deaths, 4 severe/vegetative: Table 2A). These findings are inconsistent with the trend of the relationship between SEP grades and the categories of outcome from head injury. However, an anlysis of the cause of death

TABLE 1

HEAD INJURY PROGNOSIS

SEP	Good Moderate*	Severe Vegetative*	Dead	Patients
A. Somatosensory evoked potentials†				
Grade I	36 (88%)	2 (5%)	3 (7%)	41 (38%)
Grade II	26 (72%)	4 (11%)	6 (17%)	36 (34%)
Grade III	7 (35%)	6 (30%)	7 (35%)	20 (19%)
Grade IV	0 (0%)	4 (40%)	6 (60%)	10 (9%)
B. Multimodality evoked potentials†				
Grade I	20 (90%)	1 (5%)	1 (5%)	22 (20%)
Grade II	42 (82%)	4 (8%)	5 (10%)	51 (47%)
Grade III	8 (35%)	5 (22%)	10 (43%)	23 (21%)
Grade IV	0 (0%)	6 (46%)	7 (54%)	13 (12%)

*Outcome determined >6 months after injury. Jennett and Bond classification.[6]
†Evoked potentials obtained within 48 hours of injury.

and of the period post-injury in which secondary insults occurred resolves this conflict (TABLE 2).

When patients with systemic complications such as septicemia, pulmonary, or cardiac compromise, and so on, and those who have delayed central nervous system insults such as ventriculitis, meningitis, and so on, are analyzed separately the strength of the SEP as a prognostic indicant can be more accurately assessed. No patient with a grade I or II SEP who did not develop a secondary insult died or had a poor final outcome (TABLE 2B).

Comparison of SEP with Other Prognostic Indicants

SEP obtained early after head trauma were the most accurate and reliable single prognostic indicants with 91% accuracy of prediction, 25% of predictions made at

TABLE 2

SOMATOSENSORY EVOKED POTENTIALS
EFFECT OF SECONDARY INSULTS ON OUTCOME PREDICTION

SEP*	Good Moderate†	Severe Vegetative†	Dead	Patients
A. Secondary insults present				
Grade I	16 (76%)	2 (10%)	3 (14%)	21 (31%)
Grade II	16 (62%)	4 (15%)	6 (23%)	26 (39%)
Grade III	1 (8%)	5 (42%)	6 (50%)	12 (18%)
Grade IV	0 (0%)	3 (38%)	5 (63%)	8 (12%)
B. Secondary insults absent				
Grade I	20 (100%)	0 (0%)	0 (0%)	20 (50%)
Grade II	10 (100%)	0 (0%)	0 (0%)	10 (25%)
Grade III	6 (75%)	1 (13%)	1 (13%)	8 (20%)
Grade IV	0 (0%)	1 (50%)	1 (50%)	2 (5%)

*SEP obtained within 48 hours of injury.
†Outcome determined >6 months after injury. Jennett and Bond classification.[6]

greater than 90% confidence level (TABLE 3). The SEP was the only indicant that made no falsely pessimistic errors, that is, predicted death when death did not, in fact, occur. The 9% error rate was due to overly optimistic predictions, that is, predictions of good outcome when the patient did poorly (TABLE 3). As seen in TABLE 2, those patients forecast into a good/moderate outcome group who did poorly often had delayed unforeseen secondary systemic or CNS complications. Evoked potential data were followed by age, ICP measurements, pupillary response to light, extraocular motility, and motor posturing in decreasing order of prognostic accuracy.

The combination of SEP and clinical data had an accuracy of 89% but provided the most reliable predictions of any combination of data (64% of predictions made above the 90% confidence level, TABLE 3).

DISCUSSION

We have previously demonstrated that in comatose, severe head injury patients, multimodality evoked potentials (MEP) obtained in the first few days following trauma are strong indicants of final patient outcome 6 months or more later.[1–3] In the present study, we analyzed the prognostic value of the somatosensory evoked potential alone.

TABLE 3

HEAD INJURY PROGNOSIS ALL INDICANTS

Indicants	Correct Predictions	>90% Conf. Level	Incorrect Predictions	Falsely Optimistic	Falsely Pessimistic
1. CT Data	64%	0%	36%	7%	29%
2. ICP Data	75%	0%	25%	11%	14%
3. GCS*	80%	25%	20%	13%	7%
4. Best Clinical Combination†	82%	43%	18%	9%	9%
5. Evoked Potentials	91%	25%	9%	9%	0%
6. Clinical† + CT	77%	52%	23%	14%	9%
7. Clinical† + ICP	80%	55%	20%	11%	9%
8. Clinical† + EP	89%	64%	11%	7%	4%

*Glasgow Coma Scale

†Age, GCS, pupillary response, eye movement, mass lesion, motor posturing.

The data suggest that a similar accuracy of prediction can be obtained whether SEP or the combined data of three modalities (MEP) are used to forecast patient outcome from head trauma. However, information regarding the pathophysiology of head injury and the location and extent of areas of CNS dysfunction is lost when only SEP are analyzed. For example, correlations with intracranial pressure measurements, CT scan data, operative observations, and autopsy material are more revealing with a multimodal evoked potential approach. Nonetheless, if a single modality had to be employed for early prediction of final patient outcome, our data suggest that the somatosensory evoked potential should be selected.

Somatosensory evoked potentials can provide information on the functional status of the cervical spine, brainstem and cerebral hemispheres. It is not surprising that a correlation of only the short-latency or brainstem components of the SEP and the outcome from head injury was not significant. Somatosensory brainstem potentials forecast good-to-moderate outcomes correctly in only 58% of cases when they were normal compared to the 88% accuracy of prediction achieved by the complete SEP (TABLE 1A).

Secondary systemic complications that may occur after SEP studies are recorded from comatose patients can be important variables to be considered when SEP are used as prognostic tools. Pulmonary infections, septicemia, renal failure, cardiac failure with hypotension, and so on have been identified with a deterioration of a patient's neurological condition and with increased abnormality of SEP activity.[3] Moreover, delayed or secondary complications directly affecting the central nervous system, such as meningitis, ventriculitis, mass lesions, elevated ICP, and so on, adversely affect patient outcome from head injury and depress the prognostic accuracy of SEP.[3] Our data suggest that the prognostic accuracy of SEP obtained in the first few days after injury from patients who do not have secondary insults is significantly greater than the accuracy of SEP data from patients who do develop secondary insults (TABLE 2). These results point to the importance of serial evoked potentials as well as on-line evoked potential evaluations of CNS function following severe head injury.

In contrast to CT and ICP data, SEP data performed well as a prognostic indicant when used alone. Further, compared to single clinical signs or combinations of clinical parameters such as the Glasgow coma scale or the best clinical combination of data (TABLE 3), it also did well forecasting patient outcome. The combination of clinical and SEP data provided the most reliable prognostic battery improving on both the clinical data and the SEP data considered individually. SEP data, therefore, may be especially useful for patients unable to cooperate or in whom a complete clinical examination cannot easily be obtained because of coma or barbiturate therapy and neuromuscular paralysis for ICP control. Furthermore, SEP studies in combination with clinical indicants can serve as the basis of a more reliable functional categorization of head injury patients. This is important both for family counseling and for the insights it might provide into the pathophysiology of head injury. From such information newer therapeutic approaches may be suggested.

REFERENCES

1. GREENBERG, R. P., D. P. BECKER, J. D. MILLER, *et al.* 1977. Evaluation of brain function in severe human head trauma with multimodality evoked potentials. Part 2: Localization of brain dysfunction and correlation with post-traumatic neurological conditions. J. Neurosurg. **47:** 163–177.
2. GREENBERG, R. P., D. J. MAYER, D. P. BECKER, *et al.* 1977. Evaluation of brain function in severe human head trauma with multimodality evoked potentials. Part 1: Evoked brain injury potentials, methods, analysis. J. Neurosurg. **47:** 150–162.
3. GREENBERG, R. P., P. G. NEWLON, M. S. HYATT & D. P. BECKER. 1981. Prognostic implications of early multimodality evoked potentials in severe head injury patients: A prospective study. J. Neurosurg. (in press).
4. BECKER, D. P., J. D. MILLER, J. D. WARD, R. P. GREENBERG, H. F. YOUNG & R. SAKALAS. 1977. The outcome from severe head injury with early diagnosis and intensive management. J. Neurosurg. **47:** 491–502.
5. NARAYAN, R. K., R. P. GREENBERG, J. D. MILLER, G. G. ENAS, S. C. CHOI, P. R. S. KISHORE, J. B. SELHORST, H. A. LUTZ & D. P. BECKER. 1981. Improved confidence of outcome prediction in severe head injury: A comparative analysis of the clinical examination, multimodality evoked potentials, CT and ICP. J. Neurosurg. (in press).
6. JENNETT, B. & M. BOND. 1975. Assessment of outcome after severe brain damage: A practical scale. Lancet (1): 480–484.

SHORT-LATENCY SOMATOSENSORY EVOKED POTENTIALS FOLLOWING MEDIAN NERVE STIMULATION

Jun Kimura and Thoru Yamada

Division of Clinical Electrophysiology
University of Iowa Hospitals and Clinic
Iowa City, Iowa 52242

Several studies have confirmed the presence of short-latency SEPs in man.[1-7] Different investigators have described nearly identical waveform and distribution of scalp recorded and cervical SEPs, without necessarily postulating the same neural sources for various components.

Although the negative peaks of the cervical potentials are similar in latency to the positive peaks of the scalp recorded potentials, possible relationships between the two have not been systematically investigated. In the present study, we recorded the short-latency SEPs simultaneously from the scalp and cervical regions to delineate their field distributions. Potentials recorded with cephalic and noncephalic references were compared to determine their polarity characteristics. To further elucidate the neural sources of the short-latency SEPs, we also studied the topography of SEPs recorded from 16 scalp-wide electrodes placed in accordance to the 10–20 International System with the use of tied ear reference.

Twenty normal volunteers were studied. There were 10 men and 10 women, ranging in age from 20 to 40 years (average age: 29 years). They were seated comfortably in a semidarkened, quiet room with their eyes closed. The left and right median nerves were independently stimulated in each subject. A knee reference electrode was used in 27 of 40 recordings in 20 subjects and connected ears in the remaining 13 recordings. Additional trials in most subjects included various combinations of referential and bipolar derivations. The topography of SEP was studied in 16 normal subjects after unilateral stimulation of the median nerve at the wrist.

Silver–silver chloride disc electrodes were attached to the skin with collodion, maintaining the impedance less than 3000 Ω. The median nerve was stimulated at the wrist using surface electrodes. The cathode was placed 2 cm proximal to the anode. Each stimulus was 0.1 msec in duration and the stimulus intensity was adjusted to produce a small twitch of the abductor pollicis brevis. Summation and averaging were carried out by a 21 MX Hewlett Packard Computer. The analysis time was 20 msec after the stimulus with a digitalized sampling rate of 0.1 msec. Each test set consisted of 1000 to 1200 summated responses.

To obtain clean SEPs, automatic editing was accomplished using a computer program which was written for sample reject on 5 successive equal points, either overloaded or flat. To reduce the number of rejected trials, the computer was also programmed to provide random triggering while automatically avoiding EKG artifact. This was achieved by detecting QRS complexes (or other overloaded artifacts exceeding the duration of 100 msec) and triggering the stimulus with random time delay from 0 to 200 msec following the overloaded period.

To analyze small neural responses, high amplitude resolution had to be achieved in computer averaging. If we divided the sum by the actual count, some small peaks might be eliminated by the division. To circumvent this difficulty, a computer program was developed to determine the smallest divisor that would keep the largest

0077–8923/82/0388–0689 $1.75/0 © 1982, NYAS

point within the display range. The sum was divided by this "artifical trial count" and a correction factor was applied to the computer measurement of the data, so that the amplitude displayed was the actual averaged amplitude. Typically, the divisors were 1/15th of the actual trial count. The number of rejected trials was greater for the short-latency SEP recorded with the knee reference than for the long-latency SEP recorded with ear reference because EKG artifacts were more prominent in the former.

Using a knee reference, four positive peaks were identified in the initial portion of the scalp-recorded potentials (FIGURE 1). They were designated P_9, P_{11}, P_{13}, and P_{14} according to their polarity and mean latency. With an ear reference, however, P_9 and P_{11} were not detected in most subjects. Cervical potentials recorded with knee reference consisted of an initial positive peak, P_{C9}, and three subsequent negative peaks, N_{11}, N_{13}, and N_{14}. With an ear reference, the initial peak of the cervical potentials was recorded as an upward deflection although identical in latency to the positive peak, P_{C9}, recorded with a knee reference. The subsequent potentials, N_{11}, N_{13}, and N_{14} were similar whether the ear or knee served as reference site but N_{11}, and at times, N_{14} were better delineated with an ear than knee reference. Comparison between knee and ear reference recordings suggested that the ear was active with a positive polarity for P_9 and P_{11} potentials.

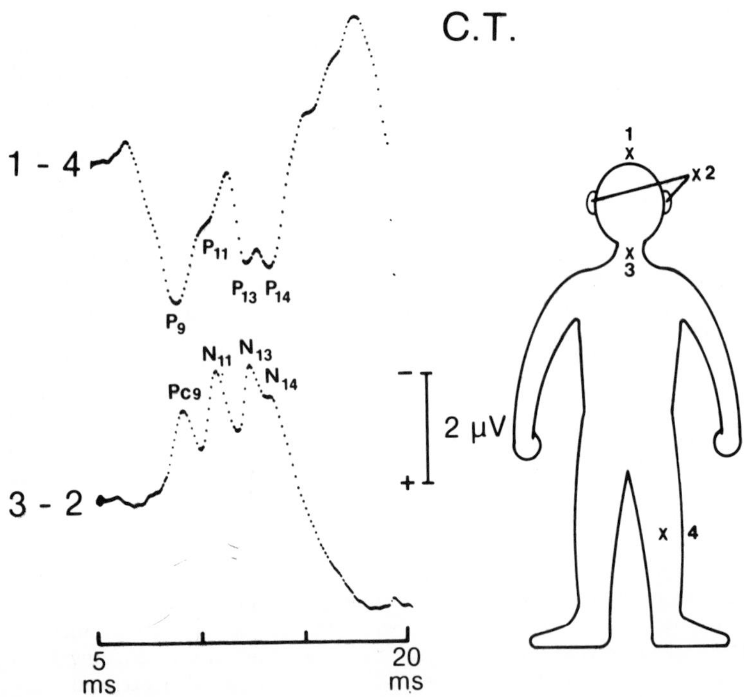

FIGURE 1. Simultaneous recording from C_Z (1) referenced to knee (4) and low cervical (3) electrode referenced to ear (2) in a normal subject. Four positive peaks, P_9, P_{11}, P_{13}, and P_{14} recorded at C_Z were nearly identical to latency to four negative peaks, P_{C9}, N_{11}, N_{13}, and N_{14} recorded at the low cervical electrode. The initial component of the cervical potential, P_{C9} is positive when referenced to the knee (not shown). (From Yamada, Kimura & Nitz.[6] By permission of *Electroencephalography and Clinical Neurophysiology*.)

TABLE 1

SHORT-LATENCY SOMATOSENSORY EVOKED POTENTIALS*

	Knee Reference			Ear Reference		
Site of Recording	Latency (msec) Mean ± SD	Latency After P_9 (msec) Mean ± SD	Number Recorded/ Number Tested	Latency (msec) Mean ± SD	Latency After P_9 (msec) Mean ± SD	Number Recorded/ Number Tested
C_Z P_9	9.5 ± 0.8		27/27			
P_{11}	11.6 ± 0.7	2.2 ± 0.3	24/27			
P_{13}	13.2 ± 0.9	3.9 ± 0.4	23/27	13.4 ± 0.7		10/13
P_{14}	14.4 ± 0.9	4.9 ± 0.4	27/27	14.8 ± 0.5		13/13
Neck P_{C9}	9.5 ± 0.7		27/27	9.5 ± 0.4		13/13
N_{11}	11.8 ± 0.8	2.3 ± 0.4	15/27	11.7 ± 0.4	2.3 ± 0.3	12/13
N_{13}	13.5 ± 0.9	3.9 ± 0.5	27/27	13.3 ± 0.5	3.8 ± 0.2	13/13
N_{14}	14.6 ± 0.5	5.1 ± 0.2	3/27	14.6 ± 0.6	5.0 ± 0.4	8/13

*After Yamada, Kimura & Nitz.[6]

The table shows the mean latency and standard deviation of each component from 27 recordings with knee reference and 13 with ear reference. The right and left sides were analyzed together since there were no statistically significant difference between the two. Cervical potentials were measured from the low cervical electrode. Although four peaks were best identified in this region, N_{11} and N_{14} were detected only in 15 and 3 of 27 recordings, respectively. The positive peaks of scalp recorded potentials, P_9, P_{11}, P_{13}, and P_{14}, were nearly identical in latency to the corresponding positive or negative peaks of the cervical potentials, P_{C9}, N_{11}, N_{13}, and N_{14} ($P > 0.2$). Similarly, the increase in latency from P_9 to the remaining scalp recorded peaks, P_{11}, P_{13}, and P_{14}, were very close to those from P_{C9} to the subsequent cervical potentials, N_{11}, N_{13}, and N_{14} ($P > 0.3$).

The first component was negative over the shoulder but otherwise distributed widely as a positive field from the neck to the scalp. Its latency was slightly less than that of the nerve potential recorded at Erb's point. The second component was best recorded at the low cervical area as a negative peak, increasing in amplitude and decreasing in latency laterally towards the site of the stimulus in some subjects. Its latency was nearly the same as an estimated nerve conduction time from the wrist to the spinal cord. The third component was most consistently recorded as a negative peak at cervical electrodes, progressively decreasing in amplitude rostrally and caudally. It was often difficult to detect over the scalp. The last component was best recorded as a positive peak at the scalp and nasopharyngeal electrode. This peak was equivocal when recorded from the cervical region unless an ear reference was used. The positivity of P_{14} at nasopharyngeal electrode, therefore, suggests that its generator site is located further caudally. Unlike N_{11} and N_{13}, N_{14} was barely detectable from cervical spine. These observations are consistent with the view that P_{14} represents an afferent volley of the medial lemniscus.[7]

Following the P_{14}, which was distributed diffusely over the entire scalp, there was a small but distinct negative peak, N_{17}, recorded bilaterally in the frontal region (FIGURE 2). In contrast, the first major negative peak, N_{19}, recorded from the central region was contralateral to the side of stimulation. The frontal negative peak, N_{17}, was also recorded consistently from the vertex and ipsilateral central electrode, and occasionally from contralateral central area preceding the major negative peak, N_{19}. The topography of N_{17} and N_{19} suggests that these are two distinct peaks of separate neural origin. In two patients, monitored preoperatively for their intractable seizures,

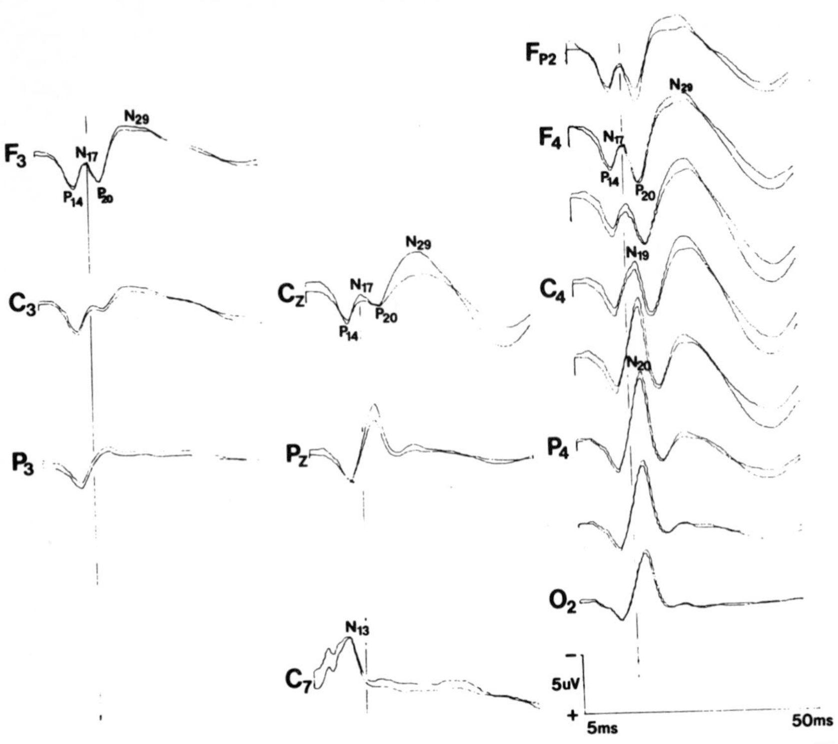

FIGURE 2. Topographic display of SEPs to stimulation of the left median nerve. Electrode placement was in accordance to the 10–20 International System with a neck electrode, C_7, placed just above the C_7 spinous process. Frontal N_{17} (F_{P2}, F_4) preceded central N_{19} (C_4) and parietal N_{20} (P_4) contralateral to the stimulus. N_{17} also appeared at the vertex (C_Z) and frontal and central areas ipsilaterally (F_3, C_3).

N_{17} was recorded from multiple levels of depths electrodes placed in the frontal lobes bilaterally (FIGURE 3).

Although N_{17} had not previously been described, recent studies by Desmedt[7] clearly showed the same negative potential preceding the parietal N_{20}. The origins of N_{17} and subsequent peaks and their field distributions remain to be elucidated. However, if P_{14} arises in the medial lemniscus and N_{20} is the cortical or thalamocortical potential, then it is likely that N_{17} is generated subcortically, as suggested by Allison at this conference.[8] The potential recorded directly from the thalamus after stimulation of the median nerve at the wrist has latency of 17 to 18 msec with positive polarity.[9,10] When recorded from the multiple levels of depth electrode, N_{17} showed neither amplitude nor latency gradient. These findings suggest that N_{17} recorded outside the thalamus may be a volume-conducted potential possibly reflecting a thalamic positive potential.

Topographic studies of the short-latency peaks[7] do not support the hypothesis of dipole relationship between parietal N_{20} and frontal P_{20}, since the latencies of the two were close but not the same, the P_{20} having slightly later onset than the N_{20}. In our present study, negative-positive peaks subsequent to P_{14} were shortest in latency at the frontal electrodes (N_{17}, P_{20}, N_{20}) showing a progressive delay toward the central (N_{19},

P_{23}, N_{32}) and parietal areas (N_{20}, P_{26}, N_{34}). This finding, however, does not suggest a "traveling wave" since N_{17} can sometimes be seen as an additional separate peak preceding N_{19} or N_{20} in the central or parietal region, respectively. The topographic specificity for various SEP components suggests the presence of a distinct neural generator in each location.

The surface-positive polarity of the short-latency SEP components results presumably because the active recording electrode is located in the approaching field of the centripetal impulse. The polarity characteristics of the short-latency SEPs led us to believe that a negative field near the generator site is responsible for the cervical potential and that the scalp recorded peak is primarily, although perhaps not exclusively, attributable to an approaching field of positivity from the same source. The presumed neural genetators corresponding to the beginning of each peak are: (1) the entry to the brachial plexus (P_9 and P_{C9}); (2) the entry to the spinal cord (N_{11} and P_{11}), (3) ascending volley of cervical cord or synaptic discharge of cuneate nucleus

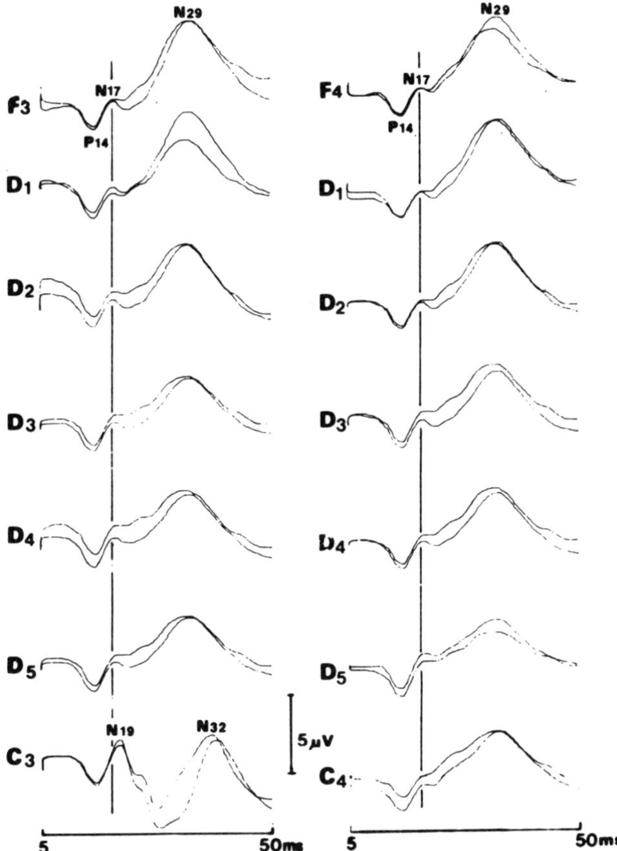

FIGURE 3. SEPs recorded from depth electrodes placed at multiple levels in the frontal lobe. The shallowest electrode was in the cortex (D_1) and the deepest lead was at the level of frontal operculum (D5). SEPs remained nearly the same in amplitude and latency regardless of the depth of recording. Both N_{17} and N_{29} appeared in the depth as well as scalp recording (F_3, F_4, and C_4), whereas N_{19} and N_{32} were primarily cortical, contralateral to the side of stimulation (C_3).

(N_{13} and P_{13}); (4) entry into the medial lemniscus (N_{14} and P_{14}); and (5) thalamic or other subcortical discharge (N_{17}). Our findings also indicate that short-latency SEPs are best delineated for clinical application by a combined recording from the scalp with noncephalic reference and from the neck with a cephalic reference.

We do not know why the traveling impulse gives rise to SEP peaks with apparent temporal relationships to its entry to the brachial plexus (P_9 and P_{C9}) and to the spinal cord (N_{11} and P_{11}). These findings indicate, however, that the presence of a standing potential does not necessarily imply a nonpropagating neural discharge such as those that occur at synapses in relay nuclei. The electrical field may alter abruptly because of change in anatomical orientation of the impulse, branching of the nerve, or alteration of the surrounding volume conductor. It is perhaps necessary to modify the traditional concept, which assumes the presence of fixed neural generators along the sensory pathways as the primary source of SEP peaks. The complex waveform of the far-field potential may be derived from dynamic interaction of afferent volleys, synaptic discharges, and possibly changes in the conduction characteristics of the surrounding tissue.

[NOTE ADDED IN PROOF: In a recent paper,[11] Desmedt and Cheron described a widespread component, N_{18}, following stimulation of the digital nerve. This negative peak, presumably generated below the cortex, was the only early component recorded at the parietal ipsilateral region after the positive far-field potentials, P_9, P_{11}, P_{13}, and P_{14}. We believe that their N_{18} corresponds to our N_{17}.]

References

1. Cracco, R. Q. & J. B. Cracco. 1976. Somatosensory evoked potential in man: Far field potentials. Electroencephalogr. Clin. Neurophysiol. **41:** 460–466.
2. Jones, S. J. 1977. Short latency potential recorded from the neck and scalp following median nerve stimulation in man. Electroencephalogr. Clin. Neurophysiol. **43:** 853–863.
3. Krichevsky, M. & W. C. Wiederholt. 1978. Short-latency somatosensory evoked potentials. Arch. Neurol. (Chic.) **35:** 706–711.
4. Chiappa, K. H., S. K. Choi & R. R. Young. 1980. Short latency somatosensory evoked potentials following median nerve stimulation in patients with neurological lesions. *In* Progress in Clinical Neurophysiology. J. E. Desmedt, Ed. Vol. 7: 261–281. Karger, Basel, Switzerland.
5. Allison, T., W. R. Goff, P. D. Williamson & J. C. Van Gilder. 1980. On the neural origin of early components of the human somatosensory evoked potential. *In* Progress in Clinical Neurophysiology J. E. Desmedt, Ed. Vol. 7: 51–68. Karger, Basel, Switzerland.
6. Yamada, T., J. Kimura & D. M. Nitz. 1980. Short latency somatosensory evoked potentials following median nerve stimulation in man. Electroencephalogr. Clin. Neurophysiol. **48:** 367–376.
7. Desmedt, J. E. & G. Cheron. 1980. Somatosensory evoked potentials to finger stimulation in healthy octogenarians and in young adults: Wave forms, scalp topography and transit times of parietal and frontal components. Electroencephalogr. Clin. Neurophysiol. **50:** 404–425.
8. Allison, T. 1982. Scalp and cortical recordings of initial somatosensory cortex activity to median nerve stimulation in man. Ann. N.Y. Acad. Sci. This volume.
9. Fukushima, T., Y. Mayanagi & G. Bouchard. 1976. Thalamic evoked potentials to somatosensory stimulation in man. Electroencephalogr. Clin. Neurophysiol. **40:** 481–490.
10. Celesia, G. G. 1979. Somatosensory evoked potentials recorded directly from human thalamus and Sm I Cortical area. Arch. Neurol. (Chic.) **36:** 399–405.
11. Desmedt, J. E. & G. Cheron. 1981. Non-cephalic reference recording of early somatosensory potentials to finger stimulation in adult or aging normal man: Differentiation of widespread N18 and contralateral N20 from the prerolandic P22 and N30 components. Electroencephalogr. Clin. Neurophysiol. **52:** 553–570.

SOMATOSENSORY EVOKED POTENTIALS TO RANDOM STIMULUS TRAINS

Robert J. Sclabassi, John K. Vries, and Daniel M. Bursick

Department of Neurological Surgery
School of Medicine
University of Pittsburgh
Pittsburgh, Pennsylvania 15213

INTRODUCTION

Attempts to characterize the input-output properties of the somatosensory system have relied heavily on the averaged evoked potential (SEP).[1,2] If the somatosensory system were linear, the SEP would provide the best mean-square error estimate of these properties.[3] However, the somatosensory system is not linear.[4] Therefore, the characterization of its features requires an analysis based on a system theoretic framework that takes into account its nonlinear properties.

These considerations led us to investigate the use of a functional power series[5] as a tool for characterizing the evoked neuroelectric activity in this system. In theory, this type of analysis, characterizes completely the nonlinear nature of this system and provides a means for quantitating the effects of stimulus interactions. Such an approach should also provide a means for predicting the response of this system to alternative sets of stimuli.

METHODS

The system under study in this paper is one in which the peripheral stimuli may be characterized as a random train of electrical events, and the observed neuroelectric activity as a continuous function of the stimulus train. The analytical and computational procedures involved have been previously reported.[6,7] We assume that the observed neuroelectric activity can be represented as a stochastic transformation in the form of the series expansion

$$y(t) = G_0(h_0) + G_1[h_1(\tau); x(t - \tau)] + G_2[h_2(\tau,\Delta); x(t - \tau)\, x(t - \tau - \Delta)] + \mathrm{E}, \tag{1}$$

where $y(t)$ is the observed activity; $[G_n]$ is a set of mutually orthogonal functionals; $x(t)$ is a discrete stimulating process where the interstimulus intervals are drawn from a Poisson distribution defined by the parameter λ; h_0, $h_1(\tau)$, $h_2(\tau,\Delta)$ are the real-valued continuous zeroth-, first-, and second-order kernels of the expansion; τ represents the time after a stimulating impulse and Δ is the separation between pairs of stimuli; E is the error that results from truncation after the second-order term. The values of the kernels are determined by cross-correlating $y(t)$ with $x(t)$ using techniques developed by Lee and Schetzen.[8]

A total of 22 subjects have been studied consisting of 11 males and 11 females, with ages ranging from 21 to 65 years. Each subject was neurologically intact with no history of previous neurological illness.

Each subject was tested while reclining in a hospital bed in a dimly lit room.

0077-8923/82/0388-0695 $1.75/0 © 1982, NYAS

During the recording session the subject was awake, with eyes gently closed. Surrounding sound levels were relatively low, and no effort was made to measure them or to remove ambient noise completely. The level of alertness was continuously monitored by observation of the background EEG. The stimulating electrodes were located over the right median nerve at the wrist, and the stimulus intensity was adjusted to produce a minimal twitch of the thenar muscles. The interstimulus intervals were drawn from a Poisson distribution with mean rates (λ) or either 2 or 15 impulses per second for each epoch. To increase the signal-to-noise ratio up to eight 60-sec epochs with interspersed rest intervals of 2 min were averaged.

For a λ of 2, 520 total stimuli were presented to each subject. For a λ of 15, 7480 total stimuli were presented. When a λ of 2 was used, the interstimulus intervals shorter than 400 msec were not presented to the subject. This produced a first-order kernel equivalent to the conventional SEP. No second-order kernels were calculated for this condition. Total data collection time was not greater than 30 minutes for either stimulus rate.

Eight channels of scalp-evoked response were detected by either needle electrodes (Grass E2B Platinum, subdermal) or disk electrodes (Grass Gold, E5GH surface electrodes) located according to the International 10–20 system with additional electrodes at the inion, and over C_2 and C_5. The scalp electrical activity was prefiltered by Grass 7P511 amplifiers with a half-amplitude bandpass of 0.3 to 300 Hz. The neck electrodes had a half-amplitude frequency bandpass of 0.3 to 1000 Hz.

The evoked EEG signals were digitized in real-time at a sampling rate of either 2.0 or 2.5 kHz per channel and stored off-line for computation of the kernels. Upon completion of these calculations the first- and second-order kernels were used to evaluate Equation 1 for a uniform stimulus train with an interstimulus interval greater than the system memory (indicated by an *). In this situation,

$$y(t) - G_0^*(h_0) = G_1^*[h_1(\tau); x(t-\tau)] + G_2^*[h_2(\tau,\Delta); x(t-\tau), x(t-\tau-\Delta)] \quad (2)$$

represents the predicted value for the conventional SEP.

RESULTS

Representative first- and second-order kernels as well as the observed and predicted SEPs are shown in FIGURES 1, 2, and 3 for subject 12. FIGURE 1 shows the SEP obtained for a λ of 2 with a minimum interstimulus interval of 400 msec (dashed lines) and the first-order kernels $[h_1(\tau)]$ for a λ of 15 with a minimum interstimulus interval of 0.4 msec (solid curves) for the electrode positions indicated. Comparing these two sets of responses it may be seen that the first-order kernel has a general decrease in amplitude as compared to the SEP and exhibits a slight decrease in latency for equivalent peaks. The same general results hold for all electrode locations examined. For λ of 2 where stimuli occurred approximately 400 to 500 msec apart this information must be interpreted under the qualification that weak temporal interactions occurred. Nevertheless, these data show excellent agreement with the previous results of other investigators.[2]

FIGURE 2 presents the second order kernels for the same subject and the same electrodes shown in FIGURE 1. The richness of the nonlinear response at P_3/F_Z (FIGURE 2A) is a function of postively increasing activity over the parietal region and simultaneously increasing negative activity over the frontal region and is seen to be the difference of the responses at P_3/A_2 and F_Z/A_2. The response recorded over the C_5 level displays little evidence of interactions, indicating that the activity at this point is being transmitted in an approximately linear fashion.

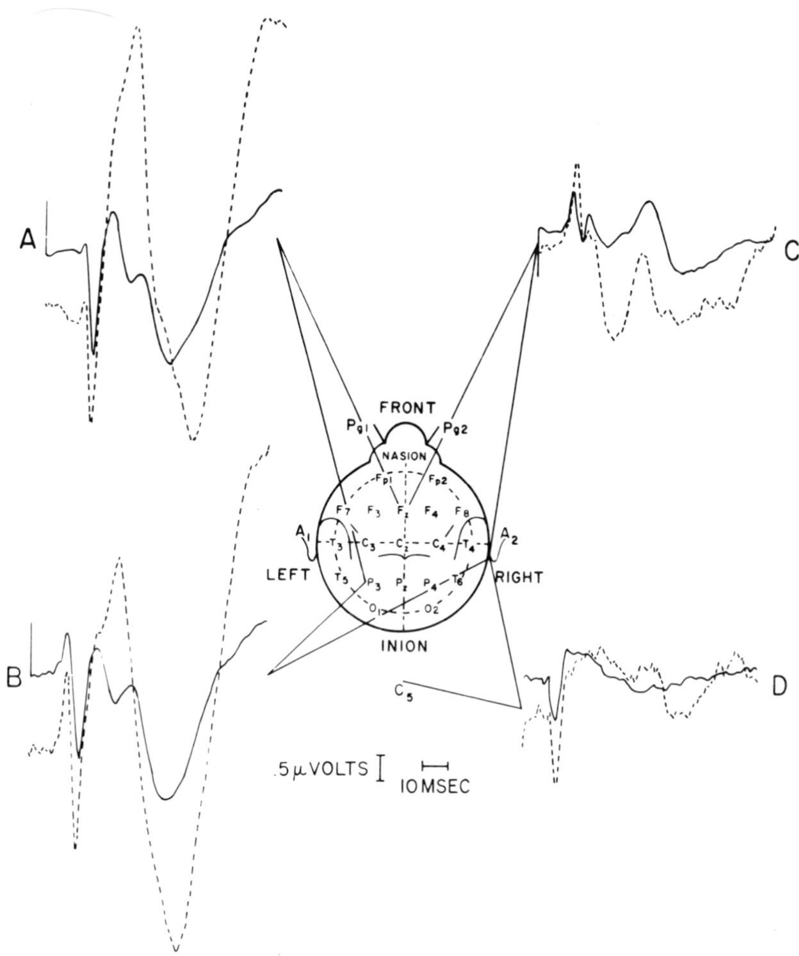

FIGURE 1. This figure demonstrates the somatosensory evoked potentials (SEPs) obtained for a λ of 2 (dashed curves), with a minimum interstimulus interval of 400 msec, and the first-order kernels $[h_1(\tau)]$ obtained for a λ of 15 (solid curves) with a minimum interstimulus interval of 0.4 msec, at electrode locations P_3/F_Z (A), P_3/A_2 (B), F_Z/A_2 (C), and C_5/A_2 (D). The ordinate is amplitude and the abscissa is time into the response. The responses are in agreement as to their form; however, the peak amplitudes of the $h_1(\tau)$'s are approximately ½ the values of equivalent SEP peaks and the peak latencies are decreased between $h_1(\tau)$'s and SEPs by approximately 1 msec on the average.

Several additional observations may be made concerning the structure of the second-order kernels. As the interstimulus interval (Δ) approaches zero, the second-order kernels take on the appearance of inverted first-order kernels at least with regards to the temporal sequence of the waves. It may also be observed with respect to this variable that the kernels demonstrate significant interactions for interstimulus intervals approaching 50 msec. An additional remarkable feature is the increasing latency to positive peak amplitudes for increasing interstimulus intervals. For the

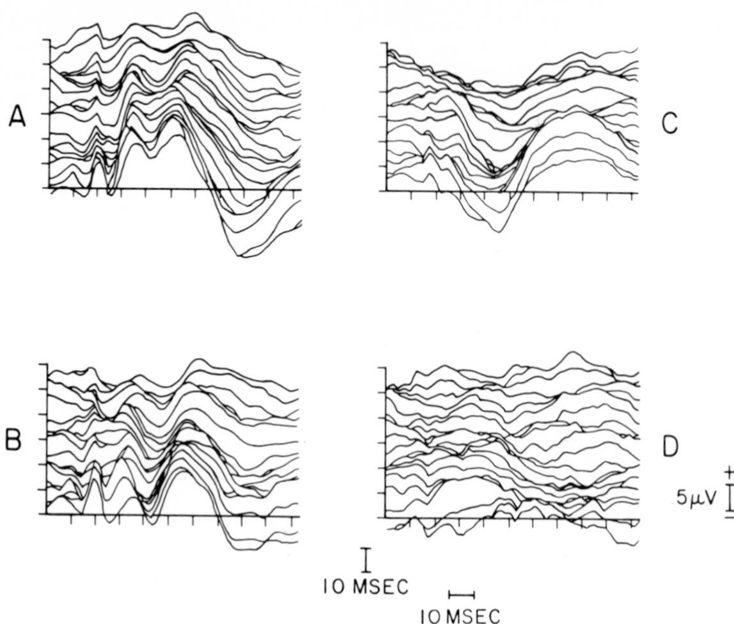

FIGURE 2. This figure demonstrates the second-order kernels $[h_2(\tau,\Delta)]$ obtained from the same subject as the SEPs and $h_1(\tau)$'s in FIGURE 1 at the same time. The same electrode locations are represented [i.e., P_3/F_Z (A), P_3/A_2 (B), F_Z/A_2 (C), C_5/A_2 (D)]. This data was obtained for a λ of 15. These figures are three-dimensional isometric plots where the ordinate is the interstimulus interval (Δ), the abscissa is the time after the second impulse in a pair (τ), and third axis is response amplitude. Note the richness of nonlinear interactions displayed in electrode locations (A), (B), and (C). (D) is the second-order kernel observed at the C_5 level and demonstrates essentially no interactions in the ascending volley at this point.

group of subjects studied so far the average shift is 0.50 msec over the range of values in which significant interactions are demonstrated.

FIGURE 3 demonstrates the SEP, the predicted SEP, the first-order functional G_1^*, and the second-order functional G_2^* for electrode locations P_3/A_2 (FIGURE 3A–D) and P_3/F_3 (FIGURE 3E–H) for the same subject whose data is shown in the other two figures. As can be seen, the nonlinear portion of the response (G_2^*) contributes a significant fraction of the total response in these electrode configurations. These predictions provided an average error at the peaks of less than 10% for the group studied.

DISCUSSION

The character of the evoked electrical activity in the somatosensory system changes as a result of previously applied stimuli. A functional power series analysis has been employed to study the nonlinear interactions occurring within this system. The results of those studies, summarized in this paper, demonstrate that this evoked activity may be represented as an input–output transformation without apriori assumptions regarding system linearity.

The first-order kernel $[h_1(\tau)]$ describes the characteristic response of the system to a single input, given that many interactive impulses have been applied to test the system, and may be considered an evoked potential to temporally interacting stimuli. It resembles the average response of the system to a single impulse, but the two are not identical unless the system is linear.

The second-order kernel $[h_2(\tau,\Delta)]$ describes the nonlinearity in a system's response to a pair of impulses separated by specified intervals, and may be likened to a generalized recovery function. It is the remainder when the ensemble average over all the impulses is subtracted from the ensemble average over impulses with a specified delay to a preceding pulse. The result represents nonlinear interactions and must be interpreted relative to the first-order kernels. If the nonlinear interactions produce a facilitory effect, then $h_2(\tau,\Delta)$ has the same sign as $h_1(\tau)$. Conversely, if the interactions are occlusive, the $h_2(\tau,\Delta)$ has the opposite sign from $h_1(\tau)$. $h_2(\tau,\Delta)$ may be zero if either $h_1(\tau)$ is identically zero or if the system is linear. The second-order kernels presented in FIGURE 2A–C demonstrate a generally occlusive interaction, while 2D shows very little evidence of interaction. The occurrence of interactions between stimulus impulses separated by as much as 40 msec, on the average, is consistent with a number of previous studies.[9-11] In Reference 11 it had been observed that early components of the SEP in man were 80% recovered at 50 msec, with minimal responsiveness at 5 to 15 msec. With interstimulus intervals less than 5 msec almost the full response returned, suggesting that paired stimuli may not be treated as such for very small intervals. This finding is also consistent with results where with small interstimulus intervals an inverted image of the first-order kernel appears in the second-order kernel. This represents an occlusive phenomenon, which is what would be

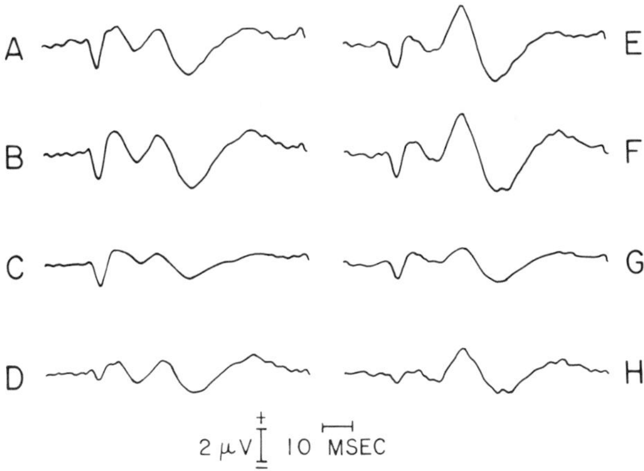

FIGURE 3. This figure presents the observed SEPs (A & E), the predicted SEPs ($G_1^* + G_2^*$) (B & F), the predicted first-order functionals (G_1^*) (C & G), and the predicted second-order functionals (G_2^*) (D & H) for electrode P_3/A_2 (A–D) and P_3/F_3 (E–H) for the same subject as the data presented in FIGURES 1 and 2. The ordinate represents response amplitude and the abscissa is the time τ into the response. These predicted values are based in $h_1(\tau)$ and $h_2(\tau,\Delta)$ obtained for a Poisson Processes defined by a λ of 15 and present the predicted SEPs (B & F) calculated as described in the text (Equation 2), for a uniform stimulus train of 1 impulse per second. The peak error between the predicted SEPs (B & F) and the observed SEPs (A & E) for all electrode locations and all subjects had an average value of less than 10%.

expected to occur as Δ decreases and which is consistent with previously reported data for bursting trains.[9]

In addition this systematic change in latency as a function of the interstimulus interval is quite likely related to the general decrease in latency observed in the first-order kernels compared to the SEPs. This may be due to a number of causes but we favor the concept that an increased rate of stimulus is equivalent to an increased effective input intensity which in turn has been shown to produce shorter component latencies in all sensory modalities.

The analytical combination of these measured functions as indicated in Equations 1 and 2 allow the expected response of the somatosensory system to alternative temporal patterns of stimuli to be predicted. For the case examined here, where the predicted response was obtained for a uniform stimulus train with an interstimulus interval greater than the system memory, the resultant waveform may be interpreted not only as the system response produced by a train of many stimuli, but equivalently as the impulse response $[h_1'(\tau)]$ of a nonlinear system to a single impulse conditioned on the fact that the system is nonlinear.[12] The resulting expression is given by:

$$h_1'(\tau) = h_1(\tau) - 2\lambda \int_0^\infty h_2(\tau,\Delta)d\Delta. \tag{3}$$

This is a powerful application of this procedure, based on the system theoretic framework in which the equations are derived, and allows quantitatively predictive experiments to be performed.

In addition to providing a more complete description of the somatosensory system and allowing predictions of responses to be made, certain other experimental advantages are obtained by use of this procedure. Due to the greatly increased number of stimuli presented per unit time, and the lack of synchrony of these stimuli with other systematic artifacts (such as EKG and 60 Hz) the signal to noise ratio of the resulting calculations is greatly increased. This latter factor also enhances the ability to obtain data under adverse conditions such as in the operating room or the intensive care unit. Finally, the second-order kernels allow the equivalent of a recovery function experiment to be performed in a matter of minutes with the added advantage that an entire response curve may be assessed rather than single points in the response.

The usefulness of this type of analysis is only beginning to be appreciated. Not only may the interactions within a single sensory system be explored, but cross-modulatory influences produced by stimulation of multiple peripheral nerves within the same sensory system and between sensory modalities may be quantified.[13]

Acknowledgments

The authors would like to acknowledge the assistance of Drs. Channing Hinman, Jeffrey Kroin, and Harvey Risch during the development of this work.

References

1. Desmedt, J. E. 1971. Somatosensory cerebral evoked potentials in man. *In* Handbook of Electroencephalography and Clinical Neurophysiology, Vol. 9. W. A. Cobb, Ed. Elsevier, Amsterdam.
2. Giblin, D. R. 1980. Scalp-recorded somatosensory evoked potentials. *In* Electrodiagnosis in Clinical Neurology. M. J. Aminoff, Ed. Churchill-Livingstone, New York, N.Y.
3. Lee, Y. W. 1960. Statistical Theory in Communication. Wiley, New York, N.Y.
4. Sclabassi, R. J., H. Risch, C. Hinman & J. S. Kroin. The nonlinear characterization of

the somatosensory system through the use of random stimulus trains. (Manuscript in preparation.)

5. WIENER, N. 1958. Nonlinear Problems in Random Theory. Wiley, New York, N.Y.
6. SCLABASSI, R. J., H. RISCH, C. L. HINMAN, N. ENNS & N. S. NAMEROW. 1977. Complex pattern evoked somatosensory responses in the study of multiple sclerosis. Proc. IEEE **65:** 626–633.
7. SCLABASSI, R. J., C. L. HINMAN, J. S. KROIN & H. RISCH. 1977. Modulatory effect of prior input upon afferent signals in the somatosensory system. Proc. 1977 Joint Automatic Control Conf. **2:** 787–795, IEEE, New York, N.Y.
8. LEE, Y. W. & M. SCHETZEN. 1965. Measurement of the kernels of a nonlinear system by cross-correlation. Int. J. Control **2:** 237–254.
9. NAMEROW, N. S., R. J. SCLABASSI & N. F. ENNS. 1974. Somatosensory responses to stimulus trains: Normative Data. Electroencephalogr. Clin. Neurophysiol. **37:** 11–21.
10. WINTER, D. L. & J. D. FROST. 1964. Recovery cycles in the lemniscal system. Electroencephalogr. Clin. Neurophysiol. **16:** 459–469.
11. ALLISON, T. 1962. Recovery function of somatosensory evoked responses in man. Electroencephalogr. Clin. Neurophysiol. **14:** 331–343.
12. VRIES, J. K., D. M. BURSICK & R. J. SCLABASSI. The use of random train stimulation and a functional power series expression to predict the somatosensory evoked potential. (Manuscript in preparation.)
13. SCLABASSI, R. J. & G. K. NOREEN. 1981. The characterization of dual-input evoked potentials as nonlinear systems using random impulse trains. Proc Pittsburgh Modeling and Simulation Conf., **12:** 1123–1130.

NATURAL STIMULI EVOKING SOMATOSENSORY POTENTIALS

A. Starr, H. Pratt,* and D. Burke†

Department of Neurology
University of California, Irvine
Irvine, California 92717

The definition of somatosensory evoked potentials has customarily employed percutaneous electrical stimuli to activate mixed nerves, such as the median nerve at the wrist, or sensory nerves, such as the digital nerves at the fingers. The use of a natural stimulus such as touch or pressure to evoke potentials has been utilized infrequently.[1–7]

An advantage of a natural stimulus is that it provides information as to the function of the receptor and its terminal nerve fibers whereas electrical stimulation of nerve trunks bypass these areas. Secondly, both the type of neural elements and the central pathways activated are relatively specific with a natural stimulus compared to electrical stimulation of nerve trunks. A major disadvantage of a natural stimulus is that complex procedures are often required for its quantification. Moreover, since natural stimuli usually activate a restricted number of neural elements, the amplitude of their evoked potentials may be less than those evoked by electrical stimulation.

Over the past few years we have had the opportunity to work on this problem and have utilized two types of natural stimuli: (1) mechanical deformation of the skin of the finger that feels like a tap,[8–13] and (2) flexion or extension of the ankle joint to activate muscle spindle receptors in the gastrocnemius or tibialis anterior muscles.[14] These natural stimuli were capable of evoking potentials at several levels of the somatosensory pathway.

Tactile stimuli elicit clear potentials at many levels of the somatosensory pathway but are of low amplitude compared to the potentials evoked by electrical stimulation of nerve trunks (FIGURE 1). Their scalp distributions are comparable[11] as are the effects of stimulus rate.[10]

Preliminary studies using tactile stimulation were completed in neonates and showed clear potentials at several levels of the pathway.[12] However, cortical potentials could only be detected in 2 of 10 newborn infants compatible with a maturational delay in cortical circuits in youngsters of this age.

Conduction times along the sensory pathway to natural tactile stimuli can be calculated and used to assess abnormalities in clinical populations.[15] For instance, in an unpublished study of diabetic and uremic patients with peripheral neuropathy, we (Starr and Pratt) found the conduction time from mechanical stimulation of the index finger to the occurrence of the evoked potentials recorded over the median nerve to be abnormal in 8 of 11 patients with uremia but normal in the 7 patients with diabetes. However, in both groups of patients proximal afferent conduction times between the wrist and axilla from electrical stimulation of the digital nerves were consistently abnormal (10/11 uremic patients; 6/7 diabetic patients). This finding would suggest

*Present address: Gutwirth Bldg., Technion, Israel Institute of Technology, Haifa 32000, Israel.

†Present address: Unit of Clinical Neurophysiology, The Prince Henry Hospital, Little Bay, N.S.W. 2036 Australia.

0077–8923/82/0388–0702 $1.75/0 © 1982, NYAS

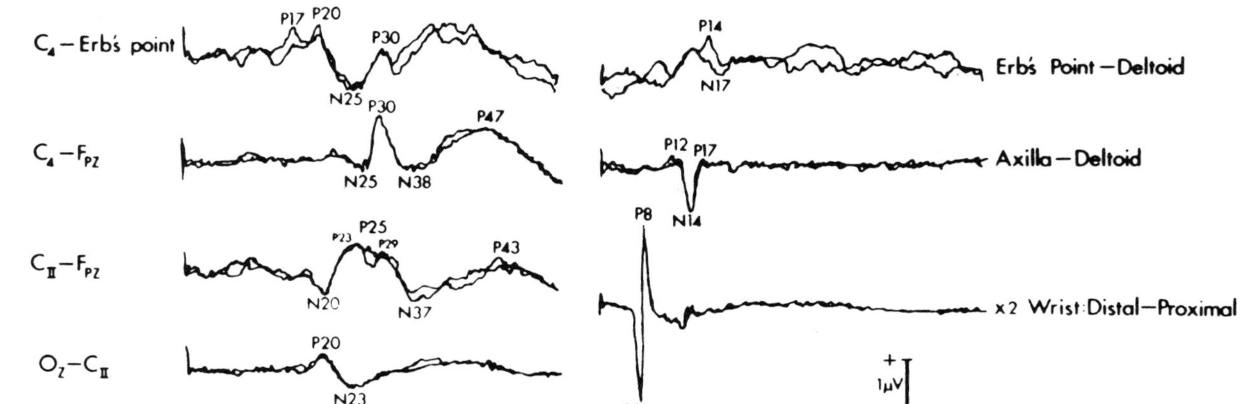

FIGURE 1. Somatosensory potentials evoked by 4.4/sec mechanical tap of the nail of the index finger. 512 stimuli comprise each average and duplicate averages are plotted. Electrodes were placed at several scalp and limb positions to define the potentials along the somatosensory pathway. The components are labeled according to their polarity (P, positive; N, negative) and latency (in msec).

that the terminal parts of the nerve fiber are particularly affected in uremic patients compared to those with diabetes.

We have recently extended the use of natural stimuli to that of muscle stretch to evoke cerebral potentials in humans.[14] We believe that muscle stretch was effective in evoking the potential by activating Ia afferents from muscle spindles (FIGURE 2). The definition of Ia afferent input to the cerebral hemispheres provides a means for the clinician to directly assess muscle spindle–cortical connections in a variety of clinical disorders.

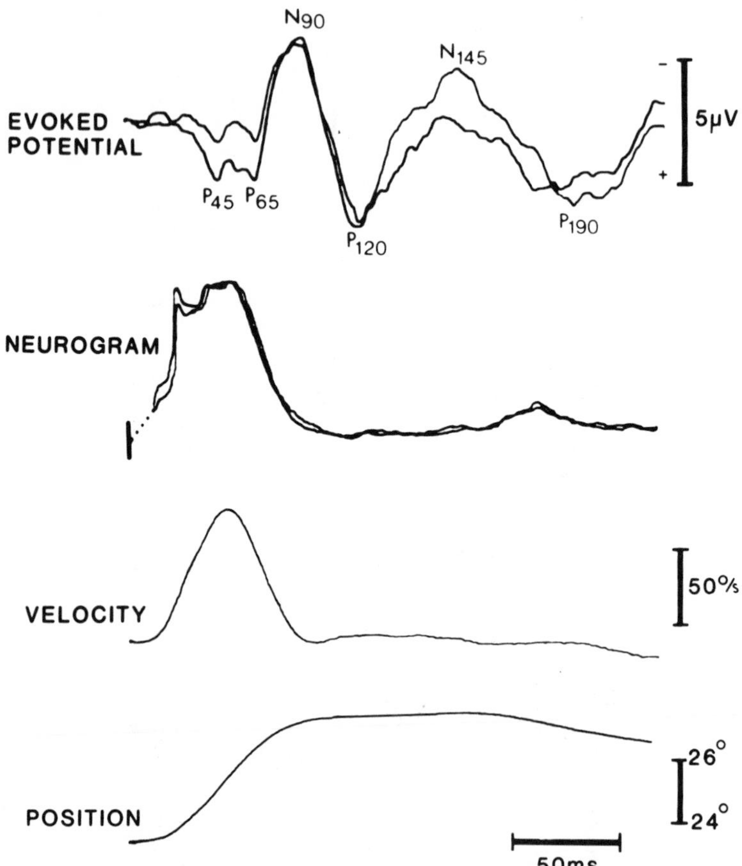

FIGURE 2. Somatosensory potentials evoked by muscle stretch. Two superimposed averages of cerebral evoked potentials (C_Z referenced to F_{PZ}: top trace), and of muscle afferent activity (integrator time constant: 0.01 sec), recorded by a microelectrode within a fascicle of the peroneal nerve innervating tibialis anterior (second trace); the velocity profile of the plantar-flexion movement of the ankle (third trace); and ankle joint position (bottom trace). The repetition rate of stretch was 1 Hz. The components of the evoked potential are labeled according to polarity at the vertex electrode (P, positive; N, negative) and approximate latency (in msec) in normal adults. Peak velocity of movement is reached at 46 msec, maximum muscle afferent activity at 37 msec, and the onset of the initial deflection of the evoked potential at 32 msec.

We propose that the use of natural stimuli to study somatosensory functions, both by behavioral and electro-physiological means, is physiologically appropriate and will result in data that can help clarify mechanisms both of normal sensory function and of clinical disorders of sensation. However, there are numerous technical problems that must be solved before such studies can be considered "routine," as are the present studies using electrical stimulation of nerve trunks. Nevertheless, techniques are now available for those wishing to analyze natural physiological approaches to evoked potential correlates of somatosensory function.

References

1. Sears, T. A. 1959. Action potentials evoked in digital nerves by stimulation of mechanoreceptors in the human finger. J. Physiol. **148:** 30.
2. Bannister, R. G. & T. A. Sears. 1962. The changes in nerve conduction in acute idiopathic polyneuritis. J. Neurol. Neurosurg. & Psychiatry **25:** 312–328.
3. Kjellman, A., E. L. Larsson & T. S. Prevec. 1967. Potentials evoked by tapping recorded from the human scalp over the cortical somatosensory region. Electroencephalogr. Clin. Neurophysiol. **23:** 396.
4. McLeod, J. G. 1966. Digital nerve conduction in the carpal tunnel syndrome after mechanical stimulation of the finger. J. Neurol. Neurosurg. Psychiatry **29:** 12–22.
5. Franzen, O. & K. Offenloch. 1969. Evoked response correlates of psychophysical magnitude estimates for tactile stimulation in man. Exp. Brain Res. **8:** 1–18.
6. Larsson, L. E. & T. S. Prevec. 1970. Somatosensory response to mechanical stimulation as recorded in human E.E.G. Electroencephalogr. Clin. Neurophysiol. **28:** 162–172.
7. Nakanishi, I., Y. Shimada & Y. Toyokura. 1974. Somatosensory evoked responses to mechanical stimulation in normal subjects and in patients with neurological disorders. J. Neurol. Sci. **21:** 289–298.
8. Pratt, H., R. N. Amlie & A. Starr. 1979. Short latency mechanically evoked somatosensory potentials in humans. Electroencephalogr. Clin. Neurophysiol. **47:** 524–531.
9. Pratt, H., A. Starr, R. N. Amlie & D. Politoske. 1979. Mechanically and electrically evoked somatosensory potentials in normal humans. Neurology (Minneap.) **29:** 1236–1244.
10. Pratt, H., D. Politoske & A. Starr. 1980. Mechanically and electrically evoked somatosensory potentials in humans: Effects of stimulus presentation rate. Electroencephalogr. Clin. Neurophysiol. **49:** 240–249.
11. Pratt, H. & A. Starr. 1981. Mechanically and electrically evoked somatosensory potentials in humans: Scalp and neck distribution of short latency components. Electroencephalogr. Clin. Neurophysiol. **51:** 138–147.
12. Pratt, H., R. N. Amlie & A. Starr. 1981. Short latency mechanically evoked peripheral nerve and somatosensory potentials in newborn infants. Pediatr. Res. **15:** 295–298.
13. Pratt, H., R. N. Amlie & A. Starr. 1979. Somatosensory potentials in humans evoked by both mechanical stimulation of the skin and electrical stimulation of the nerve. *In* Sensory Functions of the Skin of Humans. Dan R. Kenshalo, Ed. pp. 109–128. Plenum Press, New York, N.Y.
14. Starr, A., B. McKeon, N. Skuse & D. Burke. 1981. Cerebral evoked potentials to muscle stretch in man. Brain **104:** 149–166.
15. Vaziri, D., H. Pratt, J. K. Saiki & A. Starr. 1981. Evaluation of somatosensory pathway by short latency evoked potentials in patients with end-stage renal disease maintained on hemodialysis. Int. J. Artificial Organs **4:** 17–22.

Discussion of the Paper

T. Allison: Do you have data on the scalp topography of P45? One might expect that this component is generated in area 3A since this is where the muscle spindle afferents are thought to terminate.

A. Starr: Yes. This component is largest around the vertex, but our electrodes were not spaced close enough to show whether or not the distribution was asymmetric over the two hemispheres.

R. Cracco: Some of your colleagues from Australia recently provided evidence that suggests that the latency of the scalp response with mixed nerve stimulation is determined by dynamic spindle input, presumably Group 1A. Would you comment on this?

Starr: Yes, they compared SEPs of the sural nerve at the ankle with that of the posterior tibial nerve at the ankle. The latency of the posterior tibial nerve SEP was shorter than the sural nerve SEP, suggesting that the 1A afferent fiber pathways are conducted faster in the spinal cord than the pure cutaneous ones.

THE INADEQUACY OF CLICK-EVOKED AUDITORY BRAINSTEM RESPONSES IN AUDIOLOGICAL APPLICATIONS

Jos J. Eggermont

Department of Medical Physics and Biophysics
University of Nijmegen
Nijmegen
the Netherlands

An objective audiogram can be estimated basically in three different ways: (1) by using frequency specific stimuli as tonebursts and filtered clicks and determining the threshold intensities for wave V at each frequency of interest.[1,2] (2) By using a wide-band stimulus, usually a click, and deducing from the response parameters (amplitude, latency) an estimate of the audiogram.[3–6] (3) By using a place specific stimulus, i.e., combining a click with high-pass noise masking to derive narrow-band responses. The threshold intensities within each narrow-band form an estimate of the audiogram.[7]

The click stimulus is the most frequently used for evoking the auditory brainstem response (ABR), not surprisingly because the click provides the best synchronization of individual contributions to the compound response. Wave V is the component from the ABR that routinely serves in the detection of threshold. Without great modifications we may deduce its properties, with respect to audiogram estimation, from those of the action potential (AP) in the electrocochleogram. For the electrocochleogram quite a number of reports[8,9] have dealt with the relation of click threshold and AP-latency, and the shape of the audiogram. The major conclusion is that click threshold corresponds best with the audiometric threshold at 2 and 4 kHz. The correlations are usually low (< 0.6) and the slope of the regression line between AP threshold and audiometric threshold at either 2 or 4 kHz is in the range of 0.5–0.6 indicating that hearing losses in general will be underestimated.[10] The best correspondence between click threshold and the audiometric thresholds in the 2–4 kHz range can be understood if one takes into account the effect of external ear canal and middle ear resonance on the initially more or less flat click spectrum. The effective spectrum for the cochlea definitely is boosted in the 1.5–3.5 kHz range as can be deduced also from the narrow-band representation of low click intensity ABR.[11]

In dealing with click as the stimulus, an evaluation has been reported about the effect of audiogram shape on the response parameters.[12] Most studies use the click ABR as a crude but clinically convincing indicator for hearing loss in infants and adults and base this on the same latency–intensity functions that have extensively been discussed for AP studies some ten years ago.[13] Thus the present use in ABR will have the same merits and drawbacks as found earlier for the electrocochleogram.

The overall picture indicates that in using clicks a rather quick and for most clinical purposes sufficient indication can be obtained about the overall level of hearing loss for 2 kHz and up. No detailed information about the audiogram can be obtained: small localized regions with hearing loss will be missed, and diagnosing a low frequency hearing loss is impossible.[7] In other words, click ABR tells us the same about the hearing of the patients as a whispered speech test: it is a screening test and is not likely to give detailed information. In differentiating conductive from sensorineural, and mixed hearing losses, the use of bone-conducted sound versus air-conducted sound should be the major issue in ABR, yet most emphasis is laid upon the attenuating effects of an impaired middle ear upon the ABR parameters as a function of stimulus intensity. The expectation is that amplitude–intensity and latency–intensity curves shift to higher intensity values with the number of dB representing the loss in the middle ear. The latency–intensity function has attracted special attention.[5,14] There are, however, restrictions in employing this method. It has been observed that the width of the latency–intensity range for normal ears at each particular latency value is about 20 dB. The implication is that the minimum pure conductive hearing loss that can be detected on this basis will be 20 dB, and in addition the inaccuracy in the amount of conductive loss exceeding this value will also be 20 dB. An additional complication arises when wave I is absent in the ABR recording. In such cases there is no control upon the amount of wave V delay attributable to an increased central conduction time resulting, e.g., from retardation in development. Especially in children this could frequently be accompanied by a conductive hearing loss. If a brainstem abnormality is not suspected, the amount of hearing loss will be greatly over estimated.[6,15] Separation of both effects can only be made on basis of a clear wave I.

The latency–intensity functions in recruiting ears either fall within the normal range or show slightly elevated latencies at near threshold values. When clicks are used, the influence of the audiogram shape upon the latency–intensity function is marked.[16] Claims have been made that mixed hearing losses are characterized by normal latencies at high intensities and increasing difference with normal latency at lower intensities. The same shift, however, is observed in pure high-frequency sensorineural hearing loss when the audiogram has a slope steeper than 30 dB/octave.[13,17,18]

Differentiation of mixed hearing losses of up to 50–60 dB into a conductive and sensorineural component on the basis of click ABR alone to me seems mainly based on inspired guesses. Neither the conductive part, nor the sensorineural part will show a flat loss across frequency.

A way out of this problem of audiogram estimation as well as differential diagnosis, without the expense of a large increase in recording time,[7] may be the estimation of a two-point audiogram. One point is given by the unmasked click ABR threshold and the latency–intensity function and will be in the 2–4 kHz range. Another point representing the 0.5–1 kHz range may be found by presenting the click in the presence of 1 kHz high-pass noise filtered at around 96 dB/octave and determining the ABR threshold under that condition together with the latency–intensity function. This combination can distinguish between rising, flat, or falling audiograms, and may also be more suited to differentiate mixed hearing losses.

References

1. Kodera, K., H. Yamane, O. Yamada, et al. 1977. Audiology **16:** 496.
2. Terkildsen, K., P. Osterhammel & F. Huis in't Veld. 1978. *In* Evoked Electrical Activity in the Auditory Nervous System. R. F. Naunton & C. Fernandez, Eds. p. 415. Academic Press, New York, N.Y.
3. Sohmer, H., M. Feinmesser & L. Bamberger-Tell. 1972. Ann. Otol. Rhinol. Laryngol. **81:** 72.
4. Hexoc, K. & R. Galambos. 1974. Arch. Otolaryngol. **99:** 30.
5. Galambos, R. & K. Hecox. 1977. Prog. Clin. Neurophysiol. **2:** 1.
6. Mokotoff, R., C. Schulman-Galambos & R. Galambos. 1977. Arch. Otolaryngol. **103:** 38.

7. DON, M., J. J. EGGERMONT & D. E. BRACKMANN. 1979. Ann. Otol. Rhinol. Laryngol. **88** (Suppl. 57): 1.
8. ARAN, J. M., J. PELERIN, et al. 1971. Rev. Laryngol. (Bordeaux) (Suppl) **92:** 601.
9. YOSHIE, N. 1973. Audiology **12:** 504.
10. BERGHOLTZ, L. M., S. D. ARLINGER, P. KYLEN, et al. 1977. Acta Otolaryngol. **84:** 385.
11. EGGERMONT, J. J. & M. DON. 1980. J. Acoust. Soc. Am. **68:** 1671.
12. COATS, A. C. & J. L. MARTIN. 1977. Arch. Otolaryngol. **103:** 605.
13. ARAN, J. M. & M. NEGREVERGNE. 1973. Audiology **12:** 488.
14. GERULL, G., G. GIESEN & D. MROWINSKI. 1978. Laryng. Rhinol. Otol. **57:** 54.
15. STARR, A. 1977. Prog. Clin. Neurophysiol. **2:** 45.
16. COATS, A. C. 1978. Arch. Otolaryngol. **104:** 709.
17. BERLIN, C. I., J. K. CULLEN, M. S. ELLIS, et al. 1974. Trans. Am. Acad. Ophthalmol. Otol.: 401.
18. YOSHIE, N. & T. OHASHI. 1969. Acta Otolaryngol. (Suppl. 252): 71.

EFFECTS OF CLICK POLARITY ON BRAINSTEM AUDITORY EVOKED POTENTIALS IN NORMAL SUBJECTS AND PATIENTS: UNEXPECTED SENSITIVITY OF WAVE V*

Ronald G. Emerson, Elizabeth B. Brooks, Stephen W. Parker, and Keith H. Chiappa

The Robert S. Schwab Computer Facility
Clinical Neurophysiology Laboratory
Department of Neurology
Massachusetts General Hospital
Harvard Medical School
Boston, Massachusetts 02114

INTRODUCTION

Brainstem auditory evoked potentials (BAEPs) can be recorded using condensation, rarefaction, or alternating-polarity clicks. There has been no systematic study of the effects of click polarity on the BAEP waveforms in normal subjects and patients. We report 20 patients in whom the presence or absence of wave V appeared to be dependent upon click polarity. In some of these patients, a paradoxical reappearance of wave V was seen with diminution of stimulus intensity. Effects of click polarity were also studied in 45 normal subjects. Minor latency and amplitude variations were seen, but in no subject was an absence, or intensity-dependent absence, of wave V noted with either rarefaction or condensation clicks.

METHODS

BAEPs were recorded in 45 normal subjects and 600 patients using both condensation and rarefaction clicks. The normal group consisted of 25 females and 20 males with an age range of 17–54 years. The results were analyzed to determine the differential effects of stimulus polarity upon BAEP waveforms.

Details of the BAEP testing procedure and normal values are presented elsewhere.[1,2] Testing was done in a quiet, darkened room with the subject supine. Head propping was used to minimize postural muscle activity. If necessary, sedation with chloral hydrate and diphenhydramine was used. Electrode impedances were maintained below 3000 ohms.

Monaural, constant-polarity click stimuli were produced by applying a 100 μsec square-wave pulse to earphones (Telex 1470). Polarity was adjusted to produce either a condensation or a rarefaction click stimulus which was delivered at a rate of 10 clicks per second.

The hearing threshold for the click stimulus was determined for each ear individually. During recording of the BAEP, the click was 70–80 decibels (dB) above the threshold level for that ear (dBSL). The contralateral ear was masked with white noise at an intensity 20–30 dB below that of the click.

*This work was supported in part by the Edwin Minot Fund for Research in Electroencephalography.

0077–8923/82/0388–0710 $1.75/0 © 1982, NYAS

TABLE 1

ABSOLUTE AND INTERWAVE LATENCIES IN RESPONSE TO RAREFACTION AND CONDENSATION CLICK STIMULI

	Mean ± SD		Range		P Values For Sex Difference	
	Rarefaction	Condensation	Rarefaction	Condensation	Rarefaction	Condensation
Wave I latency						
M & F	1.7 ± 0.14	1.8 ± 0.16	1.4–2.1	1.5–2.3		
F	1.7 ± 0.14	1.8 ± 0.17	1.4–2.1	1.5–2.3	ns†	ns
M	1.7 ± 0.12	1.7 ± 0.15	1.5–2.0	1.5–2.0		
Wave III latency						
M & F	3.8 ± 0.16	3.9 ± 0.17	3.5–4.3	3.5–4.2		
F	3.8 ± 0.14	3.9 ± 0.17	3.5–4.1	3.5–4.2	<0.005	ns
M	3.9 ± 0.17	3.9 ± 0.17	3.7–4.3	3.5–4.2		
Wave V latency						
M & F	5.8 ± 0.22	5.8 ± 0.23	5.3–6.3	5.4–6.4		
F	5.7 ± 0.20	5.7 ± 0.21	5.3–6.1	5.4–6.4	<0.001	<0.005
M	5.9 ± 0.20	5.9 ± 0.22	5.6–6.3	5.5–6.3		
I–III latency						
M & F	2.1 ± 0.14	2.1 ± 0.12	1.9–2.5	1.7–2.4		
F	2.1 ± 0.12	2.1 ± 0.11	1.9–2.5	1.7–2.4	<0.005	<0.025
M	2.2 ± 0.15	2.2 ± 0.13	1.9–2.5	1.9–2.4		
III–V latency						
M & F	1.9 ± 0.18	1.9 ± 0.17	1.5–2.4	1.6–2.3		
F	1.9 ± 0.17	1.9 ± 0.16	1.5–2.2	1.6–2.3	ns	ns
M	2.0 ± 0.18	2.0 ± 0.14	1.7–2.4	1.7–2.3		
I–V latency						
M & F	4.1 ± 0.19	4.0 ± 0.20	3.6–4.5	3.4–4.6		
F	4.0 ± 0.18	4.0 ± 0.20	3.6–4.4	3.4–4.3		
M	4.2 ± 0.14	4.2 ± 0.15	3.9–4.5	3.9–4.6	<0.001	<0.001

*Based on 35 normals in whom waves were well defined for statistical purposes. (Also true for TABLES 2 and 3 and FIGURES 2 and 3.)
†Not significant.

Activity was recorded using a bipolar montage derived from ipsilateral earlobe and vertex, with the ground located on the forehead. If wave I was not recognizable with the earlobe electrode, a conventional electroencephalographic needle electrode was inserted a few millimeters below the skin of the anterior wall of the external auditory canal. Bipolar recordings were then obtained between the needle and the vertex. This usually helped to resolve wave I.

Amplifier gain was 500,000 times, with low and high bandpass filters set to 100 and 3000 Hz (1/2 amplitude range), respectively. Activity was recorded for 10 msec following a 0.6 msec post-stimulus delay. Generally four separate trials of 1024 responses each were obtained. Superimposed individual trials were plotted, along with the grand average. Occasionally, in cases where the waveforms were very well defined, two (but never fewer) separate trials were obtained for any given stimulus polarity. *P* values were calculated using the *t*-test for paired and independent samples.

Results

Normals

Among the 45 normal subjects, variation of click polarity resulted in minor alterations in the latency and amplitude of waves I through V. In no normal subject was wave V absent with either polarity. With rarefaction clicks the mean absolute

TABLE 2

INTER-POLARITY DIFFERENCES*

	Mean ± SD	Range	*P* Value for Rarefaction vs Condensation Clicks
Wave I latency			
M & F	0.05 ± 0.10	−0.1–0.4	<0.01
F	0.06 ± 0.11	−0.1–0.4	<0.005
M	0.03 ± 0.07	−0.1–0.2	<0.05
Wave III latency			
M & F	0.02 ± 0.12	−0.3–0.3	ns
F	0.05 ± 0.11	−0.2–0.3	<0.05
M	0.02 ± .13	−0.3–0.3	ns
Wave V latency			
M & F	0.02 ± 0.10	−0.1–0.4	ns
F	0.02 ± 0.10	−0.1–0.4	<0.01
M	0.01 ± 0.10	−0.1–0.3	ns
I–III latency			
M & F	−0.03 ± 0.11	−0.4–0.3	<0.025
F	−0.02 ± 0.11	−0.4–0.3	ns
M	−0.04 ± 0.09	−0.3–0.1	<0.01
III–V latency			
M & F	0.0 ± 0.14	−0.4–0.3	ns
F	−0.02 ± 0.14	−0.4–0.3	ns
M	0.02 ± 0.12	−0.1–0.3	ns
I–V latency			
M & F	−0.03 ± 0.11	−0.3–0.2	<0.01
F	−0.04 ± 0.11	−0.3–0.2	<0.025
M	−0.02 ± 0.10	−0.2–0.2	ns

*See notes to TABLE 1.

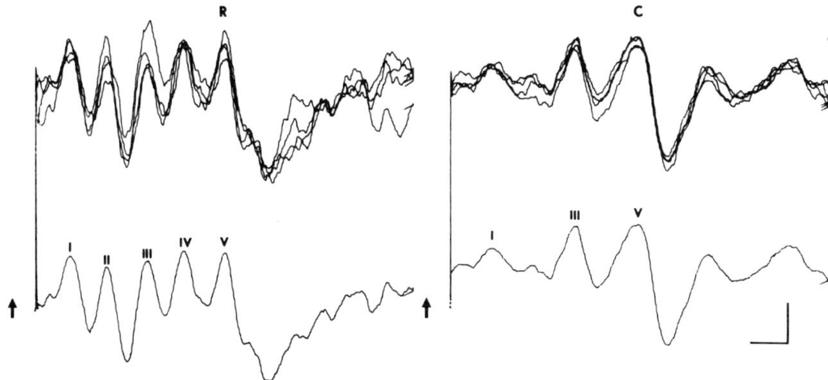

FIGURE 1. BAEPs obtained from one ear in a single normal subject showing clear waves IV and V following rarefaction clicks, and fusion of waves IV and V following condensation clicks. The left two traces show four superimposed trials of 1024 clicks each following a rarefaction stimulus with the average of the four directly underneath. The right two traces show the superimposed trials and average following condensation clicks. Derivation is the earlobe ipsilateral to monaural stimulation, to the vertex, with vertex positivity producing an upward deflection. Calibration marks are 0.25 μV and 1.0 msec. The arrow indicates the time of stimulation.

wave I latency (stimulus to peak) was significantly shortened ($P < 0.01$). Mean absolute latencies (stimulus to peak) for waves III and V were also less with rarefaction clicks, but this difference was not statistically significant. I–III and I–V interwave latencies were slightly greater for rarefaction clicks than for condensation ($P < 0.05$) (TABLES 1 & 2). Condensation clicks tended to fuse waves IV and V into a single broad peak, whereas rarefaction most often produced two distinct peaks (FIGURE 1). Wave I amplitude (positive peak to following negative trough) was lower with condensation clicks in 75% of the ears, higher in 18% and unchanged in 7% ($P < 0.01$). Wave V amplitude was higher with condensation clicks in 66%, lower in 30%

TABLE 3
AMPLITUDES*

	Mean ± SD		Range	
	Rarefaction	Condensation	Rarefaction	Condensation
Wave I				
M & F	0.29 ± 0.12	0.21 ± 0.08	0.10–0.64	0.07–0.47
F	0.31 ± 0.12	0.22 ± 0.09	0.10–0.64	0.07–0.47
M	0.26 ± 0.11	0.19 ± 0.07	0.14–0.55	0.07–0.40
Wave III				
M & F	0.26 ± 0.13	0.24 ± 0.10	0.05–0.65	0.05–0.55
F	0.27 ± 0.13	0.25 ± 0.10	0.05–0.57	0.05–0.45
M	0.24 ± 0.12	0.21 ± 0.11	0.07–0.65	0.10–0.55
Wave V				
M & F	0.39 ± 0.12	0.45 ± 0.14	0.16–0.64	0.16–0.93
F	0.42 ± 0.11	0.49 ± 0.15	0.18–0.64	0.21–0.93
M	0.33 ± 0.12	0.37 ± 0.08	0.16–0.59	0.16–0.51

*See notes to TABLE 1.

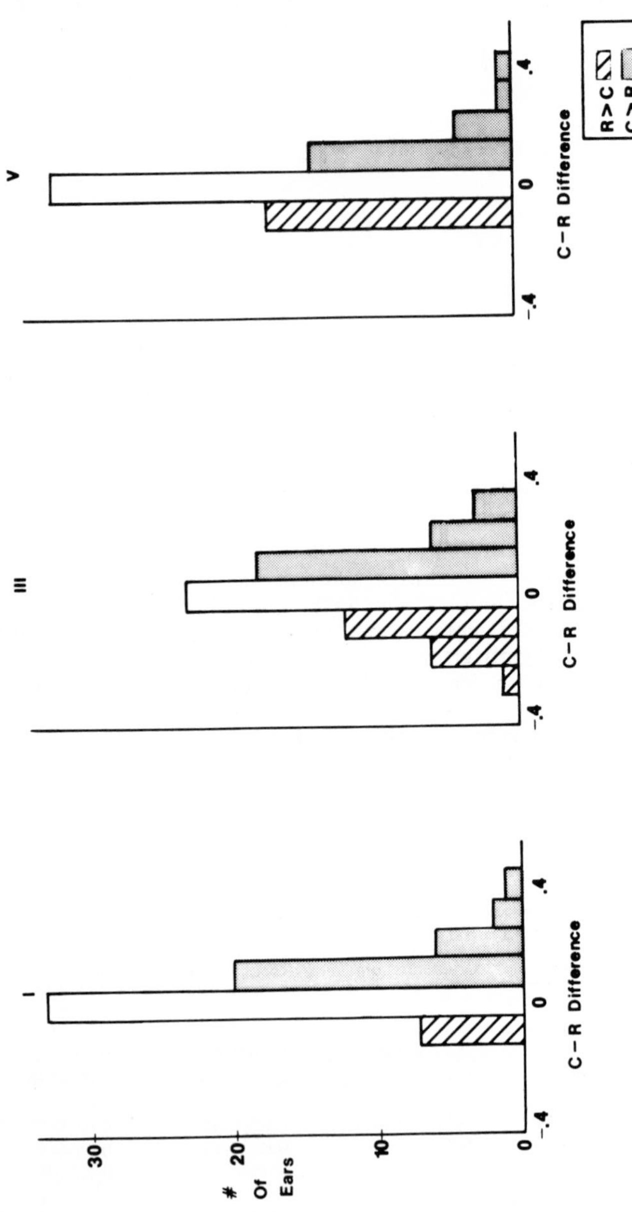

FIGURE 2. The distribution of the interpolarity latency differences for waves I, III, and V following rarefaction and condensation clicks. The horizontal axis is the latency difference (rarefaction minus condensation), in msec, with a negative value indicating a longer latency for rarefaction and a positive value indicating a longer latency for condensation. The vertical axis is the number of ears with a given millisecond interpolarity difference.

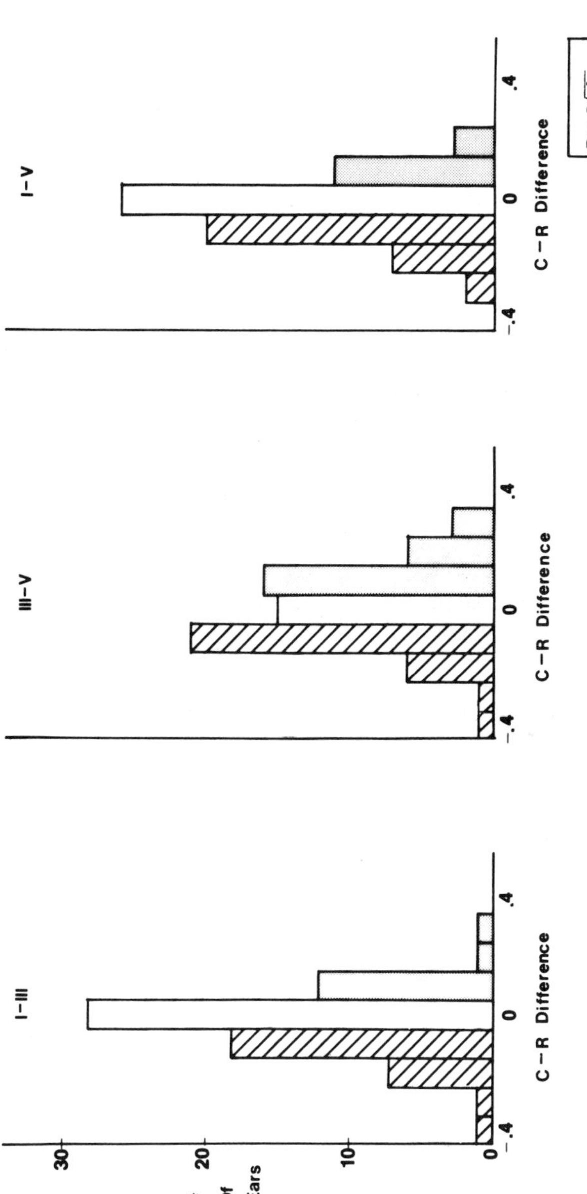

FIGURE 3. The distribution of the interpolarity latency differences for the I–III, III–V, and I–V interwave separations following rarefaction and condensation clicks. The axes are the same as those in FIGURE 2.

and unchanged in 4% ($P < 0.01$). No significant effect upon wave III was observed (TABLE 3).

Minor sex differences in the absolute wave V and the I–III and I–V interwave latencies were seen with both rarefaction and condensation clicks. Nonetheless, for both sexes, the absolute latency of wave V was shorter following rarefaction clicks and the I–III and I–V interwave latencies were shorter following condensation clicks (TABLE 1 for *P* values).

Despite significant differences in mean latencies and amplitudes of the waveforms produced by rarefaction and condensation clicks, many normals did not conform to the typical pattern, in which rarefaction clicks produced shorter absolute latencies and condensation produced shorter interwave latencies. For example, I–V interwave latencies were longer with condensation clicks in 14 ears. The distributions of interpolarity wave and interwave latency differences are shown in FIGURES 1 and 2.

Patients

Among the 600 patients, 20 were found in whom the presence or absence of wave V appeared to be dependent upon click polarity (14 females and 6 males; age range, 14–65 years). The diagnosis for these 20 patients are listed in TABLE 4. In 17 of the 20 patients (85%), wave V was seen with condensation but not with rarefaction clicks (FIGURE 4); in the remaining 3, wave V was seen only with rarefaction clicks. In all 20 cases, the recordings were otherwise normal and wave V, when obtained, fell within normal latency limits. In one case the III–V interwave latency was at the upper limit of normal. Audiometric examinations were obtained in 5 of the 20 patients. They were normal in two patients and abnormal in three (see TABLE 5). Among the patients for whom audiograms were not obtained, 8 had normal click hearing thresholds and the remaining 6 had mild losses, (mean: −11 dB; worst: −23 dB).

Seven of these patients were subjected to additional testing at decreased stimulus intensities. In all 7, diminution of stimulus intensity, e.g., from 70 to 55 dBSL, resulted in the reappearance of a normal wave V (FIGURE 5).

DISCUSSION

Maurer *et al.*[3] found that rarefaction stimulation resulted in shorter wave I latencies, longer I–III interwave latencies, and more distinct waves IV and V. Ornitz *et al.*[4,14] noted that rarefaction produced earlier absolute latencies of waves I and V. Stockard *et al.*[5] found that wave V absolute latencies did not vary with stimulus polarity, but that the absolute latencies of waves I through IV were shorter with rarefaction stimulation. Other studies have also noted shorter latencies with rarefaction clicks.[6] We also found small differences, with rarefaction stimulation generally producing slightly shorter absolute latencies and slightly longer interwave latencies.

Our observation of the selective loss of wave V with one stimulus polarity but not the other has not been previously reported. The reappearance of a wave V, clearly absent at one intensity, following stimulation at a lower intensity has not been described. In general, when stimulus intensity is decreased, the amplitude of wave V decreases and its absolute latency increases.[5,7–9] It is not known whether the sensitivity of wave V to stimulus polarity, and its paradoxical reappearance with diminished stimulus intensity, represents an abnormality or a normal variation. Since this was seen in 20 of 600 patients and in none of 45 normal subjects, it may be either a true abnormality or an infrequently occurring normal variant. Whether these phenomena

TABLE 4
INCIDENCE OF LACK OF APPEARANCE OF WAVE V

Diagnosis	Patients with Absent Wave V with Rarefaction	Patients with Absent Wave V with Condensation
Multiple sclerosis (def. prob.)*	9	2
Multiple sclerosis (poss.)*	5	1
Primary lateral sclerosis	1	0
Brainstem tumor	1	0
Head trauma	1	0
Total	17	3

*Using McAlpine's criteria.[15]

are mediated by the brainstem, eighth nerve, or cochlea remains to be discovered, although the presence of a normal audiogram in some of these patients suggests that it may be a central effect.

The effects of rarefaction and condensation stimulation on the firing rates of single acoustic nerve fibers as well as on the N1 potential of the electrocochleogram

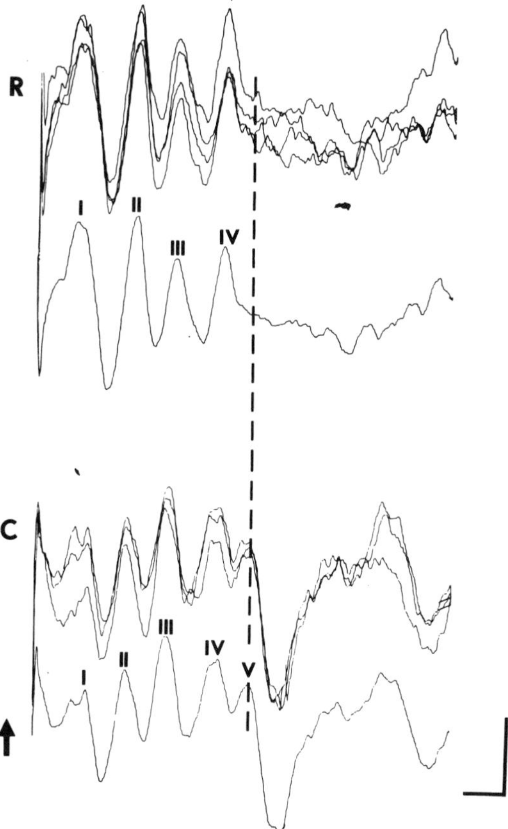

FIGURE 4. BAEPs obtained from one ear in a single patient showing the absence of wave V following rarefaction clicks. The upper two traces show four superimposed trials of 1024 clicks each following a rarefaction stimulus, with the average of the four directly underneath. The lower two traces show the superimposed trials and average following condensation clicks. Derivation and calibration marks are the same as in FIGURE 1.

TABLE 5
AUDIOMETRIC EXAMINATION FINDINGS

No. of Patients	BAER Findings	Audiogram	Speech Discrimination	Reflex Decay	Stapedial Reflexes
1	V absent with rarefaction	Normal	Normal	Neg.	Normal
1	V absent with condensation	Normal	Normal	Neg.	Normal
1	V absent with rarefaction	50 dB loss 3 kHZ & up	Normal	Neg.	Normal
1	V absent with rarefaction	Mild high tone sloping loss	Normal	Neg.	Normal
1	V absent with condensation	Flat 40–50 dB loss	Normal	Neg.	Not done

(equivalent to wave I of the BAEP) in the cat were examined in detail by Kiang *et al.*,[10] Antoli-Candels and Kiang,[11] and Peake and Kiang.[12] Following a click, a series of discharges is recorded from a single eighth-nerve fiber. The latency of the first discharge is shorter for rarefaction than for condensation clicks. The temporal relationship of single unit activity to click stimuli is markedly altered by changes in stimulus intensity and polarity, as demonstrated by the post-stimulus time histogram. Furthermore, Kiang *et al.*[12] found that the N1 latency is shorter following rarefaction clicks and that its appearance varies markedly with stimulus intensity. We observed similar effects of stimulus polarity upon the latency of wave I (FIGURE 2).

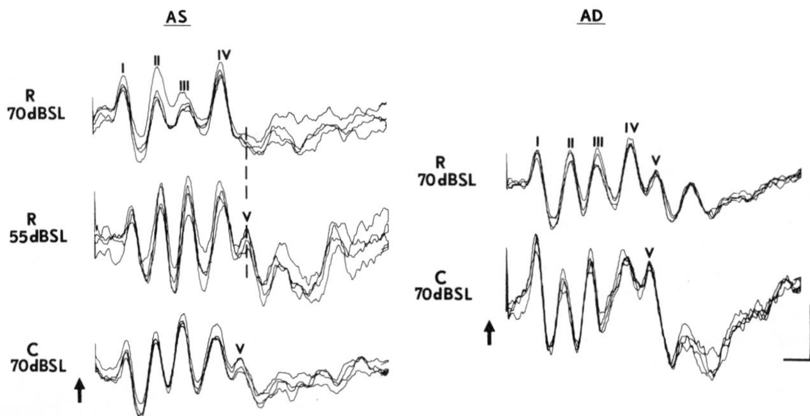

FIGURE 5. BAEPs obtained from the left and right ears of a single patient following monaural stimulation, showing intensity and polarity sensitivity of wave V. Left ear (AS) shows the lack of appearance of wave V following rarefaction clicks at 70 dBSL (top trace), however, wave V appears when the click intensity is decreased 15 dB with the same polarity (2nd trace). Wave V is present at 70 dBSL following condensation clicks in that ear (bottom trace). Right ear (AD) shows a normal ear with all waves present regardless of polarity. In all cases, four superimposed trials of 1024 clicks each are shown. Derivation and calibration marks are the same as in FIGURE 1.

In view of the sensitivity of wave I to stimulus polarity and intensity, it is not surprising that subsequent waveforms whose generation is logically contingent upon the existence of wave I are similarly sensitive. However, the relationship between the polarity dependent wave I behavior described by Kiang *et al.* and the wave V variations described in the present paper is speculative.

Since waves II through V represent structures in the auditory pathway central to the eighth nerve, alterations in the configuration of wave V may be due to a brainstem abnormality. It is equally plausible, however, that alterations in the sequential activation of hair cells in the cochlear membrane might produce a cancellation of wave V.

Levine[13] has described a derived waveform that is thought to be mediated by only central structures. BAEPs are recorded with both simultaneous binaural and monaural stimulation. The binaural waveform is subtracted from the sum of the left and right monaural waveforms. The resultant binaural difference waveform is felt to represent neural binaural interaction free from peripheral effects. The application of this technique may help to define the structures responsible for these phenomena (central versus peripheral). Ultimately histopathologic demonstration of a minimum necessary anatomic lesion would resolve the question.

Ideally, one would use both rarefaction and condensation stimulation when recording BAEPs. In the clinical setting, however, this may not be practical. We recommend that if one is to employ only a single stimulus polarity, rarefaction clicks should be used. This will produce the clearest waveforms in most cases.[3,5] Furthermore, the greatest number of cases with a polarity sensitive wave V will be revealed. Condensation can be used in those patients in whom rarefaction does not produce clear waveforms, or when wave V appears absent. The use of alternating polarity clicks as the only mode of stimulation is discouraged because of the possibility of phase cancellations.

Although we have not seen polarity-dependent disappearance of wave V in normal subjects, it remains to be demonstrated whether this is an abnormality or an infrequently occurring normal variant. At present, the clinical interpretation should reflect this uncertainty.

References

1. Chiappa, K. H., K. J. Gladstone & R. R. Young. 1979. Brainstem auditory evoked responses: Studies of waveform variations in 50 normal human subjects. Arch. Neurol. **36:** 81–87.
2. Chiappa, K. H., J. H. Harrison, E. B. Brooks & R. R. Young. 1980. Brainstem auditory evoked responses in 200 patients with multiple sclerosis. Ann. Neurol. **7:** 135–143.
3. Maurer, K., E. Schafer & H. Leitner. 1980. The effect of varying stimulus polarity (rarefaction vs. condensation) on early auditory evoked potentials (EAEPs). Electroencephalogr. Clin. Neurophysiol. **50:** 332–334.
4. Ornitz, E. M., A. Mo, S. T. Olson & D. O. Walter. 1980. Influence of click sound pressure direction on brain stem responses in children. Audiology **19:** 245–254.
5. Stockard, J. J., J. E. Stockard & F. W. Sharbrough. 1978. Nonpathologic factors influencing brainstem auditory evoked potentials. Am. J. EEG Technol. **18:** 177–209.
6. Elberling, C. 1974. Action potentials recorded along the cochlear partition recorded from the ear canal in man. Scand. Audiol. **3:** 13–19.
7. Rowe, M. J., III. 1978. Normal variability of the brain-stem auditory evoked response in young and old adult subjects. Electroencephalogr. Clin. Neurophysiol. **44:** 459–470.
8. Coats, A. C. 1978. Human auditory nerve action potentials and brain stem evoked responses. Arch. Otolaryngol. **104:** 709–717.

9. DON, M., A. R. ALLEN & A. STARR. 1977. Effect of click rate on the latency of auditory brain stem responses in humans. Ann. Otol. **86:** 186–195.
10. KIANG, N.Y.S. 1965. Discharge Pattern of Single Fibers in the Cat's Auditory Nerve. Res. Monogr. No. 35. MIT Press. Cambridge, Mass.
11. ANTOLI-CANDELA, F., JR. & N. Y. S. KIANG. 1978. Unit activity underlying the N1 potential. *In:* Evoked Electrical Activity in the Auditory Nervous System. R. F. Naunton & C. Fernandez, Eds. pp. 165–191. Academic Press, New York, N.Y.
12. PEAKE, W. T. & N. Y. S. KIANG. 1962. Cochlear responses to condensation and rarefaction clicks. Biophys. J. **2:** 23–34.
13. LEVINE, R. A. 1981. Binaural interaction in brain stem potentials of human subjects. Ann. Neurol. **9:** 384–393.
14. ORNITZ, E. M. & D. O. WALTER. 1975. The effect of sound pressure waveform on human brain stem auditory evoked responses. Brain Res. **92:** 490–498.
15. MCALPINE, D., C. E. LUMSDEN & E. D. ACHESON. 1972. Multiple Sclerosis: A Reappraisal. 2nd edit. p. 202. Williams and Wilkins, Baltimore, Md.

DISCUSSION OF THE PAPER

R. GALAMBOS: I think Dr. Chiappa has given us a very nice example of the importance of selecting one or the other of the polarities when you make these measurements.

As he knows, virtually everybody in this business has made the kinds of measurements he is talking about and have come up with very much the same sort of conclusions that he has, namely, that the rarefaction pulse is probably the better one to use if you have to use only one.

For those of you who are interested in this, initially people used mixed condensation and rarefaction clicks in order to get rid of that artifact at the front end and this was one way to do it. Then came the observation that the way we used condensation or rarefaction clicks did make a difference in the morphology of the waves.

A point that he did not stress is that the two waves that move most, although it was apparent in his recording, are waves IV and VI, whereas I, II, III, and V have a tendency not to move very much. Remember, the numbers that he gave for waves I, III, and V were on the order of 0.05 or 0.2 in his various configurations. When we tried to measure waves IV and VI we came up with, really, huge numbers such as 0.4 or even 0.5 msec between condensation and rarefaction clicks, which suggested that we may well be looking at two different systems, similar to the point that Dr. Chiappa made; one of the systems is responsible for yielding the peaks of waves IV and VI, and the other, the four remaining waves.

This, however, is just another one of those speculations, and, as he says, this is going to be a very difficult problem to parcel out.

K. H. CHIAPPA: I did not mean to suggest that no one else has studied this in normal subjects. We undertook this study because we properly got our hands slapped for not taking into account click polarity, but, I think the observation in patients of wave V absence with one click polarity and not with the other is a new finding.

A. STARR: Except your data says that if you drop the intensity then wave V appears, so, it is not absent. It happens to be absent at that particular intensity.

Is this a neurological issue or is it an end organ problem? I would bet that it is an end organ rather than a central processing problem.

CHIAPPA: The point there is that normal subjects do not show that at any click intensity.

STARR: When you get to 600 normals let me know.

CHIAPPA: I probably will never get to 600 normal subjects.

J. J. EGGERMONT: I have been intrigued by this observation of Dr. Galambos about waves IV and VI. If you have a low frequency hearing loss as in Meniere's disease, it is likely that wave VI is absent because the dominant contribution in the low frequency part is not there and the negative wave from the basal region masks the real wave VI from the basal part of the cochlea.

There is a real wave VI. There is no question about it, but normally it is 60% from the apical portion of the cochlea and the same holds for wave IV. It is actually a delayed wave III.

In this condensation–rarefaction work, there is definitely a peripheral effect. There should be about a half-wave shift, especially if some parts of the nerves have large problems as in myelinating disease so that they are not that well synchronized. Maybe other regions of the nerve are well synchronized. This may displace certain specific waves such as IV and VI. I think the likelihood of differences in ABR to click polarity derives from peripheral end organ mechanisms.

CHIAPPA: But, that myelin would, then, be central myelin and that is what we want to know anyway. Is there a lesion in central myelin? Even if it is in part of the eighth nerve, it has central myelin on it. That is what we are looking for. I call that a central effect.

COMMENT: An acoustic tumor is also a central thing. Anything peripheral to the cochlea nucleus is, for me, really peripheral and not central. That is a matter for discussion now.

CHIAPPA: The variability of IV and VI in normals is why, in neurologic applications, we stress measurements of waves I, III, and V.

TACTILE AND AUDITORY STIMULI REPEATED AT HIGH RATES (30–50 PER SEC) PRODUCE SIMILAR EVENT RELATED POTENTIALS*

Robert Galambos

Department of Neurosciences
University of California, San Diego
La Jolla, California 92093

Speech Hearing and Neurosensory Center
San Diego, California 92123

Only a few investigators have attempted to record event related potentials (EPRs) at stimulus rates above 20 or 30 stimuli per sec (Hz). Van der Tweel and Verduyn Lunel,[6] in initiating the study of what are now called steady-state visual ERPs, applied sinusoidally modulated light stimuli at rates up to about 70 Hz and recorded approximately sinusoidal responses from the scalp throughout this range. Subsequently Regan and colleagues[4] extensively explored the relationship of response amplitude and stimulus rate, identifying three ERP amplitude peaks of which one was in the 40–55 Hz region.

In somesthesia Namerow[5] applied median nerve shocks at rates up to 100 per sec or more, recording from scalp overlying the cortical projection of the stimulated limb. Substantial sinusoidal ERPs were recorded at stimulus rates above 20 Hz but no important response peak was found at a particular frequency.

In audition Campbell *et al.*[3] have reported the auditory analog of the visual steady-state experiment, similarly recording nearly sinusoidal ERPs from vertex-mastoid electrodes to rapidly-repeated clicks or tone bursts. Response amplitude declined so regularly with drop in stimulus intensity that the curve could be used to predict threshold of audibility with remarkable accuracy.

Campbell *et al.* did not examine ERPs to stimuli at rates above 32 per sec. We have recently done this and reported[2] the following main results: (1) For many subjects sinusoidal, steady-state ERP amplitudes peak at stimulus rates near 40 per sec, and for this reason we have named the phenomenon the auditory 40 Hz ERP; (2) the 40 Hz ERP has a widespread scalp distribution; (3) it is produced by a wide range of repetitive auditory stimuli, including pulsed tone bursts made up of frequencies as low as 125 Hz; and (4) the response appears to represent a consolidation, or superimposition of the several ERP waves known as the auditory middle latency response.

Our report also raises the question of whether the 40 Hz ERP is a phenomenon common to all modalities of sensation. In line with this thought we have been applying vibrotactile stimuli to fingers or toes at high stimulus rates—between about 10 and 60 Hz. Details of these studies will be reported in a publication now in preparation.

The purposes of the present report are: (1) to demonstrate separate auditory and somesthetic "40 Hz ERPs," (2) to show a way in which they interact, and (3) to discuss the possible significance of these facts.

The demonstration, summarized in FIGURE 1, involved a subject seated before a

*This work was supported by Grants NS 11707 and 11154 from the National Institutes of Health and BNS 77-14923 from the National Science Foundation.

0077–8923/82/0388–0722 $1.75/0 © 1982, NYAS

table on which an ordinary 8-inch loudspeaker had been placed about a half meter from his head. Standard electrodes on the forehead (C_Z does just as well) and each earlobe conducted his EEG to an amplifier (gain: 10^5; bandpass 10–100 Hz) connected to an averaging computer. A brief electrical pulse (one half cycle of a 250 Hz sinusoid repeated 33 times per sec for FIGURE 1) simultaneously activated the loudspeaker and triggered the computer. (As the reader will recognize, this experimental setup closely resembles—except for amplifier bandpass—that commonly used to record the ABR, or auditory brainstem response).

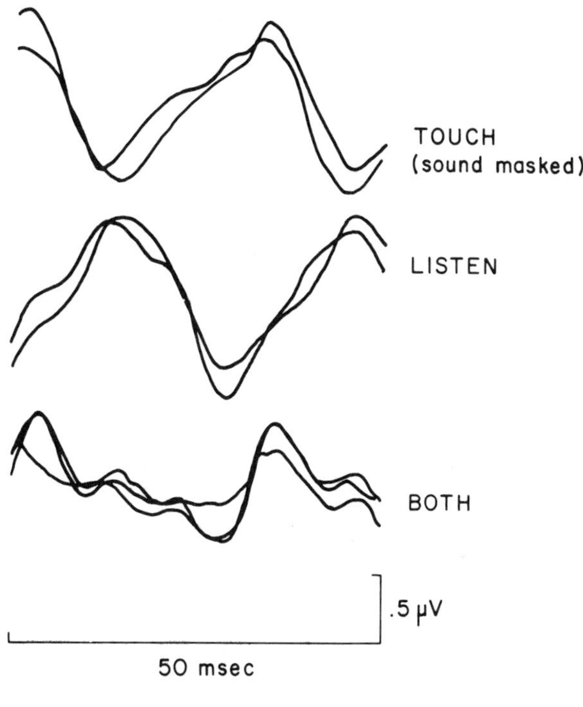

FIGURE 1. Intermodality cancellation of 40 Hz evoked response potentials. Stimulus: loudspeader motion. The brain potentials associated with touching a vibrating loudspeaker cone (top line) and with listening to the sound it produces (middle line) approximately cancel each other (bottom line) when both sensations are simultaneously experienced. The text provides further details.

In FIGURE 1 the top records were made while the subject (in this case the author) rested the index and middle fingers of his right hand upon the vibrating speaker cone. He was at this time wearing earphones through which white noise was being applied to mask the sound produced by the loudspeaker. The middle records were made while the subject, having removed the earphones, sat with hands folded in his lap listening passively to the sound the loudspeaker produced. The bottom records were made as he simultaneously listened to the sound and felt the vibration of the loudspeaker cone that produced it. Each of these conditions was repeated to produce the superimposed traces, each of which represents the average of 2000 presentations.

The major point to be made from FIGURE 1 is that both modalities yield similar ERPs when their adequate stimuli are repeated 33 times per sec. Thus, both responses approximate sinusoids, and their amplitudes (about 1.25 μV) are nearly equal (for which careful preliminary adjustments of stimulus rate, voltage applied to the loudspeaker and positions of head and fingers are mainly responsible). The large phase difference that separates them is due, we believe, to the difference in time required for the respective sensory impulses to reach their ERP generators in the brain: the ear is relatively close, whereas the fingers are distant by the amount of time required for impulses to reach the brain after conduction through the peripheral nerve and spinal pathways. Fortuitously, this phase difference is nearly 180 degrees, and so when the same stimulus simultaneously produces touch and sound sensation the bimodal ERP is much reduced, reflecting this phase cancellation as shown in the bottom traces of FIGURE 1. As we will report in detail elsewhere this inter- or cross-modality algebraic summation of ERPs (which yields reduction of amplitude in FIGURE 1) yields amplitude enhancement if the auditory stimulus (presented this time via earphones) is delayed by an amount necessary to bring the two ERPs into the same phase relationship. Furthermore, very similar rules hold for intra- or within-modality stimulation also: an auditory 40 Hz ERP produced by binaural stimulation will be larger than either monaural ERP with simultaneous (in-phase) stimulus presentation and smaller with out-of-phase. Algebraic summation phenomena like this were, I believe, first reported by van der Tweel and Verduyn Lunel[6] who showed that visual ERPs resembling those described here become larger or smaller depending on whether the sinusoidally modulated light stimuli they used were applied to the two eyes in-phase or out-of-phase.

Discussion of these results will be limited to some proposed anatomical generators of these ERPs, and to some ways they might prove useful in clinical situations.

GENERATORS

The ERPs of FIGURE 1 probably reflect activity in a brain region where afferents from the two modalities converge and produce similar effects rather than, as in the classical lemniscal systems, where one modality is anatomically segregated and separated from the other. Many such extralemniscal and polysensory regions are known to exist in the brainstem. Our current working hypothesis holds that the convergence occurs in the rostral reticular formation, perhaps in the human equivalent of the intralaminar and center median regions most recently described for cat by Buchwald *et al.*[1] Here we presume the response to input from the peripheral receptors in any modality to be a brief series of events in the 30–50 Hz range. [This assumption is particularly well supported by the auditory case where stimuli at slow rates evoke the so-called middle latency response (latency: 8–100 msec), which closely resembles a series of 3 or 4 sine waves]. When stimuli are delivered at rates within this 30–50 Hz range the waves are "driven" more or less well, just as the brain wave alpha rhythm can be more or less well driven or synchronized by light flashes presented at rates near 10 Hz.

The wave-like ERPs generated in the postulated subcortical site are further assumed to spread to the scalp electrodes by volume conduction, or, alternatively, to generate a similar response in cortex following conduction of impulses to it, or to produce currents at the recording electrodes by both mechanisms; the uncertainty on this point stems from the demonstration by Buchwald *et al.*[1] in cat that a prominent, scalp-recorded ERP component presumed to arise in cortex persists after total decortication.

Within this hypothetical polysensory region (possibly located in the thalamus) we suppose that each modality (including the visual) engages a different group of cells, not that all converge upon the same group of cells. We prefer the first alternative because explaining the algebraic cancellation of FIGURE 1 by convergence of two inputs upon the same cells would require widespread and precisely timed synaptic inhibitions that would reduce or eliminate the current flowing from the synaptic region; this seems less likely than supposing that two different cellular aggregates generate equal currents opposite in phase which simply add externally.

How these ideas will survive the experiments needed to test them is, of course, an open question. It seems clear, however, that the cellular activities responsible for generating the late cortical waves (N_1–P_2; latencies: 100 msec plus) must be different from those under discussion here. The late cortical waves show refractory periods some two orders of magnitude different from those of the 40 Hz ERPs. The late wave cellular events cause responses to drop progressively in amplitude as stimulus rate rises above about 1 per 10 sec and so are not likely to be the same ones as those generating a response having an amplitude maximum in the 30–50 Hz stimulus repetition range.

Clinical Applications

The ideal clinical test would be so innocuous, unobtrusive and trauma-free that a patient would scarcely be aware an examination was under way. It would be all over in a short time, with data collected, analyzed and presented in a form that unequivocally answers specific questions important for patient diagnosis and treatment. On several of these points the ERPs here under discussion receive high marks. Thus, to obtain one trace like those in FIGURE 1 requires a minute or two during which a patient would be inconvenienced by EEG electrodes, earphones, and a sensation of buzzing in the ears or at the finger tips. The responses recorded are robust in size, replicable, and they stand alone without need for further processing; additional analysis will require, mostly, only a ruler for measuring amplitudes.

As we argue elsewhere,[2] the auditory 40 Hz ERP, because it measures hearing thresholds so promptly and accurately in normal listeners, may prove a convenient way to estimate sensitivity at the speech frequencies in babies and children difficult to test with conventional methods. If this turns out to be the case, the auditory 40 Hz ERP will add a needed and useful tool to the audiologist's armamentarium. Whether similar specific clinical applications of the tactile ERP will develop remains to be seen. A necessary first step will be to make readily available a vibrotactile device that can be accurately calibrated and properly shielded against electrical artifacts to replace crude laboratory solutions like the loudspeaker cone of FIGURE 1. If such a device were available, physicians might compare and contrast the clinical information obtainable from the finger tip buzz with that yielded by median (or other) nerve shocks. The population of fibers stimulated by the two procedures would surely be different, with the tactile method the only reasonable one for initiating impulses from a specific peripheral sensory field. However, precisely identifying which end organs are responsible for producing responses like those of FIGURE 1 may challenge analysis: apparently those in the skin of the finger tip are involved because the major sensation experienced is of vibration there, but the joints also move, and so a contribution from excited joint receptors is also probable.

Important clinical applications for these ERPs would undoubtedly be forthcoming if it can be shown that they originate in some such extralemniscal polysensory region as the one outlined above. Existing clinical ERP tests—the ABR and the other auditory, somesthetic, and visual EPs—either deliberately measure activities in the specific lemniscal pathways or are generally interpreted as reflecting these with more or less precision. Electric responses known to originate in an extralemniscal site would, so to speak, open a window through which activities in the brainstem core could be visualized. Just as the ABR now allows observation of damage to a lemniscal pathway due to tumor, trauma, or demyelination, so a specific measure of activity at some grand junction of the extralemniscal ascending sensory pathways might permit estimates of level of consciousness, state of sleep, and the effect of drugs. If one believes behavioral states and the changes in them induced by external and internal events are mediated by brain cell activities outside the lemniscal pathways, a good place to look for them is in just such polysensory brain regions as the one apparently identified by the experiment of FIGURE 1.

REFERENCES

1. BUCHWALD, J. S., C. HINMAN, R. J. NORMAN, C.-M. HUANG & K. A. BROWN. 1981. Middle- and long-latency auditory evoked responses recorded from the cortex of normal and chronically lesioned cats. Brain Res. **205:** 91–109.
2. GALAMBOS, R., S. MAKEIG & P. TALMACHOFF. 1981. A 40-Hz auditory potential recorded from the human scalp. Proc. Natl. Acad. Sci. USA **78:** 2643–2647.
3. CAMPBELL, F. W., J. ATKINSON, M. R. FRANCIS & D. M. GREEN. 1977. Estimation of auditory thresholds using evoked potentials. *In* Auditory Evoked potentials in Man. Psychopharmacology Correlates of Evoked Potentials. J. E. Desmedt, Ed. Vol. 2: 68–78. Karger, Basel, Switzerland.
4. REGAN, D. 1975. Recent advances in electrical recording from the human brain. Nature **253:** 401–407.
5. NAMEROW, N. S., R. J. SCLABASSI & N. F. ENNS. 1974. Somatosensory responses to stimulus trains: Normative data. Electroencephalogr. Clin. Neurophysiol. **37:** 11–21.
6. VAN DER TWELL, L. H. & H. F. E. VERDUYN LUNEL. 1965. Human visual responses to sinusoidally modulated light. Electroencephalogr. Clin. Neurophysiol. **18:** 575–586.

DISCUSSION OF THE PAPER

QUESTION: As a neurologist, one of the tests one does is extinction, to look for changes in sensation. Some patients will extinguish somatosensory input from one side if you simultaneously stimulate the other side. Interestingly enough, there are patients who will extinguish across modalities. They will hear a click and they will not report to you the touch, and so it appears that you are getting to the pathophysiology.

What is the percept in the auditory modality when you do not get any electrical responses? What happens to the binaural percept?

R. GALAMBOS: Well, if you put clicks or short tone bursts at 40 per second in the one ear, you hear a certain fluttering sound. When you put it into the other ear in-phase you hear a flutter. When you put it in out of phase, then, you hear twice as much flutter.

QUESTION: There is no electrical response?

GALAMBOS: Not when they are cancelled. In the electrophysiological cancellation situation there is not a corresponding cancellation of the percept. I very carefully studied that. You can imagine my reaction when I was listening to that loudspeaker, touching the loudspeaker, looking at the tube face, and seeing there was no electric response. I was saying, well, now have I lost something in the way of sensation, but I could easily go from listening to touching. They were both there.

QUESTION: So, presumably one could get late auditory components and the early ones and not the middle ones with this paradigm?

GALAMBOS: Yes. I think there are many experiments to do on this and I certainly do not mean to convey to you the idea that I have much here except a set of observations. But I can guarantee what I have told you will work. Why it works I do not know.

T. ALLISON: To comment on that point, the fact that you are canceling extracellular potential fields does not mean, of course, that you are canceling the single unit activity that is going on there.

The really impressive thing about your 40 Hz potential is the fact that it is so large, so close to threshold. Have you compared that potential and a standard ABR in the same subjects? How much better does it do?

GALAMBOS: The standard ABR is of the order of 0.3–0.5 μV, and these responses can get up to close to 2 μV.

ALLISON: So, it may be a signal-to-noise problem.

GALAMBOS: The bandpass is different between the two measures. For the middle latency 40 Hz potentials, it extends from 10 to 100 Hz. The brainstem bandpass is 150 to 2000 or so.

If you use the brainstem bandpass, you will not see this 40 Hz phenomenon, and if you use the 40 Hz bandpass you will not see the brainstem.

ALLISON: Have you thought about using this in a neurological situation because sometimes, if you have a patient with a high-frequency hearing loss or something that may or may not be related to the disease you are interested in studying, then this might very well be a good adjunct to the standard ABR testing.

GALAMBOS: I agree.

J. J. EGGERMONT: Have you had the opportunity to derive an audiogram using this stimuli?

GALAMBOS: Yes. The interesting thing about this response is that its amplitude is largest for low-frequency tones and gets smaller and smaller with high frequencies, so that at 4000 or 5000 Hz you have, perhaps, 10% or 20% of the amplitude that you have with 500 Hz.

This electric response looks as if it is measuring how much of the basilar membrane is being moved.

EGGERMONT: That would fit very well. Actually, if you would add this to a clinical routine for estimating thresholds, it would be useful for signals up to 1 k Hz.

GALAMBOS: Yes.

H. VAUGHAN: Your suggestion that these auditory and somatosensory responses arise from a common physiological system is a very important concept. Are you stating that the topography of the auditory response and the somatosensory response is the same? It really does not look that way to me.

GALAMBOS: No. You are asking an excellent question for which I simply do not have enough information. I can say the following: You notice that at C3 we have an excellent response and that is because we are stimulating the right hand. If we were stimulating the left hand we would get a response that resembles this. However, there are territories where all three modalities seem to give very much the same response and, of course, there are territories such as C_3 and O_Z where one modality gives a much better response than the other. How to parcel out this is a question that I actually have been puzzling with for several months.

I need a very good complete mapping study of a whole variety of frequencies for all three modalities, and that is a big job. Once I obtain the data I will be able to answer the question you asked.

VAUGHAN: If one superimposes potentials that are out of phase, these will merely subtract by essentially electrical cancellations, so that one cannot necessarily infer that this algebraic or this cancellation effect is physiological. One cannot tell the difference between a physiological cancellation and a clearly electrical one in the extracellular field potential.

THE EFFECT OF CHANGE OF STIMULUS FREQUENCY ON THE PREDICTION OF AUDIOGRAMS

James Jerger and Deborah Hayes

Department of Otorhinolaryngology and Communicative Sciences
Division of Audiology and Speech Pathology
Baylor College of Medicine
Texas Medical Center
Houston, Texas 77030

In order to investigate frequency specificity of the auditory brainstem response (ABR), tone pip signals were used to elicit a response in the 10–14 msec range. The pips were 500 Hz and 2000 Hz tone bursts with a rise-fall time of 2.5 msec and no plateau (total duration = 5.0 msec). Responses to 2048 fixed-polarity tone pips presented at a rate of 17/second were averaged to define a single ABR. Recording characteristics were: (1) preamplifier gain of 50,000 to 100,000:1; (2) preamplifier filter bandwidth from 30 to 1000 Hz (slope = 6 dB/octave); and (3) epoch of 20 msec. A total of 30 normal-hearing children and adults, and 35 hearing-impaired children was tested.

Results in normal-hearing children and adults showed a clearly defined response to tone pip signals at intensity levels ranging from 20 to 50 dB above the normal hearing threshold (nHL). At levels below 20 dB nHL, responses were consistently observed in less than 50% of normal-hearing subjects.

At high signal levels ($\geq$50 dB nHL), the waveform of the response appeared as a well-defined, vertex positive peak occurring at about 8 to 10 msec following onset of the tone pip signal. The positive peak was followed by a well defined vertex negative trough. At lower intensity levels, sharp definition of the response peaks diminished, and a more gradual and rounded waveform was observed.

Latency of the ABR to tone pips in both normal-hearing and hearing-impaired subjects showed the expected inverse relationship with signal intensity. In addition, latency of the response at any given signal level depended on spectral peak of the stimulus. Latency of the ABR to tone pips systematically increased as frequency of the response-eliciting signal decreased.

Amplitude of the ABR to tone pip signals showed considerable intersubject variability with responses ranging from less than 0.20 μV to more than 1.00 μV in normal-hearing subjects.

Responses in hearing-impaired subjects were similar to those observed in normal-hearing subjects. However, agreement between behavioral threshold and ABR threshold varied depending on configuration of the pure-tone audiogram and frequency of the tone pip signal. In general, agreement between behavioral and electrophysiological threshold decreased as slope of the pure-tone audiogram increased and frequency of the tone pip eliciting the ABR decreased. In subjects with basically flat audiometric configuration (i.e., difference in behavioral sensitivity between 4000 and 500 Hz of 20 dB or less), threshold of the ABR to both the 500 and 2000 Hz tone pip was typically within 20 dB of the audiometric threshold at that frequency. In subjects with sloping audiometric configurations (i.e., difference in sensitivity between 4000 and 500 Hz of 40 dB or more), threshold of the ABR to the 2000 Hz tone pip was usually within 20 dB of the behavioral threshold at that frequency. Threshold of the ABR to the 500 Hz tone pip, however, was as much as 40 dB higher than the behavioral threshold at that frequency (ABR to tone pip predicted poorer sensitivity than actual behavioral audiogram).

These results suggest that frequency specificity of the ABR may be limited to frequencies above 500 Hz even with the use of "frequency-specific" tone pip signals.

THE CHOICE OF STIMULUS IN THE AUDITORY BRAINSTEM RESPONSE TEST FOR NEUROLOGICAL AND AUDIOLOGICAL EXAMINATIONS

Jun-Ichi Suzuki and Hitoshi Yamane

Department of Otolaryngology
Teikyo University School of Medicine
Kaga 2-11-1, Itabashi-ku
Tokyo 173, Japan

In Japan, the auditory brainstem response (ABR) has already been established as a useful test. Most of the ear, nose, and throat clinics in university hospitals are equipped with the ABR system. Neurologists, neurosurgeons, and pediatricians are also interested in the ABR. However, most of them are not familiar enough with the test yet.

In our ENT department, there are two groups of patients. One is comprised of patients with neurological disorders and the other of infants and young children with suspected hearing impairment. The purpose of the ABR test is apparently different for each group of patients. According to the data requested from the test, appropriate stimuli were studied and established in my department. Consequently, clicks are utilized for neurological examination and tone pips with a 5 msec rise and fall were selected for audiological examination.

It does yet not seem to be commonly recognized that the appropriate choice of stimulus helps to increase the data from the ABR test.

FIGURE 1 shows the ABR to a click stimulus (left) and that to a tone pip (right). The click-evoked ABR delineates clearly separated waves that represent the sequential activities in the brainstem auditory pathways.

On the other hand, response to the tone pip may seem unfamiliar to those who have been utilizing the click stimulus only. The separated five waves are missing in the tone-pip-evoked ABR and the slow component of the ABR is enhanced.

The difference of wave form is due not only to the difference of stimulus sound, but is also related to differences in the stimulating and recording system.

TABLE 1 summarizes the arrangement of the ABR system for neurological examination and audiological examination. In addition to the choice of stimulus, the important role of the high-pass filter and the time base of the display should be emphasized.

The high-pass filter, which is common for neurological and audiological examinations, has a slope of 5 dB/octave and its 3 dB roll-off point is set at 48 Hz. This filter was selected empirically for the purpose of maintaining the slow component of the ABR. The slow component, which rises positively toward the peak of wave V and falls negatively afterwards, is essential for audiological examination because it is the indicator of the threshold of ABR.

The time base of the display also influences the ABR threshold. By using a 10 msec time base, details of the response configuration are clearly visible and accurate measurement of latency is possible.

Use of a 30 msec time base results in loss of the detail. However, it enhances the slow component and makes threshold measurement easier.

In addition, the peak latency of wave five is sometimes much longer than 10 msec when the intensity of the stimulus is close to the threshold.

0077–8923/82/0388–0731 $1.75/0 © 1982, NYAS

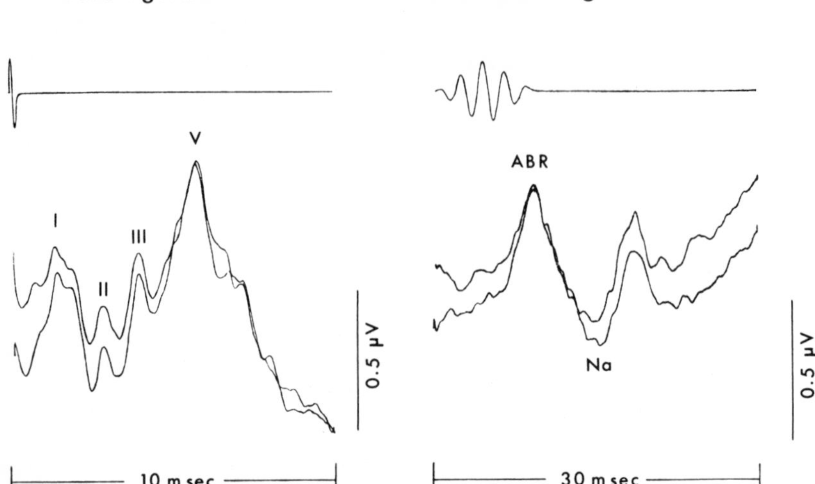

FIGURE 1. Auditory brainstem responses (ABRs) to a click stimulus (left) and to a tone pip (right). The click is generated through a dynamic earphone (TDH-39) by delivering one cycle with a 3 kHz sine wave. Intensity of the click is 80 dB above normal sensation level. The tone pip consists of a 5 msec rise and a 5 msec fall without plateau. Frequency is 500 Hz and intensity is 60 dB above normal sensation level. Note the difference of time base.

Various kinds of neurological disorders have been examined in my clinic and in the intensive care unit of Teikyo University.[1-3] Examples of ABR abnormalities are shown in FIGURE 2. These responses were recorded from comatose patients with acute brain damage and were arranged in the order of severity of the damage.

The ABR from case N.E., whose brain damage was the mildest, had a normal configuration and normal peak latencies. This was the only case who recovered.

Case A.H. showed normal configuration with prolonged latencies.

In case S.A., only wave I and wave II were distinguishable and later waves were depressed. In case S.J., a significant increase of central conduction time, i.e. the interval between wave I and wave V, was noted. These three cases had damage from the midbrain to the medulla and finally died.

TABLE 1

ARRANGEMENT OF THE ABR SYSTEM FOR NEUROLOGICAL AND AUDIOLOGICAL EXAMINATIONS

	Neurological Exams	Audiological Exams
Stimulation		
Signal	Click	Tone pip
Rate	10/sec	26/sec
Recording		
Filter	48–1000 Hz	48–1000 Hz
Display	10 msec	30 msec
Measurement	Latency & waveform	Threshold

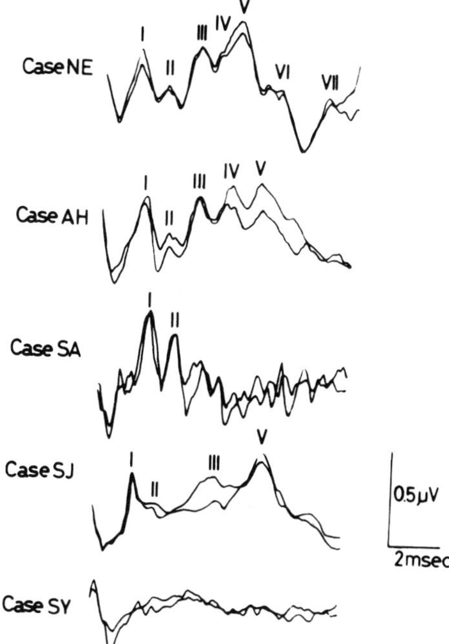

FIGURE 2. ABRs in comatose patients. Severity of the brain damage represented is in the order of arrangement. Stimulus is a click at 85 dB above normal sensation level.

No response could be observed in case S.Y. This patient fulfilled the criteria for brain death and died soon after the ABR examination.

Generally, evaluation of the ABR in neurological patients is based on the pattern recognition of the waveform and on the peak latencies.

However, it must be noted that neurological examination by the ABR test depends

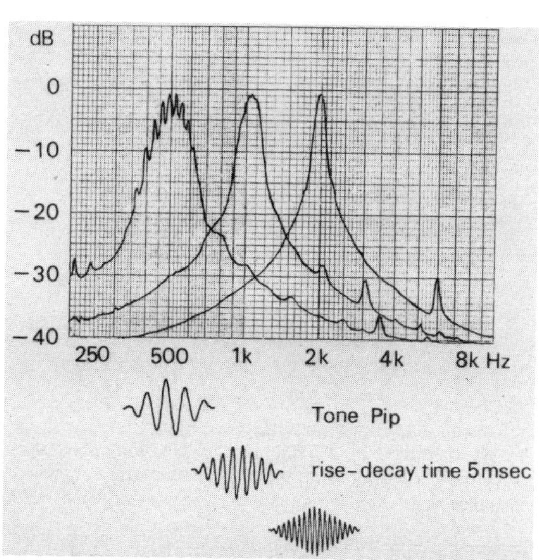

FIGURE 3. Frequency spectra of tone pips. Center frequency of these pips was adjusted at 500, 1000, and 2000 Hz. Each pip has the same shape with a 5 msec rise and a 5 msec fall but without a plateau.

upon the normal function of the peripheral auditory organ. Even if there was no response in a comatose patient, this could be due to the profound deafness. The relationship between the ABR deterioration and the peripheral hearing loss should be taken into account in the neurological examinations.

In my clinic, click-evoked ABR was utilized for assessment of peripheral hearing impairment until 1975. However, it became apparent that the threshold of click-

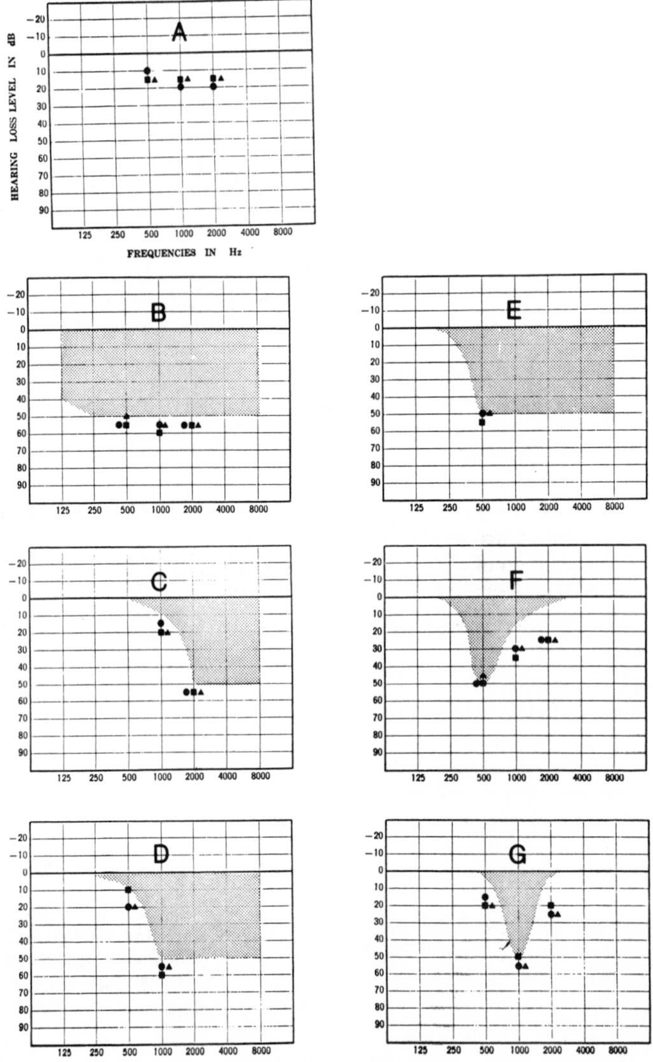

FIGURE 4. Thresholds of ABR to tone pip stimuli (500, 1000, and 2000 Hz) were plotted on the audiogram. Hearing losses were simulated by loading filtered noise. Shaded areas represent simulated hearing losses. Closed circles, closed squares, and closed triangles indicate ABR thresholds from three subjects.

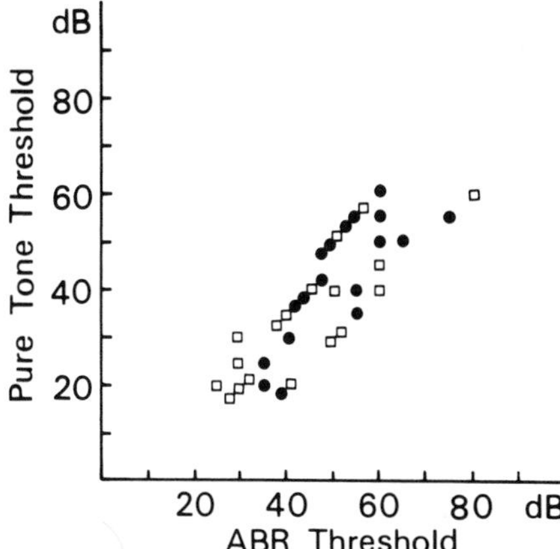

FIGURE 5. Comparison of ABR thresholds and pure-tone thresholds in cases with confirmed hearing losses. Filled circles indicate a threshold at 500 Hz, and open squares indicate a threshold at 1000 Hz.

evoked ABR only reflects hearing at high frequencies.[4] Thus, the use of tone pip stimulus was then studied experimentally[5–6] and clinically.[7]

Tone pips with a 5 msec linear rise and 5 msec linear decline without plateau were chosen as the sitmuli. This "diamond" shape is steep enough for eliciting easily recognizable response and shallow enough for maintaining frequency specificity.

FIGURE 3 shows the frequency spectra of the tone pips at 500, 1000, and 2000 Hz. These narrow spectra suggest the frequency specificity of the response to these stimuli.

The frequency specificity of the tone-pip-evoked ABR was confirmed experimentally and clinically. FIGURE 4 summarizes the results of a masking study[8] in which various kinds of hearing losses were simulated by loading filtered noise. Thresholds of tone-pip-evoked ABR from three subjects corresponded well to the stimulated audiograms.

Clinical cases with reliable audiograms were also examined by the tone-pip-evoked ABR.[7] FIGURE 5 shows the relationship between ABR thresholds and pure-tone thresholds in 18 patients.

The ABR thresholds were up to 20 dB above pure-tone thresholds in all subjects.

According to these results, prediction of audiograms in cases with unknown hearing was considered to be possible.

Five years have passed since this method was established in my laboratory. Many infants and young children have been tested by this method. Follow-up studies of these children with repeated multimodal behavioral audiometry confirmed the reliability of the predicted audiograms by the ABR.[9]

In conclusion, each laboratory should select the stimulus according to the purpose of the ABR test. Furthermore, control data subjects with normal hearing and hearing-imapired subjects should be collected for each stimulus before the clinical application.

REFERENCES

1. KAGA, K., E. KITAZUMI & K. KODAMA. 1979. Auditory brain stem responses of kernicterus infants. Int. J. Ped. Otorhinolaryng. **1:** 255–264.
2. KAGA, K., Y. TOKORO, Y. TANAKA, *et al.* 1980. The progress of adrenoleucodystrophy as

revealed by auditory evoked responses and brain stem histology. Arch. Otorhinolaryngol. **228:** 17–27.

3. Kaga, K., T. Nagai, A. Takamori, *et al.* 1981. Auditory brain stem reponses and cold caloric stimulation in comatose patients. Brain and Nerve **33:** 171–179 (in Japanese).
4. Yamane, H., O. Yamada & K. Kodera. 1976. Clinical evaluation of the auditory evoked brain stem response (BSR) in the prediction of pure tone threshold. Audiology Japan **19:** 178–186 (in Japanese).
5. Kodera, K., H. Yamane, O. Yamada & J.-I. Suzuki. 1977. The effect of onset, offset and rise-decay times of tone bursts on brain stem response. Scand. Audiol. **6:** 205–210.
6. Kodera, K., O. Yamada, H. Yamane & J.-I. Suzuki. 1978. Effects of number and interstimulus interval of tone pips on fast response. Audiology **17:** 500–510.
7. Kodera, K., H. Yamane, O. Yamada & J.-I. Suzuki. 1977. Brain stem response audiometry at speech frequencies. Audiology **16:** 469–479.
8. Yamane, H., K. Kodera & O. Yamada. 1977. Frequency selectivity of the brain stem response: Correlation between response thresholds and selective masking. Audiology Japan **20:** 143–150 (in Japanese).
9. Kodera, K. 1981. Brain stem response audiometry in infants. Japan J. Otol. Tokyo **84:** 139–145 (in Japanese).

Discussion of the Paper

Question: I have one concern regarding the filter settings during the recording of the ABR. Your high linear frequency filter was set at 1000 Hz. In assessing brain death, one of the neurological applications, we usually try not to interfere with the potentials in any way. To attenuate the portion of the components above 1000 Hz probably decreases some of their amplitudes. If we want to rule in or rule out brain death, perhaps we ought to be keeping our high-pass recording filters at 3,000 or 10,000 Hz. Otherwise, I agree with what you are saying about needing to vary what we are doing as a function of the purpose of the test. My question is, are you using a high-pass filter of 1000 Hz?

J.-I. Suzuki: Yes, we use a 1000 Hz filter with satisfactory results in assessing brain death. We would evaluate some other filter settings.

Index of Contributors

(Italicized page numbers refer to discussion comments.)